The Mammalian Radiations

John F. Eisenberg

The Mammalian Radiations

An Analysis of Trends in Evolution, Adaptation, and Behavior

The University of Chicago Press

The University of Chicago Press, Chicago 60637
The Athlone Press, Ltd., London

Paperback edition 1983
Printed in the United States of America
89 88 87 86 85 84 83 2 3 4 5 6

Library of Congress Cataloging in Publication Data

Eisenberg, John Frederick.
The mammalian radiations.

Bibliography: p. 509.
Includes index.
1. Mammals—Evolution. 2. Mammals—Behavior.
3. Adaptation (Biology) I. Title.
QL708.5.E57 599.03′8 80-27940
ISBN 0-226-19537-6 (cloth)
0-226-19538-4 (paper)

To Otto and Bernice
who attempted to start us on the high road

To Thomas J.
who became "Mr. Standfast"

And to Karl and Elise
who carry the burden as pilgrims with good nature

Ach Gott! die Kunst ist lang;
Und Kurz ist unser leben.
Mir wird, bei meinem Kritischen Bestreben,
Doch oft um Kopf und Busen bang.
Wie schwer sind nicht die Mittel zu erwerben,
Durch die man zu den Quellen steigt!
Und eh man nur den halben Weg erreicht,
Mus wohl ein armer Teufel sterben.

Goethe

Contents

Preface

I have enjoyed the privilege of working with living mammals in two rather distinct settings: in the field and in a major zoological park. I have also studied mammals in captivity or at semiliberty in my home and office and at the three major universities I have been associated with over the past twenty-two years. A dichotomy still exists between fieldworkers and those who study captive animals. There has always been a mystique about fieldwork, and this sense of romance is important in coping with its genuine hardships and privations which necessitate a discipline stemming from dedication to some ideal. Field data are the only control whereby one can determine whether captive data reflect behavioral norms or pathological deviations, and data from captive studies are often looked on with some suspicion by fieldworkers. Some caution is of course healthy, but all too often the records from years of captive study are treated as irrelevant by the true-blue fieldworker.

Since I have had one foot in each area of research—captive and field—I hope I can bridge the artificial gap between the two sets of researchers. As national parks become more managed and as zoological institutes develop breeding farms with large fenced ranges, the distinction between field and "pen" studies begins to blur. Captive studies must be designed, and the data derived from captives must be evaluated with a background of knowledge derived from the field. At the same time, a field study may be executed more efficiently if the basic data on life history from captives are already in hand.

Sections in this book lean heavily on life-history data derived from captive mammals. Where possible, these data are put into perspective with existing field data. I have tried to summarize information in tabular or graphic form rather than relying solely on narrative. This method has both advantages and deficiencies. One could hope that some of the postulates will encourage further, more refined research on life histories.

Since my approach to behavior is two-faceted (phylogenetic and ecological), the species accounts are developed with my own biases firmly in mind. Comparison of behavior patterns demands some attempt to develop detailed behavior profiles for those forms that derive from distant phyletic branching. To this end, the Monotremata, Marsupialia (Dasyuridae and Didelphidae), Edentata, Pholidota, Tenrecidae, Macroscelididae, Tupaiidae, and Rodentia will be considered in some depth. For contrast, some attention will be devoted to the extremely specialized taxa, especially the Chiroptera and Cetacea.

To bring the ecology of the various species into perspective, I will attempt to define trends in habitat utilization, mating systems, and life-history patterns. Social structure and demographic data will be reviewed where pertinent. I will not attempt to review anatomy, taxonomy, and physiology systematically, since these are as have been adequately covered by other volumes (see Grassé 1955*a*, *b*; Weber and Abel 1928; Anderson and Jones 1967; Vaughn 1978). I hope to present a summary statement about what we believe is true and what we do not know about the natural history of mammals and about how the different adaptive syndromes have evolved. This is not a task to be undertaken lightly, and I am the first to admit its difficulties. Nevertheless, I have been fortunate enough to work with a wide variety of species at first hand and thus hope to pass on a bit of what I think I know about their behavior.

Acknowledgments

Although I might be tempted to believe this volume represents my own thoughts about the course of mammalian evolution, I realize that any major ideas I may have developed are the result of a process of accretion through gaining and sharing ideas with professors, students, colleagues, and friends. There is not one thought here that I do not take full responsibility for, but if some usefulness derives from these writings then the credit must be shared with many.

Teachers and professors should be thanked because they often pointed out paths: Lena Miller, who taught that natural history could be respectable; Dr. Herbert Eastlick, who demanded and got intellectual discipline; Dr. D. S. Farner, who introduced many of us to the experiment; the late Dr. G. Hudson, who lived his philosophy of the environment; and Dr. Cynthia Schuster, who taught constructive skepticism.

The intellectual whirl at Berkeley would not have been complete without the wit and insights of Professors Seth Benson, Howard Bern, Frank Pitelka, Starker Leopold, and Frank Beach. To Peter Marler I owe a great debt because he inspired me and so many others to develop our graduate research programs as ethologists. I treasure the friendship shared with him and his wife, Judith.

Fellow students are the true matrix of education. I thank Jared Verner, Donald Isaac, William Hamilton, III, James Mulligan, Robert Kuehn, and George Hirsch for companionship and mutual instruction in the field. Mark Konishi, Jerram Brown, Keith Nelson, Edith Neal, Tom Struhsaker, Gordon Orians, Richard Root, Steve Wainwright, and Robert Behnke, together with the previous group, made seminars profitable and life as a graduate student tolerable.

I thank my senior colleagues at the University of British Columbia, Professors I. McTaggert-Cowan and W. S. Hoar. They were most encouraging during the formation of my early research plans and the development of my undergraduate training programs.

Colleagues and friends contribute in so many ways that it is impossible to thank them all for their insights. There are uncited contributions from the thinking of Ellen Franzen, Wolfgang Pfeiffer, Richard Sadleir, William Z. Lidicker, Howard Winn, Robert Ficken, Douglass Morse, and Wolfgang Schleidt, though they have never seen the completed manuscript. Special mention must be made of the debt incurred in numerous discussions with Paul Leyhausen, Valerius Geist, and Arthur Myrberg.

Certain ideas developed from key conferences sponsored by the Wenner-Gren Foundation, and I wish to thank Robert Martin, Malcolm McKenna, Irv Bernstein, Stephen J. Gould, Thelma Rowell, Matt Cartmill, and Tim Clutton-Brock for freely sharing ideas with me. I should like to thank Madhav Gadgil for inviting me to participate in the fifth Mahabeleshwar Symposium, sponsored by the Tata Institute for Advanced Studies. Madhav and Sulachana Gadgil, John Maynard Smith, John Crook, Mary Jane West-Eberhard, William Eberhard, and Robert Trivers greatly contributed to the thoughts expressed in part 4 of this volume. Professors Heini Hediger, Konrad Lorenz, Hans Peters, and François Bourlière are especially acknowledged for their inspiration and encouragement through the years.

During certain phases of fieldwork, I have had the companionship and advice of some superb friends. Theodore Grand has been a participant and friend on every major expedition, and Edwin Gould gave me the opportunity to participate in the Madagascar program—a rich intellectual experience for us both. Suzanne Ripley was the inspiration for the Ceylon Primate Survey, and Marcel and Annette Hladik were the undisputed champions of discipline and endeavor in the early Ceylon years. Melvin Lockhart made the Elephant Survey move into the realm of real life. G. G. Montgomery introduced me to the rigors of telemetry in Panama and set an example with his boundless energy. Melvin Sunquist has worked as a true "soldier" in the Smithsonian field programs. Fieldworkers who also shared ideas, hospitality, and friendship include John Seidensticker, Richard Laws, Bristol Foster, Burney Le Boeuf, John Robinson, Tom Struhsaker, Rasanayagam and Ranji Rudran, Martin Moynihan, Neil Smith, Mike Robinson, Mary Willson, George Schaller, and Stanley Rand. Recent work in Venezuela, which contributes heavily to section 16.4, was set in motion and sustained by the efforts of Dale Marcellini, Eugene S. Morton, Edgardo Mondolfi, Juan Gomez-Nuñez, and Tomas Blohm.

Ideas shared and criticized take on a special poignancy when the partner is deceased. I acknowledge my debt to the late Professors H. K. Buechner and R. F. "Griff" Ewer.

Former pre- and postdoctoral students are a special category, since they grow and become colleagues so that it is difficult to adequately pinpoint any single contribution; rather, it is a continuous unfolding. Suffice it to say, it is best described as a form of mutualism. Nicholas and Tanis Smythe shared the joys of caviomorph rodents in Panama and Chiapas; Becky Myton successfully challenged some pet hypotheses of small mammal ecology; Robert Horwich unraveled the development of gray squirrels; Lang Elliott made the chipmunk a subject of beauty and instruction. Christen and Shirley Wemmer shed light on the mysteries of small carnivore life cycles; George McKay made the Asian elephant a prominent part of our knowledge; Nancy Muckenhirn developed our understanding of the leopard; Rasanayagam Rudran pioneered the ecology of Old World primates; Wolfgang Dittus cast primate population dynamics into an evolutionary perspective. Kenneth Green deciphered the communication system of the golden marmoset, and Robert Hoage showed the detail and intricacy of the marmoset parental care system. Rebecca Field set new standards for the analysis of auditory communication; Charles Brady opened the investigation of pair-bonding in South American canids; Christine Schonewald developed the computer analysis of communication to new dimensions; Sue Wilson made the social development of young rodents take on special significance. Galen Rathbun introduced us to the Macroscelidea; Larry Collins raised marsupial studies to a new level; Ilan Golani taught us all to look more closely at mammalian interaction patterns; Michael Murphy established the art of experimentation in chemical communication; Georgina Mace introduced us to new statistical methods. Margaret A. O'Connell carried on the Venezuela small mammal project with insight and professionalism; Peter August set a new standard in critical field experiments. To all of you and those to come, my debt and affection.

Ideas and work shared with colleagues keep coming to mind. At the National Museum I am indebted to R. W. Thorington, Robert Brownell, Clyde Jones, Donald Wilson, C. O. Handley, Jr., and Henry Setzer. At the National Zoo, I thank Judith Block, our registrar; Harold Egoscue, senior curator; Miles Roberts, curator of mammals; Clinton Gray, senior veterinarian; and Dale Marcellini, curator of reptiles. I acknowledge the superb maintenance of our captive collections through the efforts of Dr. Mitchell Bush, chief of animal health; Dr. Richard Montali, pathologist; and Messrs. Eugene Maliniak, John Hough, Mike Deal, Larry Newman, Tex Roe, and Peter Beaman. Data collection and reduction were aided and sustained by Lynn Dorsey, David Mack, Donna Pyle, Ranka Sekulic, and Kent Redford. Kent kept me jumping with his intellect and enthusiasm. Where would the organization of my professional life be without Betty Howser, Gail Hill, and Tabetha Gilmore? My loyal and indispensable partner of some fourteen years of manuscript preparation remains Wyotta Holden. The illustrations are the result of the sensitive efforts of Sigrid James.

Part of the research summarized herein was supported by grants from the National Institute of Mental Health, the National Science Foundation, the National Geographic Society, and the World Wildlife Fund. My profound thanks.

I thank the staff of the National Zoological Park, its director, Dr. Theodore Reed, and its former deputy director, Ed Kohn, for the freedom to pursue my work. I am most grateful to the Smithsonian Institution for providing funds, support, and the atmosphere of inquiry necessary for writing. Special thanks are due Ross Simons; David Challinor, assistant secretary for science; and S. Dillon Ripley, the secretary of the Smithsonian, for their support and concern.

There is also a category of friendship and dialogue that defies an intellectual categorization but probably boils down to mutual support and respect. My thanks to Warren and Ghislaine Iliff, F. E. "Gene"

Wood, Jim and Dianne Robbers, Walter and Crystal Poduschka, Ludwig Franzisket and family, and Werner and Hilde Eisenberg.

Finally and by no means least, I wish to thank my close collaborators who have read parts of this manuscript and remain a source of constructive criticism and friendship: Drs. Katherine Ralls, Eugene S. Morton, and Devra Kleiman. To Devra must go a very special debt, for she has supported me in the bleakest moments of my struggles with the material. Thus, and last, to my colleague, friend, and wife, Devra—my profound thanks and gratitude.

I thank the following publishers for permission to reproduce figures: the University of California Press, figures 127 and 130; E. J. Brill, Leiden, figures 126, 135, and 136; the Smithsonian Institution Press, figures 116, 125, 128, 129, and 143.

Part 1 The Historical Perspective

1 Introduction

In 1957 I became entranced by the title of C. D. Forde's book *Habitat, Economy and Society* (Forde 1934). It came to my attention immediately after I read J. B. Calhoun's paper comparing spacing and territoriality among two inbred strains of the domestic mouse (Calhoun 1956). As the title of Forde's book suggests, there is an interrelationship among these three aspects of life-history studies. In his study the subject was mankind, and he was unable to establish consistent correlations between habitat and the two aspects of human endeavor: the formation of societies and the methods of extracting food and shelter from the environment. The dominance of learned traditions or "culture" intervenes strongly when one attempts to correlate human behavior with gross habitat categories. On the other hand, there should be some lawful relationships among these three facets of natural history studies when infrahuman mammals are studied, and recent attempts to establish correlations in humans have been more successful (Meggers 1971).

If we consider mammalian life histories, economy may be taken as a shorthand way of referring to the mode of habitat exploitation. The term habitat stands as a self-evident category referring to that part of the biome in which an animal habitually attempts to make its living. Society would then refer to the *modal social structure* of a species or a deme. By a modal social structure I mean that form of social organization most typically shown by the species, even though there will be a range of social structures that the species may exhibit over the entire range of habitats it exploits (see chaps. 31 and 32).

Now, it is well and good for Forde to undertake the analysis of the interrelatedness of habitat, economy, and society for a single species, such as *Homo sapiens*. In cross-species comparisons, however, species differences in anatomy and physiology often become paramount determinants of behavioral differences. Thus some examination of differences in anatomy and physiology for the mammalian species under investigation becomes a prerequisite for any correlations that are accepted. All mammalian species are distantly related by phylogenetic descent, stemming from a common origin, and there is a basic unity with respect to anatomy and physiology. However, when we compare two species, such as a shrew (*Sorex vagrans*) and a deermouse (*Peromyscus maniculatus*), numerous differences in anatomy become evident. Some characteristics of the shrew's anatomy suggest those of mammals that lived in the distant past. These characters we refer to as *plesiomorph*. On the other hand, the shrew has certain structural attributes that bear no resemblance to ancestral forms, and these we refer to as *apomorph* (Hennig 1966).

Comparing the shrew and the mouse, we find that the shrew has more plesiomorph characters than does the mouse. We would be inclined, then, to say it structurally resembles mammals of an earlier phylogenetic time. Using older terminology, we would say that, along with an assemblage of derived or specialized characters, the shrew carries forward into time an assemblage of conservative characters. By engaging in a series of inferences from knowledge based on the anatomy of fossil forms, we attempt to understand the present-day adaptations of living mammals in terms of descent from common ancestors. To do this, we have to reconstruct something about the past. This requires some knowledge of the literature in geology, paleontology, paleoclimatology, and paleobotany. Detailed examination of the paleontological literature can give us a feeling for faunal assemblages in the distant past, their composition, the trophic levels they exploited, and perhaps the limits imposed by morphological adaptations (see Eisenberg and Gould 1970, pp. 119–21).

The fossil record, however, has certain limitations. In most cases the presumptive lineage from a modern form into the distant past is broken. Often there are not enough intermediate fossil forms extant at recognizable geological strata to complete the linkage. Usually the fragments that are left to us in fossil forms consist of teeth or parts of the cranium. Very often a fossil species described in the literature from cranial anatomy cannot be matched with postcranial skeletal fragments. Thus we may know something about a fossil species' feeding habits as deduced from the teeth, but little of its probable mode of locomotion. In spite of limitations, the fossil record is the only record we have and, at least for mammals, tends to be better than that of any other class of vertebrates (cf. Simpson 1975).

In the phylogenetic trees and cladograms presented throughout this book, I am inferring phylogenies from fragmentary fossil evidence. Many of these cladograms could be taken only as blueprints for a phylogeny. A true phylogeny would consist of enough intermediate fossil forms at recognizable geological strata to determine a rather precise time sequence for the branching. This is a goal toward which any natural system of classification works, even if it is as yet unattained.

Of special interest in our analysis of mammalian phylogeny are those cases where a natural experiment has occurred, as where two phyletic lines of descent that branched off distantly have evolved similar body forms on different continental land masses, probably under similar selective pressures. Inferences that we develop concerning the evolution of behavior and the evolution of social systems or trophic differentiation can be tested by comparing the radiation in a phyletically distant line with the radiation under study. As I will point out subsequently, the Metatheria and Eutheria are such a set of "sister groups." Inferences and hypotheses concerning adaptation and the interrelatedness of social structure and niche exploitation in the Eutheria can be tested adequately only by turning first to the Metatheria as a control group to discern if similar trends are shown (see fig. 1).

This leads us, then, to a consideration of the comparative method and what it entails. The comparative method as approached in ecology and behavior studies involves the description of systems, whether systems of trophic exploitation, systems of interaction among members of the same species, or life history strategies in terms of demography. Once a system has been described, we can attempt to find a "sister system" in a group that is of different phylogenetic descent but has been exposed to what we assume to be the same sorts of selective forces. Then, by means of a point-by point comparison between the two systems, we can elucidate similarities and differences. Presumably similarities or convergences result either from parallel evolution based

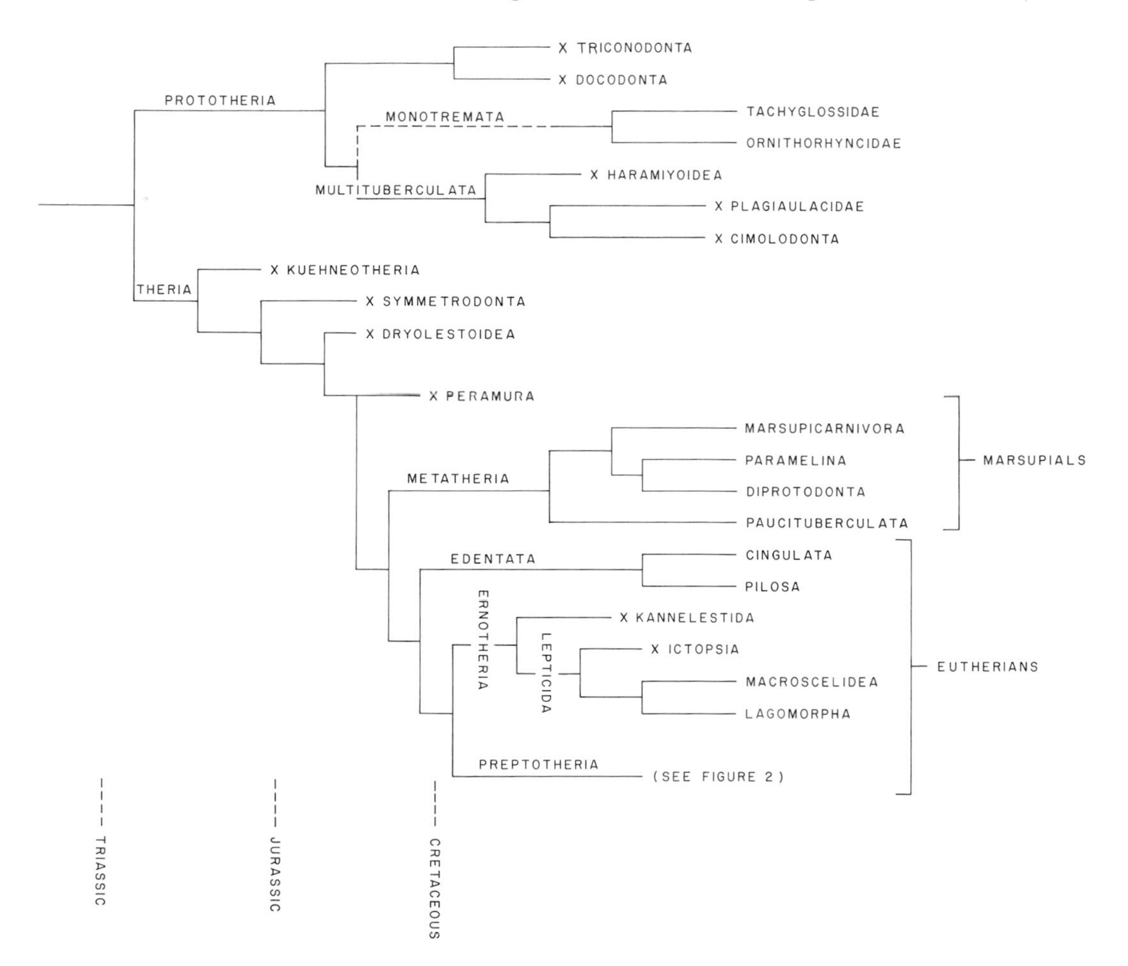

Figure 1. The early radiations of the mammals: a cladogram based in part on McKenna 1975. Note that by the late Cretaceous the marsupial and eutherian radiations were distinct and well under way. The Monotremata exist today as a remnant of the early radiation of the Prototheria. Ordinal terminology for the Metatheria from Ride 1964; other nomenclature from McKenna 1975. X = extinct taxon in this and all subsequent cladograms.

on shared genotype or, more important, from a similarity of selective forces through time acting upon somewhat different genetic substrates (Simpson 1945). Differences will result from either discontinuities in the historical development of the two lines or perhaps from inherent limitations, ultimately genomic, in the degree to which alterations in the evolving substrate can be expressed.

Whenever we delve into problems of comparison, we return to the old conundrum of how to separate homologous from analogous characters. Two structures are said to be homologous if they in fact reflect a similarity in genetic background. To state it another way, homology requires that, in the two lines of descent, the genetic substrate that determines the structure in question be nearly identical for the two forms under consideration. On the other hand, analogous characters are structural attributes that look similar in two phylogenetic lines but in fact derive from different genetic backgrounds. The process of forming analogues is phylogenetic convergence. Convergence thus occurs when two entirely different genetic stocks produce a similarity of form that does not derive from similarity in genetic background (Eisenberg 1975*a*; Simpson 1944).

2 The History of Mammals

The evolutionary history of mammals has been reviewed several times in the past seventy years. Gregory (1951) provides a lavishly illustrated outline of animal evolution, while Romer (1966) stands as the classic reference for vertebrates. Scott (1937) is still an excellent treatment of mammalian history in both North and South America. Lillegraven et al. (1980) have provided us with a beautiful update on mammalian origins and radiations in the Mesozoic. Simpson (1980) recently reviewed the history of mammals in South America. Comprehensive reviews of the evolutionary history of mammals as a worldwide survey may be found in Thenius (1969) and Thenius and Hofer (1961).

There have been many natural experiments resulting from the early separation of the various phyletic lines that we recognize today as the class Mammalia. Twenty years ago, the stability of the continents, with respect to their global position, was taken for granted.[1] The disjunct distributions of living mammals over the surface of the earth were accounted for by various theories. Land bridges were invoked, as well as notions of sweepstakes or probabilistic inoculations of land masses through the rafting of founder stocks. The very elaborate set of explanations developed by G. G. Simpson and summarized in 1965 in *The Geography of Evolution* holds up very well for explaining the present distributions of living mammals. However, this is so because most of the major continental drifting patterns, which we accept as real patterns whenever they occurred, happened so far back in time that they do not reflect much about the distribution of modern forms. The original distribution of ancient forms is a very real question.

Sometime in the Triassic, more than 250 million B.P. (before the present), the mammalian grade of anatomical structure was reached by the descendents of therapsid reptiles. Terrestrial plants were dominated by conifers and cycads. The land masses were still a connected series of plates (see fig. 2*a*). At 190 million years B.P., the Jurassic era commenced with the dinosaurs dominating the land masses. The world continent of Panagea had divided into Laurasia and Gondwanaland (see fig. 2*b*). The stage was now set for the evolution of the mammals and the radiation of the birds. The pantotheres became the dominant group of mammals, and in them and the multituberculates, symmetrodonts, and triconodonts we can discern the first adaptive radiation (see fig. 1).

1. Wegener's brilliant hypotheses (1929) were considered eccentric (Tarling and Tarling 1975).

At 135 million years B.P., in the early Cretaceous, the marsupials, multituberculates, and earliest placentals form a diversified assemblage in North America. Although mammals of the Triassic were quite small (< 120 mm) (Jenkins and Parrington 1976), postcranial skeletons from 100 million years B.P. suggest that some are now about the size of a modern opposum, *Didelphis* (Gregory 1951). The dentition suggests an insectivore/frugivore trophic level (see fig. 3*c*). The angiosperm plants have begun to flourish, and one would recognize palms, magnolias, and tulip poplars. Flowers are large and conspicuous, and fruits are large, with attractive aerials or pericarps that induce vertebrate feeding, seed dispersal, and coevolution of plants and mammals (Corner 1945). Insect pollination has become a method of passing pollen from anthers to pistils. Mammals are adapting to insects, fruit, seeds, and herbaceous vegetation as feeding substrates. Trophic specializations are beginning to unfold (see chap. 19).

By the mid-Cretaceous, western America is connected to Asia, but eastern America and Europe are adrift. Africa and India subsequently separate from Gondwanaland. The Condylarthra and marsupials are differentiated in "Euramerica" and probably in the contiguous plate that will later be South America, Australia, and Antarctica. The stage is now set for the explosive independent radiations of the Mammalia that will take place in the Paleocene (see figs. 2*c* and 2*d*).

What sorts of inferences can we make from the fossil record concerning the ecology and behavior of mammals as they evolved between 250 and 100 million years B.P.? For the most part, we are left with teeth, jaws, and skull fragments. Thus most inferences involve extrapolations from living forms. The relative abundance of fossils belonging to different orders may suggest parallels with present numerical distributions for different trophic levels. In addition, we can say something about the other vertebrate fauna that was contemporary with our mammalian fauna and, of course, about the plants that provided the ultimate feeding substrate for the animal community. (For a brilliant synthesis, see Lillegraven, Kielan-Jaworowska, and Clemens 1980.)

In the Jurassic it is safe to assume that either two therapsid lines independently achieved the mammalian grade of anatomical organization or a single stock of therapsids gave rise to two divergent lines. One line of descent gave rise to the present-day Monotremata. The second line showed many early radiations but eventually separated in the early Cretaceous to yield the Marsupialia and the Eutheria.

Let us go back to the first great dichotomy in the Jurassic—the division into the Prototheria and the Theria (see fig. 1). The prototherian line diverged into two major stocks, one branch splitting into the Triconodonta and the Docodonta. The second stock eventually divided into the Multituberculata and the Monotremata. The multituberculates were an early, successful radiation that first appeared in the Upper Jurassic, when America and Asia were still connected at their west and east coasts. Although western Europe had broken off from eastern North America, multituberculates were firmly established in Asia, North America, and Europe, suggesting their appearance in the Trias-

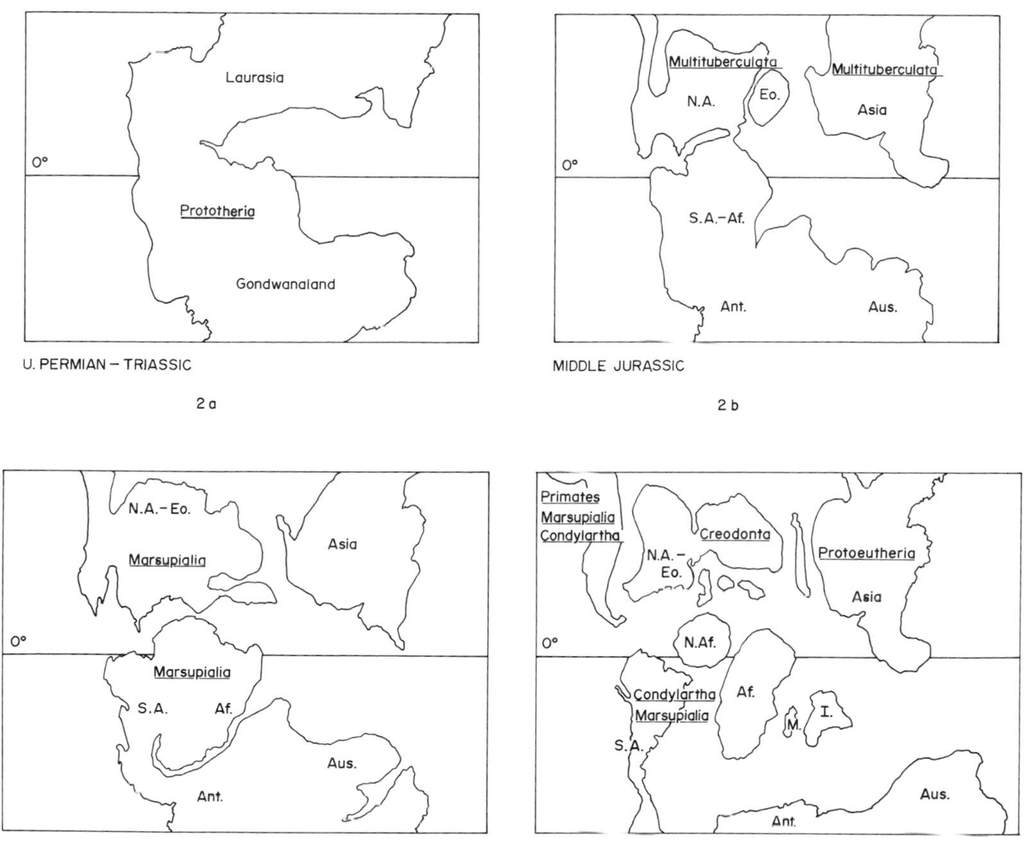

Figure 2. The breakup of the world continent Panagea, adapted from Cox 1974 and Tarling and Tarling 1975. First appearance of major taxa are indicated according to Cox 1974 and by inference from present distributions. (*a*) Differentiation of the Prototheria on the world continent (see fig. 1). (*b*) Isolation of Multituberculata in Laurasia. (*c*) Marsupials appear in the Cretaceous and are inferred to be in South America by the mid-Cretaceous. (*d*) Early Condylarthra now isolated in South America. Marsupials and monotremes could have begun movement to Australia via Antarctica.

sic. In some way the multituberculates must have passed to South America in the Jurassic and then moved on to Australia via Antarctica to establish themselves as the present-day Monotremata, but there is no fossil evidence to substantiate this hypothesis (Cox 1974). (See Griffiths 1978 for a discussion of Multituberculata-Monotremata relationships.)

Mammalian diversity perhaps became accelerated when, at 135 million years B.P., the continents began to break up, thus isolating the major stocks. At 94 million years B.P., Asia and Europe began to separate. During this interval of time, North America, South America, and Africa were somewhat connected, and Asia with part of Europe

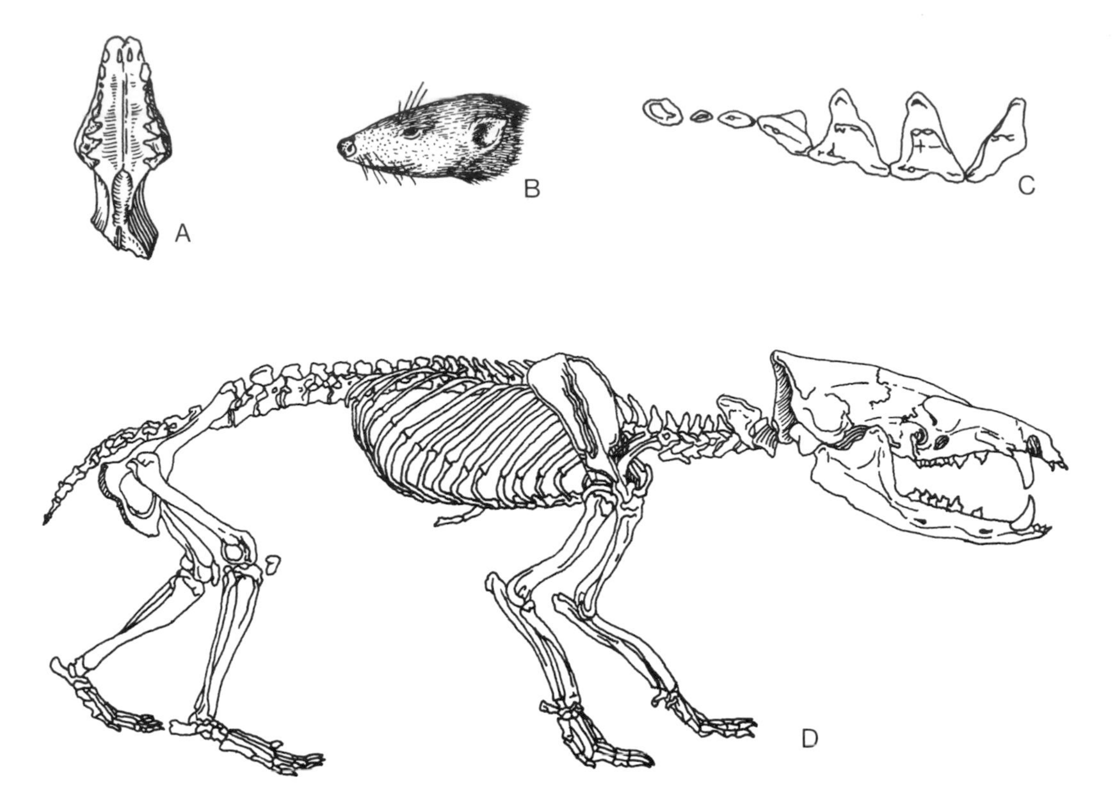

Figure 3. Skeletal anatomy of conservative mammals. Redrawn from Gregory 1951. (*a*) Ventral view of upper jaw of *Deltatheridium*. (*b*) Reconstruction of head of *Deltatheridium*. (*c*) Upper dention of *Deltatheridium*. Arrows indicate anterior and lingual orientation. (*d*) Articulated skeleton of *Tenrec ecaudatus*. The short tail of the common tenrec is an apomorphic character.

also became separate. In the mid-Cretaceous there occurred a union of the eastern part of the Eurasian land mass and North America to create Laurasia, a large northern continent (see fig. 2). During the interval when North America and Eurasia were separated, the marsupials could have become differentiated in the North American mass. We believe marsupials could have had a North American origin because many Cretaceous marsupials have been found in North America, but none have been found in Asia (Lillegraven 1974). Before North America touched Asia, some of the marsupials rafted their way to South America. In South America the marsupials may have been able to move to Antarctica at a time when it was not completely glaciated. From Antarctica they could move to Australia. As Antarctica moved to the south and became glaciated, Australia and South America became the refugia for the marsupials (see figs. 2*c* and 2*d*).

Regardless of how continental drift theory bears on the early separation of the mammalian stocks, one notes that the metatherians have an amazing unity in their basic body plans (see section 3.2). The didelphids in South America and the Australian marsupials show remarkable conformities, suggesting a monophyletic origin for marsupials (Clemens 1977).

At the time of the Eocene, 50 million years B.P., the geographic situation has become rather fixed. A filter bridge between Africa and Eurasia is still to be established, a filter bridge exists between Madagascar and Africa, and India is now approaching Asia and will crash into it to create the Himalayas. Australia is now isolated, and Antarctica is on its way south. South America is also totally isolated (see fig. 4). This provides an opportunity for mammalian stocks to develop in total isolation from each other. North America will eventually become connected with South America in the Pliocene (ca. 12 million years B.P.), and North America will be connected with eastern Asia through the Bering land bridge during the Pleistocene. (There were undoubtedly numerous contacts between eastern Asia and North America between the Eocene and the Pleistocene.)

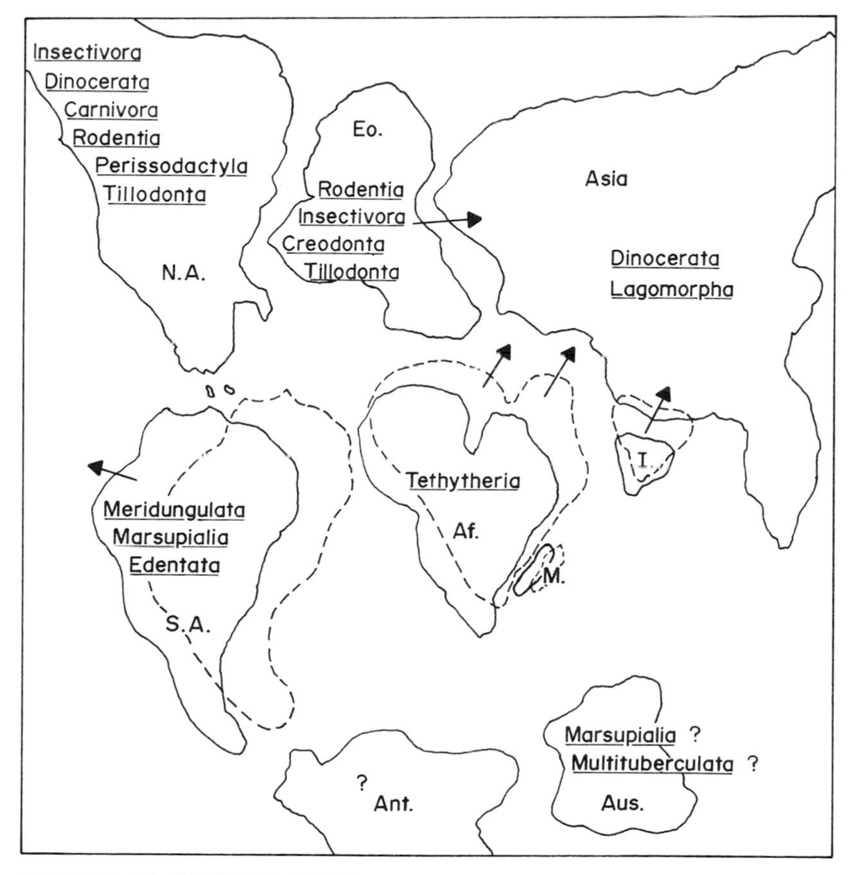

Figure 4. The continents at the beginning of the Eocene; the independent radiations can now begin. Movement of the continents during the Paleocene is indicated by dotted outlines and arrows. The presence of marsupials and multituberculates in Australia is not yet supported by fossils. The terminology of Tethytheria and Meridungulata is from McKenna 1975.

Early continental breakup and isolation set the stage for the creation of different phyletic lines of descent. This offers an unparalleled opportunity for comparison using what I call "natural experiments."

From the initial radiations, there are today some living mammals, mostly in the tropics, that in many morphological features show relatively little modification from the ancestral stocks. This could be because a widespread population originally adapted to a relatively homogeneous environment that persisted through time in those areas that underwent little modification—a process first described by Simpson (1944) as bradytely. If we can accept the existence of forms with numerous plesiomorph characters, it seems legitimate to infer that certain morphologically conservative species inhabiting niches in the tropics represent forms that are in a sense adapted to a "conservative niche." Thus a phylogenetic reconstruction of mammalian behavioral evolution should start with generalizations developed by comparing morphologically conservative forms that occupy what might be considered the niches of the Paleocene.

If the living monotremes represent the last descendents of the multituberculates, what can we deduce concerning the anatomical and reproductive specializations of the primitive prototherian? It seems safe to say that this hypothetical ancestral form had hair, laid eggs, nourished its young on milk, and may well have incubated the eggs in a pouch. These inferences are based on a knowledge of the breeding biology of *Tachyglossus* (see sect. 3.1.2). It is very likely that the ancestral therian possessed this same set of attributes. Thus, in the early Cretaceous, the ancestral "therians" gave rise to two major stocks that differed in the construction of their reproductive tracts and in their dentition. The Marsupialia incubated their eggs in an intrauterine fashion and may be thought of as ovoviviparous (Luckett 1975). The eutherian ancestral line evolved a placenta that not only provided an intrauterine nutritional source as an adjunct to the yolk sac, but also developed the necessary uterine physiology to prevent the genetically distinct fetus from inducing an immune rejection response from the maternal leukocyte system. The Marsupialia retained the pouch as an incubatorium. The eutherians solved the riddle of the maternal-fetal rejection process (Moors 1974) and thereby had the option of extending intrauterine development and reducing lactation time. Fluids from the mother's body (milk) persisted as the mode of early neonatal nutrition. (For an expanded discussion see Lillegraven, Kielan-Jaworowska, and Clemens 1980.)

3 The Early Radiations

3.1 Prototheria

Although I have been fortunate enough to have spent some hours observing the living platypuses (*Ornithorhynchus*) on exhibit at Taronga Park, New South Wales, and at Healesville, Victoria, I have not had an opportunity to work with these striking creatures. At the National Zoological Park, we have exhibited echidnas (Tachyglossidae) since my tenure began. Although the staff noted mating in 1975, we have not had a conception and rearing cycle. Aside from casual observations, I have not systematically worked with the echidna group. The text that follows is thus drawn mainly from the literature and augmented with only a few of my own observations (see plate 1).

The order Monotremata is the only surviving taxon from the subclass Prototheria. Typically, systematists divide the three living genera into two distinct families: the Tachyglossidae (echidnas) and the Ornithorhynchidae (platypuses). The affinities of the monotremes with the remaining orders are obscure (Griffiths 1968, 1978), but most system-

Plate 1. *Tachyglossus aculeatus*, the echidna. The echidna is specialized for feeding on ants and termites. Note that the nasal bones are prolonged to form a beaklike snout. Stout spines can be erected, and the echidna can roll itself into a virtually impregnable ball. (NZP archives.)

atists agree that the surviving genera represent specialized derivatives of a lineage that separated early in the evolution of the class Mammalia. The living monotremes are characterized by a horny "beak" that overlies an extremely modified set of nasal and frontal bones in the skull. The animals are plantigrade and have claws. The teeth are transitional during the development of the young; they are lost entirely in the adult spiny anteaters and are replaced by horny plates in the adult platypus. The mammary gland of the monotreme is well developed but is not organized into a single unit with a nipple. Instead, the mammary glands are formed from two "gland fields" with several separate openings from which milk droplets ooze and are licked up by the young. Griffiths (1968) presents evidence that milk is sucked by the young echidna from the pits into which it is secreted (Griffiths 1968, p. 196). It has been postulated that very young monotremes lick milk droplets from the hairs associated with the mammary ducts, but this needs to be confirmed.

The spiny anteaters, family Tachyglossidae, typically possess a pouch that is functional in the female and supported by epipubic bones. The platypus also has epipubic bones, but no functional pouch develops. The monotreme skeleton thus resembles that of the marsupial, in that epipubic bones are developed, but it also differs—for example, in having a coracoid bone in the pectoral girdle that is completely free from the scapula. Plesiomorph traits shown by the monotreme male include nonscrotal testes. Both the male and the female possess a cloaca, since the urinary tract and intestinal products are released into a common chamber. The body temperature may vary over several degrees centigrade within a given twenty-four hour cycle, though the platypus typically maintains a high body temperature during its active phase. The males of both the spiny anteaters and the platypus have a horny spur on the hind foot. In the platypus the spur is connected by a duct to a gland that secretes a mildly toxic substance.

The two living families represent entirely separate foraging strategies. The Ornithorhynchidae include just one genus and species, *Ornithorhynchus anatinus*, the platypus. The Tachyglossidae include two genera, *Tachyglossus* in New Guinea and Tasmania, and *Zaglossus* (*Proechidna*) in New Guinea. The platypus typically inhabits streams and excavates a bank burrow, where it maintains a nest. The animal forages along stream bottoms for crustaceans, annelids, and aquatic insect larvae. The manner of foraging and the life cycle have been well described by Fleay (1944), Eadie (1935), and Burrell (1927). The living platypus exemplifies a "K" selected form in that adults may live more than seventeen years, and only one or two young are typically produced each year by an adult female. By referring to table 1, one can see that, when active, its resting metabolism (as reflected by oxygen consumption) is in line with those of eutherians of comparable body size.

The monotremes all lay a cleidoic egg that is incubated for some time before hatching. This unique reproductive strategy will be outlined for two representative monotremes: *Tachyglossus aculeatus* and *Ornithorhynchus anatinus*.

Table 1 Fundamental Data for the Monotremata

Species	Body Length (mm)	Weight (kg)	Metabolic Rate[a]	Maximum Captive Longevity (months)	Encephalization Quotient	Litter Size	Litters per Year	Assumed Maximum Reproductive Span (years)[b]	Estimated Maximum Lifetime Production of Young[b]
Ornithorhynchus anatinus	300–450	0.5–2	.46	168	.944	1–2	1	8	8–16
Tachyglossus aculeatus	350–530	2.5–6	.22	600	.715	1	1	18	18
Zaglossus bruijni	450–775	5–10	—	231	—	—	—	—	—

Sources: Walker et al. 1968; Collins 1973; Griffiths 1968; Altman and Dittmer 1962.
[a]As cc 0^2/g/hr.
[b]Estimate only; based on age at maturity and average life-span.

3.1.1 *Ornithorhynchus anatinus*

The standard life-history work is by Burrell (1927). Recently P. Temple-Smith (1974) has completed a thesis that considerably augments our knowledge. Fleay (1944) described the first successful captive rearing cycle, and Strahan and Thomas (1975) have contributed to our knowledge concerning courtship and mating. Griffiths (1978) reviews the biology of both the echidna and the platypus.

The male platypus is somewhat larger than the female (Temple-Smith 1974). His sternal gland is quite active during the breeding season, and he marks with it on banks and at burrow entrances. During courtship the male marks with his cloacal glands on stones underwater (Strahan and Thomas 1975). The scant data concerning copulation in the wild suggest that only one male actively courts a female (Strahan and Thomas 1975). Courtship may be prolonged over several weeks and includes the exchange of a variety of tactile signals. The male follows the female, occasionally seizing her tail in his beak (Crandall 1958). Before intromission he may grip the skin of her nape in his beak while positioning himself above her. It is believed that the egg is laid approximately 11–12 days after copulation.

Before mating, the female enlarges her burrow system. She constructs a leaf nest, transporting the nesting material by clasping a mass of leaves in her tail. After laying the egg, she incubates it for a period believed to last 11–12 days. During this interval the femals feeds little or not at all. Since the female platypus is nervous and easily excited during the rearing cycle, much of the behavior carried out in the extensive tunnel system is unknown, yet Fleay (1944) was able to give us a partial indication of the sequence of events. The hatchling, which is about 11 mm in length, is brooded and suckled over a protracted period. At approximately 48 days after hatching, the eyes are still closed, and a light growth of hair shows. About 9 weeks after hatching, the eyes open, and at 93 days—at a weight of 430 g—the young is first seen to swim. Independent feeding and complete weaning take approximately 21 more days. Separate nesting by the mother and young occurs shortly thereafter (see fig. 5).

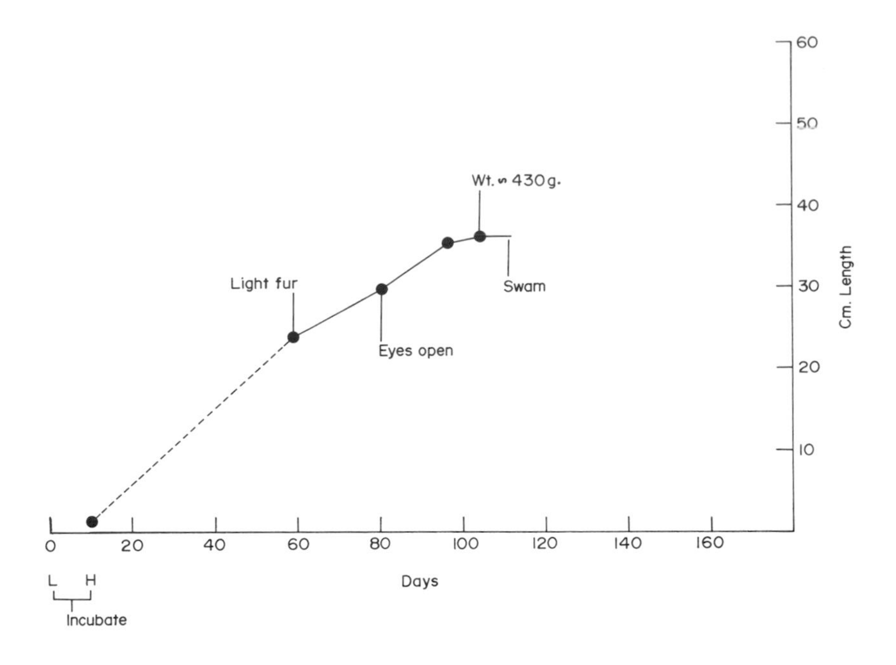

Figure 5. Growth and development of *Ornithorhynchus*. Adapted from Fleay 1944. L = lay; H = hatch.

Throughout the lactation phase, the mother remains very active, feeding prodigiously. Since the species appears to be a seasonal breeder, not more than one clutch per year could be successfully raised. The male appears to nest separately from the female outside the breeding season; in fact, the female seems somewhat intolerant of the male except during courtship and mating.

3.1.2 *Tachyglossus*

Griffiths (1968) presents a synopsis of reproductive data for *Tachyglossus aculeatus aculeatus* and *T. a. multiaculeatus*. The latter subspecies is somewhat smaller on the average, and its developmental schedule is apparently somewhat more rapid. The following account concerns the nominate race.

When a female echidna comes into estrus, she attracts more than one male. Up to five males have been noted in attendance on one female (Augee, Ealey, and Price 1975). Since Brattstrom (1973) has described the capacity of *T. aculeatus* to form dominance hierarchies, it is safe to assume that such groups of males must be organized with some priority of access to the female. Dobroruka (1960) has described copulation, which may take place with the animals on their sides. Copulation is prolonged, and the animals may become locked or tied for a time, a phenomenon that has also been suggested for *Ornithorhynchus*. The males mark with their cloacal glands on the substrate, but the functional significance of this is unclear. The female has a fixed den site during gestation and incubation.

The length of time the egg remains in the oviduct after fertilization is not precisely known, but 27 days is widely quoted from Broom (1895). During this interval the female develops a pouchlike "incubatorium" in which she deposits the egg, probably by curling herself to oppose the cloacal lips to the pouch opening. Griffiths (1968) has recorded an interval of 10.5 days from egg-laying to hatching within the pouch. Taking the growth data from Heck (1908) and Griffiths (1968), it is possible to derive a developmental schedule (see fig. 6).

The young is brooded and nursed in the incubatorium for at least two months. During this time the female maintains a high body temperature. At about 10 weeks of age, the young begins to develop spines, and the eyes are open by 12 weeks. At this time the young is "cast out" from the pouch and begins to live in the natal burrow, where it will be fed by the mother at 24- to 36-hour intervals. This absentee parental care system may continue for at least 10 more weeks. Weaning occurs at 140 to 150 days of age, and during this period the young may begin to follow the female on her excursions. *T. a. multiaculeatus* may have a developmental schedule some 4 weeks in advance of that of *T. a. aculeatus* (see fig. 6).

Since echidnas are strong seasonal breeders over most of their range, a female could not rear more than a single young per year. Aside from the mother-young association and the period of mating, there appears to be little social interaction. At the same time, field studies employing radio tracking have shown no evidence of territoriality (Augee, Ealey, and Price 1975).

The living monotremes exhibit two distinct variations in maternal care. Echidnas carry the egg and subsequent young in a pouchlike

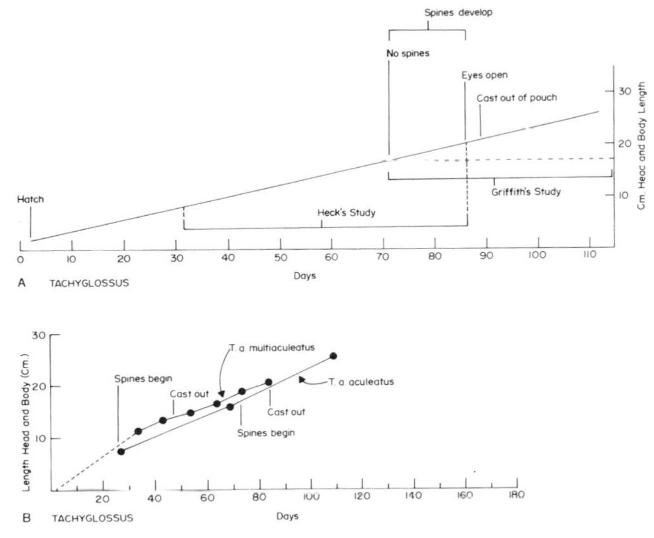

Figure 6. Growth and development in *Tachyglossus*. Adapted from Griffiths 1968 and Heck 1908. (*a*) Reconstruction of probable time course of development in *Tachyglossus*, based on a combination of data from Heck 1908 and Griffiths 1968. (*b*) Plot of growth data for two subspecies of *Tachyglossus aculeatus*, from Griffiths 1968.

structure that develops at each annual breeding season. The platypus constructs a nest in a complex tunnel system, where the eggs and young are periodically brooded. The monotremes are relatively long-lived and compensate for their low reproductive rate with an extended reproductive life. Maternal care is complex and intense.

3.1.3 The Adaptive Syndrome of *Tachyglossus*

Tachyglossus aculeatus is adapted for feeding on ants and termites. Griffiths (1968) reports that ants constitute 80 percent and termites less than 20 percent of its annual intake. Termite feeding appears to fluctuate in intensity, depending on abundance. Some species of ants (*Iridomyrmex detectus*) may be intensively fed upon at restricted seasons when the fat-rich imagoes are emerging. Over much of the echidna's natural range, the seasonal flux in mean temperature and rainfall influences its rate of movement, feeding activity, and reproduction. There may be seasonal torpor or "hibernation," when the animal draws on fat reserves for maintenance energy. Activity patterns correlate strongly with the ambient temperature, and temperature extremes are avoided (Augee, Ealey, and Price 1975). The home ranges of neighbors overlap considerably, though they forage alone. There is no evidence of territorial defense. One average home range in southern Australia was recorded at 50.6 ha (Augee, Ealey, and Price 1975). Adults showed remarkably stable home range use patterns over a twenty-five month study period.

Zaglossus feeds on earthworms and soil arthropods (Griffiths 1978). In superficial morphology it resembles *Tachyglossus,* but the anatomy of the tongue and hyoid bone is quite different.

The behavior patterns during feeding and general maintenance have been described by Brattstrom (1973), Griffiths (1968), and Hediger and Kummer (1961). The repertoire seems simple compared with that of the platypus (see following section), but this probably reflects their peculiar anatomical specializations rather than a basic simplicity of repertoire that might be thought of as accompanying their conservative anatomical features (see table 2).

Table 2 Communication Mechanisms for the Monotremata

Species	Chemical Signals and Marking	Sound Production	Tactile Behavior
Ornithorhynchus anatinus	♂ cloacal gland deposits on underwater rocks. ♂ sternal gland larger than ♀.	Shrill growl—defensive. Sucking sounds (young). Hissing.	Rubbing; crawling over—mutual. ♂ tail grasp with bill. ♂ neck grip with bill. ♂ allogrooming of ♀.
Tachyglossus aculeatus	♂ cloacal gland deposits.	Sniffs of long and short duration. Sucking sounds (young).	Pushing; butting.

As an antipredator strategy, the echidna rolls itself into a spiny, impregnable ball. Such an elaborate defense suggests a long-term selective pressure from avian or mammalian predators. The great modification of the forepaws for digging precludes their use in manipulating objects or in elaborate grooming patterns. Autogrooming is accomplished by scratching with the hind foot. The peculiar mobility of the hip joint allows the animal to reach almost the entire body surface (see Brattstrom 1973, p. 54). Brattstrom describes dust bathing in depressions in the ground, but its functional significance is unknown.

3.1.4 The Adaptive Syndrome of *Ornithorhynchus anatinus*

The platypus is well adapted to an aquatic life, having webbed feet, reduced external ears, and a fusiform body. The long claws permit efficient digging, and the pecular "bill" is an adaptation for feeding on bottom-dwelling freshwater crustaceans, worms, and other aquatic arthropods. In captivity it also eats frogs (Crandall 1958). Foraging in streams in the evening or at night and sleeping in a bank burrow during the better part of the day seems to be the normal activity pattern. The animals may move overland when streams dry up and during dispersion of subadults. In Australia there are no other semiaquatic mammals except *Hydromys* (Rodentia), and the niche appears to be uniquely occupied by *Ornithorhynchus*.

Adults seem to occupy individual burrow systems, and females are intolerant of males outside the breeding season. Fat storage and seasonal torpor occur in the southern parts of the range in Victoria (Fleay 1944), but in New South Wales seasonal torpor may be absent (Burrell 1927). When not in torpor, the platypus is very active and maintains a high body temperature. Tremendous seasonal fluctuations in weight can occur, and food intake also varies depending on the season and the stage of the reproductive cycle (Fleay 1944).

The behavioral repertoire is rather complex. Swimming after and catching prey in the water demands fine coordination, with hearing and tactile sensitivity very highly developed. Antipredator behavior consists of flight but, when confronted in the burrow, platypuses will growl and even bite (see table 2). The female transports nesting material in her ventrally folded tail. This pattern will be noted again for the Marsupialia. Burrow construction is elaborate, with nest chambers and plugs of earth (Burrell 1927).

The care of the body surface involves scratching and nibbling with the "beak." Strahan and Thomas (1975) report that before mating the female shows an increased incidence of cloacal grooming. During courtship the male has been observed to groom the female (Crandall 1958).

3.1.5 Synopsis

The living monotremes are extremely specialized for two distinct feeding niches: terrestrial myrmecophagy and semiaquatic crustacivority/omnivority. Both families show adaptations for digging and burrowing, but no special adaptations for climbing are demonstrable. The spurs of male *Ornithorhynchus,* which connect to functional venom glands (Calaby 1968), are unique to living mammals. The laying of a cleidoic egg and its subsequent incubation may be considered plesiomorph attributes. The very low rate of reproduction coupled with a rather extended longevity suggests a "K" strategy of adaptation. The relatively high encephalization quotient of *Ornithorhynchus* correlates with its semiaquatic specializations.

3.2 Metatheria

In 1949 I paid the rather stiff price of five dollars to purchase a living opossum from South Carolina. On its arrival in the state of Washington, I made my first acquaintance with a marsupial. Since that time various specimens of *Didelphis virginiana* have been guests on and off at the laboratory and at my home. Some fifteen years after my initial purchase, the metatherians had become of great theoretical interest to all of us at the National Zoological Park, and we began to assemble a collection for comparative studies in Washington, D.C. We concentrated on the Didelphidae, Dasyuridae, and Phalangeridae, but the exhibit collection also had some representatives of the Vombatidae and Macropodidae.

Our efforts led to a manual of captive management (Collins 1973) and several papers on communication and movement analysis (Eisenberg, Collins, and Wemmer 1975; Wemmer and Collins 1978; Eisenberg and Golani 1977; Golani 1976). We came to recognize the pivotal importance of the metatherians in comparative behavior studies and were all personally enriched for the efforts (see plates 2 and 3). Recently M. A. O'Connell (1979) has undertaken a study of the comparative ecology

Plate 2. *Dasyurus viverrinus,* the Australian native cat, is a medium-sized marsupial predator showing a convergence toward the eutherian civets in its body form and predatory habits. (Courtesy of Ian Eberhard.)

Plate 3. *Dasyuroides byrnei,* Byrne's marsupial mouse, is a versatile small predator of semiarid Australia. The animal is a digitigrade locomotor, and its lengthened hind foot permits quadrupedal saltation. The ornamented tail provides a visual signal during social interactions. (Courtesy of Larry Collins.)

of a set of sympatric didelphids in Venezuela that will greatly advance our knowledge of the ecology of the South American radiation.

The family Didelphidae is recognizable in the Cretaceous of Europe and North America. We of course have no way of knowing how the early didelphids behaved, but it seems safe to assume that, given a relatively unchanged anatomy, their behavior was in some measure similar to that of present-day species of the genus *Didelphis*. *Didelphis* is only one genus of the New World family Didelphidae. Together with the Caenolestidae and the Microbiotheriidae (Kirsch 1977), the didelphids make up the living remnant of the marsupials in the New World (see fig. 7). Formerly the marsupials were more diversified in North America, Europe, and South America. The northern radiations declined at the end of the Eocene but persisted in South America. With the completion of the Pliocene land bridge between North and South America, the carnivorous Neotropical Borhyaenidae became extinct (see Patterson and Pasqual 1972). The remaining living marsupials are confined to Australia, Tasmania, New Guinea, Celebes, and a few offshore islands. How this curious distribution came about has been the subject of much debate (see chap. 2), and recently Kirsch (pers. comm.) has proposed an Antarctic origin. In the absence of any fossils, the hypothesis remains untested (see Keast 1977 for a review).

Following Simpson (1945), the taxonomic scheme for the marsupials would break down as subclass Theria, infraclass Metatheria, and order Marsupialia. Ride (1964) proposed that the Metatheria be divided into four orders that would more properly reflect the antiquity of the different lines of descent. The subdivision of the Metatheria has been accepted and amplified by Kirsch and Calaby (1977) and Kirsch (1977).

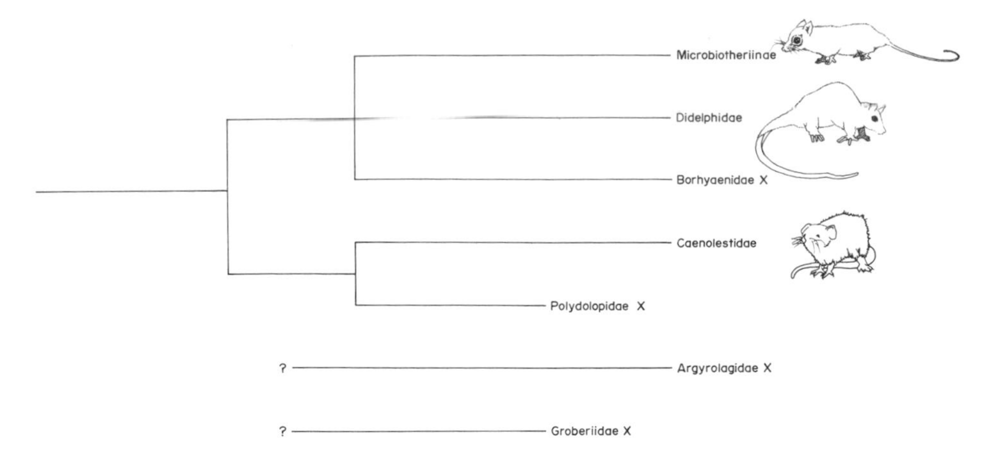

Figure 7. Cladogram illustrating the radiations of the marsupials in the New World (the Didelphoidea). Adapted from Patterson and Pasqual 1972 and Kirsch and Calaby 1977.

The current outline of living and extinct orders and families is included in figures 7 and 8.

The later radiations of the marsupials took place on continental land masses that were isolated from those more contiguous masses in the northern hemisphere. They reached a peak of diversity in South America before the formation of the Pliocene land bridge. The diversity of marsupials declined progressively through the Pliocene and Pleistocene, partly in conjunction with the movement of North American forms into the southern hemisphere.

On the island continent of Australia, with its associated large offshore islands (New Guinea and Tasmania), the marsupial radiation proceeded much longer in isolation from eutherian mammals. Bats (Chiroptera) invaded Australia rather early, and muroid rodents followed somewhat later. Until the arrival of the Australian aboriginals, perhaps some 25,000 years ago, no other eutherian mammals were present on the continent. Since the marsupials have radiated to fill a wide variety of ecological niches, and since this radiation has been accomplished utilizing the same fundamental body plan, the marsupials are the only "control" group with which we can test hypotheses about the evolution of behavior within the eutherian mammals.

The marsupials show a number of typically mammalian anatomical features, including a heterodont dentition in the adult, but they retain a number of plesiomorph features. Typically the bony portion of the palate is fenestrated and the jugal bone participates in the articulation of the lower jaw. As with the monotremes, epipubic bones are present, supposedly to support the pouch. Recently Bakker (pers. comm.) has proposed that the epipubic bones really function as sites for abdominal muscle attachment. This attachment site permits a rapid change in the volume of the abdominal cavity, thus assisting the diaphragm during periods of hyperventilation. The implication is that the whole epipubic

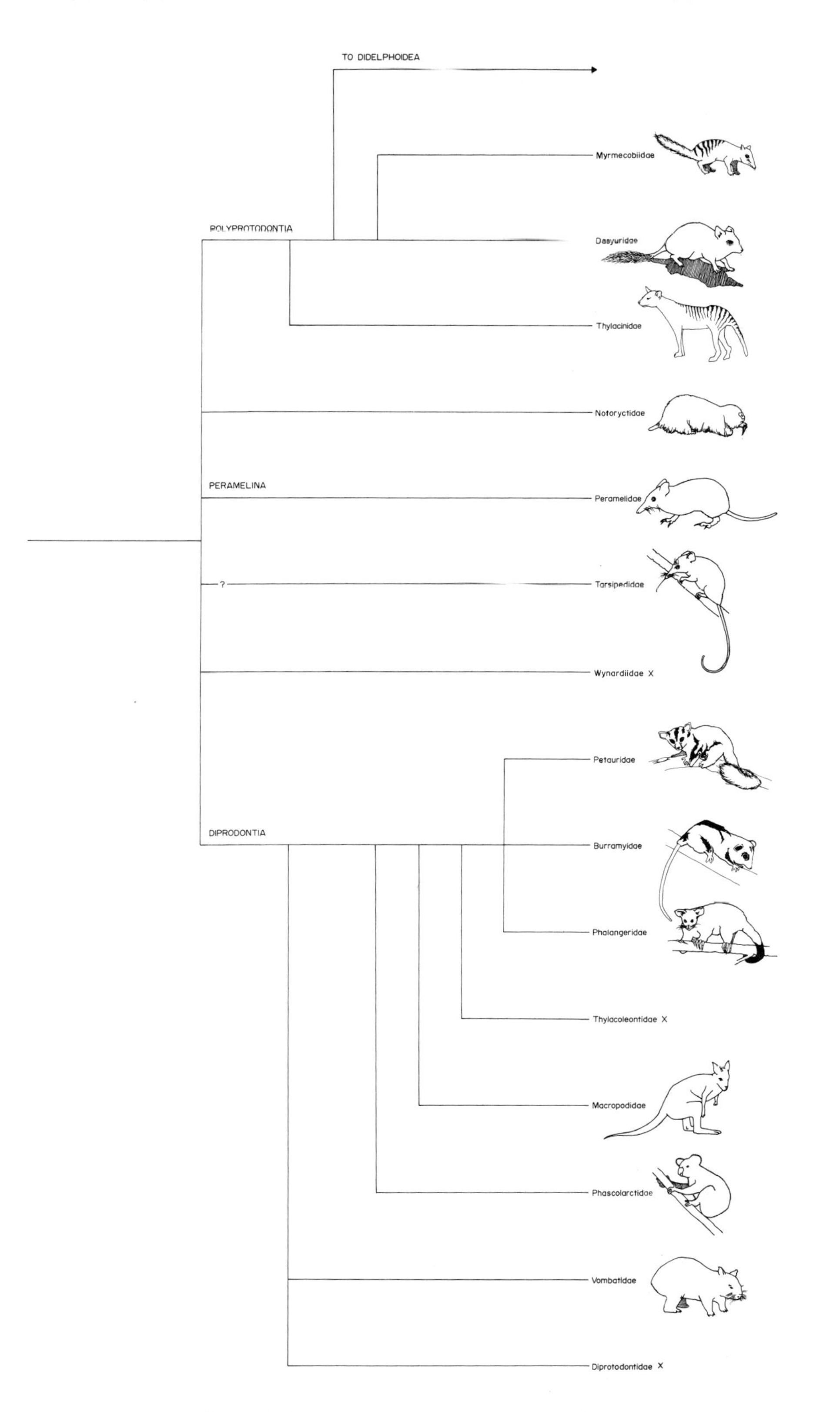

Figure 8. Cladogram illustrating the radiation of the marsupials. Living families found in Australia are portrayed (see fig. 7 for the New World radiation). The Caenolestidae are not illustrated in this figure. Adapted from Kirsch 1977.

bone-muscle complex is an adaptation for increasing ventilation rate and is not directly involved in pouch support.

Differences in the structure of the reproductive tract separate the Metatheria from the Eutheria (Tyndale-Biscoe 1973). The males have a well-developed prepenial scrotum. The structure, development, and position of the scrotum are so unusual as to suggest an independent evolution from scrotal development in the line leading to the Eutheria. The females have a double vagina connected to a bipartite uterus. Given the morphological differences in the reproductive tract, there is yet an additional morphological separation between metatherians and eutherians, though of a transient nature—the way material is exchanged between the developing intrauterine young and the products of the female's uterine glands and body fluids. The late retention of an egg membrane in the Metatheria and the development of functional, noninvasive yolk sac placenta stand in stark contrast to the eutherian mode of placenta formation. This involves the development of a trophoblast and the formation of an intimate connection between the fetal and maternal tissues, allowing for sustained and efficient transport of molecules and ions. Even in those marsupials that have developed a chorioallantoic placenta (the bandicoots), the length of time maternal and fetal tissues are conjoined is extremely short. The intrauterine duration of the conceptus in the Marsupialia is brief compared to that in the eutherians, and in general gestation in the marsupial is less than the interval between two estrous periods. The developing embryo utilizes yolk in its early development and, as a result, is born in an extremely altricial state. It appears that most families of the Marsupialia have not solved the problem of prolonging intrauterine gestation with an efficient nutrient exchange between the mother and the young (Moors 1974).

Luckett (1975) outlines the early embryology of the Marsupialia, pointing out that they are in essence ovoviviparous. The association between maternal tissue and neonatal tissue is curtailed, and Parker (1977) has suggested that the birth of marsupials may be associated with leukocyte buildup at the point of maternal-neonatal tissue contact. Moors (1974) has pointed out that, in the absence of a eutherianlike trophoblast, the embryonic marsupial may provoke an immunorejection response from the maternal tissue, thus eliminating any possibility of a prolonged gestation.

At birth the young marsupial, though very altricial, has well-developed forearms and transports itself into the teat area, which in many marsupial females is enclosed in a fold of skin termed the "pouch." A pouch in a pregnant female is not a diagnostic character for the Marsupialia, since many forms in the families Didelphidae and Dasyuridae show no pouch development. What is common in all marsupial postpartum development is that the young attaches to a teat and remains attached for a relative long period while it undergoes further growth.

3.2.1 Some Life-History Sketches for Selected Marsupials

The living marsupials present a rich assemblage of species showing a variety of feeding adaptations. In Australia almost all types of niche occupancy are demonstrable. To develop a portrait of natural history trends, I have selected five species ranging from insectivorous adaptation to a grazing, terrestrial herbivore. Table 3 serves as a guide to the literature.

3.2.2 *Antechinus stuartii*, Stuart's Marsupial "Mouse"

The following account is drawn from Marlow (1961), Wood (1970), and Braithwaite (1974) (see table 3). These small dasyurid marsupials inhabit the eastern forests of Australia from the Cape York peninsula to Adelaide. Although able to climb well, they forage to a considerable extent on the ground. They build nests of leaves in sheltered spots such as hollow trees, rock crevices, and even the abandoned nests of birds. Over large parts of its range, this species is strongly seasonal in its reproductive activity. Although animals may be maintained in captivity for more than two years, most males become senile after passing through their first reproductive season and die shortly thereafter (Lee, Bradley, and Braithwaite 1977). Males are extremely aggressive toward one another during the breeding season. The gestation period is from 23 to 27 days. The litter size is rather large, averaging between 10 and 12 young per litter. The course of development following birth is similar to that of the mouse opossum, *Marmosa* (see sect. 3.2.3). The eyes open by 62 days of age, and by 84 days all teeth have erupted. The young are fully furred and actively foraging on their own by 90 days of age. They usually become sexually mature in the following reproductive season, when they are approximately 9 to 10 months of age. It is improbable that adult males in the wild survive to a second breeding season. Female survival to a second breeding season is also improbable in the wild state but may occur in captivity. Some species of this genus, however, are adapted to reproduce over several seasons.

3.2.3 *Marmosa robinsoni*, the Mouse Opossum

This small, arboreal didelphid feeds on insects and fruit, foraging principally at night in trees and shrubs (see plate 4). The head and body length averages about 90 mm, and adult males may exceed females in size and weight by 5 percent. The basal metabolic rate for *M. robinsoni* is not too different from that of a eutherian of comparable size, but some species of *Marmosa* may show seasonal torpor (Morrison and McNab 1962). The smaller species of *Marmosa* rarely live more than 20 months. Females become reproductively senescent after 16–18 months (Barnes and Barthold 1969). Resident adult females may have a home range with a core area of 0.22 ha (O'Connell 1979). Aspects of their behavior have been summarized by Barnes and Barthold (1969) and Hunsaker and Shupe (1977).

Maintenance behaviors involve elaborate grooming patterns, including ritualized face "washing" and marking with sternal glands. Prey-

Table 3 Recent Literature on the Monotremata and Marsupialia

Taxon	Ethology	Ecology	Reproduction	General
Monotremata				Collins 1973 Griffiths 1978
Ornithorhynchus anatinus	Crandall 1958	Burrell 1927 Temple-Smith 1974	Strahan and Thomas 1975 Fleay 1944	
Tachyglossus aculeatus	Griffiths 1968 Brattstrom 1973 Hediger and Kummer 1961	Augee, Ealey, and Price 1975	Griffiths 1968 Dobroruka 1960	
Marsupialia	Hediger 1958 Eisenberg and Golani 1977		Sharman 1959, 1963	Collins 1973 Tyndale-Biscoe 1973
Polyprotodontia				
Didelphidae	Hunsaker and Shupe 1977	Hunsaker 1977		
Didelphis virginiana[a]	Francq 1969 McManus 1970		Reynolds 1952 McManus 1967	
Marmosa robinsoni (= mitis)		Enders 1935	Barnes and Barthold 1969 Eisenberg and Maliniak 1967	
Dasyuridae				Woolley 1966
Antechinus stuarti	Braithwaite 1974	Wood 1970	Lee, Bradley, and Braithwaite 1977	
A. flavipes			Marlow 1961	
Phascogale tapoatafa	Cuttle 1978			
Dasyuroides byrnei	Aslin 1974			
Dasycercus cristicauda	Sorenson 1970		Michener 1969	
Sminthopsis crassicaudata	Ewer 1968*a*	Morton 1978*a,b,c*		
Sarcophilus harrisii	Golani 1976 Eisenberg, Collins, and Wemmer 1975	Guiler 1970*a,b*	Fleay 1935	
Myrmecobiidae				
Myrmecobius fasciatus		Calaby 1960		
Peramelidae	Stodart 1977		Stodart 1977	
Perameles gunni		Heinsohn 1966	Stodart 1966*a*	
Isoodon obesulus		Heinsohn 1966	Lyne 1952	
Diprotodontia				
Phalangeridae				How 1978
Trichosurus vulpecula	Biggins and Overstreet 1978	Ward 1978 Dunnett 1956, 1964	Lyne, Pilton, and Sharman 1959 Pilton and Sharman 1962	
Burramyidae				
Cercartetus sp.	Hickman and Hickman 1960	Bartholomew and Hudson 1962	Clark 1967	
Petauridae				
Petaurus breviceps	Schultze-Westrum 1965, 1969			Smith 1973
Pseudocheirus peregrinus		Thomson and Owen 1964	Hughes, Thompson, and Owen 1965	
Schoinobates volans		Tyndale-Biscoe and Smith 1969*a,b*		

[a] Formerly *Didelphis marsupialis*.

Table 3—*Continued*

Taxon	Ethology	Ecology	Reproduction	General
Vombatidae				Wünschmann 1970
Lasiorhinus latifrons	Wünschmann 1966		Crowcroft and Soderlund 1977	
Vombatus ursinus				
Tarsipedidae				
Tarsipes spenceri	Vose 1973			
Phascolarctidae				
Phascolarctos cinereus		Eberhard 1978		
Macropodidae	Kaufmann 1974a,c			Frith and Calaby 1969
Potorous tridactylus		Heinsohn 1968		
Bettongia lesueur	Stodart 1966*b*			
Setonix brachyurus	Packer 1969	Dunnett, 1962, 1963		
Macropus giganteus	Grant 1974 Kirkpatrick 1966	Kirkpatrick 1965*a,b,c*		
M. parryi	Kaufmann 1974*b*			
M. robustus		Ealey 1967*a,b*	Sadleir 1965	
M. rufus	Russell 1970, 1973	Newsome 1965	Sharman and Calaby 1964	
Dendrolagus matschiei	Schneider 1954		Olds and Collins 1973	
Wallabia rufogrisea	LaFollette 1971			

Plate 4. *Marmosa robinsoni,* the mouse opossum. This New World marsupial is strongly adapted for an arboreal existence. The long tail is prehensile, and the pollex is opposable to the other digits. (Courtesy of Eugene Maliniak.)

catching has been described by Eisenberg and Leyhausen (1972). Reproductive behavior involves moderate courtship and a prolonged copulation period (50 min to more than 5 hr). Such prolonged intromission times are characteristic of morphologically conservative marsupials and eutherians (Eisenberg 1977).

The rearing cycle (see fig. 9) takes at least 70 days. Intrauterine development lasts 14 days, followed by a teat-attachment phase that persists for some 28–30 days after birth, or 42–44 days after conception.

The young are fully weaned at 53–60 days postpartum or 67–74 days postconception. During the interval between the end of the obligate teat-attachment phase and the end of lactation, the young pass through a "nest phase" where they remain in the nest while the female forages. During the nest phase the young begin to make excursions from the nest at about 56 days postpartum. Occasional following of the mother begins between 45 and 50 days of age.

3.2.4 Semelparous Reproductive Strategies in Small Marsupials

The article by Wodinsky (1977) concerning the hormonal basis and interrelatedness of the onset of reproduction and subsequent mortality in *Octopus hummelincki* sets the stage for this section. Mean age at first reproduction, litter size, mean number of litters per lifetime, and effective reproductive life-span are all interrelated. Let us consider those cases where an individual is "genetically programmed" to survive only one brief season of reproduction. We are then confronting the proposed phenomenon of semelparity in mammals (see Lee, Bradley, and Braithwaite 1977; Braithwaite and Lee 1979). The modal paradigm is the genus *Antechinus*. Males exhibit a life strategy as if they were semelparous, since as a rule *A. stuartii* and *A. swainsonii* males die some 3–4 weeks after becoming sexually active at about 11 months of age. This phenomenon is confined only to those species of *Antechinus* that are adapted to environments having a predictable insect "bloom" at a given season of the year. The females may approximate semelparity, but they usually live somewhat longer than the males and may pass through a second breeding season. Where else have we found

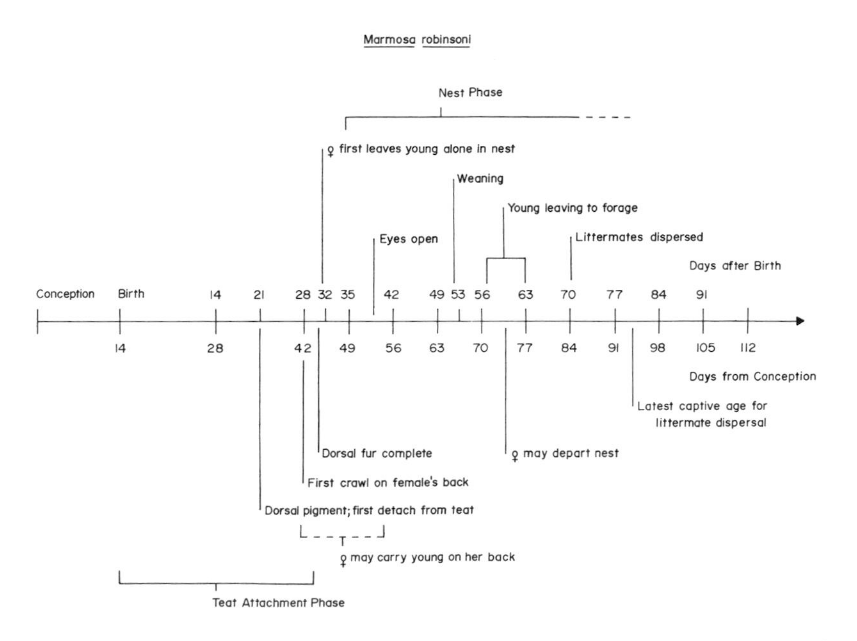

Figure 9. The course of development for *Marmosa robinsoni*. Based on both field and captive data (Collins 1973; Eisenberg, unpublished; O'Connell 1979).

examples of semelparity in mammals? I suggest that the genus *Marmosa* may show such a phenomenon in some habitats (O'Connell 1979) and that some species of tenrecoid insectivores come close to the phenomenon (see chap. 23). (Low 1978 attempts to discuss semelparity in marsupials as peculiar to marsupial reproduction, but this is somewhat misleading.)

According to Barnes and Barthold (1969), the female *Marmosa robinsoni* reaches sexual maturity at about 5 to 6 months of age. She begins to show estrous cycles with an interval of 18 to 30 days. Should she become pregnant, the entire rearing cycle from conception to weaning takes some 77 days at the minimum. Thus a female would complete her first rearing cycle at 7½ to 8½ months of age. If the habitat shows a strong seasonality in productivity, as is the case in the Venezuelan llanos, the female may have an opportunity to raise only a single litter, since her ovaries become inactive and senescent at 18 months of age. The opportunity for a second litter may occur for a few females, but in general few animals survive more than one season (O'Connell 1979). While this is not a true case of semelparity, it is an approximation of the condition and should be recognized as such.

3.2.5 *Sarcophilus harrisii*, the Tasmanian Devil

The following account is based on Guiler (1970*a,b*). Currently *Sarcophilus harrisii* is confined to the island of Tasmania, but in former times it was distributed in eastern Australia. In Tasmania the animals occupy a variety of forested habitats. They may seek shelter in hollow logs, caves, or other natural cavities, and they generally build nests by transporting leaves or grass to the shelter site. The animal is a generalized carnivore feeding on a variety of vertebrates and invertebrates, and it will also eat carrion. The young climb well, but the adults seldom climb and are more or less terrestrial in their habits. Sexual maturity is reached at approximately 2 years of age. For mammals of their size, Tasmanian devils are not particularly long-lived; captive longevity rarely exceeds 6 years.

In the southern hemisphere *Sarcophilus* breeds in March. The gestation period lasts approximately 31 days. The newborns transport themselves to the mother's pouch and remain there for the first 100 days of life. The eyes open at approximately 90 days, and from about 105 days on the young are no longer transported but remain in a nest. Weaning occurs at approximately 140 days, whereupon the young begin to forage on their own.

The home range of the Tasmanian devil is rather large, with male home ranges larger than those of females. There is no evidence for territoriality; rather, home range overlap by the two sexes is common. There does not appear to be any enduring social relationship beyond the mother-young bond. There is no sustained pair-bonding, and there is little evidence that the male *Sarcophilus* participates in parental care.

3.2.6 *Petaurus breviceps*, the Sugar Glider

The life-history synopsis is derived from Smith (1973). Sugar gliders are widely distributed from Queensland to South Australia in the forests to the east of the Great Dividing Range. They are also found on New Guinea and have been introduced to Tasmania. *Petaurus* is strongly adapted for an arboreal life and has a fold of skin, or patagium, joining the fore- and hind limbs that permits it to glide. *Petaurus* feeds on nectar, flowers, and sap, but insects may be included in its diet, as well as small vertebrates. It is thought that *Petaurus* may occupy a feeding niche similar to that occupied by lemurs of the genus *Phaner* on Madagascar. In its mode of feeding and in its presumed role as a pollinator, *Petaurus* may represent a relict animal pollinator (Sussman and Raven 1978).

Petaurus generally nests in tree holes and constructs a nest of dry leaves. A male and a female may use the same nest throughout the rearing phase. There is a tendency for a family group to occupy a single nest site, and young may remain with the parents until they are sexually mature. The gestation period is less than 27 days, and litter size ranges from one to three. The young make their way to the mother's pouch, where they remain until approximately 60 days of age. Between 60 and 70 days, the young begin their nest phase, nursing when the mother returns from foraging. The eyes open at 80 days of age. At approximately 120 days of age, the young are weaned. *Petaurus*, in common with other small phalangers, shows a tendency to develop small closed social groups based on a family (Schultze-Westrum 1965), though it is not known whether the mating system is monogamous.

3.2.7 *Macropus rufus* (=*Megaleia rufa*), the Red Kangaroo

Frith and Calaby (1969) provide the basic data for this synopsis. This large kangaroo species is adapted for grazing in the semiarid central regions of Australia. They are crepuscular in their activity cycles, generally resting during the midday heat. Red kangaroos live in bisexual social groupings that seem to be somewhat plastic in structure, with individuals joining and leaving and larger groups subdividing into smaller subgroups; however, the stable social unit within a mob tends to be an adult female, her subadult offspring, and her dependent young. Thus there is some continuity in social relationships based on kinship.

The species is strongly dimorphic. Adult males contend among themselves for dominance. Dominant males have priority access to estrous females. The young is born after a 32-day gestation period and makes its way unaided to the mother's pouch, where it attaches to a teat to undergo development for 144 days, at which time the eyes open. At 150 days of age, the young first protrudes its head from the pouch, and it begins to leave the pouch at 190 days of age. By approximately 240 days of age the young is independent of the pouch, but it continues to follow the female for some months thereafter (see plate 5).

Females become sexually mature at approximately 18 months of age; males are sexually mature at 2 years but are not sociologically mature

Plate 5. *Macropus rufus* (= *Megaleia rufa*), the red kangaroo, is one of the largest of the living Marsupialia. This bipedally locomoting grazer of central Australia has converged in social and feeding habits toward the small grazing eutherians, such as gazelles. (NZP archives.)

until about 5 years of age, when full growth is attained. The larger Macropodidae have developed the most complex social organizations found in the Marsupialia (Kaufmann 1974*a,c*). In many respects the social structure they exhibit has converged toward forms of social organization that one finds among the grazing ungulates of the orders Artiodactyla and Perissodactyla on the contiguous continental land masses.

The preceding sketches can only begin to give an impression of the life histories of metatherians. Appendixes 2–4 summarize longevity, body size, and reproductive characteristics. I have included in table

3 the recent literature on marsupial and monotreme life histories as a guide to further reading, and I have drawn heavily on the authors cited there for the summary sections that follow.

3.2.8 Behavior Patterns and Communication in the Marsupials

When a typical marsupial is compared with a eutherian, we find several obvious anatomical differences ultimately related to their behavior. The relative brain size of many marsupials is somewhat less than the brain size of eutherians of comparable body weight (see chap. 21), which may have implications for behavioral organization. The sense organs are comparable to those of conservatve eutherians. Complex social interactions in the Marsupialia must be mediated by signaling systems and sense organs in the same manner as for the Eutheria. The analysis of communication systems in the Metatheria is still in its infancy, and the relevant literature has been reviewed elsewhere (Eisenberg and Golani 1977). In this review we developed the idea that the analysis of interaction patterns in mammals could well proceed by two routes. On the one hand, interactions between individuals can be studied to ascertain the stereotyped postures, movements, and sounds that accompany certain qualitatively different forms of interaction. Examples of extreme stereotypy in sound production or posture may be referred to as displays.

On the other hand, the sequencing of motor interactions may be described from the standpoint of several defined analytical frameworks. Golani has pointed out that interaction sequences may be described in terms of how contact points change on the animal's own body as a consequence of adjustments of the partner. Golani proposed that we consider the establishment of contact and the persistence of forms of contact as the development of "joints" between two interacting individuals (Golani 1976). He points out that often the actual contact between two individuals is what is fixed in the relationship, and thus that angular relationships between interacting individuals can vary considerably. Thus "joints" and the trajectories of movement to the next "joint" can account for the positions and movements of the animals' heads, with the rest of their bodies being carried along. A behavioral "joint" can be maintained through a variety of mechanisms, the most important apparently being either maintenance of a mechanical connection or the steady maintenance of a particular sensory input without necessary mechanical input.

An alternative analytical approach to the presumptive signal systems is to classify them according to the sensory modality involved. Thus it is convenient to consider tactile input, olfactory input, auditory input, and visual input. In the review by Eisenberg and Golani (1977), we indicated how important it is that certain kinds of tactile input be maintained during the interaction forms of many of the Marsupialia. Development of contact patterns including allogrooming give rise to interactive configurations in the Marsupialia that are analogous and probably homologous to similar patterns demonstrable for the Eutheria.

The utilization of olfactory signals in the coordination of marsupial behavior is of paramount importance. One of the first analyses of olfactory communication in the Marsupialia was that performed by Schultze-Westrum (1965) in his analysis of social behavior in *Petaurus breviceps*. Since this pioneering study, other workers have begun to describe the methods of chemical deposition and the glandular areas that are of importance in scent-marking. Specialized glandular areas and marking movements are only one aspect of chemical communication. We have noted that urine, feces, and saliva are also important in passing chemical information during social interactions in marsupials (Eisenberg and Golani 1977) (see table 4).

Sound production in marsupials seems to involve the use of four basic syllable types that provide information concerning the location of a conspecific and that indicate mood, such as aggressive arousal or submissive tendencies. Vocalization patterns may be rather elaborate in a species such as *Sarcophilus harrisii* (Eisenberg, Collins, and Wemmer 1975), or the system may show little elaboration, as in the opossum, *Didelphis virginiana* (McManus 1970).

The analysis of the role of the visual channel in integrating information in the Marsupialia is still in its infancy. Some descriptive data suggest that stereotyped postures ("displays") convey information to the presumptive receiver. This is especially true when one analyzes interaction norms for the macropod marsupials. Visual input seems to be of reduced importance in nocturnal forms, a finding consistent with similar studies of the Eutheria. In comparing the communication systems of the Metatheria and Eutheria, there is great similarity in postures and configurations. The most direct comparison of interaction systems involves the study by Golani (1976) with the Tasmanian devil, *Sarcophilus harrisii*, and the golden jackal, *Canis aureus*. Golani concluded from his analysis that, if one considers only postures and interaction forms as static entities, then there is very little difference in the complexity of the repertoires of *Canis* and *Sarcophilus*. But, if one uses his approach by emphasizing the dynamics of the motor interaction sequences, then there are major differences. These involve the following: (*a*) *Sarcophilus* maintains configurations through actual mechanical contact to a greater extent than does *Canis*; (*b*) when maintaining a contact point with a stationary partner, *Sarcophilus* is more stereotyped and continues to work with its partner while in direct contact rather than at some distance; and finally (*c*) the integration of movements in a coordinated sequence seems to be more stereotyped in *Sarcophilus* and shows less flexibility than is the case with *Canis*. It is unclear from the comparison, however, whether these differences are in part due to anatomical differences when *Canis* and *Sarcophilus* are compared or whether they can be generalized to the Marsupialia and Eutheria as a whole.

3.2.9 Convergence of the Didelphidae and Eutherian Lorisiformes

The family Didelphidae presents an adaptive radiation in the Neotropics that may be compared with independently evolved radiations

Table 4 Communication Systems in the Marsupialia

A. Behaviors

						Sandbathing			Distribution of Saliva		
Taxon	Urine Dribble	Cloacal Drag	Chin Rub	Cheek Rub	Sternal Rub	Ventral Rub	Side Rub	Forehead Rub	Biting of Bark or Twigs	Mouth Wipe or Drool	Face Wash
Didelphis			+ +[a]	+						+	+ +
Marmosa	+	+ +	+ +	+					+		+ +
Sminthopsis	+	+	+[b]			+	+		+		+ +
Dasyuroides	?	+	+	+		+ +	+				+ +
Antechinus	?	+			+	+			+		+ +
Sarcophilus	+	+ +									+ +
Petaurus		+			+			+			+ +
Trichosurus	+	+	+	+	+				+	+	+ +
Macropus giganteus	+				+[c]					+	Modified
Megaleia rufa	+[c]				+[c]					+ +	Modified

Source: Eisenberg and Golani 1977.
[a]On female's neck.
[b]On female's back.
[c]Specialized, see text.
+ Observed.
+ + Strongly expressed.

Table 4—*Continued*

B. Context and Call Types

Taxon	Agonistic				Male Approaching or Following Female	Female Responding to Displaced Young	Juvenile Displaced from Mother	Special Context
	Threat or Alarm at a Distance	Low-Intensity Arousal	Intermediate Arousal	High Arousal				
Didelphis marsupialis		Hiss	Growl	Screech	Click	Click	Chirp	
Marmosa robinsoni		Hiss			Click	?	Chirp	
Dasyuroides byrnei	Chit	Hiss	Chur	Chatter	?		Chirp	
Sminthopsis crassicaudata			Hiss	Churr, squeak (juv.)	"di-di-di"	?	Buzzy call	
Sarcophilus harrisii	Bark	Growl or hiss	Whine, whine-growl	Shriek	Huff or clap	?	?	Foot-stamp
Trichosurus vulpecula	Chatter, "buck-buck"	Hiss	Loud hiss	Screech	Click	?	"Zook-zook"	Postcoital chatter (♂)
Petaurus breviceps	"Wok-wok"		Hiss		?	?		Rattle call
Megaleia rufa	Snort, "ha"		Hiss	Growl	Cluck	Soft cluck	Squeak	Alarm call, "cough," foot-stamp

of mammals in the African and Asian tropics. Many didelphids are arboreal and nocturnal, feeding upon insects, small vertebrates, and fruits. With respect to their activity patterns and habitat utilization, their niche occupancy resembles that of the galagos and lorises in the Old World tropics. The ecology of the small nocturnal prosimians has been analyzed by Charles-Dominique (1977), and aspects of prosimian behavior will be reviewed in section 14.4.

In contrast to the Didelphidae, which have numerous young per litter, the galagos and lorises produce only a single young; furthermore, they are remarkably long-lived, given their small size. If we regress the maximum longevity in months against body size (fig. 10), it can be demonstrated that the potential longevity of the galagos and lorises far exceeds that of the didelphids. The genus *Caluromys* (woolly opossums) is somewhat intermediate.

The encephalization quotient for the didelphids is also much lower than that shown for the lorises and galagos. If one again regresses brain size against body size for the didelphids and the prosimians, there is no overlap between the two sets of points. Although the slopes are similar, the intercepts are completely different (see chap. 21; Eisenberg and Wilson 1981). Given the rather low encephalization quotients (EQ values) for the Marsupialia, it is worthwhile to examine the family Didelphidae in some detail. Table 5 shows the EQs calculated for a series of didelphids. *Caluromys* has on the average one of the largest relative brain sizes, whereas *Didelphis virginiana* has a low encephalization quotient. Table 6 indicates that the antipredator responses of the Didelphidae are to some extent correlated with the relative EQ values. Those species with the lowest encephalization quotients, after

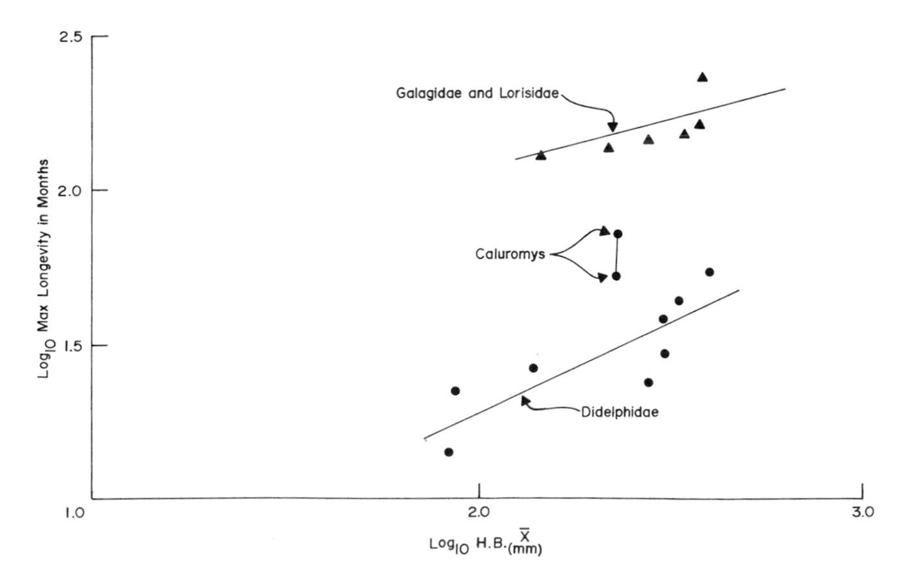

Figure 10. Longevity of didelphids and lorisiform primates plotted against mean body size. Most didelphids show rather short potential life-spans (Eisenberg and Wilson 1981).

Table 5 Encephalization Quotients for the Didelphidae

Species	Mean Body Weight (g)	Mean Cranial Volume (cc)	EQ
Caluromys philander	170	2.69	1.09
C. derbianus	294	3.84	1.04
Monodelphis brevicaudata	84	.91	.62
M. dimidiata	48	.44	.46
Marmosa cinerea	107	1.54	.88
M. constantiae	73	1.34	1.01
M. fuscata	40	.89	1.03
M. impavida	43	.98	1.09
M. mexicana	57	.99	.89
M. murina	54	1.03	.96
M. noctivaga	62	1.15	.97
M. ocellata	41	.93	1.07
M. pusilla	27	.47	.75
M. robinsoni	116	1.32	.71
Philander opossum	309	3.70	.97
Metachirus nudicaudatus	376	2.87	.65
Lutreolina crassicaudata	639	3.80	.59
Didelphis albiventris	1,265	8.39	.79
D. marsupialis	1,176	7.20	.71
D. virginiana	2,420	9.04	.53
Chironectes minimus	749	5.80	.80

Source: Eisenberg and Wilson 1981.

Table 6 Antipredator Responses of the Didelphidae

Species	Response to Being Cornered or Captured
Didelphis virginiana	Gape; hiss; growl; repose[a]
D. marsupialis	Gape; hiss; growl; bite; repose[a]
Metachirus nudicaudatus	Gape; hiss; tremble; tooth-click; rigid, paralyzed stance
Marmosa robinsoni	Gape; churr-hiss; bite
Philander opossum	Gape; churr-hiss; bite
Caluromys philander	Gape; hiss; growl-whine; bite

Note: All species first attempt to freeze or flee, depending on the initial stimulus. *Didelphis* secretes vile odors from the anal glands.

[a]Usually only if gripped by the neck or shaken; generally referred to colloquially as "playing possum."

an initial bluffing display toward a predator, generally show a paralysis if they are gripped and shaken. In colloquial speech this is referred to as "playing possum." On the other hand, those species of didelphids having higher encephalization quotients will generally show an active antipredator behavior involving outright attack and biting. This is especially true for the genus *Caluromys*.

3.2.10 Ecological Adaptations of the Marsupialia

This section considers size and trophic strategy as exemplified by marsupials. These life-history parameters also have predictive value—for example, the area of home range and the mean biomass that a population can achieve. Table 7 lists those marsupials for which enough data exist to allow some generalizations. Small, mouse-sized insectivores such as *Marmosa* and *Antechinus* may reach very high densities, but because they are small they contribute only a small biomass to the total mammalian community. Larger, generalized omnivores such as *Didelphis marsupialis* achieve moderate densities and high biomass levels. Differences in mean densities between populations of *Didelphis* reflect probable differences in carrying capacity. The herbivorous phalangeroid marsupials may exist at moderate to high densities, and they achieve rather high biomass levels. The koala, *Phascolarctos cinereus,* may achieve biomass levels reminiscent of the folivorous primates of the subfamily Colobinae (Eisenberg, Muckenhirn, and Rudran 1972). The large grazing macropod marsupials also achieve rather high densities, given their size, and are significant contributors to mammalian biomass levels.

In table 7 the home range estimates for selected marsupial species are also portrayed, and in figure 11 these data are analyzed in terms of the relationship between home range area and body size. One can see that, for their size, insectivores and carnivores have larger home ranges than herbivores, the same correlation that McNab (1963) was able to demonstrated primarily with a eutherian sample. Furthermore, it is evident that species from temperate-zone habitats have larger home ranges than species with the same trophic strategies in tropical habitats. Perhaps this indirectly reflects the differences in carrying capacity between the tropics and the temperate zone.

3.2.11 Social Structure and Habitat Exploitation

Although the data are sparse compared with those on the Eutheria, we now have sufficient information on the social structure and ecology of marsupials to attempt a synthesis. Given, then, that marsupials show habitat exploitation trends in parallel with their eutherian counterparts, I have attempted an analysis of the correlations between social structure and mode of habitat exploitation for those Marsupialia where sufficient data are available from the literature. Table 8 contains an analysis of the degree of sociability shown in three different contexts, the potential for defending four types of resources, degrees of approach to semelparity, and, finally, the temporal stability of the resources exploited. For further explanation of the analysis, see chapter 33.

There is a moderate correlation between the mobility of a species and the degree to which resources are dispersed in space. The tendency toward a semelparous mode of reproduction correlates strongly with the large litter size that in turn is associated with a high value for the intrinsic rate of natural increase. The potential for forming complex social structures or substructures is greater if the intrinsic rate of natural increase is low; therefore there appears to be a strong correlation be-

Table 7 Home Range, Density, and Biomass Data for Selected Marsupials

Taxon	Mean Head and Body Length (mm)	Unit Weight (g)	Home Range[a] (ha)		Density (no./km²)		Biomass (kg/km²)		Locality	Reference
			Core	Extended	Ecological	Crude	Ecological	Crude		
Didelphidae										
Didelphis virginiana	412	1,000	13.2	—	12	6	12	6	Kansas	Fitch and Sandidge 1953
D. virginiana	412	1,000	4.7	15.5	62	—	62	—	East Texas	Lay 1942
D. marsupialis	400	900	.43	—	132	9	119	8.1	Panama	Fleming 1972
Marmosa robinsoni	90	40	.22	—	225	31	9	1.2	Panama	Fleming 1972
Philander opossum	300	450	.34	—	65	—	29	—	Panama	Fleming 1972
Dasyuridae										
Antechinus stuarti	110	45	.20	—	1,400	—	63	—	Queensland, Australia	Wood 1970
Sarcophilus harrisii	664	5,700	129	517	11.6	—	66	—	Tasmania	Guiler 1970*a*
Peramelidae										
Perameles gunni	312	500	18	25.6	146	44.5	73	22	Tasmania	Heinsohn 1966
Isoodon obesulus	365	560	2.3	5.3	19.8	7.4	11	4	Tasmania	Heinsohn 1966
Phalangeridae										
Schoinobates volans	390	1,200	1.3	—	100	—	120	—	New South Wales, Australia	Tyndale-Biscoe 1973
Trichosurus vulpecula	295	900	.63	6.06	150	—	135	—	Australia	How 1978
Pseudocheirus peregrinus	330	1,000	.37	—	70	—	70	—	Australia	How 1978
Trichosurus caninus	300	900	—	6.3	150	—	135	—	Australia	How 1978
Phascolarctos cinereus	725[b]	11,000	<1.0	—	100	—	1,100	—	Australia	Eberhard 1978
Macropodidae										
Potorous tridactylus	350	1,200	—	—	—	17.7	—	21.3	Tasmania	Heinsohn 1968
Setonix brachyurus	537	3,000	1.6	9.6	511	—	1,533	—	West Australia	Holsworth 1967; Dunnett 1962
Macropus giganteus	1,200[b]	40,000	—	—	—	4–7	—	200	Queensland, Australia	Bell 1973
M. parryi	900[b]	12,000	71[c]	110[c]	47	3	546	36	Queensland, Australia	Kaufmann 1974*b*
Wallabia agilis	900	10,000	—	—	20	—	200	—	Queensland, Australia	Bell 1973

[a]Average of ♂ and ♀ ranges.
[b]Strongly dimorphic species; measurements are average of ♂ and ♀ values.
[c]Group (50) home range.

Table 8 Social Structure and Ecology

Species	Social Unit				Defensibility			Temporal Stability of Resources	Reproductive Rate
	Mating	Parental Care	Feeding	Refuging	Mate	Feeding Site	Shelter		
Ornithorhynchus anatinus	Transient pair	Female only	Single	Single adult	High?	High?	High	Stable	Very low
Tachyglossus aculeatus	Several ♂♂, one ♀	Female only	Single	Single adult	Low	Low	?	Seasonally predictable	Very low
Didelphis virginiana	Transient pair	Female only	Single adult	Single adult	Transient high	Transient high	High	Seasonally predictable	Very high
Antechinus stuarti	Pair, temporary	Female only	Single adult	Single adult	Transient high	High	High	Seasonally predictable	Very high
Perameles nasuta	Transient pair	Female only	Single adult	Single adult	High	High	High	Seasonally moderately predictable	Moderate
Trichosurus vulpecula	Transient pair	Female only	Single adult	Single adult	Moderate	Moderate	High	Variable over range	Low
Phascolarctos cinereus	Transient pair	Female only	Single adult	Single adult	Transient high	Moderate	Moderate	Stable	Low
Macropus parryi	Dominance hierarchy among ♂♂	Female only	Group	Group	High	Low	Low	Moderate	Low

Note: See table 58 for references and further development.

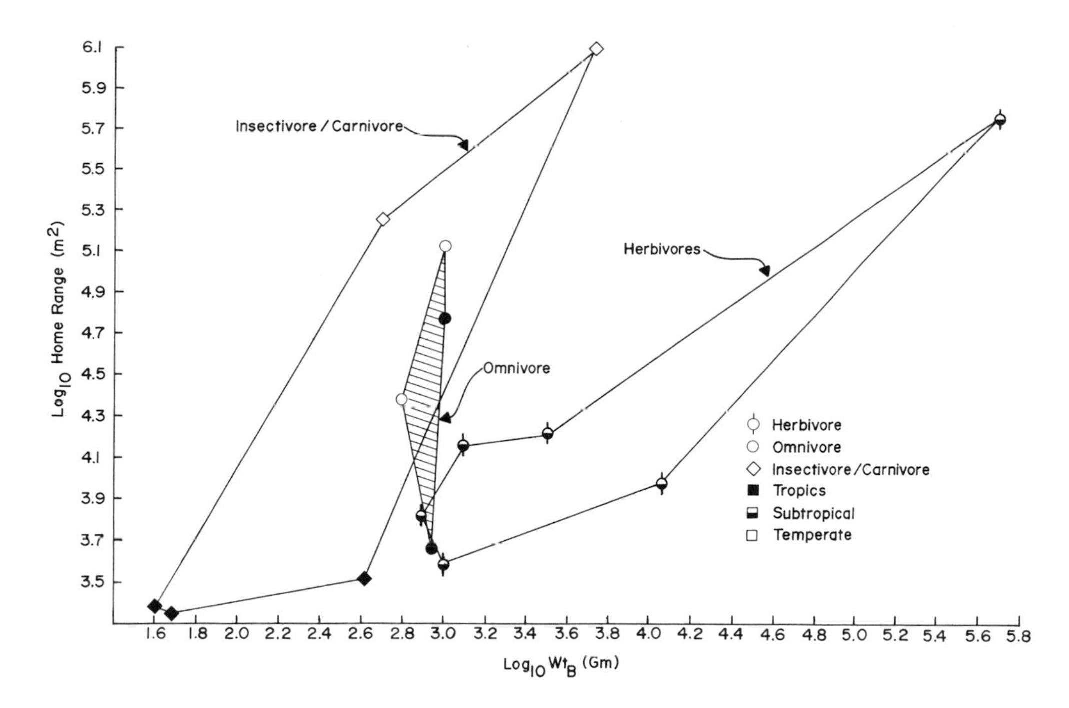

Figure 11. Relationship between home range size, body size, and trophic strategies for selected marsupials (data from table 7).

tween the degree of social complexity manifested by the species and such life-history parameters as length of life and mean percentage of time that the young associate with the parents. The same sorts of trends are demonstrable if one performs a similar analysis with a set of eutherian mammals (see chap. 33).

4 The Edentata

We now come to the eutherian mammals that are characterized by a suppression of the yolk sac placenta during their early embryology and an elaboration of the chorioallantoic placentation that allows the effective exchange of nutrients from the maternal to the fetal circulation. Figure 12 portrays the early radiation of the Eutheria, and one may note that the Neotropical Edentata and African Pholidota separated early in the late Cretaceous. The original ancestors of these two orders may have been distributed in both Africa and South America before their sepearation. The age of the edentate and pholidote lineages allows them to serve as a basis for comparison with the remaining radiations of the eutherians.

The order Edentata radiated in South America between the Paleocene and the Pliocene. The Marsupialia and Meridungulata were their only mammalian contemporaries in South America until the Oligocene. During this interval, several lines evolved that included the following

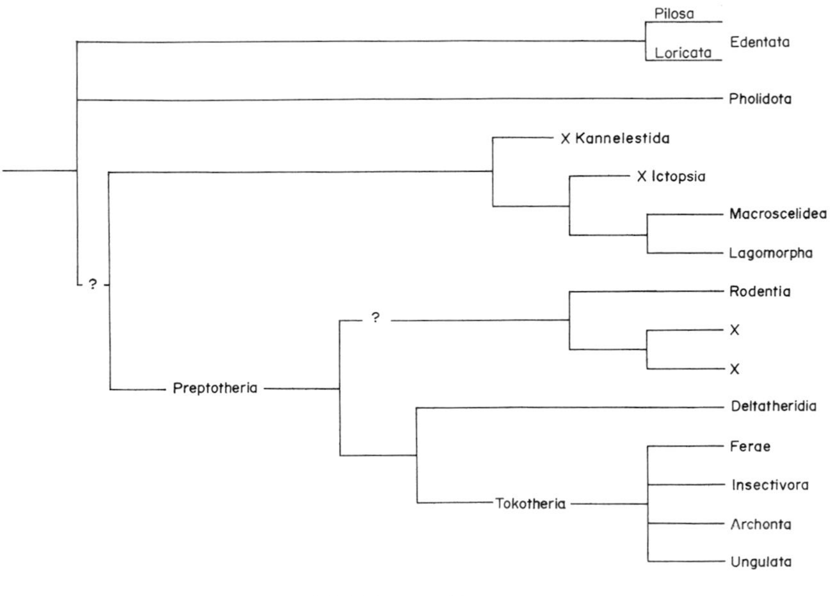

Figure 12. Cladogram illustrating the major branching of the Eutheria in the late Cretaceous. Terminology adapted from McKenna 1975.

clades: (1) a lineage of armored terrestrial grazing herbivores culminating in rather large individuals, the Glyptodontidae. (2) An arboreal nonarmored group specialized for browsing, which by the late Eocene had split into four lineages—the highly specialized arboreal Bradypodidae and three families that increased their size and became terrestrial browsers, collectively referred to as "giant ground sloths" (Megalonychidae, Mylodontidae, and Megatheriidae). (3) The armadillos, represented today by the Dasypodidae but descended from near relatives of the ancestral glyptodonts. The living family shows an enduring adaptational trend for exploiting an omnivore/insectivore or myrmecophagous niche. (4) The anteaters (Myrmecophagidae), a lineage with an ancient history of adaptation for feeding on ants and termites (see fig. 13).

The Edentata are eutherians, but they exhibit rather unusual morphological and physiological characters, some of which are conservative. It is fair to say that the Edentata tend to have rather low metabolic rates relative to many other eutherians. It must be understood, however, that anteating and folivority, as expressed by *Myrmecophaga* and *Bradypus,* respectively, are characterized in general by lower metabolic rates (McNab 1974, 1978*a*). Even allowing for these concomitants of adaptation, many edentates have slightly lower to profoundly lower rates of basal metabolism than eutherians (Goffart 1971). The testes undergo either no descensus (Bradypodidae) or partial descensus to lie in the pelvic cavity (Myrmecophagidae and Dasypodidae). The female edentates and to some extent the males of the Bradypodidae and Myrmecophagidae exhibit a urogenital sinus, which is a conservative morphological feature (Weber and Abel 1928).

Although not necessarily conservative, the tendency to develop dermal bone under an overlying epidermal scale is an ancient trait in this group that is expressed to this day in the elaborate armor of the Dasypodidae. The teeth of the Edentata lack a true enamel and consist of dentin and cementum. Herbivorous edentates evolved hypsodont teeth, whereas myrmecophagous edentates either have reduced the tooth size (*Priodontes*) while retaining a large number of teeth or in the course of natural selection have lost them entirely (Myrmecophagidae).

The living edentates show numerous apomorphic characters, which is to be expected given the long evolutionary history of adaptation to specialized feeding niches. For example, the Bradypodidae have chambered stomachs and use a process of digestive fermentation for processing the leaves that make up the bulk of their diet (Goffart 1971). The Myrmecophagidae and some genera of Dasypodidae (*Tolypeutes, Priodontes*) show adaptations for feeding on ants and termites that are convergent with similar morphological adaptations shown by anteating forms that evolved on other continents (see Griffiths 1968, appendix 1, pp. 230–46, for a review).

Although many gaps exist in our knowledge of edentate morphology, behavior, and ecology, some progress has been made in the last decade.

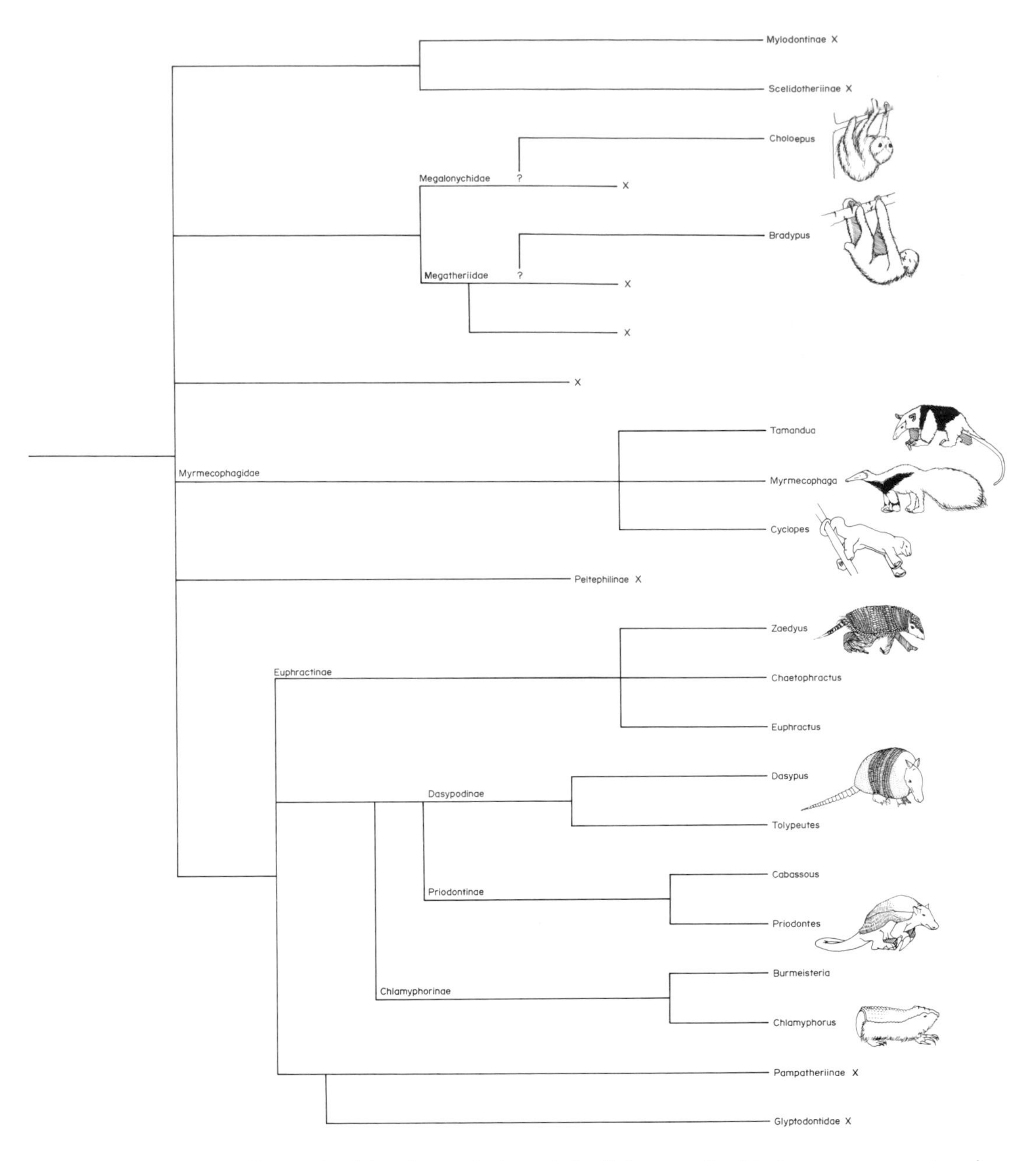

Figure 13. Adaptive radiation of the Edentata. In this figure, a separate origin for the living genera of tree sloths is portrayed. Adapted from Patterson and Pasqual 1972.

Brain evolution in the Edentata has been summarized by Röhrs (1966). Recent studies on the morphology of the Dasypodidae have been carried out by Moeller (1968). The ecology of anteaters and sloths has been studied in some depth by G. G. Montgomery, M. Sunquist, and Y. Lubin (Montgomery and Lubin 1977; Sunquist and Montgomery 1973; Montgomery and Sunquist 1974, 1975, 1978). The following discussion of behavior owes much to the efforts of these researchers.

4.1 The Natural History of the Edentata

4.1.1 Some Generalizations concerning the Maintenance Behavior of Edentates

In chapter 8 I will review the behavior patterns of the conservative mammals, and at that time I will contrast the maintenance behavior patterns of monotremes and marsupials with those of the conservative eutherians. It seems useful at this point, however, to review the variation displayed by the Edentata.

Care of the Body Surface

Scratching with the the hind foot is present in all forms studied; the forefoot may be used to wipe at the face, but there is no stereotyped "face washing" movement sequence such as is seen in some marsupials and insectivores. Scratching movements seem more highly elaborated in the Myrmecophagidae and Bradypodidae than in the Dasypodidae.

Nest Building

The mode of transporting nesting material to a burrow system appears to be unique in the Dasypodidae (Eisenberg 1961). Straw or leaves are scraped together with the forelimbs until a mass accumulates under the body. The leaves are then pressed to the venter with a forefoot while the animal hops backward to its burrow. Nesting materials are not transported by the Myrmecophagidae or Bradypodidae.

Defecation and Urination

Although the Myrmecophadidae seem to pay little attention to a specific locus for urination or defecation, the sloths and some species of armadillos offer interesting exceptions. The sloths, unlike most highly arboreal mammals, descend from the trees to defecate on the ground. *Bradypus* excavates a small depression in the ground with its stout tail before defecating. Defecation invariably occurs on the ground, often at the base of a preferred feeding tree. Montgomery has hypothesized that in the coevolution of sloths and their food trees, perhaps selection has favored the survival of sloth populations that tend to "fertilize" the trees they harvest leaves from (Montgomery and Sunquist 1975).

Dasypus may defecate without any special preparatory movements; however, it is noteworthy that both *D. novemcinctus* and *D. septemcinctus* will bury feces. In captivity *D. novemcinctus* shows locus specificity for urination and defecation. On awakening it will proceed to the "toilet" locus, sniff the substrate, dig with the forepaws, turn around, scratch backward with the hind legs, defecate or urinate or both, then again scratch backward with the hind legs, covering the excreta.

4.1.2 The Natural History of the Bradypodidae

The sloths (*Choloepus* and *Bradypus*) represent the survivors of an early radiation that adapted itself to browsing. The conversion of structural cellulose to simple sugars as well as detoxification of tannins and other secondary compounds of leaves was accomplished in both genera by evolving a chambered stomach where leaves could be fermented with the aid of protozoan and bacterial symbionts. The sloths are completely arboreal, are long-lived, and produce a single precocial young after a prolonged gestation. For a complete synopsis of the life cycle of *Bradypus,* see Montgomery and Sunquist (1975) (see plate 6).

The biology of *Choloepus* has been the object of extensive study in captivity, since, unlike *Bradypus, Choloepus* can readily adapt to artificial diets. *Choloepus* females have a gestation period of 11½ months. The interbirth interval is about 14–15 months, since the lactation anestrus is somewhat variable in its duration. The mother carries the young sloth until its weight exceeds 15 percent of her body weight. In the wild the youngster would be discouraged from clinging to the mother by her outright departure from its vicinity (see Montgomery and Sunquist

Plate 6. *Bradypus infuscatus,* the three-toed sloth, is highly adapted for climbing. It descends to the ground only to move from one tree to another or to defecate. It is one of the primary examples of extreme specialization for arboreal folivority (Courtesy of Carrie Thorington.)

1975). In captivity, however, the young may manage to cling until its mother gives birth to a sibling some 14 months later.

This brief synopsis should help to point out several important features of the parental care system. First, the male does not participate in rearing the young. Second, the female has no nest site and serves as a "movable nest" for the neonate as it matures. The young develops in utero for almost one year before birth (*Choloepus*) or for 6 months (*Bradypus*), and at birth it is almost 5 percent of the mother's body weight. The slow reproductive rate suggests a "K" selection strategy for the sloth, as will also be shown for *Myrmecophaga*.

Social interactions are subtle in sloths. Montgomery and Sunquist (1975) clearly demonstrate through radio tracking marked animals that sloths space themselves, yet do not defend territories. Spacing may be promoted by auditory and chemical cues. To date, only the distress call of the young sloth has been analyzed by Montgomery. Adults produce calls similar to those of the young, but their function is obscure. Male two-toed sloths will mark branches with secretions from their cloaca and their nostrils. This nasal secretion originates from a gland in the orbit, probably Harder's gland, and is homologous with the eye secretions of the tenrecoid insectivores (see Eisenberg and Gould 1970; Poduschka 1974).

4.1.3 The Natural History of the Myrmecophagidae

Our understanding of anteater biology has been considerably advanced in recent years by G. G. Montgomery and Y. Lubin (1977, 1978). It is obvious that the anteaters have specialized anatomically and physiologically for obtaining most of their nutritional requirements from feeding on ants or termites. The metabolic consequences of such an adaptation have been analyzed by McNab (1974, 1978*a*). McNab noted the tendency for myrmecophages to exhibit lower basal metabolic rates. As with the sloths, we find a reproductive pattern that suggests a "K" strategy. One young is produced after a prolonged gestation, and the young is well cared for by the female during its initial stages of development. The female anteater (all three genera) will carry her young when moving from one bivouac site to another, but when she forages she generally leaves the young alone in a nest. Thus the continual carrying seen in the sloth is not shown by the anteater.

Anteaters are solitary except for the parental care and mating phases. Sloths show a similar lack of complex social life. Although they are "solitary," there is no evidence to indicate that either sloths or anteaters "defend" a discrete feeding territory.[1] Home ranges show some overlap, but adults are spaced and feeding strategies suggest an individual familiarity with the terrain (Montgomery and Lubin 1977). Clearly further work is necessary to clarify the mechanisms whereby spacing among like-sexed adults is achieved.

1. The silky anteater, *Cyclopes*, demonstrates intrasexual territorial behavior in Panama (G. G. Montgomery, pers. comm.).

4.1.4 The Natural History of the Dasypodidae

Clearly the outstanding morphological feature of the living armadillos (Dasypodidae) is the possession of a peculiar series of dermal bones embedded in the dorsal epidermal matrix. An inspection of the feet of all living genera indicates elaborate claws and immediately suggests varying abilities to dig, coupled with differing capacities for cursorial locomotion. Krieg (1929) and Külhorn (1936) were able to formulate a morphological classification based on antipredator mechanisms that these morphological adaptations in part reflect (see table 9). The variation on a common theme of predator avoidance within this family parallels the set of antipredator adaptations evolved by the insectivore subfamily Tenrecinae, making use of spines rather than armor (Eisenberg and Gould 1970).

An analysis of foraging strategies of armadillos has lagged, no doubt as a result of our very imperfect knowledge of the dietary requirements of many of the rarer species. Figure 14 attempts to summarize what is currently known concerning diet for the major genera. To synopsize, the genera *Chaetophractus* and *Euphractus* appear to exhibit a generalized feeding niche and a less specialized antipredator defense (see table 9). *Dasypus* is predominately an insectivore but takes some small vertebrates. *Cabassous* and *Priodontes* are specialized myrmeco-

Table 9 Antipredator Strategies of the Dasypodidae

	Primary Strategies		
Genus	Flight	Burrowing	Rolling into a Ball
Dasypus	+ +	+	−
Chaetophractus	+	+ +	
Priodontes		+ + +	
Cabassous		+ + +	
Tolypeutes			+ +

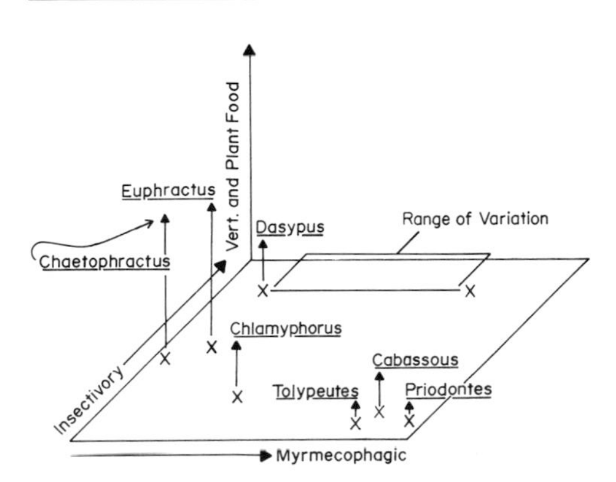

Figure 14. Feeding specializations for the genera of the Dasypodidae. Data for *Cabassous, Priodontes, Euphractus,* and *Tolypeutes* are summarized by Griffiths 1968 and Redford, in press. *Chaetophractus* and *Dasypus* data are summarized by Greegor 1974. *Chlamyphorus* data are from Rood 1970*b*. Axes are arbitrary but represent relative proportions of dietary intake.

phages with an antipredator strategy centering on their digging ability. *Tolypeutes* is a specialized myrmecophage with a highly stereotyped antipredator strategy (see table 9). *Burmeisteria* and *Chlamyphorus* are specialized for a fossorial niche similar to that occupied by the semiarid- to arid-adapted golden moles (Chrysochloridae) of South Africa (see plate 7).

Table 10 synopsizes the reproductive trends for the family. Clearly the genus *Dasypus* has the highest reproductive potential, followed by the Euphractini. The other genera (*Tolypeutes, Priodontes, Cabassous,* and *Chlamyphorus*) produce single young and suggest a "K" strategy of reproduction. An examination of the fossil record and a close comparison of the living genera suggest that *Euphractus* and *Chaetophractus* together display a conservative syndrome and that the other genera are to some extent derivative specialists (see plate 8).

The Euphractini are to some extent "plesiomorph." If the Edentata are among the most primitive living eutherians, then the behavior of a "conservative" dasypodid such as *Euphractus* has great significance for comparisons with other morphologically conservative mammals. Unfortunately, we have virtually no field data on spacing, density, parental care patterns, and activity rhythms, with the exception of Greegor's (1974) study of *Chaetophractus*. We must therefore rely heavily on captive studies. *Euphractus* females build a nest before parturition, usually in a burrow that the female constructs. After a 60–65-day gestation, the young are born with eyes and auditory meatus closed. This is far more altricial than in sloths and anteaters. The female actively grooms, broods, and suckles her young for some 60 days. The mother will retrieve a young that has been displaced from the nest. The eyes of the young open about 32 days postpartum. The young can produce both soft clicks and squeaks.

Dasypus novemcinctus exhibits delayed implantation, so the actual length of the gestation period is imperfectly known. The schedule for

Plate 7. *Chlamyphorus truncatus,* the fairy armadillo. This species has specialized for a fossorial way of life and to some extent has converged toward the feeding niche of the golden moles of South Africa. (Courtesy of Jon Rood.)

Table 10 Reproductive Patterns of the Dasypodidae

Taxon	Length of Gestation (days)	Number of Young	Number of Embryos	Time Eyes Open (days)	Reference
Dasypodidae					
Dasypodinae					
Euphractini					
Chaetophractus vellerosus	65	1–2	—	—	Greegor 1974 Encke 1965
C. villosus	60	2 ♂, ♀[a]	—	32	Beck 1972
Euphractus sexcinctus	60–65	1–3 2 ♂, ♀	1–3	22–25	Gucwinska 1971
Zaedyus pichiy	—	1–3 ♂, ♀ 2	—	—	Walker et al. 1964
Priodontini					
Priodontes giganteus	—	1–2	—	—	Walker et al. 1964
Cabassous unicinctus	—	1	—	—	Walker et al. 1964
Tolypeutini					
Tolypeutes matacus	—	1	—	22	Meritt 1976
T. tricinctus	—	1	—	—	Walker et al. 1964
Dasypodini					
Dasypus hybridus	—	—	4–8	—	Walker et al. 1964
D. kappleri	—	2	—	—	Wetzel and Mondolfi 1979
D. sabanicola	—	4	—	—	Mondolfi (pers. comm.)
D. novemcinctus	240[b] <70 (actual)	4–5	—	—	Talmage and Buchanan 1954
D. septemcinctus	—	4	—	12–18	Block 1974
Chlamyphorini					
Chlamyphorus truncatus	—	1	—	—	Walker et al. 1964

[a] ♂, ♀ indicates a mixed-sex litter.
[b] Delayed implantation.

development in *Dasypus* and *Tolypeutes* is portrayed with that of *Euphractus* in figure 15. Actually, those few living genera of dasypodids that still show multiparous births may be the sole surviving "r" strategists of the Edentata. (The genus *Dasypus* is a special case that I will discuss in the next section.)

The social life of adult armadillos is poorly understood. Pair tolerance is high for *Priodontes, Dasypus,* and *Cabassous*. At the time of parturition, females of *Tolypeutes, Euphractus,* and *Chaetophractus* become aggressive and nest alone (Meritt 1972; Gucwinska 1971; Encke 1965). Considerable home range overlap is shown by *Dasypus* (Clark 1951), and several individuals may use the same burrow (Taber 1945). Taber implies that multiple burrow occupancy may be by a family group or siblings, but Kalmbach (in Taber 1945) clearly indicates that aggregations of unrelated individuals may occur. Multiple burrow occupancy has also been noted for *Tolypeutes matacus* in the austral winter (Meritt 1972). This may be a thermoregulatory adaptation to conserve heat (see table 11).

Plate 8. *Euphractus sexcinctus,* one of the more generalized armadillos. This species has a mixed diet and can feed on small vertebrates as well as on insects. The dermal scutes provide protection against small predators. (NZP archives.)

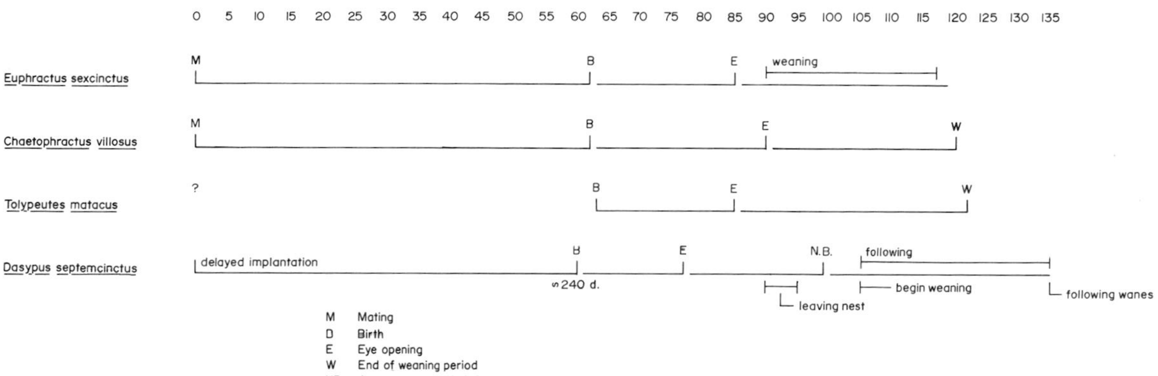

Figure 15. Development as it is understood for four armadillos. M = mating; E = eye opening; B = birth; W = end of weaning period; N.B. = first attempts by young to build nests. Data from table 10.

4.1.5 Some Theoretical Questions concerning the Social Behavior of Armadillos

Current kin selection theory predicts that the amount of "relatedness by actual genetic identity" will be directly correlated with the degree of altruism shown to relatives (Trivers 1971). Thus, siblings that are identical twins or quadruplets should show the highest mensurable altruism, or "enlightened self-interest." Only species of the genus *Dasypus* exhibit polyembryony. The female *Dasypus* is monovular (an ancestral character for the genus), but after zygote formation, mitosis takes place twice on the average before blastocyst formation and implantation. All other genera are either monovular without polyembryony or polyovular.

Table 11 Evidence for Degrees of Sociability for Dasypodidae

Species and Reference	Multiple Occupancy of Burrows	Home Range Overlap	Foraging in Proximity	Overt Aggression or Avoidance
Dasypus novemcinctus				
Clark 1951	—	+ +	♂, ♀, Subadult ♂	—
Taber 1945	— Adult + ♀ and young + littermates	n.a.	n.a.	—
Kalmbach 1943	+ + (12)	n.a.	n.a.	—
Eisenberg 1961	n.a.	n.a.	±	Mild at feeding
Tolypeutes matacus				
Meritt 1972	+ + + Winter (12)	?	?	♀ at partus?
Euphractus sexcinctus				
Gucwinska 1971	?			♀ at partus?
Chaetophractus villosus				
Encke 1965	?			♀ at partus?
Zaedyus pichi				
Meritt 1972	—		—	?

n.a. = subject not addressed in the cited publication.
+ + = strongly expressed; — = not shown.

If we could derive an actual measure of altruism for a species with naturally occurring identical twins, triplets, quadruplets and so forth, as a reproductive norm, the theory could be tested. Let us place the problem in some perspective. A hypothetical set of quadruplets born to a female *Dasypus novemcinctus* could be all males. If so, the brothers all share genes with their mother on the average of 0.50 and share *all* genes in common among themselves. Reproductive success of only *one* brother means that the genetic structure of all the brothers passes to the next generation. If some derived benefit accrues to brothers who nest together, forage together, and construct artifacts together, then even if only one brother survives as a result of communal activities, the genetic complement persists. Nonreproduction by all the siblings (resulting from the communal efforts) would clearly be maladaptive. Assuming dietary similarity and metabolic similarity, the "brothers" would smell alike, and they should to some extent perceive an identity with their own body scent when exchanging chemical signals with a sibling. Hole sharing among presumed siblings has been documented (Taber 1945), and nest building not only is a relatively stereotyped, innate pattern, but obviously is adaptive in reducing cold stress (Eisenberg 1961). It follows, then, that communal nest building by siblings could enhance survival of the group and especially of the emerging dominant brother. Does one sibling in fact emerge as the dominant in a litter? Based on weight gains in captives, a clearly dominant individual (in terms of access to food) emerged in *Dasypus septemcinctus* (Block 1974). This brings us right to the problem of mensuration. How does one realistically quantify "altruism" in an experimental context? Consider the problem in further perspective. Siblings of *Peromyscus leucopus* (not necessarily genetically identical) will nest together and

Table 12 Sound Production in Edentata

Taxon	Sounds Produced Mechanically	Exhalations (Unvoiced)	Vocalizations			
			Clear or High-Pitched Sounds		Low-Pitched "Noisy" Sounds	
Dasypodidae						
Dasypus septemcinctus[1]	Clicks	Huff and puff	Whuffling	Squeak	—	Growl
Dasypus novemcinctus[2]	—	Puff and huff	—	—	Grunt	—
Cabassous centralis[2]	—	Huff and puff	—	—	Buzz	Growl
Tolypeutes tricinctus[2]	—	—	—	—	Hum (buzz)	—
Chaetophractus sp.[2]	—	—	—	—	—	Snarl
Burmeisteria retusa[3]	—	—	—	Wail	—	—
Bradypodidae						
Bradypus infuscatus[4]	—	Hiss and puff	—	Distress call	—	—
Choloepus hoffmanni[4]	Jaw-clapping	Hiss and puff	—	Distress call	—	—
Myrmecophagidae						
Myrmecophaga tridactylus[2, 5]	—	Hiss and puff	—	Young: *Kriechen; trillende laut,* squeak, "trumpet"	—	Growl-roar
Tamandua mexicana[2]	—	Hiss and puff	—	—	Grunt	Growl
Cyclopes didactylus[3]	—	—	—	Whistle	—	—

References: 1 = Block 1974; 2 = Eisenberg, unpublished; 3 = Walker et al. 1964; 4 = Montgomery and Sunquist 1974; 5 = Schmid 1938.

derive considerable survival benefits (Howard 1948). In almost every species of mammals studied, some family members will be larger and more dominant, in some cases deriving a nutritional head start from the first day of lactation by developing a strict teat order through agonistic behavior (Ewer 1959; McBride 1963). There is no clear evidence that *Dasypus* siblings, in spite of their genetic identity, are behaving any differently from rodent siblings or other mammals during their normal ontogeny.

If *Dasypus* littermates stayed together as a unit during the reproductive cycle and formed some cooperative mating or rearing unit that was outstandingly different from those social units formed by mammalian siblings with less genetic identity, we might see the possibility of a natural experiment, but so far all field data indicate that solitary foraging and nesting by the parturient female is the rule. Male participation in parental care is apparently nonexistent, and communal rearing by sisters has not been described. Long-term studies of natural populations of marked individuals can contribute significantly to our knowledge of the survival advantage conferred on individuals through communal nest construction by siblings.

Given the above limitations on our knowledge, we can only conclude at this point that quadruplet production by *Dasypus sabanicola,*

Table 13 Home Range and Population Data for the Edentata and Pholidota

Taxon	Mean Head and Body Length (mm)	Unit Weight (g)	Home Range (ha)		Density (no./ha)		Biomass (kg/km²)		Locality	Reference
			Core	Extended	Ecological	Crude	Ecological	Crude		
Edentata										
Bradypodidae										
Choloepus hoffmanni	620	3,500	1.96	—	1.1–2.7	1.14	665	399	Panama	Montgomery and Sunquist 1975
Bradypus infuscatus	500	2,700	1.58	—	7.7–8.5	4.56	2,187	989	Panama	Waage and Montgomery 1976
Myrmecophagidae										
Myrmecophaga tridactyla	1,100	20,000	—	2,500	—	0.0012	—	2.4	Venezuela	Montgomery and Lubin 1977
Tamandua tetradactyla	560	4,000	—	375	—	0.02	—	8	Venezuela	Montgomery and Lubin 1977
T. mexicana	510	3,400	—	70	—	0.05	—	20	Panama	Montgomery and Lubin 1978
Cyclopes didactylus	160	200	—	—	—	0.37	—	10	Panama	Montgomery and Lubin 1978
Dasypodidae										
Chaetophractus vellerosus	250	2,000	1.78	4.66	—	—	—	—	Argentina	Greegor 1974
Dasypus novemcinctus	380	3,000	1.1	13.8	0.22	—	66	—	Florida	Layne and Glover 1977
	—	3,000	3.4	—	—	—	—	—	Texas	Clark 1951
	—	3,000	—	20	—	—	—	—		Fitch, Goodrum, and Newman 1952
Pholidota										
Manidae										
Manis tricuspis	400	2,800	3.5 ♀ 20.0 ♂	—	0.5	—	140	—	Gabon	Pages 1976

D. novemcinctus, *D. septemcinctus*, and *D. hybridus* is a novel way of increasing the reproductive capacity of these generalized armadillos. Clearly it suggests an "r" strategy deriving from a monovular "K" ancestor. In short, the "advanced" *Dasypus* species represent an escape from reproductive specialization.

4.2 Behavior and Habitat Utilization by Edentates

The edentates have radiated to fill a variety of feeding niches. Antipredator behavior is reflected in the structure of the carapace and limbs of the Dasypodidae, which have been molded by the activities of their principal predators for millennia. The specializations for arboreal herbivority by the Bradypodidae and anteating by the Myrmecophagidae have imposed modifications in the structure of their bodies and subsequent specializations in their behavior patterns.

Although olfaction undoubtedly plays an important role in mediating social behavior, this aspect of their ethology remains almost unexplored. Auditory communication seems somewhat rudimentary compared with that in many other mammalian taxa. What we know concerning several species is summarized in table 12.

Maternal care and sexual behavior have been described only for *Bradypus*, *Choloepus*, *Myrmecophaga*, *Chaetophractus*, and *Euphractus*. The patterns of maternal care are clearly very specialized and in conformity with feeding adaptations. Much more work is necessary before an adequate synthesis can be developed.

Table 13 presents such data on body size and population as are available for the Edentata. In figure 16 the home range size is regressed

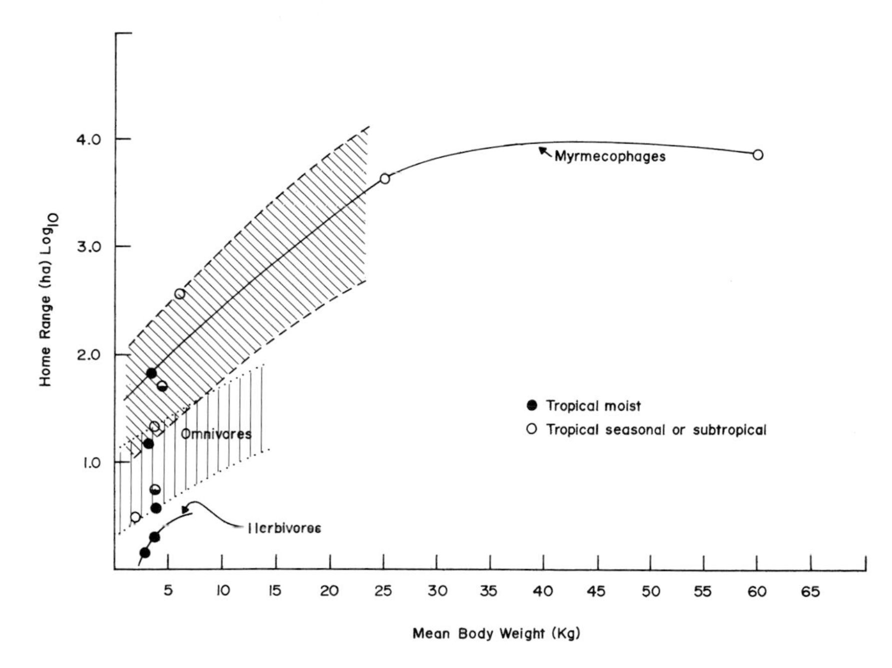

Figure 16. Home range size as a function of absolute body size and trophic strategy for several edentates. See table 13.

against mean body size. On the average, the herbivorous edentates have the smallest home ranges and the myrmecophagous edentates the largest; the omnivores have intermediate-sized home ranges. There is some suggestion that home range size is smaller in tropical forests showing reduced seasonality, but many more data will be necessary before we can assess the reliability of this correlation.

These data show trends similar to those established for the Marsupialia. Larger mammals exist at lower numerical densities than do small mammals. Herbivores have higher densities for a defined size-class than do omnivores, and they contribute high biomass levels to any given mammalian community. Anteaters exhibit home range sizes comparable to those of carnivores of the same size-class. As we shall see, the generalizations established for this ancient radiation will hold for mammalian radiations on the more contiguous continental land masses.

5 The Pholidota

Recent research by Elizabeth Pages (1965, 1970, 1972*a,b,* 1976) has clarified a great deal concerning the ecology and behavior of the Pholidota. McKenna (1975) has implied that the pholidotes are the "sister group" of the edentates. Indeed, the pholidotes share with the edentates some rather conservative mammalian features. The testes of the male pangolin do not descend into a scrotum, though they do pass through the inguinal canal when sexually mature and enlarge in a fold of the skin in the groin. The dorsal region of the body is covered with short, cornified scales that are epidermally derived. This has given rise to their common English name, "the scaly anteaters." Small hairs may grow in association with the scales. This protective epidermally derived armor is a functional analogue to the armor of armadillos but is not completely homologous. For the most part pangolins are plantigrade, with five digits, each bearing a large claw. All the living pholidotes are highly specialized for feeding on ants and termites, and they share with other anteaters numerous convergently evolved anatomical adaptations that are independently derived (for a review, see Griffiths 1968).

The living pangolins are clearly separable into two geographically distinct subgroupings, the "Asiatic" pangolins and the African pangolins. The Asiatic radiation is smaller, having only three species, two of which (*Manis crassicaudata* and *M. pentadactyla*) are for the most part terrestrial and the third of which (*M. javanicus*) is moderately arboreal with a semiprehensile tail. The African pangolins include five species with a slightly more diversified radiation. Strongly terrestrial forms include *M. gigantea* and *M. temmincki*. Semiarboreal forms include *M. tricuspus*, and *M. longicaudata* is highly arboreal with the best-developed prehensile tail.

The manner in which the ecological niches are divided among sympatric species of the genus *Manis* in Africa was elegantly described by Pages in 1970. The pangolin's overlapping scales are involved with its antipredator behavior, which includes rolling into a ball and thus presenting a surface relatively impregnable to a small carnivore (see Pages 1972*a*).

The social behavior of pangolins seems to be dominated by spacing, without the overt defense of a territory. Males stay very well spaced, with little home range overlap. They appear to accomplish this by olfactory marking involving anal glands. Adult males, when they do happen to meet in the wild, may fight aggressively (Pages 1972*a*). Pages has demonstrated that neighboring males come to know one another through occasional contact, and the responses of neighboring males toward one another are quite different from their responses toward unknown males. She points out that females show little agonistic behavior to one another, but they nevertheless appear to effect a certain amount of spacing in the wild owing to idiosyncratic patterns of foraging (Pages 1976). A male's home range generally overlaps with the ranges of one or more females, and whenever the male encounters a female's scent trail he attempts to follow her. The female will actively drive the male away if she is not in estrus or is carrying an infant. During estrus, various aspects of courtship behavior may be noted (Pages 1972*a*).

The maternal behavior of pangolins shows some remarkable parallels with that of the New World Myrmecophagidae. A single young is born, generally in a reasonably well developed state. In the species *Manis tricuspus,* the mother actively takes care of the young for the first 2 weeks during an initial nest phase. After 2 weeks of age the young pangolin begins to accompany the mother on her foraging excursions, riding on her back or at the base of her tail. This mode of transport is reminiscent of the convergently evolved infant transportation mechanisms of the Myrmecophagidae. The young pangolin's attainment of independence, beginning at 3 to 4 months of age, is carefully documented by Pages (1972*b*). Carrying young at the base of the tail has also been carefully described by Rahm (1961*a*).

The general maintenance behavior of pangolins has been described by several authors (Rahm 1961*a*). I have observed *Manis javanicus,* and these observations may be considered representative, though not necessarily typical for all species. The comfort movements of the pangolin are somewhat reduced compared with those of mammals that retain a complete coat of fur. The wrist is used to wipe the eyes and the face. Generally face wiping is performed unilaterally. The forelimb is also used to scratch under the scutes on the hindquarters and tail, whereas the hind limb is used to scratch under the scutes over the shoulders and back.

Male pangolins may urinate in a tripod stance, standing on the hind legs with the tail braced like a third foot. The stream of urine is directed forward. Highly aggressive male pangolins may also stand in a similar posture, often showing an erect penis, and attempt to strike with the claws of the forelimbs. Defecation in the pangolin, at least in *M. javanicus,* is unique. The animal arches its back while tucking its head down to the ground and standing on its hind legs in a tripod stance with the tail braced like a third foot. It then moves its head forward and backward, scraping a hole with the scutes of the head and occasionally pausing to dig with its forelimbs; it then defecates and often urinates in this hole and covers the fecal material by backward strokes with the forepaws. Though the pangolin buries its feces in a manner reminiscent of *Dasypus novemcinctus,* the movements it employs are different and indeed unique. Perhaps the burying habit has evolved independently in these two mammalian lines and does not represent derivation from a common ancestral movement.

6 Conclusions concerning the Early Radiations

Having considered the Monotremata, Metatheria, Edentata, and Pholidota, we are in a position to assess the results of the early radiations of mammals by reviewing the extant forms. The adaptations of the living monotremes are probably too specialized to give us any clues to overall evolutionary trends, but the Edentata and the Metatheria are more instructive. The Metatherian radiation took place in both South America and Australia. Many of the forms previously found in South America have been extinct since the Pliocene, but the Australian radiation remains fairly well intact. The Edentata evolved in isolation on the South American continent and were contemporary with the early marsupials and the extinct protungulate groups.

One striking feature that becomes evident from a comparison of these radiations is the extreme stability of the myrmecophagous niche (in the sense of both termite feeding and ant feeding). Table 14 illustrates the zoogeographical distribution of mammalian myrmecophagous adaptations. One notes that, in a convergent fashion, the myrmecophagous feeding niche has been filled on all major continents that have extensive land masses in the tropics. This niche is tropical in distribution, probably because colonial insects such as termites and ants are available more predictably and at greater density in the tropics than in the temperate zone. Also, it appears that the myrmecophagous niche is narrow. For any given continent, there are a limited number of species. Niche subdivision, where several sympatric species are present, is usually based on degree of arboreality. Arboreal myrmecophages predominate in tropical forests, while terrestrial adaptations occur in both forest and savanna conditions.

A number of more recently derived species from diverse orders may be considered partial and facultative myrmecophages (see table 14). Marsupials, carnivores, and elephant shrews have to some extent invaded the myrmecophagous niche. In the New World tropics, the obligate myrmecophages are all from the Edentata; in the Old World tropics, the myrmecophages are derived from the Pholidota and the Tubulidentata. In Australia, one genus of Monotremata is an obligate myrmecophage, and one genus (*Myrmecobius*) of dasyuroid marsupial has penetrated the niche in West Australia.

All obligate species were derived early in phylogenetic time. Convergence and parallelisms have been rampant as selection has shaped the morphology of the different anteater groups to increase their effi-

Table 14 Zoogeographical Distribution of Myrmecophagous Adaptations

Locality	Arboreal	Semiarboreal	Terrestrial Surface Foraging	Semifossorial	Fossorial
Complete adaptation					
Australia and New Guinea			*Myrmecobius fasciatus*	*Tachyglossus aculeatus*	
Asia		*Manis javanica*		*Manis crassicaudata* *M. pentadactyla*	
Africa	*Manis tetradactyla*	*Manis tricuspus*		*Manis gigantea* *M. temmincki* *Orycteropus afer*	
South America	*Cyclopes didactylus*	*Tamandua tetradactyla* *T. mexicana*	*Myrmecophaga tridactylus*	*Cabassous* *Priodontes* *Tolypeutes*	
Partial adaptation					
Australia			*Thylacomys lagotis*		
Asia			*Melursus ursinus*		
Africa			*Rhynchocyon* sp. *Elephantulus* sp. *Proteles cristata*		
South America					*Burmeisteria* *Chlamyphorus*

ciency for feeding on and digesting ants and termites (see Griffiths 1968 for a review).

The second major feature of the early radiations is that specialized herbivores evolved early in time. The terrestrial edentate herbivores are now extinct, but the metatherian herbivores have replicated in their habitat-utilization trends the adaptations made by their eutherian counterparts on the contiguous continental land masses. Of special interest is the convergence between South American edentates and Australian metatherians in the evolution of leaf-eating herbivores or folivores. In their metabolic, structural, and digestive processes, the sloths demonstrate a strong convergence with the more specialized leaf-eating folivores of the marsupial family Phalangeridae (e.g., *Phascolarctos* and *Phalanger*) (Bauchop 1978).

Early mammals found themselves in environments with a reasonable abundance of insects and small vertebrate prey. Together with foliage and fruits, these provided rich food sources. Mixed feeding on fruits, invertebrates, and small vertebrates probably prevailed, with stronger specializations for feeding on insects evolving early. Specialization for feeding on social insects led to the stabilization on different continents of forms feeding on ants and termites, an archaic adaptation retained in many extant conservative mammals. Exploiting trees for their fruits as well as for the arthropods associated with them could have led gradually to increased plant feeding and finally to specializations for utilizing leaves themselves as a source of energy. The culmination of this trend has been the convergently evolved leaf-feeding patterns among the primates, marsupials, and edentates, but the radiation of primates into the folivore niche occurred somewhat later (see Montgomery 1978).

7 Madagascar as a Refugium: The Radiation of the Tenrecidae and Other Madagascan Mammals

We now must consider the later radiations of the Mammalia. McKenna (1975) has published a revised classification of the Mammalia, in part, employing cladistic taxonomy (Hennig 1966). Although the results are interesting as well as useful, he has had to introduce many more subcategories than Simpson employed in 1945. Superlegions, legions, sublegions, supercohorts, cohorts, magnaorders, and grandorders have been incorporated in an attempt to reflect the phylogenetic branching. I have modified his system in developing the following discussion. The radiation of the Tokotheria occurred at the Cretaceous/Paleocene boundary and is portrayed as a cladogram in figure 17. I have indicated an affinity between the golden moles (Chrysochlorida) and the tenrecoid insectivores (Tenrecomorpha) (see Van Valen 1967), and I further suggest that this line of descent branched off very early from the stock leading to the contemporary Insectivora. I am almost tempted to give the golden moles and tenrecoids an even earlier origin, but instead I have taken the dubious step of raising each to an ordinal rank.

The Tenrecomorpha contain two living families, the Potomagalidae and the Tenrecidae, with the latter currently confined to the island of Madagascar. (The two families are often considered subfamilies of the Tenrecidae.) Madagascar has remained the repository of a variety of conservative eutherians, and some consideration of its geography and history is essential to our understanding of the unique status of the Tenrecidae in eutherian evolution.

Madagascar is the fourth largest island in the world (its area is approximately 227,760 square miles), situated in the western portion of the Indian Ocean roughly 250 miles off the coast of Mozambique. It has been isolated from Africa since the late Cretaceous. The island lies mainly in the geographically defined tropics of the southern hemisphere, between 11° 57′ and 25° 38′ south latitude. The center of the island is dominated by a high plateau that ranges between 2,500 and 4,500 feet above sea level. To the east the plateau falls off rather abruptly, and at the time of man's arrival rather extensive multistratal tropical evergreen forests occupied the east coast. The central plateau falls off gradually to the west and supports a sclerophil forest. In the extreme southwest, prolonged periods of aridity occur within each annual cycle, and the vegetation is characterized by scrub and by endemic spiny succulents.

Since their extensive occupation by man, beginning approximately 2,000 years ago, the forests of the central plateau have been degraded.

Major exterminations occurred after the advent of man on Madagascar, including that of some of the giant lemurs, the giant flightless birds (*Aepyornis*), and the giant land tortoises.

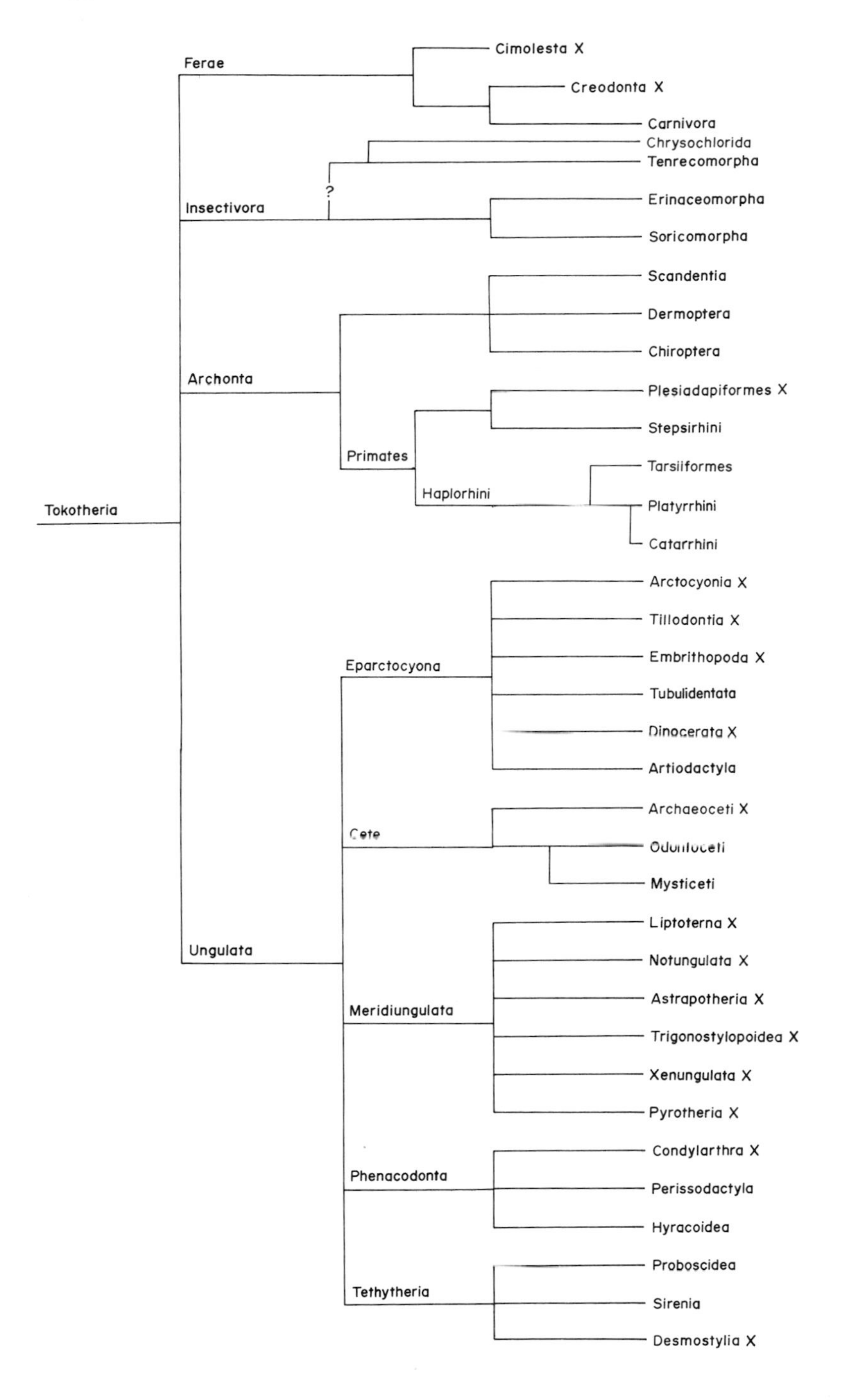

Figure 17. Cladogram indicating the later radiations of the Eutheria. Terminology according to McKenna 1975.

Six orders of mammals have been represented on Madagascar in recent times. In addition, species from four orders of mammals have been introduced by man within the past 2,000 years. At present there are eleven genera of insectivores, twelve genera of bats, ten genera of primates, ten genera of rodents, seven genera of carnivores, and one genus of artiodactylans. The ''Insectivora'' include one endemic family, the Tenrecidae, and the introduced family Soricidae. The order Chiroptera includes six families, only one of which is endemic to the island (Myzopodidae). The Carnivora include one family—the Viverridae—divided into four subfamilies,[1] three of which are endemic and the other represented by Viverricula, undoubtedly a recent introduction by man. The Primates are represented by ten living genera divided into three families, all forms being endemic to Madagascar. The native rodents constitute eight endemic genera lumped artificially into one subfamily, the Nesomyinae. The Artiodactyla were represented before man's invasion by one endemic species, the hippopotamus, *Hippopotamus lemelii,* but today the hippopotamus is extinct and the river hog *Potamochaerus,* introduced prehistorically, is widespread.

The mammalian fauna of Madagascar derives from a series of invasions that probably began in the Paleocene and ended in the Pliocene. The colonization of Madagascar by the Mammalia has been discussed by Simpson (1940) and Darlington (1957). The insectivores were among the earliest invaders, and they may have rafted across the Mozambique Channel when it was not nearly as wide as it is now. The insectivores were followed by the lemuroids, probably in the Eocene. The dramatic radiation of both lemurs and insectivores has been accomplished from stem forms that were rather conservative in their morphology.

Eisenberg and Gould (1970) compared the feeding niches of Panama and Madagascar with respect to the major genera occupying each niche. An inspection of our data indicates that the mammalian fauna of Madagascar has, through adaptive radiation, occupied most of the major feeding niches available. Of course the two faunas differ dramatically in that Panama has many archaic Neotropical elements, such as the Edentata, that are entirely absent from Madagascar.

The Madagascan mammalian fauna is derived from Africa. Obviously, during the Eocene a connection of some sort existed between Africa and Europe. Invasion of North Africa by Eocene forms apparently was rather successful, and many of these archaic forms, such as the lemurs, were able to pass to Madagascar, perhaps by rafting. Although lemuriform primates no longer exist in Africa, this was the only logical route for their colonization. Similarly, the tenrecoid insectivores show their nearest affinities with subfossil, fossil, and extant African Insectivora—for example, the Potamogalidae.

Madagascan mammals occupy feeding niches equivalent to those of continental mammals. We must conclude that, given a long enough evolutionary history and a rather unspecialized founding form, then through natural selection and adaptive radiation we can find a rather

1. The subfamilian divisions of the Viverridae are unstable; see section 13.1.

uniform occupancy of feeding niches. Thus the archaic mammalian fauna of Madagascar can serve as a test case when we examine evolutionary trends on the more contiguous continental land masses, much like the comparisons I outlined for testing hypotheses utilizing the Australian Marsupialia and the Eutheria. The mammalian fauna of Madagascar, then, represents a eutherian remnant, which can be very instructive in elucidating patterns of adaptive radiation.

The biogeography and ecology of animals in Madagascar are extensively reviewed in the volume edited by Battistini and Richard-Vindard (1972). This volume synopsizes chapters reviewing the Insectivora (Heim de Balsac), the rodents (F. Petter), the Carnivora (R. Albignac), and the lemurs (J.-J. Petter). Recent research on prosimians in Madagascar, including field studies, has been reviewed in the volumes edited by Tattersall and Sussman (1975) and by Martin, Doyle, and Walker (1974). The ecology and behavior of the Carnivora have been monographed by Albignac (1973), and the tenrecoid insectivores were monographed by Eisenberg and Gould (1970).

The founder stock for the family Tenrecidae must have been isolated on Madagascar at approximately the Paleocene. Coevolving through time with lemuriform primates, viverrid carnivores, and nesomyine rodents, the contemporary tenrecs today present us with the limits of morphology evolved within the basic body plan of a conservative insectivore (see fig. 18). The morphology of the Tenrecidae shows conservative as well as advanced characters. The presence of a cloaca in both sexes and the absence of a scrotum in the male tenrec represents a set of conservative characters. On the other hand, the very small eye and the reduction of the tail in the subfamily Tenrecinae are definite specializations. The development of spiny pelage for antipredator defense is also a specialization of the Tenrecinae.

Considering general behavior, the Tenrecidae can use simple echolocation. This capacity probably evolved very early in the history of the Mammalia as they radiated into nocturnal niches. Reproduction involves prolonged periods of courtship and attentive behavior on the part of the male. It is almost certain that most, if not all species, of the Tenrecidae are induced ovulators, which probably is a conservative feature. Copulation is prolonged in this group, a characteristic seen in other induced ovulators (Eisenberg 1977). Gestation periods are rather long in the different species, which may shorten the postpartum developmental time for young. However, the relatively longer gestations may also reflect metabolic differences between tenrecs and shrews. Litter size within the Tenrecidae shows great variation, reflecting differing selective pressures acting on the species. Litter size may be as low as two in *Microgale talazaci* and as high as thirty in *Tenrec ecaudatus*.

To promote acquaintance with this unusual group, I will synopsize the life histories of three species: *Microgale dobsoni, Setifer setosus,* and *Hemicentetes semispinosus*. The accounts are drawn from Eisenberg and Gould (1970), Eisenberg and Maliniak (1974), and Eisenberg (1975*a,b*).

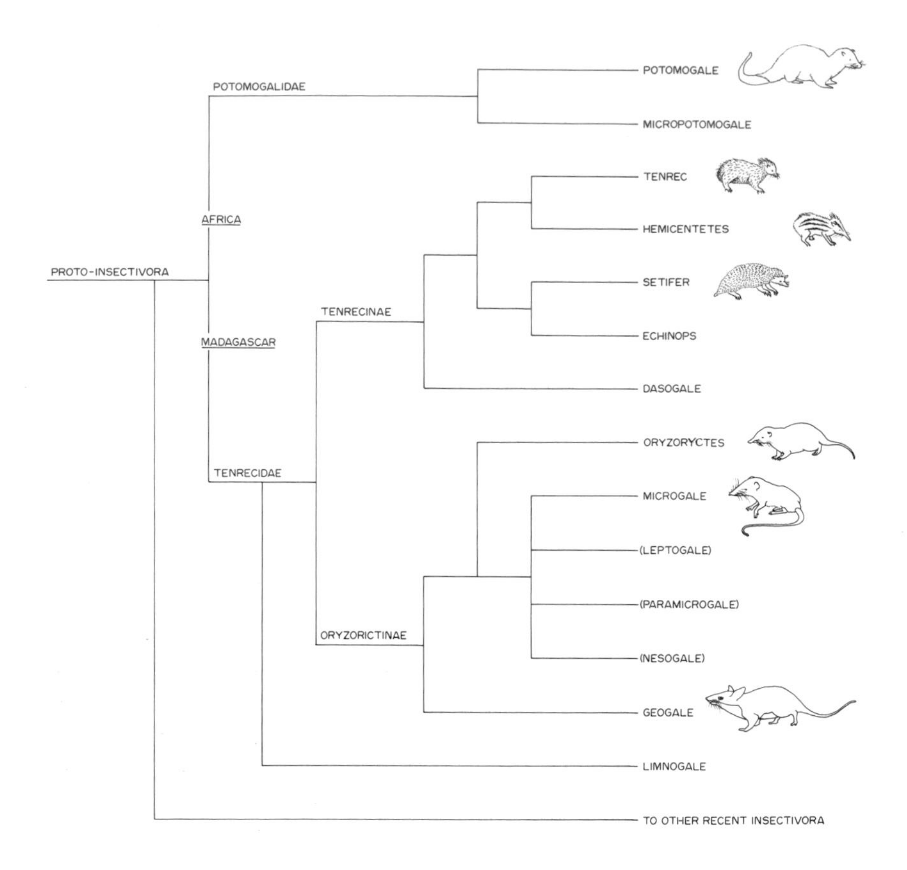

Figure 18. Cladogram indicating the radiations of tenrecoid insectivores. Parentheses indicate subgenera. Modified from Eisenberg 1975*b*.

7.1 *Microgale dobsoni*

Microgale dobsoni will serve as a model for the nonspiny tenrecid insectivores. This modest-sized insectivore, generally weighing less than 45 g, is distributed in the eastern plateau evergreen forests. It forages on the forest floor at night, seeking out the arthropods and other invertebrates that compose the bulk of its diet. Since some areas it inhabits have periods of seasonal aridity, provision is made to pass the season when insect abundance is reduced. *M. dobsoni* stores considerable quantities of fat on its body and becomes semitorpid during the months of reduced rainfall.

Litter size in *M. dobsoni* varies from one to three. Gestation is believed to exceed 57 days, and the young mature rather slowly. The young are born in a rather altricial state and do not begin to venture from the nest chamber until approximately 1 month of age, but they begin individualistic foraging soon after this. There appears to be no complex social grouping formed by species of the genus *Microgale,* but trapping data and observations in captivity strongly suggest that pair preferences may be formed and that adult pairs may live close to one another and indeed share the same home range.

Plate 9. *Microgale dobsoni,* one of the oryzorictine tenrecoid insectivores. This 50 g mammal typifies a terrestrial insectivore body plan. Compare with *Suncus murinus,* plate 24. (NZP archives.)

Plate 10. *Setifer setosus.* This tenrecoid insectivore exhibits an antipredator strategy similar to that of the continental hedgehogs of the family Erinaceidae. The quills are nondetachable. *Setifer* can roll itself into an impregnable spiny ball. (NZP archives.)

7.2 *Setifer setosus*

The dorsum of *Setifer setosus,* the greater hedgehog tenrec, is covered with stout, nonbarbed spines, and the animal can roll up into an impregnable spiny ball. Its antipredator behavior has thus converged with that of *Erinaceus,* the European hedgehog. Setifer occurs on the east coast of Madagascar, extending well up onto the plateau. The animal digs its own burrows. Depending on local climatic conditions and the relative abundance of food in the plateau area, the *Setifer* population may become torpid for a certain part of the year. On the other hand, the capacity for torpor is apparently facultative, and populations at lower elevations can be active almost all year round.

After a gestation period of nearly 60 days, the young are born in an altricial state. The eyes open at the end of 2 weeks, and from 14 days on the young animal begins to take solid food. Aside from the male-female association during the period of reproduction, the only other social grouping demonstrable appears to be the female-young unit during the rearing phase.

7.3 *Hemicentetes semispinosus*

The genus *Hemicentetes* includes two species, *H. semispinosus* and *H. nigriceps,* the latter being adapted for the eastern plateau region of Madagascar. Each species exhibits a bold striped pattern on the dorsum: black and white in *H. nigriceps,* black and yellow in *H. semispinosus*. In addition to the hair of the dorsum, *Hemicentetes* has quills that are highly concentrated on the crown of the head. The body quills are barbed and detachable, except for a modified group of quills in the midposterior region of the dorsum. This group of specialized quills serves as a stridulating organ, and the sound produced is employed in certain social contexts.

H. semispinosus is confined to the areas of rain forest below the high plateau, extending its range nearly to the southern limit of tropical evergreen forest in Madagascar. *H. semispinosus* constructs burrows; there is some evidence that complex burrow systems may be used by several generations. The animals are active predominantly at night but show a pronounced midday peak of activity when they may be found foraging on the forest floor. *Hemicentetes* may exhibit facultative hypothermia and brief seasonal torpor that need not be synchronized in the entire population. The species is specialized for feeding on soil arthropods and especially earthworms.

Plate 11. *Hemicentetes semispinosus,* one of the most specialized members of the family Tenrecidae. The quills are barbed and detachable. The striped pattern serves as a warning. This species has specialized for feeding on earthworms, with reduction of its dentition and a prolongation of the rostrum. (NZP archives.)

The gestation period is approximately 58 days. The young are rather altricial at birth but grow rapidly. The eyes open by the 8th day, and at approximately 9 to 10 days of age the young begin to explore the vicinity of the burrow entrance. From approximately 13 to 22 days of age, the young follow the mother closely while she forages, gradually dispersing to forage on their own. Contact is maintained with the mother because the young are attentive to the stridulation produced by the specialized quills on the back. Thus the foraging unit can maintain auditory contact while pursuing individual activities. This communication system apparently helps the young learn routes to foraging areas so they can return to the nest site for several days after they emerge.

Although a female approaching parturition tends to isolate herself and build a nest away from other colony members, the potential for colony formation does exist in *H. semispinosus*. The female tolerates the male after the young are approximately 2 weeks of age, and he may remain in the same burrow system with the female through the birth of her first litter. We have excavated colonies that included 18 individuals.

The Tenrecidae provide a basis for inferring the behavior patterns of early mammals. It seems appropriate at this stage to turn to the general topic of how such behavior patterns may have been organized.

8 The Phylogeny of Behavior: A Mammalian Baseline

Since behavior patterns do not fossilize (though tracks and artifacts do), a phylogenetic reconstruction tends to become an exercise that cannot yield refutable hypotheses. Rather, such reconstructions become an exercise in the "coherence test of truth" (Pepper 1961). If behavior patterns are strongly tied to a morphological feature that is preserved in fossil forms, then we are on rather secure ground. Unless the fossil record and the secondary morphological evidence are very complete, however, we still do not know whether similar behavior patterns in contemporary forms arose by convergence or by parallelism. In fact, "evidence" for homologies among behavior patterns is only inferred as the presumptive homologues are tested according to the criteria outlined by Wickler (1961). In most cases we are led to a position of some uncertainty.

Nevertheless I feel encouraged to build on the framework established by Jerison (1973), who has employed endocasts of fossil craniums to determine brain size and proportion. He has outlined a history of brain and sense organ evolution in the Mammalia that conforms in its major features with brain and cranial nerve morphology. The few postcranial skeletons that have survived from Cretaceous marsupials and eutherians permit us to infer some features of locomotion, and the teeth and jaws give us some information on methods of mastication, prey capture, and probable diet.

8.1 Communication and Sense Organs

Early mammals were nocturnal and relied on tactile sensations from vibrissae, olfactory input, taste, and hearing. The eyes were not excessively large, nor were they vastly reduced as in present-day soricoid insectivores. Kay and Cartmill (1977) outline methods of inferring relative placement and size of the eye and of relating the contribution of the vibrissae to the fifth cranial nerve in their reconstruction of *Palaechthon nacimienti*.

Hearing was probably important in detecting prey and potential predators. Vocal communication is not elaborate in living monotremes and marsupials, but the hiss, click, and growl may have been present in Cretaceous mammals. Elaborate vocalizations may have appeared in modern mammals with the evolution of social organizations more complex than those I believe were present in Cretaceous mammals.

Threat behaviors involved opening the mouth, hissing, and attempting to bite. These movements and sounds may have been employed in both inter- and intraspecific contexts.

Marking objects with glandular exudates probably involved glands associated with the cloaca or anus. Exudates from glands associated with the eyes and mouth may have been spread on the face with the forepaws (see sect. 28.3). This movement may also have served to clean the face of foreign matter. Face washing is still widespread in both metatherians and eutherians. Specialized marking movements employing other skin glands may have evolved convergently or in a parallel manner among the various Cretaceous mammals.

8.2 Comfort Movements

Homeothermy and hair probably evolved together with adaptation to nocturnal life and adaptation to cooler climates. Hair divisible into a dense undercoat and a sparse but overlying set of guard hairs is difficult to keep clean, but cleanliness is a prerequisite for maintaining its insulating properties. For this reason adequate pelage cleaning movements are mandatory. Removal of ectoparasites from fur is also a demanding task, and we may assume that as mammals and their ectoparasites have coevolved the cleaning movements of the mammals have become more elaborate.

Scratching with the hind foot is the most nearly universal method of pelage dressing in monotremes, marsupials, and eutherians. The elaborate "face wash," which may also be a self-marking movement, is present in the living Marsupialia, tenrecoid insectivores, Macroscelididae, and Scandentia. The absence of this pattern from both the edentates (Bradypodidae) and the monotremes may be a secondary loss, since both of these taxa show forefeet vastly modified for digging or clinging. A similar loss or simplification of the face-washing movement can be observed in taxa that wipe the face with a single paw, where a species has feet modified for digging or clinging.

Licking or nibbling the body surface, cloacal opening, or anus is well established for *Ornithorhynchus*, most marsupials, and the eutherians. Where the mouth and tongue are vastly modified, as in myrmecophagous mammals, the utility of the mouth as an organ for self-grooming is vastly reduced (e.g., *Tachyglossus*).

8.3 Nest-Building Movements

Building nests with leaves or grasses was probably a useful adjunct to thermoregulation. Little can be said concerning the use of tongue, teeth, and forepaws by Cretaceous mammals in preparing nesting material, but the transport of nesting materials is another question. Carrying vegetation for nesting material in the mouth seems to be the simplest method, yet many metatherians (Didelphidae, Phalangeridae, and some Macropodidae) carry nesting material with their semiflexible or prehensile tails. This remarkable tail transport method is also shown by *Ornithorhynchus*, which can bend its tail ventrad, though it would not be considered to have a prehensile tail. Thus, use of the tail is not confined to those species having prehensile tails; both *Potorous* and *Bettongia* of the Macropodidae show this behavior though such caudal dexterity would not be expected.

The living eutherians of conservative morphology do not have prehensile tails, nor do they transport nesting material with a temporary ventral flexure of the tail. Mouth transport of nesting material seems to be the general rule, but where the mouth is modified for feeding on termites or ants it is less useful for carrying leafy material. Although not obligate myrmecophages, members of the edentate genus *Dasypus* do have small mouths and transport nesting material by raking it under the body, holding the material to the ventral surface with the forelimbs, and hopping backward on the hind feet. The myrmecophagous *Tolypeutes* shows an identical movement sequence (Eisenberg 1961). Whether this curious behavior complex is convergently evolved when mammals can no longer efficiently utilize the mouth for carrying vegetation is not yet known, but *Rhynchocyon* of the Macroscelidea shows a similar movement sequence when nest-building (Rathbun 1976). Are we seeing an ancient trait carried forward through time, or is this a simple case of convergence developing out of a generalized digging movement, with a clasp and a backward hop sequence "patched in"?

8.4 Elimination

Defecation seems to require little comment, but one feature deserves elaboration. Many morpholigically conservative eutherian mammals bury their feces. Since the excavation of the hole may involve forefeet and hind feet, only the hind feet, the dorsal scutes as in *Manis javanicus*, or the tail as in *Bradypus infuscatus*, I do not feel that we are dealing with homologues in all cases. Yet the digging and covering pattern for *Tenrec ecaudatus, Hemicentetes semispinosus, Dasypus novemcinctus*, and *Orycteropus afer* (see sect. 28.4) are remarkably alike. Does this imply the ancestral eutherian buried its feces in a similar manner? What is more interesting perhaps is that no metatherian does so. Does this relate to the fact that metatherians had an arboreal origin while eutherians were terrestrial or scansorial? If so, did the early eutherians bury their feces in part of a burrow system whereas the metatherians defecated from a limb outside an arboreal nest? Although this type of speculation is unanswerable, one further point should be made. Whatever the original selective advantage for burying feces—if it occurred at all in Cretaceous mammals—we cannot easily decide whether the mammals of the present continue to exhibit an ancient trait or have developed the pattern anew in response to entirely different selective pressures.

The use of urine for marking appears to be widespread in the metatherians and the eutherians. Locus-specific urination may be an ancient method of demarcating the limits of activity.

8.5 Locomotion and Digging

In the absence of extensive postcranial material from the Cretaceous, we can say little beyond that the multituberculates were terrestrial, as were pantotheres. The early metatherians were scansorial, and many contemporary forms show a long history of arboreality (Dollo 1899; Gregory 1951). Early eutherians seem to have been terrestrial or scansorial; however, they soon invaded arboreal niches. Many anatomical convergences and parallelisms can be discerned in eutherian lines adapting to arboreal life. The problem has been ably reviewed by Cartmill (1972).

Excavating burrows by digging with the forepaws and kicking back the accumulated earth appears to be an ancient coordination pattern. It is highly developed in the monotremes, some marsupials, and most conservative eutherians. It does not seem unreasonable to assume that this trait was well developed in ancestral therians.

8.6 Capture of Prey and Mastication

Pantotheres evolved their tribosphenic molars at a time when arthropods and angiosperm plants were undergoing a rapid phase of coevolution. The first experiments in herbivority by mammals appear to have been undertaken somewhat earlier by the Doconodonta and Multituberculata, and such adaptations were to follow later in the metatherian and eutherian lines. Insectivority evolving toward myrmecophagy and specializations for feeding on the colonial Isoptera and Hymenoptera apparently began early and was carried forward into the present by *Tachyglossus* (Monotremata), *Myrmecobius* (Metatheria), and many Edentata, Pholidota, and Tubulidentata. Insectivority in a more generalized sense was carried into the present by the Didelphidae, Dasyuridae, tenrecomorphs, erinaceimorphs, Scandentia, soricomorphs, and strepsirhine primates. The insectivorous Macroscelidea and some Carnivora may well represent a secondary adaptation. The modified tribosphenic molar of the conservative Marsupialia and Eutheria allowed both crushing and shearing. Some lateral movement of the lower jaw permits a grinding action (Crompton and Hiiemäe 1969). When these animals chew, the dominant motion is in the vertical plane, and a characteristic smacking sound is produced.

Prey capture in an insectivore or a small carnivorous form involves a coordinated use of the mouth and forepaws. Pinning small prey or grasping with the forepaws is present in both eutherians and metatherians (Eisenberg and Leyhausen 1972). The mouth is employed to disable and kill the prey. In some contemporary soricoid insectivores, poison glands provide an adjunct to prey killing, but there is no evidence that this is a plesiomorph character.

8.7 Courtship and Mating

The use of tactile and chemical signals during male-female interaction is widespread in the conservative metatherians and eutherians. The complex rituals involved in the precopulatory behaviors of *Ornithorhynchus* have already been described (see chap. 3). Sound appears important in metatherian and conservative eutherian patterns of courtship, especially clicks and hisses (see chap. 4). As I pointed out in a previous publication (Eisenberg 1977), copulation in conservative mammals is usually relatively long. I have suggested that the prolonged stimuli from copulation in primitive mammals is necessary to induce ovulation in the female, and that advances over this process involve adornment of the penis, which increases mechanical stimulation and reduces copulation time, and spontaneous ovulation by the female, with concomitant high synchrony between male and female courtship patterns.

The courtship and copulation of mammals has in the main remained rather conservative. Exchange of chemical stimuli, monitoring the female's estrous condition by sampling the urine or exudates from the genital tract, and exchange of tactile stimuli are common to marsupials, tenrecs, and most specialized mammals (see Eisenberg and Golani 1977; Eisenberg and Gould 1970; Eisenberg, McKay, and Jainudeen 1971). The neck grip by the male before and during mounting is displayed by *Ornithorhynchus,* many didelphid and dasyurid marsupials, and tenrecoid insectivores.

8.8 Parental Care

In the oviparous monotremes, the parental care patterns are prolonged and rather intricate. In *Ornithorhynchus,* nest-building, incubation of the egg, and brooding and suckling the young are highly developed. The immediate ancestors of the eutherians and metatherians may have been either oviparous or ovoviviparous. In either case the same sequence of brooding and lactation must have been highly evolved. With the development of a true nipple and a pouch, the marsupials could carry the young for a considerable period before a nest phase when they became too large for transport. The early eutherians independently evolved a nipple and probably bypassed the use of a pouch, depositing the neonate in a nest at birth.

Conservative metatherians (e.g., *Didelphis*) do not show a typical retrieving response by picking up an infant in the mouth when it is displaced from the nest. Rather, they respond to the chirping call of the young by approaching and drawing it under them where it can find the pouch opening. Conservative eutherians typically respond to the call of a displaced infant by retrieving it in the mouth and transporting it to a nest site (e.g., *Euphractus, Elephantulus, Tenrec,* and *Microgale*) (Encke 1965; Rathbun 1979; Eisenberg and Gould 1970).

It appears that the common ancestors of the Metatheria and Eutheria, as well as the early derivatives of these two major radiations, had the basic communication and interaction systems between female and young that we see today in the most conservative contemporary forms. This should come as no surprise. The social system, interaction patterns, and parental care of present-day crocodilians are as advanced as those of many conservative eutherians and metatherians and probably reflect a nearly comparable level of neural organization (Hopson 1977).

Part 2

Mammalian Radiations on the Contiguous Continental Land Masses

North America has been in contact with Eurasia several times since its dissociation from East Asia in the Eocene. Most recently, the Pleistocene allowed a major faunal interchange. North America did not come into permanent contact with South America until the Pliocene. After detaching from South America, Africa enjoyed considerable isolation until more regular contact was established with the Near East in the Eocene. In spite of the intermittent contacts among the northern major continents, faunal exchange has been more consistent among them than exchanges with either Australia or South America. Perhaps I am oversimplifying somewhat by lumping the rest of my discussion of mammalian adaptation under the heading "the contiguous continental land masses," but in a relative sense these major land masses had more faunal interchanges during the Cenozoic before the Pliocene than did the southern continents of South America and Australia. Africa is somewhat intermediate in its isolation and contains much evidence of endemism.*[1] The Tethytheria (McKenna 1975) differentiated in Africa, and the Macroscelidea evolved there from an obscure ancestry.

1. Supplementary notes added while this book was in press, indicated by asterisks in the text, appear on page 503.

The Order Macroscelidea: The Elephant Shrews

The elephant shrews represent one of the unique radiations of Africa. Recently McKenna (1975) has suggested an affinity between the elephant shrews and the Lagomorpha (rabbits and hares). Certainly the lineage for the elephant shrews has been distinct for a considerable period (Patterson 1965). Ecological studies have been carried out on *Macroscelides proboscideus* (Sauer and Sauer 1971, 1972). Recently Rathbun (1976, 1979) has concluded a study on the ecology and behavior of *Elephantulus rufescens* and *Rhynchocyon chrysopygus*.

The order Macroscelidea is characterized by contemporary forms adapted for a cursorial, digitigrade, quadrupedal, ricochetal mode of progression. The eye is rather large, and most species tend to be crepuscular or nocturnal. The genus *Rhynchocyon,* however, is adapted for diurnality. The snout is prolonged, extending well beyond the nasal bones, and is rather flexible. It is apparently sensitive to touch, and the animal employs the proboscis in probing for arthropods among leaf litter and loose soil. The sounds of disturbed arthropods moving pinpoint their locations for the foraging elephant shrew. Some species, such as *Elephantulus rufescens,* feed seasonally on termites during their annual production of imagoes. The animals are not adapted for digging into termite mounds.

Elephant shrew females give birth to one or two rather precocial young. The most altricial condition at birth appears to be demonstrable in the young of the genus *Rhynchocyon*. The young rest in a crevice (*Macroscelides proboscideus*) or leaf nest (*Rhynchocyon chrysopygus*) and are nursed by the female only at 24-hour intervals. This type of maternal care system has been referred to as an "absentee parental care system" (see Eisenberg 1975*a*). When very small, the young of *Elephantulus rufescens* may be moved by the female. She employs a classical eutherian behavior pattern, carrying them in her mouth. The lactation period in the elephant shrew female is rather short. The young begin to forage independently, with infrequent nursing, 10 days after birth.

The elephant shrew has a monogamous mating system. Rathbun (1976, 1979) has demonstrated that males and females jointly occupy a territory that they defend against intrusions by like-sexed strange elephant shrews. Males defend against other males; females defend against foreign females. The young are allowed to live within the parental territory until they reach sexual maturity. Interaction between a given male and female on their territory is minimal, and one may have the illusion that the animals are in fact solitary. The male spends a great deal of time in trail cleaning. He pushes debris away from the surface of the ground with the forepaws, thus maintaining clear escape routes within the territory if any family member should be surprised by predators. Territorial marking with chemical secretions includes marking with pedal glands and sternal glands.

General maintenance behavior patterns do not differ from those displayed by other conservative eutherian mammals. Scratching with the hind foot is one of the principal methods of pelage dressing. *Elephantulus* demonstrates stereotyped face washing, using both forepaws simultaneously to comb both sides of the face. This trait has been lost in the genus *Rhynchocyon*.

Nest-building behavior in *Rhynchocyon* is reminiscent of the nest-building pattern in dasypodine edentates. Each evening an adult makes a sleeping nest by gathering leaves together with the forepaws. Leaves are swept under the body, and the animal moves backward with the leaf mass, only to pass over it and kick or scratch the leaves backward again. Gradually a small pile of leaves accumulates, whereupon the animal burrows into the center to bivouac for the night.

10 The Order Lagomorpha: Hares, Rabbits, and Pikas

The order Lagomorpha does not exhibit great diversity of species and genera. Yet before the advent of modern man it was found on all continents except Australia and Antarctica. Classically, the order is divided into two families, the Ochotonidae, or pikas, and the Leporidae, including the hares and rabbits (see plate 12). Although the lagomorphs have ever-growing incisors and a dentition superficially similar to that of the Rodentia, the incisors are unique. Each incisor has enamel on both the posterior and the anterior surfaces, and, further, the upper jaw has two pairs of incisors, the second directly behind the first.

Originally, hares and rabbits were absent from Australia and only marginally represented in South America, the genus *Sylvilagus* having invaded from North America during the Pleistocene. Europeans from the fifteenth century to the eighteenth century introduced rabbits and hares to both Australia and the southern part of South America. The pikas are more limited in their distribution and are confined to the mountainous areas of East Asia and western North America.

The young of the pikas, *Ochotona,* are born sparsely haired and with their eyes closed. The young of the European rabbit (*Oryctolagus*) are born in an even more altricial state, with virtually no hair and with

Plate 12. *Lepus nigricollis,* the black-naped hare. This Indian hare typifies the lepid body plan. Elongated hind feet and large ears make most lagomorphs easy to recognize around the world. (Photo by author.)

closed eyes. However, the young of most species of the genus *Sylvilagus* are born well furred, and the eyes are closed for only a few days after birth. The genus *Lepus* is characterized by young that are quite precocial, born with eyes fully open and possessing fur.

Ochotona can be thought of as the most conservative genus within the order. Female pikas generally bear their young in a tunnel system, in a nest constructed by the female. The European rabbit, *Oryctolagus,* digs a complicated burrow system and builds a nest of fur plucked from the mother's chest. The genus *Sylvilagus* is characterized by a modest nest, generally little more than a depression in the ground, but lined with fur plucked from the mother's chest and often covered loosely with vegetation. The precocially born hares of the genus *Lepus* rest in a shallow depression that may be partially covered with grasses. One can note here a phylogenetic tendency toward less and less complex nest construction as the young are born in a more precocial state.

The rhythm of lactation shown by female hares and rabbits is worth remarking upon. The young are generally nursed at rather long intervals, usually once every 24 hours. Hares and rabbits thus exhibit an "absentee parental care system." Picking the young up in the mouth to move them to a new place is not part of the normal female repertoire in the Leporidae. The problem has yet to be investigated in *Ochotona*.

10.1 The Family Ochotonidae

In both the New World and the Old World, pikas inhabit alpine habitats. Most species form colonies in areas of rock slides or talus slopes. During the summer months, they gather a variety of grasses and forbs and generally place them in the sun to dry out. The dried material is then stored in rock crevices as a food supply during the winter. The ecology of the pika has been studied in some detail by Severaid (1956), Broadbrooks (1965), and Kawamichi (1976) in North America, while Kawamichi (1968, 1970, 1971*a,b*) and Hagga (1960) have studied the ecology of the Asiatic species. As a generalization, pikas are diurnally active and are colonial; but within a colony individuals are spaced and appear to occupy exclusive home ranges during at least part of the year. During the reproductive season in the North American pika, a male and a female may occupy a home range jointly, although in some areas a male's home range may overlap those of more than one female.

The pikas are rather vocal. Somers (1973) has described two types of calls employed in distance communication among members of a family or among neighbors in a colony. A "short call" is employed when a territory has been invaded by a strange conspecific or when an aerial or terrestrial predator is sighted. The pika also produces a rather long, complicated call referred to as a "song." A song sequence can last up to half a minute and is produced predominantly by males. In Asiatic pikas, singing apparently is done by both males and females. Dialect variations between separate populations of pikes have been described for the "short calls" (Somers 1973). In addition to auditory communication, the pikas have a well-developed system of chemical signals. Feces are generally placed at one locus in the home range and could serve as a source of chemical information to neighbors. Pikas have apocrine glands on the cheeks (Harvey and Rosenberg 1960), and their glandular activity shows seasonal shifts, with secretions predominating during the period of reproduction. Cheek rubbing is a conspicuous form of marking.

10.2 The Family Leporidae

Although rabbits and hares may produce several types of sounds, auditory communication is less conspicuous than in the Ochotonidae. On the other hand, the system of chemical signals that rabbits employ during their social interactions is complex and has been well studied. A dominant male European rabbit, *Oryctolagus cuniculus,* has active anal glands that impart an odor to the hard fecal pellets, which may be deposited at specific dung piles (Mykytowycz 1966*a*). The size of a male's anal gland correlates rather well with his dominance status in a colony. In addition, adult male rabbits have active inguinal glands that play a role in sexual attraction (Mykytowycz 1966*b*). Harder's gland is dimorphic in its activity and size, being dependent on androgen in the male (Mykytowycz 1966*c*). Adult male rabbits have an active submandibular gland, and secretions are deposited by the characteristic "chinning" behavior, where the male rubs his chin on various objects in the environment and on conspecifics.

The European hare, *Lepus europaeus,* does not show strong dimorphism or seasonal change in the size of its anal glands, but there is seasonal activity and conspicuous sexual dimorphism in the size of Harder's glands (Mykytowycz 1966*c*). Both male rabbits, *Oryctolagus,* and hares, *Lepus,* exhibit enurination during courtship. During approaches to the female, the male may urinate directly on her, or he may urinate on her while leaping over her (Southern 1948). Enurination during courtship also predominates in caviomorph rodents (Kleiman 1974*a*).

The European rabbit forms warrens colonies of related individuals. The burrows may be used for many generations, and a given warren contains several adult females. The females may develop a dominance order, thereby restricting breeding to the top females. This is generally accomplished by restricting the use of favorable dens to dominant females for rearing their young. The males associated with a warren form a dominance hierarchy, but generally one adult male has access to all the breeding females (Southern 1948). In contrast, hares do not show such a formal social structure. This may partly be related to the failure of hares to construct complicated burrow systems that tie several generations to a fixed site.

11 The Order Rodentia

11.1 Introduction

The rodents constitute a distinctive order in the class Mammalia. Contemporary rodents show a bewildering variety of adaptations, and the subordinal classification of rodents has met with extreme difficulties (Wood 1955). Classically, rodents have been divided into three suborders: Sciuromorpha, Myomorpha, and Hystricormorpha (Simpson 1945). The artificiality of such a tripartite division has been in part resolved by Wood (1965), who refined a superfamilial classification. Some problems in the subordinal classification of rodents derive from the high frequency of convergent or parallel evolution (or both). The initial radiation of the Rodentia in the Eocene was almost "explosive," and breaks in the fossil record make it difficult to form natural groupings of the families at the subordinal level. The living families of rodents are distinct, however, and it may be wise to rely on a superfamilial or familial terminology rather than to revert to the old trichotomy of myomorphs, hystricomorphs, and sciuromorphs (see fig. 19).

The subordinal classification adopted by Simpson was grounded in the morphology of the jaws and their associated musculature. The hystricomorph rodents exhibit an enlarged infraorbital foramen with a slip from the masseter muscle passing through it to attach anteriorly on the rostrum (fig. 20). The jaw is sharply flared posteriorly and is distinct from the more gradual arc described by the lower jaw of other rodent families (see Landry 1957).*

The sciuromorph rodents typically show no enlarged infraorbital foramen, and the masseter attaches anteriorly on the rostrum and beneath the zygomatic arch (see fig. 20*c,d*). The typical myomorph has a small slip of masseter muscle passing through a slightly enlarged infraorbital foramen, but most of the anterior attachment of the masseter is similar to that described for the Sciuromorpha (fig. 20*e,f*).

Stehlin and Schaube (1951) had begun to study rodent teeth and proposed that the Sciuridae exhibited a conservative "trigonid" pattern. Their classification of the Rodentia based on cusp patterns was adopted in the *Traité de zoologie* (Grassé 1955*a*). They once again showed the New World and Old World hystricomorph rodents to be derivable from a common ancestor, a view supported by Landry (1957) using a wide variety of morphological criteria. Wood (1965, 1974) has held to the view that Old World and New World hystricomorphs exhibit convergences but were independently derived. Lavocat (1974) presents the counterargument. I have sided with Lavocat and Landry in figure 19.

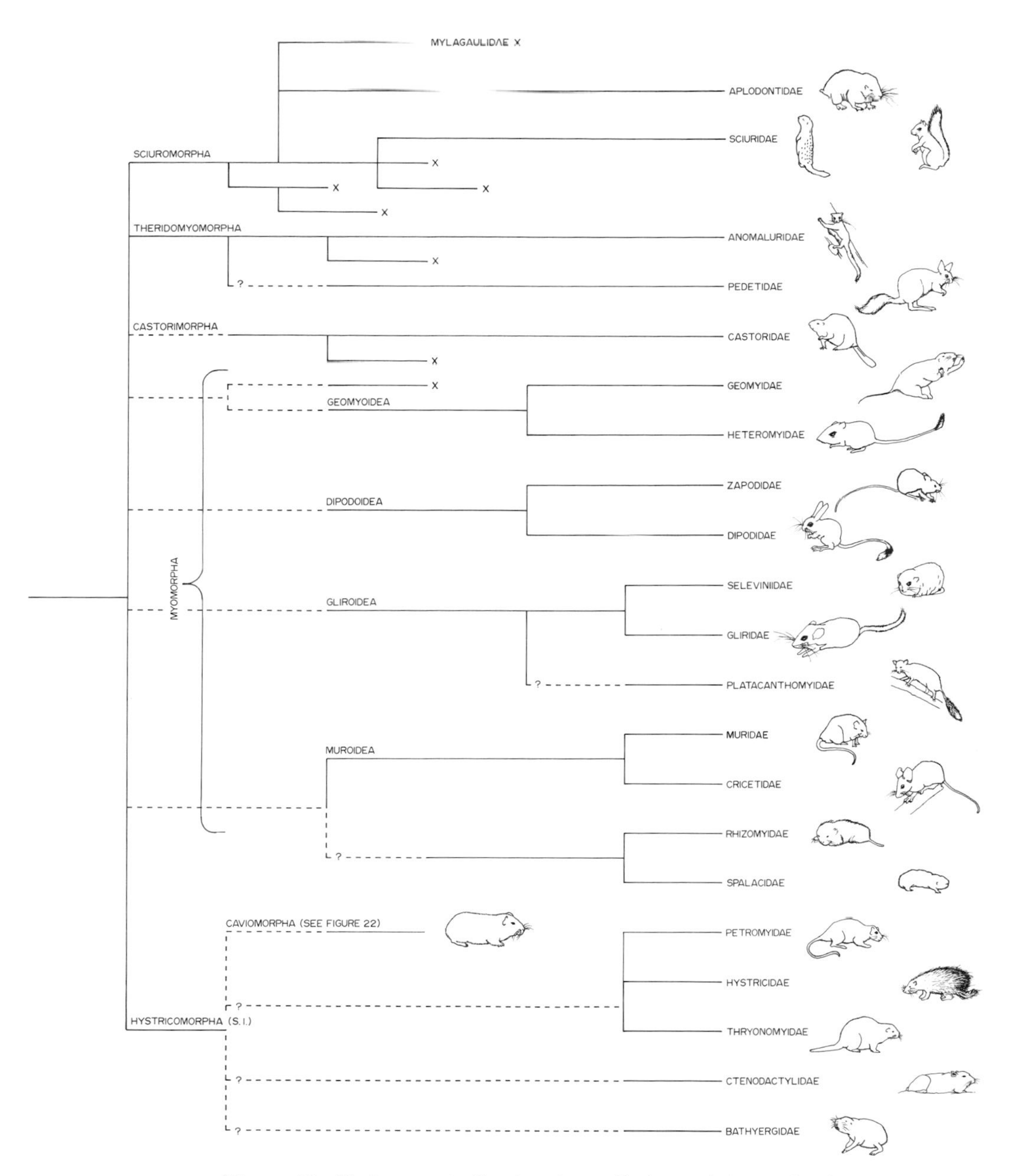

Figure 19. Cladogram indicating the radiations of the Rodentia. Scheme based mainly on Wood 1955 and Simpson 1959.

The ecology (see Snyder 1978) and the behavior (see Eibl-Eibesfeldt 1958) of rodents have been analyzed for years. Much of our knowledge of mammalian population dynamics, physiology, psychology, and psychophysics derives from pioneering studies on rodents (Munn 1950). It is difficult indeed to imagine how much of our basic knowledge concerning mammalian physiology could have been derived without their use.

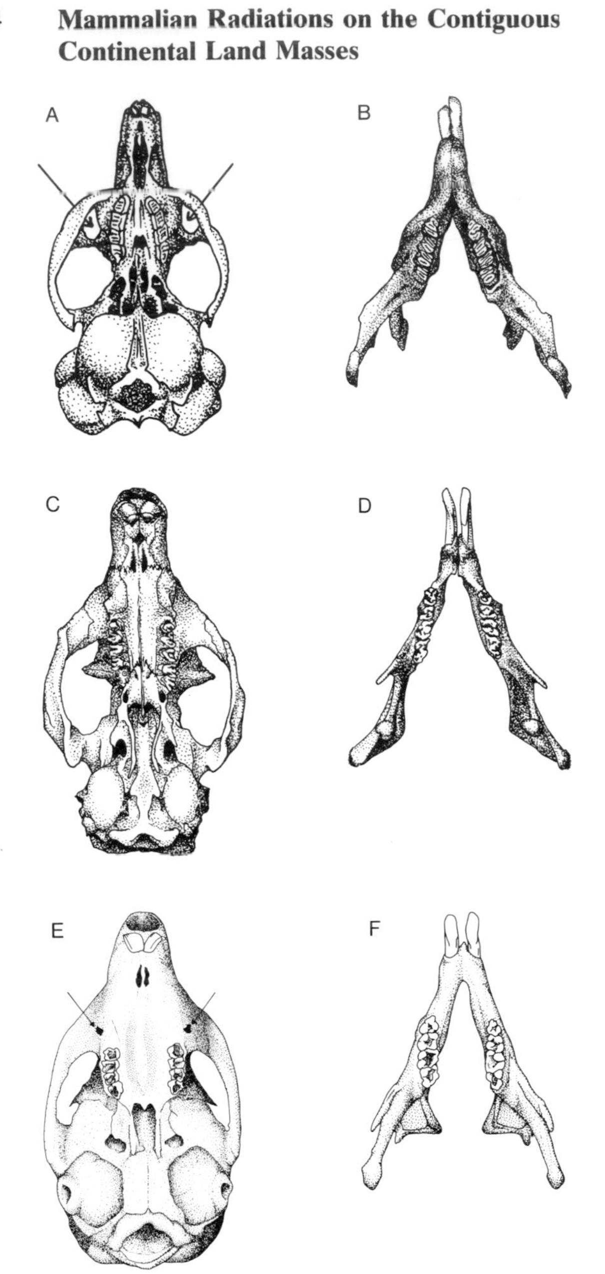

Figure 20. The three major skull types of the Rodentia. (*a*) Dorsal view of the lower jaw of *Chinchilla*. (*b*) Ventral view of the skull of *Chinchilla*. (*c*) Ventral view of the skull of *Marmota*. (*d*) Dorsal view of the jaw of *Marmota*. (*e*) Ventral view of the skull of *Peromyscus*. (*f*) Dorsal view of the jaw of *Peromyscus*. Arrows indicate enlarged infraorbital foramina for anterior attachment of masseter muscles in *Chinchilla* and *Peromyscus*.

11.2 The Sciuromorpha and Related Families

11.2.1 The Family Aplodontidae: The Mountain Beaver

The family contains only one living species, *Aplodontia rufa,* which inhabits the northwestern portion of the United States and southern British Columbia. The species has a number of conservative features with respect to its jaw musculature and is considered the most morphologically conservative of living rodents. It has a number of apomorphic characters, however, that are a consequence of becoming adapted for a fossorial life. It maintains an extensive tunnel system in which it stores food and constructs nests. The animal is strongly specialized for an herbivorous diet and does not hibernate, but rather subsists on food stored in its burrow system. The litter size averages about three young, which are born in an altricial state with the eyes closed and are relatively hairless. Breeding usually does not occur until the animal is at least two years old. The animals are not easily maintained in captivity, so that very few data concerning maximum longevity are available to us. They are of extreme interest, however, since the family has a geological history dating from the Paleocene (Anthony 1916; Martin 1971).

11.2.2 The Family Sciuridae: The Squirrels

The Sciuridae are a diverse family of rodents exhibiting a number of conservative features in the jaw musculature and dentition. There are more than fifty genera, distributed worldwide except for Madagascar and Australia (Walker et al. 1964). The family reaches its greatest diversity in terms of species numbers in Southeast Asia, though it is nearly as diverse in Africa. It radiated in South America quite recently, and nowhere in the Neotropics do the Sciuridae show the variety of forms attained in the Paleotropics.

It is convenient to consider three basic adaptations: the tree squirrels, the gound squirrels, and the so-called flying squirrels. The flying squirrels are set aside in their own subfamily (Petauristinae) and have evolved a skin membrane between the forelimbs and hind limbs that they can extend by flattening the body. In leaping they are able to glide a considerable distance. The greatest diversity of sciurid flying squirrels occurs in Southeast Asia (Muul and Lim 1978). From Asia as a presumed center of adaptive radiation, they have spread to North America. The flying squirrel niche is occupied in tropical Africa by a convergently evolved rodent family, the Anomaluridae. In Australia the niche has been filled in part by several species of the marsupial family Phalangeridae.

The tree squirrels and ground squirrels show a strong tendency for diurnality, whereas the flying squirrels are almost invariably nocturnal. In conformity with the diurnal adaptation, the squirrels have rather large eyes, and those genera that have been tested (*Sciurus, Spermophilus, Cynomys,* and *Tamias*) demonstrate an ability to discriminate spectral hues and thus may be said to have "color vision" (Jacobs 1978). Most members of the family are specialized for feeding on nuts and fruits, with arthropods as a dietary supplement (see plate 13). On the other hand, many of the terrestrial sciurids, such as *Cynomys* and *Marmota,* utilize grass and forbs and are true grazers. Some genera of tree squirrels have specialized for arthropod feeding, most notably the long-nosed squirrel, *Rhinosciurus,* of Southeast Asia.

The nut- and seed-feeding squirrels generally hoard food in burrows, in terrestrial scatter hoards, or in tree cavities, and these species are

Plate 13. *Sciurus carolinensis,* the North American gray squirrel, typifies the squirrel family. Note the large eye, which correlates with its diurnal arboreal adaptation. (Courtesy of U.S. Fish and Wildlife Service.)

often highly territorial. In contrast, the ground squirrels may exhibit colonial to communal behavior patterns, and a whole range of social tolerances is demonstrable within a single genus. As I noted previously, some ground squirrels have adapted to rather open habitats and have specialized for feeding on herbaceous vegetation, thus departing from the strong adaptation for feeding on nuts and fruits; most typical of this is the North American genus *Cynomys* (see plate 14). All the squirrels give birth to altricial young, with little hair and the eyes closed. (Illar Muul informs me that *Petinomys genibarbis* has a single, precocial young.)

Some Life-History Sketches for the Sciuridae

The eastern chipmunk, *Tamias striatus*. The genus *Tamias* is adapted for exploiting the deciduous forests of eastern North America. The adults typically defend an area immediately around their burrow system, but some home-range overlap is shown at the periphery. The burrow is assiduously defended and contains hoarded foods that are utilized not only in wakeful periods during winter torpor but also in seasons of food scarcity during the summer months. The chipmunk hoards considerable quantities of nuts and seeds. It also feeds on a variety of nonhoardable food items, such as fungi, arthropods, and even small vertebrates.

Plate 14. *Cynomys parvidens,* the Utah prairie dog, a member of the squirrel family adapted for burrowing. Praire dogs are colonial. (NZP archives.)

When a female is in estrus, adjoining males may gather on her home range and engage in successive sexual pursuits. During mating bouts a loose dominance order exists based on previous aggressive bouts among the males; however, several males may mate successively with the female during the day she is in estrus. Elliott (1978) concludes that a dominant male chipmunk may be responsible for earlier copulations during the estrous period, but it is by no means proved that the dominant male is likely to sire more young.*

Such male aggregations and mating chases have also been described for the eastern gray squirrel, *Sciurus carolinensis* (Horwich 1972) and the tassel-eared squirrel, *S. aberti* (Farentinos 1972). Exclusive mating may occur in the red squirrel, *Tamiasciurus hudsonicus,* where a male is able to maintain a competitor-free area around a given estrous female (Smith 1968).

After a gestation of some 31 days, the young are born underground in a grass-lined nest. At approximately 44 days of age, the young of *Tamias* begin to emerge from the burrow system. They are prone to follow the female on short excursions, and for a few days they form a littermate group and play together. Approximately 4 days after emergence from the burrow, the female begins to show more intolerance toward the young, frequently chasing them. The young begin to set up their own home range areas within 15 to 18 days of emergence, and they increasingly show more adultlike aggressive chasing toward one another. Juveniles that can obtain a burrow system recently vacated by an adult that has died will probably survive the winter; but winter survival depends completely on establishing ownership of a burrow and its associated cache (Elliott 1978).

The eastern chipmunk exemplifies a dispersed social system with individual territorial defense of the burrow and its immediate vicinity. This type of social system is characteristic of rodents that store large quantities of seeds and grain in their burrow system, regardless of the habitat type they exploit. A similar system is shown by the desert-adapted heteromyid rodents of the genera *Perognathus, Microdipodops,* and *Dipodomys* (Eisenberg 1963).

The yellow-bellied marmot, Marmota flaviventris. The species of the Holarctic genus *Marmota* are adapted for feeding on herbaceous vegetation. They do not typically hoard food but rather rely on accumulated fat reserves as a means of passing the winter months while in profound torpor. Nevertheless, sufficient foraging areas must be available to each individual, and aggressive defense of the vicinity of the home burrow is common. Adult females are usually well spaced, but in habitats with a high carrying capacity females may defend small areas. In such habitats a single dominant male may defend an area large enough to contain several adult females, and a polygynous mating system will result (Downhower and Armitage 1971). Young do not disperse during the season of their birth but utilize the maternal home range as a foraging area. Spring dispersal of the yearlings tends to be the rule at higher elevations (Armitage 1962).

In the Olympic marmot (*Marmota olympus*), the young may not disperse from the natal home range until their second or third year. In this species an extended family group consisting of an adult male, one or two adult females, and two generations of offspring is demonstrable (Barash 1973). This colonial structure based on a family unit exhibits a constellation of the following characters: reduced annual reproductive capacity per individual female; a slow maturation rate for the young; and a greater potential longevity for individuals. Generational overlap in one social unit sets the stage for the development of complex interdependencies and the origins of reciprocal altruism (see chap. 31).

The two brief life-history sketches cannot begin to do justice to the literature on the Sciuridae. I have summarized some recent literature in table 15 that should serve as a guide to further readings.

11.2.3 The Superfamily Geomyoidea

Two families are contained within the superfamily Geomyoidea, the Geomyidae or pocket gophers and the Heteromyidae, including the pocket mice and kangaroo rats. The superfamily is confined to North and Central America and parts of northern South America and is characterized by externally opening fur-lined cheek pouches. The Geomyidae are strongly specialized for a fossorial way of life. The Heteromyidae are highly adapted for surface foraging, generally subsisting on caches of seeds and nuts. The heteromyids are all excellent burrowers and accumulate vast food reserves in the burrow system or

Table 15 Recent Life History Studies of the Sciuridae

Taxon	Ecology	Behavior	Reproduction and Ontogeny
Sciuridae			
Sciurinae			
Sciurini			
Sciurus vulgaris	—	Eibl-Eibesfeldt 1951*a*	Eibl-Eibesfeldt 1951*a*
S. carolinensis	Barkalow, Hamilton, and Soots 1970	Horwich 1972	Horwich 1972
S. aberti	—	Farentinos 1974	—
Tamiasciurini			
Tamiasciurus hudsonicus	Smith 1968; Hamilton 1939	Kilham 1954; Smith 1968	Layne 1954
Callosciurini			
Callosciurus swinhoei	VanTien 1966	—	—
Funambulini			
Funambulus pennanti	Prakash, Kametkar, and Purohit 1968	—	Prakash et al. 1968
Xerini			
Atlantoxerus getulus	—	Poduschka 1971	—
Xerus erythropus	—	Ewer 1965	—
Marmotini			
Marmota marmota	Müller-Using 1956	Koenig 1957	—
M. monax	Hamilton 1934	Bronson 1963, 1964; Lloyd 1972	—
M. flaviventris	—	Waring 1966; Armitage 1962	—
M. olympus	Barash 1973	Barash 1973	—
Cynomys ludovicianus	Koford 1958	King 1955; Smith et al. 1973	—
C. leucurus	—	Waring 1970	Bakko and Brown 1967
Spermophilus (=*Citellus*) *undulatus*	—	Watton and Keenleyside 1974	—
S. richardsoni	—	Yeaton 1972	—
S. beecheyi	—	Lindsdale 1946	—
S. columbianus	Manville 1959	Steiner 1970, 1971, 1973	—
Spermophilus armatus	—	Balph and Stokes 1963; Balph and Balph 1966	—
S. lateralis	McKeever 1964	Gordon 1943; Wirtz 1967	—
S. tridecemlineatus	McCarley 1966	—	—
Tamias striatus	Elliott 1978	Dunford 1970	Elliott 1978
Eutamias amoenus	Broadbrooks 1958	—	—
Petauristinae			
Glaucomys volans	Muul 1968	Muul 1968	Sollberger 1943
G. sabrinus	—	—	Muul 1969

its vicinity. Defense of food caches and exclusive use of burrow systems characterizes both the Heteromyidae and the Geomyidae.

The Family Heteromyidae: Kangaroo Rats and Pocket Mice

The kangaroo rats (*Dipodomys*), kangaroo mice (*Microdipodops*), and pocket mice (*Perognathus, Liomys,* and *Heteromys*) have had a long evolutionary history in North America (Wood 1931). The family portrays among its five genera a series of adaptations to arid habitats. At one extreme *Heteromys* is confined to the moist woodlands of Central America and northern South America (see plate 15). *Liomys* is spe-

Plate 15. *Heteromys anomalus*. This heteromyid rodent exhibits one of the more conservative body plans of the family. From a generalized form such as this, the bipedal kangaroo rats evolved. (Photo by author.)

cialized for more xeric habitats (Wagner 1961) but occurs sympatrically with *Heteromys* in Panama (Fleming 1971). The genus *Perognathus* shows an enduring trend toward the exploitation of arid habitats and is conservative in many morphological features. *Microdipodops* is a Great Basin form derived from an ancestral perognathine, while *Dipodomys* shows an early independent phylogenetic origin with adaptations for bipedal, ricochetal locomotion (Eisenberg 1963; Howell 1932). The locomotor adaptations shown by *Dipodomys* have been convergently evolved by desert-dwelling rodents from other taxa: *Notomys* (Muridae), *Macrotarsomys* (Nesomyinae), *Pedetes* (Pedetidae), and *Jaculus* (Dipodidae). The phenomenon of behavioral and morphological convergence in desert rodents has been reviewed by Eisenberg (1975*c*).

Heteromyid rodents have been the subjects of studies on hearing (Pye 1965; Webster 1962); behavior (Eisenberg 1963; Kenagy 1976; Rood 1963); reproduction and development (Eisenberg and Isaac 1963; Fleming 1977; Butterworth 1961*a*); and foraging ecology (Brown 1975; Rosenzweig 1973).

The Family Geomyidae: The Pocket Gophers

This family illustrates well the morphological specializations correlated with fossorial adaptations. The eyes are reduced in size, as are the pinnae. The incisors are procumbent, and the manus is broad, with long, stout claws on the digits. The fur is short and the tail is reduced in length. Convergence in morphology with the adoption of fossorial life is shown by the Bathyergidae, Ctenomyidae, Spalacidae, Rhizomyidae, Myospalacinae, and the genera *Ellobius* (Microtinae) and *Spalacopus* (Octodontidae) (see Nevo 1979).

The New World pocket gophers represent an interesting radiation, having adapted to a variety of soil types (Miller 1964), and their specific adaptations to microhabitats have been studied in some detail (Best 1973).

Although pocket gophers tend to occupy exclusive burrow systems as adults and fight viciously if brought together outside the breeding season (Scheffer 1931), plural occupancy of burrow systems has been noted (Hansen and Miller 1959). In *Thomomys talpoides,* the greatest plural occupancy of burrows occurs in the breeding season, and the commonest combination is a male and a female. Russell (1968) reports in his study of *Pappogeomys* that the social structure may be more subtle. Although nonbreeding adults may be found in exclusive burrows, the local populations seem to be spatially segregated into demes consisting of one adult male and several adult females. The reproduction behavior and development of pocket gophers has been summarized by Andersen (1978) and Schramm (1961).

11.2.4 The Family Castoridae: The Beavers

This family contains one genus—*Castor,* the beaver—distributed in both North America and Europe. Beavers are extremely large rodents that may weigh up to 32 kg. They are adapted for aquatic life, having webbed feet, and are well known for constructing burrows near water, dams to impede water flow, and lodges in small ponds that offer them great protection from predators. Beavers are specialized for feeding on the bark, cambium, twigs, and leaves of deciduous trees. They establish colonies that generally consist of a mated pair and their various-aged progeny. The social organization of beavers has been the subject of extensive study (Bradt 1938; Hediger 1970; Richard 1967, 1968; Wilsson 1968, 1971).

11.2.5 The Family Anomaluridae: The Scaly-tailed Flying Squirrels

The scaly-tailed flying squirrels exhibit a locomotor pattern convergently evolved with that of the true flying squirrels of the family, Sciuridae. They are confined to West Africa. In common with the other forms of flying squirrels, these species habitually den in tree hollows. One genus of the family, *Zenkerella,* does not have a gliding membrane, and it also exhibits a diurnal activity rhythm. The remaining genera of the anomalurids are nocturnal, in conformity with the adaptive trend already established for the true flying squirrels. Their natural history has been summarized by Kingdon (1974*b*).

11.2.6 The Family Pedetidae: The Springhaas

This family contains one genus with two species, *Pedetes capensis* and *P. cafer.* In general morphology they resemble small kangaroos. They are nocturnal, bipedal ricocheters living in semiarid habitats. In this adaptation they express a convergence with the kangaroo rats of the family Heteromyidae and the jerboas of the family Dipodidae toward bipedal locomotion. Bipedal ricochetal locomotion has convergently evolved in rodent genera where the species is above a minimum size and forages in exposed open habitats characteristic of arid and semiarid regions (see Eisenberg 1963, 1975*c*). *Pedetes* constructs complex burrows, occasionally organized in a colony; however, exclusive occupancy by an adult or a pair appears to be the rule. The animal is adapted for feeding on vegetation and plant roots. The young are born in a near precocial state, with fur present and the eyes opening a short time after birth (Hediger 1950). The natural history is summarized by Kingdon (1974*b*).

11.3 The Superfamily Muroidea: Mouselike Rodents

The superfamily Muroidea includes the Rhizomyidae, or bamboo rats; the Spalacidae, or mole rats; the Cricetidae; and the Muridae. Figure 21 illustrates the relations between the various families and breaks down the cricetid and murid rodents to subfamilies.

The Rhizomyidae and Spalacidae may be artificially lumped in this superfamily, but the Muridae and Cricetidae clearly are closely related. So diverse are the genera, with so many clear cases of parallel and convergent evolution, that assignment to families is difficult. It is now generally conceded that the Cricetidae represent an older radiation once dominant in North Asia, Europe, and Africa which then spread to North America, where it dominated and spread to South America. The Muridae evidently branched from an early cricetinelike form and radiated in southeast Asia. They then spread to Australia, Europe, North Asia, and Africa. The African subfamilies Otomyinae, Petromyscinae, Cricetomyinae, and Dendromurinae were once thought to be in the Muridae, but recent research (Petter 1967; Lavocat 1959, 1967) suggests that these subfamilies represent early offshoots from a cricetine ancestor. The problem is well summarized by Kingdon (1974*b*).

11.3.1 The Family Cricetidae: Voles, Hamsters, and Their Relatives

This family of more than a hundred genera is distributed over the entire world with the exception of Australasia. It successively reinvaded South America at the time of the Pliocene land bridge formation. A subfamilial breakdown of this highly diverse family is necessary if we are to comprehend certain adaptive trends.

The Microtinae represent a North American and Eurasian subfamily with a tendency toward terrestrial, semifossorial adaptations; that is, the eye tends to be somewhat small, the tail reduced in length, the ear small, and the neck short. The species have adapted to feeding on herbaceous vegetation. A high degree of adaptation to cool, boreal climates may be strongly expressed in some genera (e.g., *Lemmus, Dicrostonyx, Microtus,* and *Prometheomys*). On the other hand, *Lagarus* is adapted to xeric habitats. Unique specializations for feeding arboreally on fir needles have evolved within the genus *Arborimus* (=Phenacomys) *longicaudus* (Hamilton 1962).

The subfamily Gerbillinae includes a number of genera adapted for arid and semiarid regions widely distributed throughout North Africa, the Middle East, and Central Asia (Petter 1961). In morphology and behavior patterns, certain genera (e.g., *Gerbillus*) converge strongly toward the semiarid-adapted pocket mice of the New World heteromyid genus *Perognathus* (Eisenberg 1967, 1975*c*) (see plate 16). A synopsis

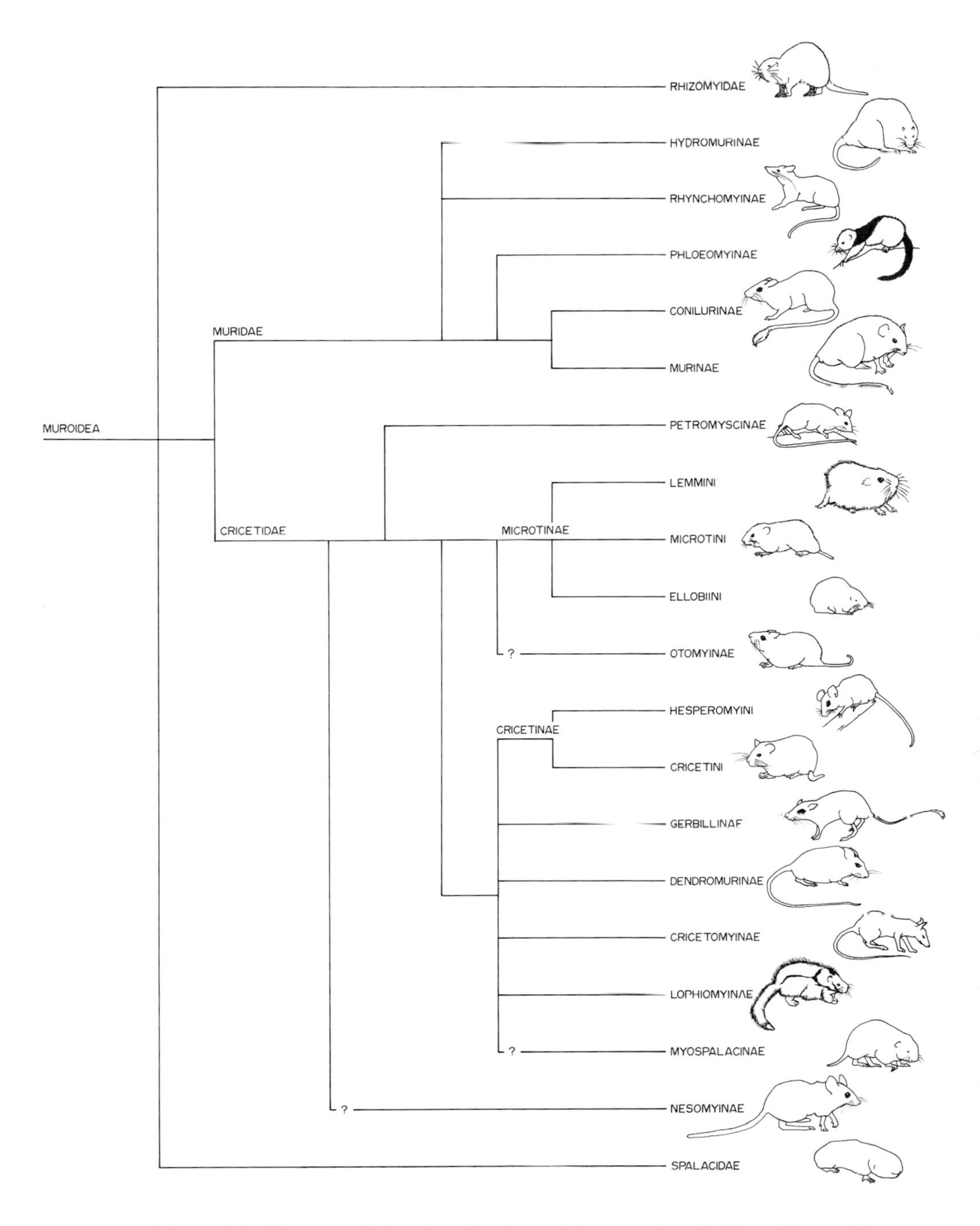

Figure 21. Diagram of the classification of the superfamily Muroidea, with modifications according to Petter 1967 and Lavocat 1959, 1967.

of recent literature on the natural history of the Gerbillinae is included in table 16.

The subfamily Nesomyinae is perhaps an artificial taxonomic category, but it conveniently encompasses the primitive cricetines of Madagascar. The range of adaptation shown by the nesomyine rodents on the island of Madagascar has been recently reviewed by Petter (1972).

Plate 16. *Gerbillus gerbillus,* the gerbil of North Africa, typifies adaptation to a desert environment by small rodents. Note the pale coat color, which serves to minimize the animal's conspicuousness when it forages on desert sands. (Photo by author.)

The Myospalacinae are a monotypic subfamily with one genus, *Myospalax,* found in China and boreal East Asia. They are strongly adapted for a fossorial life, having converged in morphology with the Spalacidae and Geomyidae. Some aspects of their biology have been summarized by Chen et al. (1966).

The Otomyinae represent a relict group of African cricetines. They are volelike in appearance and feed on herbaceous vegetation. Aspects of the ecology and behavior of *Otomys* are discussed by Rahm (1967), Davis (1972), and Kingdon (1974*b*).

The Petromyscinae (Petter 1967) are African endemics and include two genera, *Delanymys* and *Petromyscus. Delanymys* is arboreal and adapted to rain forests. *Petromyscus* is confined to the semiarid areas of southwestern Africa.

The Cricetomyinae are confined to Africa and represent a unique radiation of three genera, *Beamys, Cricetomys,* and *Saccostomus.* They possess internally opening cheek pouches that permit them to collect seeds and transport them to the nest and tunnel system. *Saccostomus* is short-tailed and hamsterlike, while *Cricetomys,* though large, has typical ratlike body proportions. The biology of this subfamily is summarized by Kingdon (1974*b*), Hanney and Morris (1962), Egoscue (1972), and Davis (1963). Ewer (1967) described the behavior patterns of *Cricetomys gambianus.*

The subfamily Lophiomyinae contains a single species, *Lophiomys imhausi.* This relict form has a unique external appearance and color pattern that may be aposomatic (Kingdon 1974*b*). In contrast to the young of most muroid rodents, newborn *Lophiomys* are precocial, fully haired, and have their eyes open.

The subfamily Dendromurinae remains as a relict, arboreal radiation in equatorial Africa. Of the three genera, *Steatomys* is the most ter-

Table 16 Literature on Desert-Adapted Rodents

Taxon	Ecology	Behavior	Reproduction and Ontogeny	Physiology
Heteromyidae	Chew and Chew 1970; Brown 1975	Eisenberg 1963	Eisenberg and Isaac 1963	Schmidt-Nielson 1964
Perognathus longimembris	—	—	Hayden, Gambino, and Lindbert 1966	Bartholomew and Cade 1957
Microdipodops pallidus	—	—	—	Bartholomew and MacMillan 1961
Dipodomys merriami	Reynolds 1950, 1958, 1960; Soholt 1973	Kenagy 1976	Chew and Butterworth 1959; Chew 1958	Soholt 1977
Dipodomys deserti	—	—	Butterworth 1961*a,b*	—
Dipodomys microps	Kenagy 1973*a*	Kenagy 1976	—	Kenagy 1973*b*
Cricetidae				
Gerbillinae	Petter 1961	Lataste 1887	—	—
Meriones unguiculatus	—	Thiessen and Yahr 1977	Kaplan and Hyland 1972	—
M. lybicus	—	Daly and Daly, 1975*a*	—	—
M. persicus	—	Eibl-Eibesfeldt 1951*b*	—	—
Gerbillus dasyurus	—	Fiedler 1974	—	—
Psammomys obesus	—	Daly and Daly 1975*b*	—	—
Muridae				
Conilurinae	—	—	—	MacMillen and Lee 1969
Notomys alexis	Newsome and Corbett 1975	Stanley 1971	—	—
Pseudomys sp.	—	Happold 1976*a*	Happold 1976*b*	—
Dipodidae	Naumov and Lobachev 1975	—	—	—
Jaculus jaculus	Happold 1967	—	Happold 1967, 1970	Kirmiz 1962*a,b*
J. orientalis	—	Eisenberg 1975*c*	Eisenberg 1975*c*	—
Allactaga elator	—	Eisenberg 1967	—	—

Note: General review volumes include: Prakash and Gosh 1975; Kirmiz 1962*a,b*; and Schmidt-Nielson 1964.

restrial. *Deomys* and *Dendromus* show numerous adaptations for climbing. Aspects of the ecology of *Dendromus* have been studied by Dieterlen (1971). The behavior of *Deomys* has been described by Kingdon (1974*b*), and Vesey-Fitzgerald (1966) and Rahm (1966) have synopsized the ecology of *Steatomys* and *Deomys*.

The remaining genera or Cricetinae proper are distributed in both the Old World and the New World and, as I noted before, have radiated extensively in the Neotropics. Almost without exception, the members of this subfamily give birth to rather altricial young. Exceptions are notable—for example, the genus *Sigmodon,* which gives birth to young that are furred and whose eyes open within 36 hours after birth.

It is convenient to divide the Cricetinae into two tribes: the Hesperomyini and the Cricetini. The latter tribe is distributed in Eurasia and Africa and includes the true hamsters, *Cricetus, Mesocricetus, Phodopus,* and *Cricetullus,* as well as the hamster-mouse, *Callomyscus,* and the African white-tailed mouse, *Mystromys*. The Hesperomyini, together with the Microtinae, occupy all the small murid rodent niches of North America. In South America the radiation of cricetines has been explosive. The adaptive trends are outlined by Hershkovitz (1962).

So vast is the literature on just two North American cricetid genera, *Peromyscus* and *Microtus*, that a synopsis is impossible. The recent literature on *Peromyscus* (see plate 17) is summarized in the volume edited by John King (1968).

In South America the cricetines were able to invade certain niches that are generally not exploited by rodents. *Scotinomys* (Hooper and Carleton 1976) in Central America and *Oxymycterus* (Dalby 1975) in southern South America occupy a terrestrial insectivore niche. *Ichthyomys, Anatomys, Rheomys,* and *Neusticomys* have become aquatic and piscivorous (see Hooper 1968; Starrett and Fisler 1970).

In an effort to guide the reader to the literature on the Cricetidae, I have included a few references in table 17.

11.3.2 The Family Muridae: The Old World Rats and Mice

The murids include more than a hundred genera and, before dispersal by man, were found in Africa, Europe, Asia, Southeast Asia, Australia, and Tasmania. They have reached their maximum diversity in the Old World tropics. Two genera, *Rattus* and *Mus,* have been distributed worldwide through the activities of man. This family shows an incredible range of adaptation in the Paleotropics that parallels the adaptations of the cricetid rodents in the Neotropics. Arboreal, terrestrial, and aquatic forms provide rich material for comparison with their New World counterparts. Murids and cricetids overlap their ranges in Europe and Asia. Once again there is a strong tendency to give birth to altricial young; however, an outstanding exception is to be found in the

Plate 17. *Peromyscus californicus,* a typical cricetine rodent. This species inhabits coastal portions of California. It is one of the largest members of the genus *Peromyscus,* exceeding 80 g in weight. (Photo by author.)

Table 17 Recent Literature on the Muroid Rodents

Taxon	Ecology	Life History and Behavior	Reproduction and Ontogeny
Cricetidae			
Cricetinae			
Tylomys nudicaudatus	—	—	Tesh and Cameron 1970
Nyctomys sumichrasti	—	—	Birkenholz and Wirtz 1965
Onychomys leucogaster	—	Clark 1962; Ruffer 1965, 1968	—
O. torridus	—	Horner and Taylor 1968	—
Oryzomys palustris	Negus, Gould, and Chipman 1961	—	—
Reithrodontomys megalotis	Whitaker and Mumford 1972	Kaye 1961	—
Baiomys taylori	—	Blair 1941	—
Neotoma fuscipes	Linsdale and Tevis 1951	—	Wood 1935
N. cinerea	Finley 1958	—	Egoscue 1962
N. floridana	Rainey 1956	—	Hamilton 1953
N. albigula	Rainey 1965	Vorhies and Taylor 1940	—
Mesocricetus auratus	—	Dieterlen 1959; Johnston 1975*a,b,c*; Murphy 1973	Rowell 1961; Beach and Rabadeau 1959
Cricetus cricetus	—	Eibl-Eibesfeldt 1953	—
Peromyscus (general)	Eisenberg 1963; McCabe and Blanchard 1950; Terman 1961	Eisenberg 1968; Doty and Kart 1972	King 1961
P. maniculatus	Blair 1940; Harris 1952; Howard 1948	Doty 1973; Eisenberg 1962	King 1958
P. polionotus	Blair 1951; Davenport 1964	M. H. Smith 1966	—
P. boyleyi	Drake 1958	—	—
P. crinitus	Egoscue 1964	—	—
P. (ochrotomys) nuttali	—	Goodpaster and Hoffmeister 1954	Layne 1960
P. californicus	—	Eisenberg 1962; Dudley 1974	Dewsbury 1974; McCabe and Blanchard 1950
P. leucopus	Nicholson 1941	—	—
P. gossypinus	—	—	Pournelle 1952
Microtinae			
Dicrostonyx groenlandicus	—	Allin and Banks 1968*a,b*; Banks 1968; Brooks and Banks 1973	—
Lemmus lemmus	Krebs 1962	Arvola, Ilman, and Koponen 1962	Mullen 1968
Neofiber alleni	—	Birkenholz 1963	—
Ondatra zibethica	Errington 1963	—	—
Microtus (general)	Krebs 1970; Krebs, Keller, and Tamarin 1969	Myers and Krebs 1971	—
M. oregoni	—	—	Cowan and Arsenault 1954
M. californicus	Krebs 1966	—	—
M. brandti	—	Reichstein 1962	—
Muridae			
Mus musculus	Anderson 1961, 1965; Brown 1953; Lidicker 1966; Southwick 1955	Crowcroft and Rowe 1963; Eibl-Eibesfeldt 1950*a*	McGill 1962; Bronson 1979
M. oubanguii	Genest-Villard 1973	—	—
Rattus rattus	Ewer 1971	—	—
R. blanfordi	Brosset 1961	—	—
R. assimilis	—	—	Taylor 1961
R. norvegicus	Calhoun 1963	Barnett 1958, 1963	Kuehn and Beach 1963; Barfield and Geyer 1972
Apodemus sylvaticus	Ashby 1967	Jewell and Fullagar 1965	—

Table 17—*Continued*

Taxon	Ecology	Life History and Behavior	Reproduction and Ontogeny
Uranomys ruddi	—	Bellier 1968	—
Micromys minutus	—	Dieterlen 1960	—
Mastomys natalensis	—	Veenstra 1958	Chapman, Chapman, and Robertson 1959
Melomys cervipes	—	—	Taylor and Horner 1970
Thammomys sp.	Genest-Villard 1972	—	Yoeli, Alger, and Most 1963
Lemniscomys striatus	—	—	Petter, Chippaux, and Monmignaut 1964

Plate 18. *Acomys cahirhinus,* the spiny mouse. This murid rodent is adapted for semiarid conditions. Its fur includes numerous spiny hairs. (NZP archives.)

genus *Acomys,* where the young are born fully furred and with their eyes open (Dieterlen 1961) (see plate 18).

The subfamily Rhynchomyinae, or shrew rats, contains a single species, *Rhynchomys soricoides,* which is specialized as a terrestrial insectivore on the island of Luzon. (The rare Philippine genera *Celaenomys* and *Chrotomys* may be closely related.)

The subfamily Phloeomyinae is perhaps an artificial assemblage of seven genera distributed in Indonesia, New Guinea, and the Philippines. Medway (1967) has studied the breeding biology of the Indonesian genus *Chiropodomys*. The evolution of arboreal herbivority in *Phloeomys* and *Crateromys* has been summarized by Eisenberg (1978).

The Hydromurinae are a specialized group of aquatic murines distributed in the Philippines, New Guinea, and Australia. The genus *Hydromys* has been studied in Australia (McNally 1960; Woolard, Vestjens, and MacLean 1978; Barrow 1964). These rodents feed on mussels, arthropods, and fish and appear to have occupied the vacant aquatic niches in New Guinea and Australia.

The Conilurinae (= Pseudomyinae) are an assemblage of Australian genera that are believed to have radiated at an early time from a common ancestor in Australia (Tate 1951). The subfamily includes seven genera: *Pseudomys, Notomys, Conilurus, Leporillus, Zyzomys, Mastacomys,* and *Mesembriomys* (Ride 1970). The remaining Australian murids (*Rattus, Melomys,* and *Uromys*), together with *Hydromys,* are believed to have been more recent or independent colonizers. The conilurine rodents exemplify a diverse radiation. *Notomys* is arid-adapted and bipedal. *Mastacomys* is adapted for cool, moist habitats and has converged in form toward the Microtinae. *Leporillus* is rat-sized and builds a stick nest and thus is convergent with the North American cricetine *Neotoma fuscipes*. *Conilurus* and *Mesembriomys* are arboreal murids inhabiting the eastern forests of Australia.

The behavior and reproduction of *Notomys* and *Pseudomys* have been analyzed in beautiful detail by Meredith Happold (née Stanley) (Stanley 1971; Happold 1976*a,b,c*). Crichton (1969) has analyzed reproduction in *Mesembriomys gouldii*. Calaby and Wimbush (1964) have conducted preliminary studies on *Mastacomys*. Natural history accounts for the Australian murids are summarized by Ride (1970).

The subfamily Murinae contains the vast majority of species and represents the most recent radiation. Two genera, *Rattus* and *Mus,* are widespread geographically. *Rattus* extends from Europe through Asia to Australia. Together with *Mus musculus, Rattus norvegicus* and *R. rattus* have been introduced throughout the world as commensals of man. The islands of the Pacific have been colonized in a similar manner and have provided interesting habitats for the study of species introductions (Storer 1962). *Mus* is found naturally in Africa as well as in Europe and Asia. Marshall (1977) has reviewed the taxonomy of this cosmopolitan genus. *Rattus* has radiated widely in southeast Asia, occupying arboreal, terrestrial, and scansorial niches. In Australia the genus *Rattus* includes only six endemic species.

The African continent has seen a vast radiation of murine rodents into semiaquatic, arboreal, and terrestrial niches. The niche subdivisions are ably reviewed by Kingdon (1974*b*). Table 17 includes a synopsis of recent literature for the Muridae and Cricetidae. Opportunities for comparative studies in ecology and behavior should readily suggest themselves.

11.3.3 The Family Spalacidae: The Eurasian Mole Rats

This family contains a single genus, *Spalax* (three genera if one follows Topachevskii 1976), which is found in the eastern part of the Mediterranean region, the Middle East, and into southern Russia and eastern Europe. It has strongly evolved fossorial adaptations and is convergent in its morphology and habits with the North American pocket gophers of the family Geomyidae. The family has been dealt with in the volume edited by Topachevskii (1976).

Speciation is rapid in the sedentary fossorial rodents, and the behavioral mechanisms preventing hybridization or inbreeding have been subjected to intensive study by Nevo and his associates (Nevo 1969; Nevo, Naftali, and Guttman 1975).

11.3.4 The Family Rhizomyidae: The Tropical Mole Rats

The mole rats or rhizomyid rodents occupy the fossorial "gopherlike niche" in Africa and Southeast Asia. There are only three genera, *Tachyoryctes, Rhizomys,* and *Cannomys,* the latter two being Asiatic and *Tachyoryctes* being African. These fossorial tropical rodents converge strongly in morphology and behavior toward the Geomyidae. Their behavior and ecology has been studied sporadically: *Tachyoryctes* by Rahm (1969, 1971), Jarvis and Sale (1971), and Jarvis (1973*a,b*), and *Cannomys* by Eisenberg and Maliniak (1973).

11.3.5 The Family Gliridae: The Dormice

Dormice constitute a small family of only seven genera, distributed from south Saharan Africa north into the Palearctic region. Most members of the family are rather nocturnal and strongly adapted for arboreality. They feed on fruits and nuts, with arthropod supplements. Some species of the family are adapted for hibernation or estivation and accumulate quantities of fat as a means of passing months of food shortage.

The behavior of *Glis glis* has been described by Koenig (1960). Reproduction and hibernation have been analyzed for *Eliomys* and *Graphiurus* (Nevo and Amir 1964; Lachiver and Petter 1969). Recently the bioenergetics of *Glis, Muscordinus,* and *Eliomys* have been compared (Gebczynski, Gorecki, and Drozdz 1972).

11.3.6 The Family Seleviniidae: Selevin's Mice

This family contains a single species, *Selevenia betpakdalaensis.* It is found in the central deserts of Kazakhstan and is thought to be an aberrant dormouse adapted for a semiterrestrial way of life. It constructs burrows and is variously active either in the day or at night (see Ognev 1963).

11.3.7 The Family Platacanthomyidae: The Spiny Dormice

Nearly allied to the dormice are the so-called spiny dormice of the family Platacanthomyidae. Two genera of spiny dormice are confined to southern India and southeastern China. Very little is known concerning the natural history of these nocturnal and arboreal rodents.

11.4 The Superfamily Dipodoidea

11.4.1 The Family Zapodidae: The Jumping Mice

The jumping mice have a distribution in the Palearctic and the Nearctic. There are only four genera, all of which are adapted to colder climates. They have never successfully invaded tropical regions. The species hibernate in the autumn and accumulate fat before hibernation. Litters are usually small. The young are extremely altricial at birth, and gestation is rather prolonged for animals of their size-class. The same tendency is exhibited by the nearly related family Dipodidae.

Ecology and life-history studies have been developed for *Zapus* (Quimby 1951; Whitaker 1963) and for *Napeozapus* (Brower and Cade 1966). *Sicista* behavior and development have been studied by Pucek (1958), Cerva (1929), and Gottlieb (1950).

11.4.2 The Family Dipodidae: The Jerboas

The family Dipodidae, the jerboas, contains ten genera with approximately twenty-five species confined to the arid portions of North Africa, the Middle East, and Central Asia. They have converged in their morphology toward the heteromyid rodents of North America and are primarily bipedal ricochetors. They show their nearest affinity with the Zapodidae, and many members of this family also are able to become torpid and store quantities of fat before hibernating. Litters are usually small, and the gestation is rather long for the mean adult body weight of the female. The young are very altricial at birth.

General comments on the ecology and systematics of the Dipodidae are included in the work of Ognev (1963). The ecology and reproduction of *Jaculus jaculus* was studied by Happold (1967, 1970). Recent summaries of Russian research on dipodid ecology are included in the review by Naumov and Lobachev (1975) (see plate 19).

11.5 Summary of the Myomorpha and Sciuromorpha

Before proceeding with a consideration of the hystricomorph rodents, it is useful to look back on some of the major convergences and parallelisms in ecology and behavior that have occurred during the radiation of the so-called sciuromorph and myomorph rodents. Table 18 synopsizes these trends. Gliding adaptations have evolved convergently in the flying squirrels of the family Sciuridae and in the family Anomaluridae. Bipedal ricochetal locomotion has evolved convergently in response to selection in arid habitats among several genera of the Heteromyidae, the murid genus *Notomys*, the nesomyine genus *Macrotarsomys*, and the jerboas of the family Dipodidae. Giving birth to precocial young is a response to a complex of factors (see part 3), but it has occurred sporadically in the families Sciuridae, Muridae, Cricetidae, and Pedetidae. Fossorial adaptations have occurred convergently on the different continental land masses when one compares the families Rhizomyidae, Spalacidae, and Geomyidae and certain genera of the Microtinae.

Plate 19. *Jaculus orientalis,* the jerboa of North Africa and the Middle East, shows one of the more extreme adaptations for bipedality evolved within the order Rodentia. (NZP archives.)

Table 18 Examples of Convergent and Parallel Evolution in the Rodentia

Adaptive Syndrome	Convergent Taxa
Gliding	Petauristinae; Anomaluridae
Bipedal ricochetal locomotion	Heteromyidae; Dipodidae; Pedetidae; Muridae (*Notomys*)
Fossorial	Spalacidae; Rhizomyidae; Geomyidae; Myospalacidae
Aquatic	Cricetidae (e.g., *Ondatra, Rheomys*); Muridae (Hydromyinae); Castoridae

Note: Exclusive of the "Hystricomorpha."

11.6 The Hystricomorpha

I will now turn to the hystricomorph rodents. The biology of this group was recently summarized in the volume edited by Rowlands and Weir (1974). Whether the hystricomorphs are truly a monophyletic suborder is still open to lively debate (see Wood 1974; Lavocat 1974; Simpson 1959). Hystricomorph rodents are typified by the production of rather precocial young. The ubiquity of this adaptation suggests a rather ancient origin for this reproductive character.

It is convenient to consider the Old World hystricomorph families separately. There are three: the Hystricidae, the Petromyidae, and the Thryonomyidae. Some authors would include the Ctenodactylidae and the Bathyergidae (see sect. 11.8).

11.6.1 The Family Hystricidae: The Old World Porcupines

The Old World porcupines include four genera distributed in the Paleotropics. The genera differ in degree of spininess; all are terrestrial. *Thecurus* is confined to Sumatra, Borneo, and the Philippine Islands. *Hystrix* extends from southern Europe through Africa, on across India, and into Southeast Asia. *Atherurus* is distributed from southern China through Southeast Asia, with a break in its range at India and the Middle East, only to be found again in East Africa and the Congo. *Trichys* is found in Borneo and Sumatra. Both *Trichys* and *Atherurus* have rather long tails (see plate 20). *Trichys* exhibits the least spininess and may be considered the most conservative member of the family. The taxonomy and distribution of Old World porcupines has been reviewed by Mohr (1965). Reproduction and growth of *Atherurus* was studied by Rahm (1962). All species give birth to precocial young.

11.6.2 The Family Thryonomyidae: The Cane Rats

The cane rats are confined to equatorial Africa. These robust rodents have been the subject of much recent research. In conformity with all members of the suborder Hystricomorpha, newborns are precocial, well-furred and with eyes open. The biology of *Thryonomys* has been summarized by Asibey (1974).

11.6.3 The Family Petromyidae: The African Rock Rats

This family contains a single genus, *Petromys,* which is adapted for semiarid, rocky habitats and is confined to southwestern Africa. The animals appear to be specialized feeders on leaves of shrubs and low trees. They are diurnally active and suggest a strong convergence in niche occupancy toward the niches occupied by the Ctenodactylidae in North and East Africa and by the genus *Kerodon* in Brazil.

11.7 The Caviomorpha

We now come to the classical New World hystricomorphs or Caviomorpha. The relationships of the family are included in figure 22.

Plate 20. *Atherurus macrourus,* the brush-tailed porcupine. This Old World hystricomorph shows extreme specialization for spininess as an antipredator mechanism. The modified quills on the tail rattle when the tail is vibrated. (NZP archives.)

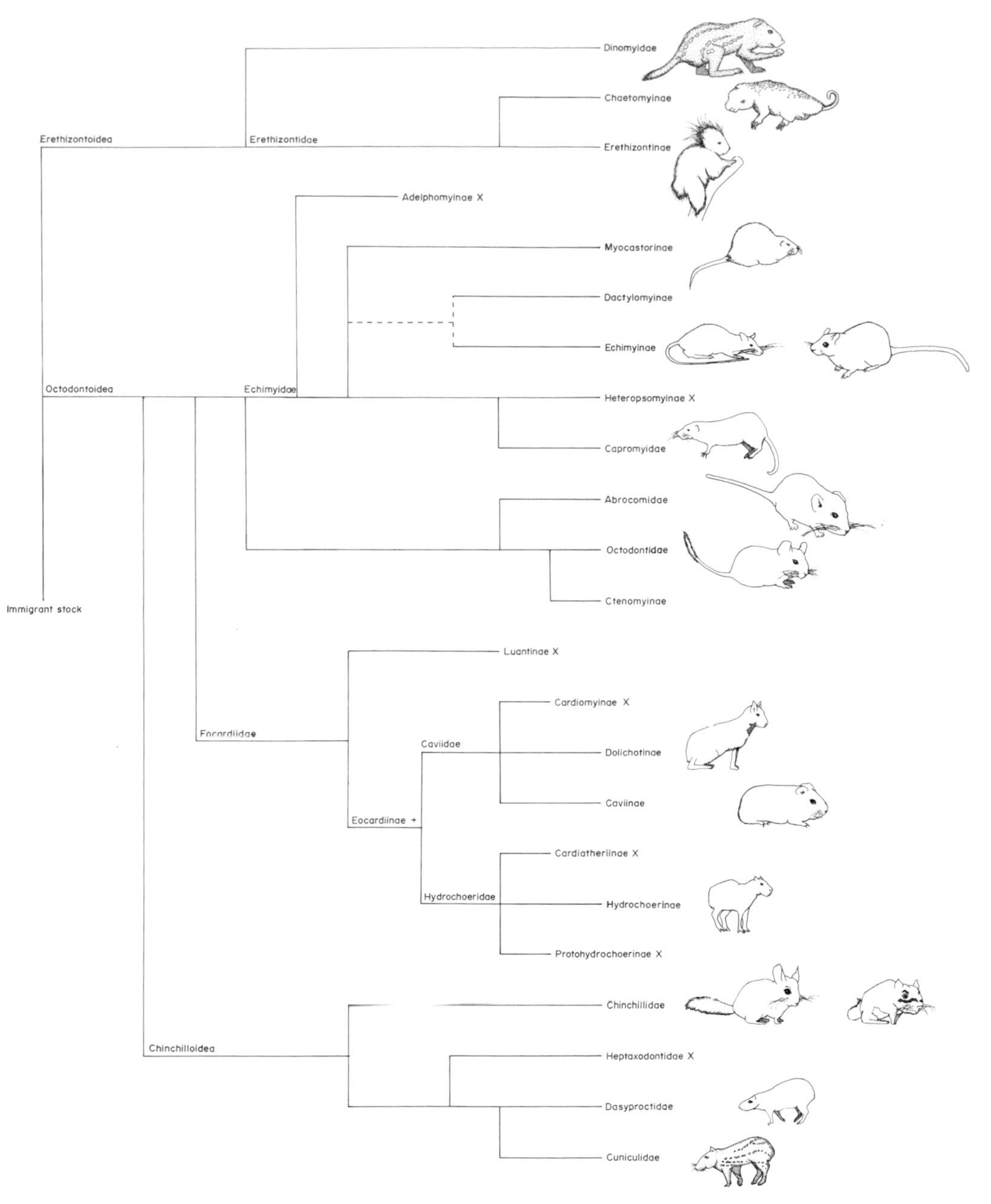

Figure 22. Diagram of the classification of the Caviomorpha. Adapted from Patterson and Pasqual 1972. Placement of Dinomyidae according to Grand and Eisenberg (in preparation).

11.7.1 The Family Erethizontidae: The Porcupines

The New World porcupines include four genera that are all remarkably similar in using spiny coats in their antipredator behavior. Two genera, *Chaetomys* and *Coendou,* have rather long tails that are not covered with spines. The other two genera, *Erethizon* and *Echino-*

procta, are characterized by short, stout tails. All are arboreal and feed on fruits, herbaceous vegetation, bark, and, in the case of *Erethizon,* cambium (Shapiro 1949). Only *Coendou* has a strongly prehensile tail.

11.7.2 The Family Caviidae: The Cavies

The cavy family includes five genera—*Cavia, Microcavia, Galea, Kerodon,* and *Dolichotis*[1] representing varying degrees of adaptation to grassland habitats. *Dolichotis* strongly converges toward the small antelopes in both its body conformity and its form of antipredator behavior (Smythe 1970*a*). *Kerodon* lives in the more arid parts of northeastern Brazil and exploits leaf-feeding in a manner reminiscent of the hyraxes of Africa; this is an example of extreme convergence. The genera *Cavia* and *Microcavia* have been thoroughly studied in the field by Rood (1970*a*, 1972), while Lacher (1979) provides a thorough account for *Kerodon.*

11.7.3 The Family Hydrochoeridae: The Capybara

This family includes a single genus, *Hydrochoerus,* the capybara. Individuals may exceed 50 kg in weight. The capybaras are semiaquatic herbivores that come ashore to graze, usually in the late afternoon or evening. They are rather sociable and live in family groups. Through fighting with other males in the breeding season, a dominant male may achieve priority mating rights with those adult females that are on his range. In body size and conformity, capybaras show a remarkable convergence with the pygmy hippopotamus. The ecology of the capybara has been monographed by Ojasti (1973).

11.7.4 The Family Dinomyidae: Branick's Giant Rat

Dinomys branickii is the only living species within this family. It shows an ancient affinity with the Erethizontidae (Grand and Eisenberg, in preparation). It may be considered a spineless porcupine confined in the main to tropical areas in the submontane regions of Colombia, Ecuador, and Peru. *Dinomys* is a selective feeder, including fruit and leaves in its diet (Collins and Eisenberg 1972).

11.7.5 The Family Dasyproctidae: The Agoutis

This taxon of cursorial caviomorphs includes three genera, *Agouti* (=*Cuniculus*), *Myoprocta,* and *Dasyprocta.** These are the pacas, acouchis, and agoutis. Some authorities dispute whether *Agouti* should be placed in this family, but I have included it pending further study. *Agouti* is a nocturnally adapted browser and frugivore. *Dasyprocta* is an obligate frugivore and is diurnal, as is *Myoprocta,* which in many respects is a smaller version of the agouti. The tail is vastly reduced, being almost absent in the paca and a small vestigial remnant in *Dasyprocta;* but the tail is short (70 mm), conspicuous, and still functional in the displays of *Myoprocta* (Kleiman 1971). To aid in under-

1. *Pediolagus* is here considered a subgenus of *Dolichotis.*

standing the cursorial Caviomorpha, I will synopsize the behavior and ecology of *Dasyprocta punctata,* according to Smythe (1978).

A Life-History Sketch for Dasyprocta punctata

The agouti is a hare-sized caviomorph rodent found in Central and South America in tropical evergreen forests as well as in semideciduous tropical forests. When not disturbed by hunting, the animals are diurnal in their activities. Their major food resource is fruits and nuts that fall to the forest floor. They are entirely terrestrial and avoid predators by fleeing. In antipredator strategy and body form, they have converged toward the small ungulates of the Old World (Eisenberg and McKay 1974). Adults have stable home ranges and defend them assiduously against like-sexed adults. A male's home range may overlap those of one or two females, but under most conditions male home ranges coincide with those of females, thus leading to monogamous pairing. The degree of defense of the home range varies throughout the annual cycle, being highest when fruit fall is lowest and somewhat relaxed when fruit is abundant. The animals scatter-hoard fruits and nuts within the home range to use as supplementary food when fruit is scarce. Juveniles may live with the adults in the defended area, but they generally disperse when they reach sexual maturity. Survival of juveniles depends on their finding a vacant home range. The gestation period, in conformity with that of other caviomorphs, is rather long, exceeding 120 days. The young is born in an advanced state with the eyes open. The female usually gives birth near a natural cavity, and the young may enlarge it and carry in nesting material. The female returns to the youngster's hiding place to nurse at intervals until the young is approximately 3 weeks of age, whereupon it begins to accompany the mother on her explorations around the home range. The young may continue to forage with the mother until it is 4 to 5 months of age. When in good condition, the female may exhibit a postpartum estrus and conceive again (Smythe 1978).

11.7.6 The Family Chinchillidae: The Chinchillas and Viscachas

The family Chinchillidae includes *Lagostomus, Lagidium,* and *Chinchilla. Lagidium* and *Chinchilla* are adapted to high montane areas. *Lagidium* is diurnal, while *Chinchilla* is nocturnal. They occupy in part a niche similar to that of the Ochotonidae of Asia and North America. *Lagostomus* is nocturnal and adapted for life on the savanna. It is colonial, living in what appear to be individually defended burrow systems. The genus is remarkable in its extreme size dimorphism—the adult male is twice as large as the female.

11.7.7 The Family Capromyidae: The Hutias

The capromyid rodents are a group of relict forms found in the Greater and Lesser Antilles. Many species have become extinct within recent times. *Geocapromys* has two extant species, one found in the Bahamas and the other on Jamaica. *Plagiodontia* is confined to Hispaniola, and the genus *Capromys,* divisible into at least four species, is confined to Cuba. The family exhibits variations on a theme toward occupying a semiarboreal folivore niche. An extreme specialization for this niche is found on Cuba in the form *Capromys prehensilis* (see Eisenberg 1978).

11.7.8 The Family Myocastoridae: The Coypu

Myocastor, or the coypu, is aquatic. Originally it was confined to southern South America, but it has been widely introduced in Europe and in North America. It has converged in morphology and ecology toward the microtine genus *Ondatra.*

11.7.9 The Family Octodontidae: The Degus

The octodontid rodents include five genera distributed from Peru south to Argentina and Chile. They are variously adapted for a terrestrial herbivorous existence. One genus, *Spalacopus,* has evolved elaborate adaptations for a fossorial life and has converged morphologically with the geomyid and spalacid rodents. *Octodon* is a diurnal genus living in coastal hill habitats in Chile (Fulk 1976). *Octodontomys* is nocturnal, adapted for the high plateau of Bolivia and Peru.

11.7.10 The Family Ctenomyidae: The Tucotucos

The ctenomyid rodents, or Tucotucos, include approximately fifty species in one genus, *Ctenomys.* They have occupied many different habitats in the southern part of South America and show great versatility in the altitudes at which they can operate. The range of *Ctenomys* does not overlap that of *Spalacopus. Ctenomys* is strongly fossorial, converging in form and behavior with the New World Geomyidae and the Old World Spalacidae (see Weir 1974; Pearson 1959).

11.7.11 The Family Abrocomidae: The Chinchilla Rats

A single genus, *Abrocoma,* with two species is confined from southern Peru to northern Chile. Typically it occupies semiarid to arid habitats.

11.7.12 The Family Echimyidae: The Spiny Rats

The echimyids or spiny rats include nineteen genera and at least eighty species. They are widely distributed in Central and South America and are the most conservative family within the entire caviomorph group. Typically the body is ratlike in form, although minor variations are shown in conformity with adaptations for a semifossorial way of life (*Carterodon*) or arboreal life (*Echimys* and *Diplomys*).

The caviomorph rodents are valuable for comparison with other mammalian taxa. The caviomorphs invaded South America at a rather early date and by the Miocene were well established. In the absence of conventional competitors found on the contiguous continental land masses, some of the caviomorph rodents evolved feeding adaptations with a concomitant morphology that reflects a strong convergence with small ungulates and lagomorphs (e.g., Dasyproctidae: *Myoprocta, Agouti, Dasyprocta;* and Caviidae: *Dolichotis*). These genera provide a fertile field for cross-taxa comparisons (Kleiman 1974*a*; Eisenberg and McKay 1974; Dubost 1968). Table 19 will serve as an introduction to the literature on the behavior and ecology of the Caviomorpha.

11.8 Families of Uncertain Affinities[2]

11.8.1 The Family Ctenodactylidae: The Gundis

The gundis include four genera distributed across North Africa down the east coast to Somalia. They are herbivores and do not hesitate to climb low trees and shrubs to feed on leaves. They prefer rocky habitats and in many respects are convergent in morphology and ecology toward the adaptive syndrome shown by the caviomorph genera *Kerodon* and *Chinchilla*. Their biology is summarized by George (1974).

Table 19 Recent Literature on the Caviomorpha

Taxon	Ecology	Behavior	Reproduction and Ontogeny
Caviomorpha	—	Kleiman 1974*a*; Eisenberg 1974	Weir 1974
Erethizontidae			
Erethizon dorsatum	Marshall, Gullion, and Scharb 1962	—	Shadle 1943, 1946
Caviidae			
Cavia	Rood 1972	Arvola 1974; Rood 1972	—
Galea	Rood 1972	Rood 1972	Weir 1974
Microcavia	Rood 1970*a*	Rood 1972	—
Dolichotis patagonum	Smythe 1970*b*	Dubost and Genest 1974	—
D. (=*Pediolagus*) *salinicola*	—	Wilson and Kleiman 1974	—
Hydrochoeridae			
Hydrochoerus hydrochaeris	Ojasti 1973	Donaldson, Wirtz, and Hite 1975	Zara 1973
Dinomyidae			
Dinomys brannickii	—	Collins and Eisenberg 1972	Collins and Eisenberg 1972
Dasyproctidae			
Dasyprocta punctata	Smythe 1978	Smythe 1978	—
Myoprocta pratti	—	Kleiman 1971, 1972*a*; Schonewald 1977	Kleiman 1970
Agouti (=*Cuniculus*) *paca*	Kraus, Gihr, and Pilleri 1970; Smythe 1970*a*	Smythe 1970*a*	—
Chinchillidae			
Lagostomus maximus	Llanos and Crespo 1952	Weir 1974	Weir 1974
Lagidium peruanum	Pearson 1948	—	—
Chinchilla laniger	Osgood 1943	—	Weir 1970
Capromyidae			
Geocapromys ingrahami	Clough 1972, 1974	—	Howe and Clough 1971
Capromys pilorides	—	Bucher 1937	—
Myocastoridae			
Myocastor coypus	Atwood 1950	—	Ehrlich 1966

2. May be related to the Hystricomorpha.

Table 19—*Continued*

Taxon	Ecology	Behavior	Reproduction and Ontogeny
Octodontidae			
Octodon degus	Fulk 1976; LeBoulange and Fuentes 1978	Wilson and Kleiman 1974	—
Octodontomys gliroides	—	Wilson and Kleiman 1974	—
Spalacopus cyanus	Reig 1970	—	—
Ctenomyidae			
Ctenomys peruanus	Pearson 1959	—	—
C. talarum	Pearson et al. 1968	—	Weir 1974
Abrocomidae			
Abrocoma bennetti	Fulk 1976	—	—
Echimyidae			
Proechimys semispinosus	Fleming 1971*a*	Maliniak and Eisenberg 1971	—
P. guairae	—	Lusty and Seaton 1978	Weir 1973
Diplomys darlingi	—	—	Tesh 1970
Dactylomys dactylinus	—	LaVal 1976	—

Note: The general reference work is Rowlands and Weir, *The Biology of Hystricomorph Rodents* (1974).

11.8.2 The Family Bathyergidae: The African Mole Rats

The mole rats include five genera distributed in South Africa east to Kenya. The family exhibits a profound adaptation to a fossorial way of life. Their ecology, burrowing behavior, and reproduction have been reviewed by Hill et al. (1957), Genelly (1965), Jarvis and Sale (1971), and Jarvis (1969, 1973*a,b*). One nude species of mole rat, *Heterocephalus glaber,* has reached the ultimate in specialization. All fossorial mammals face the problem of dissipating heat while burrowing. *Heterocephalus* has reduced its metabolic rate and does not grow body fur (McNab 1966). It seems further specialized in that it lives in small social units that are presumably extended families. The clan members cooperatively excavate burrows by forming a "digging chain" that is portrayed by Jarvis and Sale (1971, p. 486).

11.9 Concluding Statement on the Rodentia

The rodents have had a long evolutionary history, and extensive adaptive radiation was well under way in the Eocene. The most recent radiations have involved the Cricetidae in the New World and the Muridae in Africa and Southeast Asia. The contemporary sciurid and hystricomorph rodents appear to be based on earlier radiations of great stability.

Pilleri (1959, 1960*a,b*, 1961*a,b*) conducted a rather complete study of the comparative morphology of the rodent brain. Utilizing Pilleri's data, it is possible to show the advances in the development of the cerebral cortex in such large, long-lived rodents as the beaver, the paca, and some squirrels. Relative brain size also shows some interesting correlations. When rodent brain weights are regressed against mean body weight, the diurnal, arboreal sciurids stand out as having the largest relative brain size of any rodents. This strongly suggests that diurnality and reliance on input from the optic nerve is accompanied by corresponding enlargement of projection areas responsible for the integration of visual information. This development of the rodent brain in the sciurids is reminiscent of early trends in brain evolution demonstrable in primates (see chap. 21).

Rodents have shown great adaptability in specializing for different trophic strategies. While many rodents have maintained a versatility in feeding habits and may be considered true omnivores (Landry 1970), predictable trends in feeding adaptations can be demonstrated when rodent communities are compared from continent to continent. In temperate grasslands, rodents usually divide into granivores and herbivores. The granivores tend to feed on seeds and to cache quantities of food for subsistence while overwintering. The herbivores exhibit dental specializations for feeding on grasses and forbs and usually do not cache food (Baker 1971). The alimentary tract, including teeth, stomach, and intestines and cecum, may show varying degrees of specialization for processing herbaceous vegetation. The comparative morphology of the alimentary tracts of grassland rodents has been analyzed by Vorontsov (1960).

Comparative studies on rodent demography have yielded interesting insights into the consequences of trophic specializations. French, Stoddard, and Bobek (1975) analyzed the smaller temperate-zone rodents and demonstrated that the rodents that gathered grains, nuts, or seeds and cached them (Sciuridae, Cricetidae, Heteromyidae) tended to show a lower reproductive rate and a greater rate of survival. On the other hand, rodents specialized for feeding on grasses and forbs (Microtinae) tended to show adaptations for a greater reproductive capacity but to have a lowered annual adult survival. The murid rodents in their analysis were similar to the microtines, but their sample was biased toward *Rattus* and *Mus,* which are commensals of man. The fossorial rodents (Geomyidae and Spalacidae) resemble the sciurid and heteromyid pattern in their reproductive and survival strategies.

The social behavior of rodents has been well analyzed. A comparison of species-typical social structures indicates convergence in form that correlates with broad adaptive trends (Eisenberg 1967; Barash 1973). The comparative analysis of discrete social interaction forms such as the copulatory behavior of rodents has yielded much of interest (Dewsbury 1975). Dewsbury and his associates have noted that penile morphology correlates with the temporal patterning of rodent copulation patterns, and, furthermore, that prolonged intromission times and a low copulation frequency strongly correlate with monogamy.

It becomes apparent with continued analysis that demography, social structure, mating systems, and trophic strategies are interrelated. Microtine rodents display marked oscillations in population density; cricetine rodents show less dramatic shifts in density (Krebs and Myers 1974). The influences of population densities on behavior have been particularly analyzed for two muroid genera, *Microtus* and *Peromyscus.* Myers and Krebs (1971) considered the unique genotype of dispersing individuals of *Microtus pennsylvanicus* and *M. ochrogaster;* Lidicker

(1975) reviewed the role of dispersal in the demography of small mammals; Jannett (1978) discussed the formation of maternal extended families by *Microtus montanus* at high densities. Healey (1967) discussed the role of aggression by resident males and its influence on dispersal by juveniles of *Peromyscus maniculatus;* Sadleir (1965) discussed the role of aggression in population changes observable in *P. maniculatus;* Fairbairn (1978) demonstrated that subordinate dispersing males of *P. maniculatus* were genetically different from resident aggressive males. Clearly significant advances are being made toward clarifying the interrelatedness of demography, genetics, and social behavior in rodent studies.

12 The Order Insectivora

The living Insectivora are in the main confined to the Holarctic, Africa, and South Asia. Invasion of South America by insectivores occurred after the formation of the Pliocene land bridge, and only one genus, *Cryptotis,* has successfully penetrated the Neotropics. The Insectivora were never present in Australia.

Figure 23 outlines the classification of the "Insectivora." Many lines of descent are impossible to trace at this time, but I feel that the evidence favors an ancient unity among the Tenrecidae, Potomogalidae, and Chrysochloridae (Van Valen 1967).[1] I have departed from McKenna (1975) in setting these latter three families aside in their own ordinal

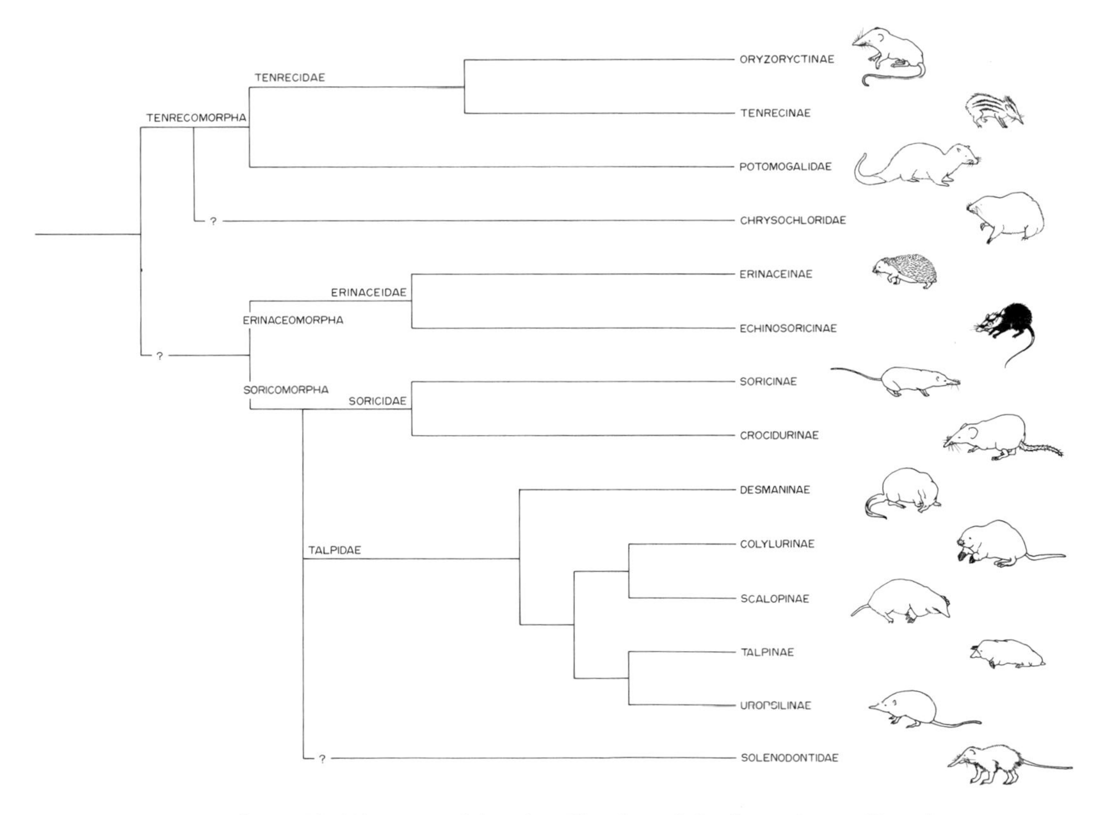

Figure 23. Diagram of the classification of the Insectivora. Based on McKenna 1975; Van Valen 1967; and Szalay 1977.

1. Van Valen (1967) united the Tenrecidae (s.i), Solenodontidae, and Chrysochloridae with the extinct Hyaenodonta in the order Deltatheridia.

grouping. In conformity with McKenna, I do not consider the Tupaiidae or the Macroscelididae to be members of the radiation giving rise to the contemporary insectivores.

12.1 The Tenrecomorpha

This ancient radiation is entirely African in its contemporary distribution. The Tenrecidae were discussed in chapter 7 (see plate 21). The Potamogalidae are confined to the wet portions of central Africa and include only two genera, both of which are strongly adapted for an aquatic life and feed on fish and crustaceans. *Micropotamogale* (=*Mesopotamogale*), which weighs up to 135 g, is the smaller of the two genera. Rahm (1960, 1961*b*) has summarized the natural history of *M. ruwenzori,* and Kuhn (1964) has discussed the biology of *M. lamottei*. The much larger *Potamogale velox,* which may reach a length of 640 mm (including the tail) and a weight of 1 kg, has been poorly studied. What little is known of its natural history is summarized by Kingdon (1974*a*) and Dubost (1965).

The Chrysochloridae include five genera that are completely adapted as fossorial insectivores. They are popularly referred to as the "golden moles." The family is distributed from east Africa to the Cape, and its members are convergent with the true moles of the family Talpidae. Some genera such as *Eremitalpa* are adapted for extremely arid habitats. Aspects of their ecology have been described by Holm (1971). The natural history of this family is poorly known, but Kingdon (1974*a*) has made an admirable summary for *Chrysochloris stuhlmanni*.

12.2 The Insectivora (sensu stricto)[2]

12.2.1 The Family Erinaceidae: The Hedgehogs

This family is divisible into two subfamilies, the Erinaceinae, or spiny hedgehogs, and the Echinosoricinae, or hairy hedgehogs. The Echinosoricinae are the most conservative members of the family and in fact

Plate 21. *Tenrec ecaudatus,* the largest living insectivorous mammal found on the island of Madagascar. Even though it has evolved a large body size, exceeding 1.5 kg, it has retained a high reproductive capacity. (NZP archives.)

2. May be broken into the ordinal groupings Erinaceomorpha and Soricomorpha.

may be the most conservative members of the order Insectivora (s.s.). There are four genera: *Echinosorex, Hylomys, Podogymnura,* and *Neotetracus*. All are confined to Southeast Asia. *Podogymnura* is endemic to Mindanao in the Philippine archipelago. The behavior and ecology of *Echinosorex gymnura* has been studied by Lim (1967) and Gould (1978*a*). Aspects of the biology of *Hylomys* have been noted by Davis (1965) (see plate 22).

The spiny hedghogs may be divided into four genera and show a range of adaptations and a wide distribution. The family ranges from the Holarctic through the Middle East to North and East Africa. *Parechinus* and *Hemiechinus* are adapted to extremely xeric habitats. All members of the family possess stout spines and roll into an impregnable ball when attacked by a predator. Recent literature on their ecology and behavior is summarized in table 20.

12.2.2 The Family Solenodontidae: The Solenodons

The family Solenodontidae contains one genus, *Solenodon,* with two species, *cubanus* and *paradoxus*. The living members are currently confined to Cuba and Hispaniola in the Greater Antilles. They possess very archaic morphological characters in conjunction with some apomorphic characters. These animals were the top small carnivores in the Antilles and, though they were probably preyed upon by boas and large owls, in general no other terrestrial form could offer them serious competition or prey upon them. They are among the largest of the Insectivora, reaching 800 g. The solenodontids show some affinities with the true shrews or soricids. Their exact taxonomic status is in

Plate 22. *Echinosorex gymnura,* the moon rat. One of the largest of the Insectivora, it may reach over 1 kg in weight. It has converged strongly in external morphology toward the opossum, *Didelphis*. (Courtesy of Christen Wemmer.)

Table 20 Literature on the Insectivora.

Taxon	Ecology	Behavior	Reproduction and Ontogeny	Physiology
Erinaceidae				
Erinaceinae				
Erinaceus europaeus	—	Poduschka and Firbas 1968; Poduschka 1968, 1969, 1977	—	Bretting 1972.
Hemiechinus auritus	—	—	Nader 1968	—
Soricidae				
Soricinae	Buckner 1964, 1966	—	—	—
Sorex araneus	Michielsen-Kroin 1966; Shillito 1963	Crowcroft 1957	Dehnel 1952	—
S. arcticus	—	Clough 1963	—	—
S. minutus	Michielsen-Kroin 1966	—	Hutterer 1976	—
S. palustris	—	Sorenson 1962	—	—
S. vagrans	Clothier 1955; Ingles 1961	Eisenberg 1964	—	—
Blarina brevicauda	Dapson 1968	Rood 1958	Pearson 1944	—
Cryptotis parva	—	Davis and Joeris 1945	Conaway 1958	—
Neomys fodiens	Borowski and Dehnel 1953	—	—	—
Crocidurinae				
Suncus murinus	—	Stine and Dryden 1977	Dryden 1968, 1969	Dryden, Gebczynski, and Douglas 1974
S. etruscus	Fons 1975	Vogel 1970	Fons 1973	Fons and Sicart 1976; Vogel 1974
Crocidura bicolor	—	—		Dippenaar 1979
C. suaveolens	Rood 1965*b*	—	Rood 1965*a*	—
C. russula	—	Vogel 1969	Hellwing 1973, 1975	—
Anourosorex squamipes	—	Mandal and Biswas 1970	—	—
Talpidae	—	—	—	Hawkins and Jewell 1962
Talpinae				
Talpa europaea	Godfrey and Crowcroft 1960	—	—	Quilliam 1966
Scalopinae				
Parascalops breweri	Eadie 1939	—	—	—
Scalopus aquaticus	Arlton 1936	—	—	—
Neurotrichus gibbsii	Dalquest and Orcutt 1942	—	—	—
Desmaninae				
Galemys pyrenaicus	—	Richard and Vaillard 1969	—	—
Condylurinae				
Condylura cristata	—	Rust 1966	Eadie and Hamilton 1956	—

dispute.[3] The behavior and natural history are described by Eisenberg and Gould (1966), Eisenberg (1975*b*), Mohr (1936), and Rabb (1959*a*) (see plate 23).

3. Van Valen (1967) has grouped the Solenodontidae with the Tenrecidae in the suborder Zalambdodonta of the order Deltatheridia.

Plate 23. *Solenodon paradoxus,* one of the largest members of the order Insectivora, reaching 800 g adult weight. This species is a relict endemic on the island of Hispaniola. (Courtesy of Christen Wemmer.)

12.2.3 The Family Soricidae: The Shrews

The family Soricidae includes the true shrews distributed in the Holarctic, Africa, and Southeast Asia. These are small terrestrial to semifossorial forms that have radiated to fill the terrestrial-insectivore niche in most of the tropical habitats with the exception of South America, where such a niche is in part occupied by cricetid rodents and small marsupials. In Australia such a feeding niche is preoccupied by small dasyurid marsupials such as *Sminthopsis* and *Planigale*. The family is classically divided into the Soricinae, which have pigmented teeth (the "red-toothed" shrews), and the Crocidurinae (white-toothed shrews) (see plate 24). Many members of this family are extremely small and are characterized by rather high metabolic rates, higher than would have been predicted from an extrapolation of the classical Kleiber curve (see chap. 17). Not all small shrews, however, show such extremely high metabolic rates, and many of the semiarid-adapted forms (Lindstedt 1980) or tropical forms have lower metabolic rates (Vogel 1976) and greater longevity (Hutterer 1977). In addition, the smallest shrews may show daily hypothermia when resting (Vogel 1974). Thus it appears that high metabolic rates within the genera *Sorex* and *Blarina* may be an overall adaptation to exploitation of temperate habitats, with a predictable yearly pulse in food supplies favorable to annual rearing, whereas the tropical shrews of the genus *Crocidura* may be geared to a slower population turnover. The whole problem needs further investigation, but the division of small shrews into "hot" shrews and "cool" shrews has provoked renewed interest in their metabolic physiology.

The biology of shrews has been studied in some detail. *Sorex araneus* may serve as an example. The young are born in an altricial state but grow rapidly. Adult breeding females may produce two litters in a given season, but the adult cohort will not survive the winter. As a consequence, the survival of the young is very high, and they breed after

Plate 24. *Suncus murinus*, the oriental house shrew. This small insectivorous eutherian typifies in body plan the insectivorous adaptive syndrome. Note the long nose, reduced eye, and prominent vibrissae. The sensory hairs on the tail are typical of the crocidurine shrews. (NZP archives.)

overwintering, only to repeat the cycle (Crowcroft 1957; Michielsen-Kroin 1966). Thus *Sorex araneus* approximates a semelparous mode of reproduction, and many other species of north temperate shrews exhibit similar demographic strategies.

Some soricine insectivores have become so adapted for fossorial life that they converge in body form toward the true moles (e.g., *Blarina*). The systematic position of some "shrew moles" (*Neurotrichus, Urotrichus,* and *Scaptonyx*) has been in flux, and some authors include these genera with the Talpidae.

12.2.4 The Family Talpidae: The Moles

The family Talpidae includes the moles, shrew moles, and desmans, confined in the main to the Holarctic (North America and Eurasia). The family extends to the Himalayan region in South Asia and west to the Mediterranean. It is not represented on the continent of Africa, nor does it extend to tropical Southeast Asia. These fossorial insectivores are adapted for spending most of their life under the surface of the earth. They differ in their overall adaptation from the fossorial rodents in that they are specialized for feeding on invertebrates and do not eat appreciable amounts of plant material.

The life history of the European mole has been analyzed in some detail (Godfrey and Crowcroft 1960). The desmans (Desmaninae) not only are adapted for burrowing but also are highly aquatic, with strongly webbed hind feet. Semiaquatic adaptations have also been evolved by some soricids, for example, *Neomys* and *Sorex palustris*. A guide to life-history studies is included in table 20.

12.2.5 General Comments on the Insectivora

Since the Insectivora carry forward in time a set of rather conservative eutherian characters, their behavior and life histories have provoked considerable interest. Poduschka (1977) has summarized the literature concerning communication mechanisms, demonstrating their reliance on auditory, tactile, and chemical cues. Gould (1965) showed echolocation in several genera of the Tenrecidae. Gould, Negus, and Novick (1964) and Buchler (1976) convincingly demonstrated echolocation during orientation by soricine shrews. Grünwald (1969) was not able to elucidate echolocation in crocidurine shrews.

The behavior of the Insectivora has been reviewed by Herter (1957). Although most insectivores have relatively small brains (Bauchot and Stephan 1966), their behavioral repertoire is complex. Maternal care patterns may be very elaborate (Stine and Dryden 1977), with complex coordination of behavior between mother and young. Caravan formation, where the young seize each other's tails and one young attaches to the female's tail, maintains a cohesive unit when a female moves nearly weaned young from a disturbed nest (Vogel 1969). This behavior appears to be mainly confined to the Crocidurinae.

The great range of size in the Insectivora has led to pioneer studies of the relation among absolute size, demography, and behavior. Hutterer (1977) has demonstrated that mean length of life is positively correlated with body size in shrews, but the Crocidurinae have average longer lives than the Soricinae for all comparable size-classes. Hutterer and Vogel (1977) have demonstrated that the mensurable characters of auditory signals vary predictably when a size-graded series of crocidurine shrew species arc compared. The mean pitch of a given vocalization type is higher in smaller shrews, and the syllable duration is shorter. The complexity of the social interaction patterns in the smaller Insectivora has only recently been appreciated (Gould 1969), and it promises to be an exciting area for future research.

13 The Grandorder Ferae

In 1945 George Gaylord Simpson proposed that the ungulates and carnivores be united in a single cohort, the Ferungulata, indicating the close affinity between the extinct orders Creodonta and Condylarthra. Van Valen (1965) has demonstrated that many of the creodont families in all probability should be united with some rather conservative insectivorouslike mammals in a new mammalian order, the Deltatheridia. McKenna (1975) united the extinct orders Cimolesta and Creodonta (=Hyaenodonta) with the order Carnivora into a single grandorder Ferae. I propose to separate the order Carnivora into two ordinal groupings, order Carnivora (=Fissipedia) and order Pinnipedia. The most primitive true carnivores, then, are the Miacidae, which are now considered of sufficiently close affinity to be included in the order Carnivora itself. The miacids obviously derive from some protoinsectivorous relative, but the connection between the miacids and the preceding ancestral forms remains somewhat obscure (McKenna 1975).

13.1 The Order Carnivora

The modern carnivores are distinguished as eutherian mammals that retain a strong tendency for a simple digestive system and a conservative dentition. The members of the order tend to feed on fruit, insects, fish, and other vertebrates. Trophic specializations have occurred, and it is difficult to consider many species as "carnivores." Some, such as the giant panda, *Ailuropoda,* are definitely herbivores. The typical dentition is 3/3, 1/1, 4/4, and 2/3 for a total of 42 permanent teeth.[1] Typically the canine tooth in the upper jaw is somewhat enlarged. Premolars tend to be tricuspidate, and molars are quadritubercular. There is a strong tendency toward the reduction in the size of the clavicle or its complete loss that probably correlates with the adoption of a cursorial mode of locomotion.

Many members of the Carnivora are plantigrade, placing the whole sole of the foot on the ground when they walk, though there is a tendency in convergently evolved cursorial forms toward a semiplantigrade condition that culminates in digitigrade locomotion typified by the Canidae, some Viverridae, the Felidae, and some Mustelidae. We are fortunate in having a superb review of the fissiped carnivores in the volume produced by the late R. F. Ewer (1973), to which I refer the reader for a full synopsis.

1. Dental formulas follow the convention of listing only half of the jaw: upper and lower incisors, canines, premolars, and molars.

The skulls of fissiped carnivores are separable into two basic types. Where the expanded tympanic bullae are composed almost entirely of the squamosal bone, we refer the species to the arctoid or canoid superfamily; this includes the families Ursidae, Mustelidae, Canidae, Ailuridae and Ailuripodidae, and Procyonidae. On the other hand, where the tympanic bullae are composed of a portion of the squamosal bone and an expanded portion of the endotympanic bone, we may distinguish the superfamily Feloidea, which includes the families Viverridae, Hyaenadae, and Felidae (see fig. 24). With the exception of the spotted hyena, *Crocuta;* the Madagascan viverrids *Fossa* and *Eupleres;* and the Madagascan mongooses *Galidia, Mungotictis*, and *Salanoia,* all carnivores give birth to rather altricial young. Prolonged maternal investment and strong parent-young bonds are characteristic of this order.

13.1.1 The Family Canidae: The Dogs

The Canidae are adapted for cursorial locomotion and for pursuing prey over long distances. They were global in their distribution, except for the Celebes, Philippines, Formosa, Madagascar, and Polynesia. The dingo was introduced by man into New Guinea and Australia, and the domestic dog has recently spread with man to all insular habitats. It is convenient to divide the family Canidae into several subfamilies, but there is considerable controversy concerning their validity (see plate 25).

Classically the Canidae were divided into three subfamilies based on number of teeth. The three subfamilies were the Otocyoninae (bat-eared foxes), with 46 to 50 teeth; the Simocyoninae (including *Speothos, Cuon,* and *Lycaon*), all of which show a tooth reduction from the basic number of 42; and the Caninae, which included the remaining dogs, jackals, wolves, and foxes with the classical tooth number. Recently Clutton-Brock, Corbett, and Hills (1976) have reviewed the whole problem and concluded that no suite of characters separates any group of genera into what could be considered a valid subfamily. The tooth reduction that characterizes the bush dog, dhole, and Cape hunting dog may indeed reflect convergence in dietary specialization rather than a special genetic affinity.

Otocyon does show an exceptionally high tooth number for a eutherian mammal, but the tooth number tends to vary from 46 to 50. The number of molars varies from a minimum of 3/4 to 4/5. This peculiarity in dental structure may be related to its somewhat insectivorous habits (Nel 1978), a tendency also observed in the Edentata when one looks at the myrmecophagous *Priodontes*. The behavior, systematics, and ecology of the Canidae have recently been reviewed in the volume edited by Fox (1975).

13.1.2 The Family Ursidae: The Bears

The bears present a contrast to the cursorial canids. The ursids are plantigrade and rather slow-moving. All members display considerable climbing ability, especially when they are young. The molar teeth are

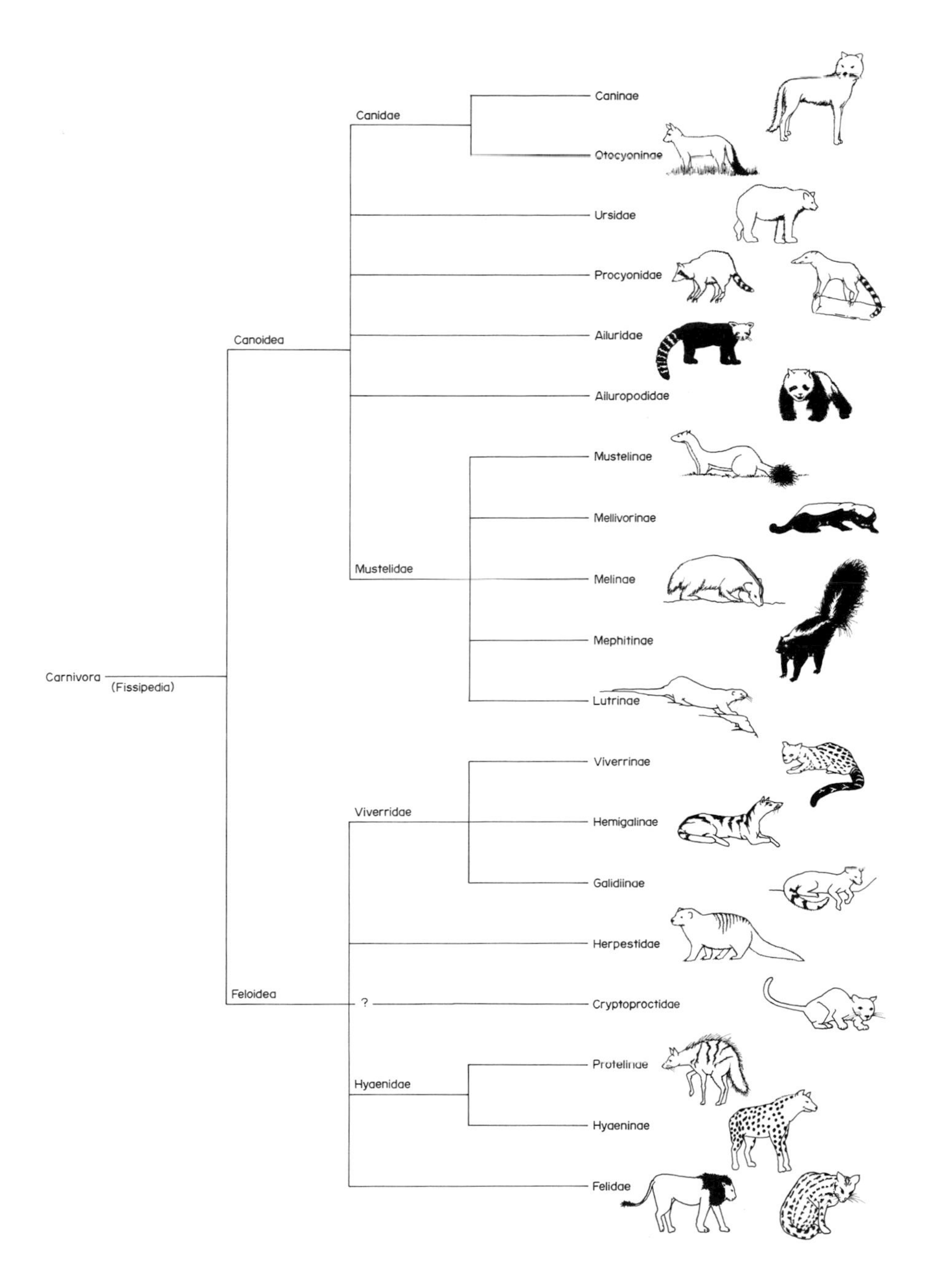

Figure 24. Diagram of the classification reflecting the phylogenetic branching of the fissiped carnivores. Based on Ewer 1973. The Galidiinae may in fact be early offshoots of the Herpestidae.

less modified for shearing and show a more cuspidate pattern, indicating a pronounced ability to crush and masticate plant substances. The tail is rudimentary in the Ursidae.

The genus *Ursus* is widely distributed in the Holarctic and includes the brown bears, grizzly bears, and the Alaska brown bear. The North

Plate 25. *Canis lupus,* the wolf. This pack-hunting canid is probably the best-studied carnivore in the world. Its life history and social structure have generated many useful hypotheses concerning the evolution of social systems. (NZP archives.)

American black bear is often separated in its own genus, *Eurarctos*. The Asiatic bears include *Selenarctos* of Manchuria, Japan, and the southern Himalayas. The Malay sun bear, *Helarctos malayanus,* inhabits Burma, Malaya, Borneo, and Sumatra and is the smallest of the living Ursidae and the most adapted for arboreality. The sloth bear, *Melursus ursinus,* is found in South India and Sri Lanka and has gone the farthest toward specialization for a myrmecophagous diet, a unique feeding adaptation in the family Ursidae. No bears now occur in Africa, and only one bear is found in the continent of South America, ranging almost exclusively in the Andes; this is *Tremarctos ornatus,* the spectacled bear (see plate 26).

13.1.3 The Family Procyonidae: Raccoons and Their Relatives

The family Procyonidae, or raccoon family, exhibits plantigrade to semiplantigrade locomotion. All members are semiarboreal. The living genera are confined entirely to the New World, if one considers the lesser panda and the giant panda to be in their own separate families. Davis (1964) concluded that *Ailuropoda* was a member of the Ursidae. Although many specialists subsume the lesser panda and giant panda under the family Procyonidae (see Morris and Morris 1966), I lean with Pocock (1921, 1928) and think that *Ailurus* and *Ailuropoda* may well be placed in their own families. If this is the case, then the family Procyonidae, sensu stricto, is entirely New World in its present distribution. It is typified by five genera: *Procyon,* the raccoon; *Nasua,* the coati; *Potos,* the kinkajou; *Bassaricyon,* the olingo; and *Bassar-*

Plate 26. *Tremarctos ornatus,* the spectacled bear, is the only bear endemic to South America. A versatile feeder, this captive is eating corn. (NZP archives.)

iscus, the ring-tailed cat. (*Nasuella,* the mountain coati, is often set aside in its own genus.)

A great variety of feeding adaptations are demonstrable in this family. *Procyon* is adapted for terrestrial foraging along stream sides for crustaceans as well as for arboreal foraging for fruit and small vertebrates. *Potos,* with its prehensile tail, is arboreally adapted for feeding on small vertebrates and fruit. *Bassariscus* is also arboreally adapted, as is *Bassaricyon,* but they have not developed prehensile tails. *Nasua,* the coati, is a versatile forager both on the forest floor and in trees. Its ecology and social structure have been analyzed in some detail by Kaufmann (1962). Its near relative the mountain coati, *Nasuella,* remains unstudied.

13.1.4 The Families Ailuridae and Ailuropodidae: The Pandas

The red or lesser panda, *Ailurus fulgens,* and the giant panda, *Ailuropoda melanoleuca,* are found in montane South Asia. *Ailurus* has a range from Nepal to Burma in the foothills of the Himalayas. *Ailuropoda* is found today only in western Szechwan. Both show convergences in the pattern of the molar teeth, since both eat considerable quantities of bamboo leaves and shoots. The crushing molar with an increased grinding surface reflects the adaptation for this dietary specialization (see plate 27). The giant panda is bearlike in its limb and body proportions, but its vocalizations and genital anatomy are quite unlike those of the true ursids. The lesser panda, with its long, ringed tail and small size, does have a superficial resemblance to the raccoon. The systematic position of both species remains to be clarified.

Plate 27. *Ailuropoda melanoleuca,* the giant panda, is a carnivore that has specialized for feeding on herbaceous vegetation. In this photograph the typical feeding posture is shown as bamboo leaves are masticated. (NZP archives.)

13.1.5 The Family Mustelidae: The Weasels, Otters, Skunks, and Badgers

The weasel family is distinguished from the preceding carnivores by the lack of an alisphenoid canal in the skull. Foot structure ranges from digitigrade to semiplantigrade. The mustelids are worldwide in their distribution, except for Madagascar and Australasia. It is possible to break them into five rather distinct subfamilies.

The Mustelinae typically have a reduced dentition and are highly specialized as carnivores. The dental formula is 3/3, 1/1, 3/3, 1/2. This subfamily includes the genus *Mustela,* which is distributed throughout the Holarctic. In the boreal north the genus is specialized into two or three sympatric species that have divided the feeding niche into different prey sizes (Rosenzweig 1966). *Martes* is also a Holarctic genus, specialized for capturing vertebrate prey in the trees. *Gulo* is a Holarctic genus of very large body size that is specialized for feeding on larger vertebrate prey and has no exact counterpart in the tropics. The Holarctic polecats (*Putorius* and *Vormela*) are strongly adapted for digging, and both genera can produce vile odors as an antipredator device.

The Neotropics have three genera of mustelines: *Eira,* which is arboreally adapted to rain forest habitats; *Grison,* which is terrestrially adapted for moist to xeric habitats; and *Lyncodon,* which is terrestrially adapted and occupies a weasellike niche in the Patagonian region.

Africa contains three mustelids adapted for semiarid to arid habitats, including *Zorilla, Poecilogale,* and *Poecilictis.* Although weasellike in form, these species bear a color pattern of black-and-white stripes reminiscent of the New World Mephitinae or skunks.

The second subfamily of mustelids is the Mellivorinae, including one monotypic genus, *Mellivora capensis,* the ratel. This species is confined to tropical Africa and adjacent parts of the Middle East. *Mellivora* is about the size of a badger and resembles the badger in its color pattern. Although a true predator on small vertebrates, it also feeds on the honey of wild bees. The symbiotic relationship between the ratel and the wax-feeding bird the honey guide, *Indicator indicator,* has been reviewed by Friedman (1955).

The subfamily Melinae includes the true badgers, which are typified by rather enlarged upper molars of the crushing type. These genera are to a greater or lesser extent adapted for digging, and many pursue their prey by burrowing. The European badger, *Meles,* which extends its range across northern Asia, typifies this subfamily, as does the North American badger, *Taxidea. Melogale (=Helictis),* the ferret badger, is long-tailed and more weasellike in its size and body shape. The genus is distributed through Southeast Asia to the Sunda Islands. Its members perhaps represent a less specialized body type than that typified by *Meles* and *Taxidea.* The three remaining genera of badgers are *Arctonyx, Mydaus,* and *Suillotaxus* the hog badgers. They are distributed respectively in East Asia, North China, and Sumatra; Sumatra, Java, and Borneo; and Palawan. These three genera are typified by a rather narrow muzzle and by specialization for feeding on soil invertebrates; in a sense they occupy an insectivore/omnivore niche. All of them have the ability to emit vile odors from the anal glands, and it has been suggested on morphological grounds that they show closer affinities to the subfamily Mephitinae than to the classical Melinae (Radinsky 1973).

The subfamily Mephitinae includes the skunks of the New World. Three genera are represented: *Mephitis, Spilogale,* and *Conepatus. Conepatus* extends its range into the Neotropics. These animals are specialized as insectivore/omnivores and have a warning black-and-white coloration. They all have the ability to emit an offensive scent as an antipredator device. As I have pointed out, this trait is shared with other mustelids, such as the hog badgers and the musteline black-and-white weasels, *Poecilogale* and *Zorilla.*

The fifth subfamily, the Lutrinae, is adapted for an aquatic life. Its members are the otters, which have webbed feet, a dense underfur, short legs, a long body, and a rather long tail. They are worldwide in their distribution, being found in freshwater streams of all continents except Australia. The genus *Lutra* extends worldwide with the exception of southern Africa. *Pteronura,* the giant Brazilian otter, reaches 7 feet in length, approximated only by the sea otter of the north Pacific coast, *Enhydra*. The sea otter is exclusively marine and feeds on mollusks and echinoids off the coasts of the North Pacific.

13.1.6 The Superfamily Feloidea

The superfamily Feloidea is characterized by the unique structure of the tympanic bullae (see section 13.1). Depending on which systematist one consults, it is divisible into either three or five families. The classical families are the Viverridae, Hyaenidae, and Felidae, but it is possible to split the Viverridae into three parts—the Viverridae sensu stricto, the Herpestidae, and the Cryptoproctidae.

13.1.7 The Family Viverridae: The Civets

As a rule of thumb, the viverrid carnivores are confined to the Old World tropics and subtropics. Ecologically they show more diversification in trophic specialization and substrate use than any other carnivore family. The niches occupied by the Old World tropical viverrids are filled in the Neotropics by the Mustelidae and the Procyonidae (see plate 28). The viverrid carnivores have not been studied in the same detail as the other families. Wemmer (1977) outlines the life history of *Genetta tigrina* and reviews the literature. The family Viverridae, together with *Cryptoprocta*, and the herpestids, *Galidia, Mungotictis,* and *Galidictis,* are the only carnivore groups to have successfully colonized and radiated on the island of Madagascar (Albignac 1973).

Most living viverrids are nocturnal. They carry forward into the present a number of conservative morphological features (Ewer 1973). Since civets may be considered to represent most closely the ancestors

Plate 28. *Viverricula indica,* the oriental civet. The viverrid carnivores show a conservative body plan. Most are strictly nocturnal small carnivores or generalized omnivores exhibiting wide adaptive radiation as a family for different niches. (Courtesy of C. M. Hladik.)

of the more modern Carnivora, they deserve more attention from students of mammalian behavior.

13.1.8 The Family Cryptoproctidae: The Fossa

Although *Cryptoprocta ferox* is obviously closely related to the viverrids, it has been separated from the stem lineage for some time. As the major carnivore on the island of Madagascar, it has converged strongly with the felid pattern in its dentition and skull shape. *Cryptoprocta* stands as a classical case of convergent evolution (see Albignac 1973).

13.1.9 The Family Herpestidae: The Mongooses

The mongooses are distributed across Africa and the Middle East to Southeast Asia. *Herpestes* has been introduced to Hawaii and the Antilles and has radiated to fill a number of terrestrial and scansorial feeding niches. Some species, such as *Atilax paludinosus* of Africa and *Herpestes urva* of South Asia, feed along streams in a manner reminiscent of the New World *Procyon*.

Some species of mongoose are diurnal and rather social. They tend to forage in a closed social group with a pronounced division of labor. The life history of *Mungos mungo* has been studied (Neal 1970; Rood 1975), and that of *Helogale* has been described by Rood (1978).

13.1.10 The Family Hyaenidae: The Hyenas

The family Hyaenidae is divisible into two subfamilies: the Hyaeninae and the Protelinae. The Hyaeninae occupy Africa, the Middle East, and Central and South Asia. They have a strongly carnassial dentition, are digitigrade, and show a reduction of toes on the forefeet. The Protelinae consist of one genus, *Proteles*, the aardwolf, found in southern Africa. The dentition is very weak, and overall they show an adaptation for feeding on termites. This is another case of a carnivore that, like the sloth bear, is specialized as a secondary myrmecophage. The genus *Crocuta* is the best known, thanks to the efforts of Hans Kruuk (1972).

13.1.11 The Family Felidae

The cat family shows the greatest tooth reduction of all the Carnivora. The dental formula is 3/3, 1/1, 3/2, 1/1, for a total of 30 teeth. The molars show the highest specialization for a carnivorous life. Typically, the claws can be retracted, except in the cheetah, *Acinonyx*. The Felidae were distributed on all continents and continental islands, except for Australia and the island of Madagascar (see plate 29).

There is some controversy concerning the number of genera in the Felidae (see Ewer 1973). Classically the genus *Felis* consists of small cats exhibiting 30 teeth. The genus *Lynx* may be set aside because of the loss of one upper premolar, giving a total dental number of 28. The genus *Panthera* is generally applied to the larger cats that show an incomplete ossification of the hyoid suspension.* This apparently allows the larynx to expand and correlates with their capacity for roaring.

Plate 29. *Panthera pardus,* the leopard. *Right,* normal color pattern; *Left,* melanistic. The leopard typifies the usual habits of the Felidae. It is a solitary hunter. (NZP archives.)

(Gustav Peters [1978] disputes this thesis and feels that the genus *Panthera* should be restricted to *P. onca, P. pardus* and *P. leo.*) Members of the genus *Acinonyx,* the cheetahs, are aberrant in many respects, not only because their claws are nonretractile, but because overall their skeleton shows extreme modification for cursorial life (Hildebrand 1961). The North American mountain lion is often set aside in its own genus, *Puma.* The Southeast Asian clouded leopard is placed in the genus *Neofelis,* while the snow leopard is conserved in the genus *Uncia.* The controversy over classification will no doubt continue, and convenience may ultimately dictate the use of the binomials. Leyhausen (1979) has provided a superb review of the behavior and range of adaptations displayed by the family.

13.1.12 The Adaptations of the Carnivora

With such an extensive literature and so many excellent monographs, I have chosen only a few to organize in table 21 as a guide for further reading. By way of summary, let me sketch the adaptations of the Carnivora by attempting to step back into the past and imagine the earliest true carnivores. The ancestor was probably a small insectivorous mammal that gradually began to specialize for killing larger and larger prey. When we recognize a true carnivore in the fossil record of the Miocene, the carnassial dentition is more highly developed; this newly evolved form was probably feeding on rodents, reptiles, and small birds.

Table 21 Recent Literature on the Fissiped Carnivores

Taxon	Ecology	Behavior	Reproduction and Ontogeny	General Review
Canidae	—	Kleiman 1967; Fuller and DuBuis 1962		Fox 1975
Canis familiaris	Beck 1973	—	Christie and Bell 1972	Scott and Fuller 1965
C. lupus	Murie 1944; Pimlott 1967; Allen 1979	Schenkel 1948; Zimen 1971, 1976	Schoenberner 1965; Field 1978	Mech 1970; Young and Goldman 1944; Murie 1944
C. latrans	Gier 1968	—	—	Bekoff 1978; Murie 1940
C. aureus	Golani and Keller 1975; Van Lawick 1971	Golani 1966, 1973	—	—
Vulpes fulva	Vincent 1958; Storm et al. 1976	Storm 1965	—	—
V. vulpes	Burrows 1968	Tembrock 1957	—	—
V. velox	Kilgore 1969	—	—	—
V. macrotis	Egoscue 1956, 1962, 1975	—	—	—
Fennecus zerda		Gauthier-Pilters 1962, 1966	St. Girons 1962	—
Otocyon megalotis	Nel 1978	Lamprecht 1979	—	—
Urocyon cinereoargenteus	Lord 1961; Trapp 1978	—	—	—
Alopex	MacPherson 1969	—	—	Pedersen 1959
Lycaon pictus	Estes and Goddard 1967; Van Lawick 1971	Kühme 1964, 1966; Frame et al. 1979; Frame and Frame 1976	Jacobi et al. 1968	—
Speothos venaticus		Kleiman 1972*b*; Drüwa 1977	—	—
Cerdocyon thous	Brady 1979	Kleiman and Brady 1978	Brady 1978	—
Chrysocyon brachyurus		Encke et al. 1970; Altmann 1972; Lippert 1973	—	—
Dusicyon culpaeus	Crespo and De Carlo 1963	—	—	—
Nyctereutes procyonoides	Bannikov 1964	Seitz 1955	—	—
Ursidae				
Ursus arctos	—	Meyer-Holzapfel 1957; Tschanz, Meyer-Holzapfel, and Bachmann 1970; Krott and Krott 1962	—	Herrero 1972
U. middendorfi	Troyer and Hensel 1964	Stonorov and Stokes 1972	—	—
Euarctos americana	Jonkel and Cowan 1971; Rogers 1977	Burghardt and Burghardt 1972	Mundy and Flook 1963	—
Tremarctos ornatus	—	Roth 1964	Dathe 1967	—
Selenarctos thibetanus	Bromlei 1956; Schaller 1969	—	—	—
Melursus ursinus	Laurie and Seidensticker 1977	—	—	—

Table 21—*Continued*

Taxon	Ecology	Behavior	Reproduction and Ontogeny	General Review
Helarctos malayanus	—	—	Dathe 1963, 1970	—
Thalarctos maritimus	Harrington 1968	Wemmer, Von Ebers, and Scow 1976; Jonkel et al. 1972	—	Pedersen 1957
Procyonidae				
Procyon lotor	Mech, Barnes, and Tester 1968; Stains 1956; Stuewer 1943	Lyall-Watson 1963; Sharp and Sharp 1956	Montgomery 1964, 1969; Hamilton 1936; Fritzell 1978	—
Potos flavus	—	Poglayen-Neuwall 1961, 1966	Poglayen-Neuwall 1973, 1974	—
Nasua narica	Kaufmann 1962	—	—	—
Bassaricyon gabbi	—	Poglayen-Neuwall and Poglayen-Neuwall 1964	—	—
Bassariscus astutus	Trapp 1978	—	Richardson 1942	—
Ailuropodidae	—	—	—	Davis 1964
Ailuropoda melanoleuca	Sheldon 1975; Anonymous 1974	Morris and Morris 1966	Anonymous 1974	—
Ailuridae				
Ailurus fulgens	—	Roberts 1975	Roberts and Kessler 1979	—
Mustelidae	—	Goethe 1964	—	—
Mustelinae	Rowe-Rowe 1978	Channing and Rowe-Rowe 1977	—	—
Gulo gulo	—	Krott 1960	—	—
Putorius putorius	—	Herter 1953; Poole 1966; Apfelbach 1973	—	—
Mustela nivalis	—	—	Heidt, Petersen, and Kirland 1968; Hartmann 1964	—
M. vison	Gerell 1967	Poole and Dunstone 1976	—	—
Martes sp.	Powell 1978	Gossow 1970	Jonkel and Weckwerth 1963	—
Eira barbara	—	Kaufmann and Kaufmann 1965	Poglayen-Neuwall 1975; Poglayen-Neuwall and Poglayen-Neuwall 1976	—
Mephitinae				
Mephitis mephitis	Verts 1967	Van Gelder 1953	—	—
Spilogale	Crabb 1948	—	Mead 1968	
Melinae	—	—	—	Neal 1977
Meles meles	Kruuk 1978; Van Wigjngaarden and Van de Peppel 1964	Eibl-Eibesfeldt 1950*b*	Neal 1958	Neal 1958
Helictis personata	—	Pohl 1967	—	—
Lutrinae	—	—	—	Harris 1968; Duplaix-Hall (in press)
Lutra sp.	Procter 1963; Erlinge 1968	—	—	—

Table 21—*Continued*

Taxon	Ecology	Behavior	Reproduction and Ontogeny	General Review
Enhydra lutra	Kenyon 1969	Fisher 1939; Sandegren, Chu, and Vandevere 1973	—	—
Viverridae	Albignac 1973	Dücker 1957	—	Dücker 1965
Nandinia binotata	Charles-Dominique 1978	Dücker 1971	—	—
Genetta genetta	Delibes 1974	Gangloff and Ropartz 1972	—	—
G. tigrina	—	Wemmer 1977	Wemmer 1977	—
Galidia elegans	—	Albignac 1969*a*	—	—
Arctictis binturong	—	Kleiman 1974*b*; Huf 1965	—	—
Herpestidae	Hinton and Dunn 1967	Rensch and Dücker 1959	—	—
Herpestes auropunctatus	Baldwin, Schwartz, and Schwartz 1952	—	—	—
Suricata suricatta	—	—	Ewer 1963	—
Helogale undulata (=*parvula*)	Rood 1978	Rasa 1972, 1973, 1977	Zannier 1965	—
Cryptoproctidae				
Cryptoprocta ferox	—	Vosseler 1929; Albignac 1973	Albignac 1969*b*	—
Hyaenidae				
Crocuta crocuta	Kruuk 1966, 1972	Watson 1965	Wells 1968	—
Hyaena hyaena	Kruuk 1976	—	Pournelle 1965	Kruuk 1972
H. brunnea	Mills 1978; Owens and Owens 1978	—	—	—
Proteles cristatus	Kruuk and Sands 1972	—	—	—
Felidae	—	—	Sadleir 1966	Leyhausen 1978, 1963, 1956; Eaton 1973*a,b*, 1974*b*, 1976*a,b*, 1977
Felis catus	Leyhausen and Wolff 1959	Rosenblatt and Schneirla 1962	Rosenblatt, Turkewitz, and Schneirla 1962	—
F. silvestris	—	—	Lindemann 1955	—
Neofelis nebulosa	—	Hemmer 1966, 1968	Fellner 1968	—
Acinonyx jubatus	Eaton 1970	Schaller 1968	Krefeld 1963	Eaton 1974*a*
Lynx canadensis	Berrie 1973	—	—	—
L. rufus	—	—	Provost, Nelson, and Marshall 1973	Young 1958
Panthera pardus	Turnbull-Kemp 1967	Muckenhirn 1972	Eaton 1977	—
P. leo	Schaller 1972	Bertram 1975	Rowlands and Sadleir 1968	Schaller 1972
P. tigris	McDougal 1977; Sunquist 1979	—	—	Schaller 1967
Puma concolor	Seidensticker et al. 1973; Hornocker 1969	—	Rabb 1959*b*	Young and Goldmann 1946

Note: The standard reference is Ewer, *The Carnivores* (1973).

As larger size was selected for, some carnivores began to feed on even larger mammals. The rise of grassland and savanna areas in the Pliocene gave selective impetus to the evolution of large cursorial grazing herbivores, and, as an adaptation for preying on these larger herbivores, selection favored larger carnivores. In addition, the grassland favored cryptic rodents and birds, so that at the same time selection was favoring the specialization of small carnivores for foraging on the smaller denizens of the grassland.

The three major lines of recent adaptation in the Carnivora that resulted in large predators include the families Ursidae, Canidae, and Felidae. The Ursidae became large but diversified in their dietary specializations. Bears feed on considerable quantities of plant parts, and, though they can be formidable predators, they have not specialized to the same degree as have the canids and felids.

Felid specializations have centered on a prey-capturing strategy involving a short rush, a leap, and a quick killing bite. One can imagine the ancestral felid as being an arboreal bird-feeder and then applying this same technique to the tallgrass savannas. This specialization culminated in the savanna-adapted lion, *Panthera leo*. Large carnivores killing large prey also open up a niche for scavengers that can be occupied by smaller carnivores, such as small canids, or by moderately large forms such as members of the family Hyaenidae.

The Canidae, on the other hand, as terrestrially adapted forms, specialized for capturing their small prey by leaping, pinning the prey with the forepaws, and administering a killing bite. As they adapted for hunting in open country, they adopted a running or chasing technique in prey capture. Cooperative predation involving a social group has apparently evolved convergently several times in the Canidae as they have adapted for killing larger and larger cursorial prey (Kleiman and Eisenberg 1973).

What, then, happened to the carnivores that remained rather small in size, such as the mustelids, viverrids, procyonids, and herpestids? These lines diversified into a number of specialists: frugivores, piscivores, and predators on small rodents, birds, and reptiles. Arboreal predatory forms developed in both the Viverridae and the Mustelidae. Exploitation of the piscivorous niche develoepd in the Viverridae and Mustelidae. Arboreal frugivority developed in the Procyonidae and the Viverridae. And in some cases there has been a reversion to an insectivorous diet, demonstrable in the Viverridae (*Eupleres*), Mustelidae (*Mephitis*), and Hyaenidae (*Proteles*).

Table 22 demonstrates the ecological specializations in the small nonfelid carnivores. It is convenient to consider convergence from the standpoint of continental areas. The insectivore/omnivore niche is occupied in North America by the mephitine mustelids *Mephitis, Conepatus,* and *Spilogale*. In Eurasia it is occupied by the mustelids *Putorius* and *Vormela,* and by the badger, *Meles meles;* in southern Europe and North Africa by *Vormela* and *Poecilictis;* in South Asia and the Sunda Islands by *Hemigalus* and *Crotogale;* in Borneo by *Diplogale* and *Hemigalus;* on Madagascar by *Galidictis;* in the Neo-

Table 22 Ecological Adaptation and Convergence in Small Carnivores

	Feeding Niche			
Geographical Area	Solitary Insectivore/ Omnivore	Piscivore	Arboreal Frugivore/ Omnivore	Diurnal Group Foraging Insectivore/ Omnivore
Eurasia and North Africa	*Putorius, Vormela, Poecilictis*	*Lutra*	—	—
North America	*Mephitis, Spilogale, Conepatus*	*Lutra*	—	—
Sub-Saharan Africa	*Ichneumia*	*Amblonyx, Osbornictis*	*Nandinia*	*Suricata, Mungos, Helogale*
Tropical Asia	*Hemigalus, Diplogale*	*Aonyx, Cynogale*	*Paradoxurus, Paguma, Arctictis*	—
Neotropics	*Lycodon, Conepatus*	*Lutra*	*Potos*	*Nasua*

tropics by *Conepatus;* and in equatorial and southern Africa by *Cynictis* and *Helogale*.

Specialization for feeding on fishes is exhibited by the subfamily Lutrinae over the whole of the major continents with the exception of Australasia, but it is noteworthy that the piscivorous feeding niche in tropical streams, in addition to being occupied by *Lutra,* has also been convergently occupied by *Cynogale,* a viverrid in Sumatra and Borneo, and by *Osbornictis,* a viverrid in West Africa.

Arboreal frugivority and carnivority have been pioneered by viverrids in Southeast Asia and the Sunda Islands, including the genera *Arctictis, Paguma,* and *Paradoxurus* (see plate 30). In Africa the viverrid genus *Nandinia* occupies a similar niche, whereas in the Neotropics the procyonid genus *Potos* occupies an identical niche. A prehensile tail has been convergently evolved in two genera of arboreal carnivores, *Arctictis,* a viverrid, and *Potos,* a procyonid.

13.1.13 Social Evolution in the Carnivora

In 1973 Kleiman and Eisenberg developed a theory concerning the origin of complex social groupings in the two carnivore families, the Canidae and Felidae, that perform group hunting and form complex, cohesive social groupings. We considered the pack-hunting canids and the prides of *Panthera leo* as two extremes in the social evolution of carnivores. We pointed out that the steps in the evolution of canid and felid societies had been different. There are now sufficient data for the black-backed jackal, *Canis mesomelas* (Moehlman 1979); the spotted hyena, *Crocuta crocuta* (Kruuk 1972); the Cape hunting dog, *Lycaon pictus* (Frame et al. 1979); the banded mongoose, *Mungos mungo* (Rood 1975; Neal 1970); and the dwarf mongoose, *Helogale parvula* (Rood 1978), to extend the discussion. Let us look for a moment, then, at the evolution of social groupings in these five African carnivores. I hope to demonstrate not only that different selective pressures have

Plate 30. *Paradoxurus hermaphroditus,* an arboreal viverrid of Southeast Asia. Contrast with the oriental civet, *Viverricula.* (NZP archives.)

achieved superficially similar social organizations, but also that the phylogenetic background has influenced the outcome.

The African lion, *Panthera leo,* the spotted hyena, *Crocuta crocuta,* and the Cape hunting dog, *Lycaon pictus,* all live in rather large, cohesive permanent social groupings. Superficially they appear quite similar in degree of sociality. The black-backed jackal, *Canis mesomelas,* lives in smaller groups, and the dwarf mongoose, *Helogale parvula,* lives in intermediate-sized bands that forage individually but as a cohesive unit.

A lion pride may include up to four adult, reproducing females. A spotted hyena clan may have even more reproducing females. The other three carnivores typically have only one reproductively active female per group. The Cape hunting dog pack may have several adult males but only one that actively reproduces. The lion pride may have one to three reproducing males, and the spotted hyena clan may have even more. Black-backed jackal and mongoose groups have one actively reproducing male. At first glance the smaller carnivores (jackal and mongoose) do not strike one as having the same degree of sociality as is represented by the social organizations outlined for the larger carnivores. Nevertheless, all five species are highly social and are usually encountered in groups. The five systems are drawn from four different carnivore families and thus represent a convergence in social structure rather than derivation from a common ancestral form that already exhibited a complex permanent social grouping.

The felid system exemplified by the lion basically consists of several females who hunt together and rear their young in common. Males that accompany a group of females have a tenure of approximately three years (see plate 31). Males acquire access to females through active competition with one another, and usually the successful competitors are two cooperating males that are probably related (Bertram 1975). The social system here does not have as its core a strong relationship between one male and one female.

As outlined by Kleiman and Eisenberg (1973), the route to sociality in the family Canidae appears to be predicated on a phylogenetically old bonding between a male and a female for the purpose of rearing a litter in common. Most of the small canids show a tendency toward

Plate 31. *Panthera leo*. Portrait of a male lion. The lions have evolved the most complicated social system among the Felidae. (NZP archives.)

monogamy, with the male provisioning the female during the lactation period, thus enabling her to rear a larger litter than would otherwise be possible. The black-backed jackal exhibits this mating system, but offspring occasionally remain with parents to aid in rearing younger siblings (Moehlman 1979). The Cape hunting dog shows a variation on this theme in that, though a dominant male and female may be the sole breeders in the pack, there may be many adult males of nearly the same age-class that are closely related (Frame et al. 1979). All adults, in the Cape hunting dog pack, male or female, participate in provisioning the lactating female and her young. The female has an exceedingly large litter, which she would be unable to raise without assistance. This may be viewed, then, as a variant on the classical canid pattern.

The spotted hyena, *Crocuta crocuta,* has arrived at sociality from a slightly different basis. Here we find traditional denning sites where young are reared by a group of females. The group of females has attached to it males that do not participate in the rearing process. The females are larger than the males and absolutely dominant over them, and selection has produced a profound genital mimicry where the female external genitalia are almost identical in appearance to those of the male. The traditional greeting ceremony of mutually sniffing the genital areas thus places the female in a position of offering the male a visual signal that he cannot discriminate from that of another male. Undoubtedly olfaction serves to identify the sexes, but the whole process whereby such genital mimicry and female dominance may evolve remains to be elucidated.

The dwarf mongoose has a society based on a founder pair that continue to breed throughout their lives. No successful breeding is accomplished by attached females or males unless a member of the dominant pair dies. Although most members of the clan are born into it, a surprisingly high number may be recruited from the outside. Recruits assist in rearing subsequent young born to the founder pair. The group forages and rests as a unit.

Foraging strategies are useful indicators of the adaptive value of social organizations. The banded and dwarf mongoose forage in semiarid savanna, using brush and scrub for cover. They feed on insects and small vertebrates, generally foraging as a somewhat dispersed group, but usually in vocal or visual contact. The group may be effective in detecting the presence of predators. The jackal is a versatile feeder and may hunt alone or in small groups. Small vertebrates form the basis of its diet, but it may also take ungulate young by cooperative hunting or scavenge the kills of larger predators.

The Cape hunting dog, spotted hyena, and lion can and do take ungulate prey larger than themselves, but they often do so by cooperative hunting strategies. This overriding trend toward hunting large prey and the advantages accruing to a social hunter seem to have dominated the evolution of these savanna-adapted large carnivores.

The basic canid pattern developed from a monogamous system where provisioning a breeding female enabled larger litters to be reared. The basic pattern of the social mongooses developed from a nearly similar situation, but the advantage of a large social grouping that persists during foraging and rest probably derives from reduced predation, especially on the young. Vigilance by all members of the group probably reduces predation, especially from raptors. Again, the foraging strategy of the group may be enhanced through greater efficiency in locating energy-rich food resources. Selection pressure favoring sociality in *Panthera, Crocuta,* and *Lycaon* probably involved the advantage of group hunting of large ungulates. Thus there has been a convergence in group size in these three carnivores that derives from increased predation efficiency. Discrete comparisons of the five social systems are presented in table 23.

13.2 The Order Pinnipedia

The Pinnipedia are adapted for an almost completely aquatic life, and all feed on fish, squid, mollusks or crustaceans. Almost all species are marine. In conformity with their aquatic adaptation, they have a fusiform body and short forelimbs. The digits are long and are enclosed in a fleshy membrane to form flippers. The tail is very reduced in length. There is a tendency for a simplified tooth structure resembling a homodont condition, which will be discussed further for the Cetacea.

Although closely related to the true carnivores, the pinnipeds have been separated for a considerable period. Although I treat them as a separate order, this is mainly for convenience and does not reflect a true phylogeny. The Pinnipedia are themselves an artificial assemblage, since they are believed to have had two separate ancestries (see King 1964 for a review). There are two superfamilies of pinnipeds: (1) the Phocoidea, including the family Phocidae, and (2) the Otarioidea, including the families Otoriidae and Odobenidae. The Phocidae probably evolved from an ancestral form similar to the one that gave rise to the modern-day otters. The Otariidae and the Odobenidae probably derived from an ancestral form closely related to the line of descent leading to the bears and canids (see fig. 25).

Let us first consider the Phocidae, or "true seals." It is possible to divide the Phocidae into the subfamilies Phocinae, Monachinae, and Cystophorinae. The Monachinae are typically divisible into two tribes: the Monachini and the Lobodontini. The tribe Monachini contains a single genus, *Monachus,* the monk seal, which is tropical in its distribution, having a representative in the Mediterranean, a species in the Caribbean (possibly extinct), and a species near the Hawaiian archipelago. The Lobodontini are divisible into four genera: *Leptonychotes, Lobodon, Hydrurga,* and *Ommatophoca*. These are antarctic in their distribution, and *Lobodon* shows a specialization toward feeding on plankton. The Cystophorinae include *Cystophora* and *Mirounga*. *Mirounga,* the elephant seal, is one of the best-studied genera, distributed in the northern hemisphere off the coast of California and in the southern hemisphere off the Falkland Islands and coastal Patagonia (see plate 32).

The Phocinae are divisible into two tribes, the Erignathini and the Phocini. The Erignathini, or bearded seals, contain a single genus and species, *Erignathus barbatus*. The tribe Phocini is divisible into two subtribes, Phocina and Histriophocina. The Histriophocina are a rather

Table 23 Correlations of Body Size and Behavior for Five Social Carnivores

	Form of Dimorphism			Forms of Dominance	Spacing between Groups	Social Structure	Adult Relationships	
	Size	Adornment	Genitalia				Male-Male	Female-Female
Felidae *Panthera leo*	♂>♀	♂ mane	Dimorphic	♂ dominates at feeding	Territorial	One or two related ♂♂ attached to a group of ♀♀; ♂♂ turnover through overt takeovers	Amicable, may be related	Cooperative, several ♀♀ may breed in a group
Hyaenidae *Crocuta crocuta*	♀>♂	None	♀≅♂	♀ dominant over ♂	Territorial	Clan of ♀♀ and attached ♂♂. ♀♀ hunt cooperatively. ♂♂ move between groups	Dominance order; no alliances	Cooperative, several ♀♀ may breed in a group
Canidae *Canis mesomelas*[a]	Nearly monomorphic			♂ and ♀ equivalent but separate roles	Territorial	Founder pair and descendents. Older young and ♂ provision adult ♀	Competitive	Founding ♀ dominant and is only successful breeder
Lycaon pictus[b]	Nearly monomorphic			♂ and ♀ equivalent but separate roles	Believed territorial	Related ♂♂ and older juveniles provision breeding ♀	Cooperative, may be related	Founding ♀ dominant and is only successful breeder
Herpestidae *Helogale parvula*[a]	Monomorphic			♂ and ♀ equivalent but separate roles	Believed territorial	Founder pair and descendents. Older young show vigilance behavior	Competitive reproductively; cooperative in defense	Founding ♀ dominant and is only successful breeder

[a] Classical monogamy.
[b] Modified monogamous system.

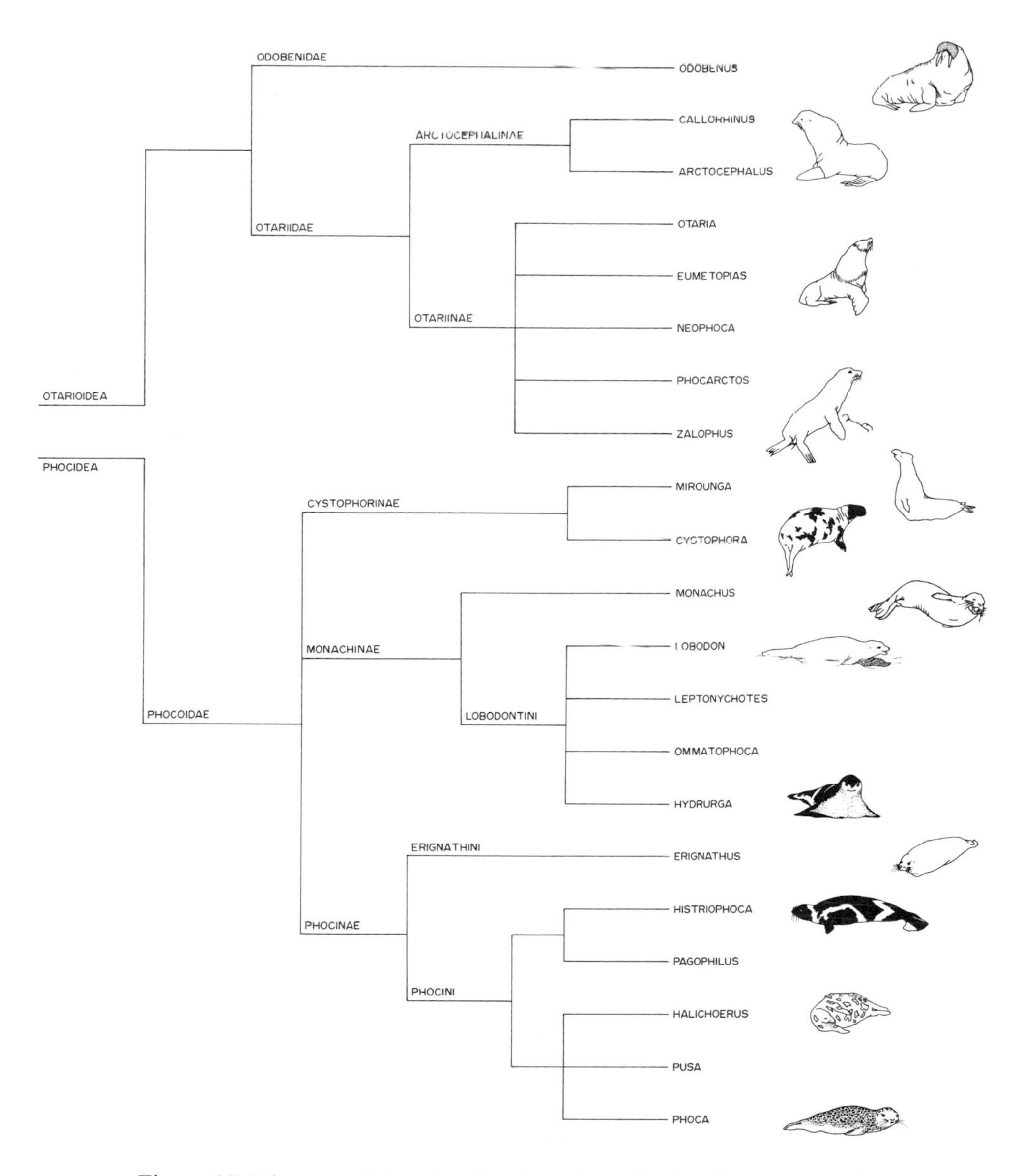

Figure 25. Diagram of the classification of the Pinnipedia. Adapted from King 1964.

small group with a polar distribution, including two genera, *Pagophilus* and *Histriophoca*. The Phocina include three genera: *Halichoerus*, the gray seal; *Phoca*, the harbor seal; and *Pusa*, the ringed seals. All three genera are distributed in the northern hemisphere. *Pusa* has two species (*sibirica* and *caspica*) that are landlocked in freshwater lakes, Lake Baikal and the Caspian Sea. *P. hispida* has several subspecies that are also landlocked in the lakes bordering the Baltic Sea.

The Otarioidea are divisible into two families, the Otariidae, or eared seals or sea lions, including the Otariinae, or sea lions, and the Arctocephalinae, or fur seals. Both subfamilies have representatives in the

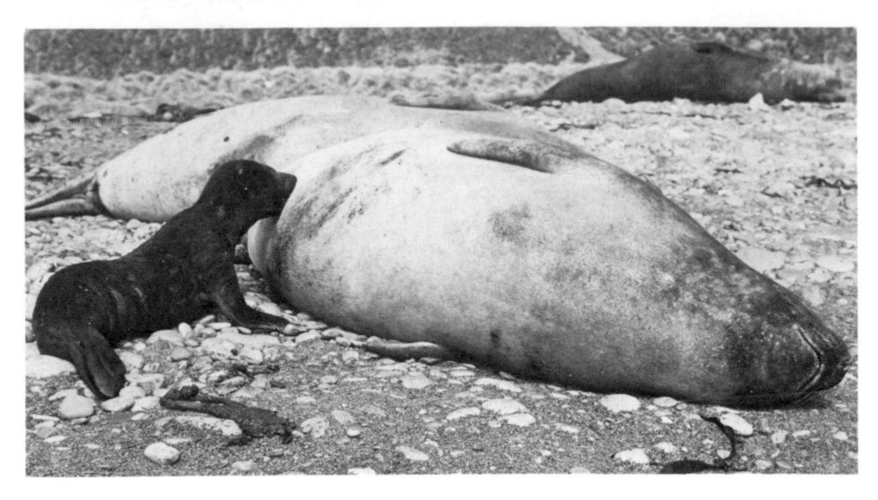

Plate 32. *Mirounga leonina,* the elephant seal. Pup feeding from mother. This species shows one of the most rapid growth rates in the class Mammalia. (Courtesy of Ian Eberhard and Brenton Watkins.)

northern and southern hemispheres. The only other family, the Odobenidae, or walruses, is uniquely distributed in the north Atlantic and north Pacific. The walruses are adapted for substrate feeding, and their long tusks are used to probe the sea floor for clams and other bivalves. In this their adaptation is unique.

13.2.1 The Natural History of the Pinnipeds

Most of what we know about pinnipeds comes from studying their life on land. We know that the California sea lion, *Zalophus californianus*, when at sea, travels in small herds while feeding. Migrating herds of *Callorhinus ursinus* or *Pagophilus groenlandicus* (= *Pusa groenlandicus)* may be observed. Aside from incidental observations, however, we know very little of what goes on when pinnipeds are at sea.

Turning first to the Otariidae, *Callorhinus ursinus,* the Alaska fur seal, is a typical example (see Bartholomew 1959). This migratory species feeds far out at sea, returning during spring to the Pribilof Islands in the Bering Sea. The males tend to arrive first and set up territories that they defend assiduously against one another. Females, when they arrive, tend to settle in groups but will not remain together unless there are more than five. Females tend to return to the same area of the beach year after year. After defending a certain portion of the beach, males will patrol their females and keep out intruding males. This promotes the male's access to females when they come into estrus. Females bear their calves on land and come into estrus shortly thereafter, whereupon they are bred by the "harem master" males.

A female can recognize her own calf when it is on the beach and will return to it to suckle. The calf grows very fast; within a month the female leaves it, and the young begins to swim and forage on its own.

This pattern of harem formation and its accompanying characteristic—that males are much larger than females—typifies the breeding behavior of otariids on land. *Arctocephalus pusilus,* the Cape fur seal, has a similar pattern of breeding behavior, but it is non-migratory and after calving it tends to remain in the vicinity of the beaches where mating takes place (Rand 1967). The California sea lion, *Zulophus californianus,* is similar (Peterson and Bartholomew 1967).

The walrus, *Odobenus rosmarus,* shows a tendency for seasonal movements. In the spring walruses are typically farther north on special beaching grounds, and in the winter they are farther south, often floating on ice floes. Breeding takes place on the ice, and this is the only time the adult male and female are together. The sexes tend to be in unisexual subgroupings outside the breeding season. A definite dominance order exists among the males, but harems are apparently not defended. The terrestrial phase in the walrus is most pronounced from April to July, when the molt takes place. A female tends to have a calf every three years. Females are very much attached to their calves, and, since they remain with them for a long period, the female rearing groups give the appearance of a group of mixed sexes. In reality, however, these are groups of females with various-aged young (Pedersen 1962; Miller 1975).

The elephant seal, *Mirounga,* is one of the few phocids that shows a harem structure similar to that reported for the eared seals (the sea lions and fur seals). *Mirounga,* however, does not defend a group of females; rather, males defend a specific area to which the females are attached. Males space themselves by assuming threat postures and giving characteristic calls (Bartholomew and Collias 1962).

Mating systems in the Phocidae are somewhat variable, but there is a much-reduced tendency toward pronounced sexual dimorphism. Dimorphism is apparent in the gray seal, *Halichoerus grypus,* and here the males do come ashore in specific places and defend territories. The topography of the beach greatly affects the size of the harem (Hewer 1960). In *Phoca vitulina* there is fall breeding, which occurs on areas where the animals haul out. Although there may be a subtle dominance order among males, there is no tendency toward harem formation (Scheffer and Slipp 1944). The leopard seal, *Hydrurga leptonyx,* is very large for a phocid and seems to be one of the most solitary of the pinnipeds. There is no tendency for females to form aggregations at the time of breeding, and males do not form harems (Marlow 1967).

To synthesize these brief examples, I should state that there are two periods when pinnipeds seem to be associated with land or ice: during the season of birth and breeding, and during molting. Both adults and juveniles molt, and at this time there is a tendency for juveniles to form a subgroup separate from the adults. Group hunting and migration favor gregariousness or school formation; however, the exact age and sex composition of foraging groups is difficult to determine. One may assume they could be either subadults of like age and sex or groups of females with their attendant young, as in the case of the walrus.

All pinnipeds are prone to return to specific inaccessible loci for molting and breeding. The inaccessibility of islands, rocky outcroppings, sandbars, and ice keeps the animals safe from molestation by terrestrial predators. There is also a marked tendency for breeding and molting to be synchronized in pinniped populations.

Social tendencies during the breeding season are manifest in two quite different mechanisms. As a general rule, the true seals or Phocinae do not form harems. Phocids that do so include *Halichoerus* in some habitats and *Mirounga*. (Boness 1979 analyzes the variation in polyg-

Table 24 Selected References for the Pinnipedia

Taxon	Natural History and Ecology	Behavior	Reproduction and Ontogeny	Physiology
Otarioidea	—	—	—	Harrison and Kooyman 1968
Odobenidae	—			
Odobenus rosmarus	Pedersen 1962; Nikulin 1940; Brooks 1954	Miller 1975; Collins 1940; Loughray 1959	—	Fay and Ray 1968
Otariidae	—	—	—	Gentry 1973
Arctocephalus phillipii	Peterson et al. 1968	—	—	—
A. pusillus	Rand 1967	—	Rand 1955	—
A. foresteri	—	—	Csordas 1958	—
Neophoca cinerea	—	Marlow 1975	—	—
Otaria byronia	Hamilton 1934	—	—	—
Callorhinus ursinus	Kenyon and Wilke 1953	Kenyon 1960	Bartholomew and Hoel 1953; Bartholomew 1959	—
Eumetopias jubatus	—	Gentry 1970	Sandegren 1970	—
Zalophus californianus	Peterson and Bartholomew 1967	Schusterman, Gentry, and Schmook 1967; Peterson 1969	Schusterman and Dawson 1968	Schusterman 1967; Schusterman and Bailliet 1971
Z. wollebacki	Orr 1967	Eibl-Eibesfeldt 1955	—	—
Phocoidea	—	—	—	Irving and Hart 1957
Cystophorinae				
Cystophora cristata	Olds 1950	Mohr 1963	—	—
Mirounga leonina	Carrick et al. 1962	Laws 1956	Carrick 1962	—
M. angustirostris	Bartholomew and Collias 1962; LeBoeuf, Ainley, and Lewis 1974	LeBoeuf, Whiting, and Gantt 1972; Klopfer and Gilbert 1966	Bartholomew 1952; LeBoeuf 1974	—
Monachinae				
Monachus schauinslandia	Kenyon and Rice 1959	—	Wirtz 1968	—
Leptonychotes weddelli	Stirling 1969	Smith 1966	Mansfield 1958	—
Lobodon carcinophagus	Corner 1972	Bertram 1940	—	—
Hydrurga leptonyx	—	—	Marlow 1967	—
Phocinae				
Erignathus barbatus	—	Ray, Watkins, and Burns 1969	McLaren 1957	—
Halichoerus grypus	—	Hewer 1960	Fogden 1971	—
Phoca vitulina	Scheffer and Slipp 1944	Wilson 1973*a,b*	Venables and Venables 1957; Foelsch 1968	Harrison and Thomlinson 1956; Hart and Irving 1959

ynous tendencies displayed by *Halichoerus* and concludes that this trait is recently derived.) The Otariidae, however, all appear to be highly polygynous; the males are quite large with respect to the females, and male-male competition resulting in harem formation appears to be the rule. Among either otariids or phocids such as *Halichoerus* or *Mirounga,* it appears that polygyny is favored whenever the species in question is confined on land for a rather long period either because the young cannot suckle at sea or because they are unable to molt at sea. This appears to be the case for *Halichoerus, Mirounga,* and all members of the otariids. In the rest of the phocines, dependence on the land or ice for molt, mating, and suckling appears vastly reduced; hence the likelihood of large aggregations of females is reduced, and the potential for harem formation becomes negligible.

The literature of pinniped natural history is extensive. Table 24 will serve as a guide to the literature and a supplement to the preceding discussion. The standard monographs and reviews include Mohr (1956), Orr (1965), Scheffer (1958), and Harrison et al. (1968).

14 The Grandorder Archonta

The Archonta represent a radiation in the late Cretaceous that gave rise to four living orders of mammals: the Scandentia, Dermoptera, Chiroptera, and Primates (McKenna 1975). McKenna has grouped these four orders because they show certain basic similarities in morphology. The fossil record is most complete for the Primates, and the three shrews (Tupaiidae) were once included in the order Primates.[1] The origin of the Chiroptera (bats) is obscured because the fossil record is imperfect, but the flying lemurs (Dermoptera) show clear affinities both with the bats and the primates. Suffice to say that it is convenient to treat the four orders as a unit, and that they probably are more closely related to each other than they are to any other contemporary mammals.

14.1 The Order Scandentia: The Tree Shrews

The Scandentia are represented today by the tree shrews of the family Tupaiidae. The skull of the Tupaiidae shows some conservative and some advanced features. The dentition shows a conservative cusp pattern on the molars, but the orbit is enclosed by a postorbital bar (see fig. 26). The digits bear claws, and the overall impression is of a squir-

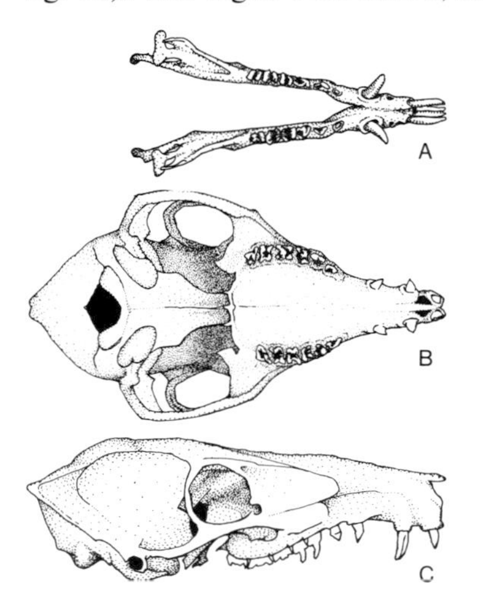

Figure 26. Three views of the skull of *Urogale everetti*. (*a*) Dorsal view of the lower jaw. (*b*) Ventral view of the skull. (*c*) Lateral view of the skull. Note the prominent postorbital bar.

1. Szalay and Delson (1979) have provided us with a comprehensive synopsis of primate evolution.

rellike mammal. Present-day forms are confined to eastern India and Southeast Asia. There are two subfamilies: the Ptilocercinae, recently reviewed by Gould (1978*a*), and the Tupaiinae, which have been the subject of several important investigations, the most extensive being that of R. D. Martin (1968).

The Ptilocercinae, or feather-tailed tree shrews, are nocturnal and are confined to the Malay Peninsula, where one genus and species still exists, *Ptilocercus lowei* (Lim 1967). The species is strongly adapted for arboreality and apparently is an insectivore/frugivore. The animals show a facultative hypothermia during the day when they are sleeping. They appear to den in tree hollows, and a social grouping may consist of an extended family. Behavior patterns are reviewed by Gould (1978*a*).

The Tupaiinae are diurnal frugivores/insectivores sheltering in tree holes at night. Some genera are adapted for feeding on the ground as well as in trees. *Tupaia belangeri* produces rather altricial young that are born in a nest, where they are left alone and fed only at 24-hour intervals (Martin 1966, 1968). The young are weaned approximately 30 days after birth. This method of parental care was termed an "absentee system" by Martin. The behavior patterns are reviewed by Kaufmann (1965) and Martin (1968). The tree shrews at one time were considered a family of the order Primates, but the weight of recent evidence (Van Valen 1965) suggests that they separated from the Primate lineage at an extremely early date.*

14.2 The Order Dermoptera

The Dermoptera, or flying lemurs, are today represented in Southeast Asia and the Philippines by one genus, *Cynocephalus*. Their affinities are difficult to determine in the absence of any complete fossil record; nevertheless, it is generally agreed that they show numerous characters in common with the Chiroptera and the Primates. Given their ancient divergence, it seems safe to treat them as a separate order. The two living species in the single genus feed on leaves, buds, and some fruits. The digestive tract is modified for processing great quantities of foliage, and the animals are extremely difficult to maintain in captivity on artificial diets. They are adapted for gliding from one tree to another and have folds of skin stretching between the forelimbs and hind limbs that may be extended during the glide. The young are altricial and are nursed for a considerable period before weaning. The major accounts of natural history and behavior are those of Chasen, Kloss, and Boden (1929), Wood (1927), and Wharton (1950).

14.3 The Order Chiroptera

The Chiroptera appear first in the Eocene as already differentiated for progression by true flight (Jepsen 1970). They are the only order of mammals specialized for true flight, with skin membranes between the fingers providing a wing surface.

In number of species, the bats are the second largest order of mammals. Obviously the phylogenetic branching leading to the bats occurred at the beginning of the Paleocene, no doubt from a stem form closely allied with those forms giving rise to the Scandentia and the

Primates. The bats radiated widely in the Paleotropics and very early invaded the Neotropics. In major activity patterns and feeding strategies, the order has remained nocturnal. All species are capable of sustained flight and thus are extremely mobile. Bat biology has been summarized in the volumes edited by Wimsatt (1970*a,b,* 1977), and the phyllostomatids have recently received special treatment in the series edited by Baker, Knox-Jones, and Carter (1976, 1977, 1978). The structure of bat communities worldwide has been analyzed by McNab (1971*b*). The trophic strategies of bats have been compared on a continental basis by Wilson (1973*b*). Their reproductive strategies have been analyzed for the Phyllostomatidae by Fleming, Hooper, and Wilson (1972). Classically, the Chiroptera are divided into two suborders: the Megachiroptera and the Microchiroptera. A scheme of their classification is included in figure 27. Figure 28 portrays the diversity in facial morphology.

14.3.1 The Suborder Megachiroptera

The Megachiroptera are conservative in certain morphological features and are confined to the Paleotropics. Megachiropterans characteristically possess a well-developed eye and large olfactory lobes. With the exception of the genus *Rousettus,* the megachiropterans do not employ echolocation, but forage only when there is sufficient moonlight to distinguish large objects. Megachiropterans are strongly adapted for feeding on the products of plants—for example, fruits, nectar, and pollen. They are specialized for several unique feeding strategies but are paralleled in their adaptations by the convergently evolved Neotropical microchiropteran family, the Phyllostomatidae.

The Pteropidae are conveniently divided into three subfamilies: the Pteropinae, the Macroglossinae, and the Nyctemeninae. The Pteropinae are specialized for feeding on fruit, whereas the Macroglossinae are pollen- and nectar-feeders, supplementing their diet with fruits. The Nyctimeninae have tubular nostrils that, by the exaggerated separation of the nares, may be of assistance in following odor trails when homing in on fruit trees (see fig. 28). The best-studied genus is *Pteropus,* which I will describe in some detail.

The Natural History of Pteropus poliocephalus, the Flying Fox

The following account is drawn from Nelson (1964, 1965*a,b*). In southeastern Queensland the flying fox exhibits a rather pronounced breeding season. This bat species is highly colonial and roosts in two situations, summer camps and winter camps. Summer camps are established in September, generally in a stand of trees that has been used for years. The sexes tend to roost separately in October and November when the births occur. For the first 3 weeks of life the young are dependent on the female not only for milk but for body heat, since they are heterothermic until approximately 3 weeks of age. They may be carried by the female when she forages during this period. At about 21 days the young may be left hanging separately in the nursery trees while the female forages.

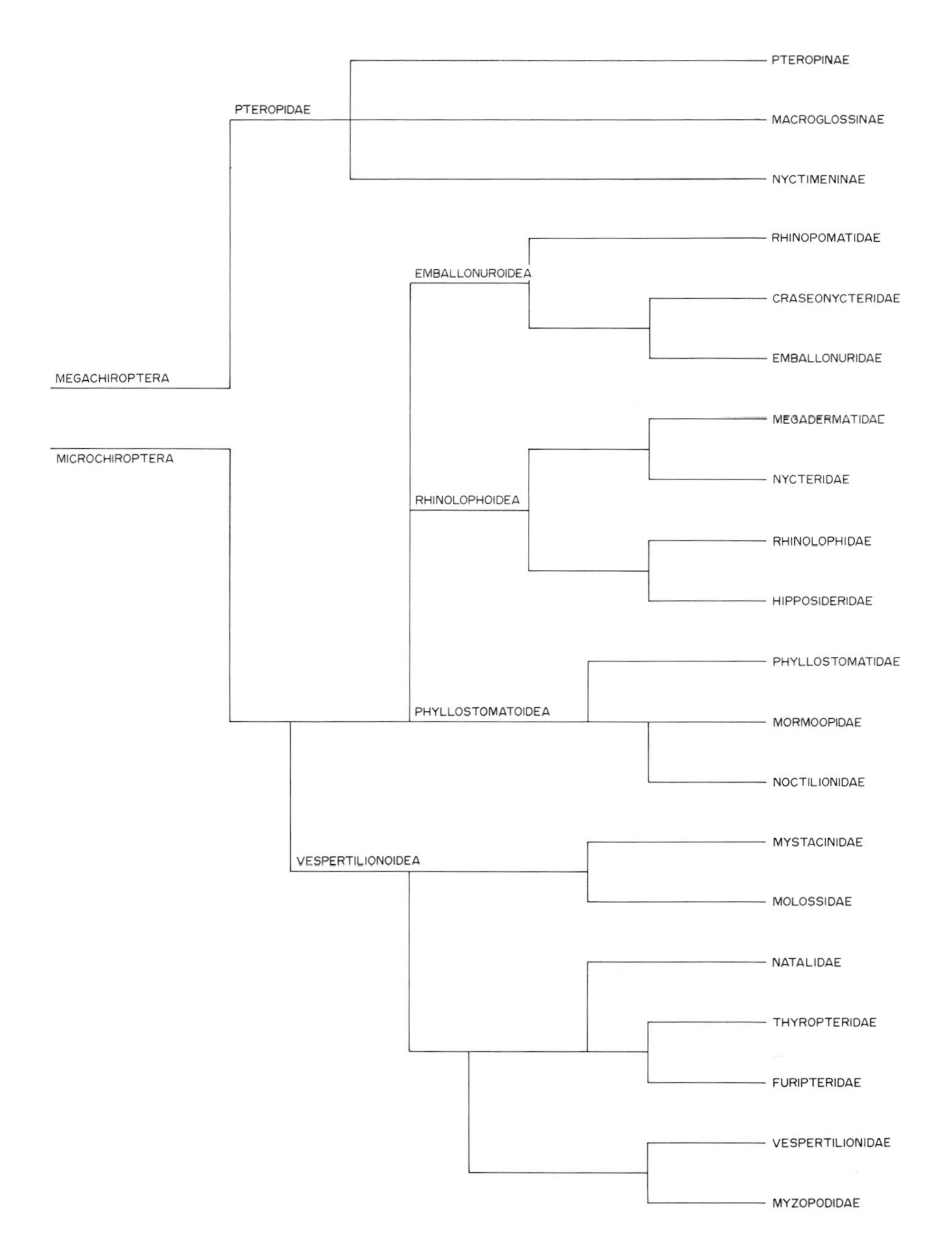

Figure 27. Diagram of the classification of the Chiroptera. Based in part on Smith 1976.

In December the males begin to occupy the nursery trees and to make attempts at pairing. Males generally try to mark a territory within the trees in order to "enclose" one or more roosting females. The males mark with their scapular glands, and fights between them are evident. Generally the growing juveniles occupy a central position in the trees, and younger males that cannot successfully compete for territory occupy the lower branches and periphery of the tree crowns.

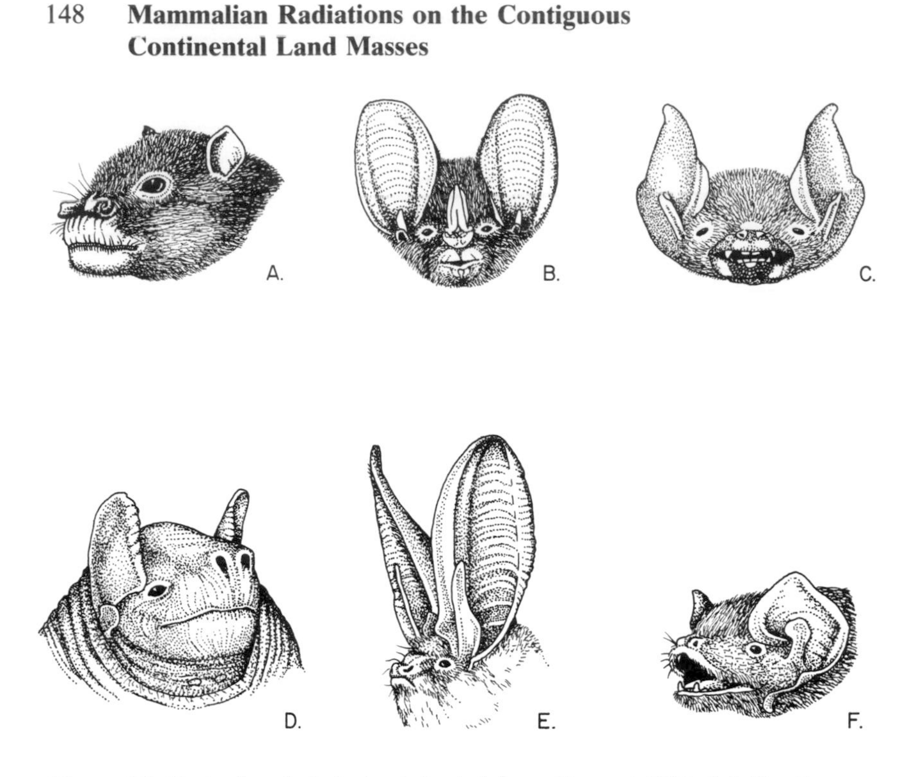

Figure 28. Portraits of six bats. Adapted from Brosset 1966. (*a*) *Nyctimene;* Pteropidae. (*b*) *Chrotopterus;* Phyllostomatidae. (*c*) *Pteronotus;* Mormoopidae. (*d*) *Cheiromeles;* Molossidae. (*e*) *Plecotus;* Vespertilionidae. (*f*) *Scotophilus;* Vespertilionidae.

The territorial males generally mate with their assembled females in March, and shortly thereafter the summer camp breaks up and the colony migrates to a winter camp. The seasonal availability of fruits greatly determines the choice of campsites during the two phases of the annual cycle.

In the Neotropics the niche occupancy of the family Phyllostomatidae has replicated the trends exhibited by the Megachiroptera in the Paleotropics. This allows one to make a number of interesting comparisons between the separately derived stocks and the degree to which they have converged in morphology and behavior. Eisenberg and Wilson (1978) have analyzed brain/body weight ratios and demonstrated that the Phyllostomatidae share with the Megachiroptera the highest brain/body weight ratios of any of the Chiroptera (see chap. 21).

In both the New World and the Old World, adaptations for feeding on nectar or pollen have been convergently evolved in the Megachiroptera and the Phyllostomatidae. A coevolution between the plants pollinated by bats and the bats adapted for feeding on them is demonstrable (Gould 1978*a*; Howell and Burch 1974; Start and Marshall 1976).

14.3.2 The Suborder Microchiroptera

Referring to figure 27, it should be evident that the Microchiroptera have undergone an advanced adaptive radiation. The paucity of the fossil record has made it impossible to construct a true phylogeny, but an analysis of morphological characters and karyotypes allows one to group the microchiropterans into discrete subgroupings that may serve as a "blueprint" for a phylogeny (Baker 1967, 1970, 1973).

Most microchiropterans have evolved a complicated echolocation system that permits them to fly on the darkest nights. The insectivorous species have refined their echolocation so that they can pinpoint insects on the wing entirely by registering sound waves reflected from their prey (Griffin 1958; Novick 1958, 1963). Communication systems involving olfaction have been analyzed by Kulzer (1962), and auditory communication between mother and young has been analyzed by Gould (1971, 1975). Communication mechanisms for the Chiroptera as a whole have been reviewed by Gould (1970, 1977).

The Family Rhinopomatidae: The Tomb Bats

This family is considered the most conservative in the suborder. There is one genus, *Rhinopoma,* with four species distributed from northern Africa through the Middle East to Thailand. The tail is very long and is only partly enclosed by a small interfemoral membrane. The species can store fat and become semitorpid during the winter, and it is insectivorous.

The Family Emballonuridae: the Sac-winged Bats

The sac-winged bats are distributed in both the New World and the Old World tropics. The relationship of the tail to the interfemoral membrane is unique in this family. The tail tip pierces the membrane and lies free on the dorsal surface; the basal part of the tail is enclosed in the membrane, which accounts for the second vernacular name for the family, the "sheath-tailed bats."

The males may have well-developed glands visible on the dorsal surface of that portion of the wing membrane that connects the tibia to the shoulder. In the genus *Taphozous,* the males have an enlarged gland in the gular region.

Bradbury and Vehrencamp (1976*a,b*, 1977*a,b*) have conducted a thorough field study of five sympatric species of Neotropical emballonurid bats. Several forms of social organization and mating strategies are found within these five species. After making a detailed analysis of their social structure throughout the annual cycle, the authors attempt to account for differences in social structure by relating them to different modes of habitat exploitation. From their comparisons of group size and territory size (which they subsume under the term social-dispersion), they conclude that population density, group size, and territory size are all limited by the spatial and temporal variations in food distribution.

Territory size is determined by the average distance between successively available food sites. Maximal group size is then determined by the minimum territory size and is related to the average richness of currently used food patches (insect aggregations). During their study, the small patches of food tended to be in the more stable habitats; thus the model predicts that temporal stability in group size and unimodal frequency of group size over space will be associated with fine-grained dispersion of resources.

In parental investment patterns, those emballonurid species using less seasonal habitats are most likely to have more than one parturition per year, and parturition periods tend to be relatively asynchronous among the females of a given group. In contrast, those species utilizing highly seasonal food supplies show one synchronous annual birth period. When food peaks come during gestation, the survival of adults will be favored over that of lactating and dispersing offspring. On the other hand, when the greatest availability of food coincides with lactation and dispersal of the young, then pregnant females suffer a high risk during gestation, but survival of the weaned young is higher. When the five species of bats are ranked by parental cost per parturition, the ranking is exactly the same as that obtained if one ranks them by either adult female mortality rates or relative seasonal abundance of food.

These four papers represent a pioneer step in the attempt to relate chiropteran social structure, mating systems, and demography to ecological parameters.

The Family Megadermatidae: The Old World Carnivorous Bats

This is a family of only three genera and five species, but it is broadly distributed from equatorial Africa across tropical Asia to Australia. With the exception of the genus *Lavia,* these bats have specialized for feeding on small vertebrates. This "carnivorous niche" is occupied in the Neotropics by *Vampyrum spectrum* and *Phyllostomus hastatus* of the family Phyllostomatidae.

The Family Nycteridae: The Slit-faced Bats

The slit-faced bats are a small group of one genus (*Nycteris*) with eleven species, distributed in tropical Africa and Southeast Asia. They are insect-feeders. The natural history of *Nycteris* is summarized by Kingdon (1974*a*).

The Families Rhinolophidae and Hipposideridae: The Old World Leaf-nosed Bats

The horseshoe bats and Old World leaf-nosed bats are closely related. Some authors relegate the families to subfamilial status. Although the Nycteridae and the Megadermatidae have folds of skin associated with the nostrils, the folding is more pronounced in the Rhinolophidae (*Rhinolophus* and *Rhinomegalophus*). In the Hipposideridae the skin folds are formed into a horseshoe pattern with associated leaflets.

Rhinolophus and *Hipposideros* emit ultrasonic pulses through their nostrils (Dijkgraaf 1946; Möhres and Neuweiler 1966); they share this characteristic with the New World leaf-nosed bats (the Phyllostomatidae). This has led to a quest to relate the possession of a "nose leaf" to some aspect of inducing directionality in the emitted pulse. The echolocation sounds emitted by *Rhinolophus* and *Hipposideros* are of a long duration when compared with those of the Vespertilionids. Their duration is remarkably similar to that of the sounds produced by the

Neotropical *Pteronotus (=Chilonycteris) rubiginosa* and *P. parnellii* of the family Mormoopidae (Novick 1963). We appear then to have a partial convergence in mode of emission and form of the signal between the Rhinolophidae and Hipposideridae, on the one hand, and the two New World Families, Phyllostomatidae and Mormoopidae, on the other.

The Superfamily Phyllostomatoidea

Three related Neotropical families, the Noctilionidae, the Mormoopidae, and the Phyllostomatidae, are subsumed under this superfamilial heading (Smith 1976). The Mormoopidae, or moustached bats, include two genera and are insectivorous; the Noctilionidae, or bulldog bats, includes one genus (*Noctilio*) with two species. *N. labialis* is insectivorous, while *N. leporinus* is strongly specialized for catching fish. The family Phyllostomatidae (New World leaf-nosed bats) contains 51 genera and more than 140 species (see fig. 28). A range of feeding adaptations can be demonstrated in the Neotropics that converges with the megachiropteran Pteropidae in the Paleotropics. So vast is the literature that a simple synopsis cannot be made. The reader is referred to the three volumes edited by Baker, Knox-Jones, and Carter (1976, 1977, 1978). Some comments must be made, however, to highlight the radiation. The subfamily Desmodontinae includes the vampire bats. The unique specialization for feeding on blood from living mammals has intrigued investigators for years. Turner (1975) has recently summarized our knowledge of *Desmodus rotundus*. In Central America vampire bats feed on the blood of domestic cattle, and where the density of domestic cattle is high the bats may become serious pests. In Turner's study, vampire bats bivouacked in hollow trees, utilizing each roost only for a few days at a time. Apparently changing roosts frequently correlates with their attempts to continue to prey upon a given herd of cattle, which may move around extensively. Turner presents some interesting evidence that vampires of one colony may defend cattle herds against intrusion by other vampires. Although this needs to be confirmed, it is extremely suggestive.

Selection of certain cattle to prey upon is demonstrable, and cattle on the periphery of the herd are most generally attacked. The young vampire bat takes a very long time to mature. Apparently, by accompanying the female on her nocturnal feeding flights, the young not only learns essential features of the prey but improves its feeding technique. The bats make a small (3 mm) incision in the skin of the prey animal and lap the blood that flows from the wound. The bat saliva contains a powerful anticoagulant that maintains a free flow of blood.

Some of the phyllostomatid bats are insectivorous and occasionally take fruit. The vast majority of species may almost be considered omnivores. The subfamilies Stenoderminae and Carolliinae are strongly specialized for feeding on fruit, while *Vampyrum spectrum* and *Chrotopterus auritus* are essentially carnivores. The subfamily Glossophaginae is specialized for nectar and pollen feeding and converges in morphology and behavior toward the nectar-feeding pteropid subfamily Macroglossinae (Gardner 1977).

Phyllostomatids also may be conveniently compared with the Vespertilionidae. A fruitful line of inquiry concerns the comparison of reproductive strategies and ontogeny. Based on an extensive analysis of growth and development in *Carollia perspicillata,* Kleiman and Davis (1978) offer the following thoughts concerning the differences in the ontogeny of behavior when the vespertilionids are compared with phyllostomatids. Although there is considerable overlap, young phyllostomatids are generally born in a more precocial state than are vespertilionids. Typically, phyllostomatid neonates have the eyes open and have a well-developed coat of fur. Phyllostomatid young are usually not deposited in large nursery groups but may be transported by the mother and either placed in a nocturnal secondary roost or perhaps carried during foraging when they are young. A distinctive crosswise position is assumed by the attached young that could be an adaptation to frequent carrying by the mother.

The Superfamily Vespertilionoidea: Small Insectivorous Bats

Seven families are included in the superfamily Vespertilionoidea. Their relationships are outlined in figure 27. Five families are relatively restricted in their distributions and have undergone little radiation.

The family Natalidae, or funnel-eared bats, contains one genus, *Natalus,* with several species distributed from Mexico to northern South America and the Antilles. The Furipteridae are also a Neotropical family of two genera and two species. The thumb is very reduced in size, thus the common name "thumbless bats." The family Thyropteridae contains one genus, *Thyroptera,* with two species distributed from Honduras to Peru. The genus is characterized by adhesive disks on both the wrist and the ankle. The family Myzopodidae contains a single genus and species, *Myzopoda aurita*. It is endemic to Madagascar and has an adhesive disk on its thumb, remarkably convergent to the condition in *Thyroptera*. The family Mystacinidae is allied to the Molossidae. It contains a single genus and species, *Mystacina tuberculata,* which is endemic to New Zealand. The remaining families of vespertilionoids have radiated greatly and will be described in some detail.

The natural history of Myotis lucifugus, the little brown bat. The Vespertilionidae represent a radiation of some 270 species, all strongly adapted for insectivorous feeding. The family enjoys a worldwide distribution in both the temperate zone and the tropics. The life history of *Myotis lucifugus* typifies the temperate-zone vespertilionid.

The following account is drawn from Humphrey and Cope (1976). The lack of available food throughout the winter imposes a prolonged period of hibernation, generally carried out in traditional caves used in common by a local population. Movement from the winter hibernating caves occurs in spring when females disperse to nursery colonies. Females may return year after year to the site where they were born to rear their own young. Nursery colonies generally tend to be established in buildings that are warm and retain a high humidity.

The young are born in a rather altricial state and are left hanging in the nursery while the females forage at night. Generally a nursery colony is composed almost entirely of adult females and their young, but in September the number of males there may increase.

In October there is generally a movement back to the traditional hibernating caves. Mating takes place in the autumn; sperm is stored by the female, and conception takes place in the spring. Only one young is borne each year by an adult female. For such small mammals, these bats have an incredibly long life expectancy, with 14 years as a maximum recorded longevity. During the summer, when the females have formed a nursery colony, small groups of males may be found roosting alone in rock crevices or tree hollows. A female may become sexually mature in the autumn following her birth. Males may take a year longer to mature.

The Family Molossidae: The Mastif or Free-tailed Bats

The free-tailed bats include twelve genera with some eighty species. They are worldwide in their distribution. These bats do not occupy the northern portions of the Holarctic, but migratory species may spend the summers as far north as 36°. These highly insectivorous bats are excellent fliers, and most species feed exclusively while in flight. The Mexican free-tailed bat serves as an example.*

The natural history of Tadarida brasiliensis, the Mexican free-tailed bat. Davis, Herreid, and Short (1962) have analyzed aspects of the ecology and behavior of the Mexican free-tailed bat. This species forms spectacular aggregations in caves at certain times during the annual cycle. As has been found with vampire bats (Turner 1975) and in the studies of Bradbury and Vehrencamp (1977*a,b*) with emballonurids, the free-tailed bats appear to fly toward their feeding grounds after emerging from the cave in discrete subgroups. There is some evidence that these groups can recognize one another. After foraging, the groups return to the cave, where they sleep in large aggregations.

The particular study by Davis, Herreid, and Short concerned summer cave use. The species is migratory in North America, and in the winter the population moves to Mexico, where the exact details of its life history are only poorly known. During summer, however, the females give birth in the caves that are traditionally used. From birth the young are able to cling alone to the cave ceiling and are never carried by the female on her foraging trips. Generally young are clustered at a specific location on the cave roof, forming a crèche or nursery. The adults abandon the cave for their southern migration at the time the young begin to forage on their own. Breeding is suspected to occur in Mexico before the northward spring migration.

Tadarida brasiliensis shows the following trends in sociality. First the lactating females form nurseries by placing their young in a selected spot within the cave. Unlike other bats that have been studied, females appear to nurse any young and do not form a specific mother-young bond. A specific bond between mother and young is common in other microchiropterans. Since the females return to these nursery spots to nurse their young, there is a spatial separation between the sexes within a cave. Second, mating in these microchiropterans takes place in the autumn or winter before the reestablishment of the traditional spring colonies in the Texas caves. The capacity to return to traditional roosting sites is highly developed in the Chiroptera and suggests no small amount of navigational ability, yet the problem has barely been analyzed.

14.3.3 Summary Remarks for the Chiroptera

Although G. M. Allen could summarize the literature for the Chiroptera in 1939, the burgeoning current research on such a diversified group does not lend itself to a brief synopsis. Topics concerning anatomy, thermoregulation, auditory physiology, nutrition, and reproduction have been adequately summarized by Brosset (1966) and in the volumes edited by Slaughter and Walton (1970) and Wimsatt (1970*a,b*, 1977). I have included some key life-history references in table 25. Some regional works have elegantly summarized life histories. I recommend for North America—Barbour and Davis (1969); Africa—Kingdon (1974*a*) and Verschuren (1957); Australia—Ride (1970); Southeast Asia—Lekagul and McNeeley (1977); South America—Goodwin and Greenhall (1961); Europe—Eisentraut (1957).

To set the stage for subsequent sections in the book, I will make a few comments on the mating systems and social structures demonstrable in this order.

Social organizations, including both mating and rearing systems, exhibit astonishing diversity when one examines the key studies for the Chiroptera. Males and females of *Pteropus giganteus* may be organized into colonies that show great cohesion in selecting roosting sites throughout the annual cycle. In *Pteropus poliocephalus* there may be a slight spatial separation within a tree or spatial separation over several kilometers between the adult sexes during parturition, so that even within one genus a range in the degree of cohesiveness in a colony is demonstrable.

Extreme separation of the sexes may be shown in the epomophorine bats. Typically, in the adults of *Epomophorus, Hypsignathus,* and *Epomops* the males roost separately from the females and emit calls that attract estrous females. The most extreme example of this tendency has recently been investigated by Bradbury (1977). *Hypsignathus monstrosus* males aggregate when they call and thereby create a true lek. Only 6 percent of the males in the lek account for 79 percent of the matings. This is in common with other lek systems, where dominant males are in the center of the group and females are attracted preferentially to them. It seems likely that an evolutionary gradient between single calling males and true leks may be demonstrated in the epomophorine bats when more species are studied (see Wickler and Seibt 1976).

One of the most common breeding and rearing systems for bats involves a separation of the sexes at parturition, with the females form-

Table 25 Life-History Studies for the Chiroptera

Taxon	Ecology and Behavior	Reproduction and Development
Pteropidae		Gopalakrishna 1969
Rousettus sp.	Kulzer 1958	Kulzer 1966
Pteropus sp.	Nelson 1964, 1965*a,b*	Marshall 1947
Eidolon helvum	Huggel-Wolf and Huggel-Wolf 1965	Mutere 1967
Rhinopomatidae		
Rhinopoma hardwickei	Khajuria 1972	—
Emballonuridae	Bradbury and Vehrencamp 1976*a,b*, 1977*a,b*	—
Rhinolophidae		
Rhinolophus sp.	Menzies 1973; Ransome 1967	—
Hipposideros sp.	Menzies 1973	—
Noctilionidae		
Noctilio leporinus	Blasdel 1955	—
Phyllostomatidae	Heithaus, Fleming, and Opler 1975	D. E. Wilson 1973*a*
Macrotus californicus	Bradshaw 1961	Bradshaw 1962
Desmodus rotundus	Greenhall 1965; Turner 1975	Schmidt and Manske 1973
Vampyrum spectrum	Vehrencamp, Stiles, and Bradbury 1977	—
Carollia perspicillata	—	Kleiman and Davis 1978
Artibeus lituratus	—	Tamsitt and Valdivieso 1965
Vespertilionidae	R. Davis 1966	Kleiman 1969; Kleiman and Racey 1969; O'Farrell and Studier 1973; Racey 1973
Lasiurus sp.	Constantine 1958	Bogan 1972
Tylonycteris sp.	Medway and Marshall 1970, 1972	Medway 1972
Plecotus sp.	Stebbings 1966	Pearson, Koford, and Pearson 1952
Antrozous pallidus	Orr 1954	Davis 1969
Pipistrellus sp.	W. H. Davis 1966	Racey 1973
Nyctalus sp.	Heerdt and Sluiter 1965	—
Nycticeius humeralis	—	Jones 1967
Myotis lucifugus	Humphrey and Cope 1976	Wimsatt 1960
M. nigricans	Wilson 1970	—
M. velifer	—	Kunz 1973
Molossidae		
Tadarida brasiliensis	Davis, Herreid, and Short 1962	—
T. condylura	—	Kulzer 1962
Eumops perotis	Howell 1920	

ing nursery groups and the males roosting separately. At the conclusion of the rearing phase, the sexes may then mingle at common roosting sites. This system is typified by a wide variety of species from many different families (see Bradbury 1977).

Some bats have developed year-round harem systems where one male associates with a group of females and assiduously defends his harem against intrusion by other adult males. This has been well described for *Phyllostomus hastatus* (McCracken and Bradbury 1977).

A few species of bats are suspected of forming a monogamous pairing system, with the male remaining near the female throughout the rearing phase and subsequent matings. Such a system may be typified by *Vampyrum spectrum,* where the male not only remains with the female through rearing but also provisions her and her young (Vehrencamp, Stiles, and Bradbury 1977).

Some bat species, however, have adapted for an almost solitary strategy, with the male and female coming together only for copulation and the female rearing her young alone. This system is characteristic of bats that typically roost on bark and remain inconspicuous by cryptic coloration. The system is typified by the North American red bat, *Lasiurus borealis*. The manner in which mating systems, social groupings, and demography correlate with the ecological specializations of the species awaits further investigation.

Reproduction in chiropterans shows a unique series of specializations, all of which are surely correlated with specific environmental adaptations (Fleming, Hooper, and Wilson 1972). Autumnal copulation with sperm storage over winter and fertilization in the spring characterizes some temperate-zone hibernating vespertilionid bats. Delayed implantation of a fertilized blastocyst occurs in *Eidolon helvum*. Variable rates of embryonic development may be shown in *Plecotus townsendii* as a result of variations in the female's body temperature (see Carter 1970 for a review). In closing, it is worth remembering that, though most microchiropterans are small compared with most small mammals, the bats have small litters (usually one young) and may have an astonishingly long reproductive life-span, exceeding 18 years in some vespertilionids.

14.4 The Order Primates

The early history of the Primates is still shrouded in some mystery. The Primate line began to differentiate in the Paleocene. By the middle of the Paleocene, several lineages can be discriminated. One leads to the Plesiadapiformes (see fig. 29). Kay and Cartmill (1977) review the early origins of Primates and point out that the ancestral form leading to the Plesiadapiformes probably was a terrestrially adapted insectivore. The Plesiadapiformes gave rise to a radiation in the late Paleocene that preoccupied the arboreal niche and became specialized for feeding on fruits and leaves. The nearest extant morphological equivalents of the Plesiadapiforme radiation survive today in the convergently evolved radiation of the Phalangeridae in Australia. By the time of the Eocene, the lineages giving rise to the extant Lorisiformes, Lemuriformes, Tarsiiformes, and Anthropoidea were established.

The living Primates present a spectrum of forms that provides a rich field for comparative studies. We have the living Prosimii, which represent variations on a conservative body plan. The lemuriform primates are preserved as a nearly intact radiation on the island of Madagascar. The lorisiform primates are distributed in Africa and Southeast Asia

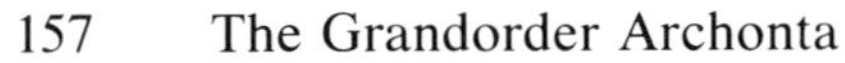

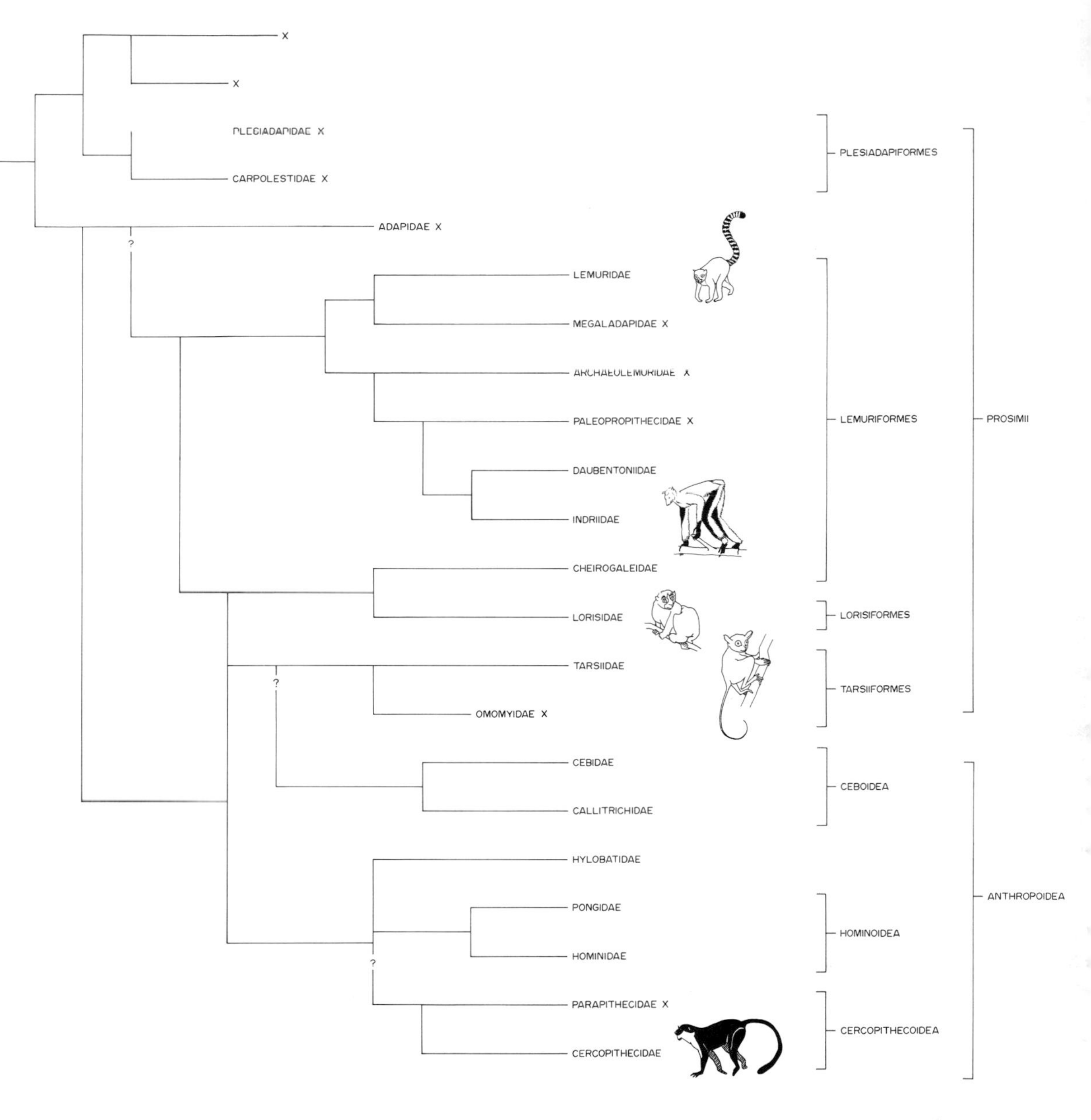

Figure 29. Cladogram illustrating the radiation of the Primates. Based on Martin 1975.

(see plate 33). The relict tarsiiform primates, which were once widespread in Eocene North America, still survive in Southeast Asia.

The Anthropoidea show two major separate radiations, the Ceboidea in the Neotropics and the Cercopithecoidea in the Paleotropics. In the Paleotropics, the Pongidae still survive in Africa and Southeast Asia as the orangutan, chimpanzee, and gorilla. Our own species, *Homo sapiens* of the Hominidae, has now distributed itself over most of the earth's surface (see fig. 30).

Plate 33. *Periodicticus potto,* the potto. This nocturnal prosimian shows some of the characteristics of the basic primate body plan. Note the opposable pollex, the binocular vision conferred by the position of the eyes, and the reduction in size of the pinna. The potto has several specialized characters, including a rather short tail. (NZP archives.)

The radiations of the Malagasy lemurs, the South American ceboids, and the Old World cercopithecids permit a rich field for comparative studies, and perhaps in no other order do we have such an unparalleled opportunity to compare behavioral and trophic convergences among stocks with respectable fossil records.

The anatomy of primates has been monographed by Hill (1953, 1955, 1957, 1960, 1962). Napier and Napier (1967) summarized the literature on living primates through the mid-1960s. The recent study of primate behavior and ecology is so active as a field of research that it seems almost futile to attempt a synthesis. The prosimians have recently been reviewed in the volumes edited by Martin, Doyle, and Walker (1974) and Tattersall and Sussman (1975). The ecology and behavior of the African Lorisidae and Galagidae have been beautifully described by Charles-Dominique (1977). Primate phylogeny has been reviewed in considerable depth in the volume edited by Luckett and Szalay (1975). Primate behavior was reviewed earlier in the volume edited by De Vore (1965), and recently primate ecology and behavior correlates have been synthesized in the volumes edited by Clutton-Brock (1977) and Michael

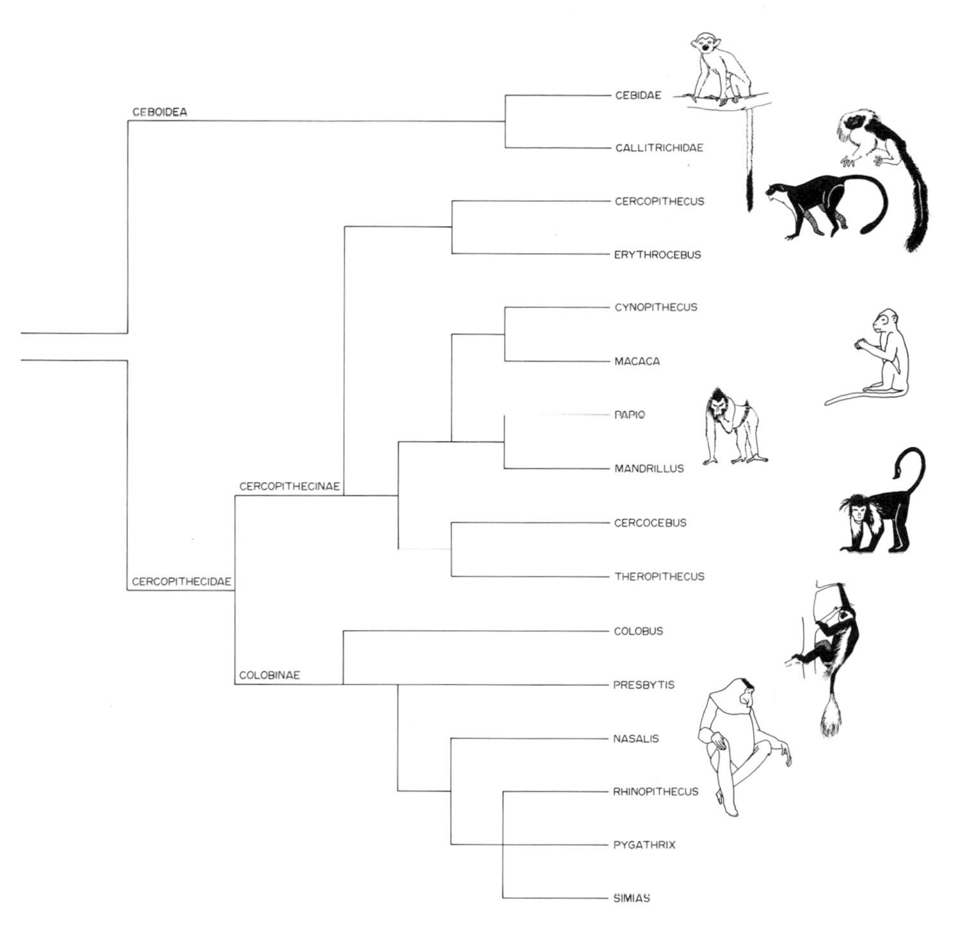

Figure 30. Diagram of the classification of the ceboid and cercopithecoid primates.

and Crook 1973). Rowell (1972) and Jolly (1972) have provided useful syntheses of primate behavior. Aspects of communication in primates have been summarized by Oppenheimer, Gautier and Gautier, Marler, and Tenaza in the volume edited by Sebeok (1977), while Altmann (1967) still remains a useful edited volume on the subject. Table 26 contains a synopsis of some of the major recent field studies.

14.4.1 The Small, Nocturnal Prosimians (Cheirogalidae and Lorisidae)

The key monograph to an interpretation of the way of life of small nocturnal prosimians is in supplement number 9 of the *Zeitschrift für Tierpsychologie* (Martin 1972; Charles-Dominique 1972). Martin's contribution deals with the ecology and behavior of the mouse lemur, *Microcebus murinus,* and Charles-Dominique considers Demidoff's galago, *Galago demidovii. Microcebus* is found on Madagascar and Demidoff's galago is found in Africa. The similarities between these two well-separated genera are remarkable and in all probability rep-

Table 26 Recent Literature on Primate Behavior and Ecology

Taxon	Reference
Prosimii	Martin, Doyle, and Walker 1974
Lemuridae	
Lemur catta	Jolly 1966
L. fulvus	Sussman 1975, 1977
Lepilemur mustelinus	Hladik and Charles-Dominique 1974
Cheirogalidae	
Microcebus murinus	Martin 1972
Indriidae	
Indri indri	Pollock 1977
Propithecus verreauxi	Richard 1974, 1977
Lorisidae	Charles-Dominique 1977
Periodicticus potto	Charles-Dominique 1975
Loris tardigradus	Petter and Hladik 1970
Galago demidovii	Charles-Dominique 1972
Tarsiidae	
Tarsius bancanus	Fogden 1974
Callitrichidae	Kleiman 1977*a*
Calllithrix jacchus	Rothe 1977
Cebuella pygmaea	Ramirez, Freeze, and Revilla 1977
Saguinus oedipus	Neyman 1977; Dawson 1977; Moynihan 1970
Leontopithecus rosalia	Hoage 1978; McLanahan and Green 1977
Cebidae	Moynihan 1976
Aotus trivirgatus	Moynihan 1964; Wright 1978
Callicebus moloch	Mason 1968; Moynihan 1966; Robinson 1977
C. torquatus	Kinzey 1977
Pithecia pithecia	Buchanan 1978
Cacajao rubicundus	Fontaine and DuMond 1977
Alouatta palliata	Milton 1978; Smith 1977; Glander 1975
A. seniculus	Neville 1972; Rudran 1979
Cebus capucinus	Oppenheimer 1968
Saimiri sciureus	Baldwin 1971; Baldwin and Baldwin 1972
Ateles fusciceps	Eisenberg 1976
A. geoffroyi	Eisenberg and Kuehn 1966
A. belzebuth	Klein 1972; Klein and Klein 1977
Lagothrix lagotricha	Williams 1968
Cercopithecidae	
Macaca sinica	Dittus 1975, 1977*b*, 1977*a*
M. sylvana	Deag and Crook 1971
M. silenus	Green and Minkowski 1977
M. arctoides	Bertrand 1969
Cercocebus albigena	Waser 1975, 1976
C. galeritus	Homewood 1978
Papio cynocephalus	Altmann and Altmann 1970
P. hamadryas	Kummer 1968
P. anubis	Packer 1979
Theropithecus gelada	Dunbar and Dunbar 1975; Crook 1966
Cercopithecus aethiops	Struhsaker 1967*a,b*
C. mitis	Rudran 1978*a*
Miopithecus talapoin	Gautier-Hion 1970, 1971
Presbytis entellus	Ripley 1967, 1970
P. senex	Rudran 1973*a,b*
P. johnii	Poirier 1969, 1970
Colobus badius	Struhsaker 1975
C. guereza	Oates 1977; Dunbar and Dunbar 1974
Hylobatidae	
Hylobates lar	Ellefson 1974
Symphalangus syndactylus	Chivers 1974

Table 26—*Continued*

Taxon	Reference
Pongidae	
Pongo pygmaeus	Rijksen 1978; MacKinnon 1974
Pan troglodytes	Sugiyama 1973; Teleki 1973; Goodall 1968; Marler and Tenaza 1977; Hamburg and McCown 1979
Gorilla gorilla	Fossey 1972, 1974; Schaller 1963

resent a conservative primate pattern carried forward into the present relatively unchanged. Both species are insectivorous but supplement their diets with fruits and in some cases pollen. Both species build either a nest in a cavity or a leaf nest on a branch (*Galago demidovii*). The newborns (one or two) are rather altricial for primates, being sparsely haired and having the eyes scarcely open. During the first seven days after birth the young are kept in the nest, and the mother leaves them on her nocturnal foraging trips. From about seven days on they are transported on the mother's back. Thus a nest phase is retained in the life cycles of these species.

The social organization of these species has been analyzed in some detail. Females with associated young tend to occupy exclusive territories, though elder daughters may settle near the mother and have a smaller home range that includes a portion of the mother's previous range. One male may be associated with one or more females, but males are apparently organized hierarchically in the extended community. Dominant males have access to females and exclude subdominant males, which then take up peripheral territories around those occupied by dominant breeding males. Social communication is maintained through vocalizations and chemical marking.

14.4.2 The Lemuriform Radiation on Madagascar

Many of the lemuriform primates on the island of Madagascar have radiated to occupy niches roughly comparable to those occupied by primates on the contiguous continental land masses. The larger species of the genus *Lemur* typically live in bands or groups reminiscent of troops of cebid and cercopithecid primates. Dietary specializations have occurred within the group, with some species strongly adapted for folivority (*Lepilemur mustelinus* and *Lemur macaco*). Other species of lemurs are mixed feeders on leaves and fruits, for example, *Lemur catta*. The lemurs of the family Indriidae, including *Propithecus* and *Indri,* exhibit two variations in sociality. *Propithecus* lives in large groups of mixed sexes, and *Indri* tends to form small, closed groups of paired adults and their various-aged offspring. Both *Propithecus* and *Indri* are strongly specialized toward folivority; it reaches its acme in *Indri indri* (Richard 1978; Pollock 1974).

14.4.3 The Superfamily Ceboidea: The New World Monkeys

In South America the earliest fossils that are recognizable as primates date from the Oligocene. This implies that the primate radiations in

South America have had a rather long history. It is reasonable to presume that a great many forms coexisted through time and to fully recognize that we are at present looking only at the end points of a broad and ancient radiation (Hershkovitz 1974). The extant ceboids are divisible into two families: the Callitrichidae (marmosets and tamarins) and the Cebidae (howler monkeys, squirrel monkeys, etc.).

The contemporary species of cebid primates exploit ecological niches similar to those occupied by primates in the Paleotropics (Eisenberg, Muckenhirn, and Rudran 1972). Even so, we find that the match between Paleotropical and Neotropical niche occupancy is not particularly good. The only nocturnal Neotropical primate, *Aotus* (the night monkey), differs in its niche occupancy from both the nocturnal lorisoids and the galagoids in the Paleotropics. In the Neotropics the feeding niche occupied by the lorisiform primates of the Old World is occupied by the didelphine marsupials (Eisenberg and Thorington 1973).

The larger, diurnal Neotropical primates do converge to some extent with the niche occupancy of the Paleotropical primates. The genus *Alouatta* occupies a niche somewhat similar to that of certain species of the Old World colobine primates. *Alouatta* has not become an obligate folivore as have some highly specialized cercopithecids (e.g., *Nasalis*). The genus *Cebus* in the Neotropics appears to occupy a niche similar to that of some species of the Old World genera *Cercopithecus* and *Macaca*. The New World genus *Ateles* occupies a feeding niche somewhat similar to that of the hylobatids of Southeast Asia. In its feeding ecology the genus *Saimiri* in the New World tends to closely resemble the West African cercopithecine genus *Miopithecus* (Gautier-Hion 1966). All the ceboid primates bear a single young, in contrast to the twinning characteristic of the Callitrichidae, with the exception of *Callimico,* which typically bears a single young.

Monogamous pairing and parental care by the male are demonstrable in the genera *Callicebus* and *Aotus* and are implicated for *Pithecia*. This suggests the possibility that monogamous pairing and shared parental duties constitute a rather ancient trait within the Ceboidea as a whole and have been carried forward and expressed as a uniform trait within the family Callitrichidae (Eisenberg 1977; Epple 1977). Social structure is highly variable within those cebid primates, which do not show monogamous pairing. In the genus *Saimiri* large groups of adults of both sexes, perhaps exceeding a hundred individuals, have been reported. Although such groups tend to subdivide into like age and sex subgroups, they nevertheless exhibit a cohesion and diversity of age-classes reminiscent of the degrees of sociality that have convergently developed in the Cercopithecidae.

The genus *Alouatta* tends to live in smaller troops with a rather fixed small home range. This is in conformity with its folivorous habits. The sexes are strongly dimorphic, and the males produce a loud, roaring vocalization that helps space neighboring troops. There is some variation in the form of *Alouatta*'s social organization when the various species are compared. *A. palliata* tends to have more adult males per troop than does *A. seniculus*. In *A. seniculus* there is a strong tendency toward a single adult male within any given troop, though each adult male may have several younger males as "satellites" (Eisenberg 1979).

The genus *Ateles* is characterized by a very labile social structure. It is strongly adapted for frugivority, though leaves constitute up to 20 percent of its diet at certain times of year. Given the frugivorous tendency, Klein (1972) has hypothesized that the extreme subdivision of the troop into small foraging units may help the members find widely dispersed fruit trees. Whatever the exact ecological significance of fractionation in *Ateles* troops during the foraging phase, it remains that, in some aspects of their social structure, they are reminiscent of the chimpanzee, *Pan* (Klein 1972) (see plate 34).

The Special Case of Monogamy in the Callitrichidae

The marmosets and tamarins have recently been reviewed by Hershkovitz (1977) and in the volume edited by Kleiman (1977*a*). Captive studies by Rothe, Epple, and Kleiman (in Kleiman 1977*a*) suggest that the modal social organization for many if not all of the callitrichids

Plate 34. *Ateles geoffroyi*, the spider monkey of South America, typifies the Old World Cebidae. The young may be carried by the mother until they are almost a year old. (Courtesy of C. M. Hladik.)

consists of the monogamous pair and their descendants of various ages. Field data from Dawson (1977) confirm the suspicion for *Saguinus oedipus*. The exact manner in which the pair bond is maintained and the frequency with which group compositions may change remains to be explored in full detail. Studies of *S. oedipus* suggest that movement between troops may be more frequent than was suspected, and that the monogamously paired condition may represent one aspect of a continuum of associations (Neyman 1977; Dawson 1977). Regardless of the controversy, it appears that in addition to monogamous pairing older juveniles may be retained in the natal group and participate in the care of younger siblings.

The detailed analysis of captive *Leontopithecus* by Hoage (1978) amply demonstrates the advantage to the reproducing female of retaining the male in a permanent paired condition and retaining her older offspring. After some varying time of carrying by the female, the male begins to carry the young. Older juveniles may also share the carrying. All of this, of course, reduces the energy drain the female would suffer if she were primarily responsible for transporting the young. In addition, older siblings and the male share food with newly maturing young. This provisioning behavior has profound significance for the survival of the most recent litter.

It appears that marmosets have undergone selection for small size. They have also increased their litter size to two young, departing from the modal single young characteristic of all Anthropoidea. Bearing two young per litter may well be a response to increased mortality through predation as a consequence of small size. Selection for small size may have resulted from niche differentiation; marmosets are capable of foraging in low shrubs and on the fine terminal branch tips unavailable to larger primates. A reproducing female can bear and raise two young if, and only if, the energy demand from lactation is reduced by the male's participating in carrying. Obviously, provisioning by the male and by older juveniles greatly assists the survival of the newest members of the group. The marmoset and tamarin monogamous system could serve as a very useful model for the analysis of cost-benefit ratios.

14.4.4 The Family Cercopithecidae: The Old World Monkeys

The living Old World monkeys are divisible into two subfamilies: the Cercopithecinae and the Colobinae (see fig. 30). The colobine primates show modifications of the gut that permit them to digest leaves. Some gastric fermentation takes place in the stomach with the assistance of microbial symbionts (see Bauchop 1978). The langurs (*Presbytis*) of south Asia and the colobus monkeys (*Colobus* and *Procolobus*) of equatorial Africa typify this radiation.

The cercopithecine primates include the macaques (*Macaca*), baboons (*Papio*), mangabeys (*Cercocebus*), and guenons (*Cercopithecus*). The genus *Macaca* occurs in isolated areas in northern Africa but shows its greatest radiation in Southeast Asia. The remaining genera are distributed in Africa south of the Sahara.

The genus *Cercopithecus* has speciated dramatically in equatorial Africa. Fourteen species (if one includes *Miopithecus talapoin* and *Allenopithecus nigroviridus*) are currently recognized. Some species are partially folivorous (*C. mitis;* Rudran 1978*a*); other species are generalized omnivores (*C. aethiops;* Struhsaker 1976). Different degrees of arboreality can also be discerned. The variation in social structure has been reviewed by Struhsaker (1969).

The genus *Cercocebus*, or the mangabeys, occurs sympatrically with *Cercopithecus*. The mangabeys have five recognized species. They are somewhat larger than the cercopithecines. The ecology of two species has been studied by Waser (1975) and Chalmers (1968).

The baboons of the genera *Papio, Mandrillus,* and *Theropithecus* are terrestrially adapted but also occur in forested habitats. Pioneer ecological studies have been carried out on the savanna-dwelling forms (see Altmann and Altmann 1970; Dunbar and Dunbar 1975; Kummer 1968).

The genus *Macaca* has radiated into thirteen species in South Asia. Some forms (*M. mulatta*) are semiterrestrial, while others (*M. silenus*) are strongly arboreal. This radiation converges with the radiations of the African Cercopithecidae in some aspects of niche occupancy. The recent literature is summarized in table 26.

The colobine primates, with two separate radiations in Asia and Africa, provide rich material for comparisons (see fig. 30). The Asiatic radiation is by far the more diversified, reflecting the longer history of large continuous tropical forest tracts in Southeast Asia compared with equatorial Africa (see plate 35). Recent literature on the colobines is included in table 26. The volume edited by G. G. Montgomery (1978) presents a current synopsis of studies dealing with folivotiry in the Mammalia, including the Primates.

Plate 35. *Presbytis senex,* the purple-faced langur. This colobine primate is specialized for feeding on leaves and has modifications of the foregut for bacterial fermentation. (Photo by author.)

14.4.5 The Family Hylobatidae: The Gibbons

The gibbons and siamangs, or "lesser apes," are currently confined to Southeast Asia, from Assam to Borneo. They are tailless and strongly specialized for an arboreal existence. The long calls of gibbons and siamangs show a remarkable stereotypy and can be used as an aid in recognizing species in the field (Marshall, Ross, and Chantharozrong 1972). Gibbons and siamangs exhibit a monogamous mating system that has been analyzed in some detail (Ellefson 1974; Chivers 1974; Tenaza 1975).

14.4.6 The Family Pongidae: The Great Apes

The living great apes include two African genera (*Gorilla* and *Pan*) and one Asian genus (*Pongo*). As the closest of man's relatives, they have received a great deal of attention, beginning with Yerkes's pioneering efforts (Yerkes and Yerkes 1929). Recently the study of man-chimpanzee communication has been pioneered by Gardner and Gardner (1969) (see Fouts and Rigby 1977 for a review). The gorilla has been studied extensively in the field (Schaller 1963; Fossey 1972, 1974). The chimpanzee has been studied extensively in captivity (Yerkes 1943) and in the field (Van Lawick-Goodall 1968, 1971). The orangutan has proved to be the most difficult of the larger apes to study in the field. It is confined to the climax rain forests of Sumatra and Borneo, and the difficulties of direct observation have been considerable (see Rijksen 1978).

Whereas *Gorilla* lives in cohesive, permanent social groupings usually led by a dominant "silver-back" male, the chimpanzee has a more amorphous troop structure. The central "clan" is capable of subdividing into subunits of which the female and her semidependent offspring appear to be the most cohesive.

The orangutan *Pongo pygmaeus* typically shows a small group size and has been described as "solitary." Schaller (1961) reviewed the status of the orangutan in Sarawak. Rodman (1973) and MacKinnon (1971, 1974) studied the orangutan on Borneo. Rijksen (1978) has completed the most recent monograph for the orangutan on Sumatra (plate 36).

Rijksen points out that the social behavior and social organization of the orangutan is to some extent similar to that shown by the chimpanzee; however, the differences may be of degree rather than qualitative. He terms the social organization of the orangutan as a "limited gregarious" life-style. It is true that as adults orangutans probably show the lowest number of social interactions of any of the great apes; however, it is not true that they are entirely without social contacts.

The adult males do not form any cohesive social grouping, and they spend a great deal of time alone unless courting females. They avoid encounters with other adult males, and to some extent their loud roaring helps them space themselves. The main cohesive social unit is the adult female and her offspring, which form a foraging unit of a mother, a young baby, and an older offspring approximately 3 to 5 years of age.

Plate 36. *Pongo pygmaeus*, the orangutan. This photo of a captive-born young and its mother illustrates the breast development of a female pongid during lactation. (NZP archives.)

Groups of adolescents form after the young leave the mother-offspring group, and this grouping may continue until puberty, when the male begins to assume the dimorphic characteristics of an adult male and the female conceives and bears her first young. Depending on the nature of the habitat, even adults can form a temporary association in certain fruit trees when fruit production is high.

It is clear that our understanding of the ecology and behavior of long-lived primates can be firmly established only after longitudinal studies of considerable duration have been undertaken. Nothing could demonstrate this process better than the case of the chimpanzee, *Pan troglodytes,* when we compare the advances in knowledge beginning with Nissen (1931) and then proceeding to Goodall (1965, 1968), Sugiyama

(1968), Suzuki (1969), and Reynolds and Reynolds (1965). By the mid-sixties there is a vision of the chimpanzee living in a loosely structured social organization with little evidence of hostility, feeding peacefully as a frugivore, and with rather free movement of individuals throughout the home range. By 1970 we learn that the chimpanzee is sometimes carnivorous, that on occasion males actively hunt small game, and that food is shared in a hierarchical fashion (Teleki 1973). By 1977 we recognize that the chimpanzee troop has a very highly organized internal structure, though all individuals may not be assembled in the same place at the same time, that the home range of a given troop is defended, that intruders from other troops may be killed, and that infanticide and even cannibalism sometimes take place (Goodall 1977).

14.4.7 Concluding Comments on Primate Socioecology

Crook and Gartlan (1966) reviewed the structure of primate social organizations and suggested a phylogenetic progression in the evolution of complex societies within classes of environmental constraints. Eisenberg (1966) independently reviewed the trends in primate social systems. A comparison of the different major phyletic stocks of primates showed the remarkable convergence in the form of social structure. Crook (1970) again reviewed the socioecology of primates, pointing up new findings and problems inherent in the classification of primate social structures. Struhsaker (1969) demonstrated convincingly that the primate species adapted to forested regions of West Africa in the main tended to exist in social groupings with only one adult male present at any given time. Subsequently, in his monograph on the red colobus monkey (1975), Struhsaker discussed the unique features of the social structure of *Colobus badius*, which, in contradistinction to the general trend for forested monkeys in West Africa, exhibits a multimale social structure.

Eisenberg, Muckenhirn, and Rudran (1972) attempted to correlate differences in foraging strategies with the form of social organization shown by various primate species. In addition, we took into account the superimposition of phylogenetic inertia on each of the radiations on the major distinct land masses—for example, Neotropics, Paleotropics, and Madagascar. Although certain general correlations could be made, obviously many problems existed in trying to discern the adaptive advantages of the varying forms of social organizations exhibited by the various primate species. Three interesting problems emerged from the reviews. The first is that variation in the form of social organization exists between separate populations of the same species. The second problem involves how troop composition and size are maintained through time. Is there a regular growth in a troop to a certain point and then a restoration to an equilibrium value? How are the sex ratios maintained within troops, and how are numbers kept within limits? Finally, there is the problem of explaining the persistence of infant death during social disruption in troops of primates, especially those troops that experience a takeover by a strange male. I will treat each of these in turn.

When we compare separate populations of the same species, we discern certain variations in home range size, ranging patterns, troop size, and social structure (Richard 1978; Rudran 1978*b*; Schlichte 1978; Glander 1978; Milton 1978). Given this variation, however, there exists a modal or central tendency toward a given form of social organization among many separate populations of a species. Differences in troop size and home range size can be correlated with differences in carrying capacity. When a population occupies a habitat where food availability fluctuates temporally and spatially, home ranges will generally be larger than for populations in habitats with more concentrated food resources and less fluctuation. That the space utilization patterns of different troops arise from peculiarities of diet and food dispersion has been amply demonstrated (Rudran 1978*b*). Although variability may be found in any mensurable parameter, be it body size or social structure, this does not preclude the utility of thinking in terms of averages or modal tendencies. However, recognizing variation is essential and necessary if we are to make progress in explaining the differences we detect when we compare populations of the same species.

The growth and regulation of troop numbers and the maintenance of sex ratios within troops is receiving increasing study as longitudinal investigations are carried out. Altmann and Altmann (1970), in their excellent analysis of baboon ecology, point out that adult males do not have a fixed tenure in any given troop. Rather, they tend to move to a new troop as they reach maturity, and they generally move to different troops after a predictable number of years of residency in any given troop. Thus it appears that the males, by their movements, introduce genetic variability, while the females show extreme fidelity to the troop of their birth. Similar trends have been noted by Dittus (1977*a,b*) in his analysis of *Macaca sinica*.

Dittus's studies are particularly instructive because they consider how numbers are maintained within limits in a given troop. During a portion of his study, which has been carried on for nearly a decade, the net growth of the entire population was zero; in short, the population was stable with its resources. His analysis of behavioral interactions during foraging demonstrated that most threat behavior occurred during foraging situations, and that recipients of threats were prevented from immediate feeding.

By analyzing threats between the age- and sex-classes, Dittus was able to demonstrate that adults dominated juveniles, which in their turn dominated infants, and that, among juveniles and infants, males dominated females of the same age-class. These behavioral polarities during foraging have a strong influence on an individual's foraging efficiency. Dittus demonstrated that adult males had the highest foraging efficiency, followed by adult females, then juvenile males, and finally juvenile females, which show the least-efficient foraging. These behavioral interactions, which affect feeding efficiency, mean that younger animals die at greater rates than adults and that infant and juvenile females die at greater rates than males of the same age-classes. Males in turn suffer a higher mortality at the transition from subadult to adult status, when they emigrate from their natal troop and compete for mates in neighboring troops.

Dittus concludes that, through socially imposed mortality, social behavior determines the age- and sex-class distribution and density within the population under study. Food resources limit the size to which any troop can grow within its home range. Larger troops tend to have larger home ranges, though differences in the carrying capacity of home ranges can affect this generalization.

Although Dittus's work with *Macaca sinica* demonstrates that males emigrate and establish themselves in troops other than their natal troop, he has recorded no cases of outright infanticide resulting from such takeover. Infanticide is yet to be described for *Macaca sinica* or, for that matter, most other species of the same genus. Yet, infanticide during takeover by alien males was recorded early on for the Hannuman langur, *Presbytis entellus* (Sugiyama 1967), and Rudran (1973*b*) recorded a similar phenomenon for *Presbytis senex*. Recently Hrdy (1977) summarized the problem of infanticide in *Presbytis entellus*, and Struhsaker (1977) has recorded occasional cases during takeovers in troops of *Cercopithecus ascanius*. Rudran (1978*b*) has recorded the loss of infants accompanying a takeover in the blue monkey, *Cercopithecus mitis*, but he did not witness the actual killing of infants. Goodall (1977) summarizes cases of infanticide, cannibalism, and hostile activities involving not only infants but other age-classes for the chimpanzee, *Pan troglodytes*, and Fossey (in preparation) documents several cases for *Gorilla gorilla*. We are here faced with an interesting situation of infant death accompanying the takeover of troops for a wide variety of primate species. Yet it does not always occur during a takeover, and in some species, for example, *Papio anubis*, *Macaca sinica*, and *Papio hamadryas*, the phenomenon has not been observed in the field, despite extensive long-term studies. The problem that remains to be clarified is in what circumstances infanticide occurs for a given species and why it is not uniformly found across all populations of the same species or when a series of species are compared. The whole problem will be dealt with in chapter 30.

15 The Grandorder Ungulata

According to McKenna (1975), at the beginning of the Paleocene the ancestral Ungulata must have differentiated into at least five lines: the Eparctocyon, Cete, Meridungulata, Phenacodonta, and Tethytheria (see fig. 17). The Cete soon became adapted to an aquatic life and have given rise to our contemporary whales and dolphins. The Eparctocyon radiation gave rise to a number of forms in North America and Asia that are now extinct. The surviving orders are the Tubulidentata, or aardvarks, consisting of a single living genus and species highly specialized for a myrmecophagous niche in Africa, and the widespread and diversified order Artiodactyla.

The Meridungulata was a radiation that became isolated in South America at an extremely early time, probably deriving from a Condylarthlike form. The Meridungulata radiated to fill most of the terrestrial herbivore niches on the island continent of South America and were contemporary with the edentate and marsupial radiations. The Marsupialia and some edentates occupied the insectivore/carnivore trophic levels, and the Meridungulata and other edentates shared the herbivore niches (see Patterson and Pasqual 1972).

The Phenacodonta radiated in North America and Europe to give rise to the newly defined Condylarthra (see p. 178) and the Perissodactyla and Hyracoidea; only the latter two orders survive in the Old World. The Tethytheria appear to be a unique radiation, perhaps African in origin, giving rise to the Desmostylia, Sirenia, and Proboscidea. The Desmostylia are extinct and probably showed their closest affinities to the Sirenia, which may in fact be derivative from them (Domning 1976). The sirenians and desmostylians early on adapted to a semiaquatic life, specializing for feeding on aquatic vegetation. The sirenians today are the only completely aquatic mammalian herbivores.

Szalay (1977) departs from McKenna and proposes a more conventional classification. The Ungulata are raised to the rank of cohort. The Meridungulata are lowered to the rank of order, but the content remains the same. The Meridungulata are included with the orders Perissodactyla, Dinocerata, Embrithopoda, Hyracoidea, Proboscidea, Sirenia, and Desmostylia under the superordinal heading Mesaxonia. Szalay then recognizes three additional superorders: the Cete, Paraxonia, and Protungulata. The Paraxonia contains only the order Artiodactyla; the Cete contains only the order Cetacea, and the Protungulata contains the orders Arctocyonia and Pantodonta. While I do not have the expertise to weigh the alternatives, I rather approve of the superorder

Mesaxonia as Szalay has defined it. Uniting the extant Hyracoidea, Proboscidea, and Sirenia seems very sensible, and further including the Meridungulata suggests that the ancestral form for all lines may have occurred in both Africa and South America before they were separated.

15.1 The Order Cetacea

The cetaceans, or whales and dolphins, have a long evolutionary history. There is good reason to believe that all members of this order had a common origin, in spite of their diversity in feeding adaptations. Fossil whales belonging to the family Protocetidae may be noted as far back as the Eocene, when aquatic adaptations were evident. The living whales may be classified into eight families, according to Hershkovitz (1966). Some authors include ten families by breaking the Delphinidae into Delphinidae, Stenidae, and Phocoenidae. Figure 31 diagrams the relationships.

The cetaceans are completely aquatic eutherian mammals. Their posterior limbs are reduced to bony vestiges embedded in muscle, or they may be lacking altogether. The anterior limbs are transformed into flippers. The posterior body is formed into a propulsive "tail" with a horizontally oriented pair of flukes. A dorsal fin is usually present, but it has no skeletal support. Although there are a few bristles on the head, for the most part the skin is naked, and there are no epidermal glands except for the inguinal mammary glands in the female. The external nares of the cetaceans are positioned dorsally on top of the head. In the male the testes are intraabdominal and do not descend

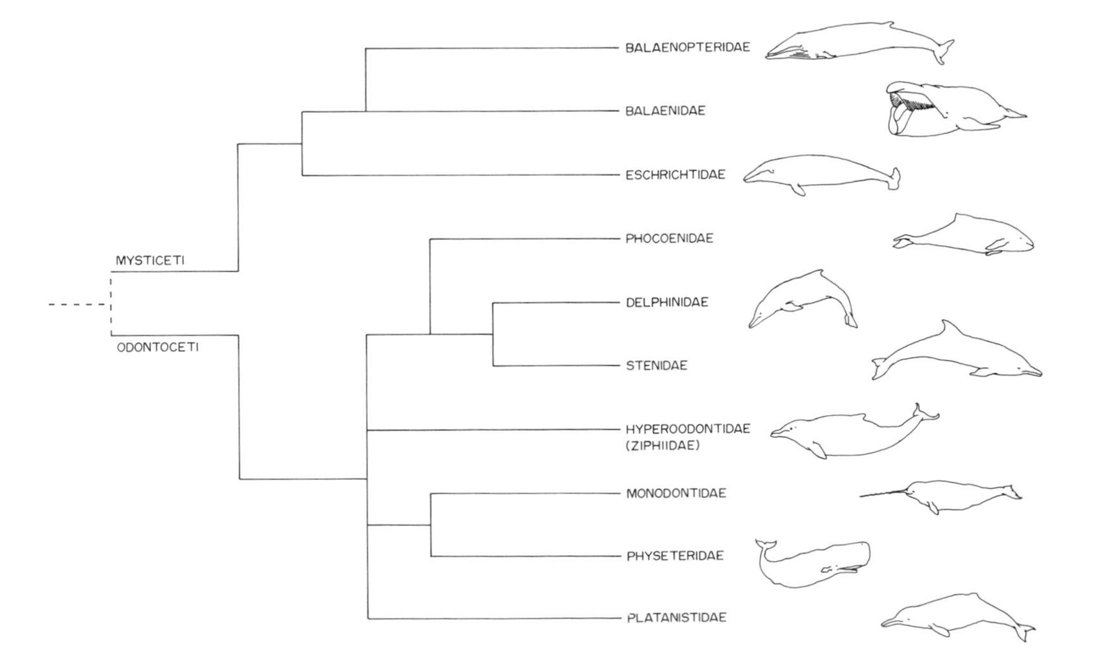

Figure 31. Diagram of the relationships of the Cetacea. Based in part on Hershkovitz 1966.

outside the body cavity. The standard synopsis of anatomy and physiology remains Slijper (1962). Burton (1973) presents a useful summary of natural history.

Broadly speaking, when one considers the cetaceans one notes two evolutionary trends reflected in the major divisions of the order into the Odontoceti, or toothed whales, and the Mysticeti, or baleen whales. In the odontocetes the teeth are retained, but there has been a reversion during the course of evolution from a complex molar cusp pattern to a unicuspid condition. All the teeth are very similar in form and size. We refer to this as secondarily acquired homodonty. In some odontocetes almost all the teeth have been lost, characteristically in the family Hyperoodontidae (=Ziphiidae). In the mysticetes the loss of teeth is complete in the adult, and epidermal plates in the roof of the mouth have been specialized as organs for straining out small crustaceans and zooplankton. The baleen whale's dominant adaptation concerns feeding upon the seasonally available zooplankton swarms.

The ancient whales, such as *Protocetis,* had a tooth formula similar to that of a primitive mammal: 3/3, 1/1, 4/4, 3/3. Feeding on zooplankton probably began at a very early stage in the evolution of the mysticete line. The extinct genus *Basisosaurus* had the beginnings of a serrated edge on its molar teeth similar to that found today in the pinniped genus *Lobodon*. This was probably an adaptation for feeding on pelagic crustaceans, which must have preceded ultimate adaptation for feeding on the smaller zooplankton. With the development of the baleen plates, tooth loss proceeded, and today only the fetal mysticetes show tooth buds. Interestingly enough, they have a triconodontid pattern.

When one examines the skeleton of a cetacean, one can discern certain trends. The skull shows a high degree of compacting in the odontocetes, and the frontal and parietal bones are shifted laterally. In the axial skeleton in both lines of cetaceans, the cervical vertebrae have become vastly shortened and in some cases are fused. The whole body of the cetacean is adapted for life in water. Beneath the epidermal layer there is generally a thick coat of connective tissue, and beneath that a sheath of fat that serves not only as a food storage organ for those species that feed seasonally, but also as an effective insulation against heat loss to the environment. The body form of the cetacean is adapted for speed in water. The fusiform shape and smooth surface reduce friction, and the capacity of many dolphins to vibrate the epithelial layer reduces the drag induced by turbulence.

All cetaceans can perform deep dives; they hyperventilate on the surface and expel some air at the moment of diving to reduce their buoyancy. During the breathing phase before diving, oxygen is bound in hemoglobin and a special hemoglobin derivative, myoglobin, associated with the muscle. Upon diving, the heart rate decreases, and oxygenated blood is shunted to the brain rather than to the muscles themselves. In addition, the medulla oblongata is relatively insensitive to CO_2 levels.

The odontocetes show a curious asymmetry in the head with respect to the nostrils. Outpocketings and blind passages leading to various distensible sacs typify the heads of many odontocetes (see Mead 1975, for a review). Furthermore, a number of odontocetes store oil in the head. Originally it was thought that the oil decreased the head's specific gravity, but the striking asymmetry of the nasal passages, in combination with the stored oil, may have something to do with beaming the sounds that are produced by moving air in the closed nasal passages during diving. These sound "beams" in turn are employed in echolocation. Although sound travels far and rapidly in water, one may recall that sound reception under water presents certain difficulties, since the flesh of the cetacean is of approximately the same density as water and thus is virtually transparent to a sound wave; of course, the skeleton is not. Bone can receive sound vibrations. But, if bone conduction is involved, the two ears would be confused about the direction of the sound. For this reason the mastoid and petrous bones are fused and lie in a pocket of connective tissue isolated from the skull itself. This allows the two inner ears to function independently of the potentially confusing bone transmission of sound by the skull itself. It is currently proposed that sound production is vital for underwater navigation and prey cpature by both mysticetes and odontocetes. The animal uses the sonar system by sending out sound pulses and listening for their echoes (Kellogg 1961). Yet navigation over long distances may involve a variety of other cues (Van Heel 1966).

There are no true vocal cords in cetaceans, and sound production is quite different in the two suborders. In the odontocetes, the pockets and folds in the nostrils, even with the blowhole closed, may allow air to move from the lungs to the air sacs, thus generating acoustical vibrations. In the mysticetes, laryngeal folds and distensible sacs that open out through the thyroid cartilage may also be employed in setting up acoustical vibrations as air moves within a closed system.

In considering sound production by cetaceans, we must distinguish between so-called echolocating pulses, or sonar pulses, and the clear tones or low-frequency sounds that are involved in social communication. The most intensively studied genus of cetacean has been *Tursiops*. The upper limit of hearing in *Tursiops* may be as high as 153 kH, which surely is an adaptation for utilizing high-frequency sounds as sonar pulses to determine the location and size of small objects.

The sounds produced during echolocation may also be employed in social communication, but there are whole classes of sounds that are clearly not involved in sonar but only in communication. The so-called songs of *Megaptera* (Payne and McVay 1971) are a prime example. The use of sounds in cetacean communication has been reviewed by Caldwell and Caldwell (1977) and Winn and Perkins (1976). The problem of human-to-cetacean communication has been discussed by Lilly (1961) and Bateson (1966).

Anatomical specializations in the cetaceans include the lack of olfactory epithelium and a consequent reduction in the olfactory lobe of the brain. Although gustatory sensitivity may be pronounced, apparently olfactory ability is very reduced or absent. The brain of odontocete cetaceans is relatively large compared with those of most mammals. Chapter 21 discusses the relative degree of encephalization demonstrable in the cetaceans. It may be noted that many species of the family Delphinidae have exceedingly large encephalization quotients.

The cetaceans have fascinated mankind over a great span of time. Systematic descriptions began with the whalers, in particular Scammon (1874). The distribution, systematics, and general ecology of the North Pacific and Atlantic cetaceans has been summarized in the volume edited by Tomilin (1967). The physiology, anatomy, and behavior of cetaceans have been recently reviewed in the volumes edited by Norris (1966) and Andersen (1969). The status of the great whales and their conservation are evaluated in the book edited by Schevill (1974). A synopsis of some recent literature is offered in table 27.

15.1.1 A Review of the Cetacean Families

In the family Platanistidae, one may note that all the cervical vertebrae are free and there is no fusion. The mandibles tend to be united for almost their full length. Typically, members of this family show a

Table 27 Ecological and Behavioral Studies of the Cetacea

Taxon	Ecology	Behavior	Reproduction
Platanistidae			
Platanista	—	Herald et al. 1969	—
Inia	Trebbau and van Bree 1974	Layne and Caldwell 1964	
Delphinidae		Saayman, Bower, and Tayler 1972; Pilleri and Knuckey 1969	—
Globicephala	—	Kritzler 1952	—
Tursiops	—	Tavolga and Essapian 1957; Essapian 1963; Tavolga 1966; McBride and Hebb 1948; Tayler and Saayman 1972	—
Stenella			
Orcinus	Baldridge 1972	Schultz 1967	Perrin, Coe, and Zweifel 1976
Monodontidae			
Delphinapterus	Kleinenberg et al. 1969	Gewalt 1967	—
Physeteridae			
Physeter	Ohsumi 1971	Caldwell, Caldwell, and Rice 1966; Gaskin 1964	—
Mysticeti			
Eschrichtidae			
Eschrichtius	Rice and Wolman 1971	—	Rice and Wolman 1971; Eberhardt and Norris 1964
Balaenidae			
Eubalaena	—	Saayman and Tayler 1973	—
Balaenopteridae			
Megaptera		Payne and McVay 1971	

long beak containing many homodont teeth, and they are strongly piscivorous. There is a tendency throughout this family for a vast reduction in the size of the eye. Most members of this family are adapted for fresh water, and they are distributed in widely different areas of the globe. The genus *Platanista (Susu)* is found in the Ganges River system. *Inia* is in the Amazon and the Orinoco. *Lipotes* occurs in the Yangtze Kiang River of China, and *Pontoporia (Stenodelphis)* is found in the Rio de la Plata. This family shows some very conservative morphological features and a uniform adaptation to estuarine or riverine habitats. It is not, however, the only family showing adaptation to fresh water, since at least three other genera from the family Delphinidae are strongly adapted for riverine systems.

The family Delphinidae includes the porpoises and dolphins. It may also be split into the Delphinidae, Phocoenidae, and Stenidae. The Stenidae include the genera *Steno, Stenella,* and *Sotalia;* the Phocoenidae include *Phocoena, Phocoenoides,* and *Neophocaena.* Most of the delphinids are marine, but the genera *Sotalia (Sousa), Lagenodelphis,* and *Orcaella* are adapted for fresh water. Some genera in this family show flat-crowned teeth and a short rostrum or no beak, with a vast reduction in the size of the dorsal fin. Typically they feed on mollusks, and they include *Phocoena,* the harbor porpoise, and *Neophocaena* of the Indian Ocean. Perhaps the best-known cetaceans, from the standpoint of behavior and physiology, are found in this family, including the genera *Tursiops, Lagenorhynchus, Globicephala,* and *Orcinus.*

The family Monodontidae shows no fusion of the cervical vertebrae and no dorsal fin. There are only two genera, *Delphinapterus,* the beluga or white whale of the arctic, and *Monodon,* the narwhal, which shows an extraordinary sexual dimorphism in that the males bear a single tooth on the anterior end of the body, twisted like that of the mythical unicorn.

The Hyperoodontidae, or Ziphiidae, are generally referred to as the beaked whales. In this family there is a pronounced loss of upper and lower teeth; only one or two remain. An evolutionary trend toward tooth reduction can be seen when we compare the five genera constituting this family. For example, *Tasmacetus* still has a considerable number of teeth, whereas *Ziphius* typically has only two lower teeth present in the male.

The family Physeteridae include the sperm whale, *Physeter,* and the pygmy sperm whale, *Kogia.*

The mysticetes are classically divisible into three families: the Eschrichtidae, Balaenopteridae, and Balaenidae. The family Eschrichtidae contains one genus, *Eschrichtius,* the gray whale. It is one of the better-studied baleen whales, being rather primitive in several features. The cervical vertebrae are still separated, there are no throat grooves, and females must return to shallow, warm water to calve. The Balaenidae are somewhat conservative, also exhibiting no throat folds. Once again, calving tends to occur in warm, shallow water. The Balaenopteridae are the best adapted to an open, pelagic existence. The cervical vertebrae are still separated, the dorsal fin is still present, and members of this family may be distinguished immediately by the presence of some forty to a hundred throat folds that allow for an extreme expansion of the throat when feeding on krill or euphausiids.

15.1.2 The Socioecology of the Cetaceans

The odontocetes have radiated into a number of different feeding niches. As a rule of the thumb, the Ziphiidae are specialized for feeding on pelagic squid. The Physeteridae are squid- and fish-feeders. The Phocoenidae are bottom feeders and include bivalves in their diet. The remaining families are more or less piscivorous.

The mysticetes do not feed strictly on the zooplankton. The gray whale (*Eschrichtius*) feeds primarily on amphipods. The bowhead whale (*Eubalaena*) feeds on both amphipods and mysids, and the right whale feeds on both euphausiids and copepods. Of the genus *Balaenoptera,* only *B. musculus* appears to be an obligate feeder on euphausiids. The other species in the North Pacific are mixed feeders on fish and euphausiids (Nemoto 1959; Nemoto and Yoo 1970).

Long-range movements and migrations of cetaceans have been well documented (Nishiwaki 1967). Some odontocetes, such as *Physeter,* the sperm whale, make annual migrations of considerable distance. Energy conservation by moving to warmer waters, especially at calving time, may be one reason. Clearly the mysticetes migrate to take advantage of the rich production of plankton during the brief polar summers, then migrate to less productive but warmer waters during the winter. Being aquatic, the great whales can transport considerable food for the off-season, stored as fat (see Brodie 1975 for a discussion).

The social behavior of cetaceans is very poorly understood, except for those species that have been studied in oceanariums and, in particular, *Tursiops*. Two useful reviews include those of Norris and Prescott (1961) and Slijper (1958). Perhaps the best procedure is to deal with some specific examples. The sperm whale, *Physeter catodon,* exhibits the following forms of social organization. Three types of groupings can be noted. There are groups of about fifty composed of females with calves, females that are pregnant, females that are lactating, and immature males less than 11 m in length. This group gives the appearance of being cohesive and travels as a unit. A second school form consists of males approximately 11 to 14 m in length. This school does not show pronounced cohesion, and individuals may be widely separated. For example, twelve individuals may be spread out over an area of 10 km^2. The third type of grouping consists of solitary males 12 to 17 m in length.

During migration, much larger groups may be composed of up to a thousand individuals. Even at this time, however, the cohesive female assemblages may be noted as distinctive subgroups. During the reproductive period, a single adult male attaches himself to the female group and defends it assiduously. This is a classical example of a polygynous system with a seasonal harem (Caldwell, Caldwell, and Rice 1966).

In contrast to this system, we may consider the California gray whale, *Eschrichtius glaucus*. This species typically shows a seasonal migration, feeding in the arctic seas between Siberia and Alaska and returning in the winter to calve in the lagoons of Baja California. Migratory movement seems to involve small cohesive groups that maintain some sort of contact with other groups of similar size. The subgroups are often recognizable as different age- and sex-classes. The only profoundly cohesive unit appears to be an adult female and her half-grown young and perhaps a calf of the year. Several males may attempt to court a female, but ultimately there is copulation only with a single male. The process of establishing dominance among males is poorly understood. Mating occurs on the calving grounds; there is no evidence of harem formation (Rice and Wolman 1971).

The composition of dolphin schools appears to be exceedingly complex. Apparently large feeding groups may be formed out of tightly cohesive subgroups. The composition of subgroups may be very stable, but the exact sexes and ages of the individuals in them is poorly understood. What is understood from close observation of *Tursiops* in oceanariums is that many bonds exist between adult individuals, especially adult females, and that mutual aid may be given to females during parturition as well as to members of the subgroup that are injured (Siebenaler and Caldwell 1956). This aid to comrades has also been shown in *Eschrichtius, Lagenorhynchus,* and *Globicephala* (Caldwell and Caldwell 1966).

15.2 The Early Ungulate Radiations

It is convenient to group the remaining ungulates in four mirorders according to McKenna (1975): the Eparctocyona, the Phenacodonta, the Tethytheria, and the Meridungulata.

Under the mirorder Eparctocyona, McKenna groups six orders (see fig. 17), only two of which have living representatives. The order Arctocyonia, as he has defined it, consists of part of the old order Condylarthra as outlined by Romer (1966). McKenna amends the order Condylarthra according to Cope (1880) to include only the Phenacodontidae and then defines a new mirorder Phencodonta, which includes the modern Perissodactyla and the Hyracoidea with the Phenacodonta (now termed Condylarthra). The order Arctocyonia includes the most conservative of the "condylarths," considered to be the stem group giving rise to modern ungulates. They appear in the Paleocene of North America.

The order Dinocerata existed from the Paleocene to the Eocene in North America and Asia. It was typified by massive size and well-developed canines as well as by bony protuberances on the head and snout. These latter forms are the uintatheres. The order Embriothopoda contains one family and one genus found in the Oligocene of North Africa. The order Tubulidentata includes only one living genus, *Orycteropus,* the aardvark. The aardvarks are of uncertain taxonomic position within the ungulates, but they clearly have affinities with the stem ungulates. Aardvarks are digitigrade and almost hairless. The molars are without enamel, and their roots are columnar; incisors and canines are lacking except in the milk dentition, and overall the group shows specialization for a myrmecophagous feeding strategy. The terminal end of each phalanx is flattened, a characteristic that suggests its protungulate affinities. Obviously, it became specialized for a myrmecophagous feeding niche very early and has remained as a living form retaining conservative anatomical characters lost in the evolution of the other ungulate groups.

The order Artiodactyla represents the most recent radiation of the Epartocyona and will be treated in subsequent sections.

The Meridungulata include the extinct South American orders that persisted until the Pliocene land bridge. The order Liptoterna began to arise as a recognizable group in the Eocene of South America. Through adaptive radiation a number of diversified herbivorous forms were evolved in the isolation of the South American continent. The height of the radiation was reached in the Pliocene and apparently ended through competition with advanced ungulate invaders from North America. Within the Liptoterna we can see a repetition of the evolution of horselike forms, from three-toed forms bearing the weight on the central toe to single-toed forms such as *Diadiaphorus*.

The order Notungulata began contemporaneously with the creodonts, starting in the Paleocene of Asia and South America and persisting in South America until the Pleistocene. Throughout the evolutionary history of this group, there was a tendency to increase in size that culminated in the suborder Toxodontia of South America, which reached massive proportions suggestive of rhinoceroses.

The order Astrapotheria was confined to South America and perhaps derived from the Notungulata. From what we can guess from its skull, it possessed a small trunklike structure reminiscent of those of the proboscideans. The Pyrotheria showed a superficial resemblance to some Tethytheria but were found only in South America. Perhaps the affinities are more apparent than real. Judging from the skull, the animals were adapted for feeding much like the present-day tapirs. The order Xenungulata was confined to South America, having molars that resemble those of the Pyrotheria, but the premolars and canines are easily differentiable.

The mirorder Phenacodonta, as defined by McKenna, includes the orders Condylarthra, Perissodactyla, and Hyracoidea. The order Condylarthra, which existed from the Paleocene to the early Oligocene in North America and Asia, was a stem group giving rise to all the Perissodactyla through adaptive radiation. Although the early condylarths were pentadactyl, each digit terminated in a protohoof. The stance was either plantigrade or semidigitigrade, and the dentition consisted of low-crowned bunodont teeth. Through time and adaptive radiation, selenodont teeth were evolved, indicating increasing adaptation for the mastication of plant food. In the advanced cursorial Condylarthra, the clavicle disappears. The Perissodactyla and Hyracoidea have living forms and will be considered in subsequent sections.

The mirorder Tethytheria, as defined by McKenna, groups two living orders, the Sirenia and the Proboscidea, with the extinct Desmostylia.

The order Desmostylia has been found in marine deposits of North America and Asia and perhaps shows some strong affinities with the Sirenia. It may have been an estuarian feeding form. The sirenians and proboscideans will be described in the following section.

15.3 The Order Sirenia: The Sea Cows

The sirenians represent an ancient adaptive radiation and are the only extant group of mammals that are both entirely aquatic and herbivorous. The hind limbs are lost totally, and the posterior part of the body is in the form of either a large paddle (family Trichechidae) or flukes reminiscent of cetaceans (family Dugongidae). The incisors are lacking in the Trichechidae but are retained in males of the Dugongidae. The rostrum of the skull in Dugongidae is deflected ventrally. The family Trichechidae contains one genus, *Trichechus,* which is distributed in fresh water and in the estuarian systems of West Africa, South America, and the Caribbean. *Trichechus senegalensis* is the West African form. *Trichechus inunguis* is found in the Amazon River system and in the Orinoco upriver from the first major cataract. *Trichechus manatus* is found in the Orinoco below the first cataract and around the Caribbean, including the Gulf coast of Florida and the Atlantic coast of Florida, almost to the Georgia border (Mondolfi 1974). The dugongs, *Dugong dugon*, are confined to the warmer waters of the Indian Ocean and the Pacific off the northern coast of Australia. All present populations are vastly reduced owing to hunting by man. The best-known species is *Trichechus manatus,* or the Caribbean manatee.

The sirenians are characterized by tooth replacement as in elephants. In short, only a few of their teeth are in use at any one time. They are specialized for feeding on aquatic plants, such as *Zostrea* and marine green algae. The dugongs feed exclusively on marine algae and do not enter freshwater systems. The lips of the dugong and manatee serve as organs of prehension for gathering vegetation into the mouth. The forelimbs are modified as flippers and are used only in stabilizing or swimming. Aspects of dugong ecology are summarized by Spain and Heinsohn (1975), Heinsohn and Spain (1974), and Heinsohn and Birch (1972).

The most extensive fieldwork has been done in Florida. The Florida manatee, in common with other members of the Sirenia, is a hindgut fermenter. A large cecum serves as a repository for microbial symbionts that assist in breaking down cellulose. In its digestive efficiency, the manatee resembles proboscideans and perissodactylans.

The manatee in Florida may undertake pronounced seasonal movements, having a summer feeding area and a winter feeding area. The north-south movement is profoundly influenced by shifts in the mean water temperature paralleling the north temperate annual climatic changes. The manatee feeds in salt water as well as fresh water, making use of a wide variety of aquatic plants. In common with other members of the Sirenia, its foraging range is restricted to water shallow enough to support the growth of herbaceous vegetation.

The extensive studies of Hartman (1971, 1979) have given us a picture of the socioecology of *Trichechus manatus*. His studies were most extensive in the season of the year when the manatee population was wintering in protected warm springs, in particular, the Crystal River of Florida. The female manatee is very much attached to her newborn young. The animals communicate by a variety of vocalizations. The mother may help the young rise to the surface by carrying it on her back. Nursing is prolonged, and a year-old young may still be suckling from the mother; however, before weaning it has begun to supplement its diet with aquatic vegetation. The female-young unit appears to be a persistent one. There is some evidence that members of the loosely overwintering group recognize one another.

When a female begins to come into estrus, she may be followed by a number of males, and at times a veritable mass mating situation can appear to occur. There is some evidence that males will butt one another, and perhaps a dominance order can be discerned. Gestation is estimated by Hartman at approximately thirteen months. The female does not become sexually mature until six to seven years of age, though this may be influenced by such environmental factors as the nutritional quality of the vegetation. Interbirth intervals in the manatee may be approximately three and a half years. It appears that the animal is a rather show-reproducing species but extremely long-lived. Captive longevity records exceed twenty-six years.

The social organization, as we currently understand it, does not appear to be very complex. The most cohesive unit consists of a mother and her offspring. Vocalizations are employed to maintain contact among members of an aggregation (Schevill and Watkins 1965). This is especially important because in many portions of their habitat the water offers only limited visibility. Where winter ranges and summer ranges are widely separated, no doubt the calf learns the route to and from such seasonal use areas by first accompanying the female. It appears that this type of early learning could be extremely important in permitting efficient use of the various patchy ranges the manatee employs at this extreme northern limit of its distribution.

Table 28 summarizes the recent literature on the Sirenia. Bertram and Bertram (1973) summarize the literature on the distribution and status of contemporary populations.*

15.4 The Order Proboscidea

The Proboscidea have had a long and spectacular evolutionary history. Beginning with *Moeritherium** in the Eocene of Africa, the following trends are evident in the evolution of this order. First, there has been a persistent tendency to increase in size by lengthening the limb bones, but not the carpals or the tarsals. The skull has become longer and the neck shorter. In many of the ancient proboscideans (Gomphotheriidae), there was a trend toward a lengthening of the lower jaw, but this is secondarily reduced in length in recent forms (Elephantidae). Second, throughout their history there has been a persistent tendency for the upper lip to be lengthened and, in the case of many extinct and all modern proboscideans, the upper lip and nostrils form a muscular trunklike organ used in feeding. Third, there has also been a trend

Table 28 Recent Literature on the Sirenia and Proboscidea

Taxon	Natural History and Ecology	Behavior	Reproduction and Ontogeny
Sirenia	Mohr 1957	—	—
Dugongidae			
Dugong dugon	Heinsohn 1972; Mitchell 1973	Spain and Heinsohn 1973; Anderson and Birtles 1978	—
Trichechidae	Bertram 1964	—	—
Trichechus manatus	Hartman 1971, 1979	Schevill and Watkins 1965; Moore 1956	Hartman 1971
T. inunguis	—	Evans and Herald 1970	—
Proboscidea			
Elephantidae			
Loxodonta africana	Sikes 1971; Buss 1961; Laws, Parker, and Johnstone 1975; Douglas-Hamilton 1972; Wing and Buss 1970; Malpas 1978	Hendrichs 1971; Buss and Estes 1971; Laws 1974; Leuthold 1976	Buss and Smith 1966; Lang 1967; Leuthold and Leuthold 1975
Elephas maximus	McKay 1973; Olivier 1978	Kurt 1974; Jainudeen, Kantangale, and Short 1972; Jainudeen, McKay and Eisenberg 1972	Eisenberg, McKay, and Jainudeen 1971; Jainudeen, Eisenberg, and Jayasinghe 1971; Jainudeen, Eisenberg, and Tilakeratne 1971

toward hypertrophy of the second incisors in the skull to create offensive weapons called tusks.

Man and the elephant have coexisted through time, and the elephant was a staple part of man's diet during the Pleistocene. At this time the mammoth and the mastodon were contemporaries with early man in both the Old World and the New World. The elephant had invaded the New World in the Pliocene, crossing the Bering land bridge and eventually passing to South America. The elephant was entirely exterminated in the New World, and man can by no means be implicated in all exterminations. Some complex puzzles concerning faunal imbalances must be solved before the extinction of the elephant in South America can be adequately explained.

The order Proboscidea has two living species, *Loxodonta africana* and *Elephas maximus*. Although each of these species is divisible into several subspecies, no complete agreement concerning the subspecific classification has been achieved. In Africa it seems clear that two subspecies can be distinguished, though they intergrade in the eastern Congo. The western race, or forest elephant, *L. a. cyclotis,* is generally smaller and possesses a smaller, more rounded ear than does the eastern "bush" elephant, *L. a. africana,* which is the most intensively studied form, found from the Sudan to South Africa. The forma typica for *Elephas, E. m. maximus,* is found on Sri Lanka. *E. m. indicus* is found from peninsular India through Southeast Asia. *E. m. sumatrensis* is found on Sumatra and Borneo.

The long evolutionary lineage of the elephant has culminated in a form that represents the ultimate adaptation for feeding upon coarse plant materials and at the same time attaining an adult size sufficient to withstand all predator attacks. (*Homo* is the exception; since the Pleistocene, elephants have been hunted by man.) The elephant calf is vulnerable to predation, and it is here that the social organization of the cow herd itself serves as a protective device as well as a mechanism for creating a social milieu in which the young elephant can mature and learn its role in adult life (see plate 37). The cow herd serves as a repository for traditional knowledge, including the routes to water holes during periods of drought, the routes to feeding grounds, and so forth. Since the adult cows undoubtedly carry the memory of habitat utilization patterns, this is a form of a living tradition. In its habitat, the elephant is never the most numerous mammalian species, but it is an ecological dominant because, in terms of biomass, it is often the greatest contributor to the cycling of plant materials within its ecological community (McKay 1973).

The teeth are highly adapted for grinding coarse plant material. The elephant is a hindgut fermenter, and the cecum is enlarged, providing a fermentation chamber that is a repository for bacterial and protozoan symbionts that assist in breaking down plant parts to simple sugars—in particular breaking the structural carbohydrate, cellulose. Because the elephant is adapted for feeding on coarse plant materials with a relatively low nutrient content, it must feed protractedly at certain times of year to make up its maintenance ration. Feeding may occupy from 70 to 80 percent of an elephant's waking hours, and in the process of selective feeding it may consume 250 to 400 pounds of wet forage in an average twenty-four-hour period.

Plate 37. *Elephas maximus*, the Asian elephant. Cow herd with young. During a slight disturbance, the young are grouped in the central portion of the herd with adult females facing outward. Compare with the antipredator strategies in plates 45 and 46. (Photo by author.)

Adult males typically forage alone or in the company of one or two younger males. Adult females are organized into cow groups that subdivide into smaller units during the day for feeding and reassemble at some point in the evening, generally at a drinking place, to reestablish contact with their herd. Although adult male elephants may occasionally sleep by lying down for some twenty minutes out of twenty-four hours, generally they sleep standing. Adult cows may lie down very little, if at all, in a twenty-four-hour period. If the cow herd does lie down, generally some cows remain standing in vigilant attitudes. Younger animals, especially calves, may often lie down to sleep. Feeding is carried out through the day and night, and drinking is generally confined to dusk or the early morning hours. An adult elephant may drink as much as 10 to 15 gallons of water. Various plant types are taken, and their amounts change with the season and the relative abundance of the individual plant species. Grasses account for a significant percentage of food intake, especially for females and young animals, but branch-feeding in the forest occupies 30 to 50 percent of their feeding time.

The Asiatic elephant utilizes areas of secondary forest, scrub, and savanna more extensively than primary or climax forest types (Olivier 1978; McKay 1973). The animal is dependent on grasses, forbs, and shrubs. Areas that show a high grass yield, such as tree savannas or interfaces between secondary forests, scrub, and grassland, are prime elephant habitat provided the distribution of water is sufficiently spread out through the area to prevent concentrations of elephants at water sources (VanCuylenberg 1977).

Since the social life of the African elephant has been discussed in a number of recent publications (Douglas-Hamilton 1972; Laws, Parker, and Johnstone 1975), I will confine my further remarks to the social organization of the Asiatic elephant, *Elephas maximus*, based on the researches of McKay (1973), Kurt (1974), Eisenberg, McKay, and Jainudeen (1971), and Olivier (1978). The initiation of a young elephant into its social unit is a gradual process. Until about four years of age, the social roles of males and females are virtually identical, but they begin to diverge in the fifth year, and from then on we can speak of a separate male role and a series of female roles. The young male may reach sexual maturity as early as 7 or 8 years of age, but this in no way implies that he has an effective role as a breeder. Seven years is considered the earliest age at which spermatogenesis may take place. The average age of puberty extends from 10 to 12 years, and its actual onset will be subject to the young male's physiological condition. In areas of poor nutrition, his growth and maturation can be retarded. Success in breeding cows seems to be limited in part by the density of other bulls that may contest for dominance as well as by the aggressive activity of older cows whom the young male cannot dominate at an early stage of his growth.

The young male Asiatic elephant does not begin to achieve adult stature and size until he is approximately 17 years of age. The years from 14 to 17 can be considered a subadult phase; it is during this interval that he begins, annually or biannually, to secrete from the temporal glands on each side of his face between the eye and the ear. The onset of secretion and the excitability that accompanies it—a condition termed *musth*—is well known to handlers of elephants. It may be likened to the rut of seasonally breeding deer and antelope (Jainudeen, McKay, and Eisenberg 1972). Although active secretory activity of the Asiatic male's temporal glands is closely associated with aggressive and sexual behavior, he may initiate breeding activity in the absence of such secretion. This activity of the glands is, however, clearly related to marking behavior, since the male will rub secretions on tree trunks.

In the African elephant both adult males and adult females periodically secrete from their temporal glands, and secretions seem to be involved in marking behavior and thus are a means of chemical communication. The sexually mature male may also secrete a second product from his temporal glands during seasonal sexual activity. This phenomenon is similar to musth in the Asiatic elephant (Poole 1979).

Throughout his years of maturation, the young male elephant contests with members of his own age-class, and a dominance order is established among males who are familiar with each other because of home-range overlap. Such dominance contests rarely result in serious injury and seem to be limited to tests of strength. Younger males tend to follow and feed in the vicinity of older males, but a fully adult male with an established position in the population's social hierarchy is generally semisolitary for the better part of his life. An adult male moves in a consistent home range, visiting various feeding and watering points on a periodic schedule. Since the cow herds or clans have larger home ranges, they may overlap with several males. During periods of local shifts in the population to preferred feeding loci, several adult males may be in the immediate vicinity of the cow herd without being in continuous attendance.

At the age of approximately 6 to 7 years young females can be considered on the threshold of puberty. The age at which a female Asiatic elephant first breeds is of course subject to a variety of environmental influences. With good nutrition, a female can probably conceive in her seventh year, which means that her first calf will be born in her ninth year. Of course, in populations existing under marginal conditions and in poor nutritional states, maturation might be delayed until ten years or older. The females remain in the cow herd of their birth, termed the clan, which is essentially a matriarchy led and coordinated by the oldest reproducing cow, who is probably mother, sister, aunt, or grandmother to most of the adult females in it.

Typically the cow herds or clans in Sri Lanka range from eight to twenty-one animals. This extended herd generally divides daily into small subgroupings of varying composition that forage within vocal contact of one another. These small subgroups reassemble at evening to drink, and they generally move from one area to another as a unit. There are differences in the composition of subgroups that are apparently based in part on small "family units." A female and her offspring constitute such a unit. Lactating females with young calves very frequently are found in the company of other females with small calves or young females nearing puberty. This leads to the formation of a "nursery unit" that may be very cohesive in its movements and that forms an effective subgroup for rearing infants separate from subgroups formed of females with half-grown young and subgroups of young males in their sixth or seventh year.

The cow herd's home range is much larger than the area used consistently by a given male; hence, cow herds will tend to pass through the home ranges of several different males. Since cows come into estrus approximately every three to four weeks when not pregnant, ample opportunity exists for a male to join a cow herd and travel with it while a female is in estrus.

The consistent presence of a large dominant male in a cow herd generally indicates that some cow is coming into estrus. Often several males may be in attendance on the periphery of the cow herd, and if a female is in estrus contests may take place among the males to reassert their dominance status. The position of the Asiatic male in a dominance hierarchy is a function of his size and his reproductive condition. As I stated before, it is not essential that an Asiatic elephant male be in musth to breed a female, but a male in musth probably has a greater opportunity to achieve a high dominance status relative to cows and other males. Males may attempt to court cows shortly before they enter their estrous period, but the intensive period of mating activity takes place during less than twenty-four hours. The courtship of elephants involves much touching, and both tactile and chemical signals seem to be involved (Eisenberg, McKay, and Jainudeen 1971). During an actual mounting sequence, the cows and juveniles may evince interest in the courting couple and form a circle around the mating pair.

After a successful mating, the female carries her calf for 22 months. After its birth, she will tend and nurse it intensively for some six months. The calf begins to experiment with feeding itself during the second half of its first year of life. Food selection is promoted by the adult's feeding habits, and the young calf can "steal" food with impunity from its mother's trunk or mouth. The calf can continue to suckle from its mother or "aunts" well into its second year, and the weaning process seems to be gradual as the youngster feeds more and more independently.

The bond between a female and her calf can become very strong. Females with small calves are extremely vigilant and circumspect in their activities. Calves may nurse from more than one female, and the presence of many lactating females in a cow herd often ensures adequate nutrition for calves, since their dependency on a single female's lactation is reduced. When threatened by danger, calves and adults will often form a circle with the adults facing outward and the young remaining between the mothers or in the center of the formation. Such a cluster or "star" formation offers adequate defense against most terrestrial predators. From this cluster, a female can charge an offending predator. Undoubtely tigers have exerted some predation pressure against Asiatic elephant calves, as do lions for the African elephant, but far and above any of the large cats man has been and is the principal predator of elephants.

My observations on Asiatic elephants show a remarkable similarity to those made on the African elephant by Laws, Parker, and Johnstone (1975), Sikes (1971), Buss (1961), and Douglas-Hamilton (1972). Most of the differences I have noted concern the annual rutting behavior of the Asiatic male, which is somewhat different from the condition shown by the African male elephant. The differences in group size and cohesiveness between Asiatic and African elephants are probably related to overall adaptations to a more forested habitat by the Asiatic elephant and open woodland savannas by the African elephant. It is thought that the social structure and herd size of the West African forest elephant may conform more to the picture outlined for the Asiatic by Olivier (1978). Recent literature for the proboscideans is included in table 28.

15.5 The Order Hyracoidea: The Hyraxes

There are three living genera of hyraxes, *Dendrohyrax, Heterohyrax,* and *Procavia. Procavia* extends from the Middle East to South Africa; the other two genera are confined to Africa south of the Sahara. The hyraxes are small, with four digits on the forefeet and three on the hind feet. The nails are flat except for that of the second digit, which is clawlike. The teeth are high crowned in *Procavia,* which grazes extensively, but *Heterohyrax* and *Dendrohyrax* have low-crowned molars and browse. The testes are intraabdominal. *Dendrohyrax* is most strongly adapted to forested habitat and, as its name implies, makes extensive use of trees, generally nesting in hollows of a traditionally used sleeping tree. Little is known concerning the behavior patterns of *Dendrohyrax,* but Kingdon (1971) has provided a useful synopsis. The males of *Dendrohyrax* employ rather characteristic calls, apparently to help space themselves in the forest. The phenomenon has been analyzed by Richard (1964).

Procavia and *Heterohyrax* have been well studied from the standpoint of ecology and behavior, both in the Middle East (Meltzer 1967) and in Tanzania and Kenya (Coe 1962; Sale 1965*a*; Hoeck 1975). The basic social organization for *Procavia* consists of a dominant male defending an area containing a number of females. Females are thus colony members, and as adults they show fidelity to a preferred habitat site, generally an outcropping of rocks. Males assiduously defend these female groups—most strongly, of course, during the breeding season. Like-aged young may assemble in small crèches, and various aspects of vigilance behavior are shown by the subadults and adults (Rahm 1964).

The hyracoids are hindgut fermenters, and the microbial flora of their cecum helps break down structural carbohydrates. Varying degrees of specialization for browsing and grazing and the width of the feeding niche have been analyzed for populations of *Procavia* where they occur independent of *Heterohyrax* (Sale 1965*a*) and for populations where *Heterohyrax* and *Procavia* occur in microsympatry (Hoeck 1975). *Het-*

erohyrax is more prone to climb trees and utilize vegetation than is *Procavia,* especially where they are in sympatry (Hoeck 1975), but late in the dry season they feed on the same plants with little interspecific strife. The two genera can use the same habitat and, without conflict, form mixed groups on rock outcroppings, so that the smaller *Heterohyrax* are often mistaken for juvenile *Procavia.*

In niche occupancy, hyracoids replicate in a convergent fashion the social organization and feeding niche of the caviid rodent *Kerodon rupestris* in northeastern Brazil (Lacher 1979). There is also a convergence in social organization and trophic strategy between the hyraxes and the ctenodactylid rodents of the genera *Massoutiera, Felovia,* and *Ctenodactylus.* As browsers, the hyracoids have a low metabolic rate (Bartholomew and Rainey 1971). They produce one or two young after an extended gestation of 7½ months (Mendelssohn 1965; Sale 1965*b*).

15.6 The Order Perissodactyla: Odd-toed Ungulates

There are three living families of Perissodactyla—the Tapiridae, Rhinocerotidae, and Equidae. All three families are basically adapted for herbivority. The number of functional toes in the Perissodactyla can vary from three to one, but the weight is generally borne on the third or central toe. In the evolution of this group there has been a trend toward reduction in toe number and increasing adaptation for cursorial, extreme digitigrade locomotion. There has been an evolutionary trend away from brachydont dentition to high-crowned hypsodont dentition in those species that have moved from browsing to grazing. In contrast to the Artiodactyla, the upper incisors and canines are retained.

The Tapiridae are confined to tropical forests in both the Old World and the New World and perhaps have the broadest dietary requirements or tolerances. The American tapirs eat a great deal of fruit and seeds in addition to browse (Terwilliger 1978). While the feeding patterns of the Malay tapir, *Tapirus indicus,* are less well understood, it appears to be strongly adapted for browsing (Medway 1974).

The Equidae (horses, asses, and zebras) have adapted to open plains habitats and are almost exclusively grazers. The rhinoceroses show mixed feeding strategies, with the African white rhinoceros, *Ceratotherium simum,* most strongly adapted for grazing. The African black rhinoceros, *Diceros bicornis,* is a mixed browser and grazer, as is the great Indian one-horned rhinoceros, *Rhinoceros unicornis. Rhinoceros sondaicus,* the Javan rhinoceros, and *Didermocerus sumatrensis,* the Sumatran rhinoceros, are at present confined to rather forested habitats and appear to be strongly adapted for browsing.

All members of the Perissodactyla are hindgut fermenters with a greatly enlarged cecum; presumably the microbial symbionts in the cecum aid the breakdown of structural cellulose and perhaps certain hard-coated seeds. The Perissodactyla produce a single young after a rather prolonged gestation period (see sect. 22.3).

15.6.1 The Family Tapiridae: The Tapirs

The Tapiridae are very little differentiated from the stem forms of perissodactyls. There are five toes on the forefeet and three on the hind feet. The teeth are bilophodont. There are four living species in one genus, *Tapirus: T. indicus* of the Malay Peninsula and Sumatra; *T. bairdii* of Central America and Mexico; *T. terrestris* of continental South America; and *T. pinchacus* of the lower slopes of the Andes. Fossils indicate that tapirs once were much more widespread and that those now occurring in Central and South America are derivative from Asia through North America (see plate 38).

Richter (1966) has summarized the behavior of the tapirs. The best-studied species, both in captivity and in the field, are *Tapirus bairdi* and *T. terrestris*. In contrast to the black-and-white *T. indicus,* the adults are uniformly colored brown, while the newborn share with the newborn Malay tapir, *T. indicus,* a pattern of horizontal yellow stripes on a brown background. Insofar as we understand it, the most cohesive social unit in the tapir is the female and her offspring. Generally before the birth of a second offspring, the older baby is driven off.

Tapirs show great fidelity to trail usage, especially at crossing points on streams and up rather steep banks. Their trails are conspicuous and are used traditionally over and over. Tapirs are excellent swimmers and generally have a wallowing area in their habitat to which they can retire. Defecation is generally in the vicinity of water, if the tapir has an option.

Generally a single young is born after a 13-month gestation. The young does not immediately accompany the female but exhibits a "hider" strategy, lying up alone in a sheltered spot while the female forages. It is assumed that its peculiar striped pattern serves to camouflage the young as do the spotted coats of newborn Cervidae (see plate 39).

Plate 38. *Tapirus terrestris,* the South American tapir, one of four surviving species of the family Tapiridae. Note the modification of the upper lip and nose to form a prehensile structure. (NZP archives.)

Plate 39. *Tapirus terrestris* young. This photograph illustrates the marking patterns of the young tapir, which confer the same camouflage benefits that young cervids exhibit. (NZP archives.)

On Barro Colorado Island there is a group of tapirs all descended from a few introduced individuals. The tapirs are generally fed at a fixed location in the evening as a supplement to their natural foraging. Under these conditions I have noted that a dominance order exists and that tapirs have a regular established use pattern at the feeding site. Females with young about 10 days old begin to bring them to the clearing, and I assume this is the time the young transfer from a hiding to a following strategy. Some key references are included in table 29.

15.6.2 The Family Rhinocerotidae: The Rhinoceroses

The family Rhinocerotidae contains four genera with five living species distributed in Southeast Asia and Africa. The African species include *Diceros bicornis,* the so-called black rhinoceros, and *Ceratotherium simum,* the square-mouthed or white rhinoceros. The Asiatic genera include *Didermocerus sumatrensis,* which typically bears two horns and is small and adapted for browsing in forests. The second Asiatic genus, *Rhinoceros,* includes two species: *R. indicus* and *R. sondiacus.* Currently *indicus* is confined to the Gangetic Plain, and *sondiacus* is known to exist only on Java.

The forest-adapted Sumatran rhinoceros, *Didermocerus sumatrensis,* appears to have habitat-utilization patterns similar to those described for the tapir. No large social groupings are formed, and the most cohesive unit appears to be a female and her calf. Very little is known concerning the life histories of these animals, except for some observations on captives at Copenhagen and Zurich and the field study of feeding habits by Strickland (1967) (see Groves and Kurt 1972).

Table 29 Recent Literature on the Perissodactyla

Taxon	Natural History and Ecology	Behavior	Reproduction and Ontogeny
Tapiridae	—	Richter 1966	—
Tapirus indicus	Medway 1974; Sanborn and Watkins 1950	—	Schneider 1936
T. bairdii	Enders 1935; Terwilliger 1978	—	—
T. terrestris	—	Hunsaker and Hahn 1965	—
Rhinocerotidae	—	Schenkel and Lang 1969	—
Rhinoceros unicornis	Laurie 1978; Gee 1953	—	Lang 1961; Buechner et al. 1975
R. sondiacus	Hoogerwerf 1970; Sody 1959; Schenkel and Schenkel-Hulliger 1969*a*	—	—
Diceros bicornis	Goddard 1968; Schenkel and Schenkel-Hulliger 1969*b*	—	Goddard 1966
Ceratotherium simum	Owen-Smith 1975	Owen-Smith 1972; Backhaus 1964	—
Didermoceros sumatrensis	Hubback 1939; Strickland 1967	—	—
Equidae	—	Klingel 1972*a*	—
Equus caballus	—	Tyler 1972	Hafez, Williams, and Wierzbowski 1962
E. asinus	Moehlman 1974	Woodward 1979	—
E. burchelli	Smuts 1975	Klingel 1967	—
E. zebra	Joubert 1972	Klingel 1968	—
E. grevyi	—	Klingel 1974	—
E. hemionus	Bannikov 1961	—	—

The Javan rhinoceros, *Rhinoceros sondiacus,* has been studied in the field by Schenkel and Schenkel-Hulliger (1969*a*). Their information on its feeding ecology supplements the earlier observations by Hoogerwerf (1970). Although Javan rhinoceroses show no cohesive social groupings besides the female and her calf, the members of a community use trails and wallows in common, and it appears that individuals recognize one another. Marking may be done in wallows, and males are prone to urine-mark in their vicinity. Dung is deposited near wallows or in specific piles at other points within the home range.

The great Indian rhinoceros, *Rhinoceros unicornis,* is the best-studied of the Asiatic species (see plate 40). Here again one has the impression that the adults in the population recognize one another. Trails are used in common, as are wallows. Males mark with urine in the vicinity of wallows and at other places in their home range, and dung piles are also utilized for marking. The most cohesive social unit again is the female and her calf. Adult males definitely have a dominance order. Although subadult males may associate with each other, adult males appear to be well spaced and to avoid one another. Severe fighting can occur between males, and they can die from wounds (Laurie 1978).

The African rhinoceroses represent two separate adaptations. The white rhinoceros, *Ceratotherium simum,* is adapted for grazing. The

Plate 40. *Rhinoceros unicornis*, the great Indian rhinoceros. This captive pair illustrate the curious folding of the skin that is diagnostic for this species. (NZP archives.)

studies of Owen-Smith (1975) demonstrate that adult males hold fixed territories in which they may tolerate subadult males but are dominant over them. Dung piles and urine are used to establish their territories. Cows and their attendant offspring (often of several ages) may move among the territories and are not necessarily confined to a single male's area.

In the African black rhinoceros, *Diceros bicornis*, the calf and mother tend to constitute the only cohesive social unit, but individual recognition of members of the same community is highly probable, and long-term associations among related females again suggest individual recognition (Goddard 1966, 1967). Nevertheless, the adult rhinoceroses tend to move singly and aggregate only at familiar points, such as wallows or water holes. The adult males are generally well spaced but exist in overlapping home ranges. Adult males may compete for an estrous cow. Marking by males again employs urine and dung piles (Schenkel and Schenkel-Hulliger 1969*b*). Further information on rhinoceros behavior has been summarized by Schenkel and Lang (1969) (see also table 29).

15.6.3 Thc Family Equidae: The Horses, Zebras, and Asses

The family Equidae has one of the best-studied evolutionary histories of any mammalian group. The Equidae were distributed in North America and Asia and are relatively recent invaders of Africa. During the Pliocene they invaded South America. For some inexplicable reason the equids became extinct in both North and South America during the

Pleistocene; they were reintroduced by Western man on his arrival in the fifteenth century.

There are numerous genera of equids described in the literature, but generally we can consider three morphoclines (see Groves 1974). It is convenient to deal with subgenera. The subgenus *Equus,* or true horses, is typified by the domestic horse (*E. caballus*) and Przewalski's horse (*E. przewalskii*). The subgenus *Asinus* is typified by the domesticated donkey (*E.* [*Asinus*] *asinus*) and the wild asses of North and East Africa (*E.* [*A.*] *africanus*). There then can be identified a series of horselike forms that tend to show morphological variations intermediate between horses and asses and to be uniform in coloration, except for a dorsal shoulder stripe. These include the onager, *E. (Hemionus) hemionus,* and the kiang, *E. (Hemionus) kiang,* of Asia. There exists a comparable series of forms that show intermediate morphological characteristics between horses and asses, but all bear striped coats to some extent. These are the zebras (subgenus *Hippotigris*), including the Grant's zebra, *E. (Hippotigris) burchelli;* the mountain zebra, *E. (Hippotigris) zebra;* and the Grevy zebra, *E. (Dolichohippus) grevyi.* The enduring trend within the recent Equidae is toward an animal supremely adapted for grazing in open country and possessing a foot remarkably adapted for locomotion over dry, hard surfaces. Their general biology is summarized in Groves (1974).

The equids exhibit a range of social organizations with two distinct forms and some intermediates. On the one hand we have a social system where the only cohesive unit is the female and her offspring. Aggregations of females may take place during short-range migration or at places with restricted resources, such as water holes. Depending on the time of the year, adult males may exhibit territoriality. Aggression by an adult male is usually against another territory-holding male. Territories are marked with dung piles and urine. Subadult males or non-territory-holding males may wander freely.

These male territories are connected with breeding behavior, since an estrous mare will not be persecuted by other males once she enters a territorial male's area. The territorial male then will breed the female. This system typifies the Equidae adapted to semiarid and arid areas, such as Grevy's zebra, *Equus grevyi* (Klingel 1974); the African wild ass, *E. africanus;* feral asses in the United States, *E. asinus* (Moehlman 1974); and the Asiatic wild ass, *E. hemionus* (Klingel 1977).

The second form of social organization is typified by the plains zebra, *Equus burchelli* (Klingel 1968), and the horse, *E. caballus* (Zeeb 1961). Here, a group of mares and their young are permanently guarded by a dominant stallion. Although the herd may move quite freely and not occupy a defended area, no other sociologically "adult" males are permitted access to the mares. The group shows great cohesiveness and displays numerous behaviors that increase social tolerance. Intermediate forms of group cohesion may occur within this classification depending on special characteristics of the habitat (Tyler 1972; Klingel 1972*a*), but in general these two types of herding behavior and mating systems are distinct. The problem has been discussed in some detail by Klingel (1972*b*), who has also reviewed communication mechanisms for the Perissodactyla (Klingel 1977).

15.7 The Order Tubulidentata: The Aardvarks

The sole surviving family, Orycteropidae, contains a single genus and species, *Orycteropus afer,* the aardvark, confined to Africa south of the Sahara. In his 1945 classification, Simpson considered it the most conservative living species of the Ungulata. As I indicated in section 15.2, the species is adapted for feeding on ants and termites. It retains its milk teeth as an adult, and termites and ants are generally chewed before they are swallowed.

Aardvarks have a large home range and appear to resemble the Neotropical genus *Myrmecophaga* in their long-range movements while feeding. Aardvarks also feed on the fruits of *Cucumis humifructus*. The seeds apparently pass through the digestive tract unharmed and germinate in the dung hills that are deposited near an aardvark's burrow (Mitchell 1965).

The social life of *Orycteropus* appears to be simplified, but the mother-young bond is strong. The gestation period exceeds 7 months. The young is about 2 kg at birth (Sampsell 1969). The development of hand-reared aardvarks has been summarized by Kingdon (1971), and he provides a useful synopsis of their natural history, while Melton (1976) reviews the literature dealing with the biology of the aardvark.

15.8 The Order Artiodactyla: The Even-toed Ungulates

The Artiodactyla represent a radiation toward ever-increasing digitigrade locomotion and utilization of plant structural carbohydrate as an energy source. The original form of the dentition was 3/3, 1/1, 4/4, and 3/3, which remains little modified in the Suiformes. In the ruminants there was an early trend toward the loss of the upper incisors and canines, while the lower canine was formed into an incisiform canine. In addition, there has been a persistent trend in selection for a cursorial mode of life, with the consequent evolution of digitigrade locomotion. The third and fourth toes are retained, and the axis of weight is borne between them. All living Artiodactyla are characterized by a peculiar tarsal modification where the astragalus exhibits a "double pulley" conformation, allowing controlled dorsoventral flexion and extension around the joints at either end. The limbs tend to move parallel to the body, and the clavicle is lost. These animals are in the main adapted for either browsing, grazing, or a mixed strategy, and all ruminants pull leaves and grass into the mouth with the tongue. The teeth show a tendency toward a selenodont cusp form, with a loss of canines and upper incisors. There is a persistent trend within the evolutionary history of the group for the stomach to become chambered, so that bacterial symbionts may break down the cellulose in the stomach rather than in the cecum (see Janis 1976). All stages toward the evolution of a grazing form with ruminant digestion may be demonstrated if we examine the families in a sequence from those exhibiting the most conservative characters to those with the most derived characters.

In the Suidae and Tayassuidae multiple births are the rule, but the Hippopotamidae and other families typically bear a single young. Twinning may occur regularly in certain species of the Cervidae, and the water deer, *Hydropotes,* is exceptional with a true litter of three to four. The young of the artiodactylans are typically born in a rather precocial state, with eyes open, natal pelage, and considerable locomotor ability.

The classification of the order is diagramed in figure 32.

15.8.1 The Suborder Suiformes: The Hippopotamus and Swine

The living Suiformes are nonruminants. The stomach is simple in the family Suidae but complex in the family Hippopotamidae, and some digestive fermentation takes place in the forestomach of the hippopotamus. The Suidae show a range of dental adaptations from a bunodont dentition in *Sus,* which is an omnivore, to a lophodont dentition in *Phacochoerus,* which grazes in addition to digging for roots and

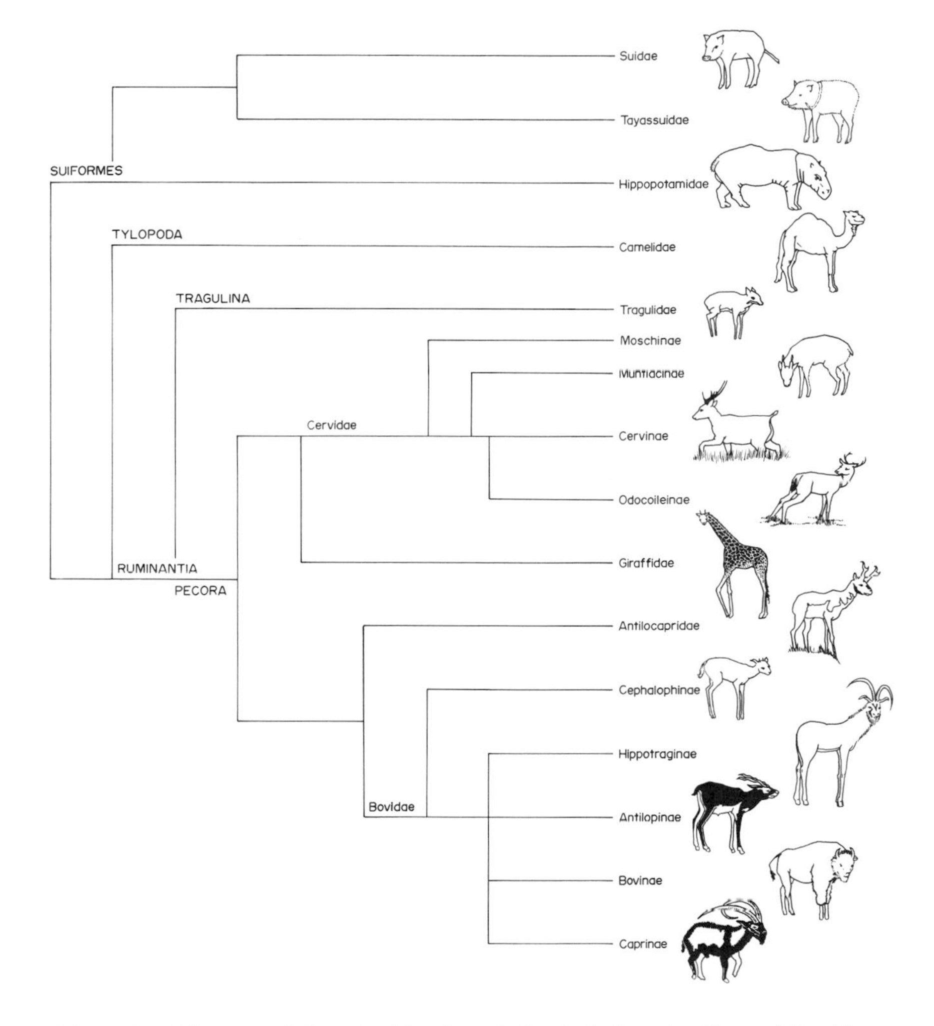

Figure 32. Diagram of the classification of the Artiodactyla. The subfamilies Cervinae and Odocoileinae correspond to the Plesiometacarpalia and Telometacarpalia, respectively.

tubers. The upper canine tooth is highly developed in the family Suidae and is used as an offensive and defensive weapon. The genus *Sus* is distributed widely in Europe and Asia but is absent from Africa south of the Sahara. Three different genera reflecting varying adaptations to open-country foraging are represented in Africa, including *Potomochoerus,* the river hog; *Hylochoerus,* the giant forest hog; and *Phacochoerus,* the warthog, which is the most completely adapted to savanna habitats. The Celebes Islands have evolved their own unique form of suid, the *Babyrousa*.

Sus scrofa is one of the best studied species both in the Old World and in North America, where it has been introduced (Grundlach 1968) (see plate 41). The adult males tend to move alone unless in association with estrous females. The sow typically constructs a nest consisting of an earth hollow lined with leaves. The leaf pile may be rather large, thus concealing the young. When the newborn are about 10 days to 2 weeks old, they begin to accompany the female in her foraging efforts. Females with young of similar age will form sounders that forage as

Plate 41. *Sus scrofa,* the wild pig. This young boar typifies the body plan of the Suidae. Omnivorous dietary habits and enlarged canines confer an ability to exploit a wide range of habitats and an active, offensive antipredator behavior pattern. (Courtesy of C. M. Hladik.)

a unit and actively repel the attacks of predators. The females in a sounder may be related. The size of the sounder appears to be a function of the carrying capacity of the habitat and the mobility of the swine. Great variation in the size of social groupings can be noted between different habitats (Eisenberg and Lockhart 1972).

In the Neotropics and Central America, a second family of suiforms occupies the omnivore foraging niche; this is the family Tayassuidae. The curve of the canine in the Tayassuidae is not so pronounced, and these animals typically have a rump gland that is used in social communication. Three species are found in part sympatric: *Catagonus wagneri,* the Chacoan peccary; *Tayassu pecari,* the white-lipped peccary; and *Tayassu tajacu,* the collared peccary. *Catagonus* was known only from fossil remains until living specimens were identified in Paraguay by Ralph Wetzel and his associates (Wetzel 1977). All species give birth to precocial young.

The three species of peccary appear to utilize different foraging strategies and exhibit dramatic contrasts in social structure. *Tayassu pecari* forms large herds of more than a hundred animals that move in seasonal patterns. This may be an adaptation for feeding on the mast fruiting palms of the Neotropics. It is paralleled by the feeding strategies of *Sus barbatus* in the Paleotropics. *T. tajacu* exhibits a lability in foraging strategy that appears to be correlated with habitat type. Band size is smaller in the Panamanian forests but may be much larger in the llanos or savanna habitat (see Kiltie 1980). Some further references on the species' biology are included in table 30.

The family Hippopotamidae exhibits a semiaquatic syndrome. The most primitive genus is the pygmy hippopotamus, *Choeropsis,* still existing in Liberia and adjacent portions of central West Africa. *Hippopotamus* was found in all major river systems of equatorial Africa. Whereas *Choeropsis* appears to be semisolitary in its habits, *Hippopotamus* is very social. The females and young form large herds that rest in shallow water or on sandbars. Males may form a distinct subgroup (Verheyen 1954). Adult males will establish territories along streams and will be totally dominant to other males who reside within such a territory. Territory holding males have priority access to estrous females within their territory (Klingel 1979).

Hippopotamus is a grazer and comes ashore at night to feed. The dietary habits of *Choeropsis* are less well understood, but one assumes that it uses a more labile feeding strategy (see table 30 for other references).

15.8.2 The Suborder Tylopoda: The Camels and Llamas

The suborder Tylopoda includes one living family, the Camelidae. In the camels a cannon bone is formed from a fusion of the third and fourth metacarpals. The orbital ring of the skull is complete, and the forestomach is chambered and digestive fermentation is highly developed. Upper canines and incisors are still present. Males have no horns or antlers. Although they are digitigrade, there is no complete fusion of most metacarpals or metatarsals. The family Camelidae is distributed in South America, North Africa, and Central Asia. The South American forms are derivatives from North American invasions during the Pliocene. The Asiatic and African genus is *Camelus,* with two species: *C. dromedarius (=arabicus)* and *C. bactrianus*. The South American genus is *Lama,* with two wild species: *L. guanicoe* and *Vicugna vicugna*. The camelids were domesticated in both the Old World and the Andean portions of the New World.

The vicuña was studied by Koford (1957) and Franklin (1974). Adult reproducing males defend a territory where a group of females lives. The territory is semipermanent, and the reproductive unit feeds within its boundaries. The territory is demarcated in part by dung heaps. The migratory camels (*Camelus*) do not show a formal territorial system, but during the rut the males fight vigorously to defend a group of females against rival breeding males (Gauthier-Pilters 1959).*

15.8.3 The Family Tragulidae: The Mouse Deer

The suborder Ruminantia is divisible into two infraorders: the infraorder Tragulina and the infraorder Pecora (see fig. 32). The Tragulina include one living family, the Tragulidae or mouse deer, characterized by no horns or antlers and by a curious aponeurosis of muscles that ossify to form a plate to which the sacral vertebrae are attached. This latter feature is unique within the Artiodactyla. The second and fifth digits are still evident, though the animal bears its weight solely on the third and fourth. The upper incisors are gone, but upper canines are present that are used in offensive fighting. There are two genera: *Tragulus* of Southeast Asia and *Hyemoschus* of Africa. These two genera represent the most conservative living Ruminantia (see plate 42).

The behavior of *Tragulus* has been described by Ralls, Barasch, and Minkowski (1975), and Dubost (1975) presents a detailed monograph on *Hyemoschus aquaticus*. Dubost points out that the nocturnal *Hyemoschus* is very conservative in its behavior patterns, showing some similarity with the Suidae. Ralls et al. point out the role of olfaction in the integration of social behavior and the males' use of the canines in intraspecific combat.

These small ruminants do not form cohesive social structures but rely on their small size and cursorial adaptations to avoid being detected and caught by predators. They typify the small, forest-adapted cursorial mammal syndrome discussed for *Dasyprocta* (sect. 11.7) and reviewed by Dubost (1968) and Eisenberg and McKay (1974).

15.8.4 The Family Cervidae: The Deer

The infraorder Pecora is divisible into the Cervoidea, Giraffoidea, and Bovoidea. The Cervoidea contain one living family, the Cervidae or deer family (see fig. 32). The Cervidae are characterized by antlers. Generally only the males bear antlers, but there are exceptions that will be discussed below. Antlers, in contradistinction to horns, are shed and regrown each year. During the growth phase, antlers are highly vascular and are covered with epidermis—the "velvet" stage. The

Plate 42. *Tragulus meminna,* the Indian chevrotain or mouse deer. This small nocturnal ungulate has a color pattern that conceals it as it lies in protected places during the day. (Courtesy of C. M. Hladik.)

second and fifth toes are vastly reduced when compared with those of the Tragulidae. The skull of a cervid typically exhibits lacrymal depressions anterior to the eye that are occupied by the preorbital glands in the living animal. Four subfamilies may be characterized: the Moschinae, the Muntiacinae, the Odocoileinae, and the Cervinae.

The subfamily Moschinae includes only one living genus, *Moschus,* the musk deer. *Moschus* has no antlers, but the upper canines in the male are greatly enlarged. Musk deer are distributed in China and the Himalayas.

The Muntiacinae, or muntjacs, exhibit a pronounced sexual dimorphism. The males have both large canines and antlers. There are typically two genera: *Elaphodus* of China, and *Muntiacus* of Southeast Asia and the Sunda Islands.

The remaining genera of deer may be divided into two groups: the plesiometacarpalians and the telometacarpalians. The plesiometacarpalians (or Cervinae) are Old World except for the genus *Cervus,* which has recently invaded North America across the Bering land bridge. In the plesiometacarpalians, the first and fifth metacarpals may still be noted at the extreme proximal end of the fused third and fourth metacarpals. In the telometacarpalians (or Odocoileinae), the second and fifth metacarpals may still be noted at the extreme distal end of the fused third and fourth metacarpals. The telometacarpalians are distributed in both the Old World and the New. Most of the New World deer except *Cervus canadensis* belong to the telometacarpalian subgrouping.*

The subfamily Odocoileinae is divisible into five tribes. The tribe Alcini is Holarctic in distribution and includes only one genus, *Alces*, the moose or European elk. The tribe Odocoileini includes the New World genera of North and South America: *Odocoileus, Mazama, Pudu, Blastoceros, Ozotoceras,* and *Hippocamelus*. This tribe has shown extensive adaptive radiation in the Neotropics since the Pliocene. Apparently this adaptive radiation has been in part a response to the extinction of many browsing ungulate forms that disappeared from the South American fauna at the completion of the isthmian land bridge.

The tribe Rangiferini is Holarctic in its distribution, including *Rangifer,* the reindeer or caribou. This is the only species of cervid in which both sexes bear antlers.

The tribe Capreolini is confined to the Palearctic and includes one genus, *Capreolus,* the roe deer. The tribe Hydropotini is the most conservative member of the family Cervidae. It consists of one genus, *Hydropotes,* currently found in China. The male does not possess antlers, and the canines are greatly enlarged.

The subfamily Cervinae, as I noted before, is in the main confined to the Old World except for the genus *Cervus,* which also includes North America in its range. The common recognized genera include *Dama,* the fallow deer; *Axis,* the axis deer; *Elaphurus,* Pere David's deer or the milu; and *Rusa,* the sambars of South Asia (see plate 43).

15.8.5 The Socioecology of the Cervidae

The Cervidae, though widely distributed in the Holarctic, show their highest diversity in the tropics of two widely separated zoogeographical realms, Southeast Asia and South America. All cervids in the areas of maximum species diversity show an enduring trend for exploiting forested habitats, and most cervids are browsers or mixed browsers and grazers. Some Cervidae have adapted to open habitats, such as the barren-ground caribou and reindeer (*Rangifer*) in the arctic tundras. The axis deer (*Axis axis*) has adapted to open, parklike habitats that show a seasonal aridity in the Indian subcontinent. *Cervus elaphus* and *C. canadensis* in many parts of their range are adapted to open parkland or high mountain meadows. The fallow deer (*Dama dama*) exploits open habitats in the Middle East. The swamp deer of the Indian subcontinent (*Cervus duvauceli*) and the South American marsh deer (*Blastocerus dichotomus*), and very probably Pere David's deer (*Elaphurus davidianus*) in the wild state, occupy marshy habitats that can be considered open (Schaller and Hamer 1978). With these notable exceptions, however, most of the species of cervids are adapted to closed forested habitats and are predominantly browsers.

Jarman (1974) established correlations between feeding strategy and social structure from an analysis of the African Bovidae. On this basis one might predict that those deer species occupying more open habitats and tending toward a grazing mode of habitat exploitation would show on the average larger herd sizes than the forest-adapted browsing species. This indeed seems to be the case. Larger groups seem to be

Plate 43. *Cervus (Rusa) unicolor.* This young sambar stag with antlers in the velvet illustrates the body plan and offensive male weaponry of the Cervidae. (Courtesy of C. M. Hladik.)

favored with the occupancy of an open-habitat, semigrazing niche, perhaps from a multiplicity of factors, but surely herd formation has been favored during selection as a form of antipredator behavior. The carrying capacity of open grazing habitats is often higher than that of forested habitats, again favoring group formation. Seasonal migrations from summer ranges to winter ranges in the temperate zone promote herding and movement as a group, well exemplified by the reindeer and caribou (Bergerud 1974; Espmark 1964).

Many of the cervid species adapted to forested habitats are small and eat fruit and seeds as well as browse. These include the Neotropical genera *Mazama* and *Pudu* as well as the Southeast Asian species of *Muntiacus*. In the Palearctic region, the small roe deer, *Capreolus,* occupies a browsing niche, but there is no ecologically equivalent cervid in the Nearctic. These small deer tend to have fixed home ranges and are well spaced. There is a suggestion of territoriality, and the problem has been analyzed in some detail for *Capreolus* (Kurt 1968) and for *Muntiacus* (Barrette 1975, 1977*a,b*). The social behavior manifested by these smaller Cervidae adapted to the forest, then, is in conformity with the analysis for the duikers of the family Bovidae as portrayed by Jarman (1974).

There are certain exceptions, however, concerning the social structure or herd size of deer adapted to open marshy habitats, since Schaller (unpublished; Schaller and Hamer 1978) offers data that *Blastocerus* (the marsh deer) lives in very small herds. The reason for this, however, may be related to historical hunting pressures or to the circumstance that the habitat of *Blastocerus* is subjected to severe annual flooding. At this time the marsh deer population may utilize traditional high-ground areas that have a very limited carrying capacity so that larger groups cannot be formed. Of course, much further research will be necessary to confirm this, but the situation is reminiscent of the small group sizes of *Odocoileus virginianus* in Venezuela, where the llanos may be flooded for several months of the year. Here again the deer retreat to high ground, and the high ground usually has a very limited carrying capacity. Larger herds are formed in *Odocoileus virginianus* in the chaparral country of southern Texas, which is not subjected to such inundation (Hirth 1977).

The rutting behavior of the Cervidae has been reviewed by De Vos, Broex, and Geist (1967). In the European red deer, *Cervus elaphus*, the North American wapiti, *Cervus canadensis*, and the reindeer, *Rangifer rangifer*, males tend to gather a harem and maintain it for exclusive mating through the rutting period (Espmark 1964; Lincoln 1971; Struhsaker 1967*c*). The caribou may show harem formation or a single tending bond between a male and a female during the rut, depending on geographical location (Bergerud 1974). The red deer, *Cervus elaphus*, may also show variations in the form of the rutting behavior (Burckhardt 1958). These variations in the capacity to form harems arc probably due to several factors. A harem is profitable to a male only if he can adequately defend it during the period when all the females will come into estrus. Thus synchrony of estrus is very important. Should females come into estrus over a prolonged period, successful harem defense will become less and less profitable. Also, a harem is possible only in those species of cervids where the females show a consistent herding tendency; thus species adapted to semiopen or open habitats with a high carrying capacity usually show larger groupings of females. The propensity to form harems is a function of the degree to which females stay together in groups and the degree to which some form of estrus synchrony is shown. Where females do not show estrus synchrony, males may develop a dominance order among themselves to determine priority of access to females when they come into estrus. This seems to be the situation in the Axis deer, *Axis axis* (Eisenberg and Lockhart 1972). A similar situation may be shown in the barasingha, *Cervus duvauceli* (Schaller 1967).

It appears, then, that some of the principles developed by Jarman (1974) for the Bovidae can be applied to the Cervidae, but much more fieldwork is necessary to formulate a clear picture of intraspecific variation.

15.8.6 The Family Giraffidae: The Giraffe and Okapi

Genera of the Giraffidae are characterized by extensions of the frontal bone that are covered throughout life by skin, forming hornlike organs. The current distribution of the family is in Africa. There are two genera,

Okapia and *Giraffa*. The okapi is confined to central African equatorial forest; *Giraffa* is adapted for open savanna country and is specialized in the length of its limbs and neck for browsing on trees (see plate 44).

The giraffes and okapis are specialized as browsers, feeding on the leaves of shrubs or, in the case of giraffes, low trees. The giraffe has invaded open country savannas that support tree growth, the ultimate adaptation for folivority in an open habitat. Herd structure is not formally organized in giraffes, but a complex social organization is developed by these large, long-lived mammals (see Dagg and Foster 1976, for a review of ecology and behavior).

15.8.7 The Family Antilocapridae: The Pronghorn Antelope

When we turn to the remaining families, we see a dominant tendency toward grazing and a versatility in utilization of habitat types. The superfamily Bovoidea includes two families, the Antilocapridae and the Bovidae. The family Antilocapridae contains one living genus, *Antilocapra americana,* which is confined to the North American conti-

Plate 44. *Giraffa camelopardalis,* mother and young. The giraffe young is able to stand within an hour after its birth and typifies the precocial infant syndrome of artiodactylans. (NZP archives.)

nent and is a Pleistocene relict. Antilocapra has horns in the male, formed by a bony core developing from the frontal bone on the skull that is covered with an epidermal sheath. The sheath is shed annually. The pronghorn is a mixed grazer and browser. Together with the bison (*Bison bison*), the pronghorn was one of the two dominant ungulate species of the high plains of western North America. Pronghorns show many behavioral convergences with the African bovine genus *Gazella*. The adult males defend exclusive breeding territories and attempt to monopolize female groups (Bromley and Kitchen 1974).

15.8.8 The Family Bovidae: Wild Cattle and Antelopes

The family Bovidae was in recent times distributed in North America, Europe, Asia, and Africa and excluded from South America and Australia. Currently, domestic livestock has been widely introduced on all the major continents. The males and often the females possess true horns that consist of a bony process of the frontal bone covered with an epidermal sheath. Usually this horn structure grows throughout life and is not shed. The bovids demonstrate a trend toward the exploitation of open habitats and profound adaptations for grazing and feeding on grasses. Although a diversity of feeding styles may be demonstrated when the group is closely analyzed, the overall adaptive trend has been for the exploitation of open habitats. The four-chambered stomach is beautifully adapted to ferment structural carbohydrate by utilizing microbial symbionts.

It is possible to divide the family Bovidae into five subfamilies (see fig. 32). The most conservative subfamily is the Cephalophinae. This includes the duikers, typified by the genera *Cephalophus* and *Philantomba*. This subfamily currently is confined to Africa south of the Sahara.

The Hippotraginae are divisible into three tribes: (1) the Reduncini, kobs and reedbucks; (2) the Hippotragini, including the sable and roan antelopes, oryx, and addax; and (3) the Alcelaphini, including the hartebeests, *Alcelaphus*, the topis, *Damaliscus;* and the wildebeests, *Connochaetes*. This subfamily is confined to Africa.

The Antilopinae are divisible into two tribes: the Neotragini and the Antilopini. The neotragines are all small, mostly browsers, and are adapted to a variety of habitat types ranging from the klipspringer, *Oreotragus*, which inhabits rocky habitats, to the dikdik, *Madoqua*, adapted to semiarid brush. The Antilopini include Asian and African forms: the black buck, *Antilope*, of India, and the widespread genus *Gazella*, found over Central Asia and across North Africa and south to the Cape.

The subfamily Bovinae includes three tribes: (1) the African Strepsicerotini includes the kudu, *Strepsiceros*, and the eland, *Taurotragus;* (2) the tribe Boselaphini includes *Boselaphus*, the blue buck, and *Tetracerus*, the four-horned antelope, both confined to the Indian subcontinent. The third tribe, or Bovini, includes Asiatic, African, and Holarctic forms. The genus *Bison* is distributed in both Europe and North America. The genus *Bos* is distributed in Europe and Central

Asia. The closely related genus *Bibos* includes the Southeast Asian wild cattle. *Bubalus* is the Central Asian water buffalo, and a derivative of it is the genus *Anoa*, or pygmy buffalo, confined to the Celebes and Mindanao. *Syncerus* is the African or Cape water buffalo (see plate 45).

The last subfamily of the Artiodactyla is the subfamily Caprinae, which is divisible into four tribes. (1) The tribe Saigini includes two Asiatic genera, *Saiga* and *Pantholops*. (2) The tribe Rupricaprini includes the so-called goat antelopes, with the genera *Rupicapra* (chamois) occupying the European montane areas. *Oreamnos*, the mountain goat, inhabits the mountains of western North America, and the genera *Capricornis* and *Naemorhedus*, the serows and gorals, respectively, are Central and Southeast Asian in their distribution. (3) The tribe Ovibovini includes two genera: *Ovibos*, the musk ox, is Holarctic, and *Budorcas*, the takin, is found in the Tibetan plateau and adjacent regions. (4) The final tribe is the Caprini, with three genera: *Pseudois*, the bharal; *Capra*, the goats; and *Ovis*, the sheep.

The ecology and behavior of some representatives for almost all subfamilies and tribes have been competently monographed (see table 30). The recent review by Leuthold (1977) synopsizes our knowledge of the African Artiodactyla. The Bovidae represent the most successful recent adaptation to open habitats by large terrestrial herbivores. Predators have acted as a selective force in forming bovids into mammals capable of developing high flight speeds and various offensive antipredator behaviors in defending their young. The precocial young may adopt either a "hider" or a "follower" strategy in their neonatal phase (Lent 1974). Followers must be so well developed at birth that they can run fast enough to keep up with their mothers.

Since female artiodactylans often form herds, there can be strong sexual selection pressures leading to male combat and active defense of estrous females. The variation in horn size and shape among the

Plate 45. *Bubalus bubalis*, the Asian water buffalo. Herd of females facing a potential predator. This defensive formation with the young animals behind or between adults typifies one antipredator option by large bovids. (Courtesy of G. M. McKay.)

Table 30 Recent Literature on the Artiodactyla

Taxon	Natural History and Ecology	Behavior	Reproduction and Ontogeny
Suidae	Ewer 1958	—	—
Sus scrofa	—	—	Grundlach 1968
Potomochoerus porcus	Sowls and Phelps 1968	—	—
Phacochoerus aethiopicus	—	Frädrich 1965	—
Tayassuidae	Enders 1935	—	—
Tayassu tajacu	Neal 1959; Knipe 1957	Schmidt 1976; Sowls 1974	—
Hippopotamidae			
Hippopotamus amphibius	—	—	Verheyen 1954
Tragulidae			
Tragulus javanicus	—	—	Cadigan 1972
T. napu	—	Ralls, Barasch, and Minkowski 1975	Davis 1965
Hyemoschus aquaticus	Dubost 1975, 1978	Dubost 1975	—
Camelidae	—	Pilters 1956	Schmidt 1973
Camelus arabicus	Gauthier-Pilters 1974	Gauthier-Pilters 1959	—
Vicugna vicugna	Koford 1957	Franklin 1974	—
Antilocapridae			
Antilocapra americana	Einarsen 1948	Bromley 1969; Bromley and Kitchen 1974	Autenrieth and Fichter 1975
Cervidae	Taylor 1956	Geist 1966*a*; De Vos, Broex, and Geist 1967	Lent 1974
Moschus moschiferus	—	Frädrich 1966*b*	—
Capreolus capreolus	Kurt 1968	Hennig 1962	Espmark 1969
Muntiacus sp.	Barrette 1975	Barrette 1977*a,b*; Dubost 1970	—
Odocoileus hemionus	Linsdale and Tomich 1953	Müller-Schwarze 1968	—
O. virginianus	Hirth 1977	—	—
Pudu pudu	—	Hick 1967	—
Mazama americana	—	—	Thomas 1975
Alces alces	Loisa and Pulliainen 1968	Geist 1963	—
Cervus canadensis	McCullough 1969; Murie 1951	Struhsaker 1967*c*	—
C. elaphus	Darling 1937	Bützler 1974	Lincoln 1972
Rangifer sp.	Moisan 1958; Bergerud 1974	Espmark 1964; Pruitt 1960	Lent 1965, 1966
Giraffidae	Dagg and Foster 1976	—	—
Giraffa camelopardalis	Foster 1966	Backhaus 1961; Innis 1958	—
Okapia johnstoni	—	Walther 1960	—
Bovidae	—	—	Lent 1974
Cephalophinae	Leuthold 1977	—	—
Philantomba maxwelli	Aeschlimann 1963	Ralls 1969, 1973, 1975	—
Cephalophus doria	Kuhn 1966	—	—
C. dorsalis	Kurt 1963	—	—
Silvicapra grimmia	Wilson and Clarke 1962	—	—
Hippotraginae	Leuthold 1977	—	—
Kobus sp.	De Vos and Dowsett 1966	Kiley-Worthington 1965	Leuthold 1967
Redunca arundinium	Jungius 1971	Jungius 1970	—
Adenota kob	—	Buechner and Schloeth 1965; Leuthold 1966	Buechner, Morrison, and Leuthold 1965

Table 30—*Continued*

Taxon	Natural History and Ecology	Behavior	Reproduction and Ontogeny
Oryx sp.	—	Walther 1965	—
Damaliscus sp.	Du Plessis 1972	—	—
Alcelaphus sp.	—	Gosling 1974	Gosling 1969
Connochaetes taurinus	Estes 1966	Estes 1966	—
Antilopinae	Leuthold 1977	—	—
Neotragus batesi	Feer 1979	—	—
Litocranius walleri	Leuthold 1971	Schomber 1963	—
Aepyceros melampus	Jarman and Jarman 1973	Schenkel 1966; Leuthold 1970	—
Gazella sp.	Walther 1968, 1969	Estes 1967	Walther 1964
Bovinae			
Tragelaphus sp.	Lobao Tello and Van Gelder 1975	—	Simpson 1968
Bubalus sp.	Tullock 1969; Ullrich 1966	—	—
Novibos sauveli	Wharton 1957	—	—
Bison bison	Meagher 1973	McHugh 1958; Lott 1974	—
Syncerus caffer	Sinclair 1977	—	—
Bos javanicus	Halder 1976	—	—
B. taurus	Schloeth 1961	Schloeth 1961	—
Caprinae			
Saiga tatarica	Bannikov 1963	—	—
Oreamnos americanus	Geist 1964	—	—
Ovibos moschotus	Tener 1965	—	—
Rupicapra rupicapra	Knaus 1960	Krämer 1969	—
Capra sp.	Schaller 1977	Schaffer and Reed 1972	—
Pseudois nayaur	Schaller 1977	—	—
Ovis ammon	Pfeffer 1967	—	—
O. aries	—	—	Banks 1964
O. canadensis	Geist 1971; Blood 1963	Geist 1968*a,b*	Berger 1979
O. dalli	Geist 1971	—	—

Note: For general reviews, see: Leuthold 1977; Geist and Walther 1974; Fraser 1968; Lamprey 1963, 1964; and Walther 1966.

Bovidae as a family is pronounced. Horns have evolved as defensive weapons against predators as well as offensive and defensive weapons in intraspecific fighting, especially among males. Often the possession of horns has been under such intense sexual selection in the male that the female is nearly hornless by comparison. To some extent the shape of the horns reflects the fighting style adopted during intraspecific combat. The whole problem has been reviewed by Geist (1966*b*).

15.8.9 Artiodactylan Socioecology in Perspective

The evolution of ungulate mating systems and the social groupings shown by various species throughout their life cycles was reviewed by Eisenberg in 1966. Since that time there is much new information concerning the adaptive nature of social structure in ungulate species. Schaller (1977) has added a great deal to our knowledge concerning the mountain goats and sheep of Central Asia. Geist (1971) synopsized the evolution of sheep society in North America. Leuthold (1977) has summarized our knowledge of ethology and behavioral ecology for African ungulates. The Calgary Symposium volume edited by Geist and Walther (1974) includes an accurate summary of the state of the art in 1971.

In 1974 Geist attempted to develop a set of theoretical postulates outlining the relationship of social evolution to ecological adaptation in the ungulates. Estes (1974) and Jarman (1974) both developed schemes to relate various phenotypic characters and feeding adaptations of the Bovidae to their social organization and mating systems. Jarman (1974) attempted to relate body size, habitat selection, and feeding style to the form of the social organization that predominates. In addition, various aspects of social organization, including the group stability, home range size, and antipredator behavior, were considered as functional correlates of feeding strategies. The mating systems in Jarman's analysis were cast into the entire ecological perspective. In Estes's analysis, mating systems became a paramount focus. Therefore the two papers are not entirely convergent but represent a parallelism in thought, with different emphases on aspects of sociality as shown by the Bovidae. Both papers point out that small ungulates that are adapted for highly selective feeding on browse and fruits and that can defend their food resources against conspecifics will tend to exist in very small social groupings, with adults moving rather independently. Such a system gives the appearance of being a "solitary" organization, but in fact a male's home range is usually coincident with those of one or two females. In some species monogamy is demonstrable (e.g., *Madoqua*); in others, polygyny (e.g., *Neotragus*).

The larger grazing herbivores adapted to open savanna generally exist in large herds that can repel predators through communal defense mechanisms. Although Jarman defines five classes of sociality, I have presented only the two extremes; the other three categories are intermediate. Whereas Jarman's analysis seems to work well if one stays within one family of the ungulates, such as the Bovidae, it does not work so well if one treats ungulates as a whole. In fact, even within the family Bovidae, Schaller found some difficulties in relating the social systems described for the Caprini to the major environmental variables that were important to Jarman's analysis of the African Bovidae. This should not be surprising, since the Bovinae are rather uniform in body plan, physiology, and feeding adaptations. Furthermore, the African bovids are adapted to a continuum of varying habitat types from forest to open savanna. Many other features complicate the situation when one tries to apply the system to an ungulate community in montane country or when one attempts to lump distantly related ungulate taxa, such as rhinoceroses, elephants, and bovids, in an attempt to derive the same correlations.

One confounding feature in a correlational system is the difficulty of determining the modal social system or modal mating system. Indeed, variation in group size and group composition can be shown within a species over a geographical range, and there may also be variations in mating patterns. Leuthold (1966) demonstrated considerable variation in the mating system of the Uganda kob, *Adenota kob thomasi*. Under modal conditions, this species may form a breeding lek among males. On the other hand, in other parts of its range, the mating system may not approximate a lek but may more nearly reflect a standard territorial system such as that shown by *Gazella thompsoni*. Variations in the structure of social groupings can be found in almost any species. One of the more remarkable studies by Hirth (1977) demonstrates that the modal social grouping changes in the white-tailed deer, *Odocoileus virginianus*, when one progresses from a forested habitat to an open grassland chaparral habitat, such as in southern Texas. Indeed, in the Texas chaparral savanna an individual male may attempt to monopolize several females in a manner reminiscent of harem formation, which is the modal mating system for the North American wapiti, *Cervus canadensis*.

In spite of such difficulties in determining modal social systems and mating patterns, for many species there is considerable uniformity over a gradient of habitat types in the form of social organization and the mating system exhibited. It is the analysis of variations, however, that will eventually be the key to determining how social organizations evolve through natural selection.

16 Trends in the Adaptation and Distribution of the Mammalia

16.1 Feeding Niches and Their Constraints

With this overview of the adaptive radiations of mammals, we can see again and again the evolution through time of similar feeding strategies. In the early Cretaceous, mammals had invaded feeding niches that involved the pursuit and capture of arthropods and perhaps small vertebrates. Judging from their dentition, the multituberculates had specialized for feeding on plant parts, probably seeds and fruits as well as herbs. At the time of the metatherian-eutherian dichotomy, the ancestral form undoubtedly was to some extent scansorial and an insectivore, very probably adapted to nocturnal foraging strategies. Subsequently the multituberculates died out on the contiguous continental land masses, and the only likely descendants of this line, the Monotremata, are now highly specialized feeders on the subcontinent of Australia.

The marsupials radiated into a variety of niches in South America and subsequently in Australia. In South America, starting from an insectivore/omnivore stem form, carnivorous species evolved, and very probably herbivorous or granivorous forms in the southern arid regions, exemplified by *Argyrolagus* (Simpson 1970). In Australia the marsupial radiation comprised a total complement of feeding strategies including insectivores, omnivores, carnivores, and various herbivorous specializations such as grazers, browsers, and arboreal folivores. The eutherians demonstrate repeated trends toward convergent niche occupancy. An early derivative of the insectivore feeding adaptation began to develop specializations for feeding on the social insects, and the potential myrmecophagous niche was filled by morphologically conservative forms independently derived in a convergent fashion throughout the Old World and New World tropics (see sect. 11.7). Insectivore/carnivore forms specialized repeatedly in different lines, persisting today in the form of the contemporary Insectivora and Carnivora. Herbivore strategies including both browsing and grazing were developed in several different lineages and persist today in the Artiodactyla, Perissodactyla, Rodentia, Hyracoidea, and Lagomorpha.

Adaptation for different strategies has had profound consequences in terms of social organization, home range size, and population density. Figure 33 shows how climate influences hatitat type, which in turn determines the range of opportunities for feeding adaptations. The trophic strategy can limit the group size, which in turn has an influence on the form of the social structure. The body size of the animal is in

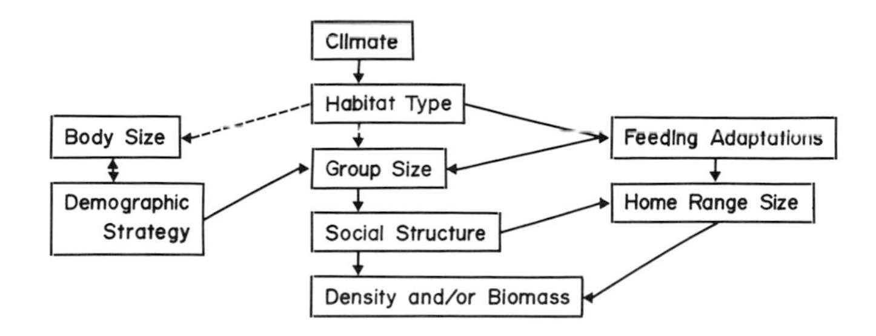

Figure 33. Diagram showing the interrelationship of environmental and organismic factors in determining social structure and density.

part determined by its feeding adaptations, and size can also influence the demographic strategy and the mating system.

Ultimately the feeding adaptation (within the constraints of the environmental carrying capacity) sets the limits on the home range size for the social group or individual (see table 31). This limit in turn determines the density or biomass that the species exhibits for a defined habitat.[1] For example, if we consider a group of carnivores and anteaters, figure 34 clearly demonstrates that the density is inversely related to home range size. To break down the correlations further, it is demonstrable that the individual weight or group weight of the species in question, if it is social, is directly related to the home range size (McNab 1963); but home range size is also a function of carrying capacity. Since plant productivity sets the ultimate limit on the biomass of animals, and since photosynthesis depends on sunlight, it follows that overall the carrying capacity will tend to be higher in temperate and tropical latitudes and lower in arctic and boreal latitudes.

The carrying capacity for herbivores or primary consumers is always higher than that for carnivores or secondary consumers. These same correlations were illustrated before in our consideration of the Edentata and the Marsupialia, two ancient radiations extant in South America and in Australia. The Edentata that are herbivores have the smallest home range for a given body weight; the omnivores and myrmecophages have larger home ranges. The Australian marsupials show the same tendency, with herbivores having smaller home ranges for individual or group weight than do insectivores or carnivores of comparable weight (see figs. 11 and 16). Obviously, then, the form of the social organization a species can show is a compromise among all the various factors that may influence it, including demographic strategy, carrying capacity, structure of the habitat, and feeding adaptation. The influences of these variables as determinants of social structure for the mammalian species will be considered in chapter 33.

1. Home range size should ultimately be related to the energetics of the population. If field data were available for the annual energy budgets of enough species, and if the carrying capacity for the habitat of the species were measurable, then annual home range area should correlate with annual calorie consumption in habitats of equivalent productivity.

Table 31 Home Range Data for Nonvolant Terrestrial Mammals

Taxon	Head and Body Length (mm)	Home Range (ha)	
		Core	Extended
Insectivora			
Soricidae			
Sorex araneus	79.3	.04–.06	
		.05	
		—	.28
S. minutus	57.8	.05–.19	
S. vagrans	60	—	3.0
Talpidae			
Talpa europaea	139	.04	
Rodentia			
Aplodontidae			
Aplodontia rufa	355	.20	1.25
Sciuridae			
Citellus leucurus			6.0
Eutamias amoenus		♂1.3–1.57	
		♀.84–1.0	
Funambulus pennanti		♂.21	
		♀.15	
Spermophilus richardsoni		.32	
S. tridecemlineatus		♂4.7	
		♀1.4	
Tamias striatus		.3–3.5	
Tamiasciurus hudsonicus	230	♂2.4–♀1.9	
Heteromyidae			
Dipodomys merriami	107	1.92	
D. microps	112	♂.93–♀.83	
D. ordii		♂1.4–♀1.3	
Heteromys desmarestianus		.56	
Liomys adspersus		.57	
Perognathus formosus	77	♂.36–♀.24	
P. longimembris	64.5	.4–.51	
Geomyidae			
Thomomys bottae		.02	
Muridae			
Apodemus flavicollis	115	♂2.2–♀1.0	
Cricetidae			
Microtinae			
Clethrionomys glareolus	108	♂.82–♀.71	
C. rufocanus		.06	.42
C. rutillus		.08	.43
Microtus oeconomus		.09	.22
M. oregoni		♂.53–♀.35	
M. pennsylvanicus		.75–1.2	
Gerbillinae			
Meriones hurrianae		.009–.01?	
Tatera brantsi		.49–.19	
T. indica		.19	
Cricetinae			
Oryzomys palustris		♂.33–♀.21	
Ochrotomys nuttalli		.26–.24	
Onychomys leucogaster	110	2.7–4.6	
O. torridus		2.7	
Peromyscus leucopus		1.26–1.82	2.80
P. maniculatus		1.66–1.89	

Table 31—*Continued*

Taxon	Head and Body Length (mm)	Home Range (ha)	
		Core	Extended
Reithrodontomys fulvescens	74	.16–.26	
Neotoma floridana		♂.266–♀.16	
Myospalax fontaniere		2.8–1.8	
Spalacidae			
Spalax		.71	
Octodontidae			
Spalacopus cyanus	150	.02	
Capromyidae			
Geocapromys ingrahami		.56	
Echimyidae			
Proechimys semispinosus	230	♂.50–♀.45	
Proboscidea			
Loxodonta africana	5,000	1,400[a] (48)	5,200[a]
Perissodactyla			
Rhinocerotidae			
Ceratotherium simum	4,000	♂170 ♀♀1,160[a] (8)	
Diceros bicornis	3,500	♂410 ♀670	♂1,580–♀1,500
Equus zebra	1,900	300–500[a] (9)	
Artiodactyla			
Suidae			
Phacochoerus aethiopicus	900	180–170	
Giraffidae			
Giraffa camelopardalis	4,000	6,200–8,500	
Bovidae			
Syncerus caffer	2,500	45,000[a] (400+)	
Tragelaphus angasi	1,425	♂130–♀40[a] (5)	♂950–♀360[a] (8)
T. scriptus	1,350	♂24–♀12	
Madoqua kirki	580	2.5–12[a] (2)	

Source: Leuthold 1977; French, et al. 1975.
[a] Group range; figure in parentheses is group size.

This brief excursion into the constraints that accompany feeding adaptations is meant to set the stage for a discussion of convergence in form and function. Feeding and substrates set limits on size and morphology. Tables 32 and 33 outline a few examples in which convergence in feeding strategy has led to a similarity in size, dentition, and digestive physiology.

16.2 The Coevolution of Species

16.2.1 Convergences in Antipredator Strategy

Body form and coloration may converge as a result of predator selection. On the other hand, there can be unique sets of solutions, such as those outlined for the Dasypodidae in section 4.1. The Perissodactyla and Artiodactyla (e.g., *Equus* and *Boselaphus*) have evolved cursorial forms in part because of selective pressure from cursorial predators. With different forms of predator pressure, the macropod marsupials evolved a bipedal ricochetal gait that has no exact counterpart in large

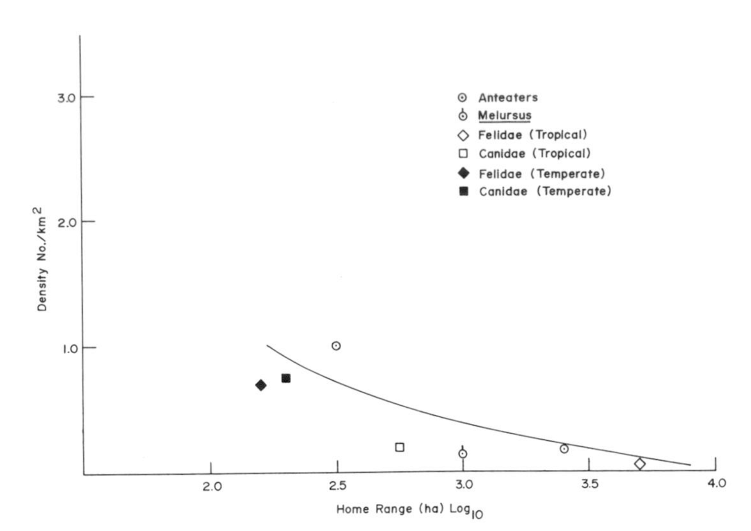

Figure 34. Plot of numerical density against mean home range size for myrmecophages and carnivores. Data are from Montgomery and Lubin 1977 and Kleiman and Eisenberg 1973.

Table 32 Nonvolant Mammalian Convergences in Feeding Strategies

Feeding Strategy	Australia	Madagascar	Southeast Asia	Africa (South of the Sahara)	South America
Arboreal folivority in tropical forests (nongliders)	*Phascolarctos cinereus*	*Lepilemur mustelinus*	*Presbytis* sp.	*Colobus* sp.	*Bradypus infuscatus*
	Phalanger maculatus	*Lemur macaca*	*Nasalis*	*Dendrohyrax*	*Choloepus hoffmanni*
		Propithecus verrauxi			
Gliding arboreal folivores	*Schoinobates volans*		*Cynocephalus*	*Anomalurus*	
			Petaurista		
Rocks and semiarid habitat folivores	*Petrogale*			*Heterohyrax*	*Kerodon rupestris*
				Procavia	
Sap- and gum-feeders[b]	*Petaurus breviceps*	*Phaner furcifer*		*Galago elegantulus*	*Cebuella pygmaea*
Nectar- and pollen-feeders[b]	*Tarsipes spenceri*[a]	*Microcebus murinus*		*Galago senegalensis*	

[a] The most specialized living, nonvolant mammal that is a pollen- and nectar-feeder.
[b] Supplemental with insects.

herbivores but is replicated in small, nocturnal, arid-adapted rodents (e.g., Dipodidae and Heteromyidae).

Convergence in form, size, and coloration is demonstrable for a series of small, cursorial mammals adapted to tropical woodlands. I refer here to the syndrome first described by Dubost (1968), amplified by Eisenberg and McKay (1974), and defined in full discussion by Rathbun (1979). Table 34 lists the genera from three orders that have

Table 33 Convergences in Fossorial Adaptation

Fossorial Herbivores

Palearctic		Nearctic		Africa (South of Sahara)		Southeast Asia		South America	
Ellobius *Prometheomys*	Microtinae	*Microtus oregoni*	Microtinae	*Tachyoryctes*	Rhizomyidae	*Cannomys*	Rhizomyidae	*Spalocopus*	Octodontidae
Myospalax	Myospalacinae	*Geomys* *Thomomys* et al.	Geomyidae	*Bathyergus* *Cryptomys* et al.	Bathyergidae			*Ctenomys*	Ctenomyidae
Spalax	Spalacidae								

Fossorial Insectivores

Holarctic		Africa		Southeast Asia		South America		Australia	
Talpa *Scapanus* *Condylura* et al.	Talpidae	*Chrysochloris* *Eremitalpa* et al.	Chrysochloridae	*Uropsilus* *Scaptonyx*	Talpidae	*Chlamyphorus* *Burmeisteria*	Dasypodidae	*Notoryctes*	Notoryctidae

Table 34 Antipredator Strategies and Morphological Convergences for Small Cursorial Tropical Forms

Asia		Africa		South America	
		Rhynchocyon	Macroscelidea		
Tragulus napu *Tragulus javanicus* *Tragulus memmina*	Tragulidae	*Hyemoschus*	Tragulidae	*Dasyprocta* *Myoprocta* *Cuniculus*	Dasyproctidae
Muntiacus	Cervidae	*Cephalophus* *Neotragus* *Madoqua*	Bovidae	*Mazama* *Pudu*	Cervidae

converged in morphology under predator selection. No better case could be made for the reality of convergence.

16.2.2 The Coevolution of Plants and Animals through Time

Many species of mammals prey upon the seeds and fruits of plants. The coevolution of this predator-prey system has been reviewed by Janzen (1971). The trees exert selective pressure on the mammalian predators by depositing toxins in their fruits, and the mammals in turn add pressure to the plants they exploit. For predation on seeds to continue, the plant must ensure itself that not all seeds will be eaten. Furthermore, for a true coevolution, the animals must act as some sort of dispersal agent for the seeds or fruits themselves. Transporting and caching seeds can create stores for the animal's use and also carry the seeds away from the parent plant (Smythe 1970*b*). Seeds may also be dispersed when they are ingested along with the fleshy pericarp. If the seeds also have the capacity to pass through the digestive system unscathed, they will generally be deposited at some distance from the parent tree. In fact, some species of seeds will not germinate unless they have passed through the digestive tract of a frugivorous predator (Hladik and Hladik 1969). The coevolution of seed predators among mammals has been ably reviewed by Smith (1975). Of special interest are his papers dealing with the seed dispersal mechanisms of conifers in the northern hemisphere and with their predators, the tree squirrels (Smith 1970).

Pollination is another mechanism whereby animals selectively influence flower structure. The flowers provide the animals with food in return for the animals' transporting pollen to the ovaries of conspecifics. In a tropical forest this pollination by mammals generally is executed by the Chiroptera. In the Old World, specialized genera of the Megachiroptera have evolved that are nearly obligate feeders on nectar and pollen. A parallel evolutionary trend is detectable in the glossophagine bats of the family Phyllostomatidae in the New World tropics. The role of bats as pollinators in Southeast Asia has been ably reviewed by Start and Marshall (1976). The situation in the New World has been reviewed by Howell (1974).

A special instance of mammal-plant coevolution has been hypothesized by Montgomery and Sunquist (1975) based on their ecological studies of the three-toed sloth, *Bradypus infuscatus*, in Panama. The sloth eats the leaves of its host plants, and most individual sloths have a modal set of trees on which they feed rather heavily. The cost to the plant is considerable, but the sloth has a pecular habit that may benefit the plant even though it has been fed upon. All sloths descend to the ground to defecate. The three-toed sloth excavates a small depression with its tail and deposits the feces in it. Since three-toed sloths tend to be associated with modal trees, it follows that defecation will usually occur at the bases of trees most heavily fed upon. The conclusion is tantalizing that the nutrients in the feces, together with the undigested portion of the leaves, are thus made available to the host tree for recycling. Given the limited availability of nutrients in tropical soils, such a condition may be strongly favored over the long run if the predated trees survive longer.

16.3 Competition between Higher Taxa

The grand ecological concepts of competition and niche have been under severe scrutiny for many years.[2] There is a prevalent notion that most contemporary niche diversification has reduced competition. A stable community should have a careful subdivision of its niches along rather strict lines. Nevertheless, the thought is inescapable that some sort of competition must be going on continually to reinforce the separation and, further, that the patterns of habitat utilization we see among members of a community today are the end results of a long-term competitive process.

Charles-Dominique (1975) outlines how birds and mammals very probably divided the feeding niches early in their evolution and maintained their separation through active competition. Birds in general occupy the diurnal, arboreal feeding niches as insectivores, frugivores, and nectar-feeders, while at night the mammals occupy the insectivore, frugivore, and nectarivore niches. While it is true that some mammals have evolved diurnality and that in turn some birds are nocturnal, the degree of separation in species of equivalent size-classes is indeed remarkable.

Cartmill (1974) has outlined how the evolution of woodpeckers has adapted them for feeding on wood-boring insect larvae and may have effectively prevented the occupancy of this niche by mammals. Mammals that subsist on wood-boring insect larvae are found in Madagascar (*Daubentonia*) and in New Guinea (*Dactylopsila*). Both areas have no avian woodpeckers. Competition between taxa reaches its highest point in granivorous ants and rodents in the southwestern deserts of North America, a story still being unraveled (Brown, Reichman, and Davidson 1979).

The most dramatic consideration of long-term trophic competition within successive mammalian faunas has been proposed by Sussman and Raven (1978). Whereas insect pollination of plants is of great age

2. In actuality, individuals compete, not taxa; however, the sense of this section is evident.

and probably was under way well back in the Cretaceous, it was inevitable that vertebrates would invade the nectarivorous niche and coevolve with the plants. At first mammals must have fed on flowers, but finally they began to act as dispersal agents for pollen as the systems coevolved. Sussman and Raven propose that marsupial pollinators may have derived from an early radiation as terminal branch-feeders in the Paleocene of North America. As the marsupials spread to South America, they occupied this niche and held it until the mid-Miocene. In North America, the plesiadapoid radiation gave rise to the true prosimians in the Eocene, which may later have invaded this niche and outcompeted some of the derivatives of the plesiadapoids. At the end of the Eocene, the bat pollinators that spread into North America from Asia probably effectively outcompeted the prosimians. The bats eventually declined in North America as climatic conditions changed at the end of the Miocene.

Bats probably developed as pollinators in Asia during the Eocene, effectively preoccupying this nocturnal niche. From there the bats spread to North America and Africa; however, bat pollinators have never been able to diversify in temperate zones. It seems possible that temperate-zone pollination could have remained with the prosimians if they had survived in a temperate climate. In South Africa, galagos persist as pollinators into the present time. The prosimians crossed to Madagascar early in the Eocene and effectively filled the nectarivorous pollinator niche in Madagascar. Subsequent invasions by three species of bats failed to displace the prosimians from this niche.

The marsupials were able to pass via South America to Antarctica and then on to Australia, where some lines established themselves as nectarivores, frugivores, and specialized pollen-feeders (*Tarsipes*). They have held this niche despite the fact that bats probably came into Australia sometime in the Miocene; but once again, as in South Africa, since much of Australia lies in the subtropics or the temperate zone, the adaptive radiation of bats has not been extensive. Furthermore, the invasion of bats in Australia has been somewhat retarded, given the distance from the main continental land masses. Thus the marsupials tend to remain pollinators in the subtropical and tropical forests of Australia, as do the prosimians in Madagascar, whereas in Southeast Asia and in South America the bats effectively preoccupy the niche. Of course, the diurnal, vertebrate pollinating function on all major continents is carried out by birds.

16.4 Mammalian Species Diversity and Niche Occupancy

With the terrestrial mammals, it is instructive to consider the number of species as a function of land area and latitude. I will attempt here an analysis along the lines pioneered by MacArthur (1970) and MacArthur and Wilson (1967). For this discussion, I will define the tropics as the area between 20° north and 20° south latitude. As a generalization, the diversity of mammalian species (in the sense of absolute number of species for a defined land area) is directly proportional to the land area surveyed, which probably reflects habitat diversity (see figs. 35 and 36 and tables 35 and 36).* Mammalian diversity

Table 35 Numbers of Mammalian Species for Selected Geographical Areas

Locality	Area (km^2)	Number of Mammalian Species	Number of Chiropteran Species	Number of Species/km^2
Madagascar[a]	589,898	92	20	1.56×10^{-4}
East Africa	1,768,450	351	98	1.98×10^{-4}
Belgian Congo (Zaire)	2,336,889	427	102	1.83×10^{-4}
South India	1,813,000	102	34	5.63×10^{-5}
Sri Lanka	65,609	83	28	1.27×10^{-3}
Northern South America	2,575,400	424	177	1.64×10^{-4}
Venezuela	912,050	304	148	3.3×10^{-4}
Canal Zone	938	128	60	1.4×10^{-1}
Colombia	1,139,155	310	122	2.72×10^{-4}
Surinam	142,700	123	62	8.61×10^{-4}
Panama	74,009	201	99	2.7×10^{-3}
Thailand	510,000	251	92	4.92×10^{-4}
Malaya	131,857	199	81	1.51×10^{-3}
Borneo	731,457	188	62	2.5×10^{-4}
Sumatra	422,527	143	47	3.38×10^{-4}
Palawan	11,785	29	9	2.46×10^{-3}
Luzon	104,688	62	33	5.92×10^{-4}
Mindanao	94,631	65	33	6.87×10^{-4}
Trinidad	4,828	94	61	1.91×10^{-2}
Puerto Rico[a]	8,866	19	11	2.14×10^{-3}
Cuba[a]	114,522	38	26	3.32×10^{-4}
Hispaniola[a]	76,462	32	16	4.19×10^{-4}
Jamaica[a]	11,424	23	17	2.01×10^{-3}
Chile	741,765	63	9	8.49×10^{-5}
Southern cone of South America[b]	3,699,000	233	34	6.3×10^{-4}
North America[c]	20,471,000	423	49	2.07×10^{-5}
California	406,120	178	33	4.38×10^{-4}
South Carolina	80,743	56	12	6.93×10^{-4}
Maryland	25,607	61	9	2.38×10^{-3}
New Mexico	314,713	137	27	5.18×10^{-4}
British Columbia	921,664	103	15	
Tasmania	67,896	28	5	4.12×10^{-4}
York Peninsula[d] (Australia)	350,000	59	30	1.69×10^{-4}
Australia	7,627,000	228	51	2.99×10^{-5}

Note: Figures are exclusive of marine mammals.

Sources: References consulted: Lekagul and McNeely 1977; Medway 1977, 1978; Handley 1966, 1976; Varona 1974; Taylor 1935; Hall and Kelson 1959; Dalquest 1948; Green 1974; Ride 1970; Ingles 1965; Meester and Setzer 1971; Cabrera 1958–61; Kingdon 1971, 1974*a,b*; Ellerman and Morrison-Scott 1951; Osgood 1943; Troughton 1967; Findley et al. 1975; Chasen 1940; Husson 1978.

[a] Includes recently (5,000 B.C.) extinct forms.
[b] Chile, Argentina, and Uruguay.
[c] Includes northern Mexico.
[d] Includes subtropical portions of Queensland.

is higher in the tropics than in the temperate zone (compare figs. 35 and 36). Diversity is higher on continents than on islands. If we divide islands into two classes—oceanic islands that are a great distance from continental land masses and continental islands that are near the major continents—then oceanic islands have a lower diversity than do continental islands, whether in the tropics or in the temperate zone. Tropical Australia is somewhat of an exception in that it shows a mammalian

Table 36 Regression Analysis for Terrestrial Mammals and Land Area

Data Set	A	B	Coefficient of Correlation	N
All mammals				
All areas	−230.46	69.53	.56	27
Tropical areas (continental)	−318.63	105.07	.91	8
Neotropics	−474.59	128.69	.84	4
Paleotropics	−687.68	169.68	.95	4
Tropical areas				
Continental islands	−495.16	114.71	.94	5
Oceanic islands	−39.16	14.97	.97	4
Temperate areas	−515.34	119.02	.90	7
Chiroptera				
Tropical areas				
Neotropics	−22.37	26.34	.97	5
Paleotropics	0.34	15.84	.99	4
Islands	−75.88	21.72	.88	9
Temperate areas				
Northern hemisphere	−49.81	13.18	.86	6
Southern hemisphere	−74.50	15.72	.87	3
Nonvolant mammals				
Tropical areas				
Neotropics	−228.84	67.21	.98	4
Paleotropics	−688.13	153.84	.94	4
Islands	−146.32	44.12	.94	5
Antilles	−19.08	6.59	.83	4
Australian region	−381.99	79.71	.95	3
Temperate areas	−498.39	112.28	.88	6

Note: Y = A + B (X); Y = number of species; X = area in $\log_{10}$ km^2.

diversity comparable to that of a continental island, such as Madagascar, or a temperate zone continental area of equal size (fig. 35).

A second interesting exception to the general trends for the tropics is South India. The area south of the Ganges and the Indus shows a reduced mammalian diversity. This probably reflects the Pleistocene extinctions and the current predominance of a single monsoon rainfall pattern with a tendency toward climax forest types that are seasonally dry and partially deciduous (see Eisenberg and Seidensticker 1967; Schaller 1966).

If we consider the temperate zone, the following trends can be determined. Intermediate continental areas such as New Mexico and California, which are almost subtropical in certain parts, show a diversity higher than northern areas, such as British Columbia or the state of Washington (fig. 36). Among the southern continents, Tasmania shows a diversity consistent with a temperate zone continental island. Chile is similar in its species diversity, which probably reflects faunal imbalances stemming from the Pleistocene extinctions.

What determines the differences in the patterns of mammalian diversity we see? First, the diversity of bat species contributes in no small measure to the numerical differences in mammalian species when we compare tropic and temperate areas. Bats are most diverse in the tropics, and we see a gradual increase in the diversity of bat species as we proceed from the high latitudes toward the equator (fig. 37).

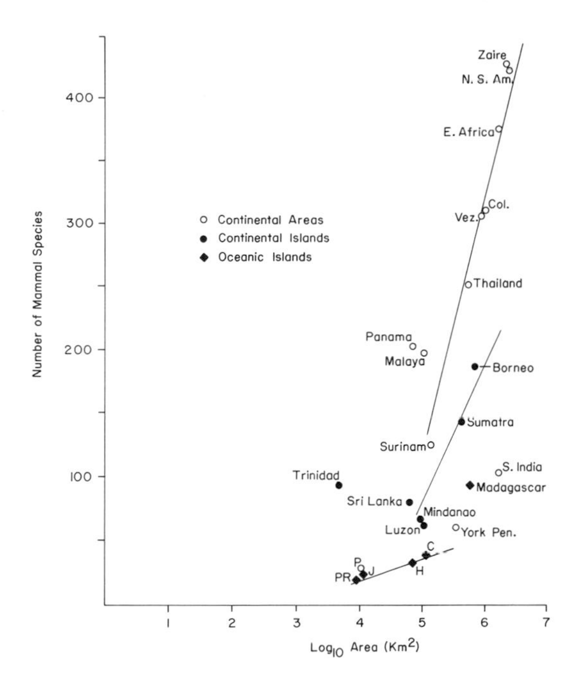

Figure 35. Regression of total number of mammalian species against geographical area. Regression lines have been drawn for mainland, continental islands, and oceanic islands. South India, Madagascar, and the York Peninsula of Australia are exceptions (see text). PR = Puerto Rico; J = Jamaica; H = Hispaniola; C = Cuba; P = Palawan.

The bat fauna of tropical Australia and South India and Sri Lanka is unexpectedly low. The bat fauna of the Neotropics is consistently higher than that of the Paleotropics, and tropical islands, both continental and oceanic, usually show a decreased diversity of Chiroptera, though in some cases the Chiroptera are the only mammalian species endemic to oceanic islands, thus reflecting their high mobility.

If bats are subtracted from the terrestrial mammalian fauna for any given region, we find the following trends. The Paleotropics show a higher faunal diversity than do the Neotropics, illustrating the circumstance that bats constitute a larger portion of the terrestrial mammalian fauna in the Neotropics than in the Paleotropics (see fig. 38). Faunal diversity still declines as we move from tropical regions to high latitudes even if the bats are subtracted, but the decline is less rapid. California, for example, shows a nonvolant mammalian diversity on the same slope as that of the Paleotropics. The southern cone of South America shows a reduced diversity of nonvolant mammals compared with equivalent areas in North America. Tropical Australia continues to exhibit a reduced diversity, as was established when I treated the total faunal complement in the preceding paragraphs. The island of Tasmania shows a depauperate mammalian fauna compared with north temperate zone areas of comparable size (see fig. 38).

Where indexes of species diversity are nearly equivalent, it is instructive to consider the faunal composition along taxonomic and

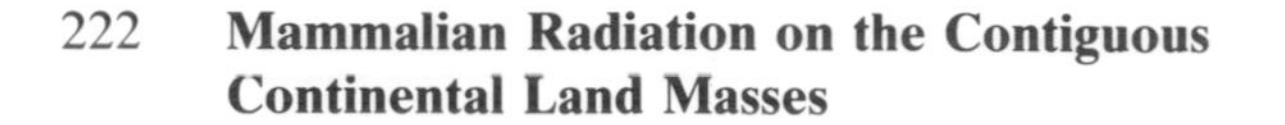

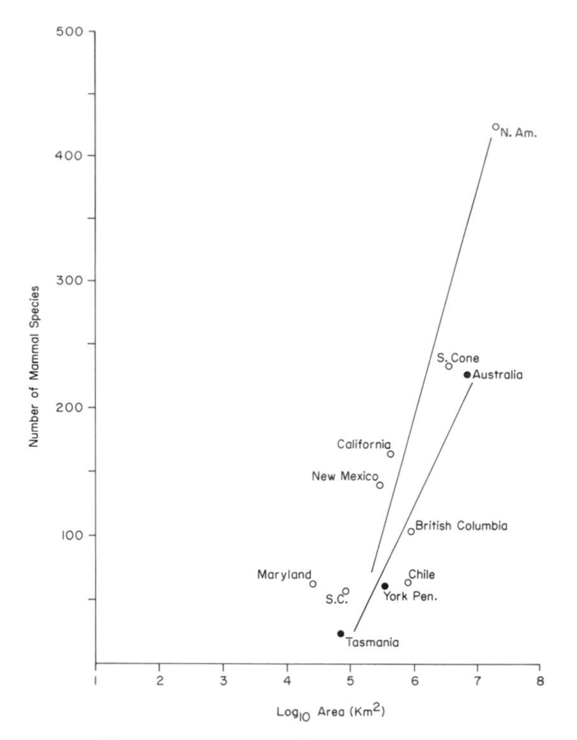

Figure 36. Regression of total number of mammalian species against geographic area for temperate zone areas. The York Peninsula of Australia and Australia have been included for comparison.

trophic lines. Table 37 compares the composition of two well-known mammalian faunas, in Malaya and in Panama. The preceding sections on the evolution of the Mammalia will explain the discontinuities, but in brief Malaya has no marsupials, edentates, or lagomorphs, while Panama has no proboscideans. In Panama insectivores have a reduced diversity, and some marsupials occupy this niche. Some of Panama's edentates (sloths) occupy Paleotropical primate feeding niches. The argument will be developed further in chapter 20. Within the Carnivora, Malaya has no procyonids, while Panama has neither ursids nor viverrids. Yet complementarity in niche occupancy is shown, for, if we compare the number of species for a set of defined trophic categories, a remarkable correspondence is demonstrable (table 38).

A further consideration of feeding niches for tropical Australia (the York Peninsula) indicates a significant deviation from Panama and Malaya. In many respects the York Peninsula resembles California in the percentage of species assigned to each trophic category. As an island, Australia has had a unique history of mammalian occupancy, and some feeding niches may well be filled by birds (see chap. 3). The temperate island of Tasmania is still more deviant in its mammalian niche occupancy (table 38), a problem considered again in chapter 20.

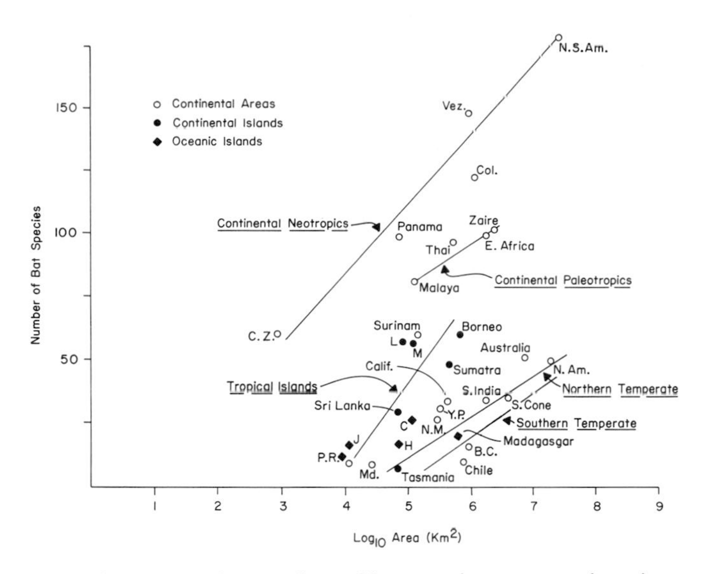

Figure 37. The number of bat species regressed against geographical area for several types of habitat. Regression lines have been drawn for the Neotropics, Paleotropics, tropical islands, and temperate zones. BC = British Columbia; Ca = California; C = Cuba; Md = Maryland; L = Luzon; M = Mindanao; J = Jamaica; H = Hispaniola; Y. Pen = Northern Queensland, Australia; P = Palawan; Mad = Madagascar.

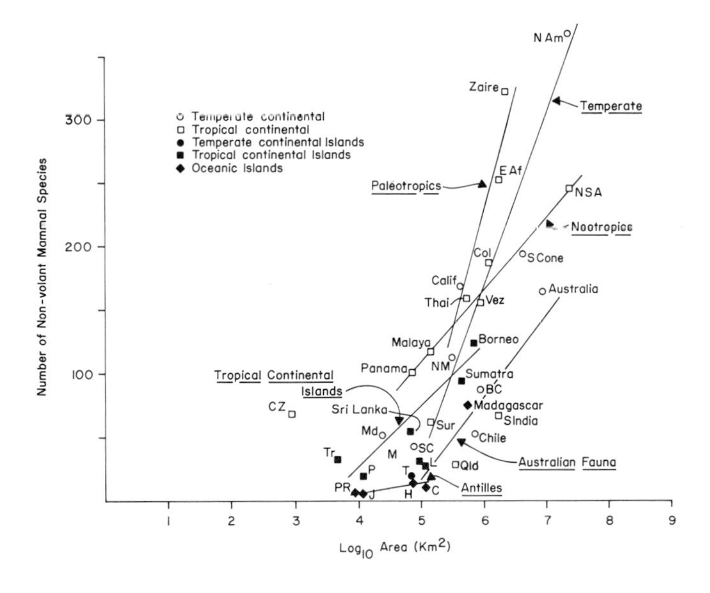

Figure 38. Number of nonvolant mammalian species regressed against total geographical area. Separate regressions have been performed for the Paleotropics, Neotropics, and temperate zones.

To complete the analysis, let us now compare the species diversity within similar habitat types for the tropical and the temperate zones. For this purpose, I will define four habitat types: montane, semiarid,

Table 37 Malayan and Panamanian Mammal Faunas Compared Taxonomically

Taxon	Malaya		Panama	
	Number of Species	% of Total	Number of Species	% of Total
Marsupialia	0	—	10	5.0
Insectivora	7	3.5	2	1.0
Dermoptera	1	0.5	0	—
Chiroptera	81	40.7	99	49.7
Primates	13	6.5	8	4.0
Edentata	0	—	7	3.5
Pholidota	1	0.5	0	—
Lagomorpha	0	—	1	0.5
Rodentia	53	26.6	48	24.1
Carnivora	30	15.1	18	9.0
Canidae	1		2	
Procyonidae	0		6	
Ursidae	1		0	
Mustelidae	6		5	
Viverridae	9+6=15		0	
Felidae	7		5	
Proboscidea	1	0.5	0	—
Perissodactyla	3	1.5	1	0.5
Artiodactyla	9	4.5	5	2.5
Total	199		199	

Table 38 Nonvolant Mammalian Trophic Analysis

Taxon	Panama		Malaya		Northern Queensland and York Peninsula		Tasmania	
	Number of Species	% of Total	Number of Species	% of Total	Number of Species	% of Total	Number of Species	% of Total
Carnivore	7	6.8	15	12.7	2	3.7	2	9.5
Piscivore	6	5.8	7	5.9	1	1.9	0	0
Myrmecophage	4	3.9	1	0.8	1	1.9	1	4.7
Insectivore/omnivore or insectivore/frugivore	14	13.7	11	9.3	15	28.3	7	33.3
Frugivore/carnivore or frugivore/omnivore	20	19.6	14	11.9	0	0	1	4.7
Frugivore/granivore	34	33.3	46	39.0	17	33.7	3	14.3
Frugivore/browser	10	9.8	12	10.2	8	15.1	3	9.5
Browser/grazer	7	6.8	12	10.2	9	17.0	5	23.8
Total	102		118		53		22	

grassland, and moist forest. I will then compare Washington (state), Maryland, California, Sri Lanka (Ceylon) and Venezuela (see table 39). My choice of areas is dictated by practical experience and by the availability of complete faunal lists.

The numbers of species in the montane zones of the states of Washington, Maryland, and California can be readily compared with those in the montane zones of Sri Lanka and Venezuela. Confining the first comparisons to montane zones reduces the complication of high bat

Table 39 Mammalian Species Numbers for Selected Continental Habitats

Habitat Type	Volant	Nonvolant	Total
Montane			
Washington State	7	49	56
Maryland	9	41	50
Venezuelan Andes	10	29	39
California Sierra	9	53	62
Sri Lanka	6	24	30
Venezuelan Andes and Coastal Range	93	86	179
California Coastal Range and Sierra Nevada	14	63	77
Arid			
Washington State	6	22	28
California (Inyo)	10	45	55
Sri Lanka (arid zone)	13	32	45
Falcon Peninsula, Venezuela	17	14	31
Grassland			
Washington State	5	20	25
California	11	31	42
Venezuela (llanos)	53	42	95
Sri Lanka (semiarid zone)	13	32	45
Moist forest			
Washington State	7	49	56
California	9	53	62
Maryland	9	41	50
Sri Lanka (lowland wet zone)	22	48	70
Venezuela (Amazonas)	91	70	161

Sources: Ingles 1965; Dalquest 1948; Eisenberg and Redford 1979; Eisenberg and McKay 1970; Paradiso 1969.

diversity in the tropics. By further focusing on the nonvolant forms, we can see that the number of species in the Venezuelan Andes and Coast Range is eighty-six; for the California coast range and the Sierra, there are sixty-three species. For the California Sierra alone, fifty-three: for the Allegheny Mountains of Maryland, forty-one; and for Washington, forty-nine. This confirms the trend of increasing diversity in nonvolant mammals as one proceeds from a temperate to a tropic zone.

Turning to the semiarid areas in the Neotropics and the North American temperate zone, we come to a different conclusion. As we proceed from the Columbian Plateau of Washington to the semiarid areas of central California, we see an increase in nonvolant mammalian diversity from twenty-two to forty-five species. If we examine the Falcon Peninsula of northern Venezuela, we find only fourteen species of nonvolant mammals (table 39). This trend probably reflects the longer history of aridity over a tremendous area in North America, which has permitted the evolution of arid-adapted granivorous rodents that contribute significantly to the faunal diversity in the arid zones of this temperate continental area. A similar conclusion was reached by Mares (1975, 1976) when he compared the recently arid "monte" of Argentina with the southwestern United States.

If we compare grassland areas among the states of Washington and California with Venezuela, we find twenty species of nonvolant mammals for Washington, thirty-one for California, and forty-two for Venezuela. This conforms to the hypothesis of high tropical diversity. Thus our conclusion for an analysis of nonvolant mammals, comparing similar life zones for temperate and tropic zones, would be as follows: In the tropics, semiaridity reduces diversity because it imposes lower productivity; but, in spite of this, the diversity in tropical grasslands tends to exceed that of temperate grasslands. In the wet tropics, the diversity of nonvolant mammals exceeds that of the wet temperate zone areas. Where a long history of aridity occurs, however, the granivore niche will be occupied by an adaptive radiation of rodents, and diversity is elevated in the temperate zone deserts relative to the smaller South American deserts.

Figures 39, 40, and 41 portray graphically the trends outlined in table 39. In figure 39 one may note that the topographically complex Cascade range of Washington State supports the greatest number of species. The arid grasslands to the east show a reduced mammalian diversity. A similar decline in diversity is shown in the transect across California as one passes to the San Joaquin Valley. On a west-to-east transect in northern Venezuela, species diversity drops to north temperate zone levels as one passes over the Andes. No such trend is shown in the lower-altitude but structurally complex Guiana Highlands (see Eisenberg and Redford 1979).

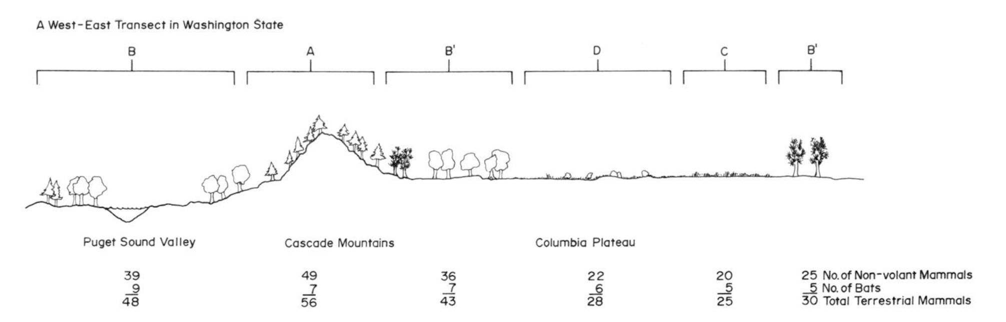

Figure 39. Diagram of species diversity for three major habitat types and six subtypes (according to Dalquest 1948) for the state of Washington.

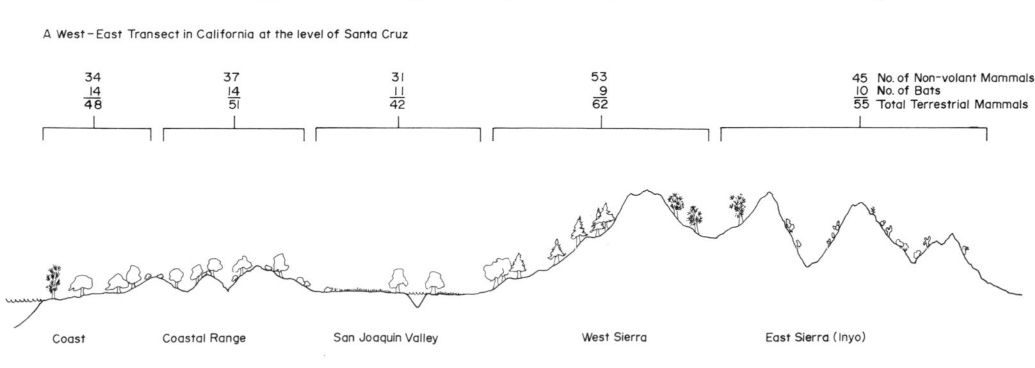

Figure 40. Diagram of species diversity for five major habitat types on a west-to-east transect in the state of California. Data from Ingles 1965.

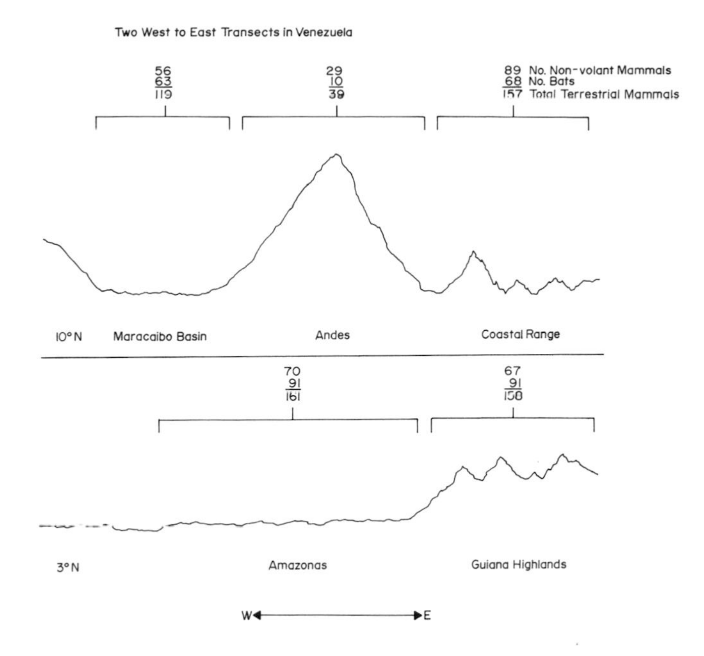

Figure 41. Diagram of species diversity for five habitat types on two west-to-east transects in Venezuela. Data from Eisenberg and Redford 1979.

16.5 The Loss of Species and the Problem of Extinctions

The preceding sections have considered the history of the mammalian radiations. Major radiations often accompany the occupancy of a new niche, either by the acquisition of new characters through natural selection or by the sudden arrival of a form that can outcompete the older, endemic species adapted to a similar niche on the new land mass. The history of extinctions and extinction rates over geological time had been reviewed by Simpson (1940, 1965). Stanley (1979) points out that over geological time rates of speciation and rates of extinction are strongly correlated within lines of descent.

In analyzing the fauna of islands that are close to the mainland, MacArthur and Wilson (1967) developed a theoretical foundation for calculating contemporary extinction rates. If one assumes that when connected to the mainland the island had the same faunal complement as the mainland has at present, one needs to know only the length of time the island has been separated to calculate the rate of species loss. When avifaunas are so analyzed, one finds that the rate of species loss is inversely proportional to the area of the island.

Oceanic islands or islands that have received random "inoculations" of species have a lower faunal diversity and have some unique types of niche exploitation by derived forms. They also tend to show more rapid rates of extinction when external perturbations are imposed, such as the arrival of *Homo sapiens*. Terborgh (1974, 1975), drawing on the data assembled by Willis (1974) concerning the loss of avian species from Barro Colorado Island in the Canal Zone, has applied the theory of island biogeography in considering wildlife preserves as habitat islands. He demonstrates that extinction rates are inversely proportional to the size of the preserved area. The extinction rates so calculated are the outcome of complex factors. Some extinctions result from processes of ecological succession with subsequent loss of habitat. Extinctions in small land areas are usually chance events where a small gene pool meets with a local catastrophe or loses heterozygosity by inbreeding and thus loses adaptability. The larger the area, the greater the probability that habitat diversity will be maintained through an equilibrium of environmental forces, and the lower the probability of random extinctions.

In table 40 I have assembled the species diversity data for three types of mammalian habitats: (1) protected habitats, such as the Canal Zone and Barro Colorado Island in Panama; (2) oceanic islands invaded by man, such as the Greater Antilles and Madagascar; and (3) continental islands that separated from the mainland at the end of the Pleistocene (e.g., Trinidad, Borneo, Sumatra, Mindanao, and Sri Lanka). According to the method developed by Terborgh, I calculated the extinction rates and have plotted them in figure 42. The continental islands fall on the same regression line remarkably well. Barro Colorado Island shows an extinction rate such as would be predicted for a continental island. The oceanic islands and the Canal Zone all have higher extinction rates deriving from the activities of man. The smaller the area, the higher the extinction rate; however, the sparser the original fauna, the more vulnerable it is to disturbance (not only by man).

It is appropriate to close this section with the warning that the radiations we have looked on with such pleasure are the current end points of a complex interrelationship among all members of a given ecological community, both plant and animal. I have only very briefly cast the mammalian radiations into the perspective of community ecology. If the organic diversity we cherish and depend on is to be preserved, we must seriously consider the necessity of preserving large

Table 40 Mammalian Extinction Rates for Selected Habitats

Locality	S_t	S_o	$1/S_o$	$S_t 1/S_o$	t	$\frac{-1/(S_t - 1/S_o)}{t} = k$	$\text{Log}_{10}k(-1)$
Sri Lanka	83	110	.0091	82.990	10,000	-1.205×10^{-6}	−5.919
Sumatra	143	199	.00503	142.995	10,000	-6.993×10^{-7}	−6.155
Barro Colorado Island	82	128	.00781	81.992	70	-1.7423×10^{-4}	−3.758
Canal Zone	128	201	.00498	127.995	70	-1.1161×10^{-4}	−3.952
Mindanao	65	188	.005319	64.994	10,000	-1.5386×10^{-6}	−5.8129
Puerto Rico	11	19	.05263	10.947	5,000	-1.8269×10^{-5}	−4.7383
Cuba	31	38	.02632	30.973	5,000	-6.4571×10^{-6}	−5.18996
Hispaniola	18	32	.03125	17.968	5,000	-1.113×10^{-5}	−4.9535
Jamaica	18	23	.04347	17.956	5,000	-1.1138×10^{-5}	−4.9532
Madagascar	86	92	.01086	91.989	5,000	-2.174×10^{-6}	−5.6627
Trinidad	101	157	.00636	100.994	10,000	-9.9013×10^{-7}	−6.0043
Borneo	188	199	.00503	187.995	10,000	-5.319×10^{-7}	−6.274

Note: S_t = Species at present time
S_o = Species at time of separation from mainland or before the advent of man.
t = Time since separation from mainland or, for Madagascar and Antilles, time since the advent of man.
k = Extinction rate.

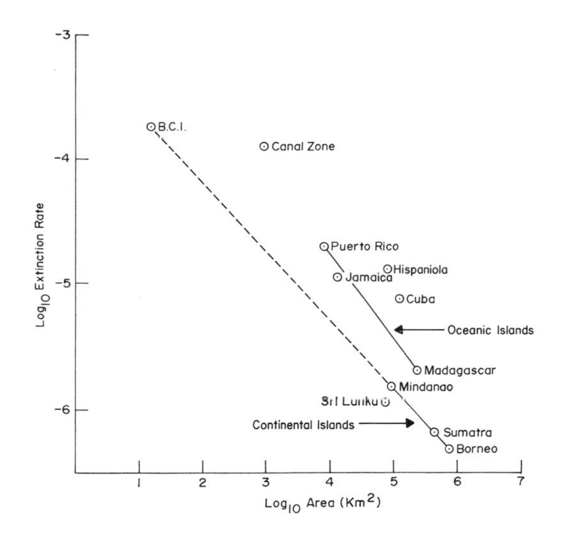

Figure 42. Regression of extinction rate against total land area for several tropical habitats. Data from table 40 (see text).

areas and intact ecosystems (see Wilson and Willis 1975). Conservation biology as a modern, scientific area of inquiry may emerge at the close of this century as a major applied discipline (Soulé and Wilcox 1980).

Part 3
Macrophysiology and Adaptation

In this part I examine physical and biological factors contributing to variations in life-history strategies. I rely almost exclusively on bivariate analysis, since multivariate statistics cannot be used because my data sets across species have too many blanks. An inspection of Appendixes 2–6 will demonstrate the problem. If enough species had all data cells filled in, then multivariate analyses would be in order. I have taken the only solution open—step-by-step bivariate analysis. A word of caution should also be offered. Many of my graphs are simple plots intended to demonstrate trends and covariance. Unless a regression or correlation has been performed, the graphic data should be taken as suggesting trends and as a convenient way to summarize information. Much more work must be done to establish the realities and the limits of variation.

17 A Consideration of Mammalian Size and Metabolic Rate

The size of a mammal's body puts a variety of limits on what it can do. As D'Arcy Wentworth Thompson (1917) pointed out many years ago, all living matter is subject to the pull of gravity and the elementary laws of geometry. For example, as any object becomes larger in a linear dimension, its surface area increases as the square of the linear dimension, but its volume increases as the cube. Since mass or weight is related to volume, weight will increase as the cube of a linear dimension, and the surface of an organism will increase as the square. This generalization means that a small organism will have a large surface area relative to its volume. As a consequence, the potential for heat loss is greater in a small animal than in a larger one. One might then hypothesize that, if the organism were endothermic and attempted to maintain a constant body temperature, there would be a higher caloric intake and a higher metabolic turnover per unit weight in a small organism than in a larger one. This is the theoretical explanation for the famous Kleiber curve (Kleiber 1961), which indicates that the basal metabolic rate of a mammal decreases as the mammal becomes larger.[1]

As an animal becomes larger in a linear dimension, its weight increases as the cube. The support structure or skeleton that bears the stresses of the organism's weight increases only as the square when considered as a cross section. Thus gravity will eventually limit the size to which a terrestrial organism can grow and still remain mobile. If we plot a frequency histogram for the average size of mammalian genera, it will be noted (see fig. 43) that the vast majority of mammals are less than a foot, or 32 cm, in head and body length.[2] Looking at the material from a human perspective, we can see that man is among the largest mammals. Most rodents are less than a foot in head and body length, as are members of the order Insectivora. Most species of the order Chiroptera are even smaller, less than 10 cm in length. Carnivores run the gamut in size, reflecting their specializations for different size-classes of prey. The aquatic cetaceans are all relatively large as a result of their buoyancy in water; most of the herbivorous "ungulates" are also relatively large (see chap. 19).

1. Metabolic rate decreases as the $-.25$ power when $\log_{10} 0_2$ consumption/unit time is regressed against $\log_{10}$ body weight. Calorie consumption per unit time increases allometrically at the .75 power on a similar double log plot regressed against mean weight.

2. Bourlière (1975) suggests that the distribution is bimodal, but my analysis indicates a skewed but unimodal form.

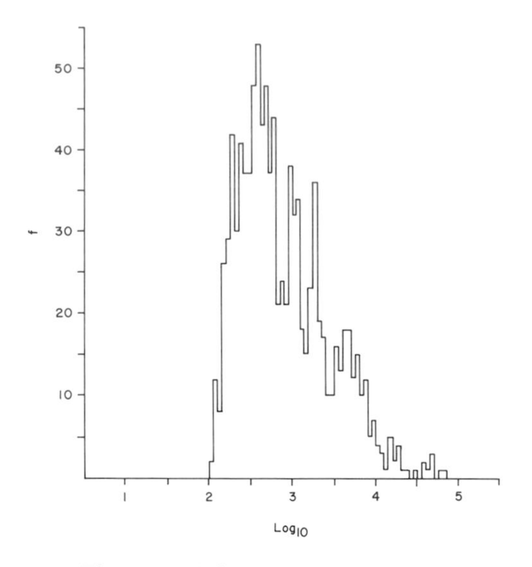

Figure 43. Histogram portraying the frequency of size-classes for the genera of the class Mammalia (abscissa, $\log_{10}$ head and body length in mm). Adapted from Eisenberg 1978.

Since most mammalian species are relatively small and, further, since most are both endothermic and relatively homeothermic, their food intake per unit of time when they are active must be relatively high. The greater potential for heat loss in a small mammal means that feeding will favor adaptation to a trophic level that gives the maximum net energy return within a minimum time for energy expended in gathering food. It should not be surprising, then, that feeding on other living animals, both vertebrate and invertebrate, as well as feeding on such concentrated energy resources as seeds, fruits, nuts, and plant storage roots or tubers predominates in mammals of these size-classes. Using plant material rich in structural carbohydrates but low in material that may be easily metabolized is an option exploited predominantly by the larger size-classes of mammals (see chap. 19).

One way to partially escape the iron law of body size and heat loss lies in developing efficient mechanisms either for insulation or for protection from extreme fluctuations in ambient temperature. Small mammals in general have accomplished this by growing dense coats and by sheltering in protected, insulated nests, thereby reducing potential heat loss. Additionally, many small mammals have slightly lowered their basal metabolic rate as a mechanism for escaping from the demand for a constant high energy intake.

A great deal of scatter is associated with the Kleiber curve, and this scatter may represent specific zones of adaptation (see McNab 1974). Some small mammals vary their metabolic rate throughout a twenty-four-hour cycle. A bat such as *Myotis lucifugus,* for example, raises its metabolic rate in the evening and maintains a high metabolic rate while flying and capturing insects. When it concludes its feeding and returns to roost, its metabolic rate will drop considerably. In addition to diel fluctuations in metabolic rate, many small mammals have adopted the technique of lowering the basal metabolic rate during part of the year by entering torpor; this torpid period is referred to as estivation or hibernation, depending on the season (see fig. 44).

Classically, the basal metabolic rate represents the amount of energy an individual expends when it is at an ambient temperature where it does not lose or gain heat excessively (the thermoneutral zone) and when it is at rest, without muscular activity. Under these strictures of definition, the metabolic rate itself is determined by measuring oxygen consumption by the whole animal. Thus the figure for basal matabolic rate is an organismic average. The various organ systems of the animal may have quite different metabolic rates, even at rest. If, for example, slices of muscle and brain tissue are removed from a living ground squirrel, *Citellus,* and placed in a respirometer and the metabolic rate

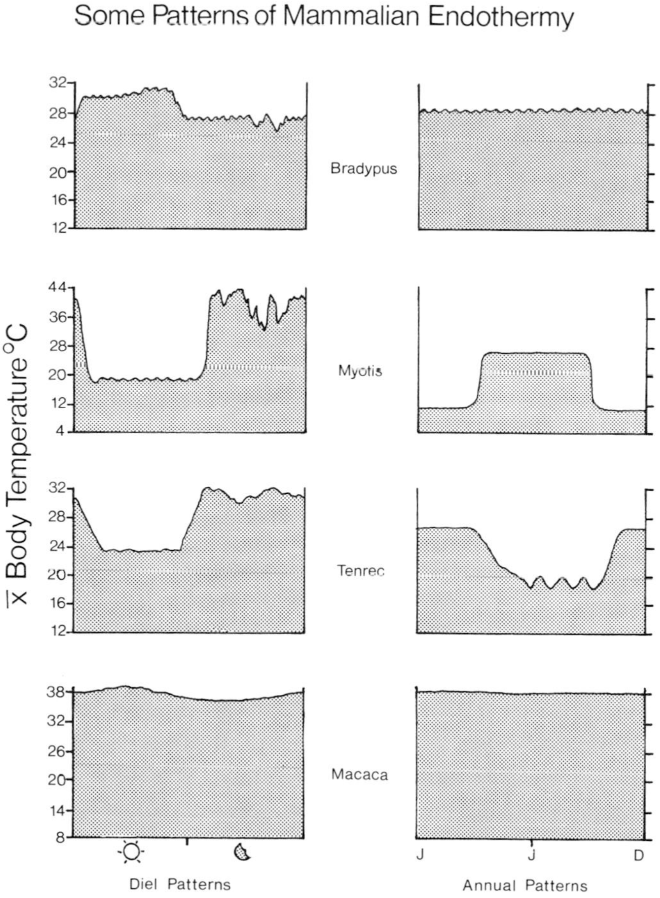

Some Patterns of Mammalian Endothermy

Figure 44. Annual and diel body temperature fluctuations for selected mammalian taxa. Note that, although *Bradypus* maintains a rather low body temperature and is an imperfect homeotherm, it shows less annual fluctuation in body temperature than does *Tenrec ecaudatus*. Both *Tenrec* and *Myotis* show diel and annual fluctuations in body temperature that relate to energy conservation. From Eisenberg 1980*a*.

of these tissues is measured, it can be demonstrated that brain tissue metabolizes at a far higher rate than tissue taken from the skeletal muscle, or even from the gut and spleen. If a hibernating mammal, such as the ground squirrel, is compared with a nonhibernator, such as the laboratory rat, the muscle tissues from *Citellus* will have a lower metabolic rate than those of the nonhibernating rat; however, the brain tissue will have a metabolic rate remarkably similar (Meyer and Morrison 1960). This indicates, then, that, a low basal metabolic rate need not imply that the metabolic activity of all of the species' organ systems, and in particular the central nervous system, is also low; indeed, quite the opposite may be true. Thus it is theoretically possible for a small organism to lower its basal metabolic rate relative to that of another species of the same size-class and still have the tissue of its central nervous system metabolizing at a high rate.

Figure 45 synopsizes some data concerning resting metabolic rates for a series of mammals. The data are taken from Appendix 5. The negative slope as a function of mean body weight is easily discerned. (In subsequent discussions, mean measurements and mean weights are recorded in Appendix 2). Four contributions to the scatter in this diagram are noteworthy. The soricine shrews typically exhibit an elevated metabolic rate when resting. The vespertilionid Chiroptera show vastly reduced resting metabolic rates as a means of energy conservation. The rodents exhibit a wide scatter in metabolic rate that correlates with specific adaptations. And fossorial rodents, which have a problem in heat dissipation, have adapted by lowering their basal metabolic rates (McNab 1974). Marsupials also exhibit great scatter, because many Australian marsupials exhibit a reduced basal rate relative to eutherians (Dawson and Hulburt 1970; MacMillen and Nelson 1969).

Figure 46 considers the same phenomenon for a series of small rodents, with one bat genus, *Myotis,* included for contrast. If we connect the mean values for *Mus* and *Rattus,* we can easily demonstrate that squirrels, some cricetines, and heteromyid rodents have lower basal metabolic rates. This correlates with a greater potential longevity for the size-class, an issue that will be dealt with in subsequent sections.

17.1 The Metabolism of Populations

Studies of the metabolism of whole populations of single species or even of ecosystems can be undertaken with great profit. By way of background, it is important to review the laws of thermodynamics. Energy tends to be conserved and it is possible to draw up a balance sheet. First, one must consider the general rule that the energy contained within the system under study is equivalent to the energy initially taken in by the system minus the energy lost in the form of both work and heat loss.

The second law of thermodynamics demonstrates that heat cannot be converted from one form to another without a loss. We must therefore consider two aspects of energy. First, the energy available to an organism for work and productivity, and, second, the energy that becomes unavailable through heat loss or entropy. Energy flow in a system can never reach 100 percent efficiency.

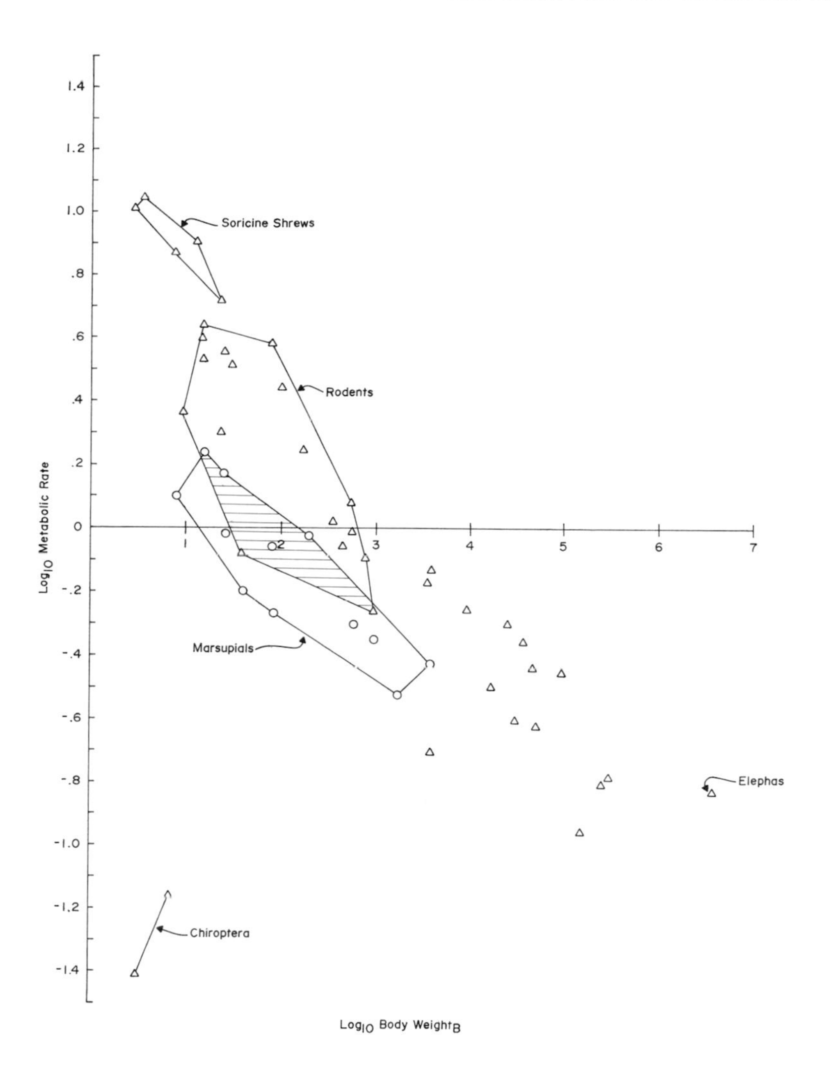

Figure 45. Plot of metabolic rate against body weight (double log plot, data from Appendix 5). Note that the soricine shrews and vespertilionid bats deviate strongly from the major slope.

If we consider the quantity of energy consumed by an individual to be I and the quantity of energy absorbed through the gut of an individual member of the population to be I_a, then the energy consumed must always be greater than the energy absorbed through the gut (i.e., $I > I_a$). The energy the organism absorbs may be diverted to producing new tissue or maintaining existing tissue. The rate of energy flow in a system has to exceed the total maintenance cost plus the total production of new material per unit time.

The average respiration per gram per hour times the weight of an average organism gives the respiration rate. If this is summed for all members of the population, this quantity can then be multiplied by 4.8, if we assume that 4.8 calories are produced per liter of oxygen consumed. This value can then be multiplied by twenty-four to give us the average respiration rate per day for the population.

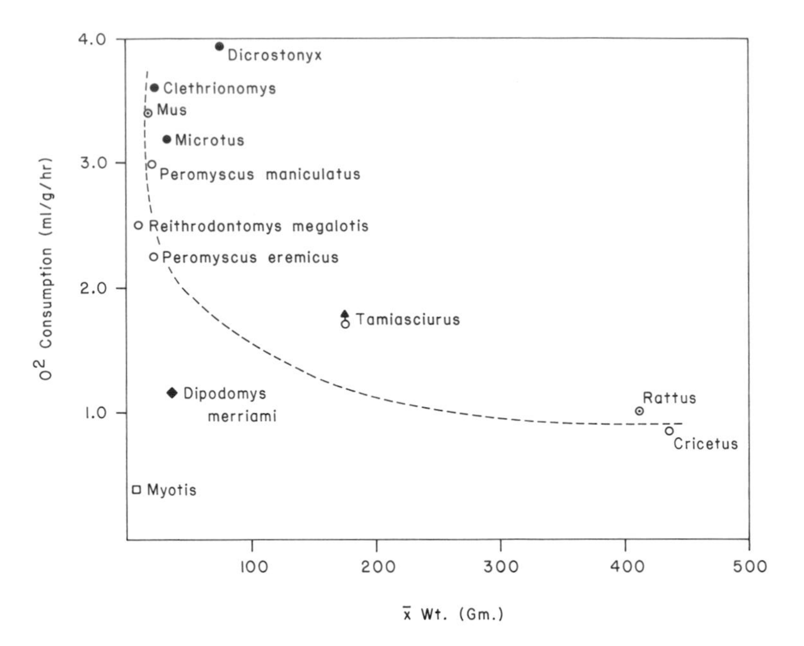

Figure 46. Metabolic rate as a function of body size for several species of small rodents. Note the low metabolic rate for Merriam's kangaroo rat (*Dipodomys merriami*) at rest. One example for the genus *Myotis* is included for comparison.

The calorie consumption for a population through a twenty-four hour period is only an initial value. This indicates the cost for total maintenance in terms of energy, but in addition we need to calculate the total production cost in terms of new tissue produced. This value can be estimated if we know the average production in grams of live tissue per specimen. This value is in turn a function of age-classes. The average production value for each age-class must be multiplied by the number of individuals within each separate age-class, and the resultant values may then be summed for the population. The final sum may then be multiplied by 1.4, which is the caloric equivalent of one gram of live tissue for an average mammal.

By calculating the two quantities—the value for maintenance and the value for production—we will obtain the rate of total energy flow for an average population. This summed value should equal the energy flow, which may be calculated by the following opposite procedure as a cross-check.[3] The reverse procedure is to measure the caloric equivalents of the food intake per unit time by the population and subtract from this the caloric value of undigested material in the feces. According to the first law of thermodynamics, these should come into balance, and the difference would represent sampling errors and entropic heat loss.

3. Of course the system is not perfect, and entropic loss of energy at each step will produce a slight inequality.

It is possible to calculate efficiencies of energy conversion for populations. The efficiency of a system is defined as the output of energy or tissue divided by the intake of energy. We may define different types of efficiencies—for example, tissue growth efficiency is defined as the potential energy used in growth divided by the amount of potential energy assimilated by the organism. Individual growth efficiencies are defined as the standing crop of potential energy divided by the potential energy used in birth and growth. Interpopulational efficiencies may be determined by comparing the efficiencies of two adjacent populations. Trophic-level assimilation efficiency may be defined by comparing, for example, the energy assimilated at the level of a carnivore with the energy assimilated at the level of an herbivore. As a general rule of the thumb, carnivores have a higher utilization efficiency than do herbivores. Thus the higher one goes in the trophic pyramid, the more efficient is the potential energy utilization. Golley (1960) determined, for example, that the weasel, *Mustela,* had an efficiency of 26.8 compared with an efficiency of 1.1 for the herbivorous *Microtus pennsylvanicus*.

Maintenance energy may be defined as the sum of the energy necessary to maintain basal metabolic rate plus both the energy cost of acquiring food and the energy cost of combating environmental stresses such as cold or heat. Energy costs can vary greatly during a given twenty-four-hour period for a given animal. Catlett (1961) gave some values for *Mus musculus*. The laboratory mouse consumes energy at the following rates in cubic centimeters of oxygen per gram per hour: when fighting, oxygen consumed rises to 18.3; when feeding, it is only at 9.5; when exploring, 15.0; when sitting, 8.0. Thus energy costs must be determined by estimating the percentage of time the animal spends in various forms of activity. Coehlo, Bramblett, and Quick (1979) provide a field example of bioenergetic analysis in their study of the energy budgets of *Ateles* and *Alouatta* in Mexico. The problem is compounded when we consider the size factor. Since heat production varies directly as the 0.75 power of the weight, a small mammal, regardless of what it does, will always use more energy per gram than does a larger mammal. One of the consequences of this is that, within any given trophic level, the smaller an animal is, the lower will be its standing crop biomass compared with the biomass of a larger species occupying the same area. Another consequence, as pointed out before, is that smaller mammals will in general tend to be at higher levels of the trophic pyramid. This brief review only highlights an extremely active area of research (Gessaman 1973).

17.2 Metabolic Rate and Reproduction

The inflexible laws of size and metabolic rate have an effect on reproductive potential. In estimating reproductive capacity, we must consider litter size, length of gestation, and maturation rate of the young. The energy drain on the female may be divided into gestation costs and lactation costs. In turn, this must be expressed in annual costs by calculating the average number of litters each female produces per year.

Reproductive potential is highly variable and tends to reflect broadly the adaptive trends within the taxonomic group under consideration. It seems that selection favors a high reproductive potential in those forms that suffer the greatest loss of offspring during an average annual cycle. This may be correlated with adaptation to ecological niches that are to some extent unstable and show great fluctuations in carrying capacity (Pianka 1972; MacArthur 1972). It is axiomatic that small mammals that maintain a high basal metabolic rate and are adapted to inherently unstable environments generally have a rather short life-span and an extremely high reproductive potential (McNab 1980). Their tendency, then, in some small mammals is to approximate a semelparous mode of reproduction (see chap. 23). A high reproductive capacity often means a large litter size as well as an adaptation toward rapid growth rates of young and an extremely short maturation time. A tendency toward semelparity is possible only if food productivity favoring the survival of the annual cohort is certain. The amount of consumable food may vary greatly from year to year, but in any given year the onset of productivity must guarantee that part of the surviving cohort will live.

Having moved from metabolism to a consideration of reproductive strategies (a topic to be dealt with at some length in the following sections), it should be obvious that basic data on life history and behavior are essential for advancement in the subdiscipline of ecological energetics and productivity. I foresee a renaissance in life-history studies as greater accuracy is attempted in the study of energy transfer within ecosystems.

18 The Influence of Body Size on Life History

A mammal's home range is usually measured in two dimensions and expressed as an area in square meters or hectares. For arboreal and scansorial species, a volumetric measurement would be more realistic (see Montgomery and Sunquist 1978). Aerial and semiaquatic forms usually feed at one locus and rest at another, so that the actual home range may assume some rather peculiar geometric patterns. If the species is migratory, then annual ranges may be connected by a transit route from one locus to another. Home ranges for solitary-foraging, terrestrial mammals usually increase as the body size (expressed either in weight or as a linear dimension) increases. When the species is a social forager, the average group weight must be employed (see Clutton-Brock and Harvey 1977). As McNab (1963) pointed out, some of the scatter displayed in the regression of home range against body weight can be eliminated if the trophic specializations are duly noted. Thus, for a given weight class, strongly herbivirous forms have smaller home ranges than do carnivores. Ultimately, however, home range size is related to carrying capacity. Arctic and boreal carnivores utilize larger home ranges than do carnivores in the temperate zone (see fig. 47). As first-order consumers, herbivores are usually sensitive indicators of plant productivity. If the herbivore in question stores little fat and does not cache food, its density may reflect annual plant productivity in a very direct way. This explains why the biomass of large herbivores parallels the physiognomic classes of vegetation when they are arranged according to the mean length of an annual drought (Eisenberg and Seidensticker 1976). Both temperature and drought duration are excellent predictors of relative plant productivity. It is in response to climatically induced, predictable shortages that food caching, fat storage, torpor, and migration have evolved in mammals. As size increases, the absolute weight of food ingested increases. This can be expressed as calories consumed, since available caloric content varies so widely among the various components of an animal's diet. The increase is negatively allometric, however, because of the decrease in basal metabolic rate for large mammals.[1]

Western (1979) has pioneered an analysis of the effect of a species' body size on its life-history strategy and ecological niche restriction. This analysis was performed for the large mammal communities of the East African National Parks.

1. Calorie consumption in homeotherms increases as weight increases to the .75 power (Kleiber 1961) in a double $\log_{10}$ plot of calories consumed per unit time against body weight.

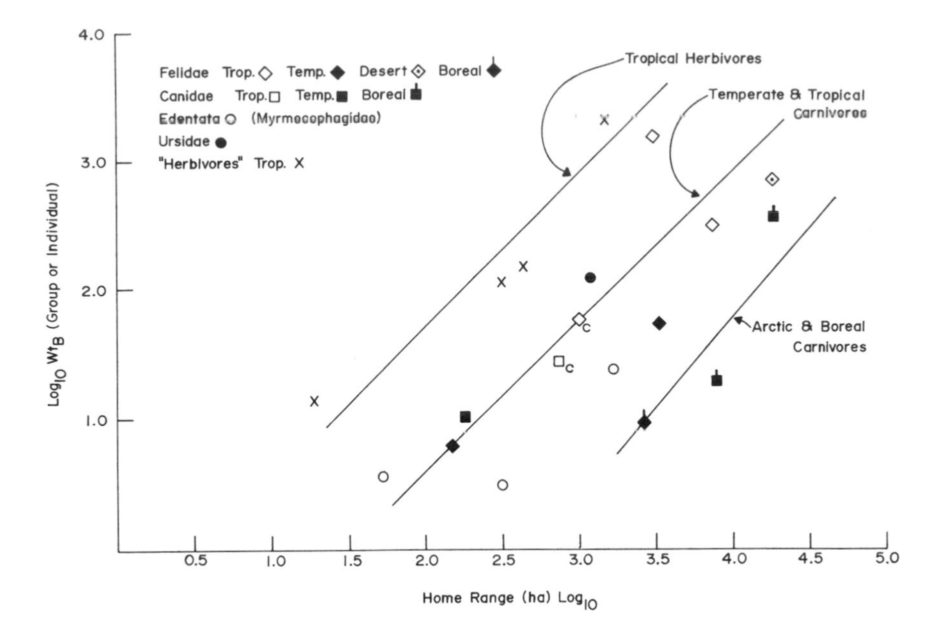

Figure 47. Plot of body weight against home range size for tropical herbivores and temperate and boreal carnivores. Data for carnivores from Kleiman and Eisenberg 1973. Data for myrmecophagous edentates are included with carnivore data for comparison. See figure 16. Lines are drawn by approximation.

Allometric relations between body size and a dependent variable are legion, but some remarkable consistencies are repeated in nature. As Jerison (1973) has so elegantly shown, brain weight increases allometrically as body weight increases. The equation may be expressed as: $Wt_{\text{brain}} = k \cdot Wt_{\text{body}}^{.67}$. Nevertheless, specific adaptations can be correlated with differences in brain weight for a given size-class. Eisenberg and Wilson (1978) demonstrated that within the Chiroptera a relatively large brain is associated with adaptations for feeding on dispersed, energy-rich food resources that are unpredictable in their spatial distribution. A further discussion of brain/body weight regressions will be undertaken in chapter 21.

Length of life is correlated with mean body size. Figure 48 presents data from diverse field studies that reflect the overall trend that the probability of dying in a given year decreases as size increases. Considerable scatter is shown, and I submit that within any size-class there can be relatively long-lived and short-lived species. Field data of this type are influenced by a variety of unpredictable environmental factors, including mortality through hunting pressures. Figure 49, which is drawn from the data of zoological parks, gives a picture of maximum attainable life-span (see Appendix 3). In general the positive correlation between longevity and body size holds very well. Certain small mammals that approximate a semelparous reproductive strategy are exceptions (see fig. 49).

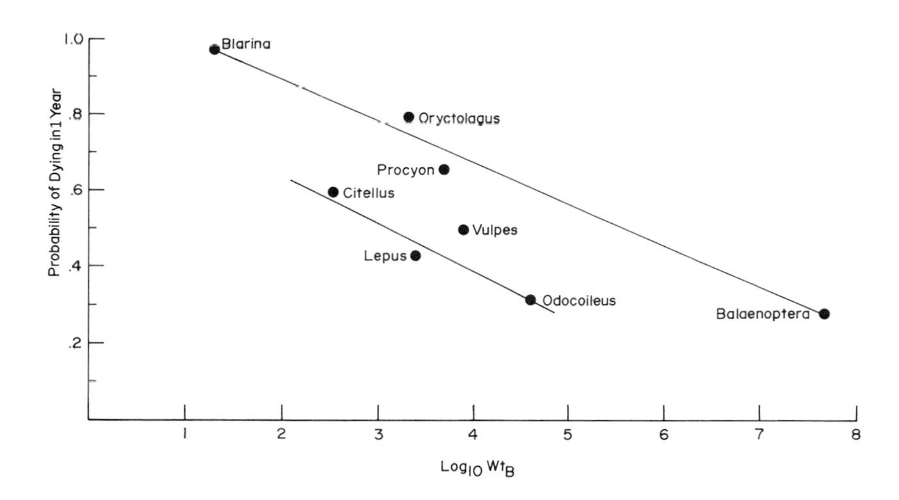

Figure 48. Probability of dying within one year as a function of body size. Lines identify species with high probability and low probability of dying over a range of weight classes. Data for *Balaenoptera* drawn from an exploited population. Data from various sources. Lines delimit upper and lower cases.

Mammals that hibernate or make diel reductions in metabolic rate often have greater longevity than species of comparable size that maintain a higher metabolic rate. The temperate zone Chiroptera are thus an exception to the rule of increasing average longevity with increasing body size. Some of the scatter in this plot can also be reduced if we note that in general marsupials have a shorter longevity than do placentals and that carnivorous mammals have a shorter potential longevity than do herbivores (see Appendix 3).

Sacher (1959) points out that potential longevity in mammals is inversely correlated with metabolic rate. As we saw before, metabolic rate is inversely related to absolute body size, but with many exceptions based on specific adaptations to conserve energy (torpor, hibernation, etc.) or adaptations for certain feeding niches. Figure 50 plots data from Appendixes 3 and 5 that demonstrate the finding described by Sacher. There is of course considerable scatter, which may reflect errors of measurement, since the measures of metabolic rate utilized in this analysis are drawn from many sources. The exceptions will be referred to again in chapter 23.

Given these facts, one can examine the consequences of increased potential longevity across species. For taxa that show both multiparous and uniparous birth patterns, there is a decrease in litter size with an increased potential longevity (see chap. 23). Also, the encephalization quotient as a species' average (see chap. 21) tends to increase for both placentals and marsupials with increased longevity. Metabolic rate is negatively correlated to some extent with the increasing magnitude of the EQ value.[2] As one might conclude from the foregoing trends, litter

2. EQ value, or encephalization quotient, refers to the relative size of the mammalian brain for a given size-class.

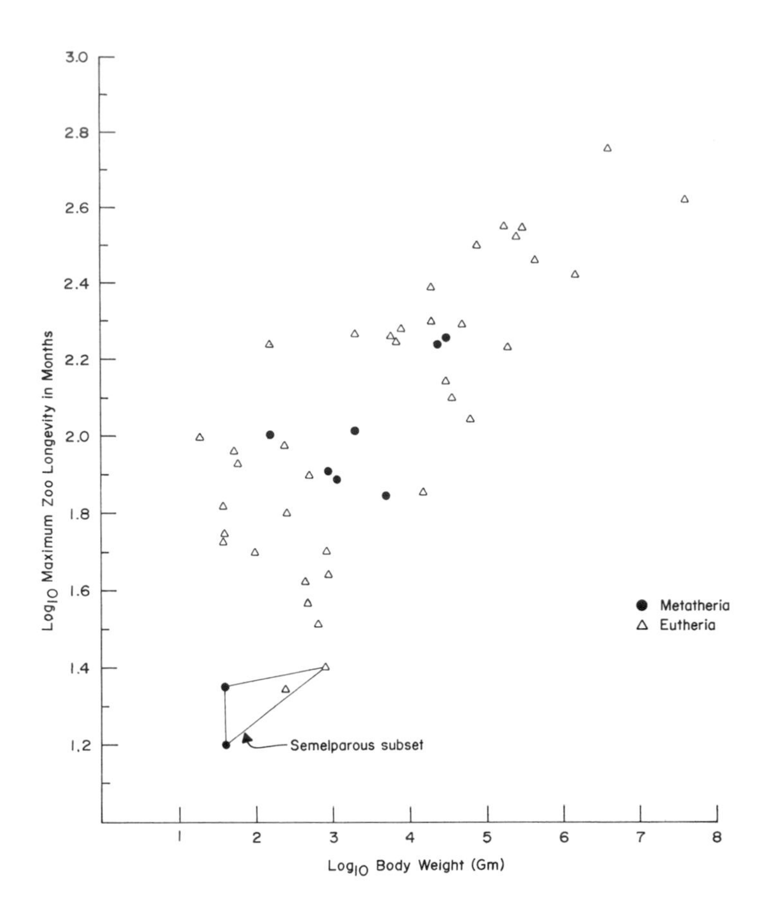

Figure 49. Double log plot of maximum zoo longevity against body size. Data from Appendix 3. Those species approximating a semelparous reproductive strategy are enclosed in a convex polygon.

size tends to decrease with increasing EQ values. Sacher and Staffeldt (1974) have demonstrated that gestation time as a rough indicator of relative developmental time is directly correlated with the cube root of the neonatal brain weight. The developmental time for a mammal is, broadly speaking, a function of absolute size and there is a general increase in developmental time with increasing adult size (see chap. 23). The suggestion is at once obvious that, regardless of absolute size, mammals with lower metabolic rates may live longer, produce smaller litters, have relatively large brains, and live a long time (see chap. 34).

18.1 Summary Statement concerning the "Escape from Constraint of a Small Body Size"

A small homeothermic mammal has a high propensity to lose heat, since its surface area is large with respect to its volume or mass. This holds only if it is essential for the organism to maintain a constant core temperature. Allowing the body temperature to fall on a diel or seasonal basis means that the basal metabolic rate can decline and energy can be conserved. The Chiroptera have exploited this alternative to the fullest in their adaptation to diverse feeding niches.

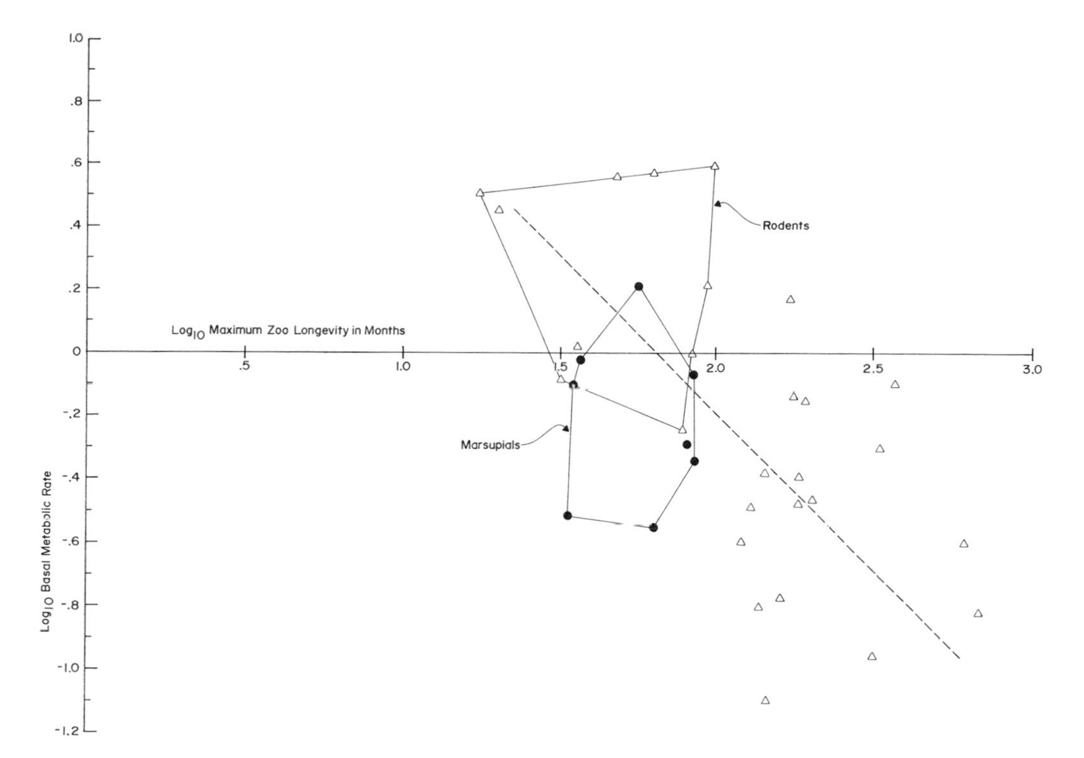

Figure 50. Double log plot of average basal metabolic rate against maximum zoo longevity. Data from Appendixes 3 and 5. A great deal of scatter is demonstrable in this plot, obscuring the negative trend. The points for rodents and marsupials have been enclosed in convex polygons. See text for a discussion of factors influencing the scatter.

Size can also influence absolute mobility, except for volants. Small nonvolant mammals are less mobile than larger ones and have higher energy costs of locomotion; so to exploit widely dispersed resources they must find a feeding "patch" rich enough to support them during a generative cycle and then defend it against competitors. If the productivity of the "patch" is temporally constrained, then the species must be able to wait out a period of nonproductivity by either storing food or entering torpor. The exceptions to the mammalian "size rules" are all species that have evolved specific adaptations to control for energy loss or to pass periods of time by surviving on cached food or on fat when their reduced mobility precludes movement to another patch (see chap. 33).

18.2 Some Reservations concerning the Use of Body Weight as a Correlative Measure

Body (*Wtb*) in mammals is related to head and body length (*HB*) by the formula $HB = K \cdot \sqrt[3]{Wtb}$. Either measure can be used in studies of allometric relationships. There is considerable scatter around the line expressed by the preceding equation, but the fit is fairly good over the range from small mice to whales. An average adult weight for a nondimorphic species is a useful value for scaling, but it presents several limitations. For example, in regressing metabolic rate or brain weight against mean body weight for a series of species, one must take care that the scatter of the points may not in part be attributable to variations in the percentage of fat, bone, or armor that contributes to the total weight. As Grand (1977) points out, the total weight of a species expresses the sum of the body's components (muscle, bone, integument, etc.), which vary among themselves in metabolic rate per unit weight and, worse yet, vary dramatically in percentage contribution to total weight from one species to the next. When using body weight from the literature, one often does not even know whether inert gut contents were subtracted. Yet the fairly good correlation between body weight and length leads one to trust that body weight is a fair indicator of relative size. Problems arise with cetaceans and pinnipeds, where fat can contribute up to 30 percent of th total weight at some seasons of the year. Mammals undergoing seasonal torpor will show less dramatic but significant variations in weight. Dimorphic species present another type of problem. When regressing brain weight against body weight, either male and female data must be treated separately or some crude average between the sexes must be calculated. Ideally, one would plot male and female data separately for dimorphic species.

19 Feeding and Foraging Categories: Some Size Constraints

19.1 Classification of Feeding and Substrate Categories

Classifying living mammals into categories according to the utilization of certain environmental substrates and according to dietary preferences is fraught with difficulty. Mammals, like all other living organisms, have a perverse tendency to defy exact classification. Consider a mammal I knew very well some twenty years ago—the dusky-footed wood rat, *Neotoma fuscipes*. In the coast ranges of California it constructs a stick nest composed of litter and detritus, and in the center of the nest it builds an earthen burrow with a side exit, often concealed, through which it can escape if the nest is disturbed. Although it forages on the surface, it climbs exceedingly well. Of course it can swim. Its diet includes seeds, insects, nuts, leaves, and even cuttings from cedar trees. In short, it is the generalized mammal so elegantly described in the essay by Landry (1970).

What does one do with an animal like this? In attempting to classify, we strive to characterize it in terms of modal strategies. We might classify it as scansorial, since it forages on the surface as well as in shrubs and trees. In terms of diet, we might characterize it as a frugivore/omnivore—a somewhat imprecise category, but one that indicates some modal tendency in its dietary preferences.

In a similar manner, if we were to try to classify the dietary habits of *Peromyscus maniculatus,* we would observe that, though it eats fruits, seeds, nuts, and insects, its modal intake over the whole year consists primarily of seeds. We might then classify it as a granivore. Thus the problem in classifying dietary specializations and substrate utilization trends is that none but the most specialized forms will conform, and many will conform only if we think in terms of modes or medians rather than the full spectrum the species can show throughout the annual cycle.

In table 41 I have constructed a matrix including eight categories of adaptations for substrate utilization, and sixteen catetories of modal tendencies in dietary specialization. Considering the substrates or mediums, category 1 may be thought of as a fossorial syndrome, exemplified by *Talpa* or *Geomys,* in which the species is adapted for several forms of soil type and conducts most of its life-support activities underground, including foraging, resting, rearing young, and mating. This

Mace (1979) has independently developed a parallel analysis of correlations among body size, reproductive strategy, brain size, and niche, using a set of small mammals (mainly rodents and insectivores). Her general conclusions parallel many of my own.

Table 41 Feeding and Substrate Matrix

		Piscivore and Squid-Eater	Carnivore	Nectarivore	Gumivore	Crustacivore and Clam-Eater	Myrmecophage	Aerial Insectivore	Foliage-gleaning Insectivore	Insectivore/Omnivore	Frugivore/Omnivore	Frugivore/Granivore	Frugivore/Herbivore	Herbivore/Browser	Herbivore/Grazer	Planktonivore	Sanguivore
		1	2	3	4	5	6	7	8	9	10	11	12	13	14	15	16
Fossorial	1	—	102	—	—	—	106	—	—	109	110	111	112	113	114	—	—
Semifossorial	2	—	202	—	—	—	206	—	—	209	210	211	212	213	214	—	—
Aquatic	3	301	302	—	—	305	—	—	—	—	—	—	—	—	314	315	—
Semiaquatic	4	401	402	—	—	405	—	—	—	409	410	411	412	413	414	—	—
Volant	5	501	502	503	—	—	—	507	508	—	510	—	—	—	—	—	516
Terrestrial	6	—	602	—	—	—	606	—	—	609	610	611	612	613	614	—	—
Scansorial	7	—	702	703	704	—	706	—	—	709	710	711	712	713	714	—	—
Arboreal	8	—	802	803	804	—	806	—	—	809	810	811	812	813	—	—	—

is not to say it never ventures on the surface or that some foraging might not involve gathering vegetation near the burrow entrance. But the dominant specialization is for a life spent almost completely under the ground. It has a number of specialized morphological features, including (1) reduction in the size of the external ear or pinna; (2) reduction in the relative size of the eye; (3) modification of the forepaws or teeth, or both, to assist in removing soil particles; (4) reduction in the length of the tail; (5) shortening of the neck; and (6) reduction of the undercoat, especially in tropical forms. In brief, the body and sense organs reflect an ability to live and forage below the ground.

Category 2 I have termed semifossorial. In this category the animal demonstrates anatomical specializations for burrowing. A burrow is typically used as a refuge, and some digging may be involved in feeding, but the animal shows considerable ability to move about on the surface. The dominant mode of life is a compromise between selective pressures for foraging and avoiding predators on the surface, and the species retains the capacity for extensive burrowing. A typical example of this category would be the pocket mouse, *Perognathus californicus,* or the badger, *Meles meles.*

Category 3 includes forms that carry out most of the life cycle in water. This includes the cetaceans and sirenians as well as the pinnipeds. Although certain members of the order Pinnipedia confine part of their life cycle to land—namely, giving birth, mating, and nursing the young—a goodly percentage of their life is spent in the water, and thus they are included in this third category.

Category 4 I have termed semiaquatic. In general, these species must spend part of each twenty-four-hour period out of the water. In small semiaquatic forms such as the water shrew, *Sorex palustris,* it is essential that the animal retreat to a dry nest to dress its pelage and conserve heat.

Category 5 is the volant specialization, where the animal can fly. This category is confined to a single order, the Chiroptera. I do not include any of the gliding forms within this specialization; they will be considered a subset of the arboreal category. Adaptation for flight may have occurred only once in the evolution of mammals, if one considers a monophyletic origin for the Chiroptera.

Category 6 includes species that have evolved specializations for foraging on the surface of the earth with minimal digging and minimal climbing. Essentially we are talking about animals that forage on the ground and use cavities opportunistically, having limited ability to construct burrows. For example, the fox, *Vulpes fulva,* essentially forages on the surface, has a limited ability to climb, and, though it digs a burrow at the time of parturition, cannot be considered fossorial in the same sense as the animals in category 1.

Category 7 includes species that show considerable adaptation for climbing but are versatile enough to spend approximately equal time in the trees and on the surface of the ground—in short a scansorial species such as the wood rat, *Neotoma fuscipes.*

Category 8 includes forms that spend most of their life in trees as well as those we might consider nearly obligatory in their arboreal adaptation, such as the sloths, *Bradypus* and *Choloepus.*

Turning now to the sixteen feeding categories outlined in table 41, I will attempt to characterize the modal feeding trends, fully realizing that variability will be obscured. Category 1 includes species that in the main prey on aquatic vertebrates and invertebrates of considerable mass (greater than 3 g), such as fish, cephalopods, and even aquatic birds. The prey is mobile and can avoid initial approaches. We may consider *Lutra* and *Tursiops* examples of this feeding specialization.

In category 2 the usual prey is a terrestrial vertebrate with considerable mobility. In short, a species that falls into this category would be generally termed a carnivore. *Panthera leo* and *Mustela erminea* are typical examples.

Category 3 includes species whose principal dietary component over much of the year is nectar and pollen. Typical examples are *Glossophaga soricina* and the honey possum, *Tarsipes spenceri.*

Category 4 includes species whose dominant food is exudates from trees, such as sap, resins, or gums. Invertebrates, fruit, or small vertebrates may also be eaten. Since some tree exudates are highly polymerized, there may be digestive mechanisms reminiscent of those in species that feed on green plant parts. A typical example is the lemur, *Phaner furcifer.*

Category 5 includes species that feed primarily upon aquatic invertebrates that have limited mobility but tend to be patchily distributed, such as crustaceans, echinoids, clams, and oysters. The chief difference between this dietary specialization and feeding category 1 is that the invertebrate prey tends to be less mobile. The walrus, *Odobenus,* is a prime example.

Category 6 includes species that feed primarily on colonial insects—in particular, ants and termites. We would characterize these as myrmecophagous. Although plant parts, other invertebrates, and even vertebrates may be included in the diet on occasion, the modal tendency is toward feeding on the Isoptera and Hymenoptera. The Neotropical anteaters and the Old World pangolins are typical specialists in this category.

Category 7 includes species that feed on arthropods but also take winged insects and thus themselves must be capable of flight; in short, aerial insectivores of the order Chiroptera.

Category 8 includes species that feed on insects by flying to particular areas and gleaning them from the undersides of leaves or from bark; these are the bats, members of the order Chiroptera, that Wilson (1973) has termed foliage gleaners.

Category 9 includes species that feed primarily on arthropods but also eat mollusks and earthworms. They may include fruit and even small vertebrates in their diets. In brief, they exhibit a modal tendency to be insectivores/omnivores. Such generalists include the badger, *Meles*.

Category 10 includes specializations for feeding upon the reproductive parts of plants that we generally refer to as fruits, primarily the pericarp or the fleshy outer covering. The seeds may or may not be ingested. In addition, these species often eat invertebrates and small vertebrates when available. I call this a frugivore/omnivore feeding category. The marmoset, *Saguinus oedipus,* typifies this adaptation.

Category 11 includes species that have specialized for feeding on fruits as previously defined and, in addition, on nuts and seeds. By seeds I mean the actual reproductive part of the plant itself, not necessarily the fleshy pericarp; in brief, the reproductive bodies of shrubs, some trees, and most grasses. Typically, these plant parts may be stored and hoarded, and this is the key adaptation for species in this category. The chipmunk, *Tamias striatus,* is a prime example.

Category 12 includes species that eat fleshy fruiting bodies, perhaps some seeds, the fleshy storage roots of certain plants, and also some green leafy material. These species tend to eat plant parts that are not readily digestible by the normal enzymes in mammalian tissues; fully utilizing the structural carbohydrate of the green leafy substances may require assists from microbial enzymatic systems. The paca, *Agouti paca,* is so categorized.

Category 13 includes species that primarily feed on stems, twigs, buds, and leaves. Here microbial enzyme systems are mandatory as a symbiotic adjunct for the digestion of structural carbohydrate (Eisenberg 1978; Parra 1978). The tree sloths (Bradypodidae) are typical members.

Category 14 includes species that have specialized for feeding on grasses. Unlike the preceding category, leafy material is less frequently included in the diet. All gradations between categories 13 and 14 may be found (see Eisenberg 1978; Hofmann and Stewart 1972). I would typify a member of category 14 as a grazer and a member of category 13 as a browser.

Category 15 includes species that live in water and feed wholly or in part on zooplankton. The great baleen whales are categorized as planktonivores.

Category 16 is occupied by only two genera in the entire class Mammalia, members of the family Phyllostomatidae that have adapted for feeding on the blood of warm-blooded vertebrates (*Desmodus* and *Diphylla*). These species are termed sanguivores.

Each cell in the eight-by-sixteen matrix in table 41 can be identified by a three-digit number. Not all cells of the matrix are occupied, because some dietary specializations are incompatible with certain forms of substrate or environmental exploitation. For example, there cannot be an aquatic aerial insectivore.

19.2 The Consequences of Specialization for Certain Habitats

It is possible, by referring to the matrix, to define certain limitations for each of the categories of substrate utilization and dietary specialization that result from the relatively irreversible changes in the animal's anatomy and physiology that take place in the process of adaptation. Let us consider the categories of substrate utilization. Table 42 gives range of sizes and frequency of distribution for mammalian genera that have specialized for certain forms of substrate utilization. (Generic "averages" have been calculated from Walker et al. [1964]. This abstract number is not referrable to a particular species but is employed as a convenience.)

19.2.1 Fossorial Constraints

Fossorial mammals fall within a very narrow range of sizes; all could be considered small. One consequence is that their potential mobility[1] is reduced (Howell 1944). Specialization of the limbs for digging further reduces surface mobility. If their feeding habits confine them to a small subset of microhabitats, then the potential for colony formation or coloniality is increased, but this is not to be confused with communal living (see chap. 31). Furthermore, extreme specialization for a particular microhabitat reduces the probability of finding a mate. Thus there may be an incipient tendency for home ranges of males and females

Table 42 Range of Mean Head and Body Lengths (mm) for Mammalian Genera Classified according to Substrate Utilization*

Adaptation	Range of Means
Fossorial	117–203
Semifossorial	88–573
Aquatic	3,050–12,491
Semiaquatic	321–1,870
Volant	58–104
Arboreal	118–508
Terrestrial	223–1,508

1. Mobility is defined as the unit of distance traveled per unit time.

to overlap and a trend toward facultative monogamy (see Kleiman 1977*b*).

A fossorial mammal has the option of performing most of its activities within a permanent burrow system that becomes the main living quarters, as does the European mole, *Talpa europa*. This confers the additional possibility of defending the area around the burrow, and thus a form of territoriality may be developed. In addition, the species can use the permanent burrow system to store the foods it gathers. There are two major feeding specializations of fossorial mammals: insectivority and herbivority/omnivority. Roots, forbs, and seeds are easily stored. For fossorial insectivores, invertebrates, such as earthworms and arthropods, may not be stored easily unless the temperature in certain parts of the burrow system remains relatively low. Nevertheless, the potential for caching food increases the probability that intraspecific defense or territoriality will be developed.

Being small increases the potential for heat loss, but a fossorial adaptation ameliorates the problem, since heat will be dissipated more slowly in a closed burrow system than if the animal were exposed to ambient fluctuations on the surface.

One difficulty fossorial mammals do have, as McNab (1966, 1974) so eloquently points out, is that in a closed burrow system not only lowered oxygen tensions but also heat dissipation can pose a problem. This suggests that the fossorial syndrome could be favored in temperate zones, where extremely high earth temperatures are not experienced. Fossorial adaptation in the tropics demands that the burrow system can be aerated by opening subsidiary entrances. Too much heat production may induce a selection for reduced metabolic rate or a loss of pelage, or both, especially in tropical fossorial mammals. As McNab has pointed out, the metabolic rate of many fossorial mammals is lower than the basal rate of surface-foraging forms, and *Heterocephalus glaber* is virtually hairless.

Reduced metabolic rate may also lengthen the gestation period relative to body size. Most fossorial mammals produce small litters of altricial young. The small litter size may reflect a very reduced predator pressure on fossorial forms.

On the other hand, if the animal is adapted to temperate zones or keeps its burrow shallow, it may not need to reduce its metabolic rate. In moles (Talpidae), the metabolic rate may stay relatively high and the animal may maintain high assimilation efficiency by remaining an insectivore or invertebrate feeder. In this case one would anticipate that the fossorial insectivore syndrome would manifest itself most strongly in the temperate zone. Of course a species always has the option of reducing its metabolic rate throughout the annual cycle and of displaying seasonal torpor.

19.2.2 Semifossorial Constraints

Adaptation for a semifossorial life appears to confer less restriction on size. Small to intermediate mammals are represented in this specialization (see table 42). Because of this range of size, the potential mobility of the larger fossorial species need not be severely restricted. Burrowing adaptations must be well developed, but surface foraging may predominate. Problems of heat dissipation during burrowing remain, but the adaptation need only be temporary, not chronic as with a species that must spend most of its life underground. The semifossorial stage, then, represents a relaxation of the restrictions found in extreme fossorial specialization. Feeding adaptations embrace great latitude, from the badger, *Taxidea,* a carnivore, to the herbivorous grazing voles such as *Prometheomys* of northern Asia. Some semifossorial forms may be rather large, such as the aardvark, *Orycteropus,* and the giant armadillo, *Priodontes*.

19.2.3 Aquatic Constraints

Aquatic specializations, in the strict sense, imply that the animal is adapted to spend the larger part of its life cycle in the water. In pinnipeds, a very small portion of the life cycle may be confined to land or ice, where the pups are born and where in some cases mating and molting take place. Aquatic mammals have certain size restrictions. No completely aquatic mammal is very small (see table 42). This reflects the problem of heat loss, which is ever present in water because the temperature is well below the thermoneutral zone. On the other hand, the largest mammals that have ever existed are aquatic. The buoyancy of water confers a liberation from the size constraints imposed on species adapted to land, and fully aquatic mammals tend to be large. The extreme increase in size displayed by some strictly aquatic forms vastly increases their potential mobility. With increased mobility comes increased speed, which may may result from predator selection.

In specialization for life in the water, the body is modified so that it offers as little resistance as possible. Adaptations for controlling buoyancy must also be developed. Some adaptations have been toward decreasing buoyancy, as in the sirenians, where the density of the bones has increased so the animals can sink easily to the bottom and forage on various forms of eelgrass and algae. Some control of buoyancy may be maintained by regulating the air retained in the lungs. In general, mammals adapted for extreme aquatic life have reduced the hair covering and achieved thermoregulation by thick layers of blubber, which serve both for food storage and as thermal insulation (Howell 1930).

The feeding strategy of aquatic forms involves using vertebrate foods such as fishes, sea birds, and other aquatic mammals; capturing mobile invertebrates such as squid or sessile invertebrates such as tunicates, clams, and other shellfish (adaptation to feeding on sessile mollusks involves foraging on the bottom—e.g., *Odobenus*, the walrus); feeding on zooplankton; and feeding on herbaceous vegetation (e.g., the Sirenia).

Feeding on the larger elements of the zooplankton has been adopted by several families of cetaceans as well as by one genus of pinniped, *Lobodon,* the crab-eater seal. Zooplankton show tremendous increases at certain seasons of the year, and, though during any twenty-four-hour period the standing crop may be relatively low, the productivity and turnover of the zooplankton while they feed on the phytoplankton is enormous. Special adaptations for feeding on this seasonally available food resource require special straining mechanisms for removing such small organisms from the water (see sect. 15.1) and the capacity to store energy in the form of blubber, since zooplankton is often available only seasonally. High areas of plankton productivity are also prime areas for cetacean feeding. Cetaceans adapted for plankton feeding thus often do not feed throughout the year, but may feed very little or even fast for as much as four to five months.

The young of aquatic forms may be born in warm water or on land. One of the keys to interpreting their movement and habitat utilization trends is to ascertain what special feeding and temperature requirements are demanded at the time the young are born. For example, pinnipeds retreat to land or ice to give birth and as refuge for the young during lactation. Some pinnipeds give birth on ice and are not constrained to move to warmer waters. Some cetaceans do move to warmer and, indeed, shallow waters to calve; this is true for the genera *Baleana* and *Eschrichtius* (Rice and Wolman 1971). For those baleen whales that calve in shallow, warm water, there is reduced opportunity to feed. During this phase the females often eat very little. Because baleen whales are large, they have a lower metabolic rate, and thus feeding need not be as frequent. Living in water allows the animal to store considerable food as fat or blubber, which not only increases its buoyancy, since fat has a specific gravity less than that of water, but also allows them to conserve heat.

19.2.4 Semiaquatic Constraints

Semiaquatic adaptation implies that the animal spends part of its life cycle on land, thus alleviating the continual loss of heat to a medium colder than its own deep body temperature. Semiaquatic forms may thus be exceedingly small, such as the water shrew, *Neomys fodiens*. Almost all semiaquatic forms are adapted to fresh water; only a few are marine. Small semiaquatic mammals feed on insects, crustaceans, and small fish. These foods require a minimum of processing and have a high caloric return. Intermediate-sized semiaquatic mammals are often fish-feeders as well. Large semiaquatic mammals may eat plants. Their larger size and reduced metabolic rate allow them to take advantage of plant foods as an energy source even though there is loss in assimilation efficiency. Typical examples include the South American capybara, *Hydrochoerus*, and the African hippopotamus, *Hippopotamus amphibius*.

Small size is possible in the temperate zone semiaquatic forms because they can conserve heat during the terrestrial phase. Dressing the pelage by grooming and combing with the hind feet takes on primary importance during terrestrial phases because it removes excess water and traps air between the hair fibers, increasing insulation. So sensitive is the pelage to disturbance that small semiaquatic mammals may die if they are not permitted to dress out their fur (Richard and Vaillard

1969). Any saponifier or pollutant in water that reduces the fur's ability to trap air during the submerged phase can be fatal to small semiaquatic mammals, since they will soon die of heat loss.

The very small, semiaquatic mammals that feed on insects, crustaceans, and small fish often adopt the solo strategy of feeding. They have a relatively large home range and may actively defend the nest area and those parts of the stream where they forage. As a result, their densities tend to be relatively low.

19.2.5 The Consequences of Flight

The volant mammals belong to a single order, the Chiroptera. Table 42 shows that the aerodynamics of mammalian flight demands that these species be relatively small. There is a definite limit on how large a bat can become and still be capable of flight. This upper limit confines bats to the smallest size classes (<800 g). Especially for bats smaller than 50 g body weight, severe heat conservation problems may be encountered, particularly for those that live outside the tropics. The Chiroptera show some amazing adaptations for the conservation of heat. During any given twenty-four-hour period, the body temperature may drop appreciably, thus conserving energy (Hock 1951). In the temperate zone, bats may hibernate for a considerable period during the winter. These metabolic adaptations have certain life-history consequences. Gestation periods for bats tend to be prolonged even if we discount such phenomenon as delayed implantation or delayed cleavage of the blastocyst. On the other hand, bats, in spite of their small size, tend to be extremely long-lived. Thus, metabolically they may be said to "mimic" large mammals. As a result, bats can afford an extreme iteroparous reproduction strategy, often producing only a single young each year (see chap. 23).

Dietary specializations of bats are profound, and most specializations in morphology are correlated with dietary specialization (see D. E. Wilson 1973*b* and sect. 14.3).

19.2.6 Terrestrial Constraints

Surface-foraging mammals with limited climbing ability and limited ability to swim span a size range unparalleled in all the other substrate adaptations, yet terrestrial mammals do not reach the extreme sizes of the aquatic cetaceans. An upper limitation on size is set by the energy and support demands of moving a tremendous mass across the ground. The smallest mammals are the terrestrial shrews, such as *Suncus etruscus* and *Cryptotis parva,* which are near the absolute lower limit of mammalian size (~2.5 g), while the largest extant terrestrial mammal is the male African elephant, *Loxodonta africana*.

19.2.7 Arboreal Constraints

Scansorially adapted mammals generally have considerable ability to climb and forage in trees, but they also forage on the ground. Here an upper limit on size results from the difficulty of climbing and moving in trees for rather heavy species. This size limitation is shown more sharply in the last class of substrate adaptations, arboreal mammals. Table 42 shows that arboreal mammals are on average about 0.5 m in length. The orangutan, *Pongo*, is an exception.

19.3 Adaptation to Environmental Stresses

To sum up, adaptations for the use of certain substrates often limit the size an animal can reach. Limitation on size or confining classes of adaptations to certain size ranges carries certain corollaries with respect to metabolic rate, potential rate of reproduction, longevity, and ultimately—as will be demonstrated later—the form of social organization. Certain size-classes have limited options in adapting to environmental stress, and it is wise at this point to review the major kinds of stresses and how size affects the way the animal can cope.

First let us consider lack of food or periods of food scarcity. A mammal can cope with food scarcity by several mechanisms. It can move, that is, change its foraging location. As I indicated before, mobility is tied to size. Potential mobility is high in larger mammals and relatively limited in smaller mammals, except for those adapted for flight. For the volant Chiroptera, there is an emancipation from the size limitation on mobility. Aquatic forms, of course, all tend to be large and thus retain a high potential mobility.

If mobility is denied a mammal during periods of food scarcity, either through size limitation or because the food scarcity is generalized over a wide geographic area, such as during temperate zone winters, then the options are to lower the metabolic rate, to draw upon food reserves, or to use a combination of both.

Lowering the basal metabolic rate and entering torpor is an option widely adopted by mammals over an immense range of sizes, from the silky pocket mouse, *Perognathus longimembris,* averaging less than 8 g, to the grizzly bear, *Ursus horribilis,* which may reach a weight of over 500 kg (see Lyman and Dawe 1960). Drawing on food reserves is also open to a wide range of mammalian forms. A large aquatic mammal may store food as blubber to draw on during times of scarcity. A terrestrial mammal may likewise store food in the form of fat. Fat may be diffused in the body tissues or localized in a specific body area (e.g., the tail). The issue is ably reviewed by Pond (1978). Storing body fat is not open to many small mammals, which must maintain high mobility, either for catching prey or for eluding predators. Such species cannot carry a huge fat store without affecting either their foraging effectiveness or their antipredator behavior. This is probably why small predators, such as weasels of the genus *Mustela,* hoard food in the winter and why shrews (*Sorex*) and moles (*Talpa*) hoard invertebrates. Surface-foraging granivorous rodents also have the option of hoarding food, with the bonus that most grains store well. They do not have to hoard only in cool weather, which limits storage time for the food of carnivores and insectivores. Surface-foraging, agile rodents that depend on speed to avoid predators thus do not store considerable quantities of fat on their bodies, but rather assemble food hoards to draw on during times of scarcity. For small mammals, the strategy of hoarding food seems to be frequently used.

The second environmental stress that many forms must meet is heat loss. This can be prevented by a variety of mechanisms, including both physiological and behavioral adaptations. Heat loss may be reduced by an appropriate microhabitat, such as a crevice or burrow. It may be reduced by morphological adaptations, such as layers of fat or thick coats of hair. It may also be controlled by a positive selection for a larger body size and a reduction in relative size of those parts of the body that dissipate heat, such as ears, tails, and limbs.

The reverse of this problem is adaptation to high temperatures where heat must be dissipated. Of course, this can be compensated for by a positive selection for a smaller body size, by the appropriate choice of a microhabitat, such as a burrow, where the ambient temperature is lower than that of the macroenvironment, and by adaptation to maintain bodily function at a higher deep body temperature. A species may evolve morphological structures such as large ears and dewlaps that rapidly dissipate heat when the blood is shunted to them. Even the color of an organism may undergo selection for increasing reflectance, thus reducing the rate at which heat is absorbed from the environment (Schmidt-Nielsen 1964; Hamilton 1973).

The third problem we should consider is water loss. Evaporative water loss is often tied to adaptations for exercise at higher ambient temperatures. The rate at which water is lost through evaporation from the lungs or mucous membranes can be controlled by appropriate microhabitat selection. This is often easier for a small mammal than for a large one. A large mammal can compensate for actual lack of water by moving to traditionally known water sources. In a small mammal, local lack of water can lead to a quick death. Small mammals are often well adapted for controlling the rate of evaporative water loss. They may have evolved discrete mechanisms for increasing the concentration of urine. This solution has been reached convergently by a number of desert-adapted rodents, including those of the New World family Heteromyidae and the Old World families Dipodidae and Gerbillidae (Schmidt-Nielsen 1964; Kirmiz 1962*a,b*).

Finally, mammals have managed to overcome chronic deficits in oxygen when adapting to certain environments. This is especially true for those forms adapted for either burrowing or an aquatic life. In aquatically adapted forms, lack of oxygen may be overcome by breathing only at intervals; during submergence, the blood is shunted to the brain and vital organs, while blood flow to skeletal muscle is reduced. The muscle can derive its energy from anaerobic respiration, compensated for during reventilation of the lungs and by periodically recirculating blood through skeletal muscle (Scholander 1940). Fossorial mammals must cope with oxygen deficits in a similar manner, though they are not under such severe restrictions as aquatic-adapted forms. As I indicated previously, a tendency toward compensation for oxygen deficits in fossorial forms may be related to the overall reduction in their metabolic rate, though this could just as well be an adaptation to control excess heat production (McNab 1966).

19.4 The Consequences of Trophic Specialization

It is now time to consider some of the broad consequences of adapting to modal feeding strategies.[2] First some general rules: It is axiomatic in ecology that the amount of biomass of living material for a given trophic level is inversely related to the number of links in the food chain (Odum 1971). This is in conformity with the second law of thermodynamics. Since plants are primary producers, deriving their energy directly from the sun, the biomass of plants for any given area will always vastly exceed the biomass of animals.

The biomass of herbivorous mammals in a terrestrial community usually exceeds the biomass of carnivores. Herbivores exploit materials near the base of the food chain, but there is a physiological problem in processing most plant material. If a species confines its feeding strategy to the by-products of plants, such as reproductive bodies (fruits and seeds) or sap, then only a small portion of the available plant biomass is tapped. To exploit the bulk of the plant biomass, the animal must adapt for processing leaves, stems, cambium, or even woody structures. This form of feeding requires vast modifications in the dental apparatus for masticating the food as well as modifications in the gastrointestinal tract for efficiently extracting the potential energy in plant structural carbohydrates (Eisenberg 1978; Janis 1976).

Obviously, invertebrates have developed many techniques for processing the energy directly from plants, and the invertebrate biomass, especially that of the arthropod component, presents a rather large energy source to be tapped in turn by mammals. It should not be surprising, then, that the numerical density of insectivorous mammals, both volant and terrestrial, can be rather high; but, since these are small mammals, their biomass will be low. On the other hand, adaptations for feeding on other vertebrates—that is, carnivorous adaptations—restrict the energy base available, and one would expect carnivores or omnivores to have relatively low numerical densities and low biomasses (Eisenberg 1980*a*). Mammalian specializations for feeding on fruit, seeds, nectar, and gum will result in numerical densities and biomasses slightly exceeding those of mammalian species adapted for feeding on vertebrates.

Thus it is axiomatic that true herbivorous mammals will, in any given area where green plant biomass is high, represent the largest biomass component of the mammalian fauna. Insectivorous and frugivorous adaptations come next in contribution to biomass, and carnivores constitute the smallest biomass component (see fig. 51) (Eisenberg and Thorington 1973; Eisenberg 1980*a*).

Let us consider the feeding adaptations in terms of their morphological consequences. One of the key features in the evolution of the Mammalia was the development of teeth that were specialized for different functions. This was a radical departure from the dental condition of reptiles. To this day we can recognize the major categories of tooth specialization in the form of incisors, canines, premolars, and molars. In fact a simple inspection of teeth will often tell us a great deal about

2. McNab (1980) explicitly reviews the metabolic consequences of certain classes of trophic specialization. His review extends the trends outlined here.

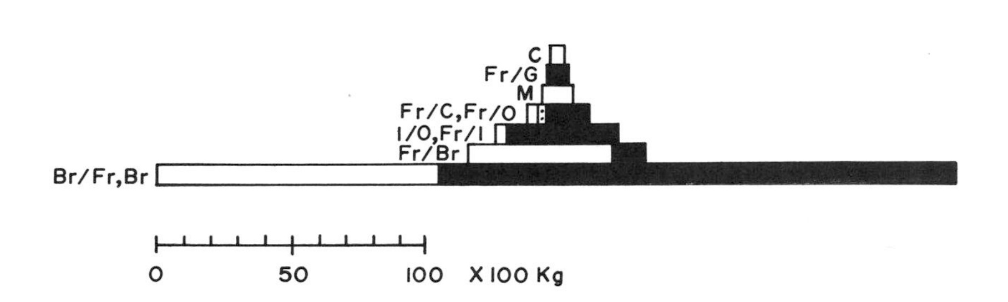

Figure 51. Typical biomass pyramid for a tropical mammal fauna (from Eisenberg and Thorington 1973). *C* = carnivore; *Fr/G* = frugivore/granivore; *M* = myrmecophage; *Fr/C, Fr/O* = frugivore/carnivore, frugivore omnivore; *I/O, Fr/I* = insectivore/omnivore, frugivore/insectivore; *Fr/B* = frugivore/browser; *Br/Fr, Br* = browser/frugivore, browser.

the dietary specializations of a species (Kay and Hylander 1978) (see fig. 52). A carnivore tends to have well-developed canines and premolars, with molars adapted for shearing. As food, animal bodies need a minimum of preparation for digestion, in contrast to plants. The more generalized a mammal's diet, the more generalized its tooth structure. Thus, an omnivore is characterized not by specialized shearing teeth, but by teeth that are somewhat low crowned and adapted for some degree of crushing. Feeding on arthropods selects for specializations in tooth structure (see fig. 53*d*) where the cusps are high for piercing but grinding surfaces are minimal.

Specialized adaptations for feeding on social insects such as ants and termites are associated with a regular morphological syndrome (Griffiths 1968). Often there is a reduction in the size of teeth or, in cases of extreme specialization, loss of teeth. The palate tends to be prolonged posteriorly; the tongue is extremely elongate; and large salivary glands are developed for coating the tongue with mucus when it is used to lap up social insects in their tunnels. In toothless forms the stomach often shows modifications for the trituration of the hard insect exoskeletons—for example, in the pangolin (Griffiths 1968).

So diversified is the use of substrate by ants and termites for nest-building that a parallel diversity may be found in any guild of continental myrmecophages. Adaptation for feeding on social insects may involve terrestrial, scansorial, and strictly arboreal foraging strategies. The interaction of social insects and their predators has been the subject of earnest study because the social insects have evolved numerous devices to repel predators, such as stings, bites, and complex methods of nest repair. Through such defense mechanisms, predation by anteaters can be tolerated without leading to the colony's extinction. The analysis of such coevolution problems is actively being investigated by E. Pages for the pangolins and by Montgomery and Lubin for the New World anteaters (see chap. 4).

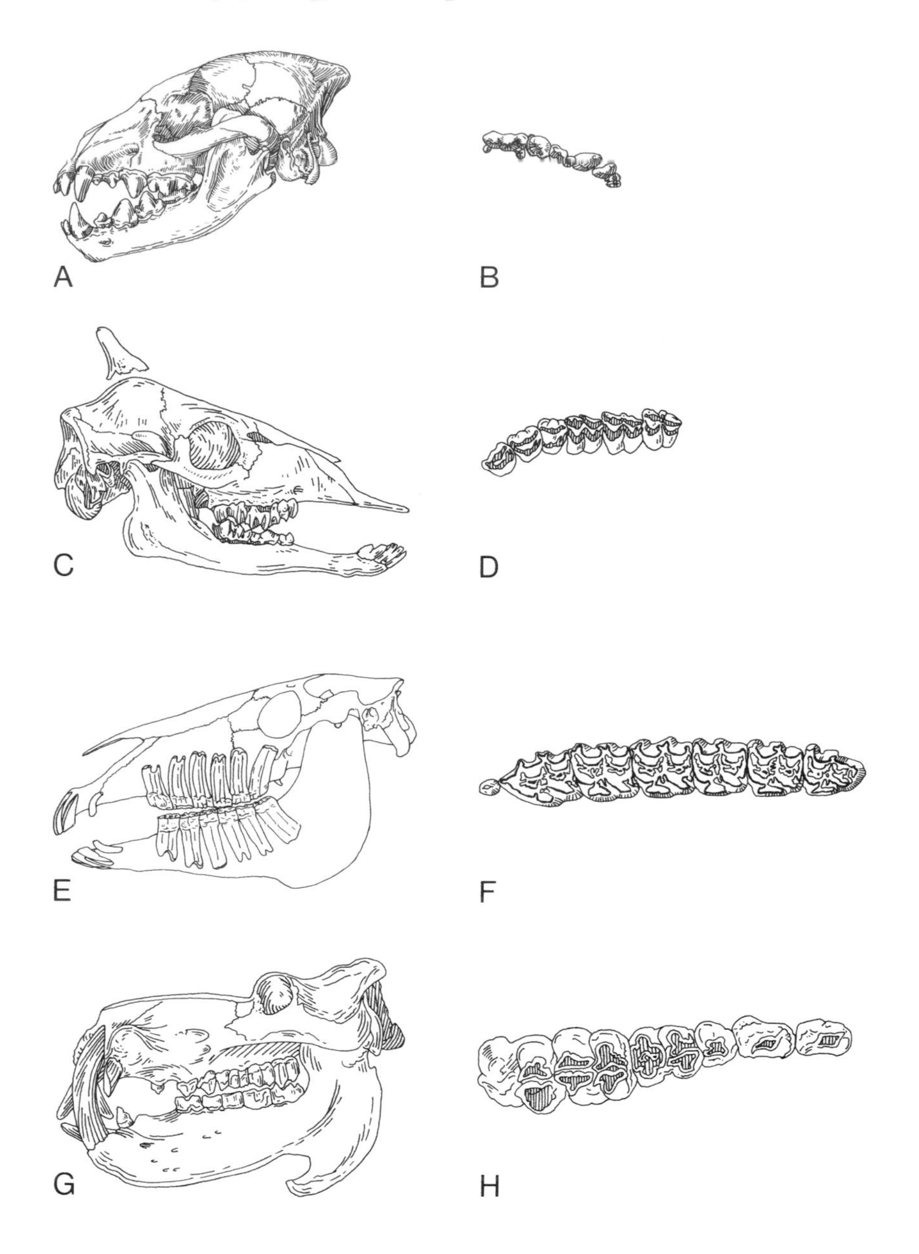

Figure 52. Comparison of molars for mammals with several disparate feeding styles. (*A*) Lateral view of *Crocuta crocuta* skull. (*B*) Ventral view of one-half of the upper dentition for *Crocuta crocuta*. (*C*) Lateral view of *Giraffa* skull. (*D*) Ventral view of one-half of upper dentition of *Giraffa*. (*E*) Lateral view of *Equus* skull. (*F*) Ventral view of one-half of upper dentition of *Equus*. (*G*) Lateral view of *Hippopotamus* skull. (*H*) Ventral view of one-half of upper dentition of *Hippopotamus*. Redrawn from Gregory 1951.

When we turn to advanced adaptation for feeding on plants, we can distinguish two broad categories of feeding strategies: mammals that are specialized for feeding on leaves (browsers), and mammals that have specialized for feeding on plants of the family Graminae, or grasses (grazers). Browsing forms, or folivores, typically show some

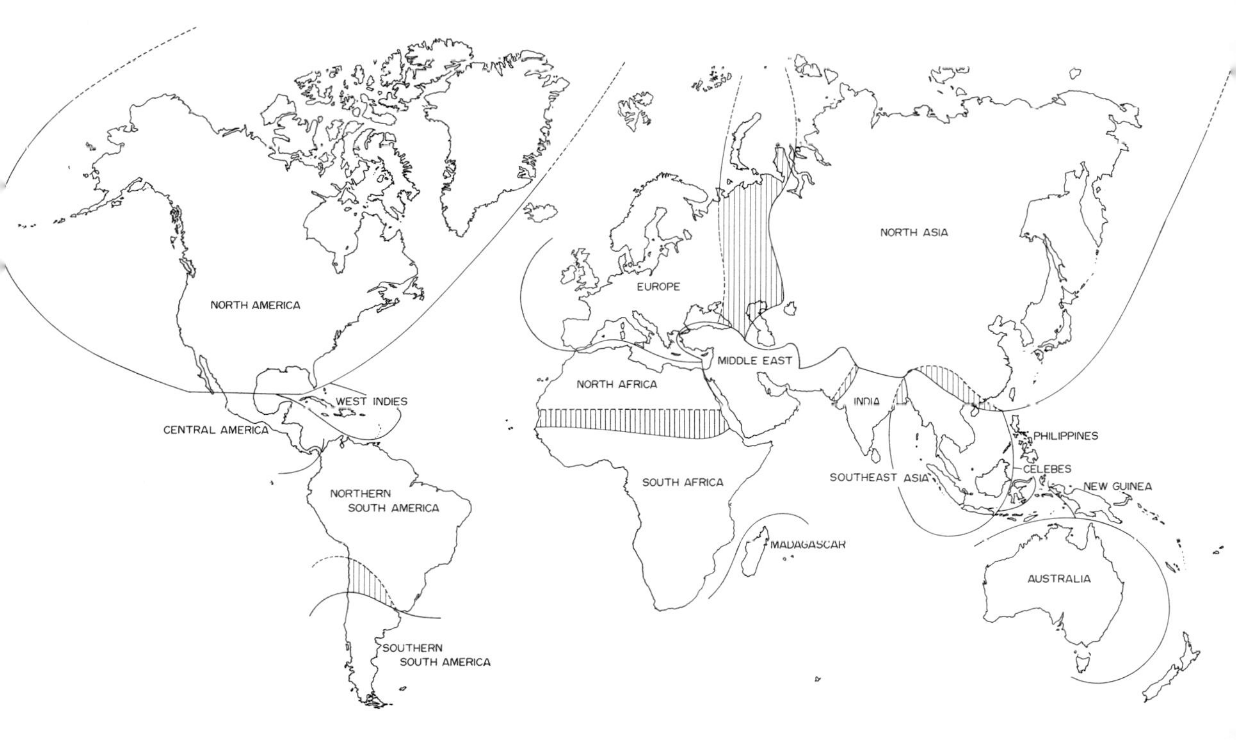

Figure 53. Zoogeographic regions as discussed in the text. *Shading* indicates regions of overlap.

modifications of the teeth for pulping leaves, but folivores often do not show the advanced specialization of their molars that can be discerned in those species adapted for grazing. The high silica content of grasses causes severe abrasion of the molars of grazers, so that they tend to evolve molars that are high-crowned and that often grow continuously throughout most of the individual's life (see fig. 52*e,f*).

Where a species is adapted for feeding on leaves or on grasses, problems of energy extraction are similar. Most of the energy in leaves and grasses is not in the form of stored starch, but is contained in structural carbohydrates, such as cellulose, which consist of simple sugars linked one to the other with what is known as the "beta linkage." No enzyme produced by the tissues of mammals can break this linkage down into simple sugars, but suitable enzymes are possessed by certain bacteria and protozoans. To tap the energy in structural carbohydrates, herbivorous mammals must evolve a symbiotic relationship with protozoans and bacteria. Special compartments in the gastrointestinal tract serve as refuge areas for bacterial and microbial symbionts that can reduce the beta linkage to simple volatile fatty acids as a first-order product.

Two general methods have been evolved within the Mammalia for providing microbial symbionts with a microenvironment. On the one hand, at the junction of the small and large intestines, a diverticulum of considerable size may be developed, termed the cecum. In this diverticulum, portions of the digesta from the small intestine may be introduced and there further acted upon by microbial symbionts. The second method involves the enlargement of the forestomach, which serves as an initial receptacle for plant parts; there they are acted upon by microbial symbionts before the material enters the first portion of the small intestine. This form of forestomach fermentation has been convergently evolved in the macropod marsupials, sloths, and Artiodactyla (see Parra, Eisenberg, and Bauchop in Montgomery 1978).

A cecum or intestinal diverticulum that serves as a locus for microbial digestion has been evolved by many rodents, the Perissodactyla, the Sirenia, the Hyracoidea, the Proboscidea, and some marsupials. One striking fact that one can readily observe is that the cecal method of microbial digestion involves an enormous variety of sizes, from extremely small mammals such as the vole, *Microtus californicus,* to elephants—almost the full range of terrestrial mammalian size-classes. On the other hand, those forms that use a portion of the forestomach for bacterial digestion appear to be confined to larger (but not the largest) size-classes. This is apparently because, if the forestomach is the locus for initial digestion, then food intake must be confined to bouts—first feeding, then forestomach digestion, then emptying the forestomach, then feeding again.[3] Apparently an extremely small mammal canot tolerate this interruption of food intake without eventually receiving too little energy to support its loss of heat. Thus the restricted, periodic nature of the rate of passage for forestomach digesters sets a lower limit on the size of animal that can adopt this strategy.

On the other hand, use of the cecum as an "assist" for dietary requirements appears widespread, especially in small mammals, which feed not only on the green parts of plants but also on such high-energy resources as seeds and fruits. In small mammals the cecum may be thought of as an adjunct and a step toward the utilization of plant structural carbohydrates, but it still permits rapid passage of food through the gut and, coupled with reingestion of feces, a sufficiently high energy return. The small mammal can be an herbivore with a low assimilation efficiency and still meet its energy budget (Parra 1978).

Why, one may ask, do some mammals that use the cecal mode of digestion also have the evolutionary option of becoming very large—larger than the Artiodactyla? The answer may be that selection can favor increased size beyond the limit for the forestomach-digesting Artiodactyla for two reasons: (1) As size increases toward that of an elephant, metabolic rate decreases, and the net demand for food of a high energy level decreases. (2) With its uninterrupted rate of passage,

3. In species that ruminate, the contents of the forestomach are regurgitated and chewed again before being passed to the second chamber of the stomach. This necessitates a further delay in the feeding process. However, remastication does reduce the food particle size, thereby aiding the subsequent digestive processes.

a giant herbivorous mammal with cecal digestion may feed on low-quality vegetation, but it must feed almost continuously. Hindgut fermenters benefit from this uninterrupted passage and from the ability to feed on low-grade forage, but large species may spend from 80 to 90 percent of their daily activity in feeding (Janis 1976; McKay 1973).

20 Mammalian Size Classes and Their Distribution According to Niche and Zoogeographic Constraints

If we analyze the size of mammals in terms of body length (exclusive of the tail), we can begin to define the average lengths for genera. By assigning genera to specific feeding and substrate exploitation categories, as defined in table 39, it is possible to make certain generalizations about body size and niche specializations. Table 43 presents the average size of a mammalian genus for each cell of the matrix as defined in table 41. The "generic averages" are arranged in descending order. According to this analysis, the mean head and body length for a mammalian genus is 619 mm. Aquatic mammals are the largest extant forms, with a mean length of 4,181 mm. Volant mammals are the smallest, with a mean generic average of 75 mm. Within each of the two extreme classes of adaptation for swimming and flight, one can observe great variation according to the trophic strategies.

If one considers trophic and substrate specialization together, there are fifty-three combinations (see table 43). The largest mammals are aquatic planktonivores—the baleen whales. They feed seasonally on the zooplankton and are thus only at the third level in a simplified food chain. In the sense of oceanic productivity, they are operating almost as terrestrial grazers. Large aquatic carnivores are a rare specialization (*Orca* and *Hydrurga*) but are intermediate in the large size category. The aquatic piscivores and mollusk-feeders are nearly the same size as the aquatic herbivores (*Trichechus* and *Dugong*), which utilize the offshore eelgrasses and algae.

The largest terrestrial mammals are included in the terrestrial browsers and grazers: Proboscidea, Perissodactyla, Artiodactyla, some Rodentia, and Hyracoidea.

Most mammalian trophic and substrate specializations grade down from 817 mm to 101 mm in head and body length. A rather sharp break occurs at 88 mm, when we encounter the semifossorial insectivores and then move on through the volant genera to the smallest size-class, the volant aerial insectivores.

Since these data are calculated on the basis of generic "averages," which in fact represent the midpoint derived from summing the mean values for the two extreme size-classes in a genus, it is useless to present ranges and attempt a statistical analysis. Table 43 should serve only as a guide and reference point in thinking about size and niche occupancy. In table 44, niche occupancies for zoogeographical areas will be compared.

Table 43 Mean Body Lengths for Mammalian Genera separated into Fifty-three "Macroniches"

Mean Head and Body Length[a] (mm)	Niche	Example
12.491	Aquatic planktonivore	*Eschrichtius*
4,875	Aquatic carnivore	*Orca*
3,204	Aquatic piscivore	*Tursiops*
3,127	Aquatic crustacivore/molluskivore	*Odobenus*
3,050	Aquatic herbivore	*Manatus*
1,870	Semiaquatic herbivore/grazer	*Hippopotamus*
1,508	Terrestrial herbivore/grazer	*Equus*
1,132	Terrestrial herbivore/browser	*Elephas*
817	Terrestrial frugivore/omnivore	*Ursus*
737	Semiaquatic crustacivore	*Enhydra*
718	Terrestrial carnivore	*Panthera*
619 =	Median Length for Mammalian genera	
597	Semiarboreal carnivore	*Eira*
585	Terrestrial myrmecophage	*Myrmecophaga*
575	Scansorial myrmecophage	*Tamandua*
573	Semifossorial myrmecophage	*Priodontes*
570	Semifossorial carnivore	*Taxidea*
565	Semifossorial herbivore/grazer	*Lagostomus*
533	Scansorial herbivore/browser	*Heterohyrax; Dendrohyrax*
508	Arboreal herbivore/browser	*Bradypus*
501	Arboreal frugivore/herbivore	*Alouatta*
498	Semiaquatic herbivore/browser	*Tapirus*
477	Scansorial frugivore/omnivore	*Didelphis*
473	Arboreal frugivore/omnivore	*Potos*
445	Terrestrial frugivore/granivore	*Dasyprocta*
427	Arboreal carnivore	*Felis marmorata*
321	Semiaquatic piscivore	*Mustela vison*
301	Scansorial frugivore/herbivore	*Coendou*
289	Terrestrial frugivore/herbivore	*Cuniculus*
277.5	Semifossorial herbivore/browser	*Lemmus*
265	Arboreal frugivore/granivore	*Cebus*
252	Semiaquatic frugivore/omnivore	*Hydromys*
248	Arboreal gumivore	*Euoticus*
223	Terrestrial insectivore/omnivore	*Chaetophractus*
221	Scansorial insectivore/omnivore	*Eupleres*
209.4	Fossorial herbivore/browser	*Ctenomys*
202.5	Fossorial frugivore/herbivore[a]	*Thomomys*
198	Arboreal insectivore/omnivore	*Metacheirus*
178	Semiaquatic frugivore/granivore	*Neofiber*
175	Semiaquatic frugivore/herbivore	*Ondatra*
165	Scansorial frugivore/granivore	*Sciurus*
156	Arboreal myrmecophage	*Cyclopes*
143.9	Semifossorial frugivore/herbivore[b]	*Spalacopus*
118	Arboreal nectarivore	*Tarsipes*
117	Fossorial insectivore/omnivore	*Talpa*
104.2	Semifossorial frugivore/granivore	*Cannomys*
104	Volant carnivore	*Phyllostomus*
101	Volant frugivore/omnivore	*Artibeus*
88	Semifossorial insectivore/omnivore	*Blarina*
81	Volant sangivore	*Desmodus*
75	Volant nectarivore	*Glossophaga*
74	Volant piscivore	*Noctilio*
66	Volant foliage-gleaner/insectivore	*Hipposideros*
58	Volant aerial insectivore	*Myotis*

[a] Mean length of generic "averages" for a defined "macroniche."
[b] Frugivore = feeding on tubers as well as "fruits" low to the ground.

Niche categorizations according to size are extremely useful if we wish to predict home range sizes. As McNab (1963) has shown, home range varies positively as a function of individual weight or group weight (Clutton-Brock and Harvey 1977). However, the home range of a carnivore is larger than that of an herbivore in the same size-class (see sect. 4.2). Several other corollaries are immediately evident: As the mean weight of the body increases, the density decreases; however, carnivore density decreases at a more rapid rate than herbivore density. Furthermore, the percentage contribution to total biomass in a community does not parallel density. To the contrary, large herbivores contribute a disproportionate percentage to mammalian biomass, while carnivores in the large weight-class contribute next to nothing (Eisenberg 1980*a*). The impact of large carnivores on an herbivore population can be considerable (Schaller 1972), but it does not even come close to limiting the density of the prey population (Sinclair 1977). The implications of this statement will be dealt with in chapter 30.

20.1 Niche Distribution on a Global Scale

Given the disparate global history of the terrestrial mammalian fauna, I felt it would be interesting to compare for various geographical regions the ways the niche matrix as defined in table 41 would be filled when the continents were subdivided according to the scheme illustrated in figure 53. Starting with North America as a standard, I looked to see what niches are discordant when we inspect north temperate Asia (see table 44). For example, in Asia niches 112 and 812 are filled, while they are not in North America. This results from the presence of *Ailurus* and *Myospalax* in Asia, with no exact equivalent in North America. Turning to Europe, niche 112 is filled, but in addition Europe has niche 802 in the form of *Genetta,* which is unfilled in northern Asia and North America. North America, on the other hand, has eight filled niches that Europe does not have, but these are replicated in north temperate Asia.

The lack of niche saturation in Europe is paralleled in temperate South America, which has ten macroniches unfilled when compared with North America, but the southern cone of South America has nine niches filled that North America leaves open. These include: 206, semifossorial myrmecophage; 112, fossorial frugivore/herbivore; 214, semifossorial herbivore/grazer; 410, semiaquatic frugivore/omnivore; 412, semiaquatic frugivore/herbivore; 516, volant sanguivore; 709, scansorial insectivore/omnivore; 810, arboreal frugivore/omnivore. These differences result from the southern extension of the Neotropical fauna into northern Argentina. I have defined North America narrowly in figure 53, and this difference would disappear in part if I included southern Mexico with North America.

Turning to the tropics, we find that Central America and Southeast Asia are almost equivalent in their niche occupancy; 209, the semifossorial insectivore/omnivore, is occupied in Southeast Asia but not in Central America; 112, fossorial frugivore/omnivore, is similar in its Asian occupancy; 614, the terrestrial herbivore grazer niche, is occupied by *Bos* in Asia but is vacant in contemporary Central America. Central America has niche 516 occupied by *Desmodus,* but the niche

Table 44 Terrestrial Macroniche Occupancy on a Zoogeographic Basis

Niche Number	North America	North Asia	Europe	Temperate South America	Southeast Asia	Tropical Africa	Tropical South America	Central America	South India	Australia
102	−	−	−	−	−	−	−	−	−	−
106	−	−	−	−	−	−	−	−	−	−
109	+	+	+	−	+	+	−	+	−	+
110	−	−	−	−	−	−	−	−	−	−
111	−	−	−	−	−	−	−	−	−	−
112	−	+	+	+	+	+	+	−	+	−
113	+	−	−	+	−	−	+	+	−	−
114	−	−	−	−	−	−	−	−	−	−
202	+	−	−	−	−	−	−	−	−	−
206	−	−	−	+	−	+	+	+	−	+
209	+	+	−	+	+	−	+	−	−	−
210	−	−	−	−	−	−	−	−	−	−
211	+	−	−	−	−	−	−	+	−	−
212	+	+	+	+	−	−	+	−	−	−
213	+	+	+	−	−	−	−	−	−	−
214	−	−	−	+	−	−	−	−	−	−
401	+	+	+	+	+	+	+	+	+	+
402	−	−	−	−	−	−	−	−	−	−
405	−	−	−	−	−	−	−	−	−	+
409	−	−	−	−	−	−	−	−	−	−
410	−	−	−	+	−	−	+	−	−	−
411	−	−	−	−	−	+	+	+	−	−
412	−	−	−	+	−	−	+	−	−	−
413	+	+	+	−	−	−	−	−	−	−
414	−	−	−	−	−	+	+	+	−	−
501	−	−	−	−	+	−	+	+	+	−
502	−	−	−	−	+	+	+	+	+	+
503	−	−	−	−	+	+	+	+	−	+
507	+	+	+	+	+	+	+	+	+	+
508	+	+	+	−	+	+	+	+	+	+

Table 44—*Continued*

Niche Number	North America	North Asia	Europe	Temperate South America	Southeast Asia	Tropical Africa	Tropical South America	Central America	South India	Australia
510	–	–	–	–	+	+	+	+	+	+
516	–	–	–	+	–	–	+	+	–	–
602	+	+	+	+	+	+	+	+	+	+
606	–	–	–	–	–	–	+	+	–	+
609	+	–	+	+	+	+	+	+	+	+
610	+	+	+	+	+	+	+	+	+	+
611	+	+	+	+	+	+	+	+	+	+
612	+	–	+	+	+	+	+	+	+	+
613	+	–	+	+	+	+	+	+	+	+
614	+	–	+	+	+	+	+	–	+	+
702	+	+	+	+	+	+	+	+	+	+
703	–	–	–	–	–	–	–	–	–	–
704	–	–	–	–	–	–	–	–	–	–
706	–	–	–	–	+	+	+	+	+	–
709	–	–	–	+	+	–	+	+	+	+
710	+	+	+	+	+	–	+	+	+	+
711	+	+	+	+	+	+	+	+	+	+
712	+	+	–	+	–	+	+	–	–	+
713	+	–	–	–	–	–	+	+	+	+
714	–	–	–	–	–	–	–	–	–	–
802	–	–	+	–	+	+	–	–	–	–
803	–	–	–	–	–	–	–	–	–	+
804	–	–	–	–	–	+	+	–	–	–
806	–	–	–	–	–	–	+	+	–	–
809	–	–	–	–	+	+	+	+	+	+
810	–	–	–	+	+	+	+	+	+	+
811	+	+	+	–	+	+	+	+	+	+
812	–	+	–	–	+	+	+	+	+	+
813	–	–	–	–	+	+	+	+	–	+

is empty in Southeast Asia. Fossorial niches seem underrepresented in the Neotropics.

Niche occupancy or nonoccupancy is of interest, but counting actual numbers of genera on a continental scale brings out additional considerations about comparability of continental mammalian faunas (see table 45). Tropical Africa contains 214 genera. Tropical South America equals this number, with a high of 219. Southeast Asia, with its much smaller area, has only 158 genera. Of the Southeast Asian genera, 36 percent are semiarboreal or arboreal, while only 26 percent are so for tropical South America and 15 percent for tropical Africa. In Africa 19 percent of the genera are volant, while the percentage rises to 26 percent in Southeast Asia and 32 percent in South America. The diversity of volant and arboreal niches in South America and Southeast Asia must surely relate to the long history of continuous tropical forest cover in these two regions.

By contrast, tropical Africa's forest has been smaller in area, and tropical Africa has had a long history of cyclic semiaridity. The terrestrial herbivore (browser/grazer) niche is replete with diversity at the generic level in tropical Africa. Twenty-four percent of the fauna falls into this category, whereas the figures are 11 percent for Southeast Asia and only 9 percent for South America. (Keast 1972, employing a different set of niche categories, reaches a similar conclusion.)

Fossorial and semifossorial niches account for 4 percent of all genera in the tropics but include 10 percent to 11 percent in Eurasia, North America, and temperate South America. Do the demands of heat dissipation restrict this niche to higher altitudes in the tropics?

In the temperate zones of Eurasia, North America, and South America, arboreal and semiarboreal mammals include some 11 to 15 percent of the genera. In this case the percentage probably reflects a lower niche diversity in conifer or temperate deciduous forests than in the tropics. The volant niche in the same temperate areas rests at 7 percent of the total genera (see sect. 16.4).

Australia, with its 96 genera, shows a pattern intermediate in niche saturation, but then the bulk of Australia is temperate and xeric. Bats include 20 percent of the genera, which is lower than tropical areas but higher than the temperate zone. Arboreal and semiarboreal forms make up 27 percent of the total genera, and this approximates the levels of Southeast Asia. Terrestrial browsers and grazers make up 14 percent of the genera, which is twice that shown by tropical South America and more nearly approaches the value of tropical Africa. One could conclude that the long history of aridity in Australia has promoted numerous grazing niches but that the marsupials' long phylogenetic history of arboreality has sustained a high level of arboreal niche occupancy even in the face of reduced forest cover.

Contemporary continental differences in niche occupancy mean that in tropical South and Central America and Southeast Asia small mammals will dominate in the fauna. Australia and tropical Africa are somewhat intermediate and resemble temperate zones in the distribution of size-classes. The evolutionary experiments in large mammals are con-

Table 45 Niche Occupancy at a Generic Level for Various Zoogeographic Regions

Region	Substrate Categories								Σ	Feeding Categories															
	1	2	3	4	5	6	7	8		1	2	3	4	5	6	7	8	9	10	11	12	13	14	15	16
Southeast Asia	3	3	5	4	42	44	24	34	159	9	10	2	—	—	1	22	2	22	27	31	14	13	4	—	—
Northern South America	2	7	4	14	69	66	19	38	219	12	9	9	1	—	8	27	9	14	48	41	19	14	6	—	3
Southern Africa	8	1	3	8	39	123	11	21	214	8	19	3	1	—	2	23	3	21	24	45	18	29	17	—	—
North America	10	19	11	5	16	51	14	2	128	10	8	2	—	2	—	11	3	16	13	24	15	16	8	—	—
North Asia	5	10	8	7	10	76	12	8	136	13	8	—	—	1	—	8	2	12	12	33	22	15	10	—	—
Australia	1	1	6	3	19	38	9	19	96	5	7	3	—	1	2	11	2	17	12	9	8	11	8	1	—
Europe	2	4	3	5	10	40	8	5	77	7	7	—	—	—	—	8	2	4	7	18	7	11	6	—	—
Middle East	1	5	—	2	17	50	6	7	88	2	9	—	—	—	—	13	3	8	9	21	10	8	5	—	—
India	1	—	3	2	22	45	9	8	90	5	6	—	—	—	1	15	1	8	15	16	6	8	8	—	—
Southern South America	2	6	7	6	5	35	8	1	70	6	4	—	—	—	1	5	—	9	11	7	14	6	6	1	1
Madagascar	—	1	—	1	15	14	7	13	51	1	3	—	1	—	—	12	1	14	6	4	5	4	—	—	—

Note: Numbers refer to numbers of genera in each category. Figures are exclusive of the Cetacea.

ducted in those continental areas that support continuous grazing areas with admixtures of low-stature shrubs for browsing. Through time and climatic change, these areas have shifted globally.

The complementarity principle of Darlington (1957) is best illustrated by closing this section with a consideration of size-classes and trophic strategies. Only nonvolant, terrestrial mammals will be considered. Given the equivalency of potential niches in two separate geographical areas with similar soils and climates, the principle of complementarity suggests that, even starting from different phylogenetic stocks, the niches will eventually be occupied by long-term natural selection. The classic example is the comparison of the Australian marsupials with the African or Asian mammalian faunas.

In figures 54 through 57, I have compared the nonvolant terrestrial mammalian faunas of California, Panama, the York Peninsula of Australia, Malaya, and Tasmania. Eight trophic strategies are considered as a function of the frequency of size-classes present as represented by generic averages.

California has many small species in the frugivore/granivore category, which reflects the explosive radiation of rodents in the arid deserts (see fig. 54). Comparing California with Panama, we note the larger range of size-classes in the browser/grazer niche of California. Panama,

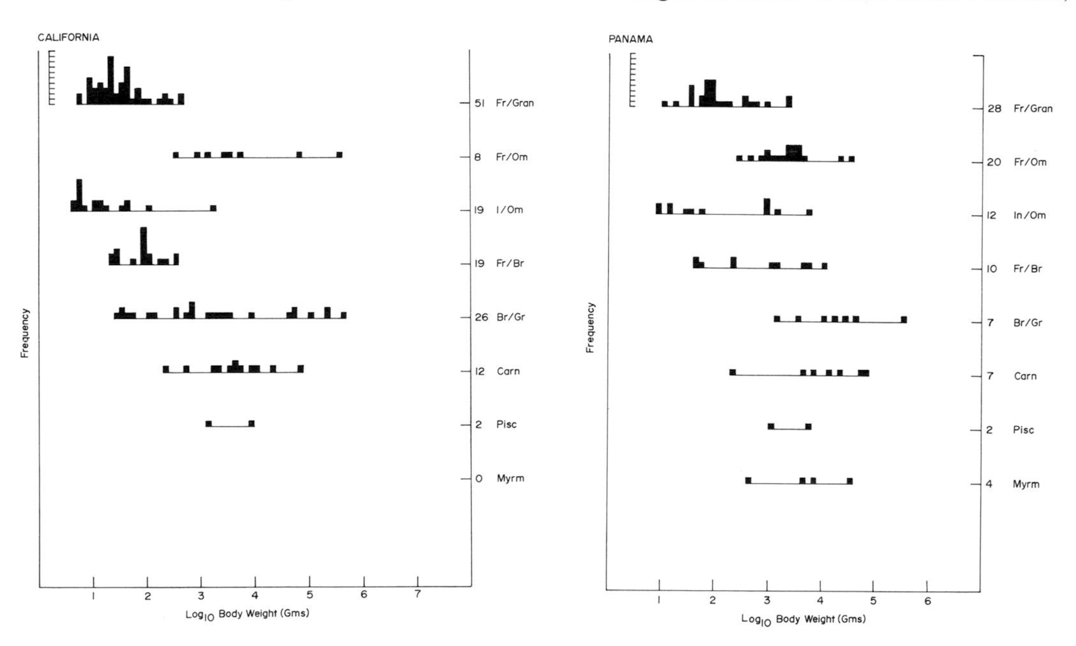

Figure 54. Frequency of size-classes according to trophic strategy: California compared with Panama. Log_{10} body weight (g) is plotted against the abscissa. The number of genera within each trophic class is given at the right of each size series. *Fr/Gran* = frugivore/granivore; *Fr/Om* = frugivore/omnivore; *I/Om* = insectivore/omnivore; *Fr/Br* = frugivore/browser; *Br/Gr* = browser/grazer; *Carn* = carnivore; *Pisc* = piscivore; *Myrm* = myrmecophage; Data from Ingles 1965 and Handley 1966.

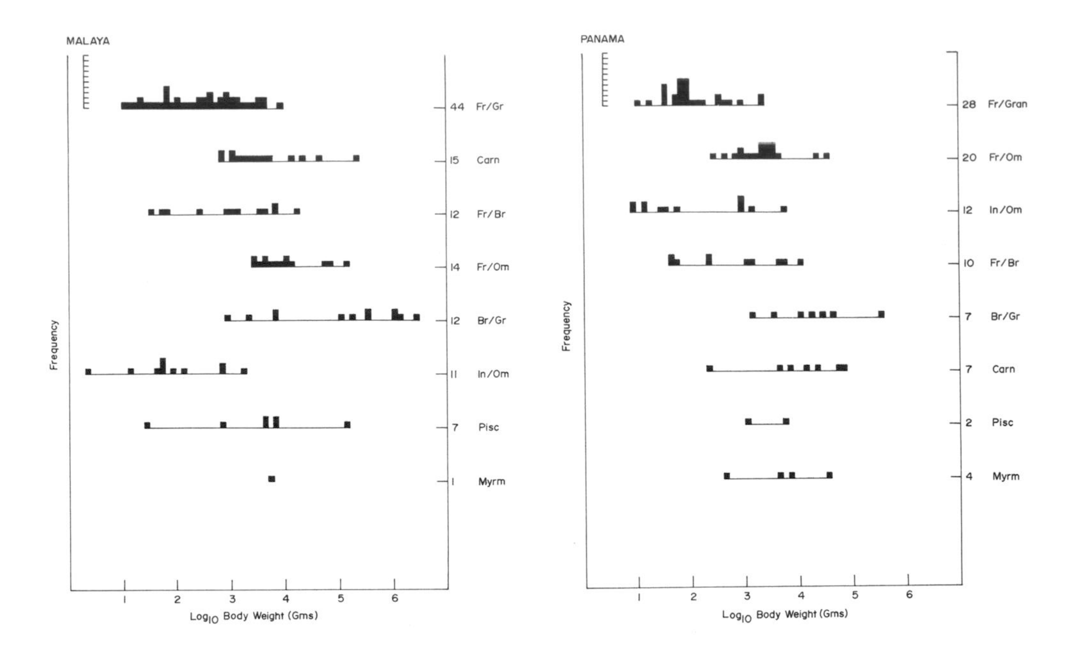

Figure 55. Frequency of size-classes according to trophic strategy: Malaya compared with Panama. Legend as in figure 54. Data from Medway 1978 and Handley 1966.

with its tropical forests, has a restricted grazing niche. The myrmecophagous niche is unoccupied in temperate California.

On the other hand, Panama and Malaya show an astonishing convergence in niche occupancy and in the frequency of size-classes in each niche (fig. 55). There are differences in that Malaya has more species of true carnivores and frugivore/granivores than does Panama. This results from the explosive radiation in Malaya of the Felidae, the scansorial Murinae, and the Sciuridae.

When we compare tropical Australia with Panama (fig. 56), we see that the frugivore/omnivore and piscivore niche is depauperate in Australia. Tropical Australia has a richer niche diversity in almost all categories than does temperate Tasmania (fig. 57).

In all major continental areas, the major trophic categories appear to be occupied; however, the true piscivorous niche is nearly absent in Australia. The myrmecophagous niche wanes as one proceeds from the tropics to the temperate zone. The mammalian frugivore/omnivore niche appears to be depauperate in Australia. Two questions may be raised: Are fruit trees relatively less abundant in Australia? Is there significant competition from diurnal psittaciform birds to preempt this niche? One may note that arboreal marsupials moved rapidly into the nocturnal folivore/frugivore niche, comparable to the hypothesized Eocene radiation of the primates in tropical areas of the Old World and the New World. Marsupials in general are nocturnal, and the noctur-

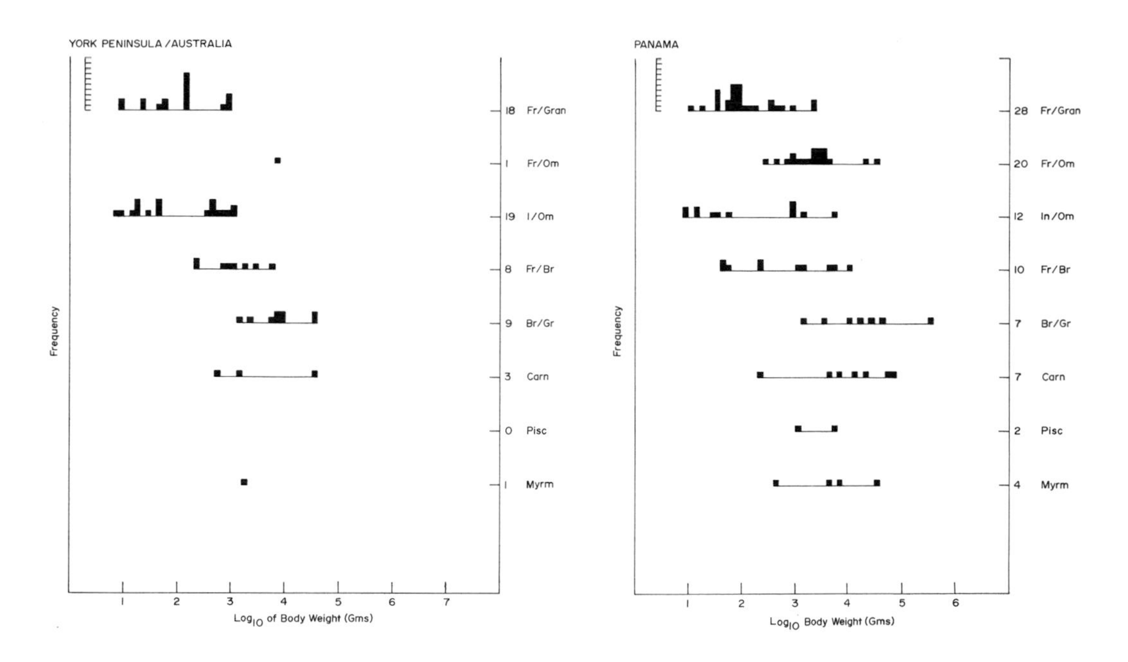

Figure 56. Frequency of size-classes according to trophic strategy: York Peninsula of Australia compared with Panama. Legend as in figure 54. Australian data from Troughton 1967.

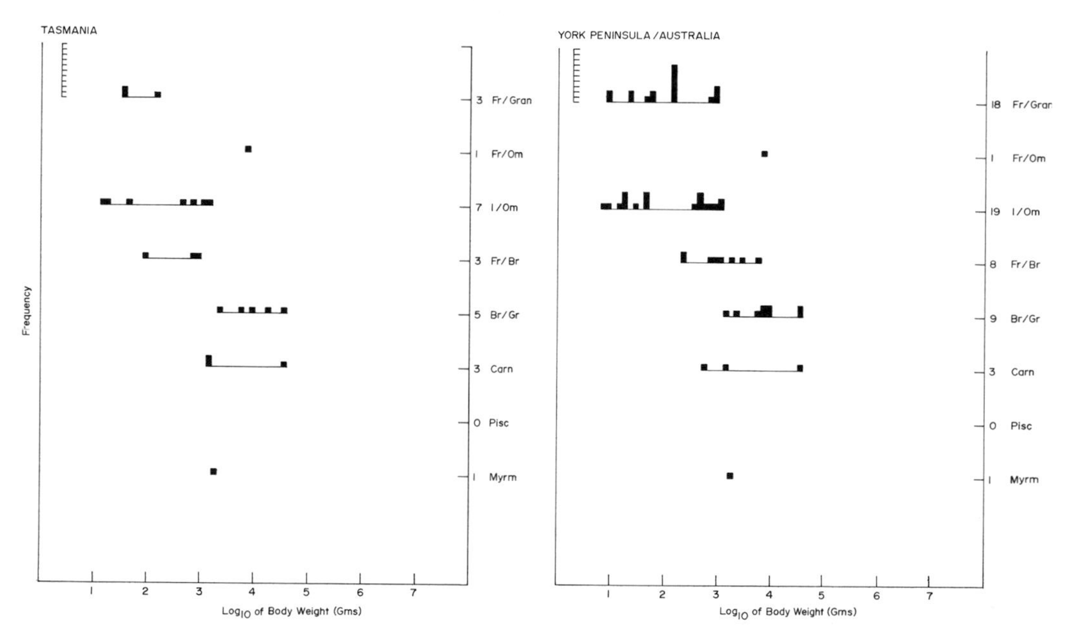

Figure 57. Frequency of size-classes according to trophic strategy: Tasmania compared with York Peninsula of Australia. Legend as in figure 54. Data for Tasmania from Green 1974; data for the York Peninsula and Queensland from Troughton 1967.

nality syndrome has dominated in the early marsupial radiations. The nocturnal frugivorous bats may have preempted the true frugivore niche and in turn may have led the marsupials to exploit a more mixed folivorous strategy.

21 The Relative Size of the Mammalian Brain

In 1973 Harry Jerison published his classic treatise, *Evolution of the Brain and Intelligence*. By an analysis of the endocasts of fossil forms, he was able to outline the changes in the relative size and proportions of the mammalian brain through phylogenetic time. Jerison reviews the literature concerning the diverse studies of brain/body weight relationships. The absolute weight of the mammalian brain (y) increases as the mass of the body (x), according to the equation ($y = bx^{.66}$).[1]

Relative brain size has increased during phylogenetic time on the contiguous continents, but the Australian mammals and the extinct Neotropical ungulates showed lower brain/body weight ratios than did contemporary forms in North America and Asia. These differences in the rate of change in relative brain size Jerison attributes in part to differences in selective pressure deriving from differences in the relative efficiencies of the predatory mammals isolated on the southern, island continents. (South America was an "island" from the late Paleocene until the Pliocene.)

Changes in the relative size of the brain's major subdivisions show interesting trends through time. Early mammals had large olfactory lobes with central projection areas from the first cranial nerve. They also had relatively large areas of the brain to process information from the seventh cranial nerve. Hearing and the chemical senses were apparently important for the nocturnal early mammals, and hearing requires much central processing, with considerable brain enlargement. Visual information in early mammals did not involve fine form discrimination until they began to invade the arboreal niche. Since much visual information can be processed peripherally (Jerison 1973, p. 265), little enlargement of those parts of the brain concerned with vision was discernible in early nocturnal mammals. Increasing use of sight and diurnality led to a parallel enlargement in the cortical and subcortical areas receiving neuronal connections from the second cranial nerve and corresponding enlargements in brain mass (Jerison 1973, p. 277). The evolution of diurnal habits further accelerated the encephalization of the Mammalia. Contemporary mammals and birds have vastly larger brains in proportion to their body weight than do reptiles and amphibians.

1. Lande (1979) analyzes the slopes for brain/body weight regressions, demonstrating the radical change in slope when such variables as ontogeny and set (genus, family, or order) are analyzed separately (see also sect. 23.4).

In contemporary mammalian taxa, a relatively large brain may result from differential increase in the size of the cortical mass, the subcortical structures, or the cerebellum. For example, the cerebellum is often relatively large in aquatic and arboreal forms. It is axiomatic that those areas of the brain that do increase in size are often related to specializations in certain sense organs and correlate with the collection, storage, and retrieval of specific kinds of data relevant to the niches the animals exploit. Pirlot and Stephan (1970) attempted to correlate the relative differences in size increase for various parts of the chiropteran brain with feeding strategies. They noted that both pteropid and phyllostomatid bats had relatively large brains and relatively large volumes of neocortex compared with equal-sized insectivorous bats. Eisenberg and Wilson (1978) followed this study with an analysis of brain/body weight ratios for the order Chiroptera. We found that, regardless of the phylogenetic relationship, bats that specialized for frugivority, nectarivority, carnivority, or sanguivority all had significantly larger brains than aerial insectivorous bats of comparable body size.

In 1973 Jerison calculated the slope and intercept for a series of mammalian brain weights regressed against body weight. The sample included very few bats. Eisenberg and Redford reassembled brain/body weight data for 547 mammalian species. The data sources are included in table 46. We balanced our sample to represent the approximate proportions of living species from the ordinal taxa and included 227 species of the Chiroptera. When the regression analysis was completed, we found that the values differed from those in Jerison's analysis: the intercept was −1.257 and the slope was 0.74.[2] On the basis of these values, we then calculated the encephalization quotients

Table 46 Primary Sources for Brain/Body Weight Data

Taxon	Reference
Mammalia	Jerison 1973; Wirz 1950; von Bonin 1937; Crile and Quiring 1940
Marsupialia	Moeller 1975
Insectivora	Bauchot and Stephan 1966; Stephan, Bauchot, and Andy 1970
Primates	Stephan, Bauchot, and Andy 1970; Bauchot and Stephan 1969
Rodentia	Pilleri 1959, 1960*a,b*, 1961*a,b*; King 1965; Compoint-Monmignaut 1973
Chiroptera	Mann 1963; Stephan and Pirlot 1970; Eisenberg and Wilson 1978; Pirlot and Stephan 1970
Perissodactyla	Edinger 1948
Artiodactyla	Oboussier 1972
Carnivora	Thiede 1973; Schumacher 1963; Radinsky 1969
Edentata	Eliot-Smith 1898; Röhrs 1966
Cetacea	Gihr and Pilleri 1969

2. Jerison (pers. comm.) has demonstrated that an exponent of 0.66 may be the most useful value for interpreting the evolution of sulci and gyri in large brains. The folding of the brain's surface increases the total brain surface area relative to its volume, thereby increasing the area for the neocortical sensory and motor projection systems.

for selected species that are included in Appendix 6. An encephalization quotient (EQ), as defined by Jerison (1973), is the ratio between the observed brain weight and the expected brain weight for a defined body weight. The expected value is that value that lies on the regression line for the entire set. Since the regression for the class Mammalia is an abstract value, all the EQ can tell one is how much the brain weight differs from an "average mammal" of a given size-class (or body weight).

Figure 58 presents the entire set of data points we employed, bounded by lines enclosing the minimum area for all points. Jerison also plotted sets by these "convex polygons." Figure 59 abstracts the values for some morphologically conservative mammals. As can be seen from table 47, the Marsupialia, Edentata, and Insectivora have similar slopes and intercepts. Figure 60 demonstrates the subsets for insectivores, marsupials, and some bats. Note that the soricine insectivores are very similar to the marsupials, but the tenrecine insectivores have brain/body weight ratios among the lowest. Small carnivores (*Mustela rixosa*) and primates (*Microcebus*) have very large brains for their body size. Figure 61 displays a similar group of subsets derived from the Rodentia and Lagomorpha. Note again the relatively large brain for tree squirrels and some hystricomorph rodents. Reference to Appendix 6 and the encephalization quotients will reflect this stratification numerically. An EQ equal to 1.0 is an "average" mammalian brain size; a value >1.0

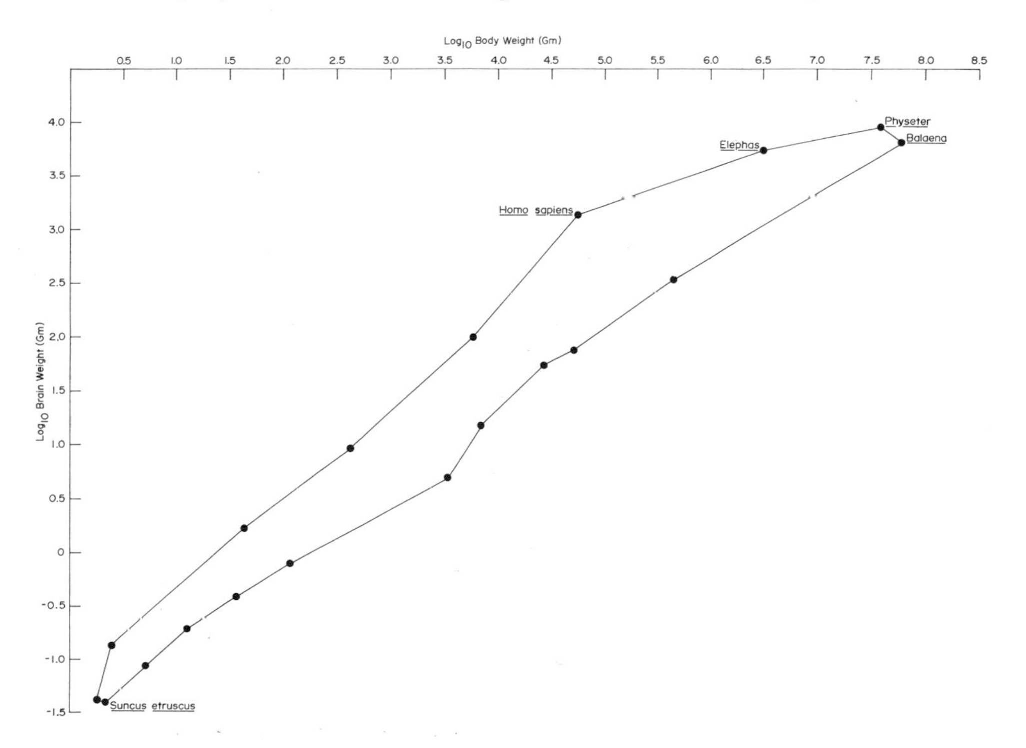

Figure 58. Convex polygon bounding the set of mammalian brain/body weight values (see text).

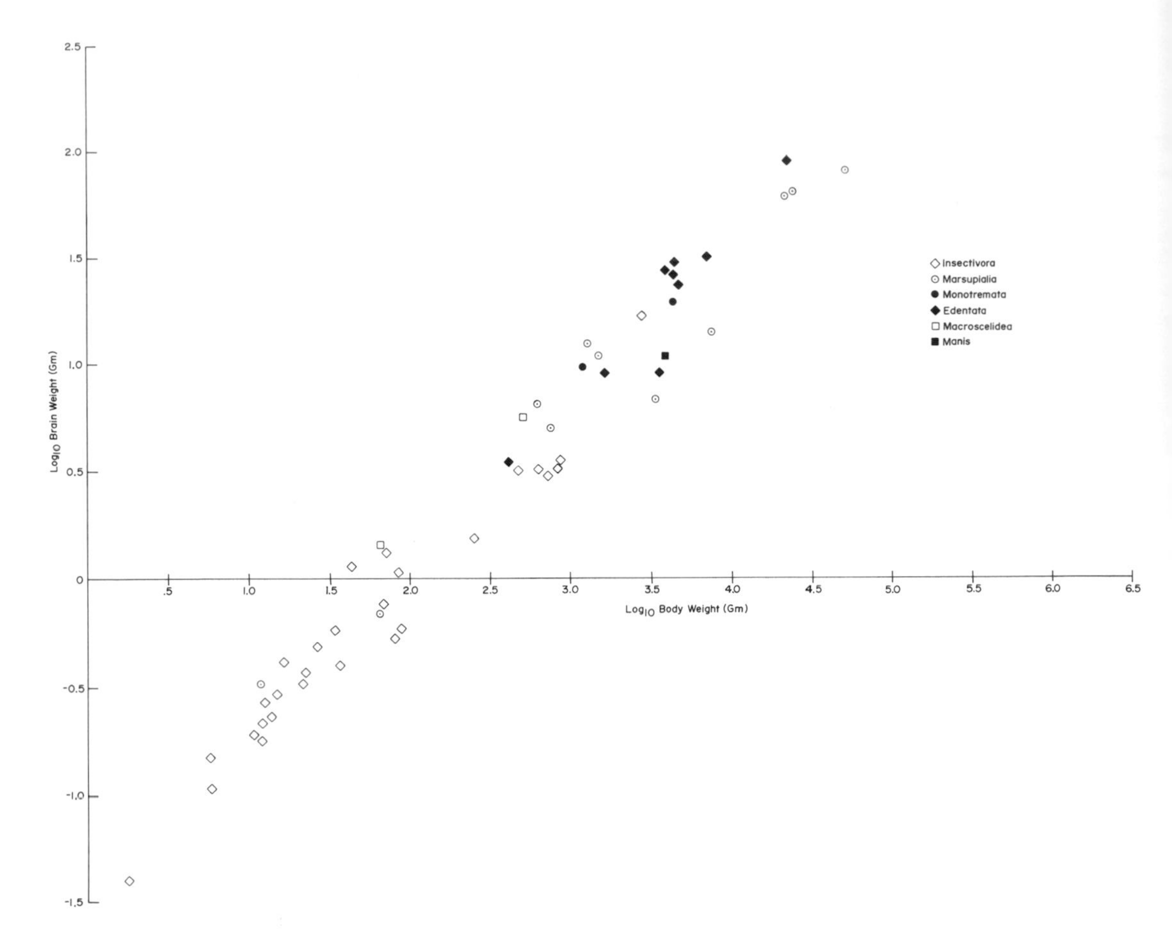

Figure 59. Double log plot of brain/body weight regression for a series of morphologically conservative mammals.

Table 47 Brain/Body Weight Regression Analysis (double $\log_{10}$ plot)

Taxon	*b*	*r*
All Mammalia	−1.26	.74
Marsupialia	−1.28	.68
Insectivora	−1.33	.68
Chiroptera	−1.34	.79
Primates	−0.64	.67
Edentata	−1.24	.72
Rodentia	−0.99	.63
Cetacea	1.96	.23
Carnivora	−0.40	.57
Pinnipedia	−0.72	.63
Perissodactyla	−0.61	.60
Artiodactyla	−0.41	.54

Note: $\text{Log } Y = b + \log X \cdot r$.

indicates a brain larger than the "average," and a value <1.0 represents a brain mass lower than the hypothetical average.

The Monotremata have respectable brain sizes, with *Ornithorhynchus* almost at the mammalian average. The Marsupialia show consid-

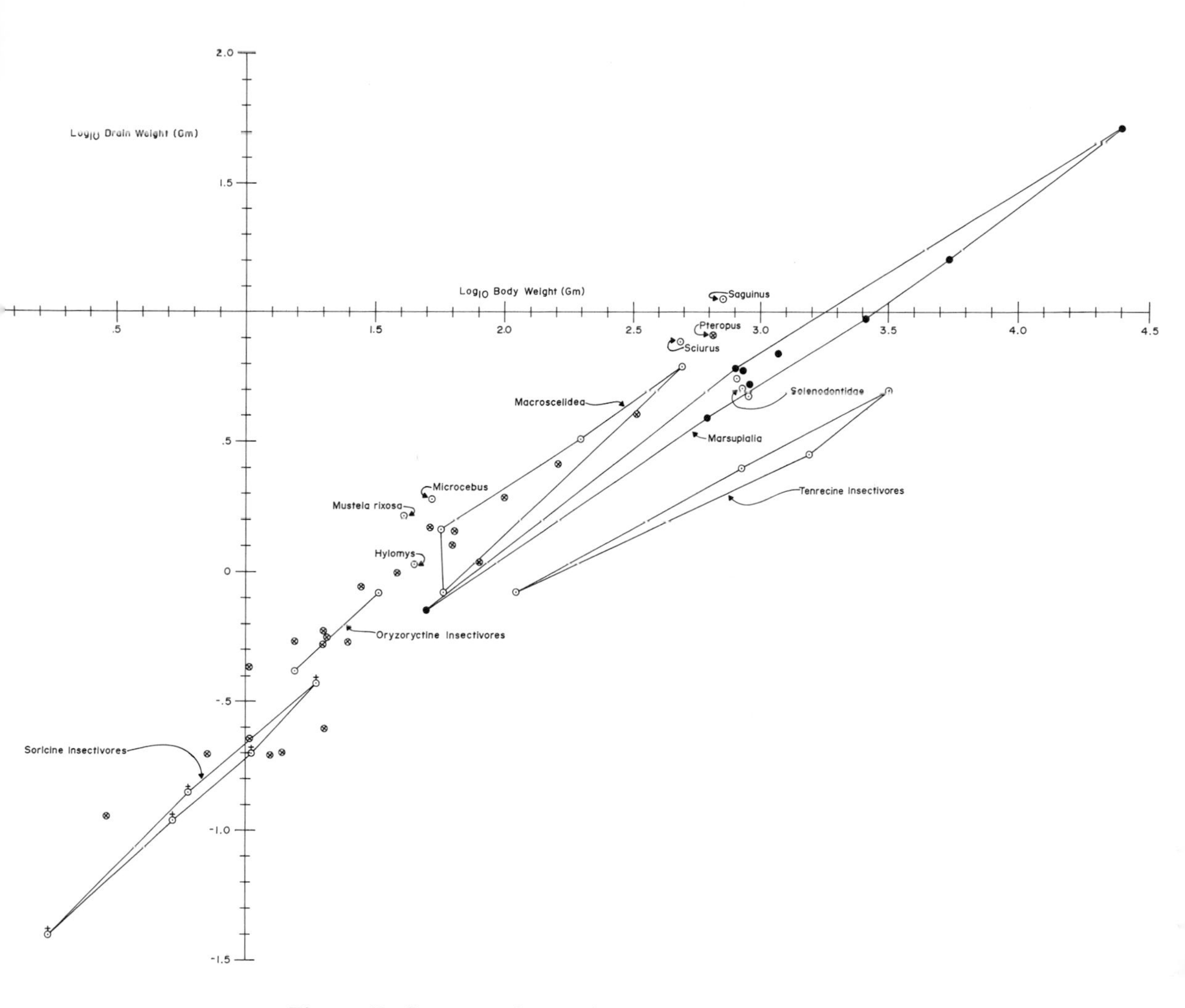

Figure 60. Convex polygons for subsets of mammalian brain/body weight values. The Marsupialia are intermediate between the elephant shrews, Macroscelidea, and the tenrecine insectivores. Several other representative genera of the Mammalia are included for comparison. The bats segregate into two groups; those having low brain/body weight ratios are as low as or lower than many soricine insectivores. Phyllostomatid and pteropid bats have elevated brain sizes, well above those of the smaller marsupials and soricine insectivores.

erable scatter. The genus *Dasyurus* has the largest relative brain size for the Dasyuridae, and *Caluromys* is the highest for the Didelphidae. The Insectivora are typically low in brain/body weight ratio, but the Talpidae are higher than average. In the Edentata, the sloths and anteaters are much higher than the Dasypodidae. The Chiroptera show tremendous diversity. The aerial insectivores, such as the Molossidae and Vespertilionidae, barely exceed the "basal insectivore" in relative brain size. The Pteropidae and Phyllostomatidae generally have EQ levels well above 1.0.

As might be expected, the Rodentia exhibit tremendous scatter, but the arboreal Sciuridae are consistently higher in EQ values. The pri-

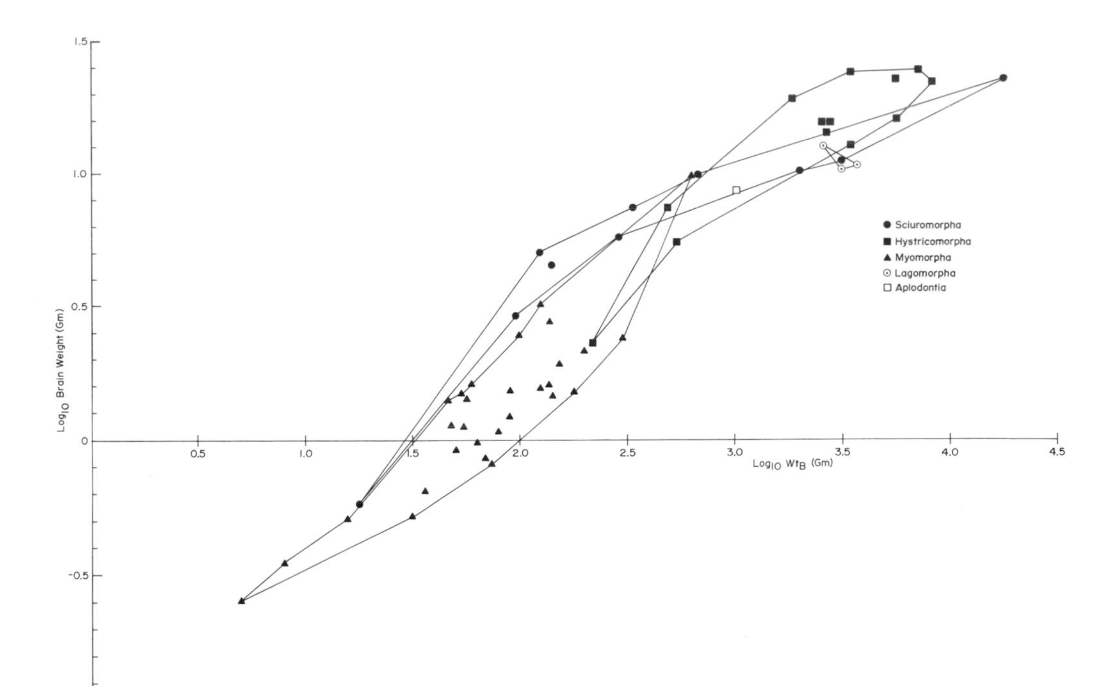

Figure 61. Plot of brain weight against body weight for a series of rodents and lagomorphs. Note that the Sciuromorpha have rather high values. The Hystricomorpha (sensu lato) show considerable scatter, but many attain high relative brain sizes compared with the Sciuromorpha.

mates are all very high, with *Homo sapiens* (at 7.0+) far above the other Mammalia. Carnivores also have high EQ values, though considerably less than the primate averages (see also table 47). The Pinnipedia compare favorably with the Carnivora. The Cetacea exhibit tremendous variability, but the baleen whales are extremely low, while the Delphinidae attain values higher than those of infrahuman primates. Proboscideans' values are comparable to those of larger primates. The zebra approaches the median value for proboscideans, but the remaining Perissodactyla and Artiodactyla show wide variations. Where sexual dimorphism is pronounced, males have very low values relative to females (see Appendix 6).

So-called average values as developed from the literature often rest on one or two specimens. Since the relative brain size varies considerably within a species depending on age and sex, the values in Appendix 6 must be taken only as a guide.

EQ values tell us nothing about what parts of the brain are enlarged relative to others, yet these ratios allow some simple correlations. Figure 62 demonstrates that high EQ values tend to be associated with an increased potential longevity for both marsupials and eutherians. Figure 63 indicates that EQ values correctly factor out any necessary relationship to body size. The same conclusions can be derived from

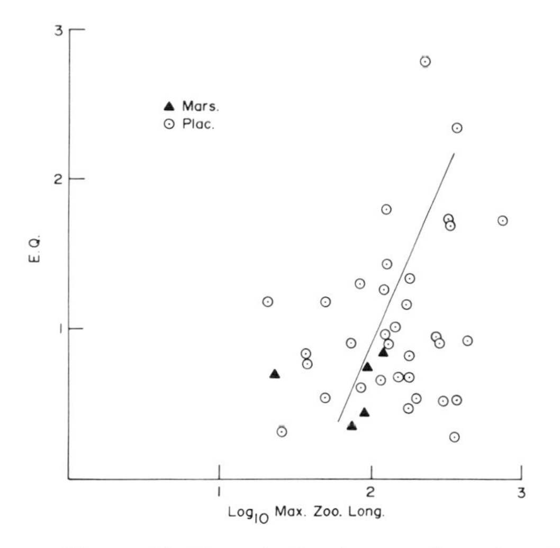

Figure 62. Encephalization quotient plotted against maximum longevity for a series of marsupials and eutherians. A weak positive correlation is demonstrable.

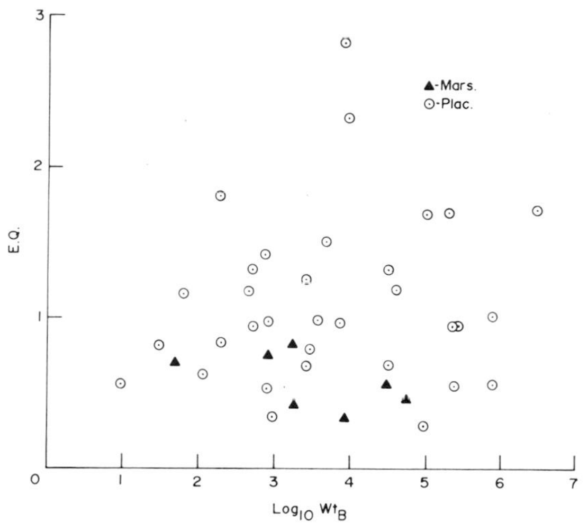

Figure 63. Encephalization quotients plotted against $\log_{10}$ mean body weight for a series of marsupials and eutherians. No correlation is demonstrable, because the EQ calculation is based on the regression line. The degree of scatter is emphasized, as are the generally low EQ values for the marsupials relative to the eutherians.

an inspection of Appendix 6. (In figs. 62 and 63, *Homo sapiens,* with an EQ value of 7.5–7.1, has been omitted.)

For broad comparisons of natural history, encephalization quotients can be very suggestive of a form/function relationship. In table 48, a series of closely related forms are compared with respect to their forms of antipredtor strategy. Passive antipredator strategies, involving rolling into a spiny or armored ball, correlate with low encephalization quotients. Active antipredator strategies, which involve directed flight and outright offensive behaviors such as counterattack, are generally correlated with higher encephalization quotients.

Table 48 Relative Brain Size and Antipredator Strategy

	Encephalization Quotient	
Taxon	Passive Antipredator Strategy	Active Individualistic Antipredator Strategy
Monotremata		
Ornithorhynchus anatinus	—	.94
Tachyglossus aculeatus	.72	—
Insectivora		
Echinops telfairi	.40	—
Setifer setosus	.45	—
Microgale talazaci	—	.77
Erinaceus europaeus	.51	—
Blarina brevicauda	—	.75
Solenodon paradoxus	—	.69
Edentata		
Dasypus septemcinctus	.79	—
Tamandua tetradactyla	—	1.452
Pholidota		
Manis javanicus	.47	—
Rodentia		
Erethizon dorsatum	.72	—
Agouti paca	—	1.09

Table 49 Relative Brain Size and Locomotion

	Encephalization Quotient	
Taxon	Locomotion in Two Dimensions	Locomotion in Three Dimensions
Monotremata		
Ornithorhynchus anatinus	—	.94
Tachyglossus aculeatus	.72	—
Insectivora		
Scalopus aquaticus	—	1.34
Galaemys pyrannicus	—	1.17
Blarina brevicauda	.75	—
Solenodon paradoxus	.69	—
Sorex araneus	.62	—
Edentata		
Euphractus sexcinctus	.86	—
Myrmecophaga tridactyla	.81	—
Bradypus variegatus	—	.95
Choloepus hoffmani	—	1.1
Chiroptera[a]		
Plecotus townsendi	—	.808
Antrozous pallidus	—	.61
Megaderma lyra	—	1.14
Rodentia		
Sciurus niger	—	1.43
Marmota monax	.83	—

[a] Chiroptera are treated separately (see Eisenberg and Wilson 1978).

If the form of locomotion is analyzed, one will note that complex locomotor patterns involving movement in three dimensions, such as arboreality or aquatic adaptations, are strongly associated with high encephalization quotients, whereas movement in essentially two di-

Table 50 Life-History Correlates of Encephalization Quotients

Low Encephalization Quotient	High Encephalization Quotient	Neural Structural Correlates
Maximum preprogramming of information	Information storage and retrieval (neocortical) based on individual experience; superimposed on preprogramming	Neocortex ↑[b]
Specialization in one or two sensory modalities	Multimodal sensory input with visual system highly developed	Subcortical projections ↑
Passive antipredator strategies	Active individualistic antipredator strategies	Sensorimotor projection ↑
Locomotion in two dimensions	Locomotion in three dimensions	Cerebellum ↑
Minimum parental association	Long period of total life spent in learning situation[a]	Neocortex ↑
Poor but ubiquitous food source	Rich but very dispersed food source	Neocortex ↑

[a] Correlates with deferred reproduction, iteroparity, and long life-span.
[b] Arrows indicate structure increased in relative size.

mensions generally is associated with a lower encephalization quotient (table 49).

Low encephalization quotients seem to involve maximum preprogramming of information in a species' nervous system and a reduced capacity to store and retrieve information; thus learning capacity is curtailed (Jerison 1973). Low EQ values correlate with specialization in only one or two major sensory input systems. Higher encephalization quotients in mammals are generally associated with sensory input from at least three modalities, with the diurnal, visual input being prominent (Jerison 1973). Low encephalization quotients are further associated with a minimum time of parental association (see chap. 34) and the utilization of an ubiquitous food source (Eisenberg and Wilson 1978). High encephalization quotients in general strongly correlate with a long potential life-span, late sexual maturation, and an extreme iteroparous reproductive strategy (see table 50 and chap. 34; see also Eisenberg 1975*a*).

22 Reproduction and Development

22.1 Introduction

All mammals reproduce sexually. After a period of courtship, the male deposits sperm in the reproductive tract of the female by means of an intromittent organ. In mammals sex is determined by two types of sperm. The male's sex chromosomes are XY in the diploid state, and his zygotes are either Y or X. Diploid female chromosomes are XX. Most litter sizes are slightly biased toward males at conception, but apparently the survival of the two sperm types may vary according to the discrete environment of the oviduct. There is a lively debate concerning the "control" females may exercise over the sex ratio at conception (Trivers and Willard 1973).

In mammals we find three quite different forms of reproduction. For example, the platypus, *Ornithorhynchus anatinus,* shows one breeding season per year, and only the left ovary of the female is functional. At the time of ovulation, the egg is approximately 3 mm in diameter. After fertilization the egg passes down the oviduct, where it receives first a coat of albumen and then a shell from various gland fields differentiated along the course of the oviduct. At the time of laying, the egg is approximately 18 mm by 14 mm. The time the egg spends in the uterus is not precisely known, but it is somewhere between twelve and twenty-eight days. After laying, the female incubates the eggs in a nest for approximately 3 weeks. This is the classical monotreme system. The system of the echidna is somewhat similar, but the egg is laid and incubated in a pouch (see chap. 3). In the monotremes there is no gestation as such. During the incubation phase the developing embryo draws upon the yolk and nutrients stored in the egg much like a reptile or bird.

The marsupials have a slightly different system. Consider the red kangaroo, *Megalia rufa*. Unless a female becomes pregnant, she has an estrous cycle every 33 days that persists throughout the entire year. Should she become fertilized, the ovum begins development in the uterus, drawing upon nutrients secreted from the uterine glands but, in the main, drawing its sustenance from a yolk sac. The membranes of the marsupial ovum are exactly homologous to those in a bird or reptile. Although a yolk sac placenta is formed briefly in the uterus, it does not transfer nutrients from the maternal circulatory system in the manner shown by eutherian mammals; rather, secretions from the uterine glands are taken up through the yolk sac. The young leaves the uterus after approximately 17 days and, by its own movements, transfers to the pouch, where it attaches to one of the teats. Lactation persists for some 6 months. Immediately after the birth of the young, the female will come into estrus again and may mate. Since she has a newborn suckling, the ovum undergoes a series of divisions but then remains in the uterus without implanting. Thus we have a phenomenon of embryonic diapause. When her nursing young decreases its suckling, the new blastocyst will implant and, after a 17-day gestation birth will occur. The previous young is now out of the pouch and will still nurse from one teat, while the newborn young attaches to the second teat and begins development (Griffiths, McIntosh, and Leckie 1972). The female will immediately come into estrus again, and development of the fertilized ovum will again be delayed as long as the newborn young nurses continuously (Sharman 1970).

A third example of the eutherian system of reproduction is the fox, *Vulpes fulva*. One estrous period occurs each year, lasting approximately 4 to 5 days. Should the female mate during this period and conceive, then after a gestation period of approximately 60 days the young will be born and nursed, and the female will not come into estrus again for another year. In the fox, howeever, intrauterine development involves the active exchange of nutrients and wastes between the fetal and maternal circulations. The young is not dependent on yolk in the egg for nutrition, and the key to this development process hinges on the evolution of the eutherian placenta (see below).

The main difference between the fox's system and the kangaroo's system is that in the fox the gestation period is longer, with active nutrient transfer from the female, and hence the young is born in a more advanced state. Furthermore, placentation in the fox allows for efficient transfer of nutrients between the maternal and fetal circulatory systems. Intrauterine development is much more complex in the eutherian mammal than in the marsupial. Since marsupials are born after a very brief time in the uterus, they are small relative to the mother's body weight. For this reason a comparison between marsupials and eutherians must take account of the marsupial young's less advanced state at birth.

The preceding examples typify the major trichotomy in mammalian reproduction. In the red kangaroo, we note a delay in development, or diapause of the embryo (Sharman 1970). This is not typical for all marsupials, nor is it exclusive to marsupials. In eutherians a variety of delays in development have been described. Many temperate-zone bats show delayed fertilization; mating takes place in autumn, but ovulation and fertilization do not occur until spring. The spermatozoa are stored and nourished in the reproductive tract of the female (Wimsatt 1960). Other Chiroptera, such as *Macrotus californicus,* exhibit fertilization upon mating, but the zygote divides slowly over a period of 15 to 16 weeks (Bradshaw 1962). This has been referred to as delayed development. A variant of this process is delayed implantation, where fertilization occurs, with zygote formation and cleavage to the blastocyst stage, but the blastocyst does not implant in the uterus but instead remains quiescent for a variable interval. Many carnivores, the roe deer (*Capreolus*), and the nine-banded armadillo (*Dasypus novem-*

cinctus) exhibit this phenomenon (see Asdell 1964). It has also been described for the convergently evolved fruit bats, *Eidolon helvum* (Pteropidae) and *Artibeus jamaicensis* (Phyllostomatidae), by Mutere (1967) and Fleming (1971), respectively.

A variation of delayed implantation may be shown by some rodent species that have a postpartum estrus and, after mating and fertilization, do not experience implantation for 3 to 4 days. The delay is controlled by the stimulation and lactation induced by the first litter (Svihla 1935). The adaptive significance of delayed implantation or delayed development cannot be explained by a unitary selective pressure but must be examined case by case (see Asdell 1964). Further consideration of ovarian cycles, testicular cycles, and environmental regulation of reproduction is undertaken in Sadleir (1969) and Austin and Short (1972).

Placentation in the Eutheria allows for an efficient transfer of nutrients between the maternal and fetal circulatory systems. The placenta has both maternal and fetal tissue components. Amoroso (1958) defined placental types for the Eutheria based on differences in the embryonic origin of the fetal portion and the tissue layers separating the maternal and fetal circulatory systems. In eutherians the chorion and the allantois fuse to form a plate of cells that either is in apposition to the uterine wall or invades it to fuse with the maternal tissue. A series of five placental types may be defined, based on the degree of tissue loss on both the maternal and the fetal sides. At one extreme is the so-called epitheliochorial placenta of ungulates, where the fetal and maternal blood vessels are separated by epithelial and connective tissue on both the maternal and the fetal portions of the placenta. At the other extreme is the hemoendothelial placenta, where the endothelial cells of the fetal blood vessels are actually bathed in the blood of the female circulatory system. Obviously the latter type of placenta is a derived characteristic that maximizes the efficiency of transfer between the fetal and maternal circulatory arcs (see Luckett 1975).

22.2 Intrauterine Development and Length of Gestation

Leaving aside the various forms of delay in development for any given size-class, a longer time in the maternal uterus should result in a fetus that is larger and better developed at birth. This generalization would hold if placental efficiencies in nutrient transfer were equivalent and if the metabolic rate of the female remained high. Racey (1973) has demonstrated that bats of the genus *Pipistrellus* extend gestation by arresting fetal development when the females are deprived of food and enter torpor.

When we compare two eutherian species that do not show gestational increase resulting from torpor, the intrauterine growth rate of the young is not necessarily the same. For example, the intrauterine growth rate of the pig shows a dramatic increase in rate between the 90th day and the 108th day of intrauterine time. In the human and the rat the rate of growth is relatively uniform and shows no such spurt just before parturition (see fig. 64). In caviomorph rodents, the egg may undergo a long period of trophoblast development before extensive division of the blastocyst itself. The extremely long gestation period of some cavio-

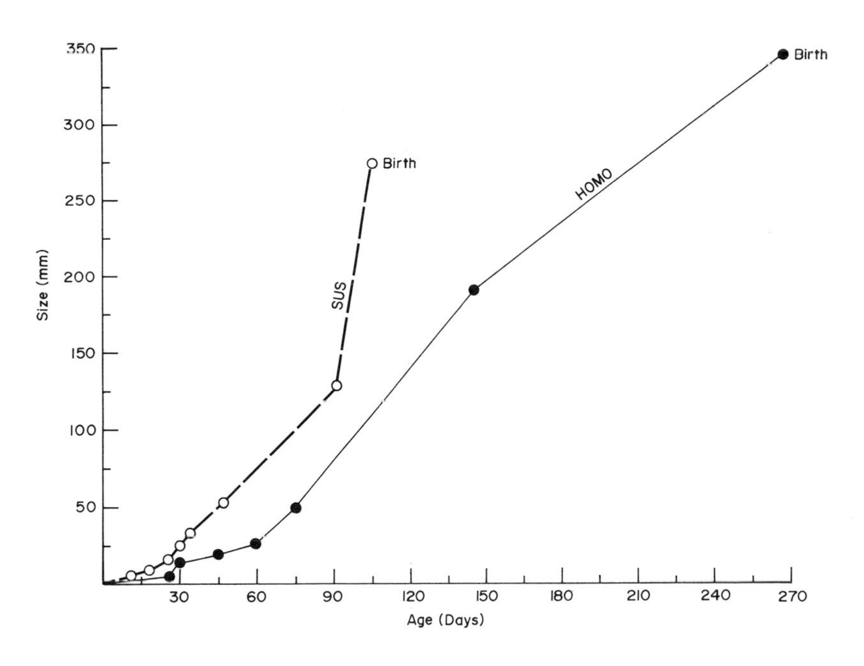

Figure 64. Intrauterine growth rate for *Sus scrofa* and *Homo sapiens*. Note the rapid growth rate for *Sus* between 90 and 105 days. Data from Altman and Dittmer 1962.

morph rodents is an outcome not only of the production of large, precocial young but also of the rather long percentage of intrauterine time devoted to the elaboration of the trophoblast (Weir 1974).

Huggett and Widdas (1951) studied specific fetal growth rates in eutherian mammals. The cube root of fetal weight, when plotted against time, gives a linear growth form. Fetal growth must be considered in two phases, however. There is an initial phase of growth with a very low rate followed by a dramatic linear growth phase with a much higher rate (see also fig. 64).[1] Frazer and Huggett (1974) showed that the specific fetal growth rate varies widely within the Eutheria. Whereas primates and insectivores have low rates (0.01–0.02), seals and cetaceans have very high rates (0.08–0.58).

Although in marsupials gestation is brief, the blastocyst does not implant (the Peramelidae are an exception), and trophoblast formation does not take place, apparently there is some nutritive transfer of material from the female's uterine glands to the conceptus. Frazer (1977) has demonstrated that fetal growth rates in marsupials range from 0.03 to 0.08, well within the range of some eutherians. By comparing fetal growth rates of mammals, birds, and reptiles, Frazer concludes that metatherians, eutherians, and birds have all developed high fetal growth rates compared to those of reptiles. The high growth rates appear to be correlated with the evolution of endothermy.

1. Van der Merwe (1978) has performed an excellent analysis of fetal growth for *Miniopterus schreibersi*.

Kihlström (1972) regressed gestation length against body weight for 208 eutherians and found that gestation length (G) increased as body size (W) of the adult female increased, according to the simple allometric expression $G = aW^b$. On a double log plot, when orders were considered separately, the value for b (0.17) was quite similar for all terrestrial mammals. The intercepts were dramatically different, and Kihlström was able to correlate the low intercept values with an endotheliochorial placenta and the higher values with a hemochorial placenta. He suggests that placenta type somehow correlates with gestation length. In an effort to analyze the problem, I have assembled reproductive data for eutherians and marsupials in Appendix 4 and performed a series of analyses.

If we take a series of mammals and plot gestation time as a function of body weight, we notice a general trend toward increased gestation time with increased adult body size (Kihlström 1972). However, the condition of young at birth is highly variable from one species to another, since some young are born with their eyes closed and no fur whereas others are born with eyes wide open and able to run about, are well furred, and have the ability to thermoregulate. These two classes of young are referred to broadly as altricial and precocial. To compensate for this difference in condition at birth, we may add to the gestation time the time from birth until the eyes open. Thus, altricial forms come more in line with their precocial relatives. If we regress the average body size of the adult against gestation time plus the time to eye opening, including both eutherians and marsupials, we can demonstrate a general tendency for developmental time to increase as body size increases. There is still a great deal of scatter; some of this scatter results because eye opening itself is not an absolute indicator of equivalent stages of development (Portman 1965) (see figs. 65, 66, and 67).

The marsupials now fall well within the limits of eutherian developmental time because we have compensated for their short gestation period by adding the time to eye opening. The scatter is instructive if we analyze the positions of the higher taxa.

Figure 65 shows that the order Cetacea has a fast rate of intrauterine development. This is especially true for the baleen whales, which can compress gestation time into a brief 12 months. For their size-class, the baleen whales exhibit the highest intrauterine growth rates (Frazer 1977). The fissiped carnivores have rapid rates of development, although there is much scatter. Lagomorpha and elephant shrews exhibit the fastest rates for small mammals. The Chiroptera and Perissodactyla have among the slowest rates of development for the Mammalia.

In figure 66, the Rodentia are portrayed with the entire set of development times for the Mammalia. As one might expect, the scatter is considerable in this diversified order. The Caviomorpha exhibit slow rates of development. I will discuss this further in a subsequent section. Figure 67 demonstrates the rapid developmental rates for the Artiodactyla and the low rates for the Primates. Primate developmental times are similar to those of the Chiroptera. The Edentata are intermediate

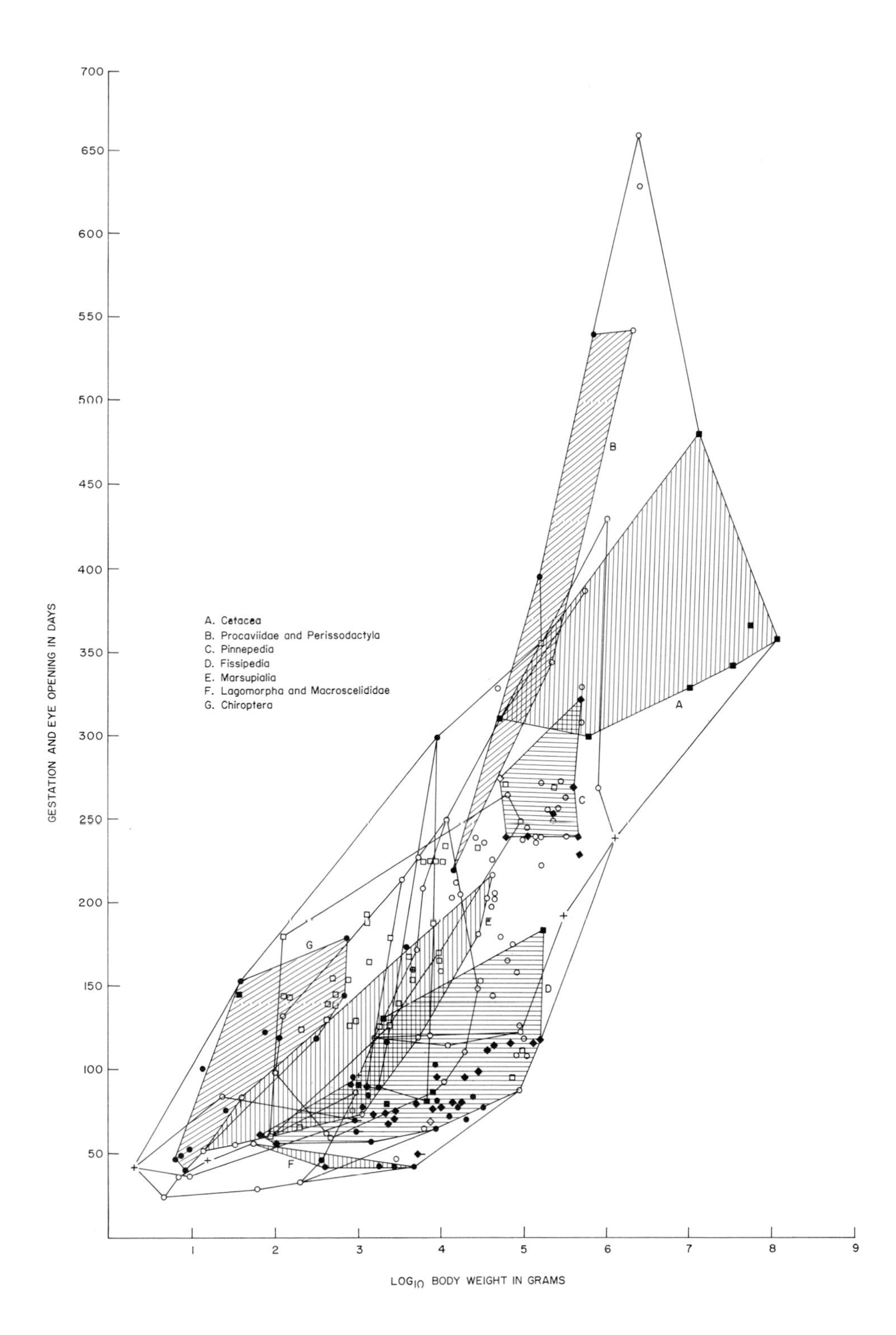

Figure 65. Plot of gestation plus eye-opening time against body weight; weights are expressed in $\log_{10}$ (g). Limit of the set of points is bounded; specific taxa are outlined (see figs. 66 and 67). Note that the Procaviidae, Perissodactyla, and Chiroptera have rather slow rates of development. The Lagomorpha and Macroscelidea have extraordinarily rapid rates of development, as do some members of the Cetacea. Note that the Marsupialia, relative to the Eutheria, are intermediate.

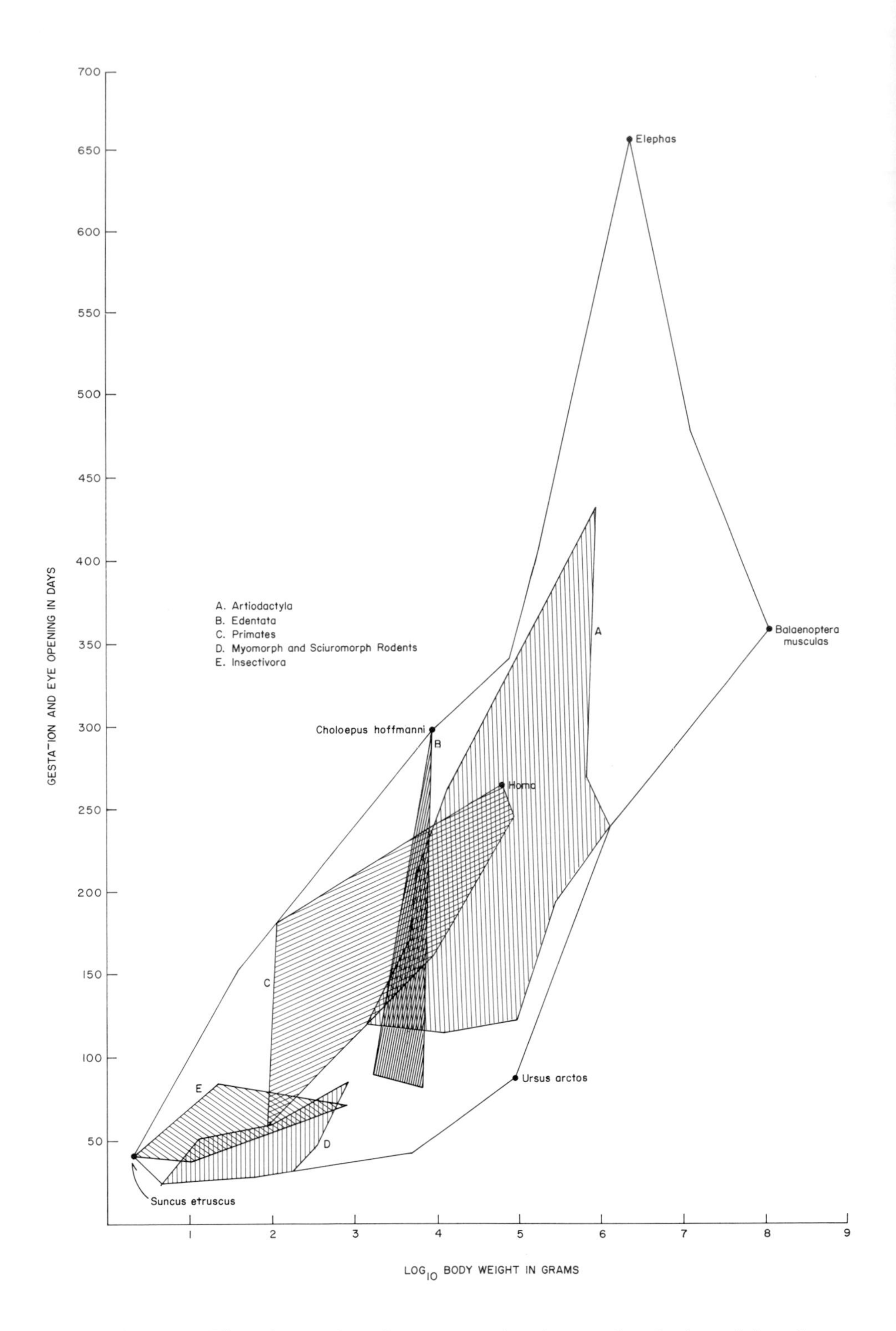

Figure 66. Plot of gestation plus eye-opening time against body weight; plotting as in figure 65. The myomorph and sciuromorph rodents have relatively rapid developmental times compared with most caviomorph rodents.

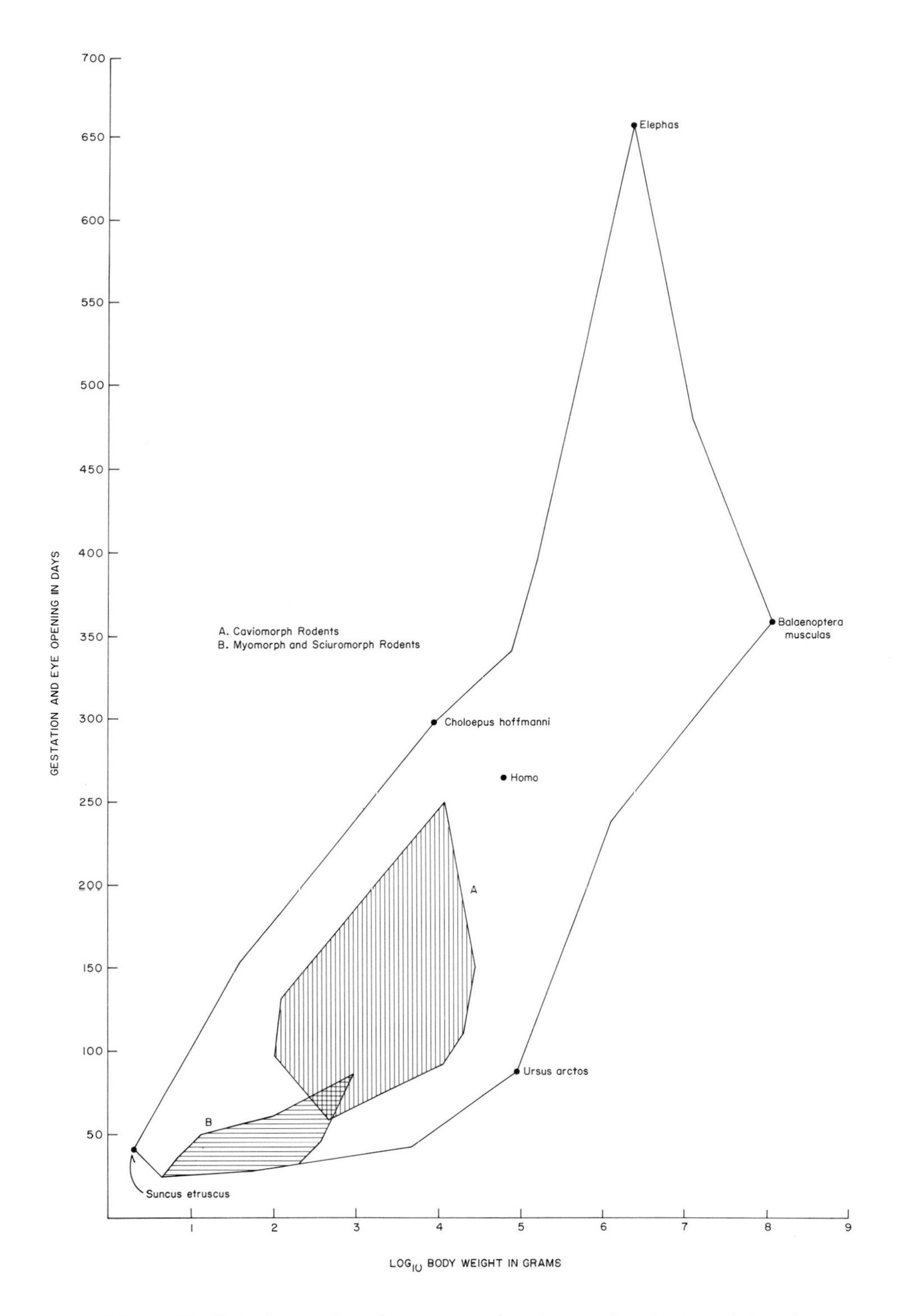

Figure 67. Plot of gestation plus eye-opening time against body weight; plotting as in figure 65. Note the relatively slow developmental rate for the primates. Some edentates have even slower developmental times (see text).

in developmental time, and soricine insectivores show rapid rates of development.

Table 51 protrays the results of a regression analysis of gestation plus eye-opening time against mean maternal body weight. The trends shown in the preceding figures are outlined here for orders and families. Note the correspondence in both slope and high intercept values for the Megachiroptera and Primates. These two taxa also have very high brain/body weight ratios and conform to the analysis by Sacher and Staffeldt (1974), who proposed that the duration of gestation bears a strong allometric relation to neonatal brain size. Marsupials correspond well with the values derived for the Bovidae. In general the remaining sets of mammals analyzed show similar slopes and intercepts.

Regression analyses of average values for higher taxa with a great range of size-classes yield rather high slopes. One may note in table 51 that taxa with a small size range (e.g., Rodentia: Muridae and Microtinae) show rather shallow slopes. Furthermore, lower taxa that are (by inference) genetically similar show very small slopes when development time is regressed against body size (see fig. 68). The same general trends derived from the broad analyses are still apparent but are not as dramatic. The long developmental time for some Chiroptera and some small carnivores relative to rodents and insectivores in the same size range is readily apparent.

Table 51 Log_{10} Gestation and Eye-Opening Time (Days) Regressed against Log_{10} Mean Body Weight (g) of a Sexually Mature Female

Taxon	*A* Intercept	*B* Slope	Coefficient of Correlation	Number of Species
Marsupialia	1.58	.13	.73	13
Insectivora				
Soricinae	1.33	.26	.86	4
Crocidurinae	1.43	.14	.65	3
Chiroptera				
Pteropidae	1.85	.12	.72	5
Phyllostomatidae	1.57	.37	.99	3
Vespertilionidae	1.45	.29	.99	5
Primates	1.81	.12	.77	33
Cebidae	1.82	.12	.86	7
Rodentia				
Sciuridae	1.76	.016	.13	13
Muridae	1.41	.09	.82	9
Microtinae	1.41	.06	.76	8
Carnivora				
Canidae	1.47	.10	.73	9
Felidae	1.48	.11	.85	20
Perissodactyla	1.49	.20	.88	4
Artiodactyla				
Suidae	1.67	.10	.40	4
Cervidae	1.73	.13	.66	12
Bovidae	1.54	.16	.78	30

Note: $Y = A + BX$.

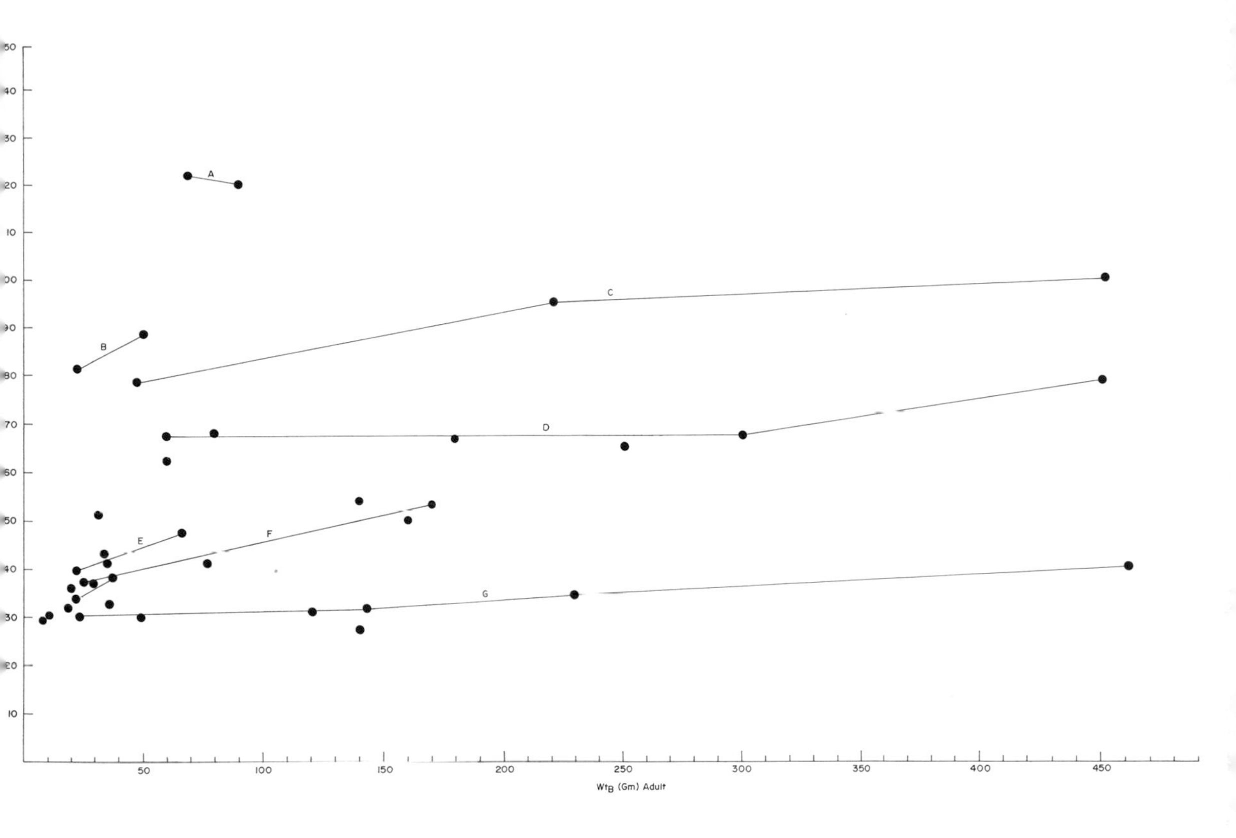

Figure 68. Plot of gestation plus eye-opening time against mean body weight for seven mammalian families. Note the extended developmental time for the Chiroptera and the striking difference between the Megachiroptera and Microchiroptera. *A* = Pteropidae; *B* = Vespertilionidae; *C* = Mustelidae; *D* = Sciuridae and Dipodidae; *E* = Heteromyidae; *F* = Gliridae; *G* = cricetine rodents.

22.3 Some Contributions to Variability in Developmental Time

General trends among higher taxa are of interest, but they pose many more questions. Figure 69 analyzes the relationship between body size and gestation for the caviomorph rodents. All forms give birth to relatively precocial young. Note that the families seem to lie on their own regression lines. The Caviidae show a rapid rate of intrauterine development. The Dasyproctidae are intermediate. The Capromyidae and the Chinchillidae have very long development times, as do the octodontoids, but the latter exhibit a very steep slope. *Erethizon* and *Dinomys* have prolonged intrauterine development times. I propose that we are looking at different adaptative sets. Since placentation and degree of precociality in the young are similar, then perhaps differences in metabolic rate contribute to the variation in development time. The fossorial octodontoids have low metabolic rates (McNab 1966), as do the folivorous erethizontids and capromyids (McNab 1978*a*). In a similar manner, the slow rate of development for some crocidurine shrews relative to the soricine shrews may result from the fact that the cro-

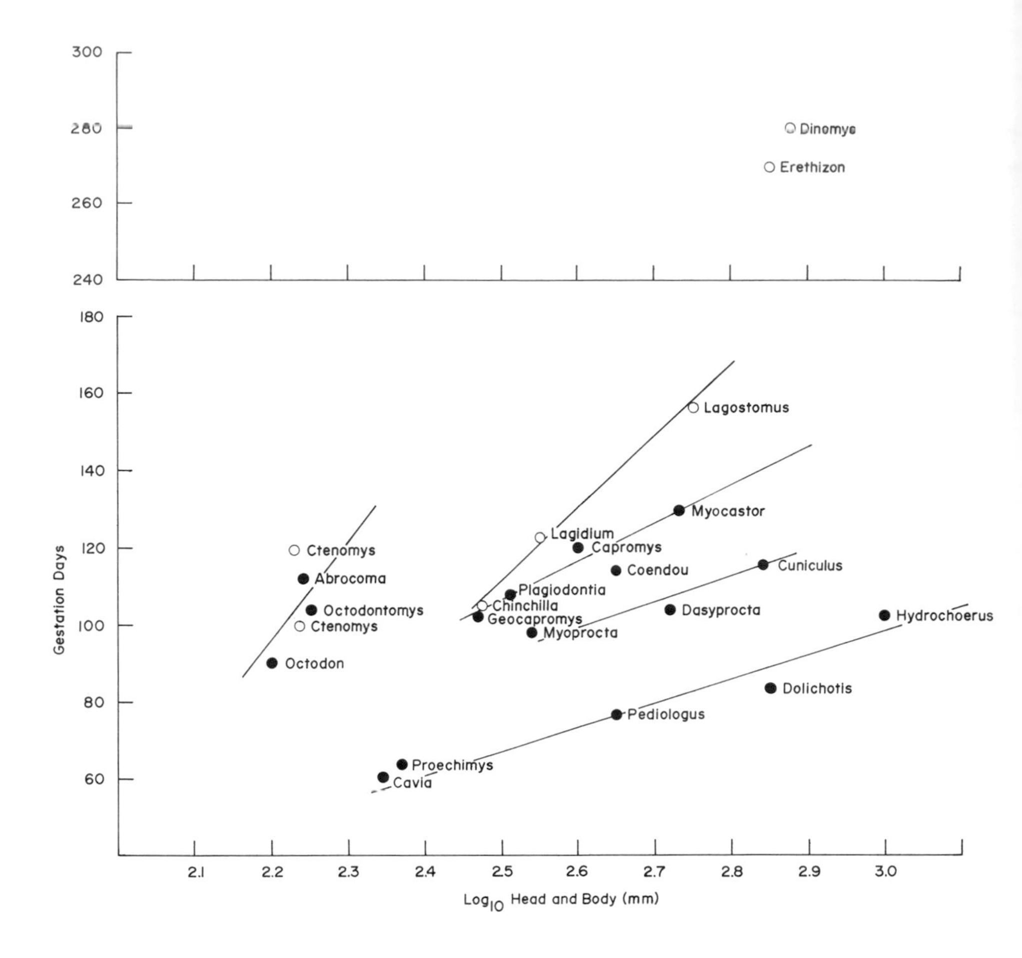

Figure 69. Gestation periods for caviomorph rodents, plotted against $\log_{10}$ head and body length in mm. From Kleiman, Eisenberg, and Maliniak 1979.

cidurines have lower basal metabolic rates than do the soricines (see fig. 70).[2] Yet this cannot be the whole story, as an inspection of the scatter in figure 70 will clearly show. Some of this scatter may be the result of differences in the size of the young at birth, since the same degree of altriciality is not expressed in these species. But, more important, there may be differences in the degree of encephalization.

Figure 71 portrays data for a series of chiropterans and insectivores that are all born in an altricial state. Neonatal weight at birth is plotted against gestation. The larger the young at birth, the longer the gestation, but the larger the adult encephalization quotient, the longer the gestation time for a defined size-class. Gestation time or developmental time from conception to eye opening seems to be a function of the absolute size of the female, the encephalization of the young, the rel-

2. In his review of the evolutionary consequences of trophic specialization, McNab (1980) points out that, within closely related taxa with a wide range of trophic adaptations, some consequences of lowered metabolic rate are longer generation time, lower litter size, and a slower growth rate of the young.

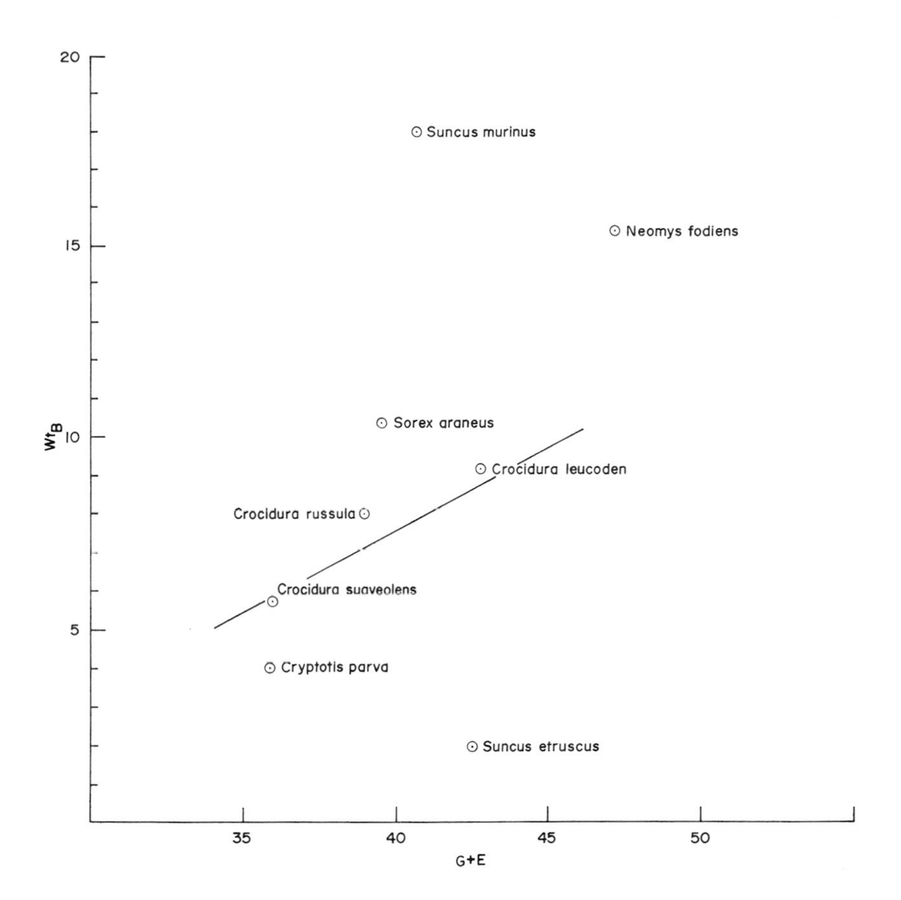

Figure 70. Body weight (g) plotted against gestation plus eye-opening time for a series of soricoid insectivores. Line connects the crocidurine shrews.

ative metabolic rate of the female, and perhaps the structure of the placenta.

I now introduce the further complication of litter size in making comparisons among species. For example, gestation time and developmental time to some criterion, such as eye opening, may vary as does the total weight of the fetal mass. In figure 72 I have plotted mean adult weight of the female against the gestation plus eye-opening time for a series of closely related cricetine rodents. The general trend is expressed, but the scatter is considerable. If, however, we plot the total weight of the average litter at birth against the mean weight of the adult female, we note a rather close linear relationship with a shallow slope (see fig. 73). If we then divide the mean weight of the litter at birth by the mean weight of the female and plot this against gestation plus eye-opening time, we find a tight negative covariation (see fig. 74). Litter size (or mass of the conceptus) at birth profoundly influences developmental time.

To reiterate: (*a*) larger species show a trend toward longer developmental times; (*b*) larger species tend to produce a larger young at birth or a larger total weight of neonates; and (*c*) as the mass of the

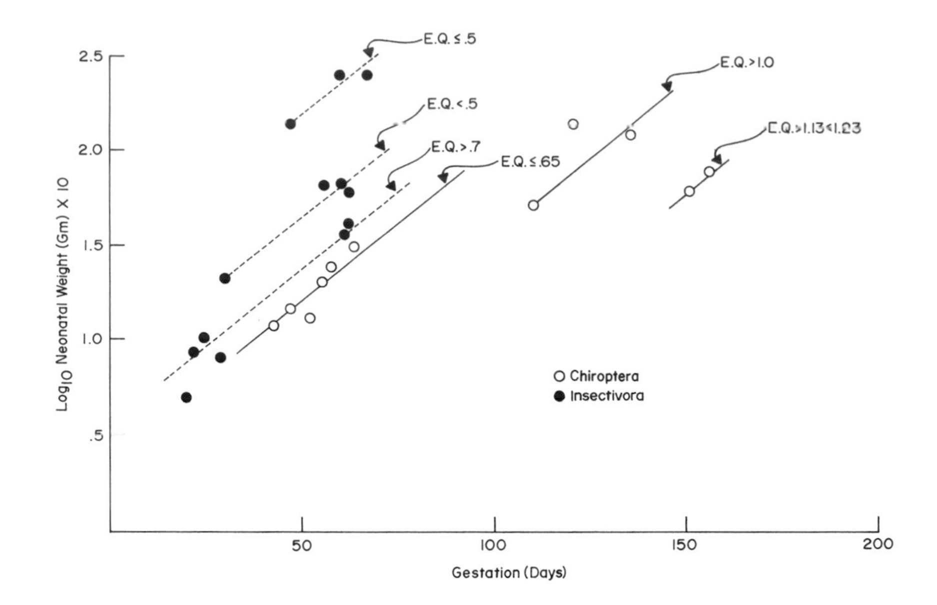

Figure 71. Log_{10} neonatal weight (g) plotted against gestation length in days. Regression lines for equivalent encephalization quotient values have been drawn through the points. Larger encephalization quotients correlate with longer developmental time.

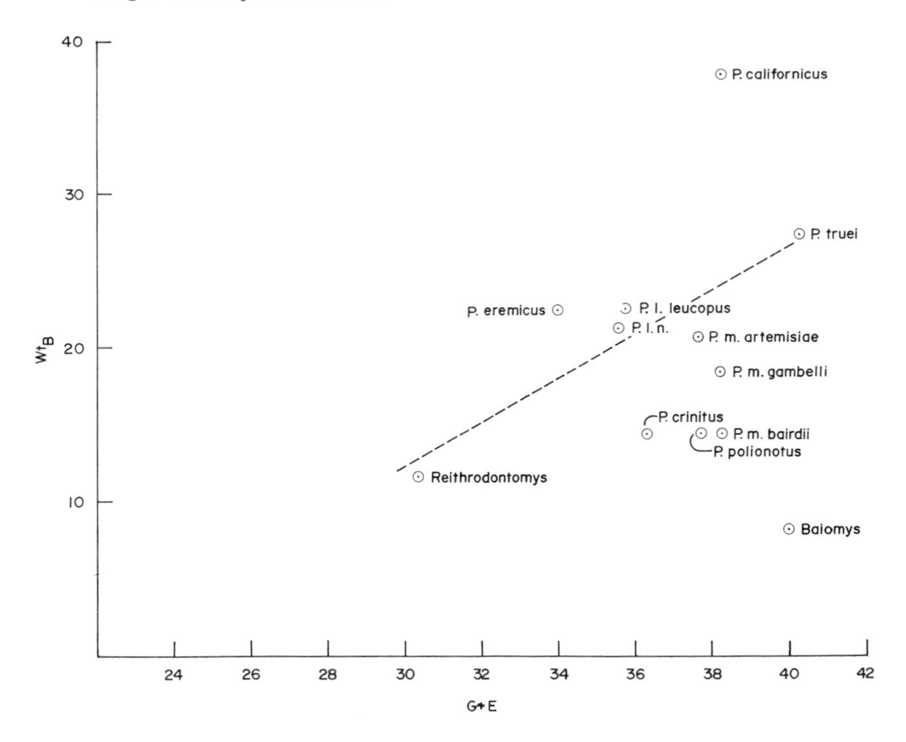

Figure 72. Adult body weight (g) plotted against gestation plus eye-opening time (days) for a set of closely related cricetine rodents. There is a weak positive covariation but considerable scatter.

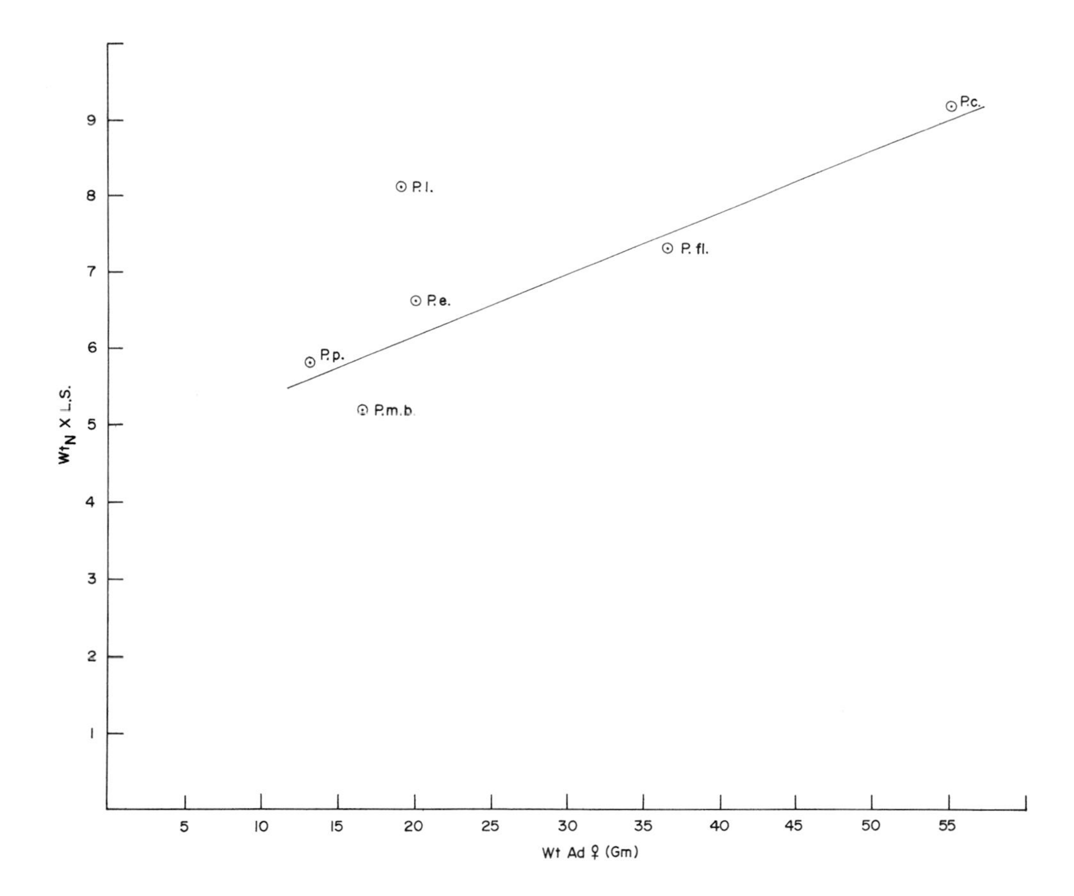

Figure 73. Mean weight (g) of the neonate times mean litter size plotted against average body weight of an adult female for six species of *Peromyscus*. *P.l.* = *P. leucopus; P.p.* = *P. polionotus; P.m.b.* = *P. maniculatus bairdii; P.e.* – *P. eremicus; P.fl.* = *P. floridanus; P.c.* = *P. californicus.*

young decreases relative to the female's mean weight, developmental time may increase. This last result seems paradoxical, but it happens because species with a larger mean size have slower rates of development and large species have the lowest ratio of fetal mass to mean adult weight. This was demonstrated when Leitch and his co-workers (Leitch, Hytten, and Billewicz 1959) examined the relationship between maternal and neonatal weights for eutherian mammals. They compared species that gave birth to several young with those that gave birth to only a single young. The multiparous species were treated by multiplying the average weight of the single young at birth by the mean litter size. They then regressed the total weight of newborn young against the average maternal weight and obtained a linear relationship showing a negative allometry. As the maternal weight increases, the total weight of the newborn young also increases; but, as the mother gets larger, the weight of the young relative to the mother's weight decreases by a constant amount.

In a second analysis, Leitch et al. expressed the total newborn weight as a percentage of the maternal weight. As maternal weight increases,

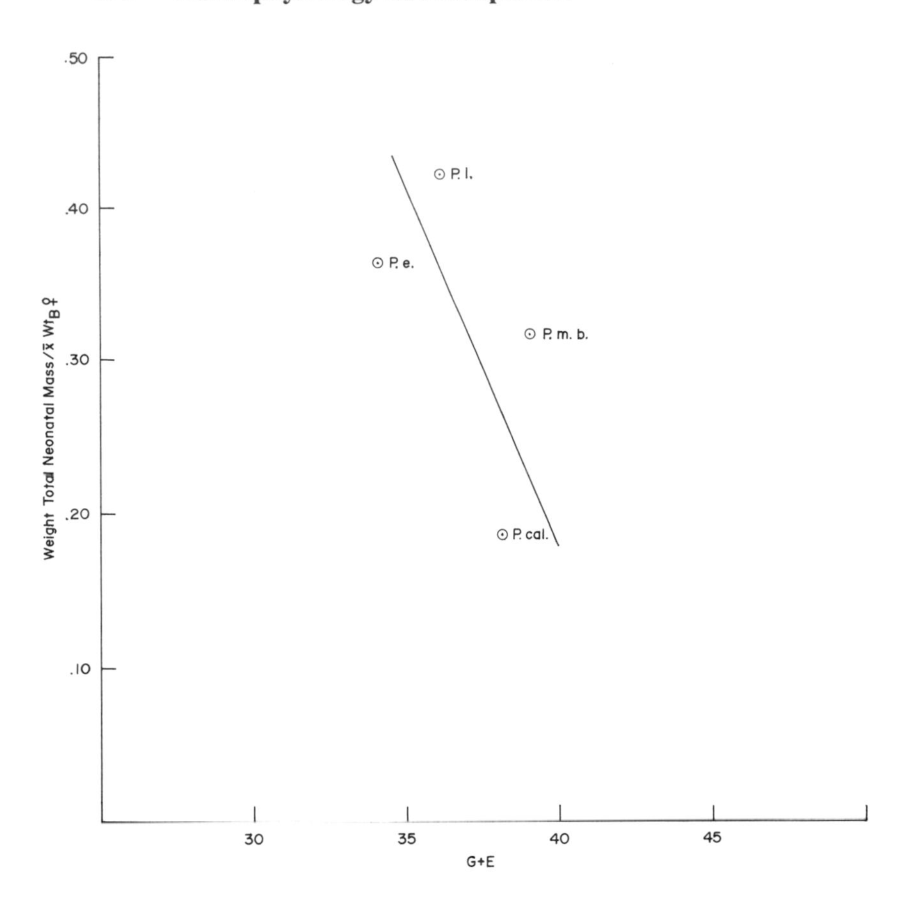

Figure 74. Ratio of total weight of neonatal mass (g) divided by mean weight of the female plotted against gestation plus eye-opening time (days) for four species of *Peromyscus*. Symbols as in figure 73.

the percentage of the maternal weight of the newborn young decreases. Thus, the large whales and elephants give birth to a relatively small young compared with small rodents or shrews. As a result, the total mass of the conceptus at birth correlates with gestation time, as one might expect, but the correlation masks many variables.

Table 52 gives the result of regressing total neonatal mass against gestation in days. Note that the Insectivora have a slightly negative slope. This results from the relatively rapid developmental cycle for *Hemicentetes* and *Tenrec,* both of which produce very large litters. Figure 75 graphically portrays the data. Once again, the relatively slow development times before birth are demonstrable in the Chiroptera and the Primates.

The preceding analyses of gestation and neonatal mass have included forms that give birth to young at quite different stages in their total develomental cycle. I have of necessity confounded the species that bear altricial young with those bearing precocial young. In my next analysis I will consider development from the measure of gestation

Table 52 Log$_{10}$ (Neonatal Weight in Grams × Mean Litter Size) Regressed against Gestation (Days)

Taxon	*A*	*B*	Coefficient of Correlation	N
Rodentia	.83	.01	.82	78
Sciuromorpha	.63	.02	.84	18
Myomorpha	.59	.02	.36	33
Caviomorpha	1.53	.01	.54	24
Primates	.47	.01	.81	31
Insectivora	−.30	.03	.73	18
Artiodactyla	1.19	.01	.75	15
Carnivora	1.78	.01	.37	49
Pinnipedia	2.84	.005	.53	10

Note: X = Gestation; Y = neonatal maximum; $Y = A + BX$.

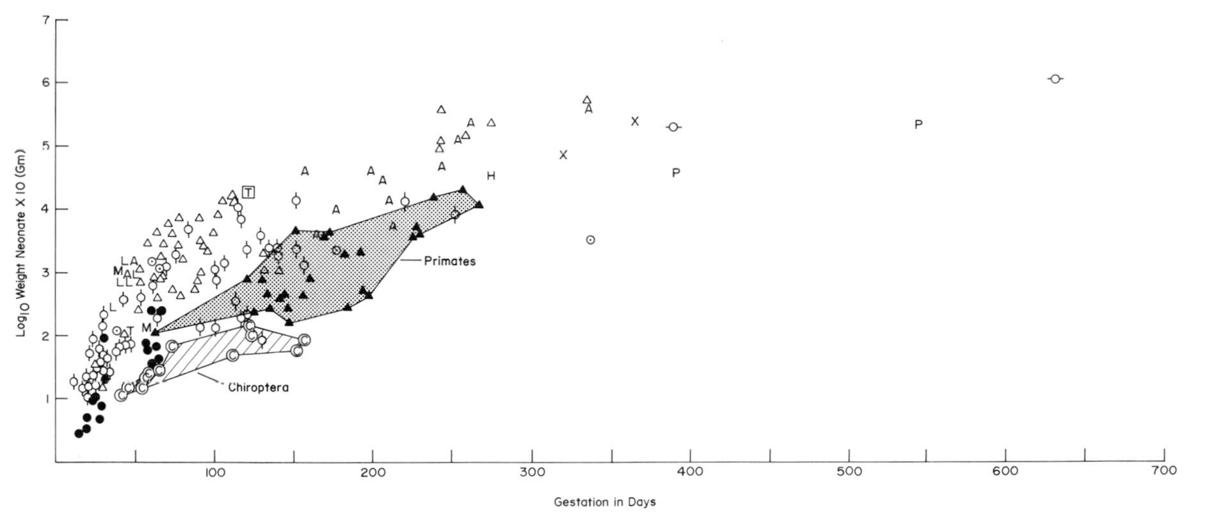

Figure 75. Log$_{10}$ neonatal weight (g) plotted against total gestation time in days. See figure 102 for explanation of symbols.

plus the mean duration of lactation and analyze this value in terms of the size of the mother.

If we plot the value for gestation plus lactation time against the ratio of the mass of the conceptus at birth to the mean adult weight, we find a negative relationship. Figure 76 portrays the data for the Marsupialia, while figure 77 integrates the marsupial curve with that of some selected eutherians. The data in figure 77 suggest that the larger mammals have smaller young relative to their body size but take a longer absolute time to complete the rearing cycle. The data in figure 76 for the Marsupialia indicates considerable scatter but, given the small size of the neonates and the relatively short gestation time, perhaps there are more confounding variables than we we can measure, given the state of the art.

If we then plot gestation plus lactation time against the mean weight of the adult female on a double log plot, we once again find the cor-

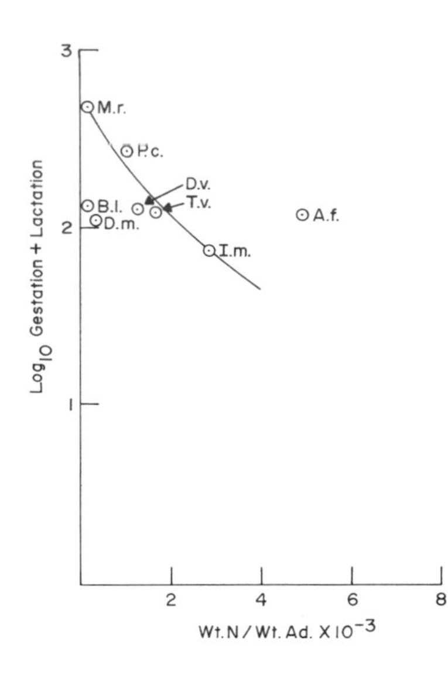

Figure 76. Log_{10} gestation plus lactation in days plotted against the ratio of total neonatal weight (g) divided by mean weight (g) of an adult female. Marsupials only are graphed here. *Mr = Megalia rufa; Im = Isoodon macrura; Tv = Trichosurus vulpecula; Bl = Bettongia lesueuri; Af = Antechinus flavipes; Dm = Didelphis marsupialis; Dv = Didelphis virginiana; Pc = Phascolarctos cinerea.*

relation of longer developmental time for larger mammals (see fig. 78). The Chiroptera and Primates clearly have the longest developmental cycle for their size-classes. The Insectivora have the shortest cycle. The tenrecoid insectivores are intermediate, as are the Marsupialia. for their size-classes, the great whales have the shortest overall developmental time. Table 53 presents the numerical analysis.

22.4 Lactation

Specialized glands for producing nutritive secretions for young animals are defining characters for the class Mammalia. The mammary glands are functional only in the female, though homologues of the glands are present in the male. The early nutrition of the young invariably falls solely to the female. In monogamous species such as marmosets or some canids, the male may be involved in provisioning the young after their initial growth; but in no case has natural selection favored the production of milk by males.

Milk very possibly has two functions—first, nutrition, and, second, passing antibodies to the young to confer passive immunity to diseases during early development. The Marsupialia possess a choriovitelline placenta throughout the intrauterine phase of development. Little exchange of material between the fetal and maternal circulations is possible, and, though the fetus may be able to absorb secretions from the lining of the uterus through the chorion and the yolk sac, it is quite certain that passive immunity is conferred not through the placenta but through milk. Passive immunity transfer through milk in *Setonix* occurs throughout the 180-day lactation period (Buss 1971).

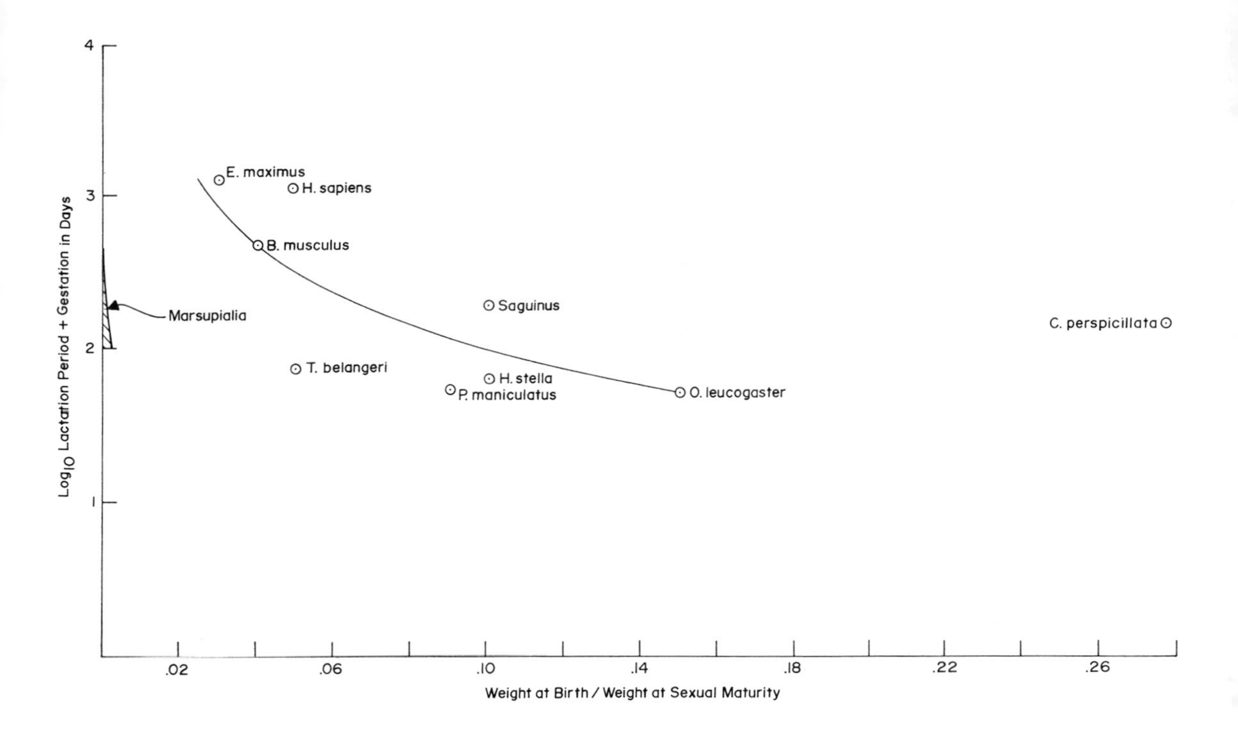

Figure 77. Log_{10} gestation plus lactation in days plotted against the ratio of total neonatal weight (g) divided by mean weight (g) of a female at sexual maturity. Marsupials are graphed with eutherians. Line eye-fitted.

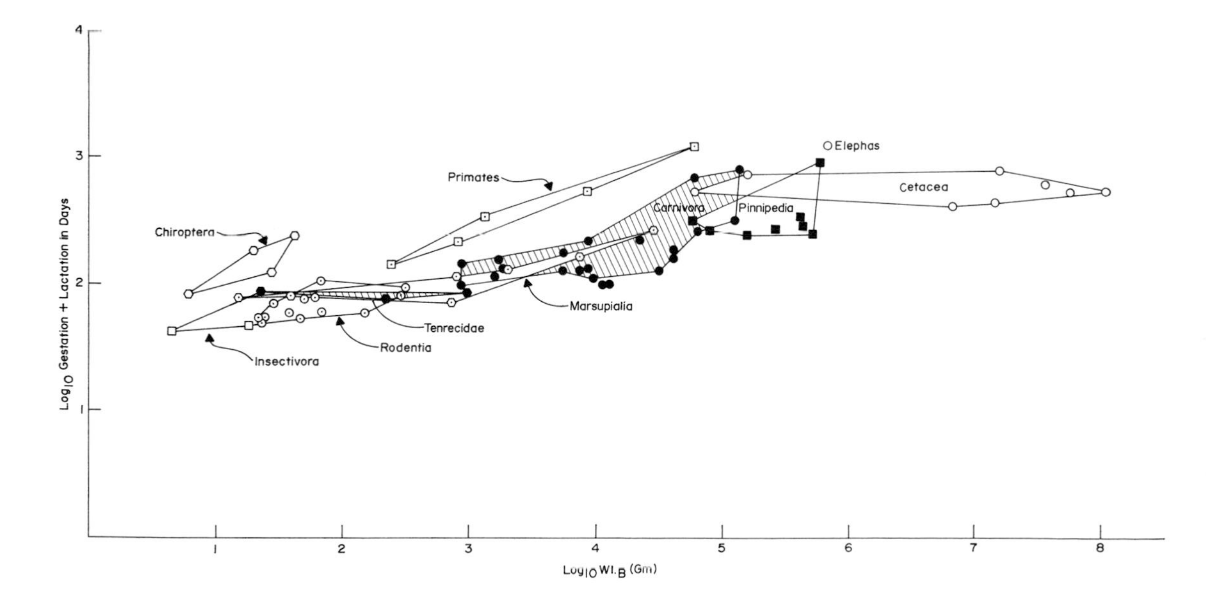

Figure 78. Log_{10} gestation plus lactation in days plotted against mean body weight (g) of an adult female.

Table 53 Log_{10} Gestation Plus Lactation (Days) Regressed against Log_{10} Body Weight in Grams of a Sexually Mature Female

Taxon	*A* Intercept	*B* Slope	Coefficient of Correlation	Number of Species
Marsupialia	1.70	.10	.70	6
Insectivora				
Soricidae	1.63	.05	.66	3
Primates				
Ceboidea	1.71	.26	.90	3
Rodentia				
Sciuridae	1.63	.11	.41	3
Cricetidae	1.32	.30	.50	7
Carnivora				
Canidae	2.01	.01	.11	5
Felidae	1.39	.21	.73	7
Artiodactyla				
Bovidae	1.51	.21	.88	4

Luckett (1975) points out that in the embryogenesis of placentals a choriovitelline placenta is transient and that the defined chorioallantoic placenta is characteristic of all eutherians. There are three types of chorioallantoic placentas according to Luckett, and the three classes reflect varying numbers of tissue layers separating the maternal and fetal circulation (see Glossary). The epitheliochorial placenta characteristic of artiodactylans, pangolins, whales, perissodactylans, and strepsirhine primates has at least two major tissue layers interposed between the maternal and fetal circulations. The endotheliochorial placenta has only the endothelium of the maternal circulation separating it from the chorion of the fetal placenta. The hemochorial placenta characteristic of haplorhine primates, most rodents, elephant shrews, and vespertilionid bats has the chorion of the embryo in contact with the blood of the female.

It is interesting to note that all eutherians studied to date having an epitheliochorial placenta also transfer antibodies thus conferring passive immunity through the milk. Some mammals with endotheliochorial placentas or hemochorial placentas pass part of the passive immunity through the milk and the rest of it prenatally from the female circulation directly to the circulatory system of the young. Some species with hemochorial placentas pass almost the entire maternal antibody complement from the maternal circulation to the fetal circulation and pass none or little of it in the milk. These include the guinea pig (*Cavia*), the domestic rabbit (*Oryctolagus*), the rhesus monkey (*Macaca*), and the human (*Homo*) (Buss 1971). These data suggest that the milk was at first not only nutritive but also a means of inducing passive immunity in the young, and that this primitive function has been lost with the progressive evolution of the hemochorial placenta.

The nutritive function of milk is of course the simplest concept to grasp. Milk has been analyzed for a wide variety of species (Jenness 1974), and its composition is known to vary throughout the lactation cycle; hence the nutritive quality of milk during early lactation can be quite different from the nutritive quality of milk at weaning. Oftedal

(1978) presents some well-taken caveats with respect to the validity of the sampling procedures presented and summarized in the literature on milk composition. Some of the data from the literature must be accepted with great caution, especially if they were based on a single sample. Figure 79 indicates that there is great variation among species in the protein, carbohydrate, and fat constituents of milk.

Some species of mammals produce milk with an exceedingly high fat content. This is characteristic of seals and whales. Fats may allow rapid weight gain, but one of the more important components for tissue growth is protein. Some years ago, Bourlière pointed out that mammals having rapid postnatal body growth generally were species with a high protein content in the mother's milk. If this is so, why, then, has natural selection not promoted the highest quality of protein nutrition in milk to ensure the most rapid growth of the young? Surely this must be related to specific adaptational constraints of the species in question. For example, the diet of the elephant (*Elephas maximus*), consisting in a large measure of structural carbohydrate, may make it difficult for a female to mobilize sufficient protein reserves to ensure rapid postnatal growth of the young. Thus selection surely has effected a compromise with a long intrauterine development period and a protracted lactation

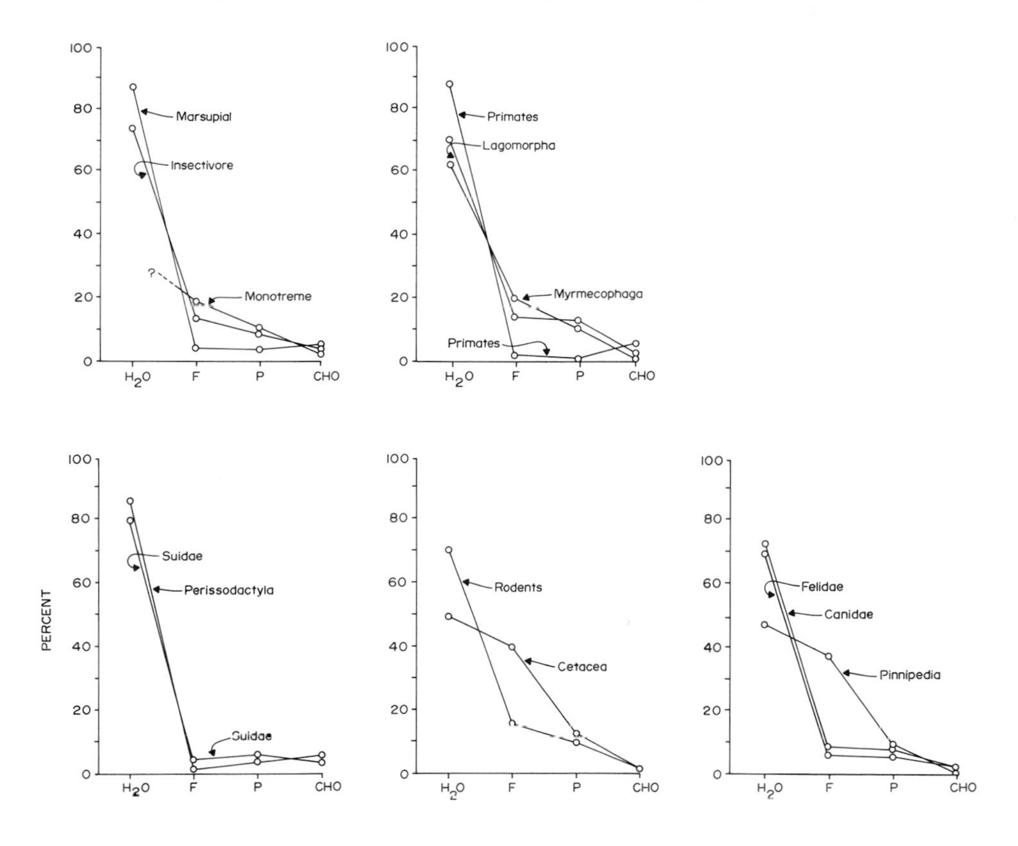

Figure 79. Composition of various mammalian milks. Data from *International Zoo Yearbook*, vol. 4, 1964. *F* = fat; *P* = protein; *CHO* = carbohydrate.

period, in conformity with the energy constraints on the female elephant, given her specific feeding adaptations.

In the previous section I considered the relationship between lactation time and mean body size of the reproducing female. Figure 80 portrays the relationship between duration of nursing and relative size of the weaned young. Larger mammals tend to have rather long lactation times but are weaned at a much smaller body size relative to adult size than are small mammals. In fact, some small mammals (e.g., soricine shrews) may be weaned at a weight approximating or exceeding subadult mean weights.

22.5 Postnatal Growth

I have little to add after the monumental efforts of D'Arcy W. Thompson (1942) and Sir Julian Huxley (1932, 1972). The formula $Y = b \cdot X^a$ or $\log Y = b + \log X(a)$ was first generalized by Huxley to describe allometric growth. The alterations in the relative growth of the parts of animals' bodies through ontogenetic and phylogenetic time have received exhaustive treatment (Gould 1977). In this section I will discuss some insights derived from an inspection of growth curves from selected mammalian studies. I am most indebted to Kent Redford for his insights and labors in this section.

Growth rate after parturition is much better understood than intrauterine developmental time. If we take a series of small mammals and plot the percentage of adult weight achieved over time, we note that in general the smaller the mammal the more rapidly it attains adult weight (fig. 81). This somewhat parallels what we would anticipate based on what we know of the mother's metabolic rate (and presumably

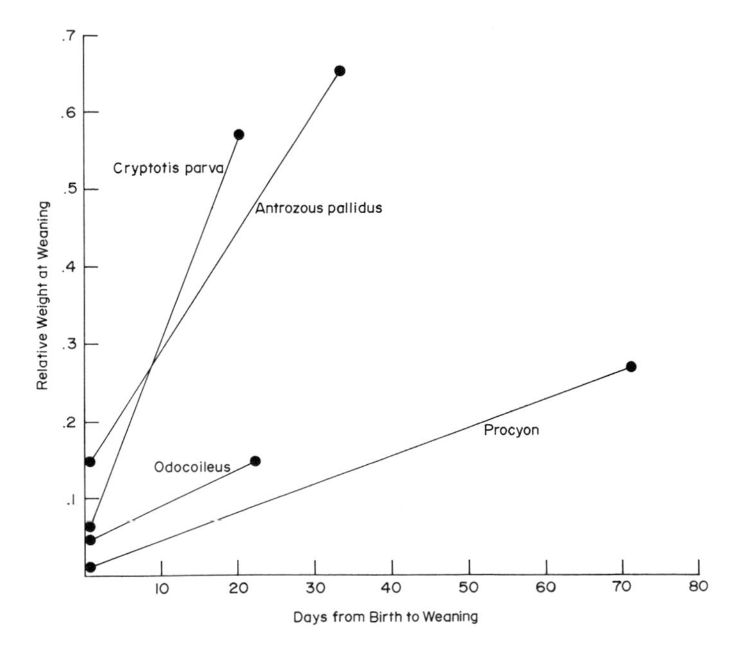

Figure 80. Relative size at weaning compared with adult weight plotted against the duration of heavy lactation for four species of mammals.

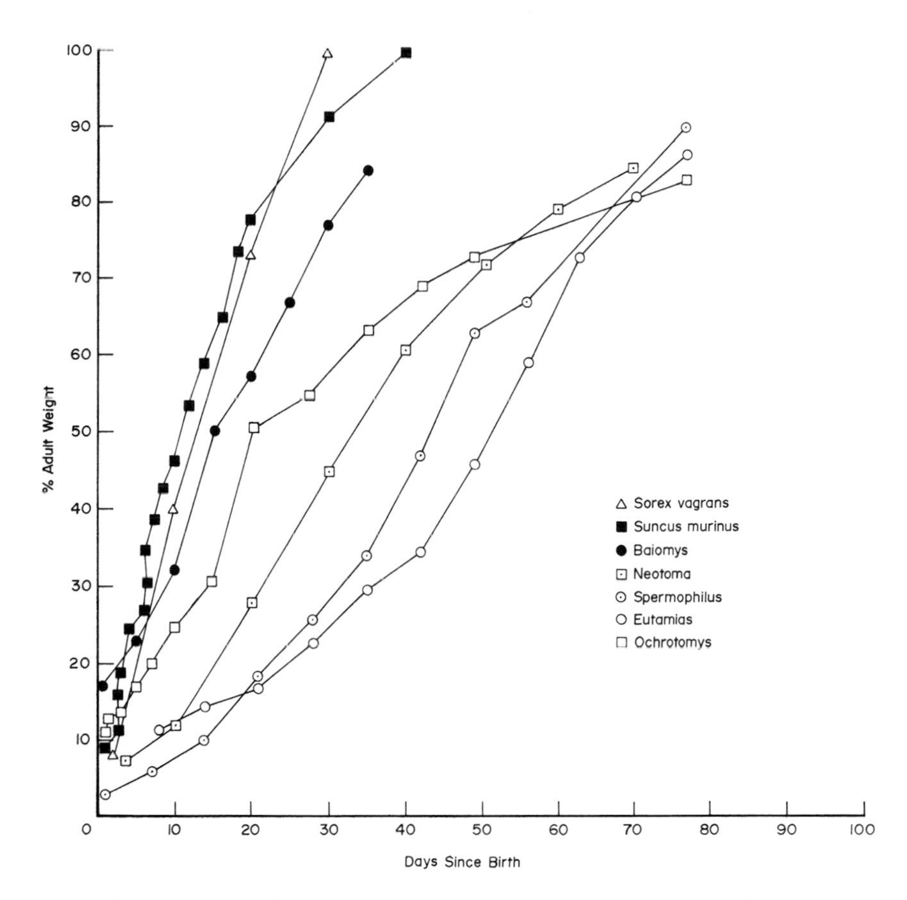

Figure 81. Relative growth of newborn small mammals graphed as percentage of adult weight attained as a function of days since birth. *Sorex vagrans* (Forsyth 1976); *Suncus murinus* (Dryden 1968); *Baiomys* (Hudson 1974); *Neotoma* (Martin 1973); *Spermophilus* (Morton and Tung 1971); *Eutamias* (Hirshfeld and Bradley 1977); *Ochrotomys* (Linzey and Linzey 1967).

her lactation level), which decreases as a function of the animal's adult size. Postnatal growth rates may be integrated with postulated prenatal growth rates as in figures 82 and 83. In general, postnatal growth is much more rapid than prenatal growth. If the growth rate of the young reflects a corresponding energy drain on the female, then the total lactation period could be a greater energy burden than is gestation (Millar 1975, 1976). (Randolph et al. [1977] analyze energy costs for *Sigmodon* and conclude that the energy cost of lactation may be twice that of gestation.) As a general rule, small mammals such as the Etruscan shrew (*Suncus etruscus*) and the wandering shrew (*Sorex vagrans*) develop much faster than larger mammals such as the laboratory rat (*Rattus norvegicus*) or the chipmunk (*Eutamias amoenus*). However, there is great variability. For example, chipmunks and squirrels consistently show slower developmental rates than one would predict from their weight. This is probably because they have a lower basal metabolic rate than do muroid rodents (see fig. 45). The mouse opossum, *Marmosa robinsoni,* and the striped tenrec, *Hemicentetes,* show rather

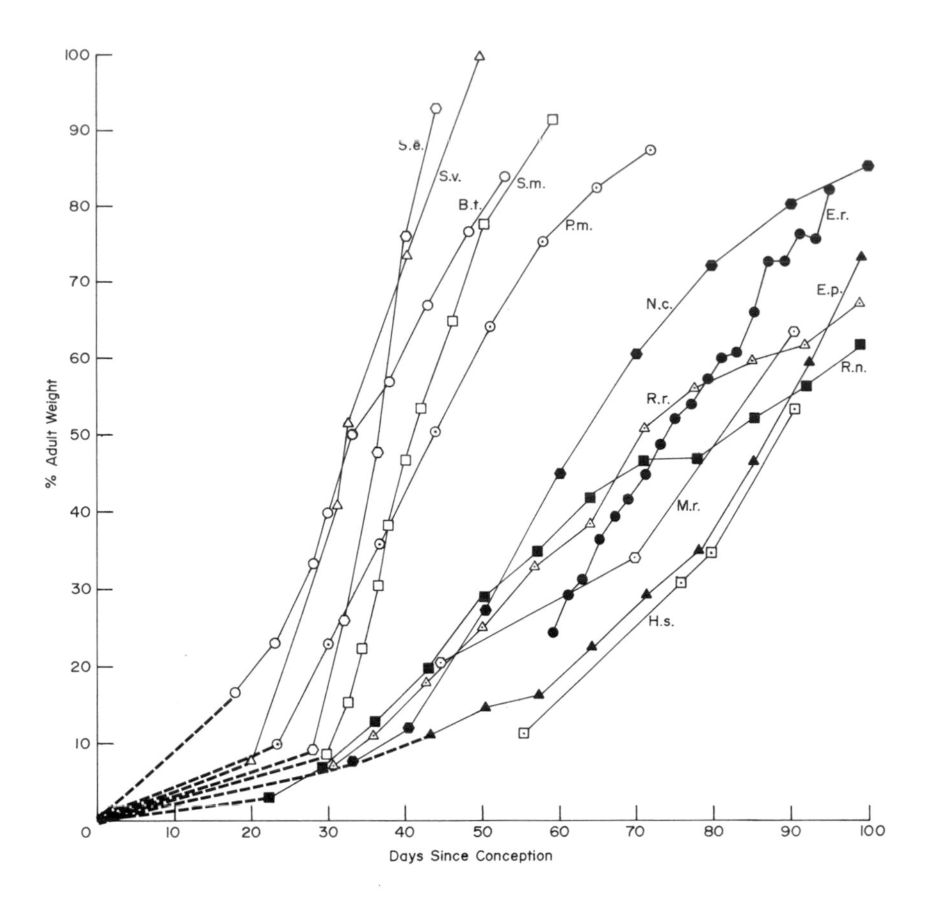

Figure 82. Relative growth of newborn small mammals graphed as percentage of adult weight attained as a function of days since conception. *Broken line* = hypothetical intrauterine growth. *S.e* = *Suncus etruscus* (Fons 1973); *S.v.* = *Sorex vagrans* (Forsyth 1976); *B.t.* = *Baiomys taylori* (Hudson 1974); *S.m.* = *Suncus murinus* (Dryden 1968); *P.m.* = *Peromyscus maniculatus* (Linzey 1970); *N.c.* = *Neotoma cinerea* (Martin 1973); *E.r.* = *Elephantulus rufescens* (Rathbun 1979); *E.p.* = *Eutamias panamentinus* (Hirshfeld and Bradley 1977); *M.r.* = *Marmosa robinsoni* (Eisenberg unpublished); *H.s* = *Hemicentetes semispinosus* (Eisenberg and Gould 1970); *R.r.* = *Rattus rattus* (Altman and Dittmer 1962); *R.n.* = *Rattus norvegicus* (Altman and Dittmer 1962).

slow developmental rates for their body size. These probably reflect fundamental differences in the basal metabolic rate as determined by McNab (1978*b*) and Eisenberg and Gould (1970). The same sort of general trend can be shown if the percentage of adult head and body length attained is plotted against time (fig. 83). Here, however, the position of the curves is shifted to the left, since adult body proportions are attained much more rapidly than average adult body weight.

When plotting developmental rates in this manner, care must be taken to indicate the time of weaning. Forsyth (1976) has elegantly demonstrated that in shrews weaning may occur when the animals have attained a weight slightly higher than their average weight as subadults. This apparently is an adaptation reflecting the young shrew's ability

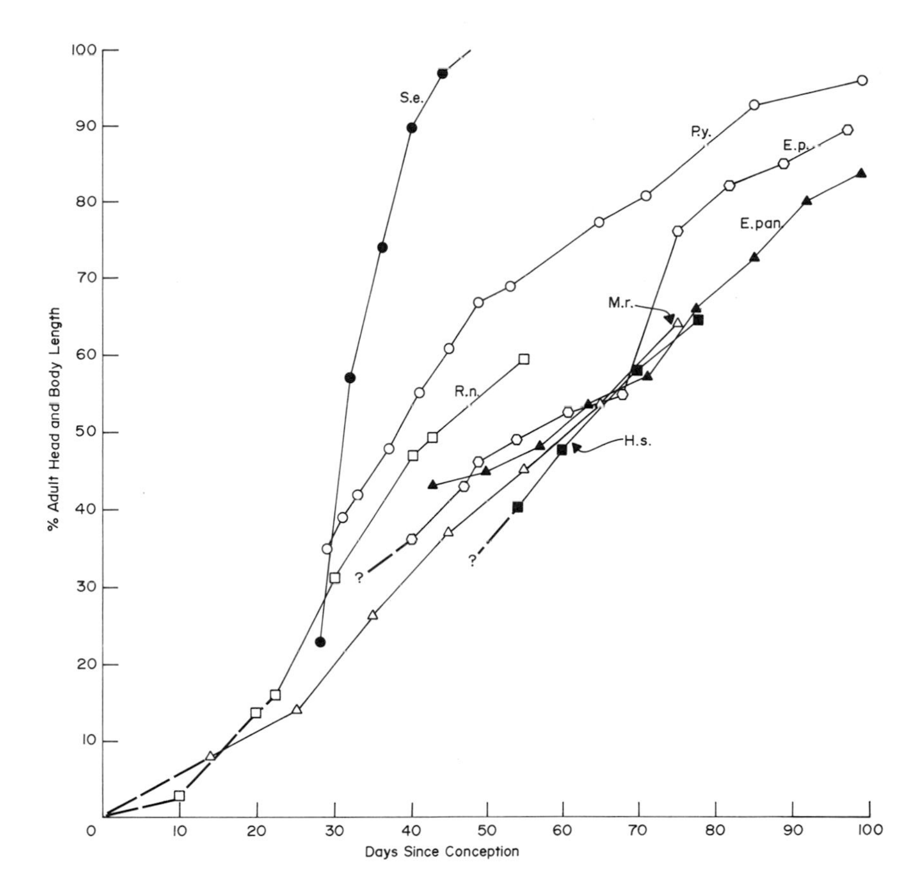

Figure 83. Relative growth of newborn small mammals graphed as percentage of adult head and body length attained as a function of days since conception. *S.c.* = *Sorex cinereus* (Forsyth 1976); *P.y.* = *Peromyscus yucatanensis* (Lackey 1976). Other abbreviations as in figure 82.

to store fat for use as an energy reserve after weaning, when it must rapidly learn to hunt for itself. A similar phenomenon is shown during the weaning of young phocine seals. Of course, in neither case does the weight at weaning exceed the mean adult weight.

One should be able to relate postnatal growth rates to such variables as the nutritional quality of the milk, metabolic rate of the female, and metabolic demands of the infant's tissues. Sufficient data are not present for a synthesis, but intriguing questions can be posed.

Kleiman and Davis (1978) point out that the postnatal growth of vespertilionid bats is greater than that of comparable-sized phyllostomatid bats. Jenness and Studier (1976) present data on milk composition for such chiropterans as have been studied. *Glossophaga soricina* has a low-energy, low-fat milk and develops rather slowly. The association time between the mother and young is prolonged. *Glossophaga* has a complex foraging strategy and a very high brain/body weight ratio. Does it follow, then, that selection for a milk with high nutritive content is relaxed in *Glossophaga* as part of a positive selection for a lower energy drain on the female and a prolonged mother-young bond?

Similar questions could be asked with great profit when we study closely related species that differ in their rearing strategies and the growth rate of the young. Perhaps prolonged lactation is one way to sustain the bond with the female when the young must learn complicated modes of habitat exploitation (see chap. 34).[3]

3. Case (1978) has performed an exhaustive analysis of postnatal growth patterns in mammals and birds. Some of his conclusions are similar to mine. He rightly points out that the selective factors regulating postnatal growth rates are different from the factors determining the degree of precociality at birth.

23 Reproduction and Life History Strategies

23.1 Reproductive Capacity and Body Size

The reproductive capacity of a female that reaches aduluhood is the potential number of young she could produce over her reproductive life. The lifetime productivity for a female depends on her reproductive life-span, the mean interbirth interval, and the average litter size. These three factors vary according to both absolute body size and metabolic rate of the species. Figure 84 illustrates the reproductive capacity of three species of *Peromyscus*. The largest species has the longest developmental cycle and the smallest litter size. Its lifetime average production of young is the least, whereas the smallest species can produce the most young. Field data from McCabe and Blanchard (1950) and

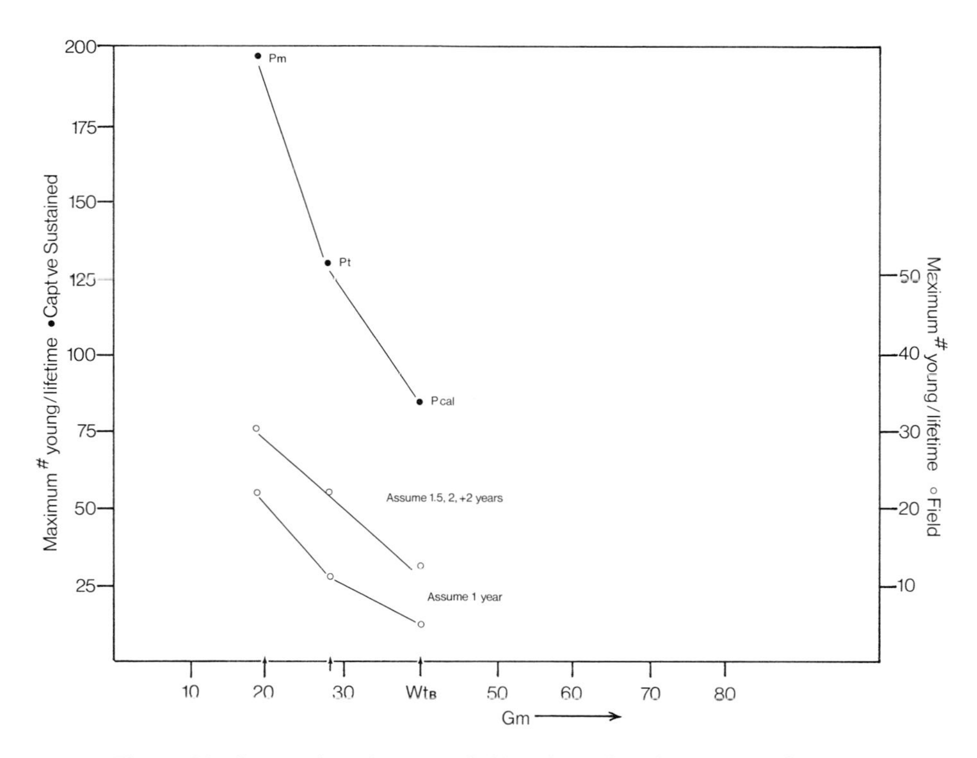

Figure 84. Comparison between field and captive data concerning the number of young produced per female for three species of *Peromyscus*. Captive data from Egoscue, Bittmenn, and Petrovich 1970; field data from McCabe and Blanchard 1950. The field data are presented with respect to two different mortality schedules. *P.m.* = *Peromyscus maniculatus*; *P.t.* = *Peromyscus truei*; *P.cal.* = *Peromyscus californicus*.

captive data from Egoscue, Bittmenn, and Petrovich (1970) are graphed together and yield the same trends. I will graph theoretical lifetime productivities for a variety of species to establish the general trends. The data are from Appendix 4. Where captive data are employed, they will not be mixed with field data. The trends should be equivalent whether based on field or captive data.

The Marsupialia offer an interesting starting point for our analyses. Utilizing captive data, I have plotted longevity against body size and demonstrated a positive trend (fig. 85). In figure 86 I have broken the data points into two sets that suggest that trophic strategy, to an extent, correlates with potential longevity. Herbivores[1] in almost all size-classes appear to have a longer potential life-span than equivalent-sized carnivores or insectivores.[2] If we plot mean litter size against longevity, there is a negative slope but a great deal of scatter (fig. 87). Note that some didelphid and dasyurid marsupials have extremely large litter sizes and a short life expectancy. Some tenrecoid insectivores show a similar demographic strategy (see chap. 7 and sect. 23.1), while the monotremes exhibit a long potential life-span and a low litter size.

Table 54 represents the regression of a portion of the data points in figure 87. The negative slope of the entire set is clearly indicated, but the scatter is considerable. Clearly we are dealing with two data subsets: the short-lived dasyurids and didelphids with larger litters, and the long-lived herbivores with their smaller, modal litter sizes.

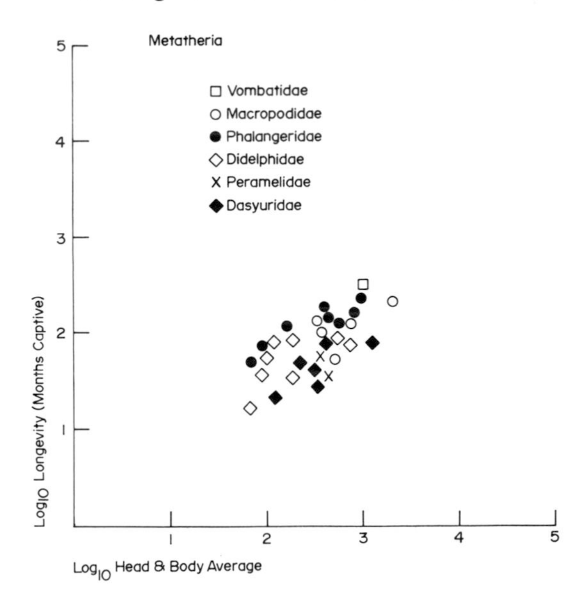

Figure 85. Double $\log_{10}$ plot of maximum longevity in captivity (months) against mean head and body length (mm) for six families of marsupials.

1. By "herbivores" I mean foregut fermenters or large hindgut fermenters. I exclude the microtine rodent herbivores.

2. Mace (1979), in her analysis of small-mammal demography, points out that herbivorous rodents such as microtines have shorter life-spans than do granivorous cricetine rodents. This results from the fact that many microtines are extremely "r" selected relative to cricetines and in no way detracts from the general conclusion presented here (see chap. 34).

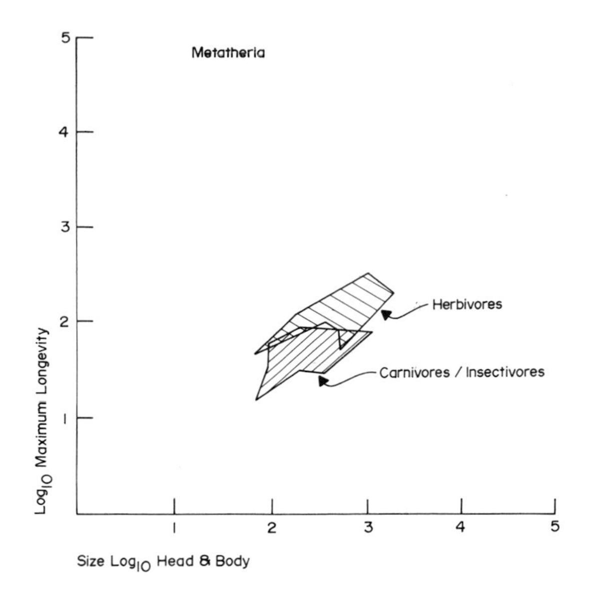

Figure 86. Longevity as a function of body size and trophic strategies. The points in figure 85 are represented as two convex polygons reflecting two broad trophic strategies. The herbivores tend to be longer-lived than the carnivores and insectivores.

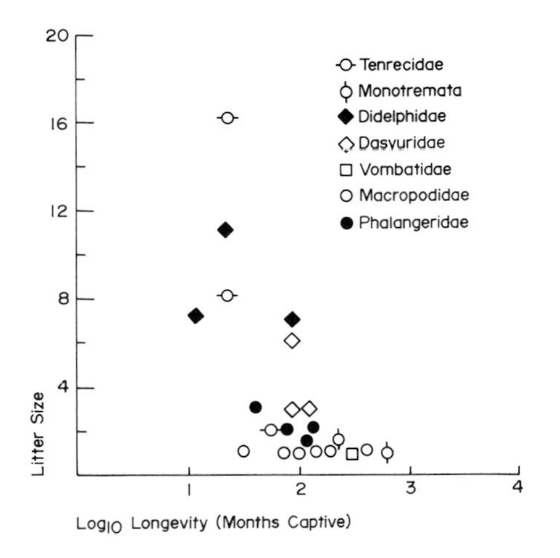

Figure 87. Mean litter size plotted against mean longevity (months expressed as $\log_{10}$). Tenrecoid insectivores are graphed with monotremes and marsupials.

Since longevity correlates positively with absolute body size, it should not be surprising that litter size regressed against body size yields a negative slope. Figure 88 portrays the data points, and table 55 presents the regression analysis. Once again part of the scatter can be resolved by separating out the long-lived herbivores, with their low metabolic rate. Figure 89 portrays this division, which reflects the same trends outlined in figure 86.

Table 54 Regression of Litter Size against Log_{10} Mean Longevity in Days

Taxon	*A* Intercept	*B* Slope	Coefficient of Correlation	Number of Species
Marsupialia	22.37	−5.64	−.51	14
Didelphidae and Dasyuridae	14.33	−2.40	−.29	6
Phalangeridae and Macropodidae	1.29	.01	.002	8

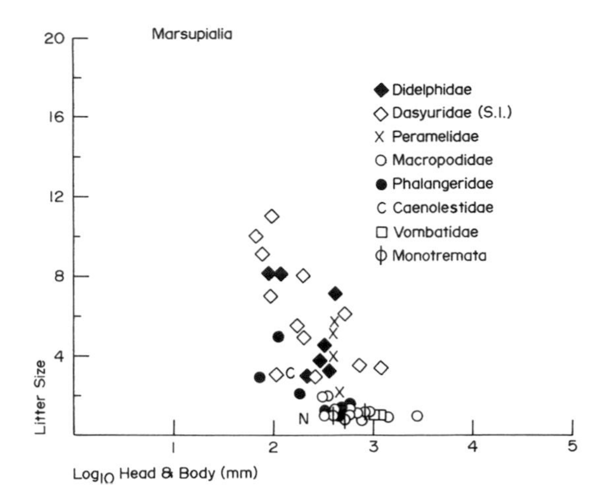

Figure 88. Mean litter size plotted against body size (mm) expressed as log_{10} for selected monotremes and marsupials.

Table 55 Australian Mammals: Litter Size Regressed against Log_{10} Mean Body Weight in Grams of an Adult Female

Taxon	*A* Intercept	*B* Slope	Coefficient of Correlation	Number of Species
Didelphidae and Dasyuridae	13.34	−2.56	−.90	6
Phalangeridae and Macropodidae	2.43	−3.48	−.75	6

How then does the lifetime productivity of a female relate to body size? Figure 90 portrays a series of marsupials and the monotremes based on the captive data in Appendix 4. In general, larger species have a lower lifetime production of young, but the range is not as wide as one might anticipate. Most females will produce between ten and eighteen young in their reproductive life-spans. Some short-lived species will produce up to twenty-four young in a season, but these forms begin to approximate a semelparous reproductive strategy (see sect. 3.2). Figure 91 divides the marsupials into herbivores and insectivore/carnivores. There is considerable overlap, but the small insectivorous forms can extend into semelparity. One may recall the generalizations

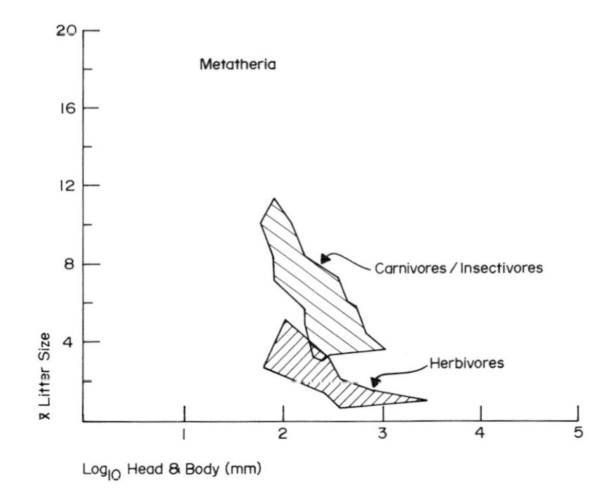

Figure 89. Data from figure 88 plotted as two subsets according to trophic strategy. Note that herbivores tend to have smaller litters than do carnivores and insectivores.

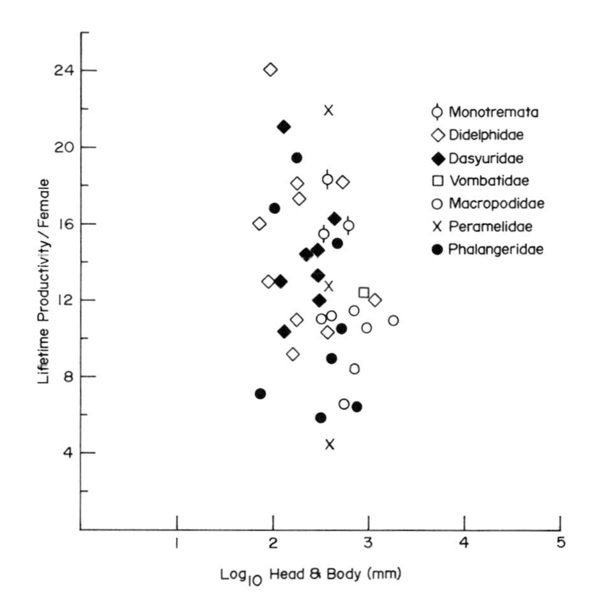

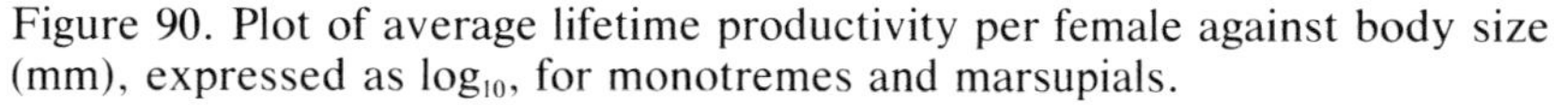

Figure 90. Plot of average lifetime productivity per female against body size (mm), expressed as $\log_{10}$, for monotremes and marsupials.

concerning trophic strategies and ease of energy conversion as outlined in chapter 19.

In anticipation of the next section, I have graphed in figure 92 the lifetime productivity for a set of marsupials and a sample of eutherians as a function of mean body weight.[3] Eutherians alive today exceed the living marsupials in size, but the convex polygons show an amazing overlap. The meaning of this correlation will be developed further in the next few paragraphs.

3. Body length (L) and body weight (Wt) are related by the equation $Wt = K(L)^3$.

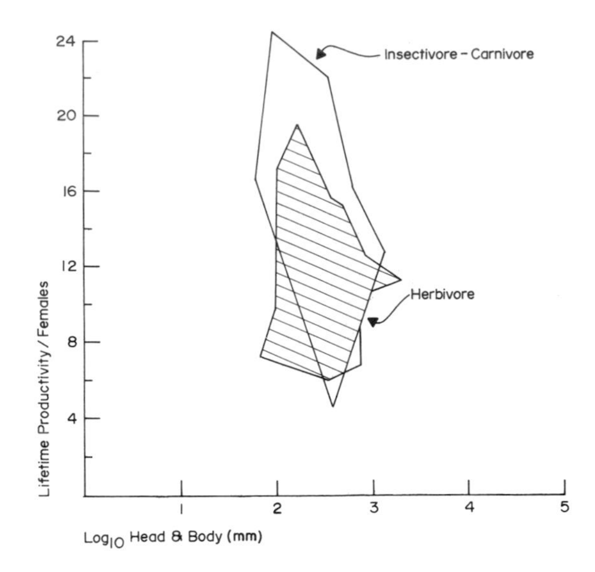

Figure 91. Data from figure 90 redrawn as two trophic subsets. Note the near identity.

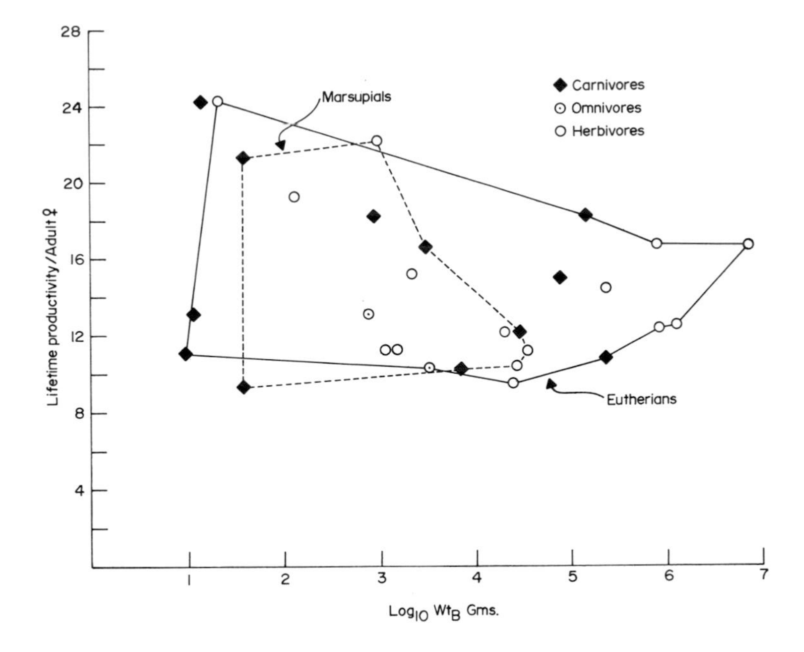

Figure 92. Mean lifetime productivity per adult female plotted against body size (mm), expressed as $\log_{10}$. Two sets are portrayed: marsupials and eutherians. Note the near identity.

The living Edentata represent an early offshoot from the eutherian line (see chap. 4). In figure 93 I have graphed mean litter size against longevity. Note again the subset of the genus *Dasypus,* which has a short mean life-span and a high litter size. In figure 94 the estimated lifetime productivity of females is regressed against body size. The

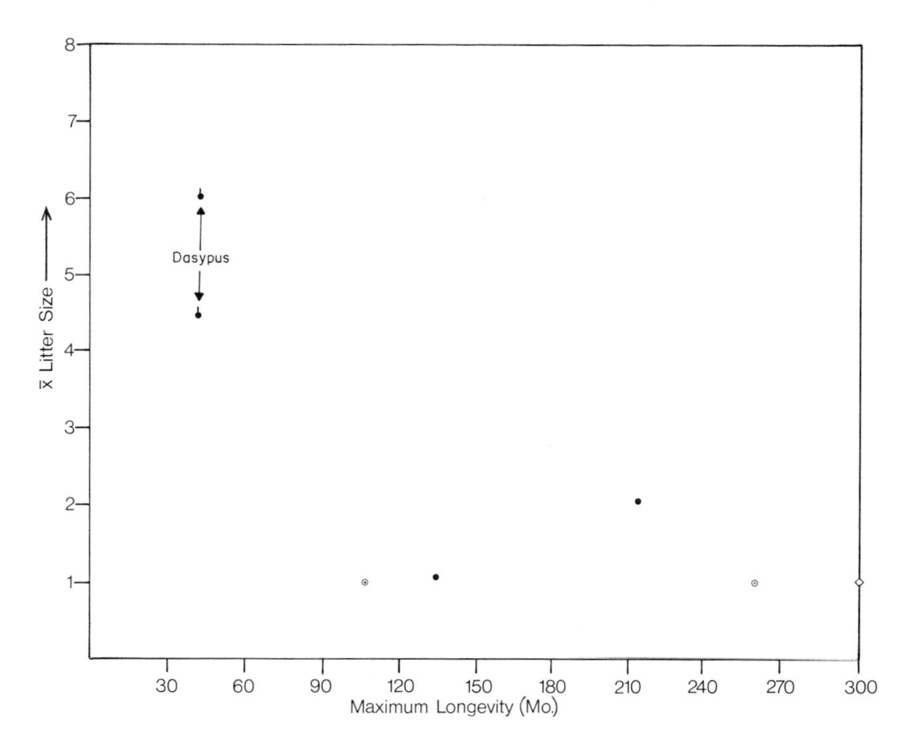

Figure 93. Mean litter size plotted against maximum longevity for a series of edentates. Note the large litter size for the dasypodine armadillos. ● = Dasypodinae; ● = Chaetophractinae; ⊙ = Bradypodidae; ◇ = Myrmecophagidae.

iteroparous species with small litter sizes are clumped between the limits of eleven and sixteen offspring per female, while the rather short-lived *Dasypus* has the potential of twenty-four young.

The same trends will be outlined for insectivores and rodents. In figure 95, selected insectivores, rodents, and a few marsupials are analyzed with respect to litter size and longevity. There is an inverse correlation. Here we have sufficient data to go a step further. Rather than settle for correlating litter size versus longevity, I have plotted the litter-size data for insectivores and rodents against the mean maximum age of the female at her last reproduction. The scatter in figure 96 is vastly reduced, and the negative slope is clearly accentuated. In this figure I demarcate the semelparous reproducing forms with a line. Note that the semelparous reproductive option is limited to certain small size-classes. Figure 97 portrays the maximum number of young an "average" female would be expected to produce in her lifetime as a function of body size. *Sorex minutus, S. araneus, Hemicentetes semispinosus*, and *Tenrec ecaudatus* are eutherians that have opted for a nearly semelparous strategy with eighteen to twenty-seven young per female. Of course, the line is arbitrary. The vast majority of mammals tend toward twelve young per lifetime. What I have outlined here for marsupials and small eutherians is only a beginning. Demography is

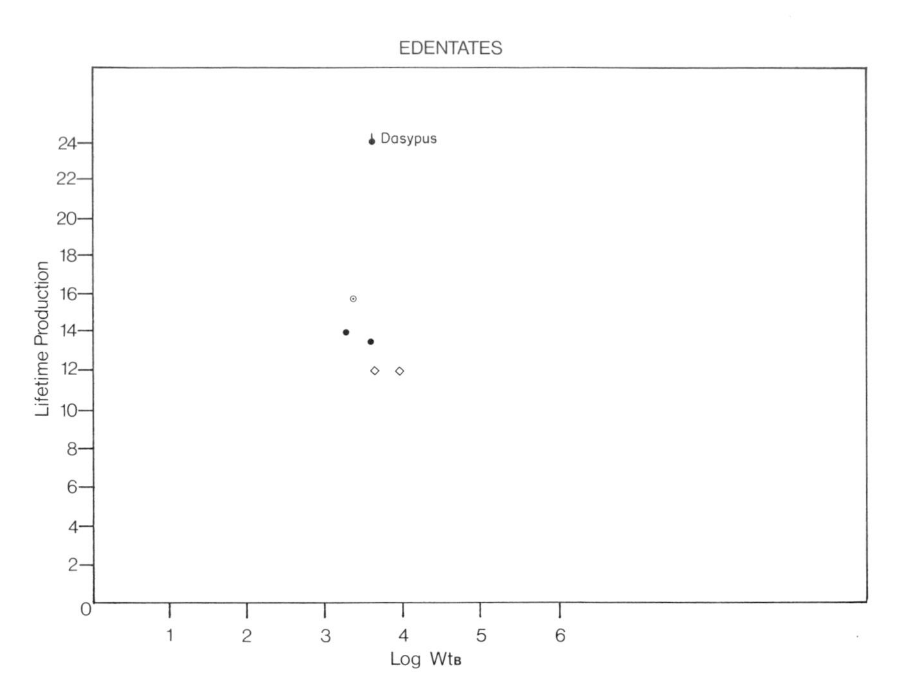

Figure 94. Lifetime productivity per adult female expressed as a function of body size for selected edentates. Note that only the relatively short-lived dasypodine edentates are aberrant. Symbols as in figure 93.

a guide to understanding adaptive syndromes. We badly need more field data and life tables derived from field studies.

The life table is basic to an understanding of reproductive strategies. One must sample a population at biologically meaningful intervals, X_1, X_2, X_3 . . . , and at the end of each interval calculate the number still living (l_{x1}, l_{x2}, . . .). If we consider only females in the population, we can calculate for each time interval the birthrate per female (m_{x1}, m_{x2}, m_{x3} . . .). By summing over all intervals, we can obtain the recruitment (R_0) for the cohort of the population under study ($R_0 = \Sigma l_x m_x$). Each organism in the founder population theoretically must replicate itself in the next generation. If this happy case took place, the population would be stable and not growing. The differential survival of different genotypes results from natural selection upon phenotypes, and indeed the "fitness" of uhe parental genotype is determined by this difference in survival of offspring. Stable evolving populations occur in time, but individual "fitness" may vary greatly.

One can derive several other parameters from a life table. One of the most important is the intrinsic rate of natural increase (r_1). This value reflects the rate at which a population can grow,

$$\frac{dN_1}{dt} = r_1 N_1,$$

and the variability in r_1 reflects the variability in generation time when species are compared. In nature the unrestricted growth of the popu-

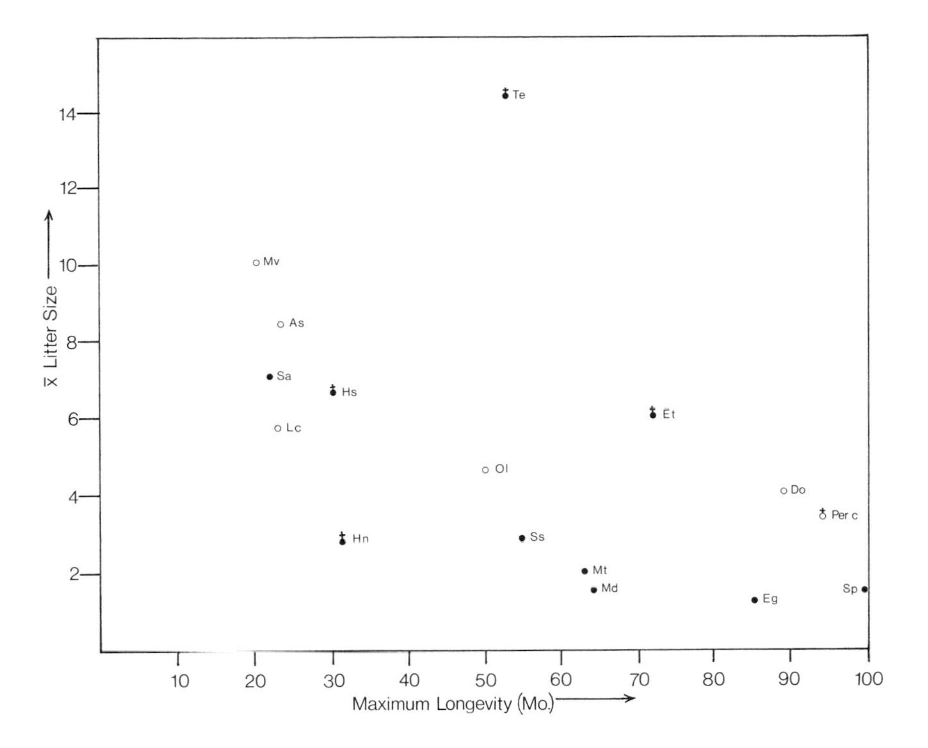

Figure 95. Mean litter size plotted against maximum longevity for selected insectivores, rodents, and marsupials. Note that the semelparous reproducing marsupials and insectivores form an elevated subset. Data in part from Egoscue, Bittmenn, and Petrovich 1970. *Mv* = *Marmosa robinsoni; As* = *Antechinus stuarti; Sa* = *Sorex araneus; Hs* = *Hemicentetes semispinosus; Lc* = *Lagurus; Hn* = *Hemicentetes nigriceps; Ol* = *Onychomys leucogaster; Ss* = *Setifer setosus; Et* = *Echinops telfairi; Mt* = *Microgale talazaci; Md* = *Microgale dobsoni; Do* = *Dipodomys ordii; Pc* = *Peromyscus californicus; Eg* = *Echinosorex gymnura; Sp* = *Solenodon paradoxus; Te* = *Tenrec ecaudatus*.

lation is determined by the carrying capacity (K), so that the actual change in numbers for a population is limited and expressed by the equation:

$$\frac{dN_1}{dt} = r_1 N_1 \left(\frac{K_1 - N_1}{K_1} \right).$$

Since many small mammals reproduce over only a year or two at best, one can often graph annual productivity from field data as a function of size. Fleming (1975) has done this and developed a graph remarkably similar to the lifetime productivity graph shown in figure 90. Fleming has further shown that for small mammals the annual probability of survival is inversely related to the annual productivity of the species.

The whole problem of graphing such field data is difficult. While it is true that survival rate is inversely related to reproductive capacity in small mammals (French, Stoddard, and Bobek 1975), one must be careful of the units in which survival rate is expressed. If one calculates

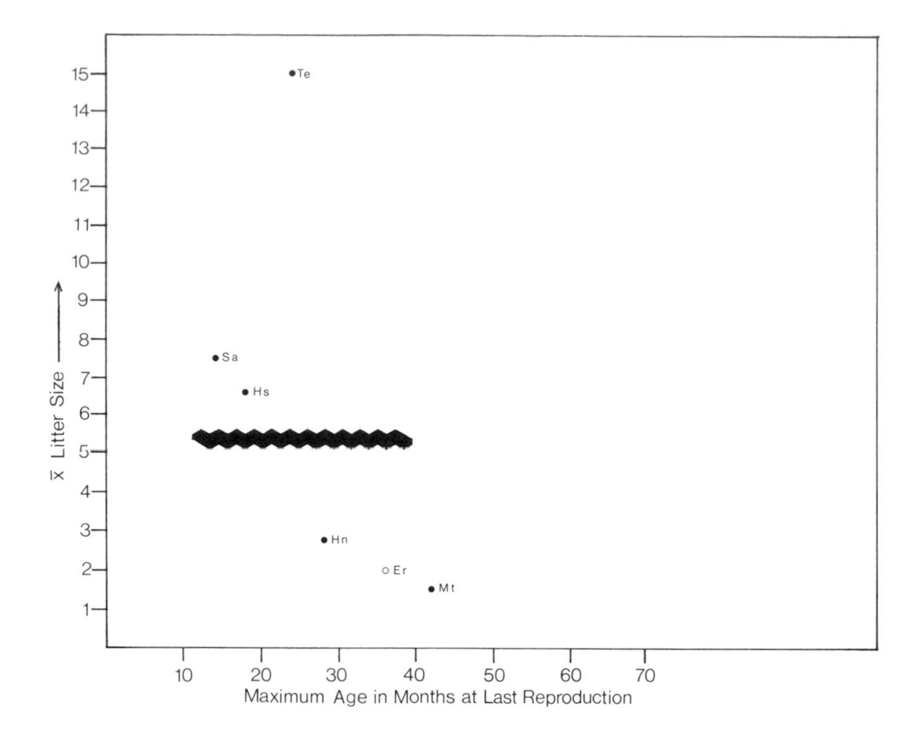

Figure 96. Mean litter size graphed against maximum age at last reproduction for selected insectivores, rodents, and marsupials. Top black line separates the semelparous subset from the remainder. Abbreviations as in figure 95.

annual survival rate for a cohort, shrews of the genera *Sorex* and *Blarina* have higher values than many small rodents. This happens because soricine shrews do not usually breed in the year of their birth but instead replace the nearly semelparous reproducing adult cohort and breed the year following. This adaptation of soricine shrews results in a high survival rate for the first year, but the entire cohort is dead within eighteen months of birth (Crowcroft 1957).

The data from French et al. suggest that some small mammals will have a high intrinsic rate of natural increase relative to larger mammals, and this is indeed the case. The Chiroptera, with a small body size and low metabolic rate, are an exception.

Small mammals have the option of producing large litters, an option apparently not retained by any large mammal. However, when we compare lifetime productivities of large and small mammals, we find near equivalence for a great many species over the whole spectrum of size-classes (see the previous sections).

It is axiomatic that the intrinsic rate of natural increase (r) tends to decline as the species' body size increases (Fenchel 1974). This in part reflects the well-known inverse correlation between metabolic rate and body size (Kleiber 1961) or the direct correlation between longevity and body size (Sacher 1959). All of these correlations may be brought together in the general rule that the interval between generations in a

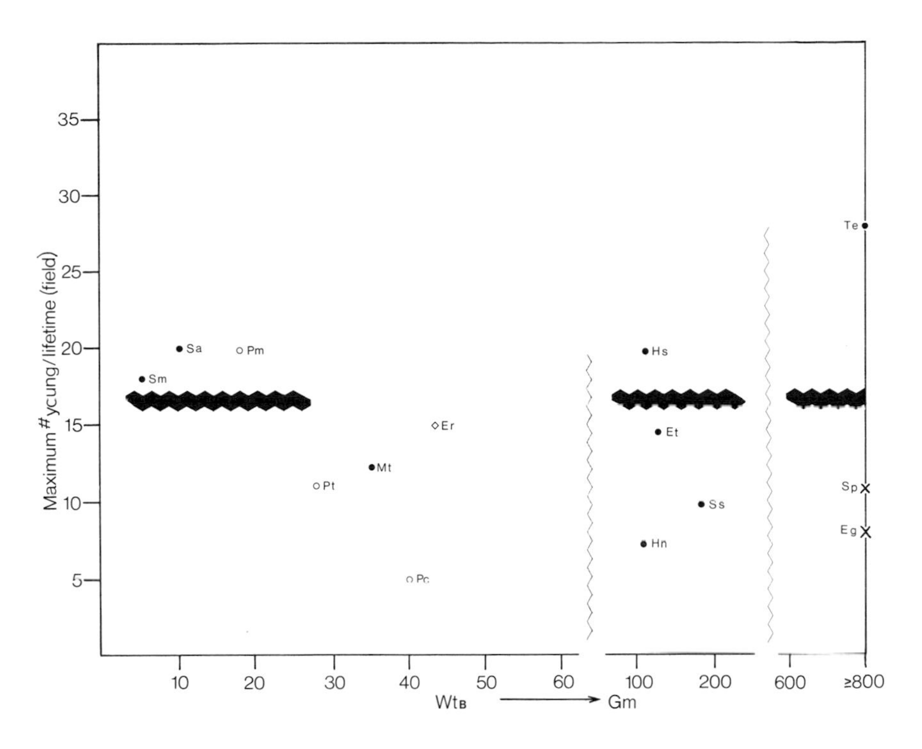

Figure 97. Maximum number of young per lifetime per adult female expressed as a function of mean body size. Data points as in figure 96. Black lines separate those species tending toward a semelparous reproductive strategy. Abbreviations as in figure 95.

population is positively correlated with mean body length for the species (Bonner 1965). Finally, it may be stated that the probability that a subpopulation of a given species will become extinct within a specified time interval is inversely correlated with body length or body weight (Rabinowich 1975). Of course the key phrase in the preceding sentence is "a specified time interval." It means, in effect, that extinction in a subpopulation of large species will take longer.

23.2 Some Demographic Exceptions in Insular Environments

When the class Mammalia is analyzed as a set of individual data points representing species averages for some mensurable phenomenon, then a number of correlations can be generated after the dependent and independent variables are designated.

As we have seen, litter size is a rough indicator of a species' reproductive potential. One could happily conclude that litter size should vary inversely when regressed against either body weight or potential longevity. The argument might then be extended to ask if a relation exists between adult body size and litter size. Figure 98 demonstrates that a mammal larger than 630 g in body weight generally has one young per litter. A mammal whose body weight exceeds 10,000 g almost always has one young per birth. Thus, multiple births are common in smaller mammals, but there are exceptions. The Chiroptera tend to be

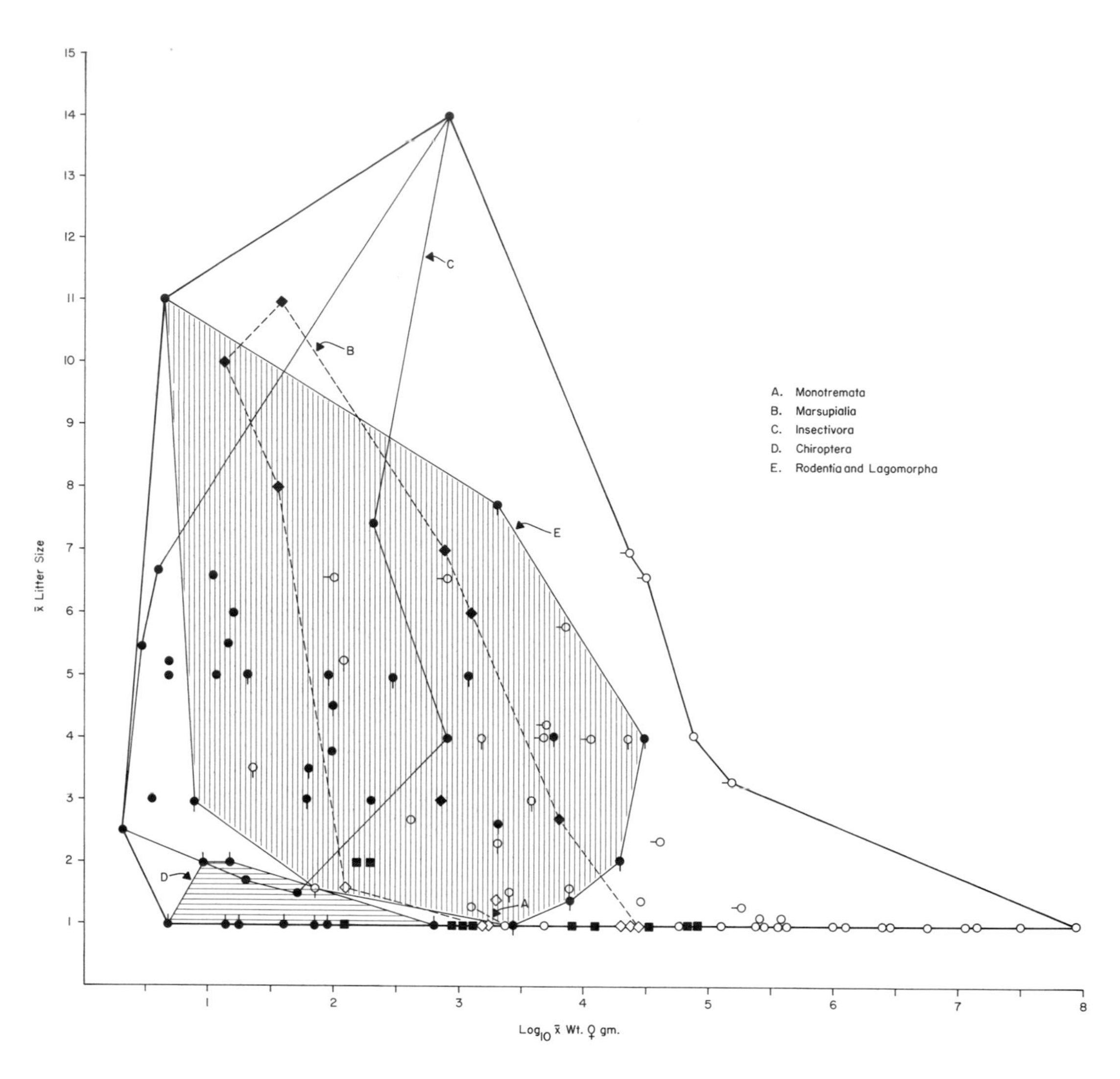

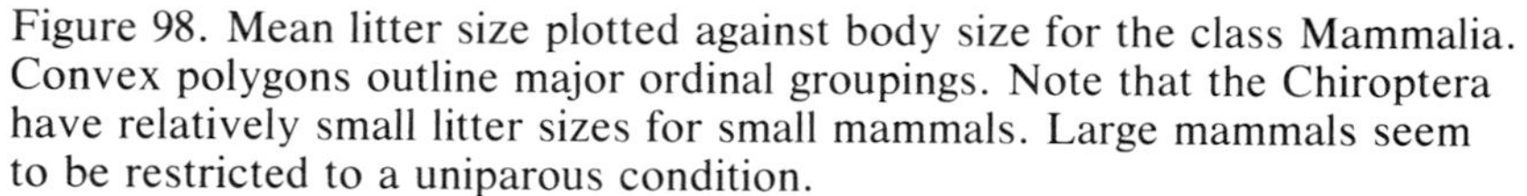
Figure 98. Mean litter size plotted against body size for the class Mammalia. Convex polygons outline major ordinal groupings. Note that the Chiroptera have relatively small litter sizes for small mammals. Large mammals seem to be restricted to a uniparous condition.

very small—most species are less than 80 g—but they almost always have a litter of less than two.

These generalities are comfortable after the Chiroptera have been removed, but certain nagging exceptions remain. The mammalian fauna of Madagascar allows an entrée into the "world of exceptions." Excluding the Chiroptera, figure 99 demonstrates that low litter sizes characterize several species that average less than 100 g in adult body weight as well as those species exceeding 1,000 g. Table 56 portrays the regressions. Three genera of the Tenrecidae falling in the middle weight range are characterized by moderate to extremely large litter sizes.

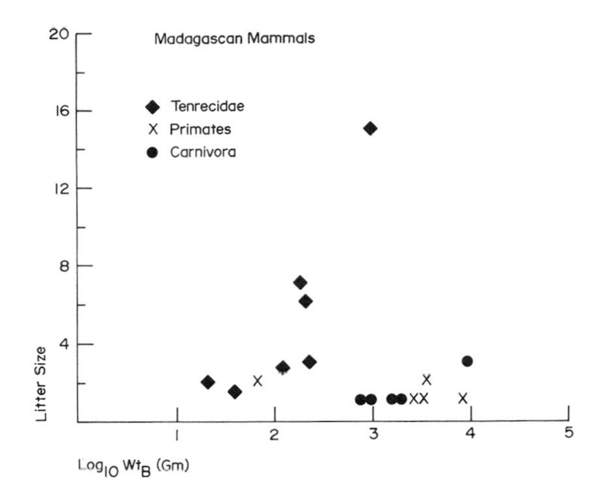

Figure 99. Litter size as a function of body size for a set of Madagascan mammals. Note the tendency for reduced litter size over a broad range of body size. Some tenrecoid insectivores approximate a semelparous strategy.

Table 56 Madagascan Mammals: Litter Size Regressed against Log_{10} Mean Body Weight in Grams of an Adult Female

Taxon	*A* Intercept	*B* Slope	Coefficient of Correlation	Number of Species
Lemuriform primates	3.25	−.67	−.99	3
Carnivora	−4.79	1.91	.94	5
Tenrecine insectivores	−14.64	9.33	.86	7

When examined from the standpoint of trophic strategy, utilizing a body length plot (fig. 100), the insectivore/omnivore Tenrecidae show a trend toward increasing litter size as their body size increases. The frugivore/granivore Rodentia show a decrease in litter size as their body size increases. The primates show no discernable trend, though the insectivorous, small *Microcebus* has two young. The viverrid carnivores have a low litter size (mode of one) in the smaller size-classes.

These are not typical of the trends discerned for continental faunas. Figure 89 portrays an analysis for marsupials where a litter size/body size correlation for continental areas was established. If carnivores and insectivores are separated from herbivores, then in both sets there is a general trend toward decreased litter size with increasing body size. The slope is greater for the carnivore/insectivore trophic set, and the intercept is much higher.

Given the peculiarities in the litter size/body weight correlations for the mammals of Madagascar, surely the longevities of some small Madagascan mammals must be relatively great. We can estimate theoretical lifetime total production of young for females that reach adult status if we know (*a*) age at first conception, (*b*) maximum age at fertile conception, (*c*) average litter size, and (*d*) maximum age before senility. Unfortunately, only nine species of Madagascan mammals are suffi-

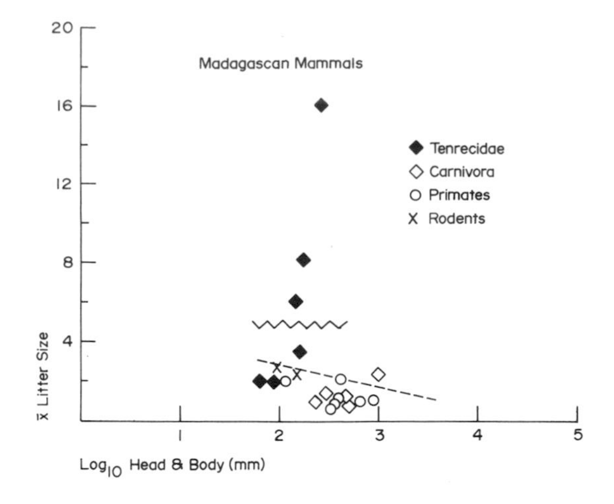

Figure 100. Mean litter size plotted against head and body length in millimeters, expressed as $\log_{10}$, for a series of Madagascan mammals. The trends in the previous figure are replicated.

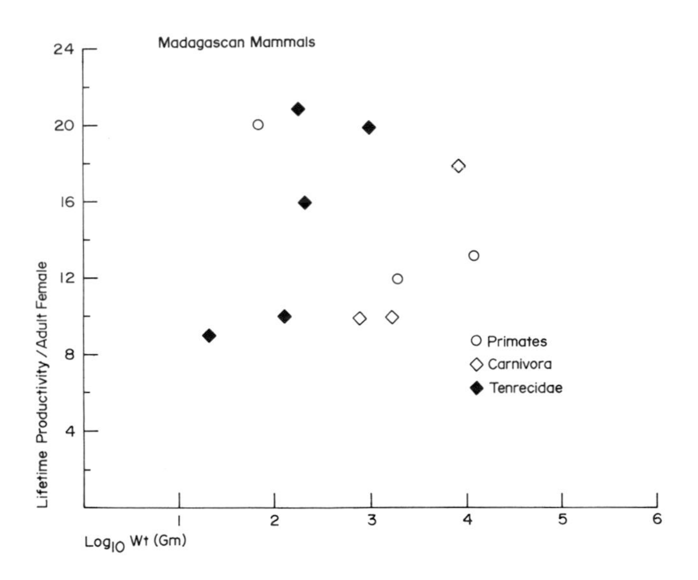

Figure 101. Estimated lifetime productivity per adult female as a function of body size for a series of Madagascan mammals. Note that two subsets can be discerned (see text).

ciently well known to permit such analysis, but figure 101 portrays the subset as a function of size. Between nine and twenty-one young appears to be the lifetime average productivity, and there is little correlation with size except that in the extremely large size-classes the value is about twelve. Those species that produce very large litters approach a condition of semelparity (one litter per lifetime; e.g., *Tenrec ecaudatus*). This implies, then, that low litter size should correlate positively with longevity and that relative longevity, though a function of absolute size, must show a wide scatter (see fig. 96).

Figure 92 has demonstrated that the "lifetime" productivity per adult female is poorly correlated with size for marsupials and eutherians. The lifetime productivity ranges between nine and twenty-four young, but the modal values clump at about twelve. The data suggest that a lower limit of some nine young per adult female seems to be fixed and that between 10 and 1,000 g adult body weight considerable lability in total productivity can be shown. At the higher weight- or size-classes, the range in total productivity is less, and this may reflect the restriction on litter size and lengthened interbirth interval that is a function of the decreased metabolic rate that accompanies size increase.

Since carnivores and insectivores show a higher litter size relative to herbivores (see fig. 89), we may conclude that as a rule of the thumb herbivores live longer than carnivores over a range of body sizes. For marsupials and placentals this seems to be the case (figs. 89 and 92), but no such "clean" trophic separation can be shown for the Madagascan mammals. Small viverrid carnivores from Madagascar seem to have as long a reproductive span as do their herbivorous compatriots within the same size-class.*

What has shaped this unique departure from the continental averages? The first question one might ask concerns the selective forces that have been molding the Madagascan mammalian fauna. Apparently the colonization of Madagascar by mammals proceeded in a series of stages, with the Tenrecidae arriving either before or at the same time as the strepsirhine primates (see chap. 7). As noted by Eisenberg and Gould (1970), the tenrecs were able to undergo an adaptive radiation unparalleled by insectivores on the contiguous continents. The viverrid carnivores arrived later and began to place selective pressure on the tenrecs, to which the tenrecs responded by developing numerous antipredator devices in addition to those evolved to protect them from raptors and snakes. The larger tenrecs, such as *Tenrec ecaudatus*, may have increased their reproductive rate as a direct response to this pressure, since cryptic coloring and small size were denied them because before the viverrids arrived they were the top terrestrial carnivore. *Tenrec ecaudatus* may have enjoyed the same status on Madagascar that *Solenodon* does on Cuba and Hispaniola; both species of *Solenodon* still have a low reproductive rate.

The foregoing may seem like a rather tight theory, but it still does not explain how the third large unarmored insectivore, *Echinosorex gymnura*, manages to survive in Malaya and Borneo with a modal litter size of one. Perhaps the relatively high encephalization quotient of *Echinosorex* is one correlate that allows this morphologically conservative mammal to retain a low litter size and long life-span while living in a Malay rain forest with 198 other species.

23.3 Some Factors Favoring the Production of Precocial Young

If one considers eutherians, then the production of precocial young by larger mammals depends in part on the predictability of food availability. Periods of food shortage can be compensated for by food storage, laying on fat, torpor, or mobility. Thus the predictability of the resource base is more important than actual periods of reduced primary productivity. How did I arrive at this conclusion?

A large body size favors the production of large, precocial young. Large size usually correlates with a mean increase in life-span and an extended time for reproduction. This in turn makes possible iteroparity, with a concomitantly reduced litter size. Large size may mean a decreased basal metabolic rate, and, if the trophic specialization (such as herbivority) further reduces metabolic rate, then the gestation will be relatively long.

If the species is very large, the relative weight of the fetus at term will be small compared with the mother's weight. It will be economical for the mother to produce a large baby while still retaining mobility. The mother will make a relatively greater total energy investment during gestation, but the investment per day may be low. The low daily investment can override perturbations in resource availability while the mother passes through the long gestational phase. The young will be able to flee and to feed itself at a relatively earlier age, and the female's lactation investment can be substantially reduced.

23.4 Brain Development and Length of Gestation

The results of Sacher and Staffeldt (1974) have implied that a long gestation and a moderately precocial condition at birth correlate with a rather high adult EQ (see chap. 22). This notion has also developed from the observation that higher primates have high EQs and are born in a rather precocial state after a moderately long gestation (i.e., the eyes are open; they cling; they have fur). It has also been noted that mammals born in an extremely altricial state (naked, blind, with limited mobility) generally do not exhibit high EQ values as adults. Figure 102 shows EQ values plotted against a five-point developmental scale for the neonate. Both low and high adult EQ values are associated with a degree of precociality at birth.

It is true that an adult EQ of 2.6 and above tends to be associated with an intermediate level of precociality (stage 3.5); but *Homo sapiens*, with an EQ of ~7.0, is graded at stage 2.9 in development at birth. In short, the mammal with the highest EQ is a bit on the altricial side (see S. J. Gould 1977, pp. 365–75, for an expanded discussion of the neotenic nature of *Homo*). Some cetaceans and most primates are precocial (stages 3.5 to 5.0) and also show high EQ values as adults (2.0 to 3.75). A few carnivores are rather altricial (stage 2.4) and have high EQ values as adults (~2.0). In general the points are quite scattered, and no useful trend is evident.

Why bother with the question? Sacher and Staffeldt (1974) noted that longer gestations correlate positively with higher neonatal brain weight. The argument could run that since brain growth rate during gestation is even more rapid than general somatic growth, then a longer gestation will achieve a relatively larger brain. Further, Huggett and Widdas (1951) demonstrated that primates showed a relatively long gestation compared with other mammals when the log of the weight at birth was regressed against gestation. Figure 75 duplicates this plot with a wider array of mammals than was employed by Huggett and Widdas. The scatter is tremendous. The best that can be said is that some bats, primates, and perissodactylans have a long gestation relative to the

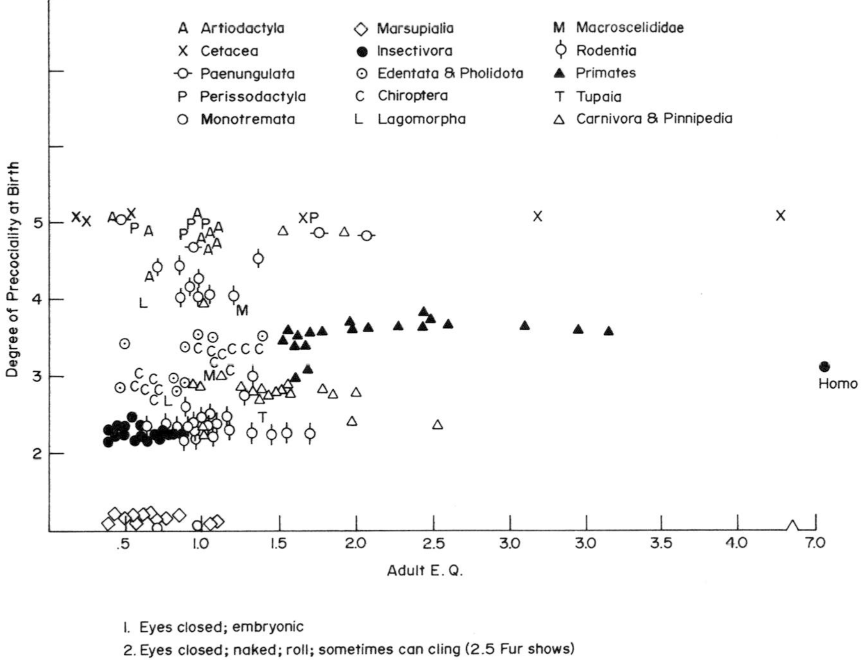

Figure 102. Degree of precociality at birth as a function of encephalization quotient. Note the lack of correlation.

mass of the litter at birth. The data imply that macroscelidids, lagomorphs, some rodents, some cetaceans, and a few carnivores produce a large neonate after a relatively short gestation.

It is known that, for some mammals, brain growth is very rapid during gestation (Holt et al. 1975). It is also known that *Homo* continues a rapid brain growth rate for some time after parturition (ibid.). How many other mammals show comparable rates remains to be demonstrated. The few that have been studied do not show a continued high rate of brain growth after birth (the Marsupials are excluded here).

I believe, given the discussion in section 22.3, that intrauterine somatic growth rates vary widely among the major mammalian taxa. This variation derives from differences in efficiency of the placental transfer mechanisms, metabolic rate of the gestating female, and various other factors. Although somme authors assume that the rate of brain growth during gestation is constant, the data base is small. Given the differences in placenta form, physiology, and efficiency as well as species differences in basal metabolic rate, perhaps it is an error to plot all mammalian taxa on one graph. A far more practical measure would be a plot of related species that show some unity in physiology and placenta morphology. Figures 71 and 103 portray a separation of three orders: Insectivora, Chiroptera, and Rodentia. In figure 71, soricoid and tenrecoid insectivores are combined. Although there are two ex-

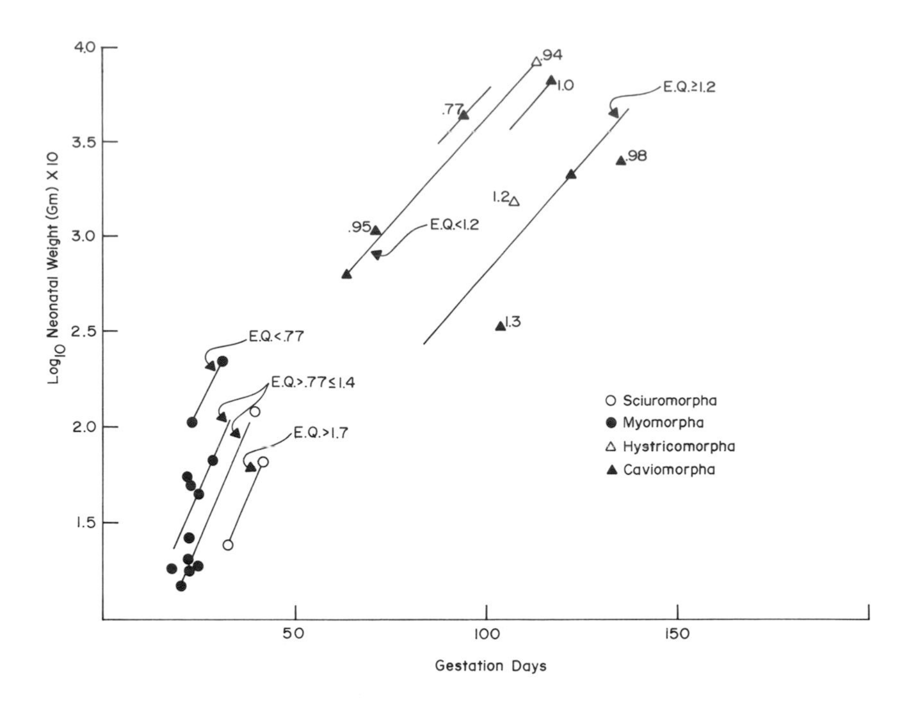

Figure 103. Gestation as a function of mean neonatal weight at birth. The effect of encephalization is demonstrable. Myomorph rodents have relatively rapid developmental times and low encephalization quotients. Some sciuromorphs have extremely high encephalization quotients and longer developmental times for their size-classes. Similar tendencies are demonstrable for hystricomorph and caviomorph rodents, which are much larger at birth than are myomorphs and sciuromorphs.

ceptions for the Insectivora, in general higher adult EQ values correlate with a relatively longer intrauterine time with respect to the fetal mass at birth. The bats plotted on the same figure show a clear separation into fast development/low EQ and slow development/high EQ. Figure 103 portrays the Rodentia for which I have data. Although the Caviomorpha clearly have a different pattern from the sciuromorphs and myomorphs, the same trends are shown as in the Chiroptera. Note that there are again two exceptions. These plots suggest that we may give some credence to the idea that a relatively longer intrauterine life correlates with a higher adult EQ, but metabolic differences may produce a striking increase in gestation time with little or no advance in EQ.

I do not offer a unified theory for mammalian development, but I suggest that useful comparisons are best made within a set of closely related species. Even in such comparisons, one will note exceptions that are worthy of special study. The development of the brain is as much subject to heterochronic change as any other mensurable character.

Larger mammals that give birth to relatively precocial young do not always form a social structure where long-term learning and a relatively large brain are necessary to survival. We need far more data dealing with measures of the length of time young learn from parents and older siblings. If we graph the percentage of the total life-span spent in a learning situation against EQ—and I include only species where I feel intuitively confident concerning the "time spent in a learning situation"—then figure 104 results. The Asiatic elephant and some primates appear to be orders of magnitude apart from the rest. Note that all species graphed produce relatively precocial young. Simple unitary explanations for the advantages of this will not suffice. Let us turn to the opposite question: Why produce altricial young?

23.5 The Advantages and Disadvantages of Producing Altricial Young

The production of young by both marsupials and eutherians may be divided into two phases: an intrauterine phase, or period of gestation, and an extrauterine phase, or period of lactauion. In marsupials the lactation phase is relatively long. Both phases demand energy from the female to sustain the growth of the offspring. Growth rate, whether fetal or postpartum, is influenced by the size of the adult. Smaller mammals tend to attain adult proportions in a shorter time than do larger mammals. The rate of fetal growth is more rapid in the second half of pregnancy than in the first. The postnatal growth rate is slow initially, then fast, and then slows again at the time of weaning. Let us consider methods of measuring energy demands on the mother, both during gestation and during lactation. Referring to table 57, we can calculate a value β that measures the g/day gained by a litter during

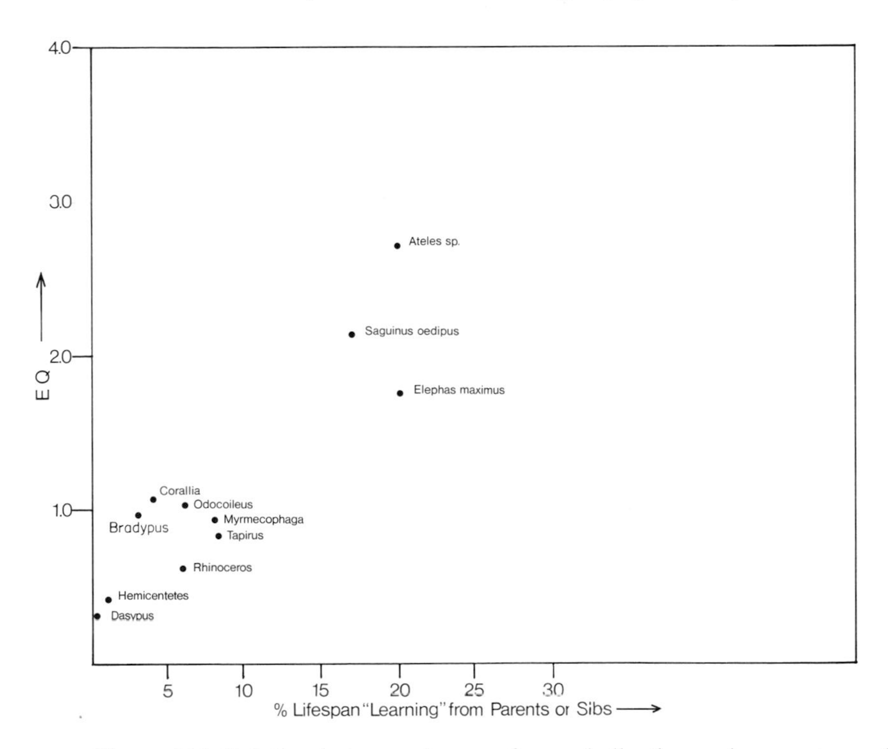

Figure 104. Relation between degree of encephalization and percentage of life-span spent in the company of parents or older siblings. There is a weak positive correlation.

Table 57 Weight Gains of Young per Unit Time for the Second Half of Gestation and the Dependent Lactation Phases

Taxon	β Value	ν Value
Marsupialia		
Marmosa robinsoni	.14	1.15
Didelphis marsupialis	.24	5.24
Insectivora		
Tenrecidae		
Tenrec ecaudatus	13.3	32.5
Microgale talazaci	.24	1.0
Hemicentetes semispinosus	1.3	9.0
Echinops telfairi	1.1	7.9
Soricidae		
Cryptotis parva	.14	.87
Suncus murinus	.26	1.44
Edentata		
Dasypus novemcinctus	3.0	6.2
Euphractus sexcinctus	6.7	9.5
Choloepus hoffmanni	2.2	6.5
Chiroptera		
Antrozous pallidus	.10	.30
Carollia perspicillata[1]	.08	.21
Pipistrellus pipistrellus[2]	.06	.07
Primates		
Callithrix jacchus[3]	.76	1.98
Saimiri sciureus[4]	1.06	2.64
Pongo pygmaeus[5]	9.09	15.75
Gorilla gorilla[5]	16.06	22.35
Scandentia		
Tupaia belangeri	.87	4.67
Lagomorpha		
Sylvilagus aquaticus	8.7	19.8
Rodentia		
Mystromys albicaudus	.99	2.83
Neotoma albigula	1.14	2.83
N. cinereus	3.15	11.3
Ochrotomys nuttali	.475	1.12
Peromyscus californicus	.693	1.13
P. crinitus	.55	1.2
P. maniculatus gambelli	.596	1.83
Sigmodon hispidus	2.84	10.24
Clethrionomys glareolus	.76	2.05
Dipodomys merriami	.50	1.22
D. nitratoides	.55	1.7
Perognathus californicus	.695	1.73
P. longimembris	.378	.64
Tamiasciurus hudsonicus	1.51	4.79
Proechimys semispinosus	1.80	8.8
Dasyprocta punctata	5.0	24.64
Hydrochoerus hydrochoeris	74.7	296.4
Octodon degus	1.5	13.2
Carnivora		
Potos flavus[6]	2.01	13.65
Felis silvestris	2.85	13.0
Panthera pardus	15.63	211.31
Mirounga leonina	333.2	6827.0
Ailurus fulgens[7]	6.0	26.7
Canis latrans	36.5	173.3
Mustela nivalis	.27	5.14
Proboscidea		
Elephas maximus[8]	288.0	1442.0

Table 57—*Continued*

Taxon	β Value	ν Value
Artiodactyla		
Choeropus liberiensis	50.4	527.0
Bos taurus	253.6	377.0
Rangifer tarandus[9]	49.9	729.0
Saiga tatarica	67.8	350.0

Note: β value = average gain in weight for fetal mass during the last half of gestation (g/day).
ν value = average gain in weight for young or litter during sustained lactation (g/day).

Sources: Millar 1976, NZP records, and (1) Kleiman and Davis 1978; Kleiman 1969; (3) Hearn and Lunn 1975; (4) Hendrickx and Houston 1971; (5) Gijzen and Tijskens 1971; (6) Poglayen-Neuwall 1962; (7) Roberts and Kessler 1979; (8) Reuther 1969; (9) Kelsall 1968.

the second half of gestation. We can further define a value ν that expresses the average rate of gain by a litter or single young in g/day during the interval from birth to weaning. The values β and ν only approximate the energy demands on the female, since her transfer efficiency is far less than 100 percent. Nevertheless, net weight gain by the fetus or neonate reflects a net loss in energy for the female. This rather crude approximation allows me to use data from a wide variety of species. Precise calculations of the bioenergetics for artiodactylans are included in Moen (1973).

In figure 105 I have graphed β and ν as a function of the mean weight of the mother. As I predicted in the previous sections, the rate of absolute weight gain per day increases as the mother's size increases whether we measure late gestation or early lactation. The absolute gain per unit time is more rapid during lactation, but there is great scatter and overlap.

If we divide β and ν by the mean weight of the mother (*Wtb*), we correct for the size bias and can plot either β/*Wtb* or ν/*Wtb* against the mean weight of the species. If we consider β/*Wtb*, we find that the gain in g/day/g of mother's weight decreases as mean adult weight increases. This relation should approximate the decline in metabolic rate associated with increase in body size, which is to say that the conceptus of a small mammal gains weight more rapidly per gram of body weight of the mother during the second half of gestation than do large mammals. There is still much scatter, indicating that even within this set there are slow gestators and fast gestators in a given size-class (see fig. 106).

If we graph β against ν, we find that almost all mammals show maximum weight gains and hence maximum relative energy input from the female during lactation rather than gestation, but the scatter is considerable. Further, those species that tend to maximize energy input during gestation also produce a relatively large litter weight at birth relative to the mother's weight. This latter correlation has nothing to do with the degree of precociality at birth (see fig. 107).

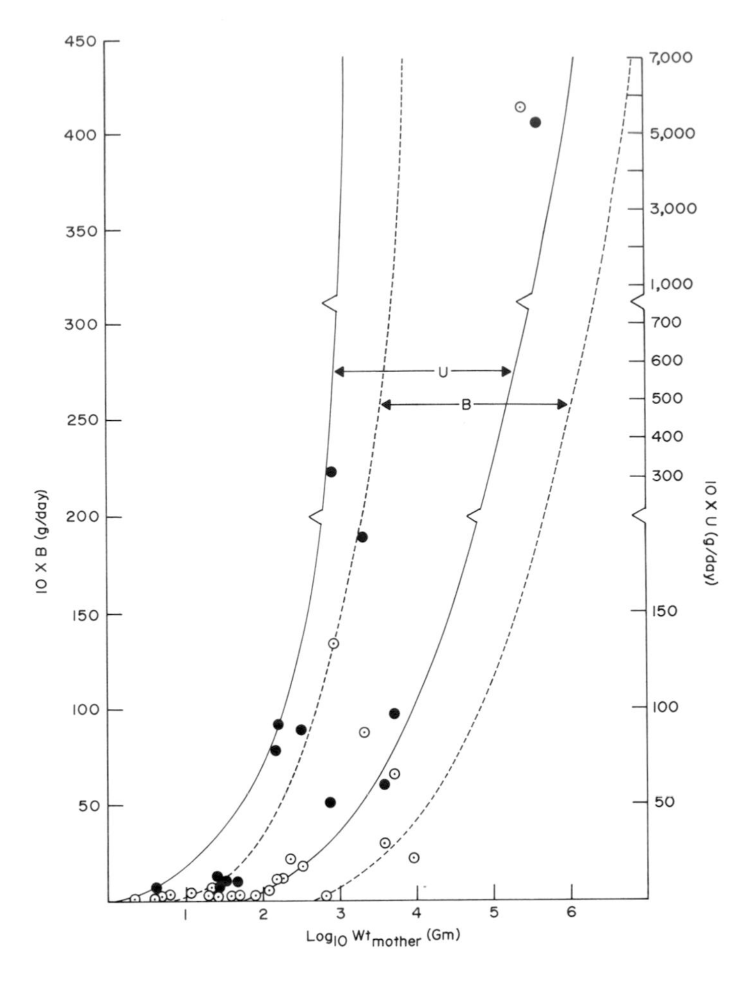

Figure 105. Weight gain (g/day) during lactation and weight gain (g/day) during gestation as a function of mean body size of the mother. Note that in general weight gains during lactation exceed weight gains during gestation.

Large mammals with a relatively long reproductive life-span can produce small litters (1–2) of extremely precocial young. The weight of the young relative to the mother's weight will be low, and the relative energy cost per day to the mother during gestation will be low. If a small mammal opts for a low litter size (1–2) with the production of semialtricial or precocial young, it must increase its potential reproductive span by either increasing its life-span or decreasing its interbirth interval; and, if the latter course is taken, it must decrease the relative cost of lactation by shortening the lactation period even more than would a large mammal. The net result is a tremendous input of energy per unit time during the second half of gestation. One would assume that the resource base would have to be very predictable (see fig. 108).

Figure 109 outlines the options open to a mother for terminating a rearing cycle, given two extreme conditions: the production of extremely altricial young, or the production of rather precocial young.

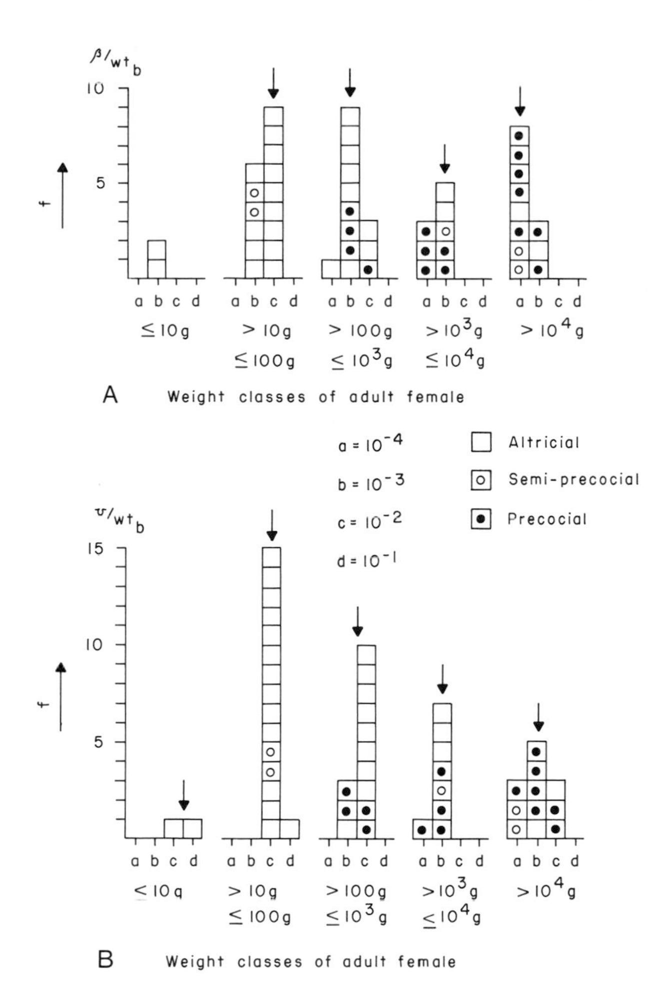

Figure 106. (*A*) Weight gain (β) during gestation divided by the mean body weight of an adult female plotted according to mean adult female body-size classes. Note that in general larger mammals exhibit a lower weight gain by the fetus relative to the mother's mass than do smaller species. (*B*) Weight gain (ν) during sustained lactation divided by the mean body weight of an adult female plotted according to mean adult female body-size classes. Note that the same trends hold as determined for relative weight gains during gestation, but the gain per unit time of the young is more rapid. Letters *a* through *d* refer to the values for β or ν. See table 55 for actual values.

During pregnancy, if an adverse set of environmental circumstances occurs, a female can resorb her embryo(s) in the early phases of gestation. In the latter phases of gestation she can abort and devour the litter, thus recycling the energy. In practice, however, large herbivores that bear precocial young are not equipped morphologically to reduce a large fetus to an acceptable food substrate, so an aborted fetus is lost as potential recycled energy. Even after giving birth, the female who bears altricial young has the option of killing and eating her young in response to adverse environmental circumstances. The large herbivore

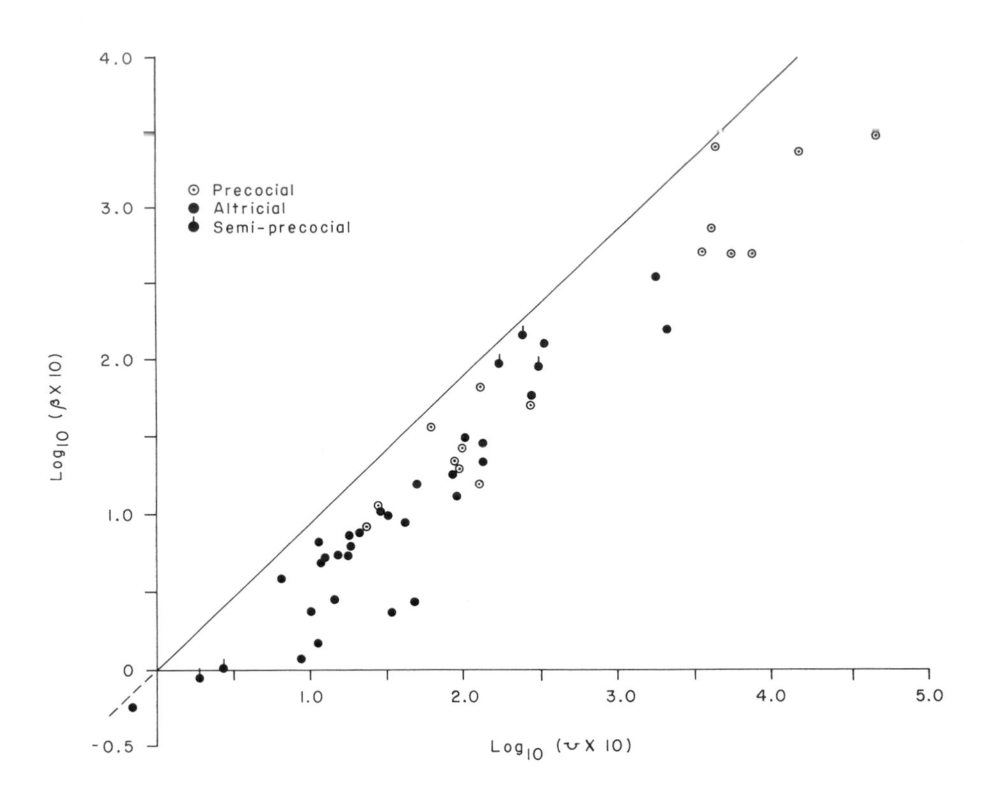

Figure 107. Litter weight gains per day during gestation (ordinate) plotted against weight gain per day during lactation (abscissa). Weight gains per day for all species are greater during lactation. Precocial forms exhibit the same trends as do altricial forms.

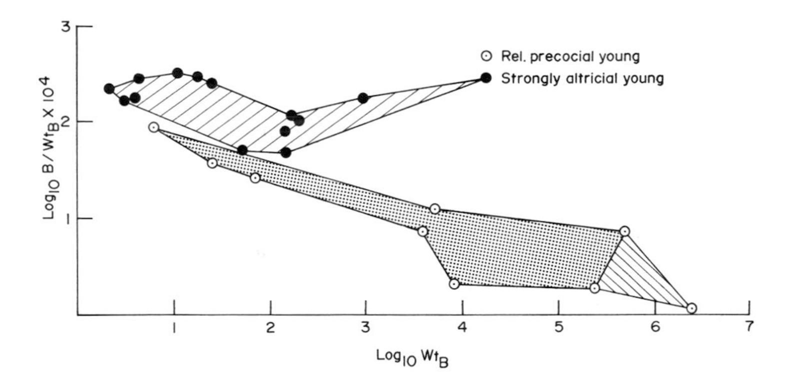

Figure 108. Weight gains per unit weight of female during gestation plotted against mean body weight of adult female. Note that species producing precocial young invest less in growth per day per unit body weight than do those species bearing altricial young.

with a precocial young can only desert it, with total loss of the reproductive investment.

To summarize the costs and benefits of bearing altricial young versus bearing precocial young, let us note the following points. Most female

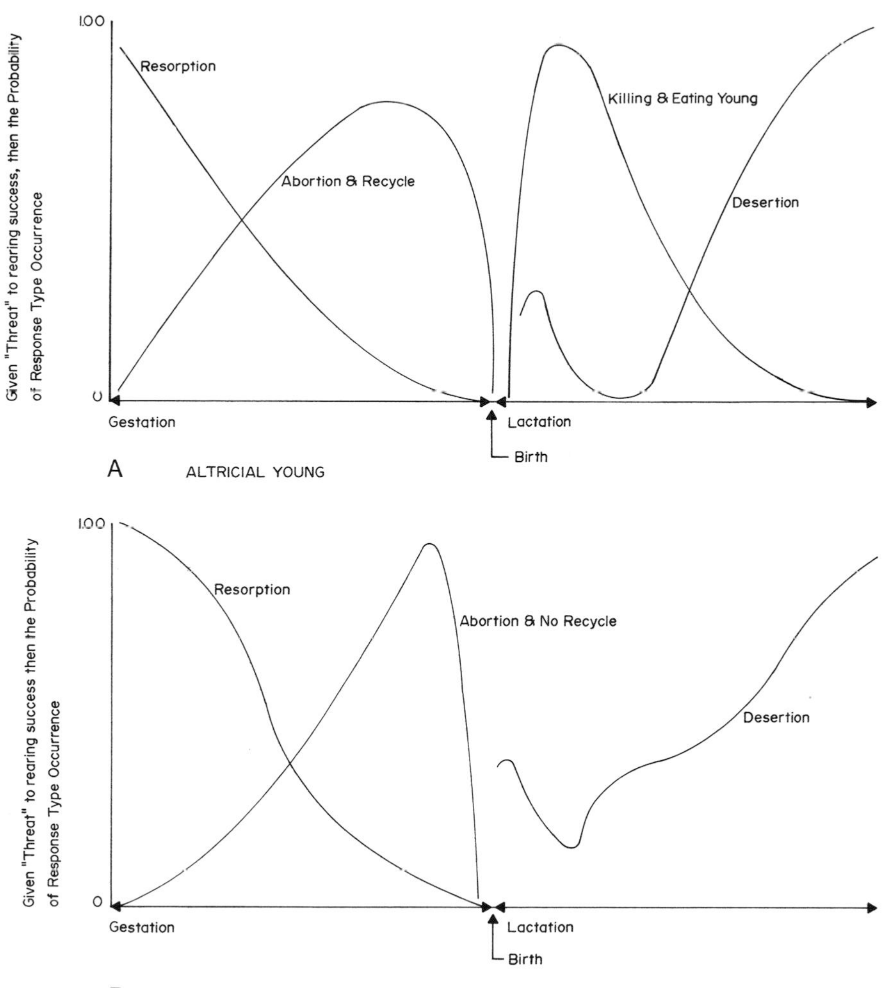

Figure 109. Possible responses of females to loss of young during gestation or lactation. (*A*) Options open to females bearing altricial young. (*B*) Options open to females bearing precocial young.

mammals that bear altricial young can "recycle" the young at many points in the gestation and rearing cycle if conditions become adverse. In small mammals that produce large litters of altricial young, the female assumes a large share of the energy cost in rearing the young relative to her lifetime energy needs. The only way to reduce the cost per litter for a female is to reduce the litter size by prolonging gestation and reducing the rate at which the fetus grows, thereby reducing the energy demand per unit time.

If precocial young are produced by a mammal that can no longer feed on an aborted fetus (e.g., ungulates), then the cost is very high if an offspring is lost; however, if the young survives the relative proportion of parental care may be reduced, and the young bears the major burden of survival by learning to feed at an early age and by developing appropriate antipredator behaviors a short time after birth.

A small mammal that opts for a brief reproduction period in its lifespan and a large litter of altricial young may elevate its metabolic rate during gestation to shorten the actual gestation time, or it may retain a brief gestation and produce extremely altricial young. During lactation, the female may either lactate at a rather constant but slow rate, with slow neonatal growth, or elevate lactation to shorten the time of parental care. Marsupials are bound into a short gestation, given the peculiarities of their yolk sac placentation (see sect. 3.2). Eutherians have an option of extending gestation. When eutherians extend gestation to produce a large neonatal mass (with or without precocial young), the lactation period is usually shortened. If it is shortened, it may be because the young are more precocial, or because the young are altricial but the greatest energy is placed in the lactation phase to terminate the rearing cycle early. Which strategy is adopted appears to depend on a complex trade-off in energetics that cannot yet be fully understood.

23.6 The Evolution of Parental Care Systems in Mammals: A Reconstruction

Imagine two therapsid reptile stocks in the Triassic. One type was reasonably long-lived, and the female produced small clutches of large-yolked eggs but had the potential to produce many clutches in her lifetime. The female incubated the eggs and was a relative homeotherm during this phase of the reproductive cycle. Upon hatching the young were well developed and could forage within a short time.

The second type of therapsid was characterized by females that produced a large clutch of eggs that were smaller and had less yolk. These eggs were also incubated by the homeothermic mother, and the young were somewhat altricial at hatching. At some point the descendents of the altricial stock evolved a mechanism for nourishing the young by secretions from modified sebaceous glands on the venter of the female. The amount of yolk could be further reduced, and "lactation" became fixed as a means of feeding altricial young.

Once lactation was fixed as a trait, the species could diverge and undergo further "r" selection or "K" selection, favoring either large or small litter sizes; but the small-yolked egg, short incubation time, and subsequent lactation phase remained inextricably tied together. One could conceive of this as the "monotreme stage." A pouch to allow the female to become a mobile incubator was probably added independently many times in the course of evolution (e.g., *Tachyglossus*, the echidna).

The greatest single advance in the derivation of the Marsupialia and Eutheria from the pantotheres was the evolution of a teat through which the milk ducts opened. In the Marsupialia, the young could then attach to the teat, and the female would not be bound to a nest with helpless infants. The pouch was probably evolved later at several points in the marsupial history. In a final abandonment of the early nest phase with eggs and incubation, the marsupial stock evolved ovoviviparity (Luckett 1975) and the phenomenon of giving birth to extremely altricial young. The nest phase became mandatory later in the rearing cycle only when the weight of the litter imperiled the foraging efficiency of the female.

The eutherians escaped from the syndrome of producing extremely altricial young by the development of the trophoblast—the embryonic contribution to the placenta. The trophoblast allows fetal development to proceed with efficient transfer of nutrients and dissolved gasses between the fetal and maternal circulations. In addition, the eutherian placenta prevents an immunorejection response from developing by preventing fetal blood proteins from mixing with the female circulatory and lymphatic system (see chap. 22). The trophoblast must have evolved under selective pressures that promoted a reduction in the relative length of the lactation phase and the production of slightly more precocial young. Portman (1965) has suggested that the "primitive" eutherian neonate was intermediate between the extremes of altricial neonate shrews (Soricidae) on the one hand and precocial rodents (Caviidae) on the other.

The original placenta type may have been epitheliochorial, as Luckett (1975) suggests. (For an alternative hypothesis, see Starck 1956). Nevertheless, the conclusion is inescapable that, once the trophoblast evolved, there occurred a rapid divergence in placental morphology. All evidence from extant forms indicates moderate to great conservatism in the form of placentation and fetal growth rate in the diverse eutherian lineages. A close comparison of reproductive patterns could indeed shed more light on mammalian phylogeny.

Did the original eutherian stock as it diverged from the pantotheres possess an ancestral form that was ovoviviparous, like that of a marsupial, or did the progenitors of the eutherians evolve a nipple and then pass from an oviparous stage to viviparity with the development of the trophoblast? If eutherians passed through a marsupiallike stage of ovoviviparity, then the retention of young in the uterus increased the female's investment in the gestational phase and decreased her relative investment in lactation. This would have been accompanied by a slight reduction in litter size and indicates a moderate "K" selection. If the eutherians moved directly from oviparity, it suggests equally an influence of "K" selection, since gestational investment still increases with a decrease in lactation investment. One cannot decide between the alternatives. It does tell us, however, that the further evolution of the Eutheria requires "r" selection forces with a reversion to increased altriciality and increased relative investment in lactation with an increase in litter size.

24 Conclusion to Part 3

As I have explored "Macrophysiology" and the correlations between body size and metabolism, relative brain size and development, I have speculated on evolution and natural selection. The reader may wonder if I intend to expand on the mechanisms of evolution. The great saga of biological discovery in this century was the identification of the nature of the "gene." The way the DNA of the nucleus was organized to encode all information necessary to construct and replicate a cell; the way mutations occurred; and the primacy of the genetic material as the determinant of growth, development, and form have all been unraveled and consolidated in my lifetime.

Of course, population genetics and an understanding of how the genotypes that underlie phenotypic expression can change over time by drift, selection, and mutation is the backbone of modern evolutionary theory. Nevertheless, these topics have been elegantly reviewed (Mayr 1963; Rensch 1959; Dobzhansky 1970). I have assumed, in writing this section, that the reader is versed in the basic theory. Furthermore, Wilson's recent synthesis (1975) of population biology, genetics, and behavior sets the stage for the next section, and I will not reiterate the basic framework now.

The recent synthesis by Stanley (1980) should offer considerable food for thought to students of mammalian evolution. Gradual change in morphology within a line of phyletic descent is the fundamental theme of classical evolutionary theory, but the branching off of divergent lines has remained a puzzle. It is now generally conceded that divergent splitting off of new lines accounts for more evolutionary change than a large number of gradual transitions within a single lineage (Bush 1975). Rapid change and divergence probably results from the fixation of new genetic traits within small, isolated populations, which can diverge rapidly from the founder stock in response to new selective pressures.

Evidence from karyotypic studies of rodents indicates a high level of inter- and intraspecific diversity (Patton 1970). In some families and genera of rodents, primary diversification in response to new habitats has been recent. Some taxa, such as the bats, have passed through an earlier diversification stage, and according to the "canalization model" of Bickham and Baker (1979) bats have within their "aerial adaptive zone" reached a stage of relative karyotypic stability.

Attempts to calculate evolutionary rate based on a study of variation in structural proteins have run into several difficulties. It is now recognized that structural proteins change slowly, and thus genetic divergence as measured by protein divergence has limited value for interpreting the rate of phylogenetic branching and net morphological change. The imporatnce of regulatory genes and of epistatic interactions among genes poses new problems. Alterations in growth rate, body proportion, and the temporal sequences of ontogeny may be controlled by simple alterations in gene position. The importance of regulatory genes is now widely recognized (Wilson et al. 1977).

Genetic concepts and the application of genetic theory to the expression of behavioral phenotypes will be deferred until part 4. In this section I hoped to establish an overview of classical zoological concepts applied to the class Mammalia. If I have sparked some interest in the reader, then the task has been a success.

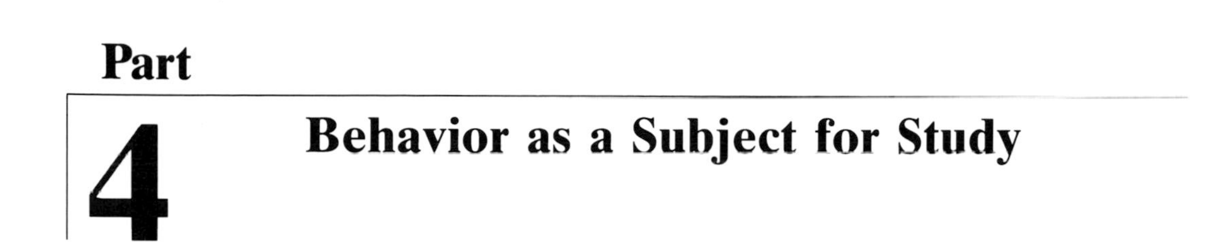

Part 4 Behavior as a Subject for Study

25 Introduction

The Nobel laureates for biology and medicine in 1973 were Konrad Lorenz, Niko Tinbergen, and Karl von Frisch. These three pioneers in the development of ethology exemplify three different general approaches to the problems of behavioral description and analysis. Von Frisch, in his elegant analyses of bee behavior and fish hearing, used an experimental approach. He conditioned the animals to various stimuli and was able to determine through their reactions the limits of perception in several of their sensory input systems (von Frisch 1954). Niko Tinbergen preferred to study animals in their natural environments. Through a cunning series of experiments in nature, he not only was able to demonstrate what aspects of the social and physical environment an animal was attentive to and responded to, but, by comparing closely related species, was also able to infer the adaptive nature of the behavior pattern under study (Tinbergen 1951). Konrad Lorenz preferred to work closely with captive animals in an attempt to unravel what parts of the behavioral repertoire required discrete learning and what parts of the repertoire seemed to develop through the processes of embryogenesis and maturation independent of specific environmental input (Lorenz 1961).

Lorenz and Tinbergen developed many ideas from their evolutionary and comparative approaches to the study of behavior. One of the cornerstones was the recognition that behavior was adaptive and ultimately the product of natural selection. The genetic endowment of a species set limits on the stimuli to which it was able to attend. The manner in which motor coordination patterns are carried out in situations specific to the survival of the organism tends to be predictable. Such behavioral sequences, which appeared to develop without imitative learning or specific environmental input, were referred to as innate. One need not go into the artificial controversy over whether a behavior pattern is innate or learned, but those interested in a historical review should see McBride (1971) and Lehrman (1970).

Intensive research on animal behavior in the 1940s and 1950s concentrated on the nature of animals' responses to defined stimulus situations. Greater and greater precision was developed in the description of movement patterns. By suitable experimentation, one could infer the function of specific movement sequences. Since fish and birds were classically used for the studies, emphasis was on visual signaling systems, and only gradually were responses to olfactory cues and auditory cues analyzed. The recognition that fishes employed chemical cues as well as auditory cues in their intraspecific communication came early, but the genuine analysis of these sensory input systems was deferred until after the close study of movement patterns had run its course. (The historical development of ethology has been summarized by Klopfer and Hailman 1967.)

When ethologists first investigated mammals, the concentration was again on the analysis of movement. A tradition of movement analysis had existed for some years (Muybridge 1887, reprinted in 1957). Study centered on the development of techniques (cinematography) that would allow the temporal resolution of mammalian movements during different types of locomotion. Basic descriptions of movement patterns led to correlations between behavior (locomotion) and anatomical structure (Howell 1932, 1944). Comparison of skeletal structure and locomotion patterns has led to more sophisticated analyses of movement (Hildebrand 1961) and direct inferences concerning the locomotor patterns of fossil forms (Gambaryan 1974; Sukhanov 1974). The description of movement patterns during an individual animal's maintenance activities finally led to comparisons between species with respect to urination and defecation (Altmann 1969), drinking behavior (Schoenholzer 1959), and mastication (Crompton and Hiiemäe 1969).

In a parallel fashion, aspects of mammalian social behavior were also being described, and the serious origins of behavioral description may be found in the works of L. H. Morgan (1868) and F. Lataste (1887). Movements associated with the deposition of chemical secretions, with vocalizations, and with an animal's reproductive behavior were described for a variety of species. The finest summary of classical ethological technique as applied to mammalian behavior is contained in *The Ethology of Mammals* by the late R. F. Ewer (1968*b*). Patterns of dyadic interaction have been analyzed in ever-increasing detail, culminating in the research of Golani and Mendelssohn (1971) and Golani (1976).

Early on, it was recognized that the general maintenance behavior of mammals as well as the forms of their social interaction often involved rather stereotyped components or subcomponents. Stereotypic vocalizations, marking movements, postures, and configurations between two interacting individuals were generally described by a human observer in the course of developing a behavioral inventory or ethogram for a species. Once the configurations, postures, movements, and sounds had been defined, the patterns or units could be analyzed from the standpoint of their temporal regularities or irregularities. The temporal patterning of behavior and how this patterning was influenced by the social and the physical environment became the basis for further analytical procedures (Marler and Hamilton 1966, pp. 25–72).

Always, however, the ethologist is concerned with bringing the interacting animals back into the context in which they evolved. He constantly attempts to see the total behavioral repertoire as the product of natural selection. To be sure, in a developmental sense, the behavioral repertoire is the outcome of ontogenetic processes whereby the developing organism continually interacts with its microenvironment, thus shaping its behavioral capacities. Nevertheless, the final product of development is usually an organism finely attuned to a specific environmental context that in turn is the outcome of millennia of natural selective processes on the parental stocks, giving rise to an organism that is "adapted" to the normative or average environmental context in which it finds itself (see J. L. Brown 1975).

In the analysis of behavior it is convenient to categorize behavioral acts functionally, according to the ends to which we believe they are geared. Such a classification might include such categories as pelage dressing, acquisition of food, storage of food, burrow construction, and mating behavior. Interaction is also defined by the outcome of types of social interactions; again, function is implied in the formation of categories (e.g., parental care behavior, mating behavior, agonistic behavior, and courtship behavior).

The description of mammalian interaction patterns has been fraught with technical difficulties. One usually records movements and sounds, knowing full well that in mammals chemical secretions are also involved in the coordination of activity (Eisenberg and Kleiman 1972). Where the deposit of chemical secretions, urine, feces, and so forth, is accompanied by easily noted postures, then the use of "chemical signals" becomes built into the description; but, where odor traces are passively carried on the body of the animal, their significance and function are often overlooked. Interaction is also dynamic, with a temporal dimension, and by definition involves at least two individuals. Thus the description of behavior is often best accomplished with an assist from a visual recording device such as videotape or cinefilm.

Since it is assumed that, during social interactions, the interactants are exchanging stimuli among themselves, all interactive behavior by definition involves communication. The exchange of signals is often categorized in terms of the presumptive receiving sense organ—for example, olfactory signals, visual signals, auditory signals. Exchange of signals between interacting partners became the primary focus for the earlier attempts at the analysis of social behavior. Most research on mammalian communication has been recently summarized in the book edited by Sebeok (1977). The conceptual foundation of communication analysis has been explored by Smith (1977).

Although the analysis of animal communication has had great popularity, it contains a number of built-in assumptions. First, one assumes that there are a designated sender and receiver. A message passes between them, and the human observer assumes that its "meaning" for the presumptive receiver can be interpreted by noting its reactions. One assumes also that the "mood" of the presumptive sender can be discerned. By an opposite approach, one assumes that the message content can be assessed from the standpoint of the sender by an approach that also employs motivational analysis.

Golani (1976) has proposed an alternative paradigm for use during the initial analyses of interaction forms in mammals. His key assumption is that the sole "goal" of the presumptive sender may be only to maintain levels of homeostasis with respect to sensory input. This approach to interaction analysis reduces the number of initial assumptions.

One of the most difficult goals in the study of mammalian interaction processes is to analyze the forms of social interaction for the different age- and sex-classes of the life cycle. We may aspire to a description of the form and nature of interactional norms for a species in a defined habitat throughout an entire generation for both sexes. Much descriptive research so far has concentrated on only a small aspect of the social life of a species—for example, mating behavior, temporal patterning of copulation, mother-neonate interactions, or male-male aggression. What we are left with, then, is a series of fragments. And, because we have to deal with fragments, we often compare among species only those "pieces" of the behavioral repertoire that we have at hand.

Volumes of "comparative" studies exist that look at only maternal care patterns or courtship behavior. This has led in turn to systems of functional classification for those aspects of social behavior for which there are sufficient data to permit comparison across species. Of course, if we have knowledge of mating systems and rearing systems, it is possible to capture the major events of a life history and to compare them across species. Indeed, in my earliest attempts at an analysis of how social structure is adaptive, I did exactly this (Eisenberg 1966).

A preliminary classification for mammalian social systems could be based on: (1) the degree to which the social group in question shows cohesion through time; (2) how many adult individuals are contained within the group; (3) the degree of genetic relatedness among the adults; (4) the size of the group itself; and (5) how group size is maintained at some level of stability with respect to the space it utilizes. Once having formulated this woefully inadequate classification system, one can begin to look for the way the system reflects adaptation for particular foraging strategies within defined habitat types; in other words, to elucidate how the social structure itself is an adaptation for living in a certain manner in a defined environment.

The key to the foregoing approach to social behavior is the recognition that the social system (or systems) a species shows throughout its life cycle is itself an adaptive mechanism and may be analyzed in the same way one would describe and analyze any phenotypic character. The goal for the human observer is to elucidate the way the social system functions in the natural history of the species. How natural selection shapes social systems through time becomes a derivative question and could lead one to develop hypotheses concerning the function of social systems in a natural context. Once such hypotheses have been developed, we can make predictions and draw comparisons to test the predictions. This is where the phylogenetic approach becomes so important, because one can envision the different lines of mammalian descent evolving separately on different continental land masses and adapting in a convergent fashion to similar environmental situations.

The ultimate question, then, is, whether such comparisions of independently developed radiations can confirm our hypotheses concerning the adaptive nature of the social organization itself. It therefore follows, if one considers that the social structure is an adaptation, that somewhere along the great chain between the gene and the overt expression of behavior there must be a true linkage. In other words, the propensity to show certain kinds of behavior under certain conditions is rooted ultimately in the genetic code. This guiding principle should not be oversimplified to say that behavior patterns are innate, that learning plays no role, and so on; rather, I mean that the form of the social structure exhibited by a population in question is the outcome of an interplay between genetic predisposition and the actual environmental circumstances in which the organism finds itself.

26 The Genetic Basis for Behavior Patterns

That genes influence physical development, which ultimately expresses itself as discrete, quantifiable behavior patterns, there can be no further doubt. Experiments on crickets, frogs, birds, and dogs have indicated that the genes affect numerous behavior patterns (see Fuller and Thompson 1962; Scott and Fuller 1965; Huber 1962, 1963). Occasionally a discrete pattern, such as a cricket's song, may be measured and the variations studied when various hybridization experiments are run. In most cases, however, the genetic influence on a behavior pattern is subtle and requires very careful quantitative procedures.

From the standpoint of the behavioral geneticist, the completely developed adult organism has reached its finished state of organization through a process of organic development based on a specific genetic composition, hereafter referred to as the genome. At all points in the course of development, the genome reacts with a specific cellular environment. The genetic makeup of the individual dictates the nature of protein synthesis in the cells of the organism and thus the number and form of the enzyme systems, which influence the rate of chemical reactions, which in their turn influence the rate of cell division and differentiation, which ultimately results in the differentiation of tissue types in a predictable order. The differentiation of the complete organism and its rate of growth is under genetic constraints. Thus any physical trait characteristic of a whole organism is the product of an interaction between a specific genome and a partially known environment.

Geneticists often speak of the heritability of a trait. Heritability (h) is mathematically defined in terms of the observed variance of the numerical expression of the trait under study. The variance is attributable to either variations in the genotype (g) or variations in the environments (e) of the individuals in the study population. Thus, if the environment is held to as near a constant set of conditions as possible, all variation in a trait should be the result of variations in the genotypes present in the population: $\left(h = \dfrac{\partial(g)}{\partial(g) + \partial(e)}\right)$. If all exhibited variation within a population of organisms is caused by the genome, then phenotypic correlation between siblings would be 0.50 (assuming this is not an inbred line with a heritability of near zero). Defined in this way, heritability is not an attribute of a trait but an attribute of a population.

To a certain extent, almost all phenotypic characters (morphological or behavioral) show a component of heritability. Behavior geneticists are always discussing the net component of variability in a population contributed by the genetic makeup of the population. They are not asking questions concerning innate versus learned behaviors. They are assessing the extent to which the variation observed in a population is the contribution of the genetic makeup or a contribution of variations and perturbations in the environment. When the heritability in a population falls to a very small value, that is to say, when selection is no longer effective in changing the behavior of a population, then one assumes that all the minor variations observed in the population are caused by environmental perturbations and are no longer manifested by the genetic variation as such.

Genetic control over the external appearance of a behavior pattern is of pivotal importance when one is considering the evolution of behavior. Konrad Lorenz (1961) developed a basic conceptual framework. He assumed that all modes of behavior that are relatively invariant in their appearance could be treated as structures in the sense that a morphologist considers a given physical structure as a static entity. The relatively invariant motor patterns an animal exhibits as components of its behavior were considered to be the result of nerve-muscular relationships laid down in accordance with the genetic makeup of the individual by the known embryological laws of development. Examples of such relatively invariant movement patterns included movements used for autogrooming in rodents; the fixation, chase, and killing bite of the ferret; the pitch and phrase length of a chaffinch song; the cry of an abandoned young mouse; or the fanning of a male stickleback at its nest. Such motor patterns may be described in terms of their form and temporal patterning. Since such motor units were assumed to have a genetic basis, they could be compared from one species to the next. The comparisons could be made in the same way that morphological structures are compared to discern homology. Theoretically, behavioral analogue could be distinguished from homologues (see Eibl-Eibesfeldt 1970 for a complete theoretical treatment).

Behavioral traits thought to have a strong genetic component influencing their ultimate expression can then be analyzed in a natural context to determine their adaptive function. One is now in a position to pose the question, How does this behavior pattern aid the survival of the organism? Although this implies that the behavior has a purpose, the assumption need not be considered the result of teleological thinking. The behavior is purposeful because it results from neurophysiological mechanisms in the progeny of forms which have survived to reproduce while in competition with other forms.

The outcome of reproductive success is the survival of the parental genotype (in sexually reproducing forms, half from the father and half from the mother). Trait variability exists in all animal populations, and reproductive success depends on an organism's adjustment to environmental variables. The ability to adjust varies, as does the genotype. The living organism is thus "adapted" in that it is the end point of natural selection on successive parental genotypes. The individual organism then is in a feedback situation with the environment. The organism is goal-oriented; that is, it seeks out stimulus input relevant to its survival, and the degree to which it achieves successful adjustments to environmental variables will in the long run determine whether its genotype survives. It is the survival of its genotype that ultimately determines the characteristics of the population in the following generation.

Now it is all very well to consider a simple pattern of behavior, such as head tossing during courtship, and compare head tossing over a series of related species. It is another thing to consider the whole process of courtship, including head tosses, chasing, attempted mounts, as a macrounit of behavior. In considering such macrounits of courtship, Golani and Mendelssohn (1971) have demonstrated that variation exists among individual pairs of jackals (*Canis aureus*) during courtship. Here the principle of idiosyncratic variation can be introduced. During courtship in jackals, interaction rate and to some extent even the form of the interaction can vary from one pair to the next. Such variations can be measured and described. Comparison of macrounits of behavior thus must allow for individual variation. But despite the reality of variation one still discerns an overall pattern of jackal courtship that is quite different from the courtship, let us say, of a giraffe.

This brings us to the question, How does selection act? In attempting to study animal behavior patterns in a natural context, and giving full appreciation to the fact that variation does occur, one may be led into a trap concerning a definition of adaptation. A behavior pattern may be displayed in a context where it is atypical or even maladaptive. The most important question one can ask is whether one has correctly interpreted the normative function of the behavior pattern. Many idiosyncratic variations can be thought to be adaptive, but only a long-range study of who survives in the next generation can tell us whether the particular complex of behaviors we have seen is advantageous. One often questions the adaptive nature of a behavioral phenomenon that seems inappropriate because we are looking at only one context, not at the whole range of contexts in which the behavior may be displayed. Such a case, I believe, has arisen recently with the problem of how infanticide may have been selected for in natural populations (see chap. 30).

27 A Consideration of Biological Rhythms and Temporal Patterning

In this chapter I shall explore some aspects of periodic behavior shown by living organisms and attempt to relate such periodic events to the selective forces that have probably shaped the presumed genetic basis underlying the behavior. To do this, I will make passing references to the presumptive function. Numerous studies have confirmed that living organisms, plant and animal, multicellular and unicellular, have intrinsic rhythms that can manifest themselves independent of external regulators (see Harker 1964). Thus, we may take as a *given* the demonstrated presence in many organisms (but not all) of an endogenous rhythm of activity that approximates a diel cycle or twenty-four-hour period. The literature is filled with elegant experiments demonstrating the presence of diel rhythms. In the main, they measure some gross aspect of activity rather than the discrete units of behavior. Almost invariably the organism has the capacity to "set its biological clock" by some environmental factor—generally the natural daylight cycle—so that, in the words of Aschoff, light can act as a *Zeitgeber* (Aschoff 1958). The concept of a biological clock was an important advance in understanding animal navigation. This work was pioneered by Gustav Kramer with migratory birds (Kramer 1952) and led to the sun compass hypothesis, proved for many vertebrates and invertebrates.

While much attention has been given to twenty-four-hour periodicities, lunar periodicities or tidal rhythms have also been demonstrated in marine organisms, and annual periodicities have been shown in certain invertebrates—in particular the edible snail, *Helix pomata*. These longer rhythms are of great interest to biologists, especially annual cycles, which have attributes of endogenicity. Reproductive rhythms in migratory birds have been subjected to intense investigation; in particular, the white-crowned sparrow (*Zonotrichia*) has been studied by Farner and his colleagues. Although Farner's group concludes that light in appropriate dosages is essential to "drive" the neuroendocrine secretory system, the presence of a refractory period at the end of the reproductive cycle confers a semblance of periodicity to the annual cycle, albeit the appropriate dosages of light are necessary to "fuel" the neuroendocrine mechanism (Farner 1965). Of course, an organism must live long enough—that is, longer than a year—for natural selection to shape any of its physiological mechanisms toward annual periodicity.

Let us return to the more classical framework—namely, periodicity in activity shown by an animal essentially isolated from external influences. De Coursey (1961), working with flying squirrels (*Glaucomys*), and Kavanau (1962, 1967), in research on the diel rhythms shown by mice of the genus *Peromyscus,* pioneered in this area. Undeniably *Peromyscus* shows an endogenous rhythm in activity that approximates a twenty-four-hour cycle, and in the main this activity is confined to hours of darkness. In the absence of any light cues, the animals will maintain a twenty-four-hour rhythm, though they will gradually drift out of phase under constant conditions when cut off from an environmental *Zeitgeber.*

In the early sixties Donald Isaac asked, To what extent are discrete units of activity organized within the normal activity period under conditions of environmental isolation? By a series of recording devices, he was able to measure the periodicities of feeding, drinking, wheel-running, grooming, elimination, and sleeping in a laboratory population of *Peromyscus maniculatus.* In the isolated condition, the gross patterns of activity rhythm persists, and the organization of activity units within bouts of wakefulness shows some properties of order. For example, drinking almost always follows feeding, and grooming the face almost always follows drinking or feeding; but, apart from this, chains of activities with a predictable relationship between the units do not exist. In short, at any point in time there is almost an equal probability that the actively moving animal will do any of the major acts open to it. Ordering of activity into chains of predictable, dependent events cannot be demonstrated, with the exception of contingency between feeding and drinking and between ingestion and grooming (Isaac 1966). These results are consistent with results obtained by Nelson (1964) in another type of analysis, where he measured the dependency of discrete acts upon one another during the courtship activity of male glandulacaudine fish. Although the foregoing has little to do with rhythmicity, I wish to point out that long chains of statistically dependent acts found in a courtship bout of glandulacaudine fishes show the same types of uncertainty as is demonstrated in the analysis of the activity patterns of an isolated *Peromyscus.*

Periodicity during interaction is demonstrable if one looks not at the dependence of small units, but rather at functional blocks of behavior. Some years ago I ran an extensive series of dyadic encounters utilizing a wide range of mammalian species, but, in particular concentrating on rodents, insectivores, and small marsupials (Eisenberg 1969). The nature of the periodicities one uncovers using the encounter technique is of course a function of how long one allows an encounter to run, the locus of the encounter, the experience of the participants in the dyad, and the nature of pair-formation in the species. Obviously a knowledge of natural history is important in designing a meaningful encounter program. Essentially, each visually distinct unit of behavior is defined and named. This behavioral inventory is then used in formulating a precise, quantitative description of interaction forms throughout an open-ended, staged encounter.

One way of plotting the data may be seen in figure 110, where each symbol refers to a particular behavioral act and where the temporal patterning, in the case of the male's behavior, is demonstrable along

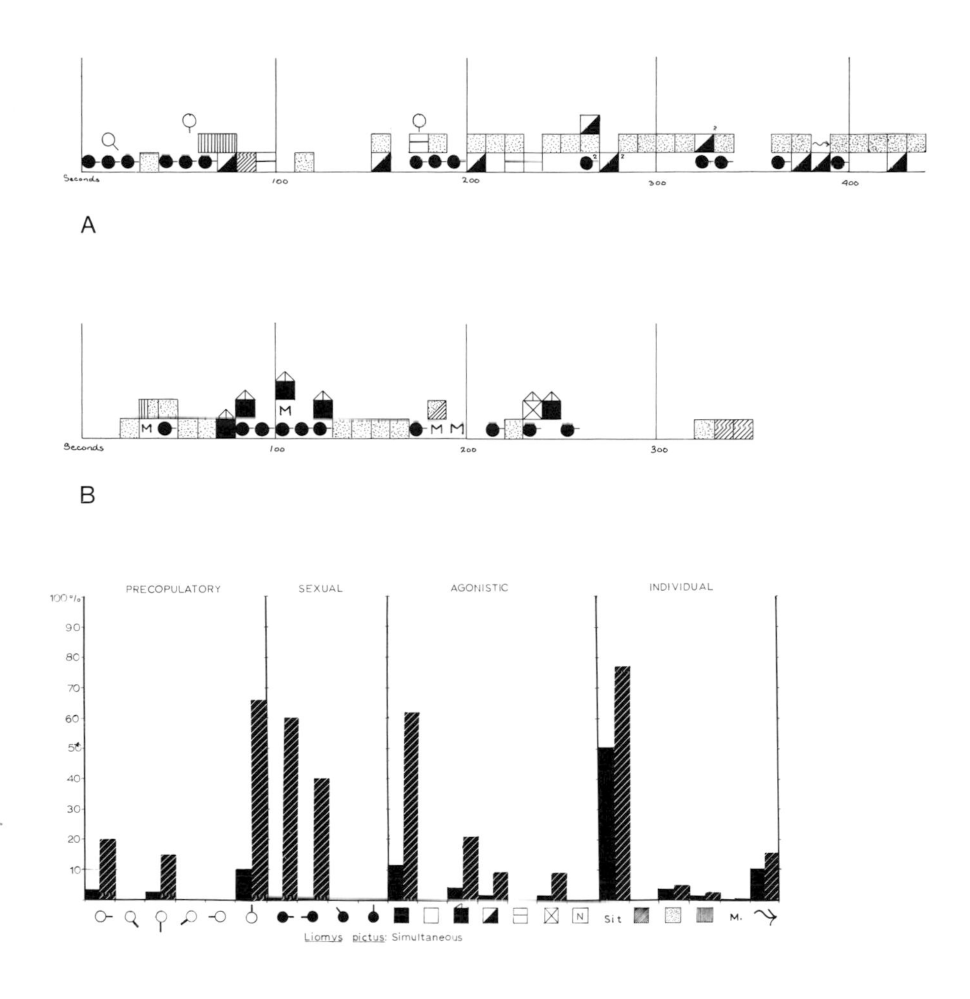

Figure 110. Temporal patterning of male courtship behavior. The symbols represent different behavior patterns as recorded within 5-sec time periods for a male in a male-female encounter. Each line represents data taken on a different day through the time course of estrus for a female. (*A*) *Dipodomys panimintinus*. (*B*) *Perognathus californicus*. (*C*) Behaviors are classified into functional groupings as in figure 111. Each individual act is then summed and expressed as a percentage of the total time for all behaviors within a given functional category. *Crosshatch* = male-initiated behaviors. *Black* = female-initiated behaviors.

a time line. A five-second period is taken as the basic time unit; hence, when three acts are stacked up within one interval, it means that three acts occurred within five seconds. Without going into definitions for the discrete behavioral units, the top line represents an encounter when the female was receptive to the male, and the bottom line represents an occasion when she was unreceptive. It should be immediately obvious that the quality of the two interactions is quite different. In figure 111 this becomes more apparent. I have set up the data in a slightly different fashion to demonstrate that, before estrus, interaction is primarily agonistic and that during estrus there is a shift to the expression

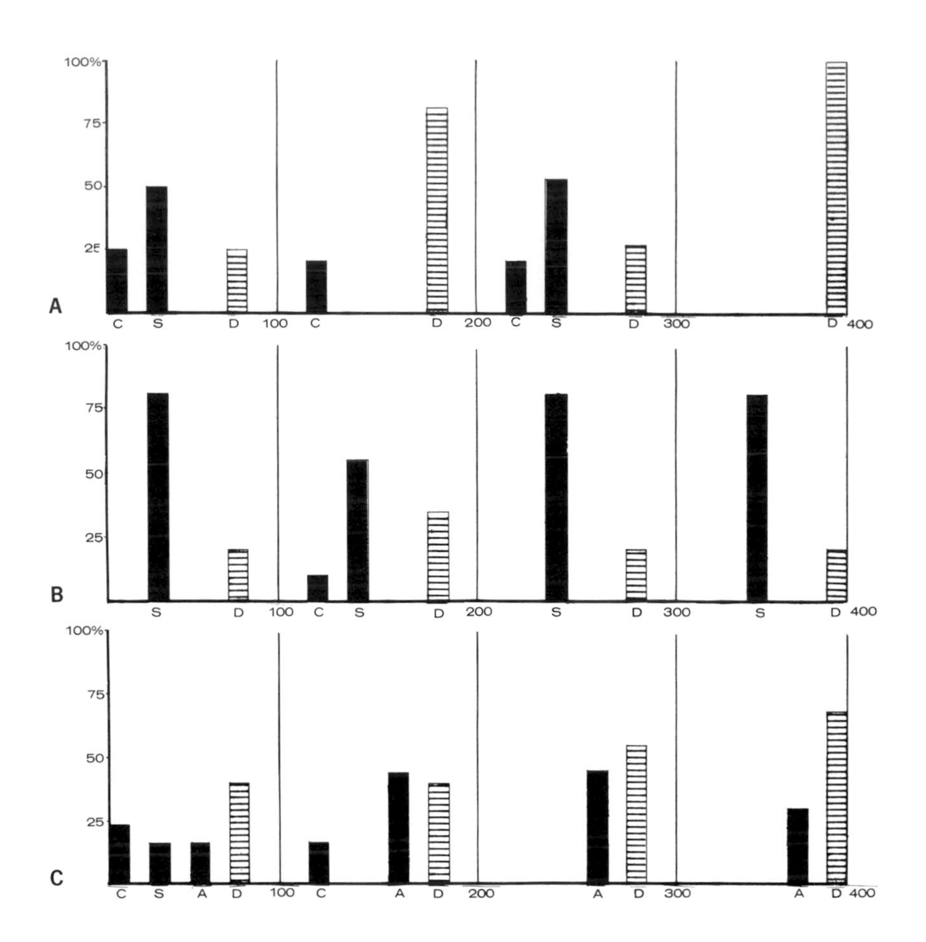

Figure 111. Percentage of total behavior for each consecutive 100-sec period shown by a male toward a female through the time course of her estrus. Line *A* represents the day before estrus, line *B* the day of estrus, and line *C* the day after estrus. (*C*) contact-promoting behaviors; (*S*) sexual behaviors; (*A*) aggressive behaviors; (*D*) displacement behaviors. Note the tendency for aggressive behaviors to occur with greater frequency on the day following estrus and, through the course of the encounter, to increase in frequency relative to contact-promoting and sexual behaviors. From Eisenberg 1967.

of sexual behavior. After estrus there is a return to the original encounter type. This "macroperiodicity" is controlled by the female's reproductive condition, which in turn exhibits properties of endogenicity. Hence the predictable structuring of the form of dyadic interaction in this species has both an endogenous basis, as determined by the female's reproductive condition, and a basis that is entirely dependent on the unique identity of the interacting individuals as well as on the circumstances of the encounter.

A predictable interactional "form" can be determined for dyadic heterosexual encounters within a given species. Differences between species are readily apparent when one chooses species that represent contrasting social types. For discussion, we can consider solitary, dispersed species and communal species. Communal species generally show a predominance of contact-promoting behaviors during male-female encounters regardless of the estrous condition of the female. An encounter can be resolved through contact-promoting behaviors without recourse to overt fighting. On the other hand, solitary species, even in heterosexual encounters, may demonstrate considerable amounts of agonistic behavior before establishing contact. This is especially true of encounters between strangers when the female is anestrous (Eisenberg 1967).

The temporal patterning of encounters may vary greatly across taxa. For example, in a study of the Asiatic elephant (*Elephas maximus*), staged encounters between males and females gave rise to the temporal patterning shown in figure 112. Note the longer time scale for single actions; note also that the form of the interaction is controlled by the estrous condition of the female (Eisenberg, McKay, and Jainudeen 1971). As a rule of the thumb, the time course of an interaction will vary directly with the size of the mammal. One must use extremely short time periods in analyzing dyadic encounters between individuals in species characterized by small size.

Variables of size or metabolic rate must be taken into account when species are compared. For example, in their analysis of humpback whale (*Megaptera*) songs, Payne and McVay (1971) not only demonstrated the long duration of syllables and intervals that compose a song, but also showed that the song itself may take more than ten minutes to perform. Contrast this with the syllable and interval durations in a marmoset (*Saguinus*) long call, where many syllables are less than 0.1 second and the entire long call may take less than thirty seconds (McLanahan and Green 1977).

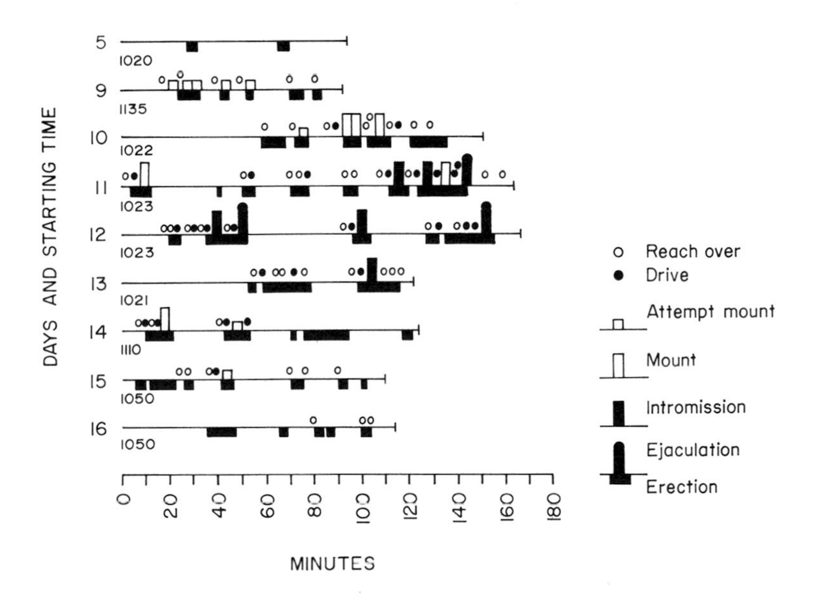

Figure 112. Temporal patterning of male sexual behavior through the estrous period of a female. Subjects were *Elephas maximus*. Estrus in the female occurred on days 11 and 12. From Eisenberg, McKay, and Jainudeen 1971.

Paradoxical are species with a very labile body temperature that fluctuates during the diel cycle (*Tenrec ecaudatus, Hemicentetes semispinosus*). These species require a "warm-up" period before sustained activity. In trying to develop measures of interaction rates for tenrecs, suitable allowances must be made for activity levels to reach modal operating levels.

Courtship itself may be organized into periodic episodes. For example, consider the repertoire of the male acouchi, *Myoprocta pratti,* as described by Kleiman (1971). When placed with a familiar female who is anestrous, the male will show episodes of courtship followed by maintenance behaviors. As Chris Schonewald has shown, these episodes may be separated by rather regular intervals (e.g., ~11 min), but the interval duration appears to be predictable only for a designated dyad (Schonewald 1977).

The organization of copulatory behavior may exhibit temporal regularities. For example, among species of rodents, it is possible to establish a species-specific rhythm for the male of mounting and thrusting, and number of mounts necessary to achieve ejaculation. The results are admirably summarized by Dewsbury in his recent review (1976). After an ejaculation, however, the male generally becomes refractory for a predictable interval preceding renewed courtship. This postejaculatory interval (PEI) is more often than not species-specific or strain-specific, and it leads to a characteristic decline in sexual activity as the male approaches satiation. The PEI is not constant within a bout but lengthens on successive ejaculations until it becomes very long and essentially all reproductive activity ceases. Such species-specific periodicities have been reviewed several times and need not be reworked here.

Another aspect of rhythmic properties in behavior is auditory communication. The sounds animals produce can be classified into units in the same way a behavioral repertoire can be broken into units. A given syllable type then, as so defined, can be delivered with different temporal patterns.

Stridulation by *Hemicentetes semispinosus* deserves some mention. A group of specialized quills when rubbed together produces bursts of sound that coordinate the movements of family groups (Eisenberg and Gould 1970). Stridulation rate has an average periodicity, but it can be modulated depending on ambient input. It is to the rate of modulation that conspecifics respond. As Gould (1971) points out, rate modulation is also a common method of signaling in vocalization—the analogue is represented in figure 113.

Consider calling by howler monkeys, *Alouatta*. Howler monkeys may call in response to a variety of environmental stimuli, such as rain and thunder, and in response to the calls of another howler troop. Yet, at extremely dense populations, an entire subpopulation or deme may exhibit synchronized calling at dawn; in effect, a dawn chorusing. Light here clearly regulates the onset of calling, and a periodicity is thus shown as a result of the flux in incident illumination. Similar periodicities controlled by the amount of ambient light can be shown in the

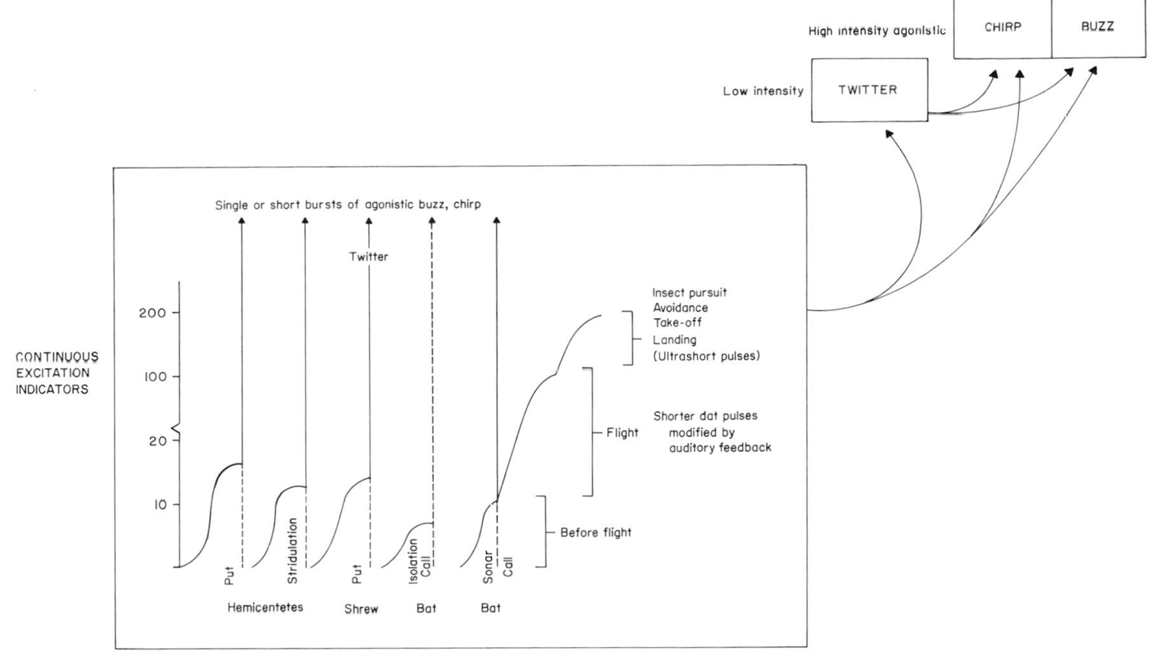

Figure 113. Formation of different call types by the reduction of intervals separating discrete sounds. Simple clicks may function in several ways. Clicks may be employed for orientation in space or for communicating position to a partner. In the scheme above, the echolocation pulses of bats are derived from basic orientation clicks considered to be a primitive character in nonvolant mammals. The rate of production of clicks can be considered an excitation indicator when perceived by a presumptive receiver. Other syllabic forms exhibit similar patterns of increasing rate of production. Fusion of syllabic forms may give rise to new syllable types. From Gould 1971.

calling of leopards, both wild and captive (Eisenberg and Lockhart 1972).

In addition to these rather gross periodicities, one can examine the structure of syllable chains themselves as a form of behavior. By plotting the interval between syllables against the duration of the syllables when they are uttered in phrases, we can get some idea of the range in variation in calling rate, which is a form of rhythmic behavior. In figure 114 I illustrate barking, showing qualitatively different kinds of barking and their tendency toward a certain regularity in rhythm of delivery. Similar extrapolations can be made with other call types, and in figure 115 I demonstrate through fusion of syllables the construction of a new syllabic category and the production of a new call as a result of shortening intervals between very brief vocalizations.

In figure 116, I indicate by separating syllable classes how some of these classes may be correlated with a given motivational state in the animal itself, the motivational being inferred from the context of the calls. In this way, then, vocalizations as one aspect of behavior can be

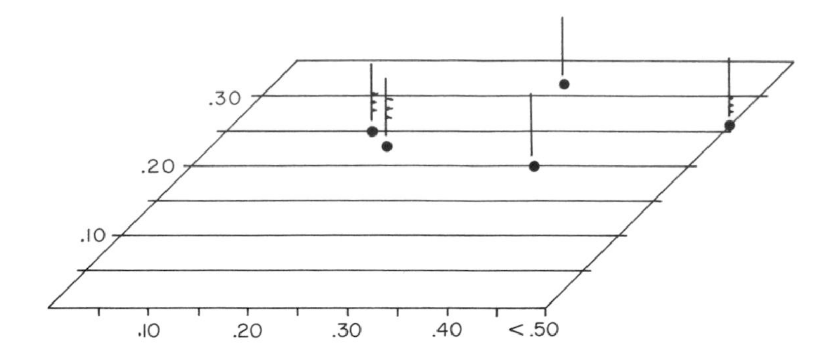

Figure 114. Temporal patterning of barks by *Ateles belzebuth*. The intervals separating the consecutive syllables are plotted on the ordinate; the abscissa indicates the average syllable duration in seconds. Each sample point includes the average of at least five. The points represent numerical values. The form of the syllable is indicated by the distribution of dark bands on the vertical line.

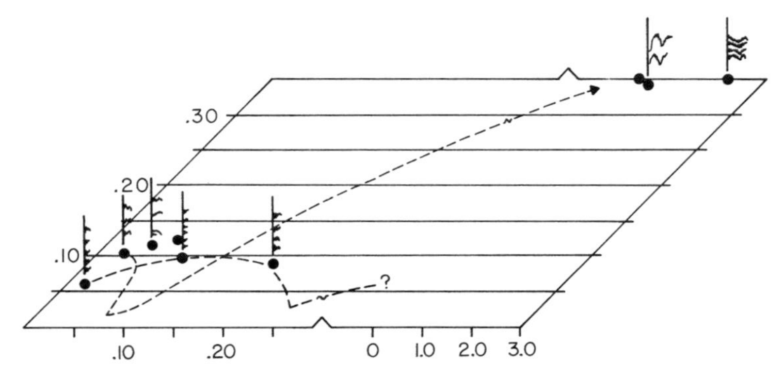

Figure 115. Temporal patterning and transformation within the whinny series by *Ateles fusciceps*. Scales as in figure 114. Syllable type VI, the whinny, may transform into syllable type XII by fusion. Broken lines indicate known syllabic transformation.

shown to be organized into sets that are correlated with moods. Within a mood, we can deduce general rules for the genesis of syllabic types (Eisenberg 1976).

Natural selection, however, can shape the form of calls in the same way it can produce different forms of nonvocal behavior. Certain classes of calls, such as long, loud calls, may be under severe selectional constraints on form and temporal patterning. Furthermore, loud, long calls may even be constructed so as to promote synchrony in their performance. For example, the long call of the mantled howler monkey (*Alouatta palliata*) has a distinctive pitch modulation in its midportion. Since howler monkeys tend to call in choruses, and since the function of the call is to space adjacent troops, then it follows that a troop that can call louder than another may maintain a more effective foraging radius with less energy expended in direct confrontation with neighbors. Thus, one could hypothesize that the pitch shift acts as a temporal marker that permits synchrony in calling by several animals so that they can synchronize amplitude modulation and thereby increase the

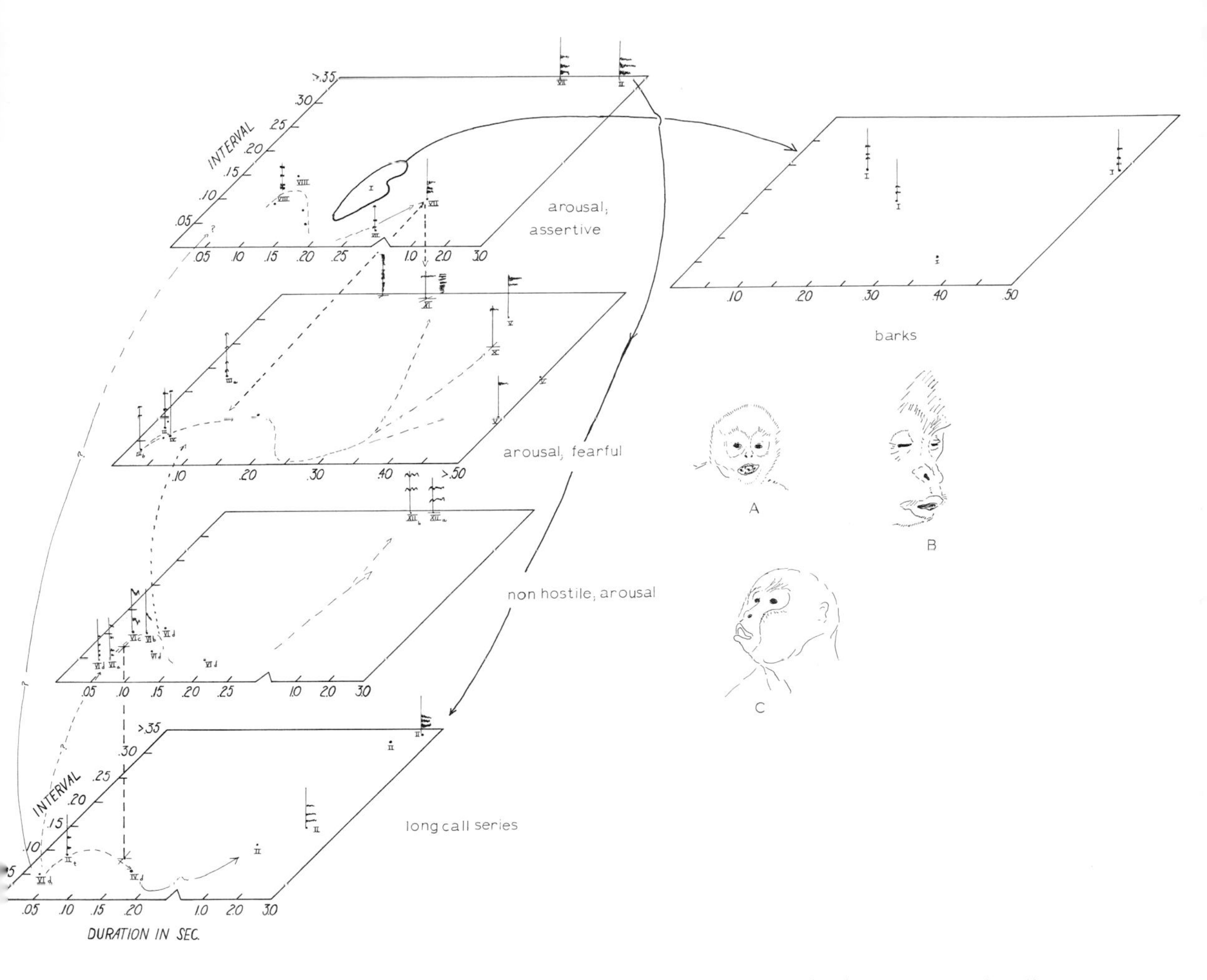

Figure 116. Call transformations and syllabic similarity, a composite diagram derived from analyses of syllabic transformations in the vocal repertoire of *Ateles fusciceps*. The roman numerals refer to the discrete call forms as portrayed in figure 115. Calls believed to be indicative of the same general mood are portrayed in the same plane. Broken lines indicate pathways of syllabic transformations. The facial expressions are generally correlated with certain moods: (*A*) expression during growl; (*B*) expression during a squeak; (*C*) expression during a long call. From Eisenberg 1976.

total volume of sound output. Although I am not implying that any endogenous periodicity is completely demonstrated in the form of vocalizations, I do hope that, in one small aspect of behavior, such as vocal repertoire, we can see the same shaping hand of natural selection (Eisenberg 1976).

28 Description and Classification of the Behavioral Elements Common to Generalized Terrestrial Mammals

In this chapter I will present a functional classification of the behavioral elements found in almost all generalized mammals. This is a derivative of the outline in part 1. One must remember that a functional classification implies certain adaptive consequences inherent in the acts in a given category; hence, the category of orientation movements subsumes those behaviors whose survival value is related to their helping the animal gain information about its position in space. In utilizing a functional classification we very often are describing apparently determinate sequences that result from similar causal mechanisms within the organism; but there is no necessary correspondence between a functional classification and a causal classification. Thus a functional classification may tend to categorize behaviors into physiologically related groups. We cannot assume, however, that all behavioral components included in a functional category are under the control of the same discrete physiological mechanism or drive center (Hinde 1970). Causal and functional classifications may coincide at times, but it has been well documented that functionally related acts need not be controlled by the same classical drive center, and that they wax and wane in intensity, very often independently of one another (Hinde 1970). Furthermore, some of the patterns discussed under a functional category are specific in the physiological sense to several general motivational circumstances or several different causal sequences.

When discussing discrete behavioral acts in this functional manner, I will have occasion to point out that some of the acts have been the evolutionary precursors of more ritualized patterns. I will discuss the consequences of this ritualization at greater length in the next chapter, which deals with the functional category of communication processes in mammals.

28.1 Positions Assumed during Sleep and Rest

The major postures and patterns animals exhibit during relaxation or the physiologically defined condition of sleep are reviewed here. The phylogenetic origins of sleeping and resting postures have been discussed by Hassenberg (1965). There are three basic positions assumed by an animal at rest: the first is the curled position, described by Hassenberg as the loose curl, where the long axis of the body is bent into a semicircle or, in a modified form, into almost a complete circle. The dorsoventral axis of the body may be oriented in the vertical plane, and the animal will then bear most of its weight on its hind feet, forepaws, and perhaps on the rostrum or even on the top of the skull. An alternative to this posture is with the dorsoventral axis on the horizontal, the animal thus reclining in a semicircular posture on its side. In either position the tail is frequently wrapped around the body. If the animal rests with the dorsoventral axis vertical, the tail tends to be wrapped around the feet; if it lies on its side, the tail may more or less follow the contour of the body itself.

The second category of rest positions includes the stretch position, where the long axis of the body is more or less in a straight line. This posture may be assumed on the side, on the back, or on the ventrum. In general it exposes more of the body surface to the air and is often assumed when the animal is attempting to increase its surface area so as to radiate excess body heat. Conversely, the curled posture is mostly assumed when an animal is attempting to conserve body heat.

A third category may be termed the sitting posture, where the animal keeps its dorsoventral axis aligned vertically and its head off the substrate, bearing part of its weight on the forelimbs. In this posture the bulk of the weight is generally borne by the hind feet, the hind limbs being flexed and the full soles placed flat on the ground. This is the typical sitting posture of generalized mammals, and it is not until we look at higher primates that we find most of the weight being borne on the ischia or, in arboreal forms, on the femurs of the hind legs. In the specialized taxa, such as ungulates and aquatic forms, the basic postures are vastly modified.

28.2 Locomotion

The subject of mammalian locomotion has been reviewed in part by Muybridge (1887), Howell (1944), and Gambaryan (1974). The more specialized forms of primate locomotion, including brachiation, have been considered in the reviews by Cartmill (1972) and Tuttle (1975). For locomotion on a plane surface by a generalized terrestrial mammal, the basic gait involves a diagonal pattern of limb coordination, whether the animal is moving forward or backward. This means that the right front limb is in synchrony with the left hind limb and the left front limb is in synchrony with the right hind limb. This diagonal locomotion has a number of subvariants, so that the animal may proceed with two feet, three feet, or four feet on the ground at a time. If it speeds into what is known as a gallop, then for a moment in each sequence of the gait the animal may have none of its feet on the ground, but may tend to bound slightly. I do not wish to divide these gaits into subtypes; I only wish to point out that, in the more conservative mammals—especially in the Insectivora, where the shoulder and pelvic girdles articulate in a less specialized fashion—the movements of the forelimb and hind limb tend to be accompanied by a slight twisting motion of the trunk musculature in the horizontal plane. This is seen in its extreme form in the more primitive tetrapod reptiles and amphibians, but the characteristic sinuous swinging of the trunk is still displayed to a slight extent in some of the more conservative Insectivora and Carnivora. This conservative trunk movement is most strongly displayed in swimming, for the same diagonal pattern of limb coordination is maintained and the sinuous side-to-side motion of the trunk becomes more pronounced.

It is fair to say that in the more specialized cursorial forms of mammals, with the pelvic and pectoral girdles more vertically aligned into the body axis, the forearms and upper legs are aligned more nearly parallel to the dorsoventral plane of the body. The weight is borne in a more vertical manner, with the thrust moving down through the girdles, perpendicular to the horizontal axis. Thus there tends to be less lateral movement of the trunk during forward progression.

A second form of locomotion on a plane surface, not involving diagonal limb coordination, is so-called quadrupedal saltation, where the forelimbs and hind limbs strike the ground alternately and the trunk moves up and down in the vertical plane, being contracted and extended alternately as the forelimbs and hindlimbs touch the ground. This form of locomotion is also seen in conservative mammals when the animal increases its speed. It is found in such forms as *Solenodon, Echinosorex,* and in many of the conservative insectivores, but it is not developed in forms such as the tenrecoid genera *Setifer* and *Echinops.* In such specialized Insectivora with spiny coverings, the locomotion pattern tends to remain conservative, being crossed extension or diagonal in form.

Although almost all conservative mammals can assume a bipedal stance and even walk bipedally for a step or two, bipedal saltation or progression by bounding on the hind limbs with the forelimbs not touching the ground is evolved only in more specialized forms. A rather detailed discussion of the evolution of bipedal saltation in both arboreal and terrestrial forms is presented by Howell (1932).

Jumping and contracting the trunk musculature and suddenly extending while thrusting with the hind feet is developed in most generalized mammals but, again, apparently is lost secondarily in the heavier insectivores, which rely on a spiny covering for protection. One may assume that the ability to leap was present primitively in the original mammalian progenitors.

Locomotion in a vertical plane, or climbing, consists of two phases: ascent and descent. The conservative mode of coordination for ascending a limb or a tree trunk consists of the diagonal pattern. Where the trunk surface is too large for the animal to grasp, one must assume that the primitive animal had claws that allowed it to grip the rough surface of the tree trunk. A second method of climbing may be employed, especially on small branches or by those forms with rather well developed claws. This alternates forelimbs and hind limbs and is a variant in the vertical plane of the quadrupedal saltation previously described (see Jenkins 1971).

Descent poses slightly different problems, and primitively mammals apparently had a rotational ability in both the ankle and hip joints and in the forearm. This allowed the hind foot to rotate so that the animal could descend headfirst. Headfirst descent may still be seen in the opossum, but it is an attribute of vertical, arboreal locomotion that has been lost in many forms that have become digitigrade and thereby concomitantly exhibit a tendency to fix the movement of the ankle joint in one plane or at the most in two directions in a single plane. In some instances an animal that has become slightly digitigrade and has lost the ability to rotate its hind foot around the ankle joint may secondarily readapt to an arboreal life. In such cases the animals frequently do not exhibit a headfirst descent on a tree trunk but must back down, using the reverse of the coordination pattern employed in the ascent. The wrist appears to be more conservative in most forms and is the last of the limb joints to lose its flexibility of movement in the evolution of digitigrade locomotion. It seems then that in backward descent the flexibility of the grip in the wrist is retained, while the hind legs act merely as brakes. For an expanded discussion of these two modes of descent, refer to Leyhausen's paper concerning cats of the genus *Leopardus* (Leyhausen 1963).

It is apparent that all generalized mammals could swim. During swimming, most of them employ the same diagonal limb coordination used on a plane surface. In addition, the trunk musculature oscillates slightly from side to side. In the more specialized aquatic forms, the trunk serves as the primary propulsive agent, and the limbs are used for steering or for fine adjustment of movement. The trunk in this case tends to oscillate in a vertical plane, and entirely new adjustments must be made in the epaxial and hypaxial musculature, especially in the caudal region. Analyses of these forms of locomotion have been made for Cetaceans (see Slijper 1962; Lang 1966)

Variations on the conservative mammalian pattern of diagonal limb coordination with trunk movement may be seen all the way through the mammalian orders that are specialized for digitigrade locomotion. Digitigrade locomotion confers greater speed; a full discussion of its evolution and the morphological consequences is contained in the article by Howell (1944).

28.3 Autogrooming

The movement sequence used in care of the body surface has been reviewed for two orders of mammals, the Rodentia and the Lagomorpha, by Bürger (1959). In almost all conservative mammals that have undergone little specialization, the forelimbs are intimately involved in grooming. Since in generalized mammals the forearm has some rotational ability and the fingers have some ability to grasp, the forelimbs are employed extensively in cleaning movements. In addition to the forepaws, the mouth (including the tongue and teeth) and the hind feet are employed. Frequently these organs are used in a rather orderly sequence, the degree of order being in part a function of the animal's motivational state. A reasonable analysis of this, with some motivational considerations, has been accomplished by Ewer (1967) for *Cricetomys*. The head itself and the movements of the neck musculature are also involved in autogrooming, since in wiping the face the forelimbs may be moved less than the head. In effect, the head is drawn through the forelimbs during face washing. With these introductory comments, let us look at a typical sequence.

Typically, in conservative mammals, the snout or the side of the face may be wiped with the forepaw. The head may or may not be moved slightly during these wiping motions. In addition to this simple snout wipe with the forepaw, the animals show a variation where they sit upright and, while opening the mouth slightly, move both forepaws simultaneously down on either side of the face, then bring them back up behind the ears and repeat the stroke. On the downward stroke, the wrists and to some extent the digits comb through the fur of the nose and pass over the vibrissae, apparently freeing the nose and associated sense organs of dirt or foreign matter. At the bottom of the downstroke, however, saliva is frequently smeared on the wrists and to some extent the forepaws. At times these members may even be licked (see fig. 117).

At first it was assumed that this slight moistening of the fur merely aided in cleansing; now, however, an additional and quite different function may be ascribed, since the animal's saliva is spread upon its head by this maneuver. Since it is well known that the scent of human saliva is helpful when dogs track humans, it is entirely possible that the saliva of other mammals also carries a great deal of chemical information. If so, then a primary grooming movement may also be related to communication, since the animal's head would by this means become impregnated with its own saliva (Eisenberg and Kleiman 1972). The forepaws not only are used to wipe the face but may also wipe the sides of the body and the flank and, in addition, are often used to hold the hind foot while its toes are cleaned with the teeth and tongue.

In addition to cleaning movements with the forepaws in which the mouth is involved, the mouth itself may be employed separately. With the tongue, the animal may clean the fur of the underside and the genitalia, and the teeth may strip through the fur of the ventrum and the posterior parts of the body and the toes of the forepaws and hind paws to remove foreign particles.

Cleaning the toenails with the tongue and teeth involves a rather rigid stereotyped movement of extending the hind foot and raising it until the mouth can be brought into contact with the digits. Finally, the last stereotyped cleaning movement involves scratching the fur with the hind foot. This pendular scratching movement, directed to various parts of the body, but generally behind the ears and over the shoulders and head, is common in most generalized, conservative mammals.

In addition to these classic autogrooming movements, several other body movements may be involved in cleaning or dressing the pelage. In most conservative mammals, these occur when the fur is wet and the body is rubbed in the substrate, generally dry soil, to speed drying. One movement is writhing, a sinusoidal contraction of the trunk musculature, which may be accomplished while the animal lies on its side but is more characteristically done on the ventrum or back. A variant of this includes extending and flexing the trunk musculature while the animal lies either on its side or on its ventrum. In conservative mammals that have adapted to arid environments and that exhibit an increased secretion from the sebaceous glands, the writhing movement and the

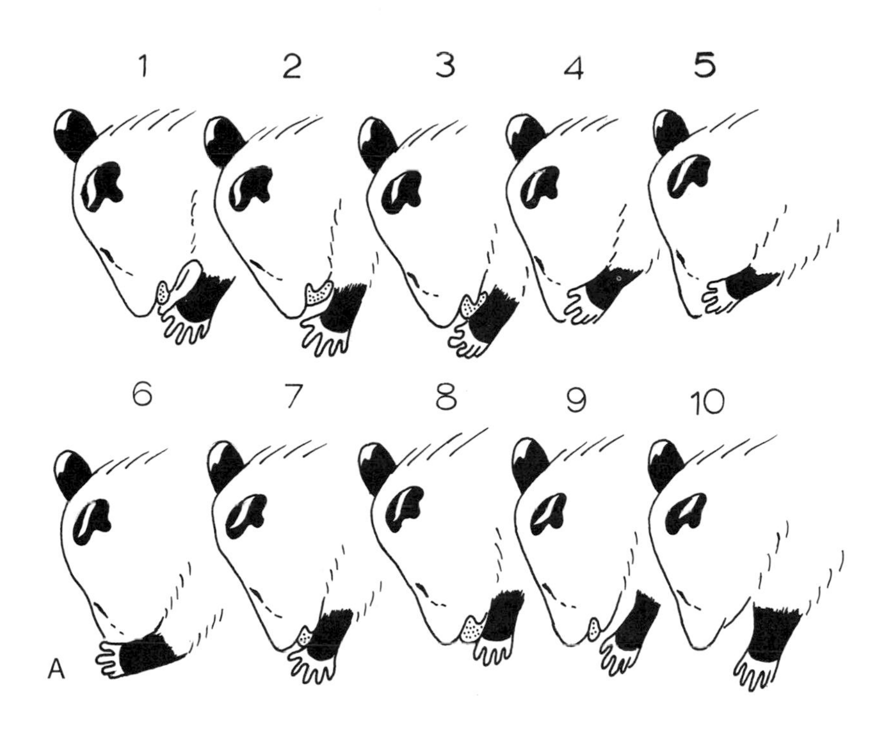

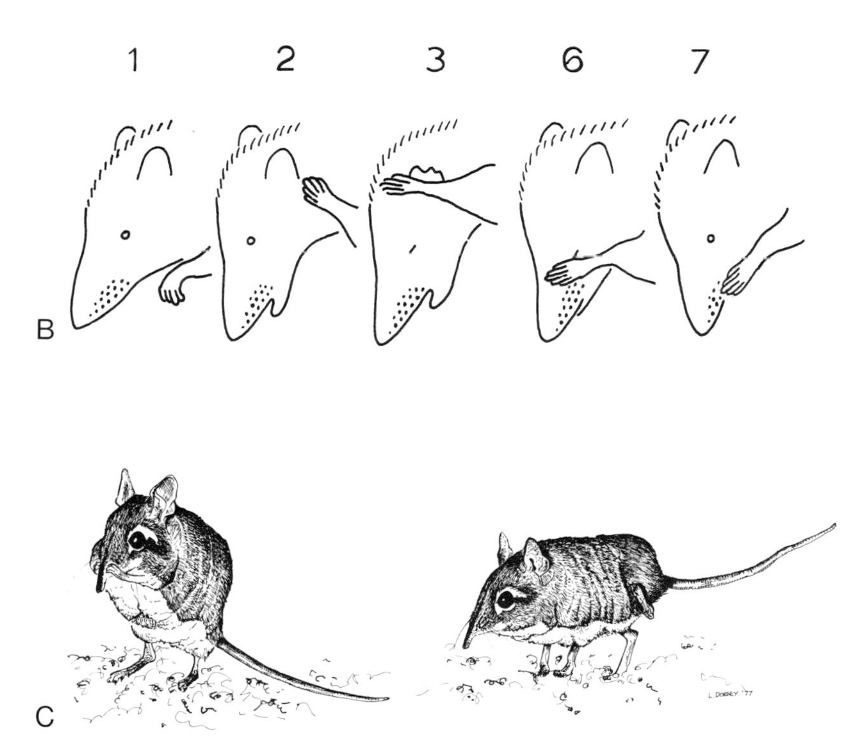

Figure 117. Sequential form of washing by (*A*) *Didelphis virginiana* and (*B*) *Microgale talazaci*. Note the tendency for stereotyped forepaw movement with licking in *Didelphis*. The frame-by-frame analysis of washing in *Microgale talazaci* does not permit the differentiation of licking, but the corners of the mouth are rubbed. (*C*) represents two types of comfort movements in *Elephantulus rufescens*, washing and scratching with the hind foot. Drawings *A* and *B* by C. Wemmer. Numbers refer to frames of film at 18 frames per second. Drawing *C* by L. Dorsey (Rathbun).

extending and flexing movements may be combined into a predictable sequence that we term sandbathing. Sandbathing has been independently evolved in many rodent families (see Eisenberg 1967) and also in the desert-adapted Marsupialia. As described in my earlier studies, it has the dual function of dressing pelage and, in addition, leaving behind odoriferous traces that serve a chemical communicatory function, so it may thus be considered a marking movement.

Another functional category of comfort movements includes rubbing various parts of the body on a surface. Rubbing is a functional term that covers several discrete movements. The more frequent ones are described here. An animal may lower its head and, while moving forward, press its nose to the substrate, thus dragging the side of the rostrum through the soil. The same thing may be accomplished by extending or flexing the torso while lying or standing near an object. In a similar manner, the side of the face or cheek may be rubbed on the substrate or on suitable objects. The chin may be rubbed by extending and flexing the neck while resting the chin on an object. Occasionally the side of the body may be rubbed, either on the substrate by extending or by scrambling forward with the limbs while lying on the side, or more rarely while standing and alternately extending and flexing the limbs of one side of the body while leaning against an object.

A separate form of rubbing, termed the perineal drag, involves depressing the perineal region on the substrate while moving forward. This involves a slight flexure of the lumbar vertebrae and at the same time a sustained flexion of the hind limbs. It may originally have been a movement for removing fecal material adhering to the anal region, but, in almost all present-day mammals, it spreads glandular exudates and is a form of marking behavior as well.

Shaking, the vigorous extension and flexion of the trunk musculature in the horizontal plane while standing in one position, is a primitive pattern of pelage dressing present in all conservative mammals.

Included under the category of comfort movements are such basic vertebrate patterns of behavior as the stretch, which includes the extension of the trunk musculature, with varying forms of extension or flexion of one pair of limbs; the yawn, which involves gaping with the mouth and extreme extension and flexion of the masseters and temporals; the cough, a physiological response to low-intensity stimulation in the bronchi; and the sneeze, a violent exhalation that occurs in almost all air-breathing vertebrates.

28.4 Elimination

Movements concerned with urination or defacation tend to be stereotypic for the Mammalia (Altmann 1969). Conservative mammals typically defecate while in an immobile position, with slight flexure of the hind limbs and contraction of the dorsal musculature adjacent to the lumbar vertebrae. The posture is frequently adopted for a protracted period. Urination, in the conservative sense, typically involves a similar stance, with a slight flexion of the hind limbs, though in some conservative mammals urination may occasionally accompany forward movement in crossed extension locomotion, in which case the animal exhibits some slight flexure in the hind limbs and dribbles urine as it moves.

In both urination and defecation, the secondary function of chemical communication is inherent. As a consequence, rather elaborate movement patterns have occasionally been evolved to accompany these elimination responses. Since the inherent chemical communicatory value of urine or feces will persist as long as they do not become dry, many primitive mammals bury their feces, leaving a slight mound of earth to mark the spot. Such a mound of fresh earth would be conspicuous to another of the same species or even a different species and, if sniffed, the odor of the buried urine or feces would remain strong for longer than if the material were deposited in the open. On the other hand, burying the urine or feces consumes energy, and selection has not always favored the burying response. To cite one example, the common tenrec, *Tenrec ecaudatus*, approaches a spot, generally a traditional one, by backing up to it. It scratches with its hind limbs, assumes a typical defecation posture, defecates and occasionally urinates, then kicks back before moving on. This typifies the pattern generally employed by these nonburying forms, and the net result is a small mound of turned earth in a traditional spot containing the feces. On the other hand, *Tenrec ecaudatus* may also dig to bury its feces in a locus near its burrow, or it may defecate at the entrance to the burrow and not bury the feces at all. It seems that there is less advantage in burying the feces near the burrow, since the burrow itself is marked by the owner's scent. Moreover, it may be that leaving the burrow to defecate during the day makes the animal vulnerable to attack, so that it does not waste time or energy in moving far from the burrow or in digging a hole.

28.5 Breathing

In all mammals breathing is necessary to exchange gaseous material within the lungs. The phylogenesis of breathing and the theory concerning its motivation have already been developed by Spurway and Haldane (1953). In short, the rate of breathing depends on the particular workload the animal is performing and, in addition, on the level of arousal of the autonomic nervous system, being thereby a concomitant of change in emotional state. Exhalation and inhalation are the sources of all vocalization phenomena, the evolution of which will be discussed in a separate section.

28.6 Drinking

A phylogenetic survey of drinking behavior in mammals has been proposed by Schoenholzer (1959). Most conservative mammals drink by lapping water with the tongue. The act of ingesting milk in primitive mammals, which also involves tongue movements, is a complex subject that begins with an analysis of the behavior of the neonate (Prechtl and Schleidt 1950). Sucking behavior and teat attachment are distinct from lapping water from a pool.

28.7 Feeding

Ingesting food involves two acts: manipulating the food and reducing it to particles small enough to swallow. Most conservative mammals, including *Marmosa*, *Didelphis*, and some of the tenrecoids, are able to hold bits of food in their forepaws and bite pieces from them. With larger pieces of food, the forepaws are generally braced against the food, and, by seizing the food with the teeth and jerking the head from side to side, small pieces are torn off that can then be chewed and swallowed. The acts of biting, tearing, and chewing food vary from one type of conservative mammal to the next depending on their dentition (Hiiemäe and Ardran 1968; Crompton and Hiiemäe 1969, 1970). The swallow reflex is uniform for the Mammalia.

Gathering food involves several aspects. First there is selection. This is in part a function of rather stereotyped responses to various stimulus inputs and is modified by early experience. In general, the selection of food may be promoted through the young animal's association with its mother (Eisenberg and Gould 1966). The gathering of food in conservative mammals also involves catching prey. Since the utilization of plant foods is a later specialization, many of the behavior patterns employed in food gathering by herbivorous rodents or ungulates are refinements or complex derivatives of the basic pattern of prey capture exemplified in the more conservative mammals.

In catching prey the mouth and forepaws are typically employed. The forepaws are used to seize or pin a small prey object. The mouth has the dual function of seizing and killing. In conservative mammals, the mouth is typically the most important organ for seizing and killing small prey, but the forepaws may be used to immobilize the prey, either by pinning it to the substrate or by clutching at it, depending on its size and nature.

Killing the prey by biting it and shaking the head slightly from side to side is a common pattern in primitive mammals, and the general strategy with a small prey object is to deliver a series of bites. With a larger prey object, separate bites are often delivered without holding the prey in the mouth until the prey becomes weakened. In the conservative mammals that frequently prey on moderate-sized organisms, there is often a tendency to bite at the anterior end. This appears to develop as a consequence of aiming the bite slightly in advance of the moving object, so that in general it tends to coincide with the anterior end. Through learning and various refinements, this biting technique may become specialized so that the bite is delivered at a very vulnerable point. This is further amplified in the article by Leyhausen (1965) and by Eisenberg and Leyhausen (1972). Shaking the head from side to side at the moment of biting and releasing the prey at the end of a shake are common conservative patterns to be found in *Didelphis, Solenodon, Echinosorex,* and *Tenrec.*

Typically the mouth is used to pick up and hold a small prey object, and mouth transport is the most frequent mode of moving prey from place to place. The forepaws are generally used for holding it during ingestion. Conservative mammals may eat prey where it is killed or may move it to a more sheltered spot by carrying it in the mouth or, if the prey is large, grasping it with the mouth and dragging it. If the prey is to be eaten in a burrow, it may be dropped outside and then hauled in by alternating movements of the forepaws, scratching it into the den, or pushed with the forepaws, sometimes aided by the snout. If the prey is buried, digging movements are used to remove the earth, and the dead prey is dragged or pushed and patted into the hole, then covered by movements of the forepaws and occasionally the nose.

28.2 The Construction of Artifacts

Conservative mammals generally construct two types of artifacts:[1] burrows in the earth and nests made of plant material. In burrow construction the basic movement is the digging pattern. Alternate movements are made with the forepaws until a pile of earth accumulates under the animal, which is then kicked backward by simultaneous or alternating movements of the hind limbs. Frequently during the digging phase the animal may turn and push the earth with its forepaws and chest; the nose may also be used to scatter earth from side to side. The forepaws and the body may be used unspecifically to mold and pack the floor and walls of the tunnel system.

Nest building includes several functional categories: the assembly of materials (generally dead leaves or bits of plant material that are picked up and transported in the mouth); their deposition by dropping them in the burrow or nest site; and a series of movements involving pushing with the nose or alternate or simultaneous pushing with the forepaws while turning the body to mold the herbaceous material into a cuplike structure. More refined movements may involve picking up pieces of the herbaceous material in the mouth and moving them to the sides of the body or raising them overhead with up-and-down movements of the head. These movements, as displayed by the shrew, *Sorex,* serve to make a roof over the cup, thereby creating a chamber (Eisenberg 1964; Crowcroft 1957).

28.9 Orientation to the Environment, both Animate and Inanimate, and Associated Protective Responses

Orientation within a familiar environment appears to be a function of gradual exploration from a fixed site. Use of olfactory, solar, and even electromagnetic sensory information has been postulated for mammals that home when displaced. Migration to a destination never before seen by a maturing individual may depend on experience with adults or even involve innately programmed mechanisms in response to environmental cues. A stepwise theory of vertebrate orientation and migration mechanisms is elegantly developed by Baker (1978) and will not be reviewed here.

During orientation to the environment, the receiver manipulates its sense organs so that the energy inputs to them are either increased or decreased. Varying degrees of eye closure may be displayed in response to varying strengths of light. In a similar manner, varying degrees of eye closure can be elicited by any strong change in the background illumination, since the eyelid closure not only regulates the amount of light entering the eye but also is a protective response. There are, in all the sense organ control mechanisms, implied categories of responses that serve different purposes from specific orientation to the environ-

1. Although a strict definition of *artifact* implies an object constructed by *Homo,* I propose to broaden the definition to include structures made by all animals. The special topic of tool use is not treated here, but the reader is referred to Beck (1980).

ment and may be subsumed under the function of protective responses, as Andrew so ably explained in his paper of 1963.

The ears of conservative mammals, such as *Didelphis*, *Echinosorex*, and *Marmosa*, are extremely mobile. They may be partially folded to reduce auditory input or to protect the ear from damage, or they may be extended. Parenthetically, in *Marmosa* and *Didelphis* and many of the marsupials, the two ears may be moved independently and are undoubtedly useful in localizing moving objects in space (see Eisenberg and Leyhausen 1972).

Orientation to the environment also involves the tongue. Animals frequently lick objects, thereby gaining some chemical information. The tongue is also used to moisten the rhinarium, which may itself be involved in the direct reception of chemical stimuli or may serve as an organ to which chemical stimuli adhere; the continual licking of the rhinarium, then, serves the dual purpose of keeping it moist and at the same time imparting some chemical information to the tongue receptors themselves (Eisenberg and Kleiman 1972).

The vibrissae are a means of orientation to the environment. First we may note the differential position of "sensory hair clusters" (see fig. 118). Primitively, the vibrissae of mammals are placed in several defined areas: (1) Vibrissae occur near the end of the rostrum on the upper lip, where they are termed the mysticial group. These vibrissae often vary in length, yielding different functional subgroups (see Eisenberg 1963). (2) The superorbital vibrissae give information about objects adjacent to the eyes. (3) The genals give information about objects next to the cheek. And (4) The mentals give information about objects immediately beneath the chin. The mysticial vibrissae are not fully under muscular control in conservative mammals, and information about the position and distance of objects relative to a primitive mammal's face appears to be mostly a function of the vibrissae type activated or deformed. In the Rodentia, as Andrew (1963) rightly observed,

Figure 118. Location of vibrissae and sensory hairs on a typical conservative mammal. 1 = mysticial, 2 = superorbital, 3 = genal, 4 = mental, 5 = antibrachial, 6 = calcaner.

the vibrissae in general enjoy much greater mobility, with a rich neuromuscular control system. The different subgroups of mystícial vibrissae can in fact scan the environment. In the Insectivora the vibrissae, though capable of some movement, are relatively fixed; very often the nose has become elongated and movable so that tactile information is fed in from the movable "proboscis" as well as from the more fixed vibrissae. This is especially true in a form such as *Solenodon*.

During orientation to the environment, there may be special breathing movements. For example, strong exhalation may occur, apparently to clear the nasal passages. Very rapid breathing, accompanied by searching movements (switching the head from side to side over the substrate) or movements of the vibrissae and a lifting of the head to test air, is seen in animals attempting to localize chemical and tactile cues (see fig. 119).

The movements of the body and head during orientation to the environment or to a partner are an important class of responses. These include the category of turning toward an object and its opposite, turning away. Turning toward an object may develop into an approach where the animal moves toward the source of stimulation and then may grade into a rush at the object or a leap upon it. These could be interpreted as motivational variants of the same process of "movement toward." Conversely, turning away includes further categories that are motivational variants of avoidance and involve moving away or ultimately fleeing. In turn, such responses can lead into true protective responses, such as leaping, climbing, or other similar locomotion variants (see fig. 120). There can be defined a category of ambivalent behaviors, in the sense of Tinbergen (1951), where the animal alternates between turning toward and turning away (approaching and avoiding). There are classical "intention movements" within this ambivalence category that involve raising the forepaws or various alternate tensing of the muscles when the animal is indecisive about approaching or avoiding.

The next subdivision under orientation to the immediate environment involves contact with objects by the vibrissae or by the nose itself (the rhinarium) by inhalation and exhalation, thereby promoting the transfer or exchange of chemical information. Contact may be accomplished by various parts of the body, including the forepaws, the side, and so

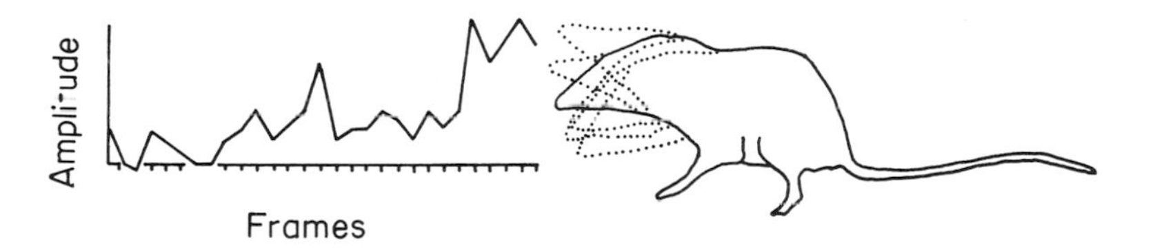

Figure 119. Head-bobbing movements associated with testing the air by *Microgale talazaci*. All data from 16 mm cine film at 18 frames per second. Drawing by C. Wemmer.

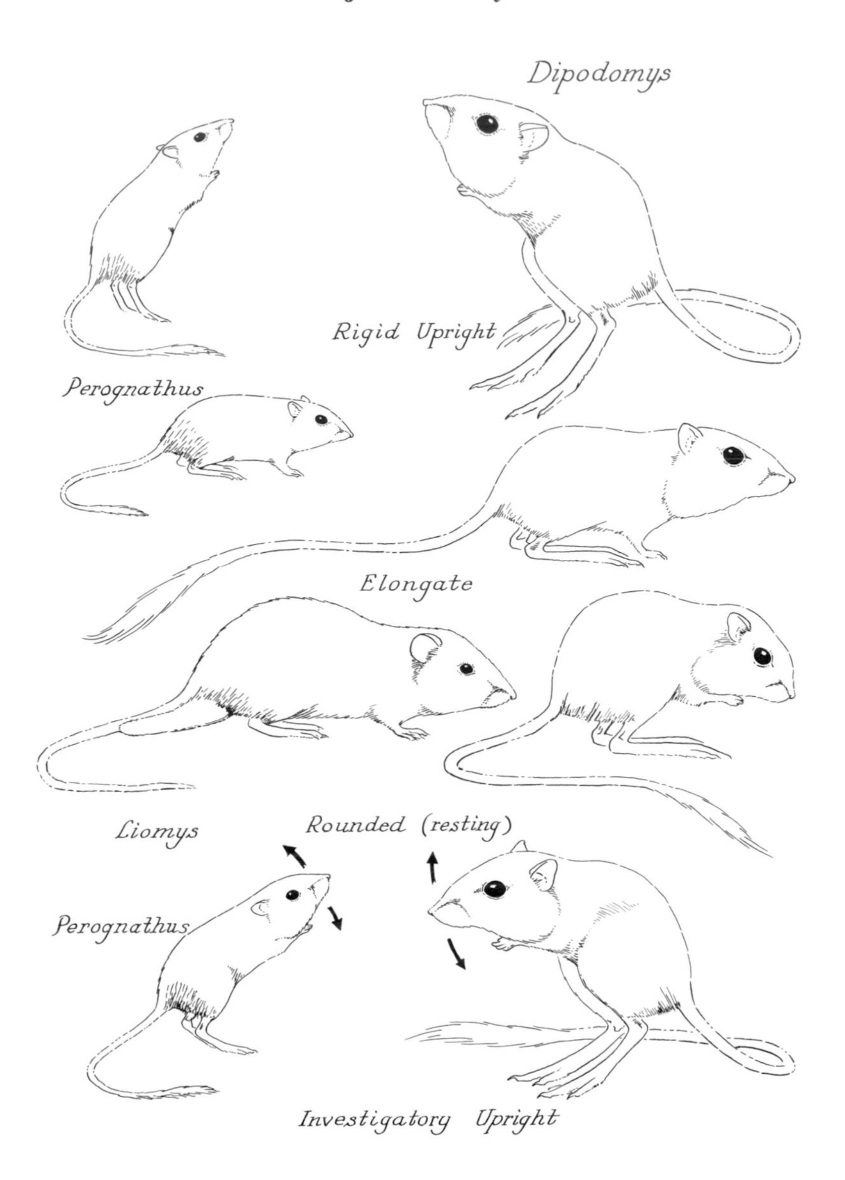

Figure 120. A comparison of the major postures assumed by an isolated rodent while exploring a novel area. The bipedal genus *Dipodomys* is contrasted with the quadrupedal genera *Perognathus* and *Liomys*. Note that the modification of posture in the bipedal *Dipodomys* is a simple exaggeration of general posture trends exhibited by the quadrupedal genera. From Eisenberg 1963.

on. The approach to a stimulus source may also be considered from the standpoint of body contour. The body may be horizontal with all four paws on the substrate, horizontal with its contours rounded where the animal has flexed its trunk muscles, or elongated where the trunk muscles are extended. Elongated postures and flexed postures may be alternated when the animal is ambivalent, or the animal may become immobile, sometimes referred to as freezing (Eisenberg 1962). In this separate functional category the adaptive advantage is assumed to be the inconspicuousness resulting from the cessation of movement. In contradistinction to the horizontal variants are the classes of postures known as uprights. Almost all generalized mammals can stand up on their hind legs (Ewer 1968). This may involve an upright posture with rounded contours or a tense, upright posture. Again, as I explained for horizontal postures, these are probably motivational variants, and ambivalence may be shown in upright postures as well as in horizontal postures.

The contours of the body are also influenced by the degree of piloerection, itself the result of massive sympathetic discharge from the autonomic nervous system (Eisenberg and Gould 1970). Extreme piloerection may increase the apparent size of an animal. All these trends outlined above may be named and defined as different postural variants, as has been done in previous publications for the Rodentia (see Eisenberg 1967) (see figs. 120 and 121).

28.10 Communication Mechanisms

The exchange of information during an encounter is the basic mechanism for regulating interaction. When discussing interaction, it is convenient to classify the potential communication mechanisms in terms of the major sense organs involved. It is wise to remember, in most of the mammals we consider, that no single sensory modality is divorced from the others. In perhaps no other form of vertebrate are the sense organ inputs more integrated than in the Mammalia. Visual cues are used in conjunction with olfactory and auditory cues, as well as with tactile cues.

28.10.1 Visual Information

In conservative mammals the role of vision in form discrimination is undoubtedly reduced. The primary emphasis thus appears to be on auditory and chemical inputs. Nevertheless, the body, the face, the tail, and the pelage may to some extent be involved in the transfer of information to a partner. In almost all conservative mammals, the open-mouth gape, with exposure of the pink mucous membranes, display of the teeth, and concomitant exhalations, forms a potential visual, chemical, and auditory signal during threat. The position of the ears will influence the contour of the face, whether they are extended forward, laid back, or folded, and this can have an inherent communicatory function (see Wemmer 1977).

Conservative mammals have very little ability to retract the corners of the mouth or to purse the lips. But in the more highly evolved mammalian forms such as the primates or ungulates, the form of the mouth may be altered by the position of the lips. Covering the teeth with the lips has evolved as part of an integrated signal system (see Andrew 1963).

The degree of eye closure is primarily a defensive mechanism, as I noted before, but it may also influence the conformation of the face, especially if the eyelids are colored the same as the rest of the face; then the black, virtually nonreflecting surface of the eye may be exposed or covered.

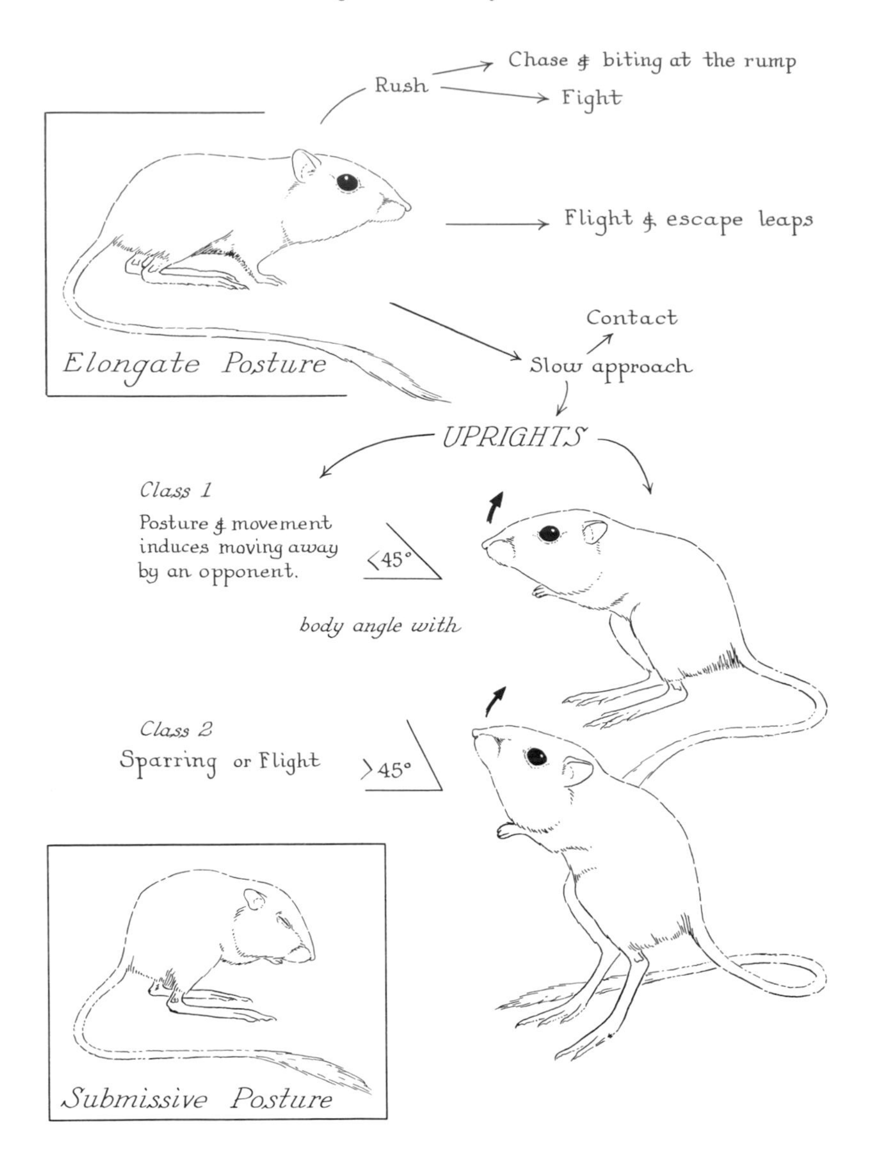

Figure 121. Postures and sequences of activity exhibited by *Dipodomys* during an encounter. The submissive posture is the antithesis of the elongated posture and is generally followed by withdrawal or disinclination for contact. From Eisenberg 1963.

The tail is an extension of the body, and in most conservative mammals it serves as an orientation point during following by a conspecific. This may be most easily observed during the driving phase of sexual pursuit or when a young animal first begins to accompany its mother on her foraging expeditions. For this reason, as the visual ability of a mammal evolves the tail often becomes marked in a fashion that renders it conspicuous not only to conspecifics but also to predators; thus the predator's sight is diverted away from the animal's body and toward the end of its tail. This is especially true of desert rodents such as the jerboas (*Jaculus jaculus*) or kangaroo rats (*Dipodomys*) (Eisenberg 1975*c*).

Piloerection, or increase in the apparent size of the body by fluffing of the hair, is an ancient visual signal (Andrew 1963). General trembling of the body and twitching of the tail often accompany extreme autonomic arousal and have independently in a parallel and convergent fashion become ritualized into signals, either visual or auditory, as when the tail rattles against the substrate (Kleiman 1971).

An upright posture displaying a light-colored ventrum also conveys inherent visual information. These postures and movements, then, constitute the major patterns of behavior associated with the exchange of visual signals.

28.10.2 Chemical Information

There are two general sources of chemical information. One involves the deposit of secretions from glandular areas, the other the deposit of products of elimination—urine or feces. Since sebaceous glands always occur in association with hair, each hair itself becomes a potential contributor to the total field of chemical information. As the animal's body brushes against the substrate or undergrowth, it leaves chemical traces. As I noted before, special hairs on the chin and cheeks, above the eyes, and on the tip of the snout that receive tactile input may by their associated sebaceous glands evolve into specific organs for depositing scents through special movement patterns. This has occurred over and over again within the Mammalia. Hence it is not uncommon to find that conservative mammals wipe the cheeks, the chin, or the rostrum on specific points in the environment (Eisenberg and Kleiman 1972).

The glands associated with the external genitalia and the anus have also been employed in marking behavior in a parallel fashion by many mammals. The movement pattern is widely distributed throughout the Mammalia, and its genesis may easily be found upon observing the conservative mammals (Wemmer 1977). The products of elimination are, as discussed before, often deposited in a semiritualized fashion. They may be placed in a consistent locus or they may be buried, thus conserving the odor and at the same time creating a slight artifact or mound that attracts the attention of a conspecific (see fig. 122).

The glands in the vicinity of the eyes may also help identify individuals. The mouth, with the associated buccal glands, is likewise a source of chemical information, as are the ceruminous glands in the ears. One can easily see that, since all these glandular areas have information to transfer, the contact-promoting postures or the postures assumed during contact must expose the nasal area to these points. Indeed, as we shall see in the next section, this is the case (see also Ewer 1968*b*).

Something that has not often been considered is the inherent communicatory value of saliva. During the open-mouthed threat, the breath itself can carry chemical information, as I have noted in other publications (Eisenberg and Gould 1966, 1970). The ritualized patterns for distributing saliva, especially on spiny forms that have reduced seba-

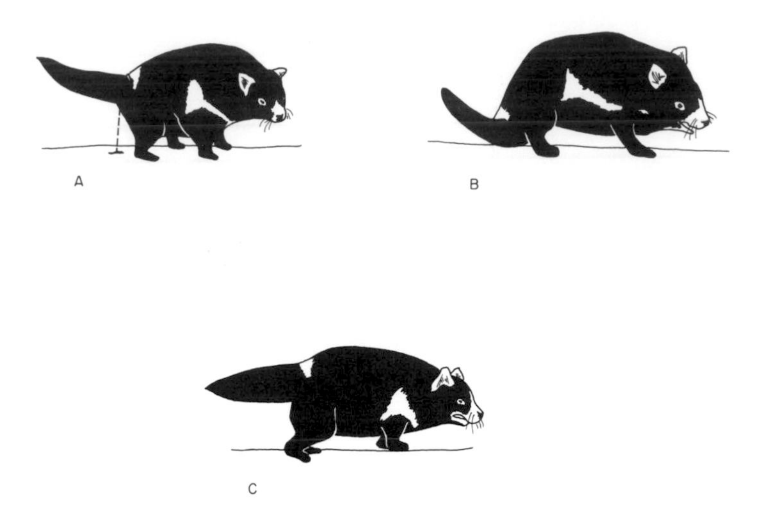

Figure 122. Various forms of chemical and visual communication in *Sarcophilus harrisii*. (*A*) urination by the female; (*B*) marking by dragging the anogenital area on the substrate; (*C*) broadside display to a partner. Note the tendency for the hair of the tail to be erected during a broadside display. From Eisenberg, Collins, and Wemmer 1975.

ceous glands, as in *Echinops* and *Erinaceus*, make saliva marking on the body an important ritual (Poduschka 1970). Indeed, as I pointed out earlier, the face-washing behavior itself may not necessarily be related to cleaning so much as to impregnating the head with the odor of the animal's saliva. The complexity of the mammalian chemical communication system has been reviewed by Eisenberg and Kleiman (1972), Müller-Schwarze and Mozell (1977), and Doty (1976).

28.10.3 Tactile Information

In tactile communication, the tail may be the organ first encountered by a conspecific. It may be used to maintain contact, for example, in the female-young unit of some species of shrews, where several animals seize each other's tails in succession (Zipellius 1972). The limbs themselves may be used to achieve or maintain contact by hanging on or gripping. And, finally, the specific body points containing glandular areas may be the sites of contact. The nose, with its associated vibrissae, is the primary organ for perceiving or initiating contact and for determining the extent of contact (see Eisenberg and Golani 1977; Ewer 1968*b*).

28.10.4 Auditory Information

Audition is the most complex form of communication in conservative as well as advanced mammals. Auditory signals may be produced by movements of the animal's body itself, such as clicking the teeth, or stamping the feet, or by even more general movements that have some conditioned information content for the recipient, such as sounds of eating, drinking, and moving along the substrate. A common sound produced by breathing is the hiss, an unvoiced expiration or inspiration or a combination of both. Another common sound is a click produced by the tongue, teeth, or lips (Wemmer and Collins 1978). Andrew (1963, 1964) noted that clicks may occur in a novel context and have an inherent function in echolocation as well as communicating an animal's position to a conspecific.

Vocalizations may be analyzed from the standpoint of repetition rate, syllable duration, amplitude of the sound, and distribution of energy with respect to frequency within the syllable. (Pye and Sales 1974 have reviewed the special problem of ultrasonics.) A syllable type common in conservative mammals is the clear tone or squeak, which on a sound spectrograph may have a linear form, a chevron form, or a sliding form (see fig. 123). Common to most conservative animals is the shriek, a noisy, nonharmonic sound carrying energy at the higher frequencies, and finally there is the growl or grunt, a noisy sound with few harmonics and with energy concentrated at the lower levels. The syllable form in and of itself does not always have the same meaning to a recipient. It appears that the function of the vocalization is related less to syllable form than to such physical parameters as repetition rate, pitch, and amplitude. Indeed, when different species are compared, it is the repetition rate and amplitude that appear to form a related series rather than the discrete syllable itself (Eisenberg, Collins, and Wemmer 1975).

In 1977 E. S. Morton refined several observations on the structure of vertebrate auditory communication systems and developed a theory to account for the remarkable convergence in signaling systems of birds and mammals. As Collias (1960) pointed out, evolution has produced

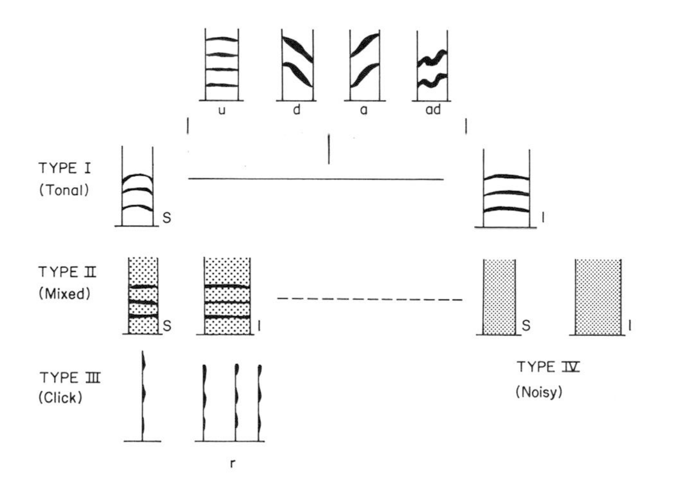

Figure 123. Four basic syllable types commonly produced by marsupials. Type 1 is characterized by clear tones that may be unmodulated (*u*), of descending modulation (*d*), of ascending pitch (*a*), or highly modulated (*ad*). The syllables may be short (*s*), less than or equal to 0.6 seconds, or long (*l*), greater than 0.6 seconds. There are eight possible type 1 sounds. Type 2 is a mixed syllable, tonal in structure but with the superimposition of noise. Type 3 is a brief syllable less than 0.02 seconds, which may characteristically be repeated. Type 6 may be short (*s*) or long (*l*) but is essentially unharmonic.

a convergence in many of the auditory signals that vertebrates produce in hostile and appeasement contexts. During aggressive interactions, vertebrates often emit low-frequency sounds and, when one partner (or both) is frightened or preparing to flee, they can produce purer and more tonelike sounds. This appears to be a general rule across taxa. A convergent use of harsh, low-frequency sounds by aggressive or hostile animals may have resulted from selection for aggressive defense of resource areas, where a low or harsh sound implied that the defending animal was larger than it really was. By the same token, a higher-pitched sound may convey to the receiver that the sender is smaller or a juvenile. This accounts for a "reversion" in appeasement situations to sounds characteristic of infants or young animals.

There is a further ramification from these observations. Here we have a built-in tendency for deception. In other words, although some forms of signal exchange may convey unambiguous information, such as the location or mood of the presumptive sender, signals may also carry an implicitly false message. In the cases Morton has been discussing, the sender appears to offer a signal implying it is either larger (aggressive signal) or smaller (appeasement signal) than it really is. The whole problem of deceit in signaling among vertebrates deserves careful and rigorous treatment.

29 The Interaction Systems of Generalized Mammals

29.1 Problems in the Analysis and Comparison of Mammalian Interaction Systems

As a prelude to understanding the interaction of group members, one can focus on a single dyad within a group and record its interaction forms over time. To isolate the variables underlying the shifts in an interaction form or a rate of interaction, artificial encounters may be set up between the various age- and sex-classes (Schloeth 1956). To compare the behavior patterns of solitary and social species, it is imperative that we deal with restricted classes of interactions (e.g., male-female, female-young, sibling group). Such a baseline of interaction-classes permits us to gain some insight into species differences and perhaps assess the underlying causes of such differences. Dewsbury and his associates (Dewsbury 1975) have obtained remarkable success by restricting observations to just one class of interaction (males' approaches to estrous females) in the family Cricetidae. Within this class of interaction, Dewsbury has focused upon copulatory behavior and has related mount duration to penile morphology.

This brings us to an obvious point. When we compare encounter behavior across taxa, the limit of the form of an interaction is ultimately determined by the organism's anatomical structure. The structures of effector organs, sense organs, and neuromuscular systems, as well as the brain, all contribute to the movement potential of the species or individual. Differences in morphology and resultant physiologies have derived from the process of natural selection and thus have a history of development through phylogenetic time. This history can be approached through studying fossil remains (bones, endocasts, dentition) and through comparing living forms that represent varying degrees of corresponding morphological specialization (see Jerison 1973).

Returning to qualitative differences, the axial skeleton of *Elephas* is constructed in such a way that certain movement patterns are restricted when compared with those of a wolf, *Canis lupus*. In like manner, the sense organs of a shrew (*Sorex vagrans*) determine what type of stimuli emanating from a partner it can process (Eisenberg 1973). Figure 124 clearly indicates that the kinds of displays utilized in an interaction will probably be related to the sense organs most highly developed for perception. Thus soricoids do not rely on visual display, but they have a finely graded vocalization system for conveying information to conspecifics.

To reiterate, then, in comparing social behavior across mammalian taxa one could start with a dyadic analysis. By controlling the experience of the interactants, as well as the age- and sex-classes compared,

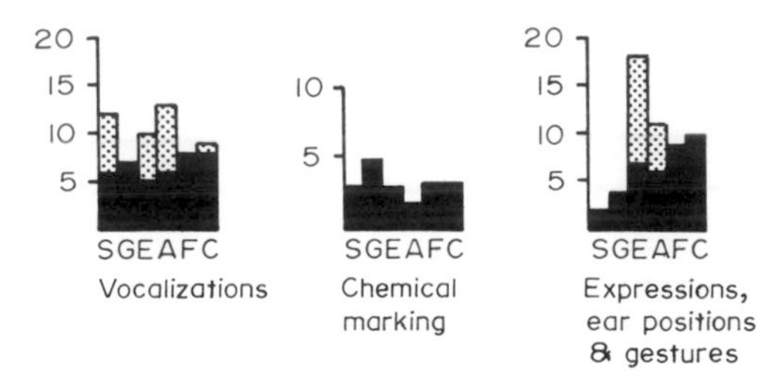

Figure 124. Comparisons of communication repertoires for selected species of mammals. *E* = *Elephas maximus; G* = *Genetta tigrina; A* = *Ateles geoffroyi; S* = soricoid insectivores; *F* = *Felis cattus; C* = *Canis lupus. Black* = fundamental patterns; *stippled* = discernible graded elements if noted. Note that the repertoires differ from species to species with respect to the communication channel, but the total repertoire tends to exhibit some constancy. From Eisenberg 1973.

one could establish baselines or "norms" of interaction. Interaction form will vary with the effector and sensory capacities of the species in question. After correcting for differences in the rate of interaction resulting from size or metabolic differences, a few fundamental rules of interaction could emerge that characterize mammalian encounter patterns. The next question is, Do these rules apply to macrobehavior, or can we apply them to subsystems of the gross repertoire?

29.2 Social Interactions

In an earlier section I discussed orientation to the macroenvironment and indicated several movements that involved the primary sense organs and several concomitant postures. As I pointed out, these same movements and postures show up again in the context of discrete orientation to a social partner or a conspecific; however, in this case the basic orientation movements and protective responses may have become exaggerated through the process of ritualization (Tinbergen 1956) and thereby may function to communicate intent to the partner. We need not go through these responses again except to demonstrate that the exaggerated response forms may serve a display function (Ewer 1968*b*).

Initial contact and contact-promoting behaviors have been discussed in a comparative sense by Schloeth (1956). The most common contact postures demonstrated by conservative mammals include nose to nose, nose of one to anal-genital region of the other, and a mutual nasogenital exploration that may be termed circling (figs. 125 and 126). These three postures seem to enhance the exchange of chemical information. They may have a further function either in promoting the toleration of contact between two individuals or in breaking off contact (see fig. 127).

After contact is established, a compatible interacting dyad may perform allogrooming. We may classify allogrooming according to the outcome of the interaction: (1) There are allogrooming behaviors exhibited by the mother to the young, which generally involve cleaning. The mother licks the anogenital area and, indeed, often the whole body surface. This behavior is shown by both marsupials and eutherians.

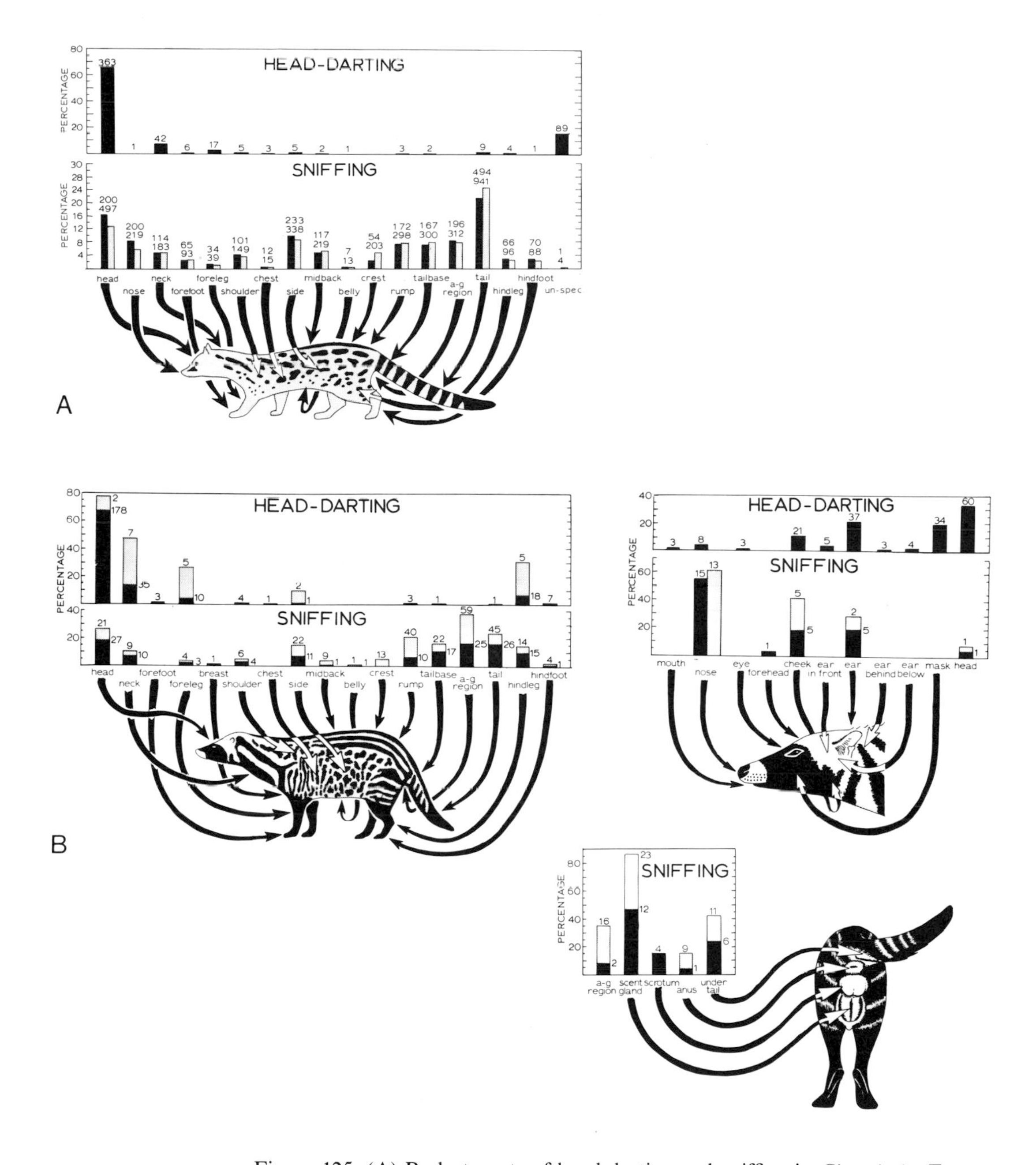

Figure 125. (A) Body targets of head-darting and sniffing in *Civettictis*. From Wemmer 1977. (B) Positions contacted during head-darting and sniffing for *Genetta*. From Wemmer 1977. Numbers above bars in the head-darting graphs and the top set of numerals above bars in sniffing graphs indicate number of observations for all the targets. Bottom numerals above bars in sniffing graphs indicate total duration in seconds of sniffs directed to each target. *Solid bars* indicate frequency percentages based on total number of all observations for targets. *Shaded bars* = duration percentages based on total duration of sniffing to all targets.

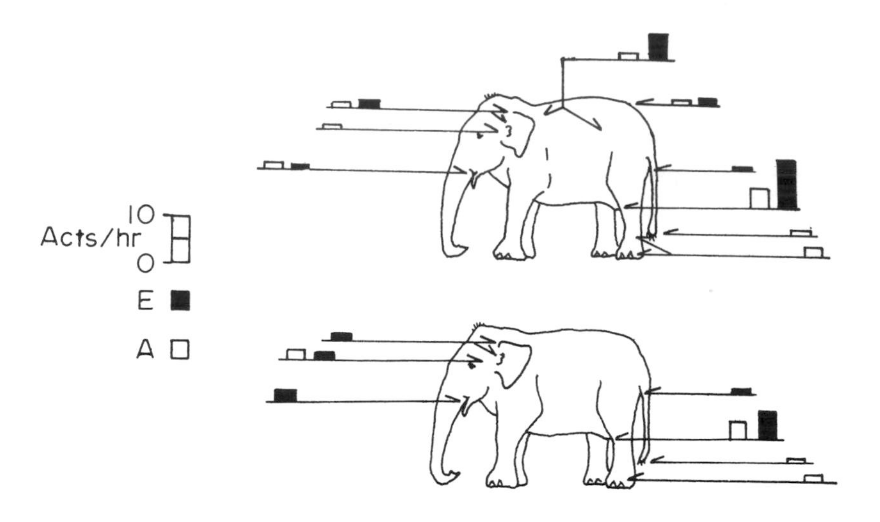

Figure 126. Positions contacted during an encounter between a male and a female *Elephas maximus*. Numbers indicate average number of contacts per hour. *Top diagram* = number of contacts shown by male to female; *bottom diagram* = average number of contacts per hour shown by female to male. *Black bars* = average contacts during the female's estrous period; *white bars* = average number of contacts per hour during diestrus. From Eisenberg, McKay, and Jainudeen 1971.

(2) A similar allogrooming behavior pattern is involved in sexual interaction between a male and a female. (3) A third form of allogrooming is involved in a dominance situation. It may occur in a sexual context, but in some mammals grooming may grade over into aggressive grooming as an assertion of dominance. Therefore, while the term allogrooming can be described with respect to the movements involved, one must remember that functionally and causally several different outcomes may be involved.

In the case of a mother grooming her young, we refer primarily to a stimulating and cleaning function. In the sexual encounter, we are implying the furtherance of contact and exchange of chemical stimuli and perhaps both an increased stimulation and tolerance of contact for the partner. In a dominance situation we are implying, as before, that not only is the behavior promoting contact, but at the same time one animal is being conditioned to submit to another's touch without an outcome of sexual behavior. Finally, as a further extension of this last category, in aggressive allogrooming the aggressor is manipulating another individual by grooming, without injuring it as would happen if it discharged its full propensities toward aggression. The recipient is tolerating and submitting to manipulation by another individual. In the course of such grooming, the recipient may in addition be marked with secretions, urine, and saliva from the "groomer."

In an alternative, descriptive classification for allogrooming, we refer to the organs employed in grooming and to the movement pattern itself. Allogrooming typically involves the mouth and tongue. It is directed to the anogenital area, the face, and those areas that produce glandular

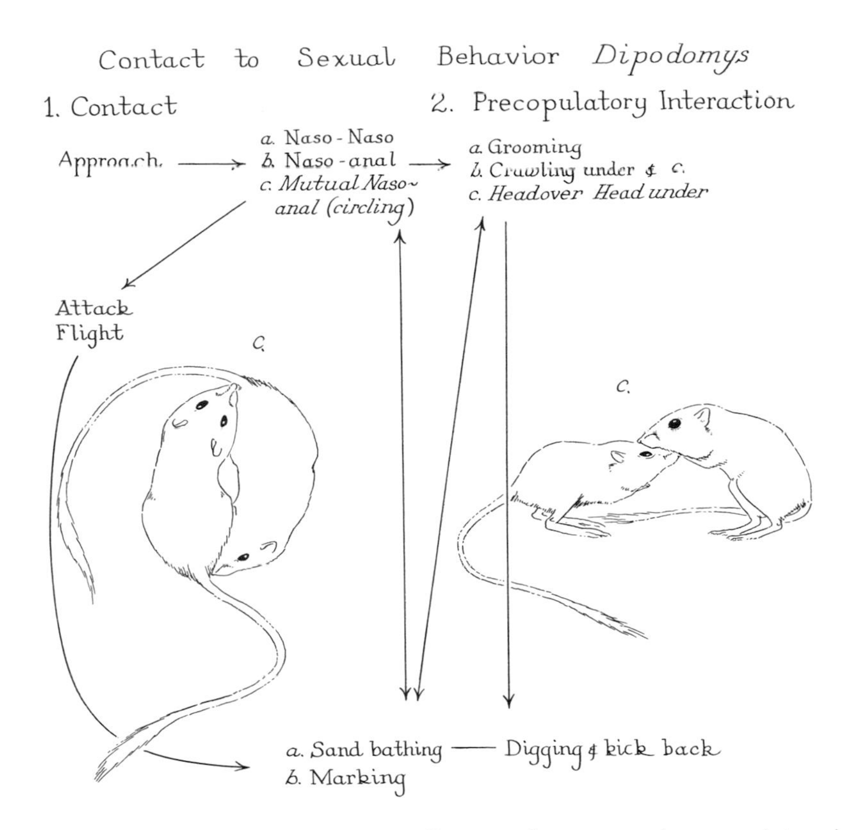

Figure 127. Generalized sequence diagram of contact and precopulatory interactions for *Dipodomys*. From Eisenberg 1963.

secretions (the cheeks, the eyes, the corners of the mouth, and the ears in conservative mammals).

29.3 Courtship and Sexual Behaviors

The male patterns of sexual behavior are basically the same for all mammals. There is a phase of orientation to the female and pursuit. The female may flee from the male, but not so fast as to escape him entirely if he follows persistently; this is termed driving. On the other hand, the female may actively incite the male if he slows his responses. Driving behavior may be interspersed with bouts of contact and contact-promoting behavior. In the sexual act, the male typically mounts the female while gripping her anterior to the pelvis with his forearms. He may also grasp with the paws. In conservative mammals a neck bite is involved. Usually the male grasps the female's skin directly behind the head. Copulation involves intromission of the penis and thrusting, usually with sustained friction movements while mounted on the female, until ejaculation occurs (see figs. 128, 129, and 130).

Dewsbury (1972) has developed a classification scheme for male copulatory behavior. His system takes account of whether the male and female tie or lock during intromission, whether thrusting occurs, and whether intromission or ejaculations or both occur in multiples during a mating sequence.

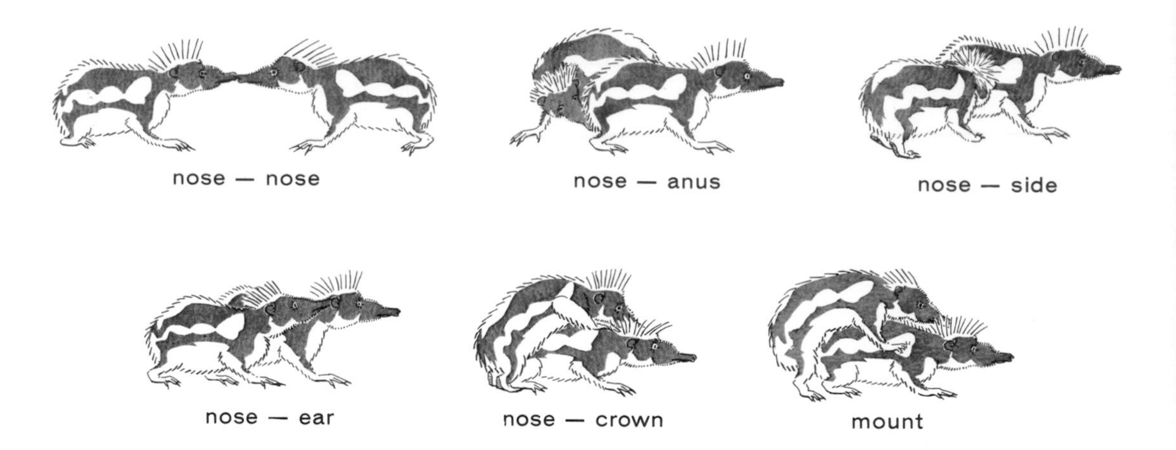

Figure 128. Example of a mating sequence for *Hemicentetes nigriceps*. Note that the contact configurations involve the basic patterns of touching nose to nose and nose to cloacal area. Glandular areas at the ear and nape are also predominantly involved in the precopulatory interaction. From Eisenberg and Gould 1970.

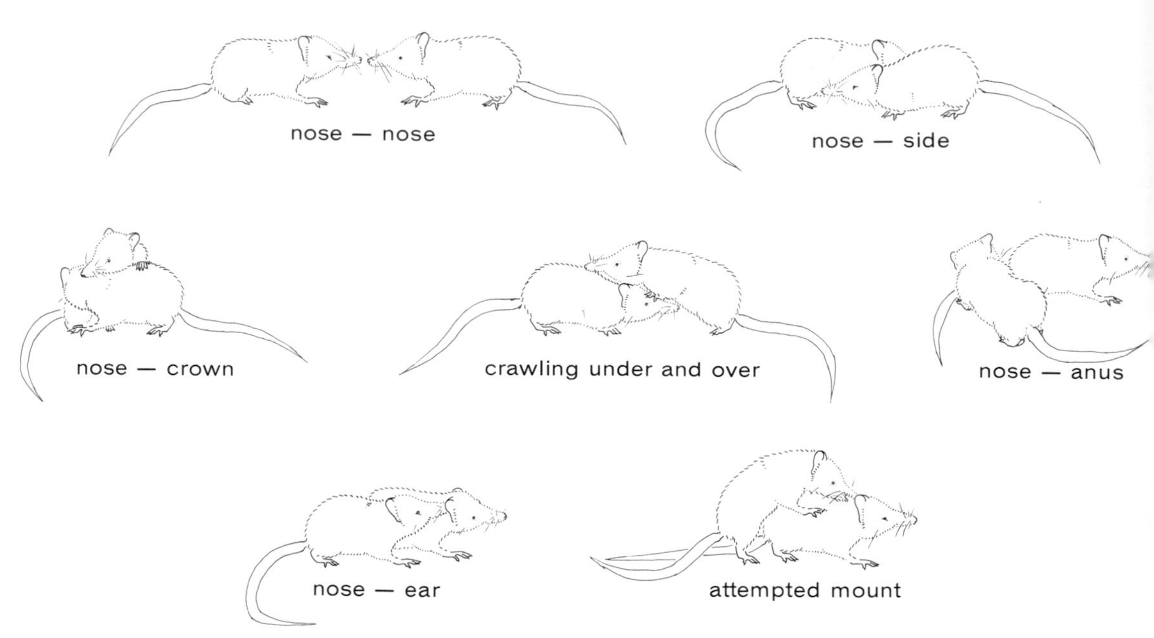

Figure 129. Precopulatory and copulatory behavior in *Microgale dobsoni*. The typical configurations displayed during initial contact include nose to nose, nose to side or groin, and nose to crown. Crawling over and under probably serves to deposit scent upon one another from the cloacal region. Compare with figure 128. From Eisenberg and Gould 1970.

Duration of intromission is highly variable in the class Mammalia. Ewer (1968*b*) speculated on the remarkably long mounts with intromission shown by the marsupials (see fig. 131). Eisenberg (1977) speculated that long intromission time may be a conservative pattern in the Mammalia, using vaginal stimulation to ensure that ovulation will oc-

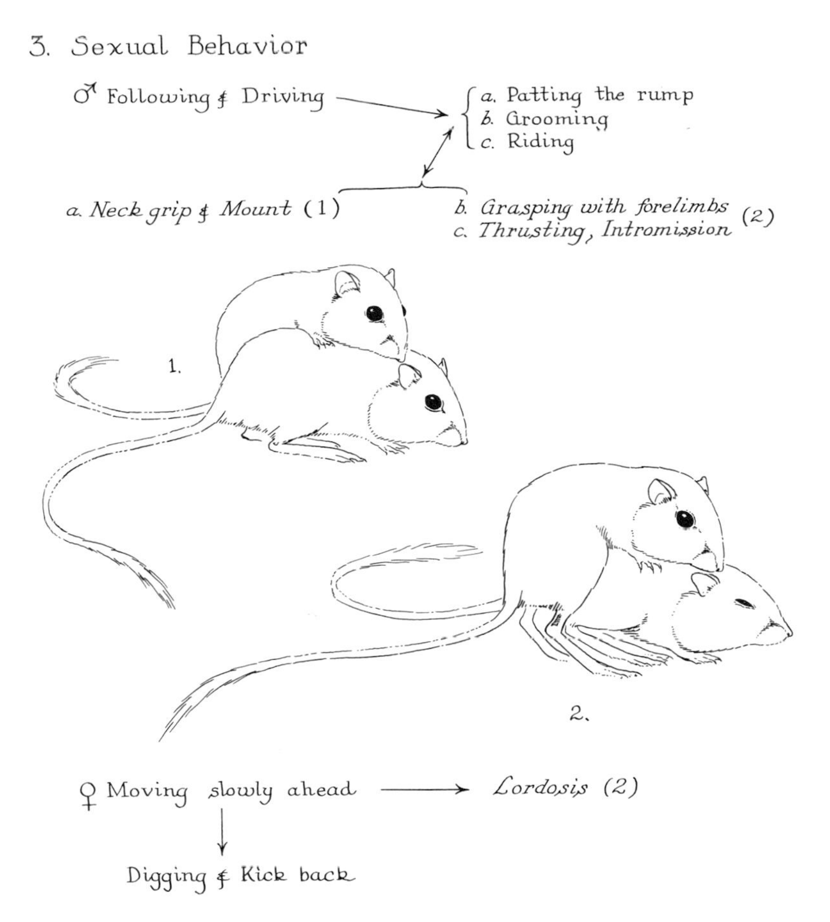

Figure 130. Sexual behavior in *Dipodomys*. Compare the fundamental similarity in the terminal postures leading to intromission in the bipedal *Dipodomys*, a rodent, with the two preceding examples from the Insectivora. From Eisenberg 1963.

cur. Morphologically conservative insectivores have long intromission times (see fig. 132). Kleiman (1974*a*) points out that in caviomorph rodents brief and few intromissions before ejaculation correlate with spines on the penis, implying that with penile adornment the female can be stimulated to ovulation with a minimum intromission time. Dewsbury (1975) similarly notes that in cricetine rodents long intromission time and a copulatory "tie" correlate with a relatively greater diameter of the male's penis.

Figure 133 portrays copulation time courses for selected primates. No clear pattern related to phylogeny is evident, but the selective pressures reducing copulation time are as yet unclear. We know that female mammals may be artificially divided into induced and spontaneous ovulators (Asdell 1964). In the former case, the female requires tactile stimulation to trigger ovulation and the development of an ovarian corpus luteum. In the latter case, the female will ovulate regardless of the discrete aspects of her interaction patterns. The dichotomy is

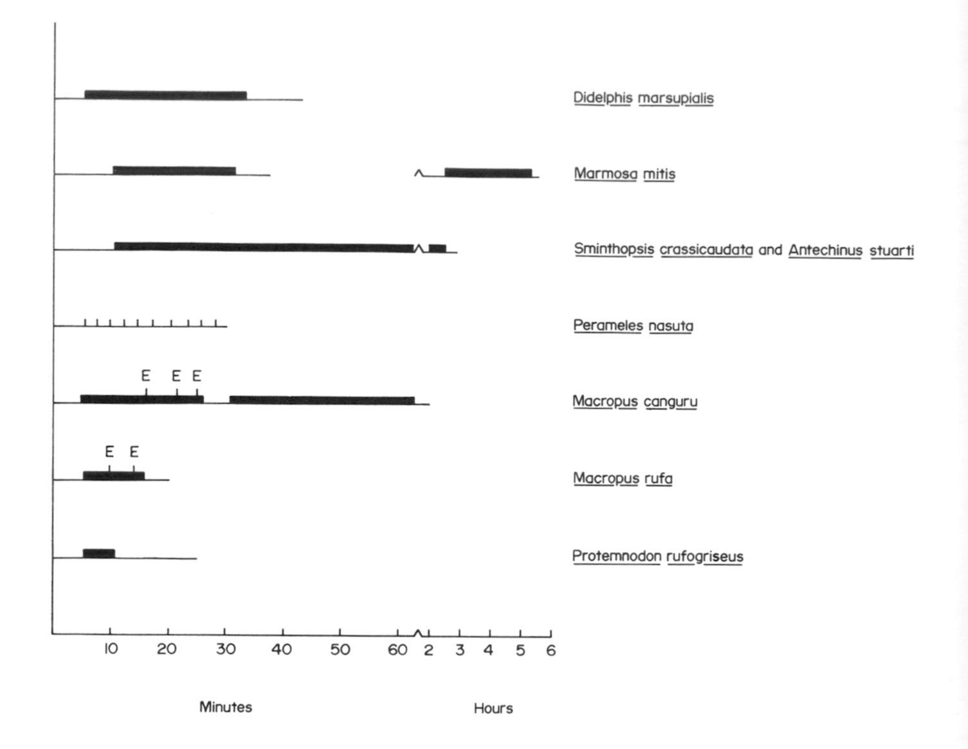

Figure 131. Temporal patterning of copulation for a series of marsupials. Abscissa is in minutes to 60 and then hours. Note the long period of intromission in the Marsupialia. E = ejaculation.

artificial in that intermediate grades can be discerned. The fact remains, however, that copulation and deposit of sperm in the female's reproductive tract must be synchronized with ovulation. Given that the option of sperm storage is closed (some vespertilionid bats are exceptions), then synchrony between ovulation and insemination are indeed critical. Vaginal stimulation of the female by the male is one way to induce ovulation or to prepare the uterus for implantation. The conservative pattern may have involved long intromission time for species that copulated in a secure place (e.g., a burrow). Copulation in open places where the couple could be exposed to predation or harassment may have caused a selection process to reduce intromission time by diverse mechanisms: (1) penile adornment, (2) spontaneous ovulation and efficient chemical signals to synchronize mating, or (3) reducing the length of individual intromission time but retaining many intromissions in a series to ovulation.

The typical female patterns involve moving ahead of the male, participating in contact-promoting behavior, and often reversion to partial male patterns of behavior and inciting the male (proceptive behaviors). Female immobilization with the assumption of a lordosis posture before mounting by the male involves raising the tail and depressing the lumbar region in conservative mammals. At the termination of a bout of copulatory activity, both male and female usually revert to autogrooming patterns involving licking the genital area.

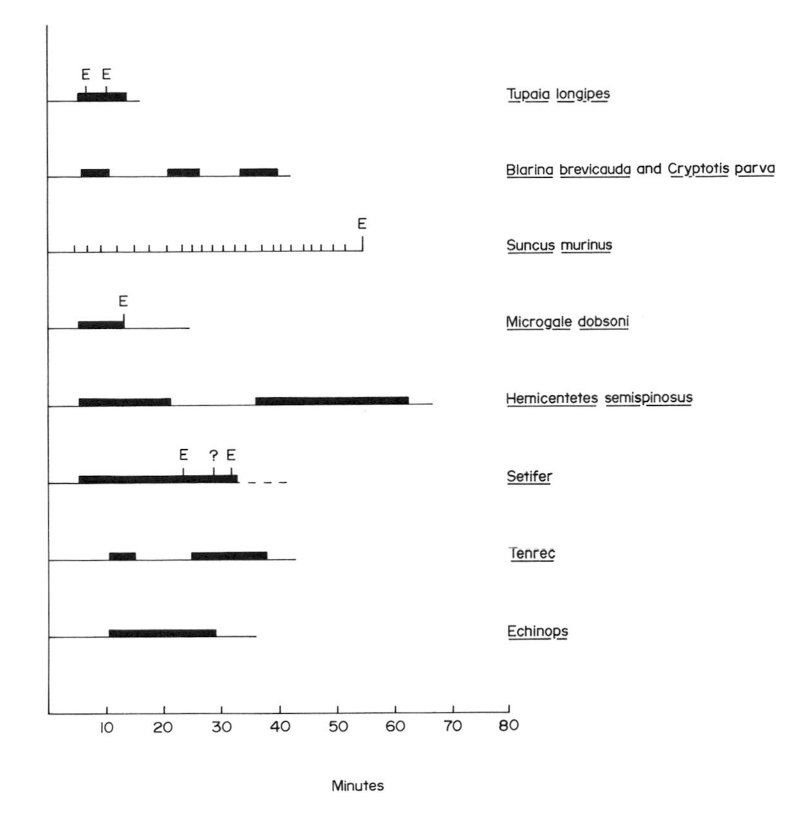

Figure 132. Temporal patterning of copulation for some insectivores and the tree shrew. Abscissa is in minutes. Note the very brief intromission times for *Suncus murinus*, considered to be a specialized derivative from the basic long intromission pattern. E = ejaculation.

Social interactions, whether they are contact-promoting, sexual, concerned with initial orientation, or of an offensive or defensive nature, frequently involve the exhibition of marking patterns. Marking patterns are highly varied, and I have discussed their derivation from the simple movements of elimination. It is sufficient at this time to point out that marking by dragging the perineal region (thus depositing glandular exudates upon or in the immediate vicinity of a partner or in part of the territory) is a basic conservative pattern. The distribution of saliva by biting at parts of the substrate and the distribution of urine and feces are also ancient traits for demarcating territory and establishing dominance. The process of courtship and mating frequently involves an elaborate exchange of olfactory stimuli (Ewer 1968*b*).

29.4 Offensive and Defensive Behavior Patterns

Within this functional category we are apt to find some of the more important modifications of behavior varying in a species-specific manner. It is here that special movements have evolved that accompany the acquisition of such antipredator structures as spines, bony scales, and epidermal plates (see sect. 32.6). In a similar manner, skeletal modifications permitting greater speed, as outlined by Howell (1944), will manifest themselves in the diverse forms of locomotion. The evo-

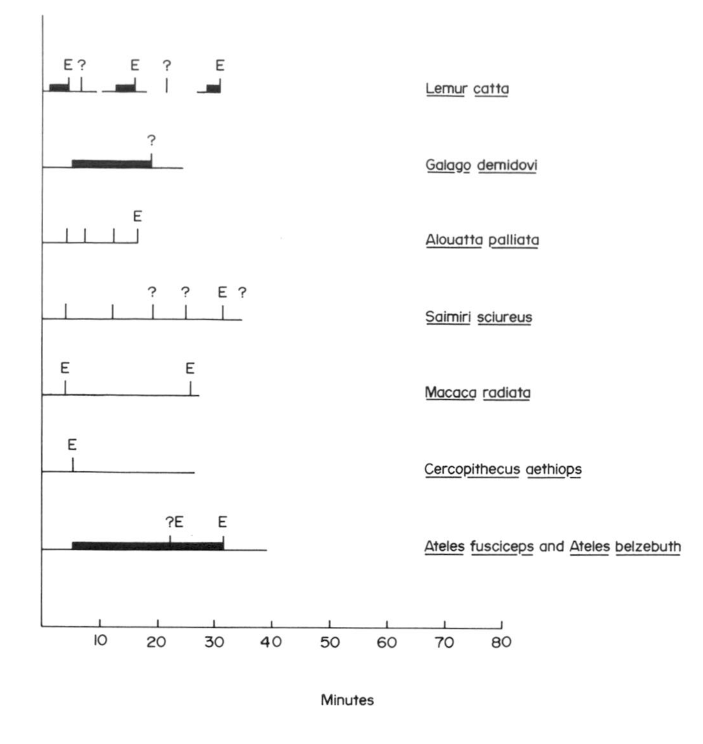

Figure 133. Temporal patterning of copulation for a series of primates. Note the extreme variability in mount duration with ejaculation. The pattern displayed by *Ateles fusciceps* and *Galago demidovii* is considered to be conservative. E = ejaculation.

lution of digitigrade progression as a method for increasing speed in avoiding predators was paralleled by the evolution of digitigrade locomotion in the cursorial predators to permit more efficient prey capture (see sect. 13.1.12). The offensive and defensive category contains many motivational variants, and indeed these reflect themselves in the different behavioral patterns that are adapted for an interspecific context on the one hand and for an intraspecific context on the other.

Let us consider a functional classification. As I mentioned in section 28.9, "Orientation to the Environment," freezing or immobility is a behavioral and postural variant frequently exhibited in mammals, and it has an antipredator effect by making the "actor" inconspicuous. Immobility or freezing is generally associated with one of the horizontal or upright postures; however, immobility may also occur in the curled or resting posture. An animal may assume a curled posture when under attack by withdrawing the head and sense organs so they are less vulnerable. This curling into a ball has become highly ritualized when the concomitant development of protective armor or spines has occurred (see fig. 134). In its most conservative form, passive withdrawal from conspecifics or other species is seen in hunching the body, folding the ears, and closing the eyes while submitting to contact. The origin of the submission posture is then in the withdrawal of the sense organs.

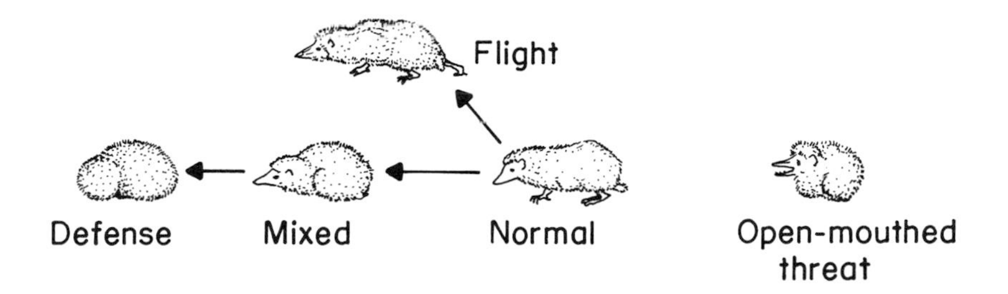

Figure 134. Examples of antipredator behavior displayed by a spiny tenrec, *Setifer setosus*. From Gould and Eisenberg 1966. Note that the ultimate form of defensive behavior involves rolling into a ball. The spines serve to deter predation by small carnivores.

In the presence of a conflicting drive state where the animal wishes to withdraw and at the same time stay, or when the animal cannot escape and attack is blocked for one reason or another, passive withdrawal appears to be the only option.

The functional categories of withdrawal and protective responses lead us to the threat function, where an animal intimidates an opponent by producing a strong stimulus contrast (in the sense of Andrew 1963). In one of the typical threat postures, the animal holds its body in a horizontal axis facing the opponent and presents a sudden contrast to the preceding activity by arching its back, accompanied by piloerection and opening the mouth. The open mouth will generally be pink or reddish in color and the teeth white, usually contrasting with the basic body color. The arched back and the erection of the hair make the attacking animal appear somewhat larger. An animal threatening more confidently may turn its side to its opponent, thus presenting a still larger image (see fig. 122).

Paradoxically, a generalized mammal may also adopt a threat posture while on its back, where it may react by arching the back while slightly stiffening the limbs and gaping. More elaborate threat postures while on the back have been highly ritualized in certain more specialized mammals, such as the rodents and carnivores. The upright posture, which derives from an orientation movement, generally causes a marked stimulus contrast by presenting a lighter colored ventrum to the opponent. In addition, the apparent size of the opposing animal is increased by the body elevation (fig. 135).

Forms of attack may develop from these threatening postures, including a direct frontal approach or an approach by sidling and shoving with the side or shoulder. During a frontal approach, a direct bite may be delivered, accompanied by a toss or a shake of the head. In the quilled forms that tend to roll into a ball during an extreme defensive posture, a variation on the bite may appear, including a toss of the head or a head buck that effectively drives the crest spines into an opponent's exposed body.

Rhythmic pushing away with the forelimbs or hind limbs may occur during fighting, and there may be many variations, from gentle pushing with one forepaw to outright simultaneous pushing with both forepaws

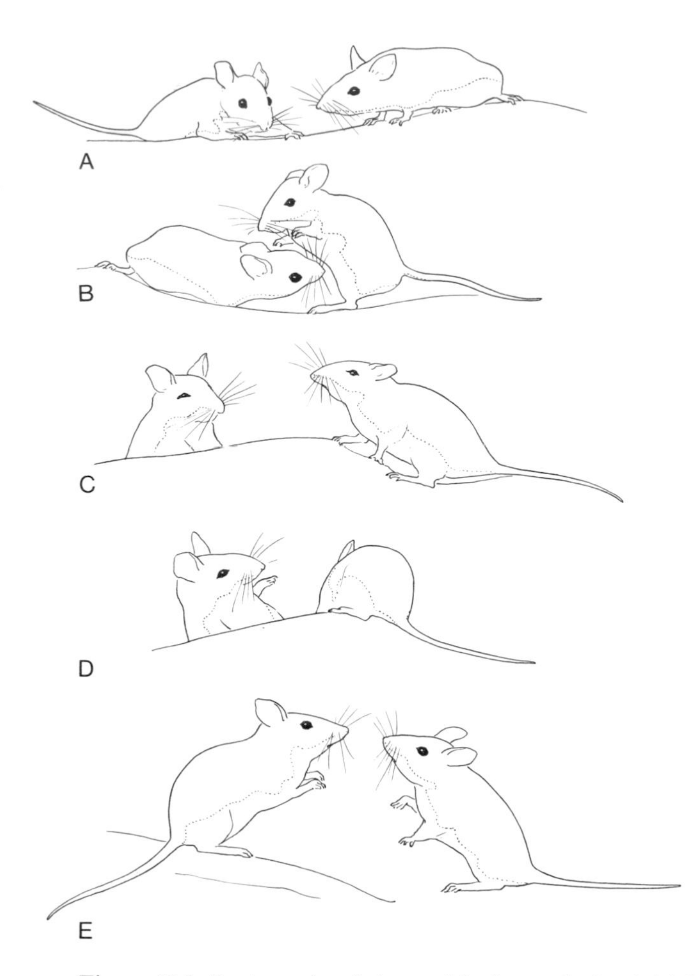

Figure 135. Postures involving ambivalence during initial encounters in the deer mouse, *Peromyscus maniculatus*. (*A*) Left animal elongates tail with slight withdrawal; right animal extended toward the other with slight flattening of the ears. (*B*) Right animal exhibits an upright posture while warding off the opponent with the forepaws. (*C*) Left animal shows a slight withdrawal while the right animal shows an incipient upright position. (*D*) Left animal is warding off with its forepaws while the animal at the right turns away, presenting its side. (*E*) Mutual upright postures. From Eisenberg 1962.

and the hind feet against an opponent. During running away, the animal may tend to kick back, and, finally, a true fighting or locked position may be adopted, where two animals grasp each other with forefeet and hind feet and attempt to bite one another while rolling and jockeying for position (fig. 136). These functional fighting categories may involve various forms of ritualization, but those I have outlined here are almost universally present in an unritualized or semiritualized form in all conservative mammals.

Many threat postures do not develop into outright attack. The inhibition of biting is an ancient trait and is displayed predominately during interactions by carnivores. Carnivorous marsupials display a

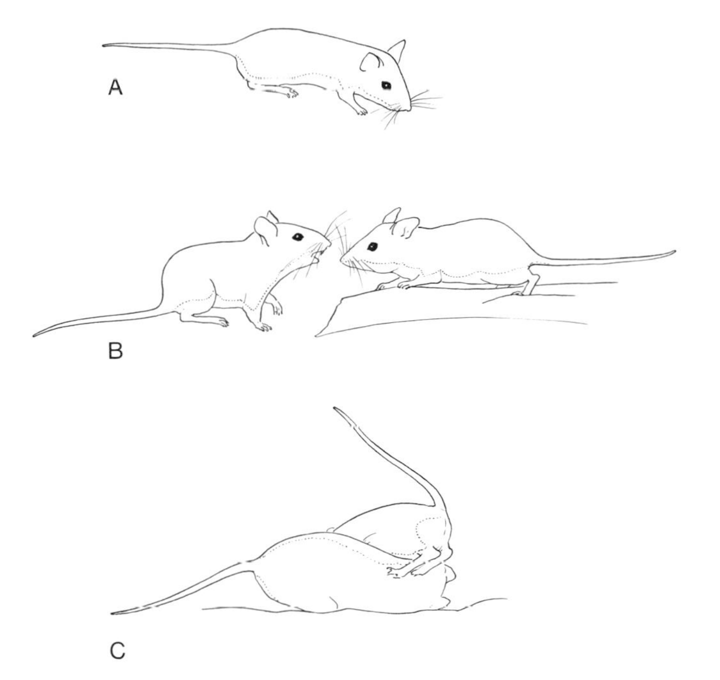

Figure 136. Sequences demonstrating threat and fighting in the deer mouse, *Peromyscus maniculatus*. (*A*) Elongate posture. (*B*) Left animal shows incipient threat and upright, indicated by the raised forepaw; right animal shows the elongate posture but the ears are laid back. (*C*) Fighting; the animals are locked together with their body axes at right angles. From Eisenberg 1962.

remarkable convergence in their threat and bite inhibition rituals when compared with eutherian carnivores (Eisenberg, Collins, and Wemmer 1975) (see figs. 137 and 138).

29.5 Parental Care Behaviors

Behaviors associated with parturition in the female generally include an increased tendency to groom the nipples and the genital areas. This may precede parturition by several days (Rosenblatt and Lehrman 1963). At parturition, it is customary for metatherian mammals to assume a position with the body axis vertical. The young moves from the vulva to the pouch unassisted by the mother. Upon entering the pouch, the young attaches to a teat and remains attached for some time. The mother attends to the young and to the pouch lining by licking them frequently throughout the long period of maturation.

The transport and retrieval phases of marsupial maternal behavior do not begin until the young emerges from the pouch and begins to move independently. Here, an interaction system is set up between the neonate and the mother whereby calls of a displaced neonate induce the mother to approach. The young then climbs onto the mother's body or into her pouch or teat area unassisted (Eisenberg and Golani 1977).

It is customary for a eutherian female to adopt a crouched posture during parturition. At birth she licks the neonate carefully. In conservative eutherians, if the mother transports the young, she generally carries it in her mouth. The female eutherian adopts a particular crouch posture over the young during nursing, but as they mature she may lie

Figure 137. Sequence of ambivalent interaction between a male and a female *Sarcophilus harrisii*. (*A*) Female shows open mouth while male presents broadside. (*B*) Male turns toward female. (*C*) Mutual uprights. Compare and contrast with basic postures illustrated in the preceding figure.

on her side while the young nurse (see Slijper 1958 for a detailed review).

Around the time of parturition there may be an increased propensity to build a nest, especially in conservative marsupials (such as *Marmosa* and many of the dasyurids that have no true pouch), even though the young are attached to the teats. Complex nest construction also characterizes the eutherian mammals that have altricial young; nest building is often extremely important to ensure adequate thermoregulation. In didelphid and dasyurid marsupials, as well as in the conservative eutherians, there is an increased tendency for the female to defend the nest site during lactation and rearing of the young.

29.6 Comments on the Parental Care Systems of Mammals

At this point we must discuss the similarities and differences found in the Marsupialia and the Eutheria. As was so admirably stated by J. D. Biggers (1966), there is no reason to assume a priori that marsupials represent a more primitive evolutionary level than placentals. All recent research indicates that highly complex mechanisms have evolved within the Marsupialia in both the male and the female. We thus cannot look upon the marsupial mode of reproduction as a simple phenomenon or as necessarily an example of an aggregation of primitive traits. (For an admirable review, see Renfree 1980.)

The marsupials are characterized by a rather short gestation period and a rather prolonged period after birth when the young is helpless

Figure 138. Defensive interaction in *Sarcophilus harrisii*. (*A*) Shouldering into the female as she defends the entrance to the nest box. (*B*) Turning away. (*C*) Offering the cheek to the approaching partner. (*D*) Offering the cheek to the female during the completion of a wrestling bout. From Eisenberg, Collins, and Wemmer 1975.

and completely dependent upon lactation for up to six months. Since the young of a marsupial are born in an extremely altricial state and the period of dependency is prolonged, the major adaptation to ensure survival and appropriate care of the young has been the evolution of the marsupium, or pouch, in which to enclose the neonates. Although the presence of a pouch is diagnostic of the order Marsupialia, many species within the order do not possess a true pouch (see sect. 3.2).

The following specialized trends in marsupial reproduction may affect the form of maternal care: (1) The evolution of a chorioallantoic placenta in the family Peramelidae has certainly had consequences for the degree of preparturient nutrition of the young (Stodart 1977). (2) One does find a trend in the more specialized, larger forms toward producing a smaller number of young, culminating in the single young of the koala (*Phascolarctos*), the wombat (*Vombatus*), and the larger macropod marsupials (see chap. 22). (3) There is a tendency to prolong the gestation period, especially in the family Dasyuridae; however, this seems to have little effect in reducing the period of infant dependency (see chap. 23). (4) The phenomenon of delayed implantation, which is highly evolved within the family Macropodidae, appears to have little effect on the form of maternal care but rather is an adaptation to guarantee a sustained, uninterrupted production of young concomitant with the reduced litter size so prevalent in the larger macropods (see chap. 23).

Within the eutherian orders, one observes convergent trends toward smaller litter sizes. The more derived species of the eutherian orders typically exhibit both a reduction in the number of young and larger young or increased neonatal mass at birth (see chap. 22). Correlated with the reduction in litter size has been the tendency toward prolonged gestation periods and more precocial young. Indeed, one may define the conservative condition of reproduction in the Eutheria as a short gestation period, a large litter size, and altricial young (see chap. 22). The conservative condition in the Marsupialia probably included no delayed implantation, a brief gestation period, and extremely altricial young, with the only functional placental system being a yolk-sac placenta.

Among the consequences of the marsupial mode of reproduction, the most important one from our standpoint is the altricial condition of the young. This is matched by a similar condition in primitive eutherians, but, as I pointed out, it is departed from within the evolution of the eutherian orders as a whole. Yet in no case do the existing eutherian orders, which produce altricial young, exhibit as prolonged a period of postpartum dependency as do the Marsupialia.

A developmental classification system based on the work of Williams and Scott (1953) for *Mus musculus* includes four developmental categories: (1) the neonatal period, during which the animal is relatively helpless and the major sense organs are not functional; (2) the transition period, during which the major sense organs are developing; (3) the period of socialization, where the youngster begins to move about and to perceive through the major sense organs relevant data concerning its mother and its littermates; and (4) the juvenile period, when the young animal is moving about independently of its mother and siblings.

Let us consider rather a discrete comparison between an insectivore, such as *Hemicentetes semispinosus*, and a didelphid marsupial, such as *Marmosa robinsoni*. At parturition, the female oppossum exhibits a characteristic posture, squatting on her heels with her head lowered, licking often at the pouch area. She remains with the long axis of her body somewhat vertical. As the young are born, they move to her teat area by their own action. The female *Marmosa* does nothing to aid the movement of the young to the teats other than to hold her position.

Most small insectivores assume a crouched position at parturition, expelling the fetus from the vagina in conjunction with strong contractions by the uterus. The young are immediately capable of some movement. The female generally turns to lick each one as it appears. Note here that in the insectivore there is a characteristic body posture during parturition, but, because it is not necessary for the young to orient to the pouch, the rigid vertical posture is unnecessary. Further, the eutherian mother pays a great deal more attention to the neonate itself, licking it and manipulating it with her tongue, and even at this time she may pick it up in the mouth and carry it from spot to another, though generally the young is shifted with the mother's nose.

During the neonatal and transition phase of the development, *Marmosa* young are intimately associated with the mother, attached to the teats. Thus at the transition period, when the sense organs become functional, the mother's body is their primary environment, and, indeed, they never detach from her until much later in the development. Although the mother may retire to a nest, the nest has no strong valence for the young. In an insectivore, such as *Hemicentetes,* the mother pays a great deal of attention to the young, licking them while lying on her side or over the young, crouching over them to permit them to suckle, and, if disturbed, even picking the young up in her mouth and transporting them to a new location. The mother must leave the young alone in the nest while she forages. The young become conditioned to each other at the time their sense organs become functional. They also explore the nest site and its environs while the mother is absent.

During the "socialization period" the young become capable of some locomotion and begin to move about and interact. Their sense organs are now almost completely functional. In the marsupial female we find an interesting pattern of interaction with the litter. The young are now transported on the mother's body, but they still attempt to go into the pouch when possible.* At times they may remain in the nest while the mother forages. The young are now getting so large that they cannot all crowd into the pouch; therefore they must nurse while outside. They are becoming conditioned to the nest without the female. At this time the mother still serves as a primary orientation point. She not only comes to nurse them, but she may carry them when she leaves the nest or permit them to ride on her body. The mother assumes a stronger valence than the nest itself. At this time the young may occasionally become detached. When they are detached and uncomfortable, the youngsters emit a high-pitched, chirping cry that causes the female to approach them and stand over them until they climb back onto her body.

In *Marmosa* there are no ritualized retrieval movements by which the female picks up a young and transports it to a nest, because, in a sense, the mother is the nest. Rather, there is a very stereotyped reaction whereby the mother responds to the cry of the young, approaches, and stands near them, permitting them to "board" her. Some species of *Marmosa* will pick up their young in the mouth and carry them back to a nest in response to their squeaking (see also Eisenberg and Golani 1977).

If we consider *Hemicentetes,* we find that the mother does carry the young in her mouth during times of danger or when she moves to a new nest site. She responds to the distress call of a young by picking it up in her mouth. Throughout postnatal ontogeny the young are building up positive associations with each other and with the nest site, since the mother tends to forage independently while they are helpless. Gradually the young will begin to follow the mother as she forages, and a new type of reaction is established where the mother becomes a primary guiding stimulus that the youngsters must keep up with and follow on her nightly foraging expeditions (Eisenberg and Gould 1970).

In the marsupials, the stage of following by the young is somewhat abbreviated. The female transports the young in most circumstances, and when they become too large for her to carry they begin to move about separately. Although some young often do follow the mother, the period of attachment is somewhat reduced in the sense of a "following" relationship. This is not, of course, the case in macropod marsupials, where the relationship persists longer. From the late socialization period and into the juvenile stage of development, eutherians and marsupials come to resemble each other more and more (see fig. 139).

In summary, in conservative eutherians, such as the "insectivores," the female generally has a nest in which the young are born; they are nourished there and kept warm by the mother, but the mother leaves for foraging expeditions during the neonatal and transition stages of the young. The mother is capable of mouth transport and can move the young to a new nest site, though this generally occurs only in emergencies. The young of conservative eutherians soon become attached to their nest site and their siblings as well as to their mother. The mother and siblings are associated at the nest site, and only later on, during the following phase, will they be associated in another context.

The marsupial female, on the other hand, is in a sense the "primary nest," since the young undergo their neonatal and transition stages while accompanying her wherever she goes. It is only when the young reach the stage of socialization that they begin to approximate the eutherian condition, with the nest attaining more importance as a source of conditioning stimuli.

In the primitive didelphid marsupials, it is customary for the young, during their early socialization stage and at the end of their transition stage, to cling to the mother and be carried by her. In conservative eu-

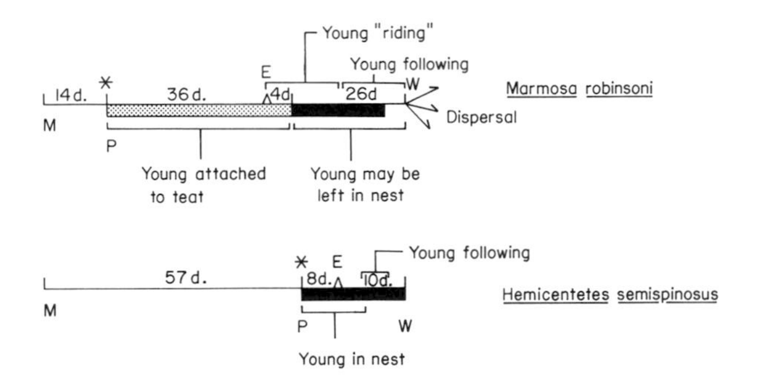

Figure 139. Time course of parental care behavior shown by a marsupial, *Marmosa robinsoni*, and a eutherian, *Hemicentetes semispinosus*. Note that the prolonged gestation of *Hemicentetes* reduces proportionately with the time of lactation relative to the marsupial. This is one of the major differences in the expression of parental care behavior between marsupials and eutherians. *M* = time of mating; *P* = parturition; *E* = eye opening by young; *W* = time of weaning; *black bar* = nest utilization phase. From Eisenberg 1975*a*.

therians, however, clinging by the young is not general. The young mature rapidly, and they would be too unweildy a load. Instead, when they are mature enough, they tend to follow the mother. This stage of following can occur later in the conservative marsupials, and it is brief.

In advanced or specialized eutherians, we see many departures from the conservative condition. In the prosimian primate *Microcebus,* one finds the origin of a specialized maternal transport mechanism that in some ways approximates that of marsupials. As the young *Microcebus* develops for the first few days, they are left in the nest while the mother forages. As the littermates become slightly older, in the stage of "initial socialization," they are carried on the mother's back. This specialization departs from the conservative eutherian method I outlined for *Hemicentetes.* In *Microcebus* it shows two stages: at first the young remain in the nest while the mother forages, and later they are carried. In other advanced primates the parent carries the young on its body, and in even more advanced primates the young cling to the mother's fur at birth.

The tendency within many advanced, specialized mammalian orders (Cetacea, Pinnipedia, Artiodactyla, and Perissodactyla) is to produce young that are more and more precocial and a smaller litter size (see chap. 23). Very precocial young are capable of following the mother after a very brief period of initial socialization that is in some cases comparable to imprinting in precocial birds (see Lent 1974 for a review).

Since most conservative mammals are nocturnal, the following response has to be mediated by cues different from those in diurnal forms or those having eyes better adapted for form vision. Perhaps the simplest cue is a tactile one, where the young attempt to maintain contact with the mother, generally with their noses touching some part of her body. Odor trails can also be used in conjunction with touching. Such a system of tactile contact is very common in small shrews, such as *Sorex,* but a number of specializations have evolved. In the genus *Crocidura,* the white-toothed shrew, the young form caravans, each seizing the tail of the mother or of the sibling in front (Zippelius 1972). It is well known that in some rodents and fewer insectivores the youngster at first maintains contact with the female by clinging to her teats as it follows. This arises out of the general nursing response and is exemplified in *Solenodon* (Eisenberg 1975*b*).

Of course the young may maintain contact with the mother through other senses. In many species the mother leaves behind a trail of secretions from her feet, or the young may be able to smell her directly as she moves ahead of them. Yet another method is for the mother and the young to produce corresponding sounds that allow them to maintain contact. This is exemplified by the specialized stridulating organ in *Hemicentetes* (Eisenberg and Gould 1970).

In the more specialized forms, vision predominates, and visual cues may take precedence in coordination. In most mammals, probably all the senses that are anatomically developed for reception are used in maintaining contact; but, as we have seen in the specialized cases cited, some may be used more than others.

Since the following response and the association of the young with the mother, is so universal—in eutherian or marsupial, generalized or specialized mammal—the female must have some important adaptive role aside from primary nutrition of the young. This surely at least involves conditioning the young to a choice of foods and the selection of a feeding site (see Elliott 1978).

29.7 Play Behavior

Play behavior is widespread in the Mammalia (Meyer-Holzapfel 1956). Play has been described in a rather sophisticated manner by Müller-Schwarze (1968) for mule deer and by Wilson and Kleiman (1974) for rodents and carnivores. Bekoff (1972) has analyzed the implications of metacommunication in the course of play. Some species show play only briefly during their maturation (Elliott 1978), whereas other species play well into adulthood (Loizos 1966).

Some forms of play in young mammals appear to result from partially activated drives that do not lead to consummatory acts (see fig. 140). Other forms of play seem to be related to acts necessary to adult survival, such as antipredator behavior (Wilson and Kleiman 1974), predatory behavior (Eisenberg and Leyhausen 1972), or intraspecific aggressive behavior (Poole 1966). Clearly the common theme in these studies is that the object or goal of a play sequence is not related to a conventional stimulus-specific consummatory act as it does in the similar adult sequence in a situation related to immediate survival. Yet play may in mammals have a strong relevance for survival later in life.

Play may be carried on alone or with a partner. Playing with a partner may involve a complex signal system to signify that the intent is play

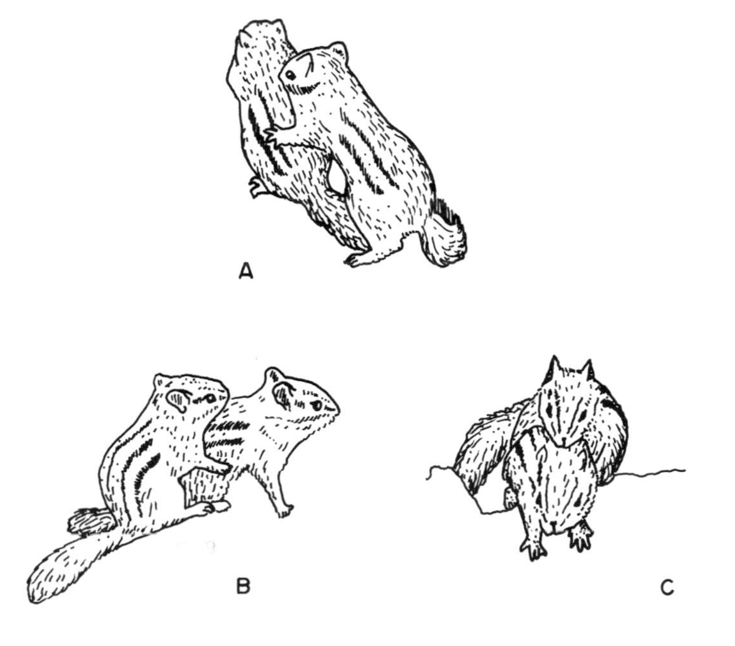

Figure 140. Juvenile chipmunks (*Tamias striatus*) engaged in interaction that simulates adult sexual behavior. From Elliott 1978.

rather than serious interaction (Bekoff 1972). One aspect that begins to emerge from studies of play is that animal species with rather high EQ values tend to spend a significant portion of their early development in play. An arithmetic correlation is not yet possible, but the suggestion is inherent in the published works.*

30 Reproductive Failure, Social Pathology, and the Special Case of Infanticide

30.1 Introduction

In view of my review of parental care and development, let us consider the breakdown of maternal responses. From recent critiques (Hrdy 1976), it appears that a definition of terms is in order before I proceed. Webster's dictionary defines pathology as "the branch of life science that deals with the structural and functional changes caused by any departure from the normality of mental and physical function." I myself define normality as "the exhibition of a phenotypic trait within the environmental context for which primary selective forces have shaped it, the outcome of the trait's expression being maximal, inclusive fitness."

Clearly, some cases of reproductive failure that allow the individual to repeat the attempt with a higher probability of success may be considered highly adaptive. This is analogous to the situation where (in an anthropomorphic sense) a presumptive parent (generally the female) deems the outcome of rearing dubious and terminates the rearing process, thereby giving herself a second chance at reproduction (see chap. 23). This is adaptive only if terminating the pregnancy in question will increase the individual's probability of bearing viable offspring in the future. For species that are seasonal breeders and have adopted an approximation to a semelparous reproductive strategy, the option is nearly closed, unless they can fit two litters into one season of reproduction. The problem for the researcher is to differentiate the forms of reproductive failure that may in fact have been selected for from those forms that are truly pathological.

In a previous publication (Eisenberg 1967) I documented examples of reproductive failure that derived in part from an enforced spatial proximity of individuals under artificial conditions. This enforced proximity could well be termed "crowding." Figure 141 shows how reproductive failure can occur and demonstrates that different species of rodents may vary in their sensitivity to crowding. I concluded from this work that one could classify rodent species into categories of sensitivity to the proximity of conspecifics. Their relative sensitivity to crowding seemed to reflect natural dispersion patterns (Eisenberg 1969). In brief, a species that showed a communal social organization in nature and that had behavioral mechanisms for coping with increased contacts with conspecifics would show a high tolerance for crowding in captivity. Reproductive failure would occur much later than in a species preadapted to a dispersed social system in nature.

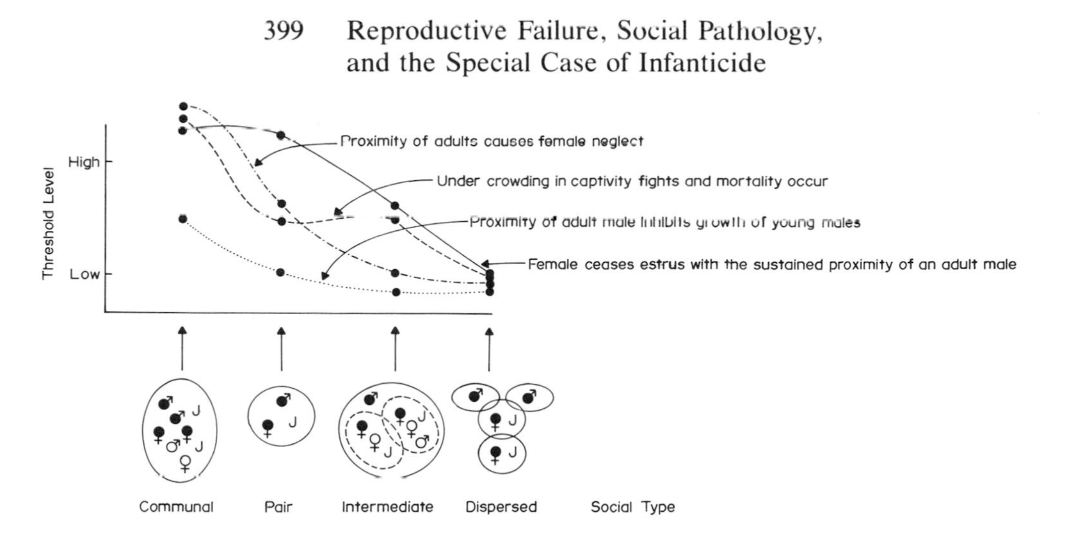

Figure 141. Diagram illustrating differential sensitivity by mammalian species exhibiting different social organizations with respect to four forms of social pathology induced by crowding. Species that tend to live communally have high thresholds for the exhibition of the following reproductive failure syndromes: cessation of female estrus, inhibition of growth of young males, fighting, with resultant mortality, and neglect of young by the female. Eisenberg 1967.

More sophisticated work involving species that habitually form pair bonds, such as the elephant shrew, *Elephantulus rufescens,* the golden marmoset, *Leontopithecus rosalia,* and the acouchi, *Myoprocta pratti,* conducted by Rathbun (1979), Hoage (1978) and Kleiman (1972*a*), respectively, indicates that reproductive failure through outright nonconception can occur when members of a pair are incompatible. In short, through courtship, opposite-sexed individuals would normally be able to entrain their activities so that mating could occur; but if the period of entrainment is disrupted or if the motivational tendencies of the two animals are somewhat out of step, successful mating may not occur for some time.

Reproductive failure can also derive from deficits in early experience. For example, if there is maternal deprivation in the early ontogeny of *Macaca mulatta,* then sexual and maternal behavior at maturity will often be deficient (Harlow and Harlow 1965). Hoage (1978) has amply demonstrated that in the golden marmoset insufficient early experience with younger siblings will cause deficits in an adult's parental care patterns. Such deficits in experience may even result in infanticide.

30.2 Infanticide in Captivity

Numerous examples of infanticide can be described from captive studies. In the African Cape hunting dog, *Lycaon pictus,* I know of a case where a female killed and ate her own young and on a second occasion merely killed the young. Both instances occurred when the female had been isolated from the male. It is now well known from the studies of Kühme (1965) and Van Lawick (1971) that the male and other pack members provision the female during her lactation period (see sect. 13.1). Clearly, separation from the mate in captivity does not provide the female *Lycaon* with the appropriate stimuli for successfully rearing her young. A similar case has been described with the bush dog, *Speothos venaticus,* in the Frankfurt Zoo (Jantschke 1973). The disturbance of maternal care patterns by deprivation of a mate need not lead immediately to the killing and eating of an infant, but the female may in fact fail to rear by merely neglecting the young. Maternal neglect is a common cause of infant death, even though it does not qualify as "infanticide" by a strict definition.

Bears are often difficult to breed successfully in captivity if the female is not isolated from all external disturbance. In this case leaving the male with the female would be the worst possible thing to do. In fact, in nature the female bear must often guard her young against the male's cannibalistic tendencies. Even if the male and the female are separated, the female may neglect or devour her young if she is subjected to noises or strange odors and sounds. Apparently the female bear is preadapted to anticipate the maternal environment of total solitude that would obtain under natural conditions when she gives birth in her hibernating den. If these conditions are not met in captivity, infanticide will result, through neglect or through outright killing.

In some cases a captive male's harassment of his own or others' young has led to infant death. In the early years of our attempts to breed the black spider monkey, *Ateles fusciceps robustus,* the male would often harass females in about the second week after she gave birth, especially if we isolated the female for any time and subsequently returned her to the group. We had far better success if we let the female give birth within the group and removed the male for a time. As the male grew older he became much more parental, and when he reached approximately eleven years of age his harassment was negligible and he no longer had to be removed.

A male and a female of the white-cheeked gibbon, *Hylobates concolor,* were left together through parturition, whereupon the male began to harass the infant. In a species preadapted for monogamy (Ellefson 1974), this seems extraordinary; but the male was rather young, which leads one to believe that sexual maturation is not equivalent to sociological maturation and that perhaps some rearing deficits in the history of this captive-reared young male explained his totally maladaptive behavior. In a subsequent mating some three years later, the male was left with the female and engaged in no such harassment.

Several females and one reproducing male of Matschie's tree kangaroo, *Dendrolagus matschiei,* have been kept as a colony. Persistent harassment of females with young in approximately the third month of pouch development was observed only after the colony had grown from one male and three females to one male, five females, and a juvenile male. This harassment by the adult male is clearly atypical, and one wonders if the proximity of a maturing subadult male contributed to the pathological behavior. Probably in nature an adult male tree kangaroo is not so continuously associated with a stable reproducing female group, and perhaps his harassment could be attributed to crowding.

30.3 Infanticide: Is It Adaptive?

To consider infanticide in the context of reproductive failure, we must differentiate among cases where the female kills her own young, the female kills another's young, the mother causes infanticide through neglect, the male kills his own young, or the male kills another's young. For each of these cases of killing or neglect, we should note whether the young are eaten.

Many species of mammals that bear altricial young exhibit maternal cannibalism of part of the litter. The mother may eat young that are born dead or weak. The general conclusion is that killing and eating such young is an adaptive process immediately recycling nutrients from the female's own body and making the most efficient use of the material in producing milk for the more viable surviving littermates. Clearly, infanticide of this type is not related to reproductive failure and could be considered highly adaptive.

Mallory and Brooks (1978) have written an interesting paper concerning the killing and eating of young by the lemming, *Dicrostonyx*. In their experiments, males killed the unrelated young of a strange female, but males introduced to a previous pair mate had reduced infanticidal tendencies or none. They argue that the response of males to the young of strange females derives from the males' strong post-fertilization competition for mates. A female whose young are cannibalized by a strange male and who then is mated by him will exhibit a shorter gestation than does a female who mates postpartum but rears her litter. This further suggests that males and females have conflicting reproductive strategies. These observations will be discussed in the light of primate research below.

When we observe a behavioral trait such as infanticide, we must immediately ask whether it has in fact been selected for. Perhaps the propensity to attack smaller conspecifics has been selected for to allow the larger individual greater access to food or other limited resources. The propensity to attack a smaller conspecific could lead to attacking infants, subadults, smaller adults, or females or be generalized to attacking strangers that exhibit inferior or submissive behaviors. This attacking trait then could be positively selected for, but a counter-selection would have to act to prevent indiscriminate killing of conspecifics, kin, or one's own offspring. Thus, if food or resources become limiting as a result of an increase in population density, one might expect increased aggression against smaller conspecifics, which might manifest itself as outright attack.

On the other hand, in infanticide and cannibalism the trait selected for may be a specific protein "hunger." The expression of this trait could lead to cannibalism of nonkin as well as to hunting and eating other species. In protected populations where mortality is low, infanticide might then manifest itself under crowding as a compromise behavioral phenotype, but the expression of the trait may not necessarily result from a specific selection for infanticide.

30.4 Infanticide in Primates

A topic of recent interest is the cause of infant deaths in troops of primates. Since primates had been considered highly social mammals with rather high encephalization quotients, preadapted to live a complex social life, the discovery of infanticide and cannibalism came as a shock to primatologists who had hoped to describe "peaceable kingdoms" in nature. It appears that infanticide is not common in those primate populations that have been studied, but when it occurs it can be dramatic. Goodall (1977) discusses the problem in the chimpanzee. Struhsaker (1977) describes it in the red-tailed monkey, *Cercopithecus ascanius*. Rudran (1973*b*) describes it for the purple-faced langur, *Presbytis senex,* at the time of troop takeovers by new males. Hrdy (1977) gives an exhaustive analysis of the phenomenon in *Presbytis entellus*. Rudran (1979) offers a description and interpretation of the process in the red howler monkey, *Alouatta seniculus*.

The field data usually consist of observations that infants disappear when a troop is taken over by a group of males or a single male. These "takeover males" formerly lived on the periphery of an established troop, before displacing a resident breeding male. The assumption, based upon the observed absence of the infant, is that it has been killed in the course of the troop takeover. Often one does not know how it died, but sufficient evidence, as established by Suzuki and Hrdy for *Presbytis entellus,* leads one to feel that the infant has actually been killed and has not died from maternal neglect. Male infanticide is by no means proved for all cases of infant disappearance at the time of troop takeover. In those cases that have been witnessed in primates, the infant is usually killed by a male; however, in the chimpanzee, Goodall reports cases of infanticide by females. Where the infant has been killed outright through an attack by a male, it generally is not eaten in *Presbytis* and *Cercopithecus,* and the killer is presumed not to be the father. In the cases of the chimpanzee infanticide, however, whether by male or female, cannibalism has sometimes been noted.

Based on her long-term study of *Presbytis entellus* at Abu, Hrdy argues that male takeovers of an established troop, with displacement of the resident male, generally result in the death of infants less than six months of age. Attempts by pregnant females to "deceive" the male include "false estrus" and a willingness to copulate subsequent to the takeover. Attempts may be made to protect infants through coalitions of females. Hrdy argues that these traits are phylogenetically ancient and represent an adaptive compromise of individual attempts to maximize the fitness of resident females by protecting their young and an individual attempt by the takeover male to maximize his fitness by inducing an early estrus in the female through infanticide. By killing a nursing infant, the takeover male shortens lactation, thus speeding the onset of estrus. Whether this is in fact a phylogenetically ancient trait deeply engrained genetically within the langur or whether the phenomenon of false estrus as described by Hrdy has another explanation remains debatable. What is difficult to contend with in Hrdy's book is that there are many cases of troop takeovers in primates where there is no indication that infanticide takes place. These instances may result from species-specific differences, troop-specific differences, or a composite of individual differences, or they may be a function of the time when the observations were made.

One thing is absolutely clear. If the information Hrdy presents concerning the langurs at Abu is compared with the data for *P. entellus* elsewhere (see Hrdy 1977, table 3.4, p. 31), and if we plot the mean troop size of *P. entellus* against the mean home range, we see that most cases of infanticide in langur populations occur when the troop size is moderately large (between fifteen and thirty-five individuals) and the home range size is extremely small. A high level of infanticide among langurs also occurs where the troops are in habitats with little or no predation pressure. Finally, langur troops existing at extremely low densities show a low frequency of infanticide or none.

The preceding evidence strongly suggests to me that the occurrence of infanticide correlates with special demographic circumstances. It appears that infanticide per se has not had a long history of selection pressure, but that it is the outcome of selection for other traits—for example, attacking smaller conspecifics and attacking nonkin. In fact, the data suggest that, under normal conditions of predation pressure and when crowding does not result from the compression of home ranges, infanticide is very rare indeed. Rudran (1979) suggests that the frequency of infanticide in howler monkey populations increases as the population density increases. He further points out that infanticide can be a potent force contributing to the mortality rate in populations that are near the environmental carrying capacity. I note in his articles an echo of the findings on rodent populations summarized by Southwick (1955).

31 The Structure of Social Organizations

31.1 Social Organizations as "Phenotypic" Characters

How does one describe a society? A social organization or group may be defined by spatial criteria. When observed for varying periods of time, the members of a population are generally found to exhibit different types of dispersion in space. When spacing behavior is actively maintained with respect to resources, it is expressed through threat displays and is negatively reinforced by agonistic encounters (see fig. 142). One could assume that such spacing mechanisms have evolved to partition resources within the environment. Spacing patterns vary between species with a broad range of trophic adaptations. We are usually viewing a selective compromise between the ability to defend an area and the benefit to the defender of investing energy to control a discrete resource.

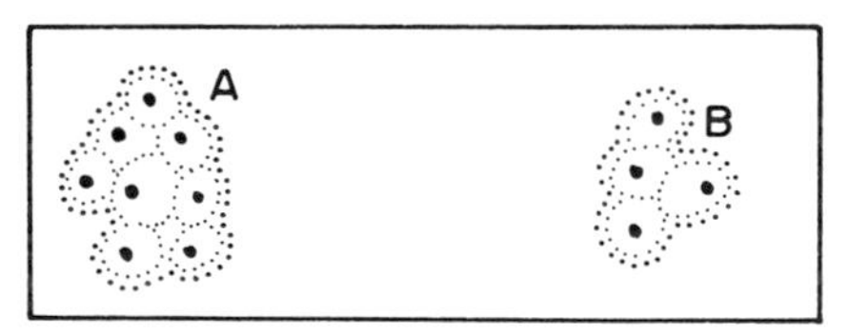

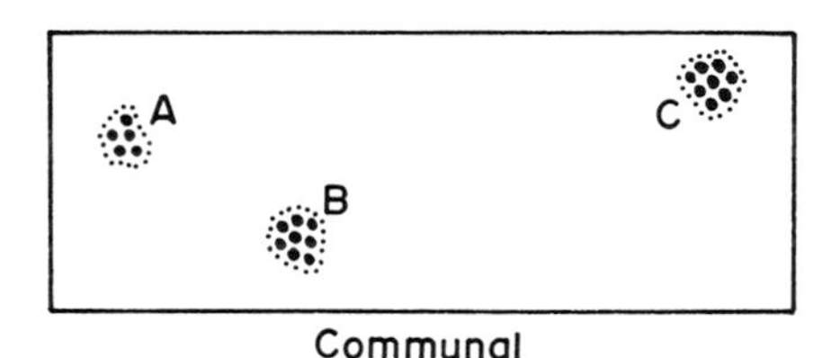

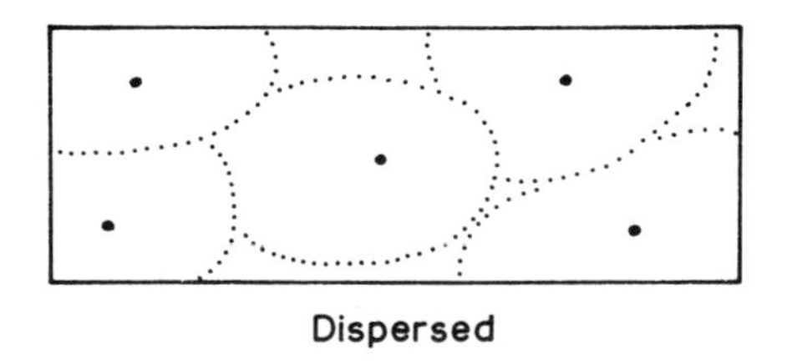

Figure 142. Three possible forms of distribution in space exemplifying three different social systems. *Black dots* represent individuals. *Dotted lines* indicate boundaries of individual living areas. *Dashed lines* demarcate group territories. From Eisenberg 1977.

In general, each adult individual of a population has a center of activity, and in some species the area of activity or home range also coincides with a defended area. With individual defense of an area, the subjects will be spaced out, each with an area of exclusive use. On the other hand, a population may be organized into groups; if areas of exclusive use are still present, they are shared by the group members. The individuals then appear to be clumped.

Some form of spacing among like-sexed adults is the rule for many mammalian species. Even within a communal group, there is a hierarchical order of precedence in the use of scarce resources. Of course, social groupings might be quite mobile and move over wide areas, and no static resource defense pattern need be exhibited. Then again, a species may alternate at different times of the annual cycle between defense of a restricted area and movement with little overt defense. If one examines the way space utilization varies through the life cycle of an individual, some species seemingly generate a bewildering number of variations.

One can thus conclude that spacing behavior may be phasic. Indeed, even in species that exhibit the most exclusive land tenure patterns, there is a time during the life history when an intimate association is set up between adults of opposite sexes and mating takes place (see Eisenberg 1966, p. 15). To understand the phasic component of social behavior, we must consider how static spacing mechanisms operate over time. Generally, such phasic changes are studied by examining interaction rates and spacing patterns among specified age- and sex-classes through a complete annual cycle. Species can then be compared with respect to the temporal patterning of interaction rates.

"Cohesion" refers to a species' mechanisms for maintaining proximity even when moving. Special signals, such as "contact calls," help coordinate group activities. Even in permanent groups, there is spacing between individuals. Among individuals living in a more or less cohesive group, social structure may be described in terms of rates and types of interaction. The data are usually presented as sociograms (fig. 143), which can be compared for populations of the same species or across taxa. This type of comparison often yields fruitful insights into the different interactive processes that underlie intragroup spacing patterns. It is often possible to demonstrate quantitative differences between species in affiliative or agonistic behavior. When one finds variation between populations of the same species, one can seek its causes. Differences or similarities between species can help us understand how selection shapes rates of affiliative and hostile interaction as the social unit evolves its particular modes of environmental exploitation (Crook 1970).

31.2 The Genetic Basis of Social Behavior

We assume that vertebrate social structures are manifestations of natural selection and reflect adaptations for exploiting specific environments (Crook 1970). One of the more popular subunits for discrete analysis within any given social structure is the parent-young unit. Here in the fundamental unit of reproduction certain forms of interaction must occur if the rearing cycle is to complete its "purpose." In

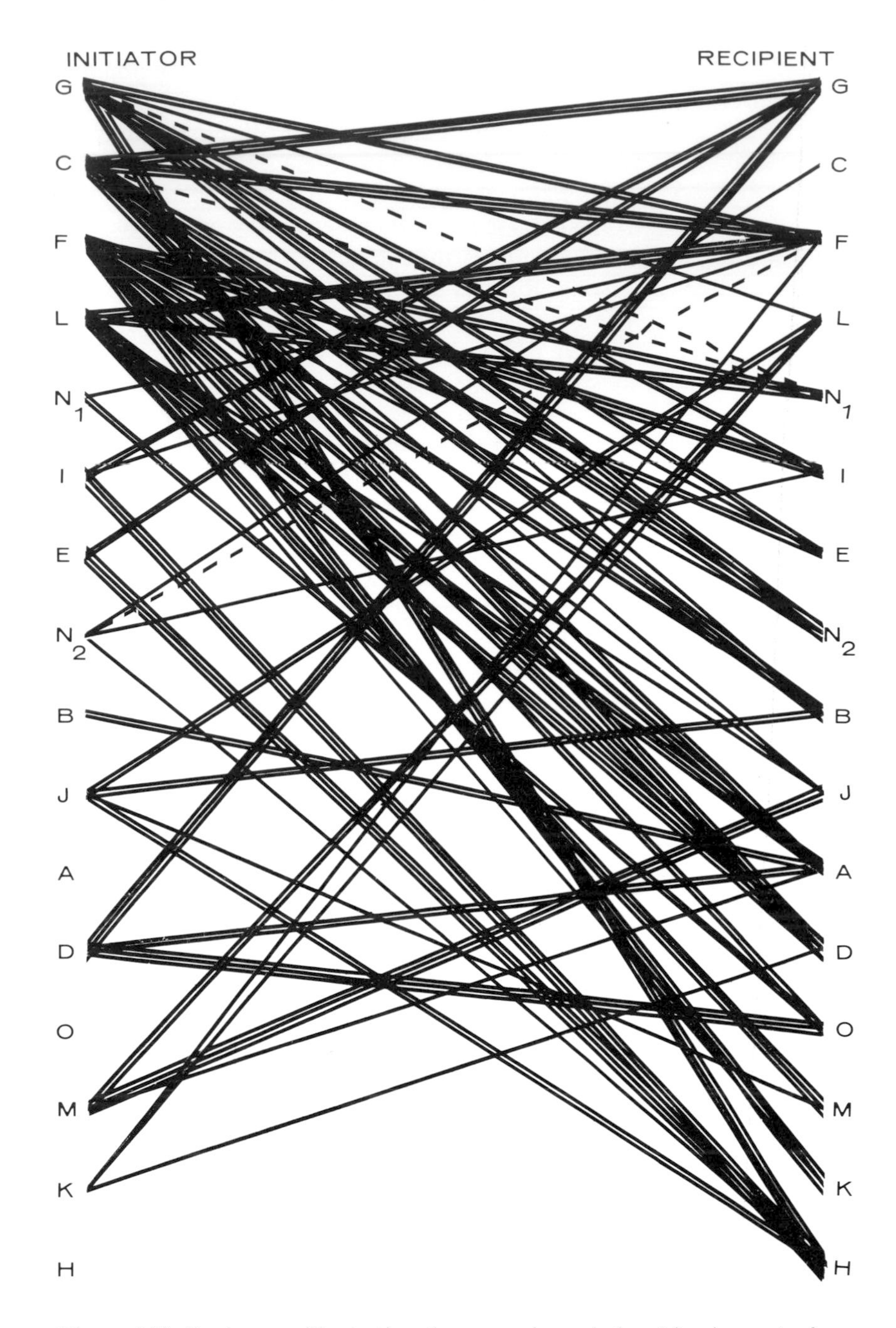

Figure 143. Sociogram illustrating the grooming relationships in an *Ateles geoffroyi* colony. Letters refer to the individual specimens. Left column indicates the groomer, right column designates the recipient of grooming. The number of lines indicates cumulative duration of grooming. From Eisenberg and Kuehn 1966.

mammals, all young are at first fed secretions from the mammary glands. Milk is the "cement" of the young mammal's first social relationship (Wilson 1975). The energy parents (generally the mother in mammals) expend in rearing the next generation is easily explained in terms of classical genetic paradigms. Since one defines fitness in terms of the reproductive success of the next generation, any behavior pattern contributing to the future success of offspring obviously increases the fitness of the parental genotype.

Fisher (1930) based his theoretical model of natural selection upon the life table. For any given breeding population, a life table can be constructed yielding information on life expectancy, the probability of reproducing in any time interval, and the number of offspring during various segments of the cohort's life-span. A breeding population is stable in numbers if the individuals it comprises leave at least one offspring each to replace them as breeders when they die. If the average recruitment per individual is less than one, the population is declining; if the recruitment rate is greater than one, the population is increasing. Genes ultimately affect survival and reproduction; hence any advantage gained by an organism with a given genotype will be conserved as an increase of its progeny—or, extending this thought, "fitter individuals leave more progeny." Relative fitness is thus defined in terms of the individual's contribution in numbers to succeeding generations.

Although Fisher's model is useful in predicting the spread of gene frequencies through time, it will run into difficulties when the trait in question is a behavioral mechanism that actually decreases the bearer's "fitness." In an effort to explain the evolution of neuter, nonreproducing castes in social insects, W. D. Hamilton (1964) proposed reconsidering the classical idea of fitness. By redefining fitness to include survival of gene frequencies rather than only progeny survival, and by proposing that behavioral mechanisms that preserve gene "frequencies identical by descent" in related individuals could themselves be selected for, Hamilton set the stage for Dawkins's book *The Selfish Gene* (1976).

The new model could account for an increase in average "inclusive fitness" for a population of related individuals. This new concept of fitness would then obey the same rules applied to Fisher's classical definition. One could now proceed to define how "altruism" evolves in populations composed of closed, multigenerational families. The probability of altruistic acts between individuals could in fact be correlated with the coefficient of relatedness between two individuals or among members of a group (Trivers 1971; Wilson 1975).

Hamilton also dealt with the fixation of behavioral traits that may be considered selfish or, in extreme cases, "spiteful" (Hamilton 1970, 1971). With the simplest of models, he was able to demonstrate how extreme forms of spiteful behavior, if genetically controlled, would soon lead to the extinction of small populations. Mildly selfish behavioral traits could be genetically fixed and, over the long run, would benefit individuals when buffered with alternative "altruistic" mechanisms. The whole theory and its applicability to studies of animal behavior and parental investment has been dealt with ably by Wilson (1975) and Trivers (1972, 1974) (see fig. 144).

Although kin selection is often invoked as the mechanism producing seemingly altruistic acts, several other mechanisms may underlie the expression of altruism. Indeed, the identification of such mechanisms and their evolutionary history should be a priority. Trivers (1971) outlined a theory of reciprocal altruism whereby an apparent individual loss of fitness to a partner may be compensated for by a later gain in fitness when the partner in turn performs an altruistic act. When this concept is broadened to involve nonrelatives, we approach the condition of mutualism, where two individuals cooperate because the mutual benefits outweigh the initial cost. Alternative views of animal cooperation and their origins are imperative now, when kin selection is being invoked everywhere to explain social processes.

One enduring controversy concerning the genesis of seemingly altruistic acts concerns parental control or manipulation of offspring. Some offspring help rear younger siblings at a reduction in the older offspring's fitness. Does this result from long term selection for an increase of inclusive fitness, or does it result from selection on the parent to reduce the fitness of some offspring so as to ultimately increase the fitness of the parental genotype by helping younger siblings survive to the age of reproduction? (See Alexander 1974.) Alexander offers alternatives to an evolutionary model of altruistic behavior derived from kin selection based solely on coefficients of relatedness. He demonstrates that some cases of altruistic behavior by progeny may in fact be forced by the parents (i.e., parental manipulation) because, by denying reproductive status to their first offspring (by withholding adequate food), the parent may guarantee the ultimate survival of subsequent offspring. The whole set of concepts has introduced a revolution in the studies of the form and rate of social interaction (West-Eberhard 1975).

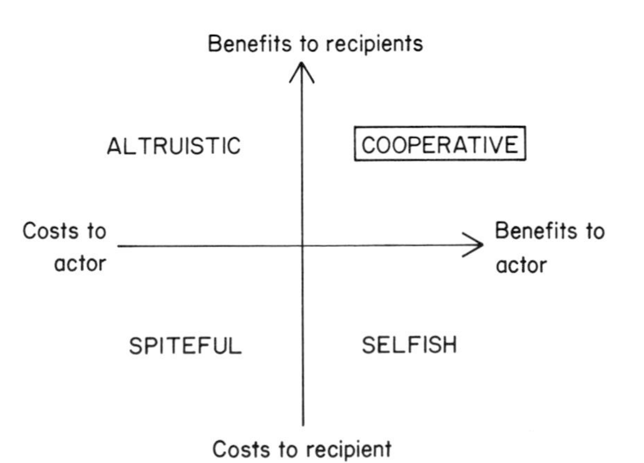

Figure 144. Relation between costs and benefits to donors and recipients that defines the terms "cooperative," "altruism," "spite," and "selfish" interaction forms. Adapted from Trivers (pers. comm.).

Parental manipulation of offspring, the evolution of altruism, and the utility of coefficients of relationship for predicting the form and rate of social interaction have been the subject of recent popular research on social organizations. Of course, sexually reproducing organisms such as mammals must form pairs for breeding, and the degree of relationship between a given male and female is usually small. Again and again we have seen that, even in the most socially cohesive species, one sex tends to disperse, thus promoting outcrossing rather than inbreeding (Harvey, Greenwood, and Perrins 1979; Greenwood, Harvey, and Perrins 1979). In some cases females rather than males appear to disperse (e.g., *Gorilla,* Harcourt 1979; *Lycaon,* Frame et al. 1979), but in most mammals males disperse.[1] Inbreeding coefficients seem to be maintained at moderately low levels in mammalian populations, and high inbreeding is usually accompanied by high infant mortality (Ralls, Brugger, and Ballou 1979).

Social organizations can be classified according to the form of affiliation during the mating phase: Is the pair relationship brief? Is the pair monogamous? And so forth. An adequate description of social structure requires an integrated understanding of the mating system and the parental care system and of their permanence through time. I shall approach the problem of classifying social structures in the last section. But before we pass on to this we must face squarely the reality of phenotypic variation.

Behavioral traits vary between individuals in strength and in frequency. A social organization (or a subset of a social structure, such as a mating system or a rearing system) can also vary depending on a number of factors (Leyhausen 1979, pp. 3–5). Part of the variation in social organizations of populations within a species may in fact result from genetic variation. Phenotypic variation may also result from a population's specific nongenetic adaptation to unique features of its environment. Thus, to generalize about the form of the mating or parental care strategy or the cohesiveness of the group over space, one must examine enough individuals or subpopulations to grasp the range of variation in the parameters under study. Once the range of variation has been described, one can think of behavioral "norms" or averages. By the same token, one can consider an abstract "normative social structure" for the normal range of environments in which a species may find itself. We should be careful to exclude those social structures that clearly are maladaptive reactions to peculiar environmental circumstances.[2]

1. In species that are strongly polygynous, male dispersal may involve elevated male mortality. Survival curves for the two sexes may have profoundly different shapes (Ralls et al. 1980).

2. Hendrichs (1978) approaches the same problem with a unique set of definitions. He theorizes that social stress within a population increases when the frequency of social contacts falls either above or below some species-specific optimum value. He defines an idealized population structure for a species as the "tonus organization." In such an idealized organization the optimal scope for interaction form and rate is realized.

If, for example, a unique variant in social structure is discovered but the population dies out because of the particular trait exhibited, then surely this behavioral response to the environment reduces fitness. The appearance of maladaptive traits should not muddle our thinking concerning the normal adaptation of the species to its average range of environments.

Another point to bear in mind is that a social structure is not static. In a cohesive social grouping of interrelated individuals, the size of the group and its age and sex composition will vary slightly over time. Only by studying social structure through time can one ascertain how much the size of the group changes and whether it tends toward a stable figure (see Dittus 1977*b*). Recognizing variation should not make one despair and say things like "It is impossible to define a social organization for a species." The task may be difficult, but even with a great deal of variation the normative social organization could still be defined. After all, the study of variability in morphological characters has proved extremely productive in elucidating specific adaptations and understanding the processes of evolution itself. We are making a beginning at a comparative science of animal sociology, and that is not easy. Comparing social structures within and between species frequently generates very useful insights. At best it may involve comparing analogues rather than homologues.

32 A Classification of Mammalian Social Organization

A mammal's social life may be divided into phases. A species may be highly social at one season of the year and not at another, and these social phases are often connected with definite functional categories of the animals' life's activities. In attempting to classify mammalian social organizations, it is convenient to consider four phases in terms of the degree of sociality shown: (1) the mating system, (2) the rearing system, (3) foraging behavior, and (4) the refuging phase, including not only diel sleeping but also hibernation, estivation, overwintering, and so forth. In examining all these life phases, it is useful to treat the social organization of males as one subset and that of females as another. The combined social structure can be viewed as an outcome of adjustments between the male and female subsets. In the following sections I consider terrestrial, nonvolant mammals. If sufficient data were available, one could approach the correlational problem for aquatic mammals and bats (see Eisenberg 1966 for an earlier attempt, and sections 14.3 and 15.1).

32.1 Mating Systems

Male mammals are generally not involved in direct parental care activities (see sect. 23.6). Males can increase their fitness by attempting to mate with many females, and as a result there is a strong tendency toward polygynous mating systems in the Mammalia.[1] For this reason, I have developed this section from a male perspective.

Let us consider the major variations of the mammalian mating systems. The first type involves a number of neighboring males who, on the basis of previous encounters, are ranked alpha, beta, delta, and so on, so that a vague dominance gradient is usually demonstrable. This group of neighboring males, either singly or together, but in an uncoordinated fashion, seeks out neighboring females at the time they come into estrus. Typically, both sexes are spaced in nature, though home ranges may overlap to some extent. We can discern two variants, one system exhibiting semiexclusive adult home ranges regardless of sex, and the other showing considerable home range overlap between opposite-sexed adults. Males begin courting a female, and priority of access is determined in part by the male hierarchy. Some female "choice" may be involved. In any event, preferential access to females is achieved by the dominant male, but the female may mate subsequently with more than one male. Though more than one male may

1. See Maynard-Smith (1978) for an extended discussion of the evolution of sex and its consequences.

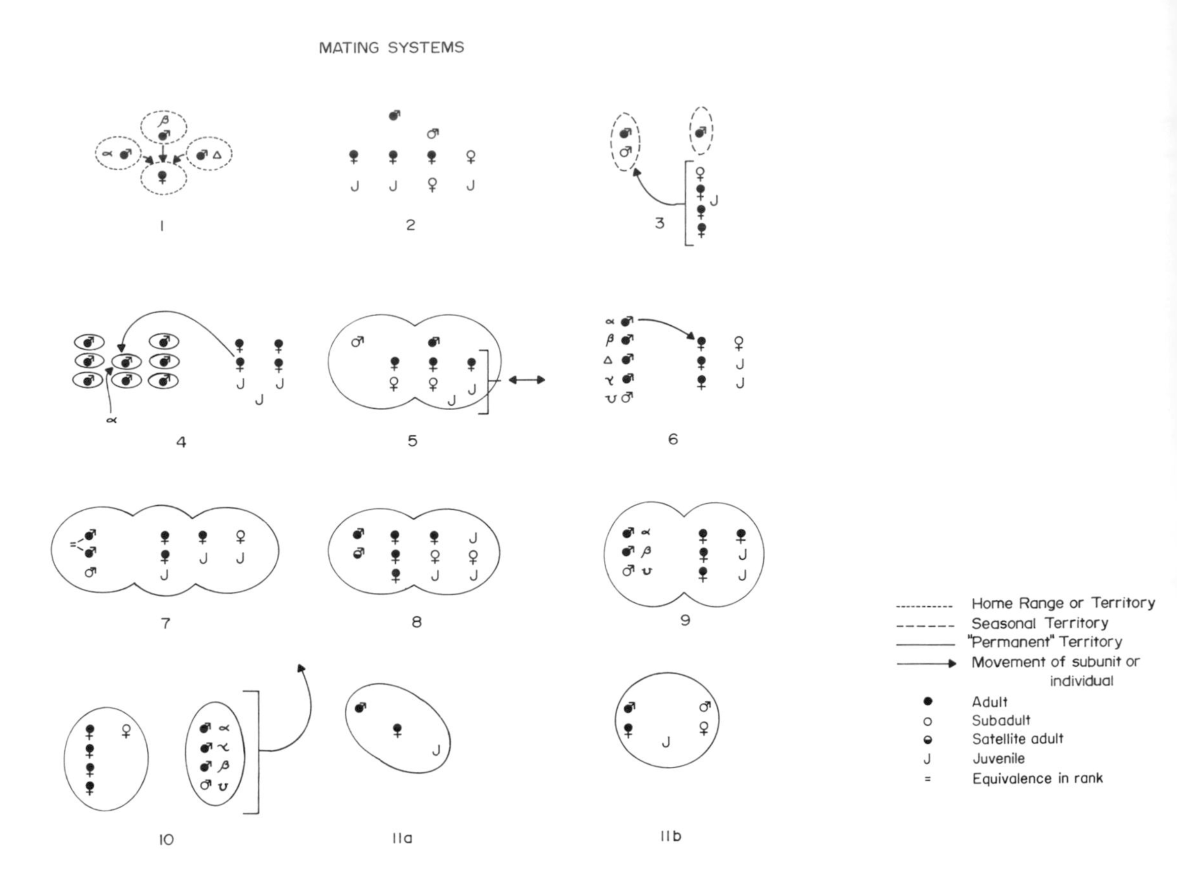

Figure 145. Diagram of mating systems as outlined in the text (see text for descriptions of systems 1 through 11). ● = adult; ○ = subadult; ◒ = satellite adult; *J* = juvenile; equals sign = equivalence in rank.

sire a litter, priority access usually ensures that the dominant male sires most of the offspring.[2] This type of system often involves a mating chase, and a typical example would be the eastern chipmunk, *Tamias striatus* (see Elliott 1978).

In the second mating system an adult male and occasionally a subadult male typically move about in a loose subunit. The adult male defends a group of females year-round. The male's defense is not necessarily restricted to one location. This typical polygynous harem system is typified by Burchell's zebra (*Equus burchelli*).

In the third system a single adult male and several satellite subadult males seasonally defend a spatial territory. Females may be attracted to the location of the male territory, and the adult male will assiduously defend them against rivals while they are on his defended area. The male will not follow the females when they wander, and, indeed, the cohesive female group may end up on a neighbor's territory. Such a system is typified by the red deer (*Cervus elaphus*) and the North

2. Where several males mate and conceive young, we are faced with a mixed polygynous and polyandrous strategy. We may number this variant 1*b* (see tables 58 and 59).

American wapiti (*Cervus canadensis*) (see Lincoln 1971; Struhsaker 1967*c*).

The fourth system is a variant of the preceding, where individual males establish and defend territories in a traditional area and are highly synchronized in their breeding season. The territories are extremely small, and no attempt is made to keep a group of females within the territory. Rather, females visit the male territorial grounds individually and generally mate with the male in the central position. This involves an active choice by the females and a rank order among males, since position on the territorial grounds, as well as the manifestation of territoriality, determines mating success. This system is referred to as a lek, and the classical example is the Uganda kob, *Adenota kob* (Buechner 1974).

The fifth mating system involves male defense of a territory over a period of several years. The territory is large enough to support a group of females. Although females are not necessarily restricted to the male's territory, they spend most of their time there, and the mating success of a territorial male is higher than that of a nonterritorial male. Although nonterritorial males may be tolerated within a territory, they do not breed. The classical example of this is the white rhinoceros (*Ceratotherium simum*), as studied by Owen-Smith (1975).

The sixth system involves a rank order of males, including both adults and subadults, where the top-ranking male tends to have priority access to females. The male group shows some cohesion, though this may break down at the time of mating. If the female group to which they attach themselves shows close synchrony in estrus, then the alpha male does not do all the breeding. On the other hand, if the synchrony is reasonably loose, the alpha male may do most of the breeding. This point is very important in considering a system where access to females is determined not on territoriality, but on a known rank order among a semicohesive group of males. Typically this is shown by the Rocky Mountain bighorn sheep (*Ovis canadensis*) (see Geist 1971).

In the seventh system typically two males exhibit near equality. There may be an attendant subadult male, and these males cooperatively protect their access to a group of females. Little dominance can be discerned among males, and the success of two males associated with a female group is greater than that of a single male. It is strongly suspected that the two males may be related. This system typifies the lion (*Panthera leo*), as described by Schaller (1972) and Bertram (1975).

The eighth system involves a single adult male, usually with an attached subadult satellite male, who defends an area containing a variable number of females, though the modal system consists of a single male and several females occupying a fixed defended area (e.g., *Cebus capucinus*) (Oppenheimer 1968).

The ninth system involves a male subgroup that is ranked hierarchically. The top male has some preference of access to estrous females. The male group controls access to a given female group. Permanent spacing is often effected through territorial defense. The male group may be age-graded, and this is typically demonstrted by the red howler monkey (*Alouatta seniculus*) (Rudran 1979). This system may be an outgrowth, in an ontogenetic sense, of system eight.

In the tenth system a ranked group of males may attach themselves in a semipermanent fashion to a female group. Priority of access to females is determined by the rank order of the males, but little cohesion is shown in the male group. A male and female briefly form a pair bond that dissolves after mating. What characterizes this system is that the females are dominant over the males and the males act as a satellite subunit to the female group. To date the only species known to typify this system is the spotted hyena (*Crocuta crocuta*) (see Kruuk 1972).

The eleventh system, monogamy, requires an extended description. Kleiman (1977*b*) reviewed the occurrence of monogamy in the class Mammalia. She defines monogamy as mating exclusivity and admits that exact proof of mating fidelity remains to be demonstrated for most species; however, enough field and captive data are on hand to let us guess at those species that may show exclusive mating over several breeding cycles. She defines two forms: type 1, or facultative monogamy, and type 2, which is obligate.

Facultative monogamy usually occurs when a species exists at such low densities that the home range of a male overlaps only that of a single female, whereas obligate monogamy occurs when aid from the male and often from older young or fraternal siblings of the alpha male is necessary to permit the female to rear her litter. This obligate, restricted system usually occurs when the carrying capacity within the home range of the group will not allow more than one female at a time to rear a litter.

Given these two types of monogamy, one typically finds a number of associated morphological and physical characters. First, the adults of monogamous species generally show little sexual dimorphism in either morphology or behavior. Furthermore, sociosexual interactions, although prominent during pair-formation, are relatively infrequent once the bonded period is established. In mammals showing obligate monogamy, the sexual maturation of the young usually is delayed while they are with the parents; this promotes breeding only by the parental stock. Very frequently the older juveniles help rear their young siblings, and usually the adult male also helps by a variety of mechanisms. The process has been most closely analyzed in captivity for some canids (Brady 1979) and marmosets (Hoage 1978).

Mating systems for well-studied mammalian species are presented in tables 58 and 59. Regardless of the variation I have outlined and diagramed (see fig. 145), the systems fall into three fundamental groups. (*a*) Systems 1*b* and 7 are polygynous in the sense that a given male conceives young with more than one female. Furthermore, since a female may mate with several males during a given estrous period, and since the females of the species so characterized are multiparous, we simultaneously have a form of polyandry. If the females were uniparous, then this polyandrous condition would not pertain unless one considered a series of successive breeding cycles. (*b*) Systems 11*a* and 11*b* are monogamous, and the parental investment of the male tends to be tied to a single female (Kleiman 1977*b*). (*c*) All other systems are essentially polygynous, with the male maximizing his fitness by mating with as many females as possible.

32.2 Sexual Dimorphism and Polygyny

As I noted above, monogamous species tend to be monomorphic. This observation leads to the whole question of sexual dimorphism. In many mammalian species adult males and females differ in size. Usually the male is larger than the female. Ralls (1976) analyzed cases where adult females are typically larger than males and speculated upon the selective processes that might produce such differences. Since the male usually is the larger, explanations for sexual dimorphism in mammals usually center on sexual selection where male combat determines the direction of dimorphic traits. Since female mammals provide the early nutrition of the young through lactation, and since females make the primary parental care investment, males could increase their fitness by mating with more than one female, thereby spreading their genes among a larger number of females. Females then become the limiting resource for males, and males would actively compete for access to females. Such male-male competition could lead to selection for larger and stronger males who would win more fights, thereby promoting the size dimorphism and providing a link between polygynous mating and larger size in males. This is especially true where males compete within a restricted time frame and can recoup energy expenditure during an off-season. As a word of caution, however, year-round harem maintenance may be so energetically costly that males cannot afford to direct energy into maintaining large weapons (horns, tusks, or antlers) or large size. Such permanently territorial or harem-holding males may be either monomorphic or smaller than females (Raedeke 1979).

By analyzing cases where females are larger than males, Ralls pointed out that different selective pressures may act on females where it is advantageous for them to reach a larger size. In many species of bats the female is larger, perhaps to improve flight efficiency by reducing the weight of the young relative to her own body mass. In mysticete cetaceans, the increased size of the female might be explained as a way to offset the energy drain caused by the rapid growth of the fetus and by lactation. Selection may also promote larger mothers if large size maximizes their reproductive efficiency by producing larger babies, which presumably have a higher survival rate.

Ralls (1977) considered strong sexual dimorphism in mammals where the male is larger in the light of theories of size dimorphism in birds. She points out that most models for the evolution of polygyny and subsequent sexual dimorphism are derived from studies of galliform birds and notes that when the male mammal helps rear the young sexual dimorphism tends to be vastly reduced. The reverse, however, where there is small male parental investment, is not necessarily a good predictor either of sexual dimorphism or of polygyny. An intriguing factor that she identifies is that polygyny has evolved more frequently in larger mammals, and that the larger a mammal is the greater the relative degree of size dimorphism. Sexual selection is thus shown to be inadequate as a total explanation not only for dimorphism but also for polygyny. Data on relative size dimorphism are included in tables 58 and 59 for fifty-nine species.

32.3 Rearing Systems

Rearing of the young is, in its early phases, tied to lactation and thus inevitably to the female. Rearing systems are then most easily classified by the role of the female. The following ten rearing systems can be discerned:

1. A single adult female living on a home range that is more or less exclusive and raising her young alone. The classical example for this is the red squirrel, *Tamiasciurus hudsonicus* (Smith 1966).

2. An adult female who accomplishes most of the maternal care tasks by herself, though her home range is contained within the territory of a male. This may approximate "facultative monogamy" as defined by Kleiman (1977*b*). It is typified by the elephant shrew, *Elephantulus rufescens* (Rathbun 1979).

3. A cohesive female unit that rears young in common, though there is minimal mutual support between females (*Gazella dorcas*). A male may be seasonally attached to the group.

4. In a variant of the preceding, females and their attendant offspring of various ages show some cooperation in antipredator behavior and perhaps in foraging. Males may (*Cebus capucinus*) or may not be attached to the group, but if they are they do not significantly take on parental care burdens (*Elephas maximus*).

5. A single adult male and a single adult reproducing female with various-aged offspring that cooperate in various ways to share the costs of rearing the offspring (e.g., *Canis lupus*). This is obligate monogamy as defined by Kleiman (1977*b*).

6. A variant of the preceding where offspring of previous years may be retained within the family group for a protracted period and do not breed in this context. The "helpers" contribute to the rearing of young siblings. This could also be termed a form of obligate monogamy (e.g., *Saguinus oedipus*).

7. A group of females that show some cooperation in foraging but are primarily responsible for the individual rearing of their own young. A group of males may be attached on a semipermanent basis. This rearing system is typified by the lion, *Panthera leo*.

8. A group of breeding females who are approximately equal in rank, with an attached subgroup of ranked males who do not significantly contribute to the early rearing of the young. This is typified by the savanna baboon, *Papio anubis*.

9. A cohesive group of nearly equivalently ranked females with an attached subgroup of males who are age-graded and who do not directly contribute to the initial rearing of young animals. This is typified by the spider monkey, *Ateles geoffroyi*.

10. A cohesive group of males and females. The subset of males does not directly aid the females, and the females do not directly aid one another. This is typified by some of the larger herding ungulates, such as the Cape buffalo, *Syncerus caffer*.

Tables 58 and 59 and figure 146 summarize the rearing systems displayed by some fifty-nine well-studied mammalian species. The ten rearing systems can be reduced to four categories based on the degree of aid the female receives from other adults. (*a*) In the simplest case, the female rears her progeny with no aid from other adults, system 1. (*b*) Where females form groups but the only benefit derives from individual responses to predators that may indirectly alert others, we have the classic example of the "selfish herd," systems 3 and 10. If females are long-lived and form groups that have traditional foraging patterns, other benefits can accrue from grouping (system 7). (*c*) Where the females are related, the stage is set for kin selection and some altruistic acts that increase the inclusive fitness of female clan members (systems 4, 8, and 9). Bringing the male into the system to provide indirect aid to the female and young (system 2) is a variant that becomes more pronounced (*d*) in species showing obligate monogamy (system 5), and, finally, when nonreproducing kin members actively provision the reproducing female we have reached the acme of assistance (system 6).

32.4 Foraging Systems

We can distinguish at least eight foraging units based on the degree of cooperation among group members. In the first foraging system, we consider a male alone defending an area for his exclusive use or use by a female and young. The second form includes a cohesive group of

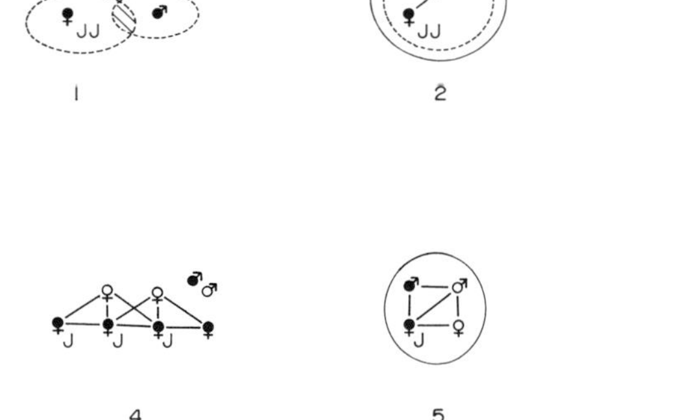

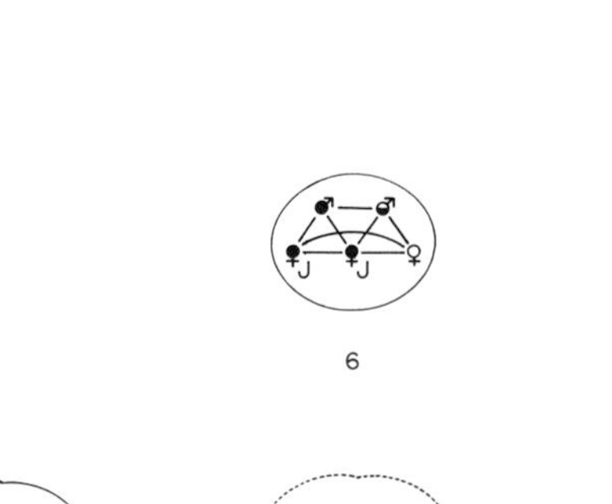

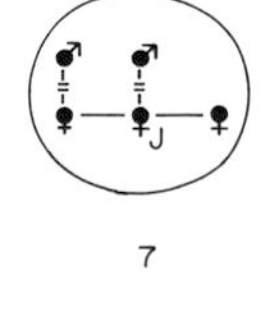

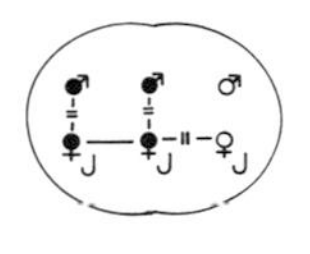

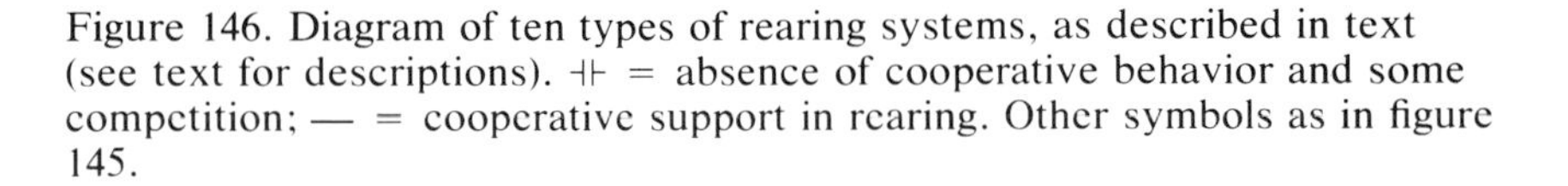

Figure 146. Diagram of ten types of rearing systems, as described in text (see text for descriptions). ⊣⊢ = absence of cooperative behavior and some competition; — = cooperative support in rearing. Other symbols as in figure 145.

males that feed individually. The third variant is a male and a female jointly occupying a home range that may constitute a territory, but their foraging habits are individualistic and they show no cooperation. The fourth variant is the female alone or with young foraging exclusively in a defended area. The fifth system is a varying number of females that forage individualistically with little mutual aid; varying subsets of males may be associated with the group but do not contribute directly to feeding efficiency. The sixth type is a group of females with a varying number of associated males, but the females show some coordination in their foraging activities that provides mutual aid. The seventh system is a variant of the preceding system where mutual assistance may be given in rearing juveniles but the adults and subadults do little to aid one another in foraging. This form of foraging and the following type are closely tied to monogamous mating systems. The eighth foraging unit is a mixed group of males and females that feed individually but cooperate in provisioning young animals (see fig. 147).

Ultimately our eight foraging systems can be reduced to three major forms: individualistic foraging where at the most juveniles are allowed to feed near an adult (systems 1 through 5); systems where sharing between two adult females can result in increased survival of their offspring (system 6); and systems where food is actually transferred from adults and subadults to juveniles (systems 7 and 8). Such provisioning of young is characteristic of monogamous species and has been noted for canids and marmosets (see Kleiman 1977*a*).

In foraging strategies, certain amounts of competition are almost always discernible. The key trend appears to be an increase in assistance to a female and her dependent progeny. Assistance may come from older offspring, fathers, and relatives of the father or mother. It is in the analysis of foraging strategies that the degree of cooperation appears to correlate most strongly with coefficients of relatedness (Malcolm 1979). Foraging involves not only actual provisioning of the young but also some form of cooperative antipredator behavior that frequently correlates with relationship by descent (Sherman 1977; see

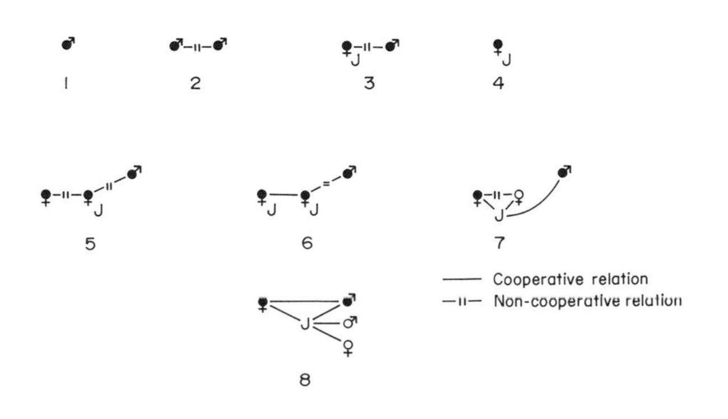

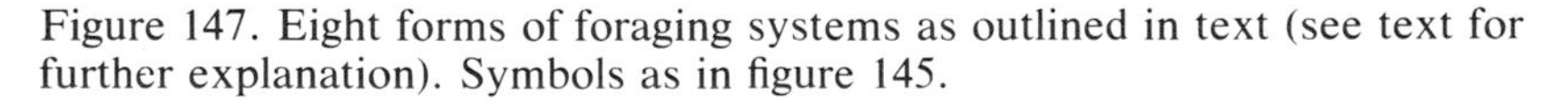
Figure 147. Eight forms of foraging systems as outlined in text (see text for further explanation). Symbols as in figure 145.

sect. 32.6). If the hypotheses concerning the increase in fitness as a function of selection acting on gene frequencies "identical by descent" has any validity, it is in an analysis of foraging and antipredator strategies that the theory will be confirmed or rejected.

32.5 Refuging Systems

The refuging systems may show a classification parallel to that of the foraging systems. This classification scheme is portrayed in figure 148. Refuging strategies tend to be strongly phasic. In the simplest case, the rearing requirements for the young may be so drastically unlike the conditions necessary for adult survival that in polygynous species male and female refuging areas may be radically different (see the life history of *Myotis lucifugus*, sect. 14.3.2). In monogamous species with direct male parental care, the refuging place will be shared by the family (example 2, fig. 148). In a permanently defended harem, the refuging place may hold a male and several females. In species exhibiting a more or less permanent social life, a single refuging area may be spatially subdivided by the age- and sex-classes (examples 4 and 5, fig. 147). Such segregation may even be shown within a single tree when primates rest for the night (Dittus 1977*b*) (see fig. 148).

32.6 Predatory Behavior and Social Structure

Although mammalian populations are ultimately limited by scarce resources such as shelter and food, predators apply a constant and significant pressure on most species. Falcons prey on bats at twilight, snakes feed on young mice in nests, Cape hunting dogs attack wildebeests, and even elephant calves are occasionally taken by tigers. Predators shape color patterns, morphology, and behavior in a variety of ways. Warning cries and their evolution through kin selection have received much attention in recent years, and a convincing case for kin selection has been made for Belding's ground squirrel (Sherman 1977). Some forms of social behavior are shaped by predators, though not all

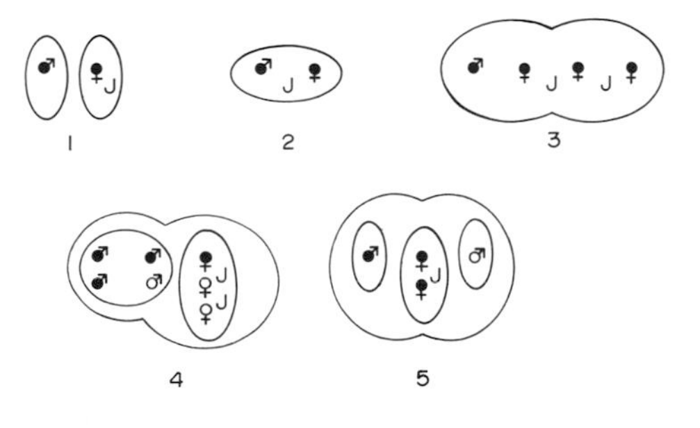

Figure 148. Five types of refuging systems: (1) Separate refuging areas for the adult male, adult female and juveniles. (2) Shared refuging system for adult male, adult female, and juveniles. (3) Shared refuging system for adult male and polygynous group of females with young. (4) Separate refuging systems for a female rearing unit and a male subgroup, though the two subsystems may be considered a single social system in terms of interaction form. (5) A variant of the previous refuging systems with separate refuging by males and females. Symbols as in figure 145.

forms of antipredator behavior evolve by kin selection. Smythe (1970*a*) discusses the idea that the "pursuit invitation" strategies demonstrable in cursorial caviomorph rodents and ungulates are a result of individual selection. In these cases a predator is alerted through a specific signal that it has been discovered by the intended prey; further stalking is thus discouraged and the predator realizes sustained pursuit would be fruitless.

Forming a herd while feeding in the open is common in some species of ungulates. Herd formation may reduce vulnerability to predation for individuals in the center, and selection for the herding tendency may involve individual selection for a mildly selfish trait (Hamilton 1971). On the other hand, active defense of young against predators by a parent or a kin group may involve true altruism, and such traits are simple derivatives of kin selection or individual selection (see plate 46).

Defense against predators may involve a long process of selection for protective structures such as spines. The evolution of spines as an antipredator mechanism has been traced within the tenrecoid insectivores (Eisenberg and Gould 1970). Cryptic behaviors are generally the product of predator selection. Indeed, they may involve selection that disfavors the formation of social groupings and thus contributes to a solitary foraging strategy by mammals that have convergently evolved to exploit similar habitats (Eisenberg and McKay 1974). The relationship between social organization and antipredator behavior is often very subtle and exceedingly complex. Indeed, the form of antipredator behavior for a given prey species may alter depending on age, sex, topography, and, most important, the predator itself. For example, Axis deer (chital) are not vulnerable to predation by jackals (*Canis aureus*) as adults but are extremely vulnerable as fawns. An adult female chital may face down or distract a jackal pair to protect her hidden fawn. Adult chital are vulnerable to predation from the leopard (*Panthera pardus*), and females are more vulnerable than males. Yet

Plate 46. *Sus scrofa*. This sounder of females with their collective young illustrates a highly adaptive antipredator strategy. Note the position of the young in the center of the moving group. Active antipredator behavior on the part of the females and subadult males serves to protect the vulnerable newborn from attack by predators. (Courtesy of C. M. Hladik.)

a small herd of female chital may attempt to mob a leopard if the animals are in the open and the leopard is moving away from them (see fig. 149). Calculating a "vulnerability quotient" for a prey animal thus must involve at the least a consideration of maturation phase, habitat, and predator.[3]

This leads us finally to the other side of the predation question—the influence of habitat type and prey on the hunting strategies open to a predator. Kleiman and Eisenberg (1973) compared the carnivore families Canidae and Felidae and suggested that, while phylogenetic inertia is demonstrable in both lines, group hunting is a more prevalent strategy within the Canidae than within the Felidae. They suggest that the deep-seated tendency for monogamy in the Canidae, where the male provisions the female during the rearing stage (Kleiman and Brady 1978), sets the stage for selection to promote a hunting strategy based on family group pursuit of cursorial prey. Highly structured social groups that hunt cooperatively are characteristic of *Cuon alpinus* (Johnsingh 1979), *Canis lupus* (Mech 1970), and *Lycaon pictus* (Malcolm 1979) (see plate 47). The influence of predation strategy on group formation

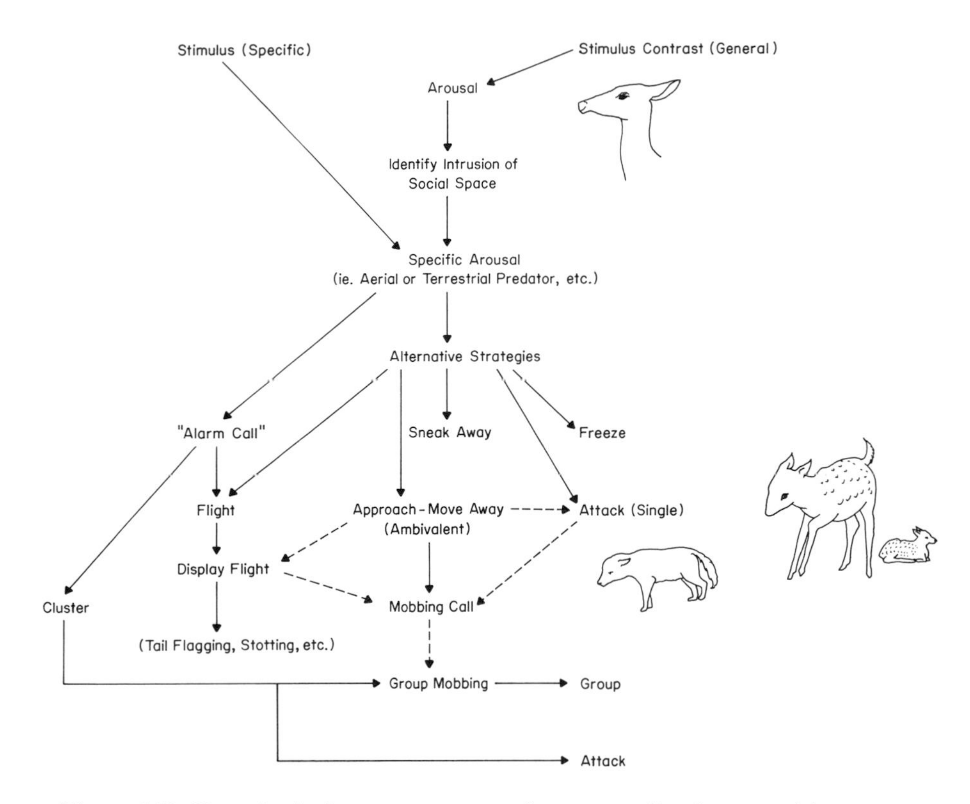

Figure 149. Hypothetical response system for a generalized mammal in response to a predator attack. Not all options are shown by all mammals. From Eisenberg and McKay 1974.

3. The preceding discussion is based on observations in Sri Lanka (Eisenberg and Lockhart 1972). The vulnerability of prey species may be quite different in peninsular India, where a different guild of large predators is present (Johnsingh 1979).

Plate 47. *Cuon alpinus*, a pack-hunting canid of South Asia, typifies the major predator for ungulates of the subcontinent. (Photo by A. J. Johnsingh.)

is classified and demonstrated in the diagrams of figure 148 and discussed in section 32.3.

32.7 Summary

A perusal of tables 58 and 59 will indicate several major trends among the fifty-nine mammalian species I have selected for comparison. Polygyny is the dominant mating system (82 percent) and of this the polygynous/polyandrous variant accounts for 12 percent. Obligate monogamy is rare but accounts for 12 percent. Facultative monogamy involves three cases. In rearing strategies, the tables include twenty-eight species where the adult female rears her young alone. Matriarchal organization with some aid in rearing among females accounts for 10 percent of the set; 12 percent of the species are characterized by direct aid from the male or from the male and older offspring or relatives. Foraging strategies involving essentially individualistic patterns account for 82 percent of the species, but 18 percent either show tolerance for young (10 percent) or directly provision young (8 percent).

The monogamous extended families show the clearest cases of cooperative rearing and food sharing. Members of matriarchal groups give an intermediate level of aid to offspring other than their own. Actual group defense against predators is also an attribute of matriarchal organizations or monogamous families. Mutualism (reciprocal altruism) or altruism through kin selection is implied.

Table 58 Some Correlations between Social Systems and Dimorphism for Monotremes and Marsupials

Taxon	Social Systems				Antipredator Strategies	Size Dimorphism	Social Complexity Rating ( )[d]	Reference[a]
	Mating	Rearing	Foraging	Refuging				
Monotremata								
Ornithorhynchus anatinus	1a	1	1, 4	1	Individualistic[b]	♂ > ♀	2	Burrell 1927; Griffiths 1978
Tachyglossus aculeatus	1b (?)	1	1, 4	1	Individualistic	♂ ≃ ♀	2	Griffiths 1968
Marsupialia								
Marmosa robinsoni	1a	1	1, 4	1	Individualistic	♂ > ♀	2	O'Connell 1979
Didelphis marsupialis	1a	1	1, 4	1	Individualistic	♂ > ♀	2	O'Connell 1979
Antechinus stuarti	1a	1	1, 4	1	Individualistic	♂ > ♀	2	Wood 1970
Dasyuroides byrnei	1a	1	1, 4	1	Individualistic	♂ ≥ ♀	2	Aslin 1974
Sminthopsis crassicaudata	1a	1	1, 4	1	Individualistic	♂ ≃ ♀	2	Ewer 1968*a*
Sarcophilus harrisii	1a	1	1, 4	1	Individualistic	♂ > ♀	2	Guiler 1970*a,b*
Perameles nasuta	1a	1	1, 4	1	Individualistic	♂ ≃ ♀	2	Heinsohn 1966
Trichosurus vulpecula	1a	1	1, 4	1	Individualistic	♂ ≃ ♀	2	Dunnett 1962
Petaurus breviceps	2	1, 2	1, 4	2	Individualistic but warning cries	♂ ≃ ♀	6	Schultze-Westrum 1965
Phascolarctos cinereus	2	1	1, 4	1	Individualistic but warning cries	♂ > ♀	2	Eberhard 1978
Vombatus ursinus	1a	1	1, 4	1	Individualistic	♀ > ♂	2	Wünschmann 1970
Potorous tridactylus	1a	1	1, 4	1	Individualistic	♂ ≃ ♀	2	Heinsohn 1968
Dendrolagus matschiei	2	1	1, 5	1	Individualistic	♂ > ♀	9	Collins, unpublished
Macropus giganteus	6	3	5	n.a.	Individualistic but "selfish herd"	♂ > ♀	15	Kaufmann 1974*b*
Megaleia rufa	6	3	5	n.a.	Individualistic but "selfish herd"	♂ > ♀	15	Kaufmann 1974*b*
Macropus parryi	6	3	5	n.a.	Individualistic but "selfish herd"	♂ > ♀	15	Kaufmann 1974*b*

[a] All species except *Ornithorhynchus*, *Perameles*, *Phascolarctos*, and *Macropus parryi* were also studied in captivity by the author and associates.
[b] Individualistic includes female defense of her own young.
[c] n.a. = category not applicable for that species.
[d] See text for scaling.

Table 59 Some Correlations between Social Systems and Dimorphism for Eutherians

Taxon	Social Systems				Antipredator Strategies	Size Dimorphism	Social Complexity Rating	Reference
	Mating	Rearing	Foraging	Refuging				
Edentata								
Bradypus infuscatus	1a	1	1, 4	1	Individualistic[a]	♂≃♀	2.0	Montgomery and Sunquist 1978
Tamandua tetradactyla	1a	1	1, 4	1	Individualistic	♂≃♀	2.0	Montgomery and Lubin 1977
Pholidota								
Manis tricuspis	2, 1a	1	1, 4	1	Individualistic	♂≃♀	2.0	Pages 1976
Macroscelidea								
Elephantulus rufescens	11a	2	1, 4	1	Individualistic	♀>♂	3.0	Rathbun 1979
Rodentia								
Cynomys ludovicianus	5	3	5	3	Group warning	♂>♀	6.0	King 1955
Tamiasciurus hudsonicus	1b	1	1, 4	1	Individual warning cries	♂≃♀	2.0	C. Smith 1968
Tamias striatus	1b	1	1, 4	1	Individual warning cries	♂≃♀	2.0	Elliott 1978
Castor canadensis	11b	5	7	2	Group warning sounds	♀>♂	4.0	Wilsson 1971; Morgan 1868
Microtus pennsylvanicus	1b	1	1, 4	1	Individualistic	♂>♀	2.0	Getz 1962; Madison 1978
M. ochrogaster	11b	5	1, 4	2	Individualistic	♂≃♀	3.8	Thomas and Birney 1979
Peromyscus polionotus	11b	5	1, 4	2	Individualistic	♂≃♀	3.8	Blair 1951
P. maniculatus	1a	1	1, 4	1	Individualistic	♂>♀	2.5	Eisenberg 1962
Dasyprocta punctata	11a	2	1, 4	1	Individual warning cries	♂≃♀	2.8	Smythe 1978
Dolichotis patagonum	11a	2	1, 4	1	Individual stotting display	♂≃♀	2.5	Dubost and Genest 1974
Insectivora								
Sorex araneus	1a	1	1, 4	1	Individualistic	♂≃♀	2.0	Crowcroft 1957
Talpa europaea	1a?	1	1, 4	1	Individualistic	♂≃♀	2.0	Godfrey and Crowcroft 1960
Hemicentetes semispinosus	8, 1a	3	1, 4	5	Individualistic	♂>♀	4.0	Eisenberg and Gould 1970
Carnivora								
Canis lupus	11b	5	8	1 Rearing den	Group defense	♂>♀	6.7	Mech 1970
Nasua nasua	5	4	4	1, 4	Group defense	♂>♀	6.0	Kaufmann 1962; Russell 1979
Panthera leo	7	7	6	n.a.[b]	Some group defense by ♀♀	♂>♀	6.4	Schaller 1972; Bertram 1975
P. pardus	1a, 2	1	1,4	n.a.	Individualistic	♂>♀	2.0	Muckenhirn and Eisenberg 1973

Table 59—*Continued*

Taxon	Social Systems				Antipredator Strategies	Size Dimorphism	Social Complexity Rating	Reference
	Mating	Rearing	Foraging	Refuging				
Lycaon pictus	11b	6	8	n.a.	Group defense	♂≃♀	7.5	Frame et al. 1979; Malcolm 1979
Crocuta crocuta	10?	8	7	n.a.	Group defense, individualistic	♀>♂	6.8	Kruuk 1972
Ursus americanus	1a	1	1, 4	1 Rearing den	Individualistic	♂>♀	2.0	Rogers 1977
Primates								
Microcebus murinus	1a, 2	1	1, 4	1	Individualistic	♂>♀	2.5	Martin and Charles-Dominique 197
Saguinus oedipus	11b	6	8	2	Group defense	♂≃♀	7.0	Dawson 1977
Cebus capucinus	8	4	6	5	Group defense	♂>♀	12.0	Oppenheimer 1968
Alouatta seniculus	9	9	6	5	Group defense	♂>♀	12.0	Rudran 1979
Macaca sinica	9	8	6	5	Group defense	♂>♀	18.0	Dittus 1977*b*
Hylobates lar	11b	6	8	2	Group defense	♂≃♀	5.0	Ellefson 1974
Pan troglodytes	6	3	5	4, 5	Group defense	♂>♀	12.0	Goodall 1968
Gorilla gorilla	8	4	6	5	Group defense	♂>♀	12.0	Schaller 1963
Proboscidea								
Elephas maximus	6	4	6	n.a.	Group defense	♂>♀	>20.0	McKay 1973
Perissodactyla								
Equus burchelli	2	3	5	n.a.	Group defense	♂≃♀	8.0	Klingel 1967
Ceratotherium simum	5	1	4	n.a.	Individualistic	♂>♀	3.2	Owen-Smith 1975
Artiodactyla								
Sus scrofa	5	4	4	1 Rearing den	Group defense	♂>♀	8.0	Eisenberg and Lockhart 1972
Cervus canadensis	3	3	5	n.a.	Individual alarm calls	♂>♀	8.3	Struhsaker 1967a
Adenota kob	4	3	5	n.a.	Individual alarm calls	♂>♀	>20.0	Buechner 1974
Gazella thomsoni	3	3	5	n.a.	Individual stotting display	♂>♀	8.0	Walther 1968
Syncerus caffer	6	10	5	n.a.	Group defense	♂>♀	>20.0	Sinclair 1977
Ovis canadensis	6	3	5	n.a.	Individual alarm signals	♂>♀	8.0	Geist 1971

[a] Individualistic includes female defense of her own young.
[b] n.a. = category not applicable for that species.

33 Toward a Definition of the Selective Forces Producing the Different Grades of Sociality

The classification of social systems was outlined in the previous chapter. The social organization I describe for a species is an integration of the various functional phases outlined there. Four possible subsets of social systems shown during a life cycle are portrayed in figures 145–46. In a fully social species showing great cohesion, the same grouping system may obtain throughout the life history. The organization of individuals in space and time results from the forms and rate of interactions among them, which may be represented as a sociogram. Variation in interaction rate and form, once age- and sex-class differences have been corrected for, can result from ontogenetic differences or from differences in the genome (Bekoff 1977). In practice it is difficult to factor out the causes of this variation.

When species are compared, differences in social organization are often evident. We can think in terms of "modal" social organizations that reflect a population's adaptation to a "typical" environment. Social organizations thus appear to evolve, but natural selection is acting on the individuals that compose the society, and if social organizations change over time it is the result of selection on individuals in the succeeding cohorts of the population. (For a fully developed argument for group selection, see D. S. Wilson 1980.) An alternative to this model is to reduce the real complex that is selected to the functional units of the genome (the replicators in the sense of Dawkins 1976). For our purposes, however, the society appears to evolve because selection acts on the bearers of the genome—the individual reproducing members of the population. In mammals the evolution of sterile, permanently nonreproducing castes is not yet demonstrable, so that in a sense one can assume that all individuals have the potential at birth of becoming reproducing members of the population. Of course in practice the different genetic endowments confer different degrees of "fitness" for an environment, and this equality is never realized.

Given that we can define modal social organizations for a species, I have sought correlations among behavior, morphology, the resource exploitation system, and the distribution of resources in space and time. To find replication in the patterns, I propose that we compare the Metatheria and the Eutheria, two lines of descent that represent a natural experiment (see chap. 2).

The rating of sociability (Σ) is based on the average-sized social unit of interacting individuals. It is somewhat arbitrary. Two is set as the minimum and represents the basic "solitary" system—male and female during mating, and the female-young unit. Four to eight is generally set for an extended family, while higher numbers represent mean troop or herd size. The scalings are indicated in the last data columns of tables 58 and 59.

Tables 60 and 61 compare the mobility of the various species, their reproductive strategies, the tendency for reproduction to be seasonally synchronous, the defense of resources, and the temporal and spatial distribution of resources. My definitions require further explanation.

To help correlate the selective forces molding the social systems, I have developed a simple rating for the defensibility of resources—high, moderate, or low. The resource categories are refuge area, feeding area, caching area, social helper, and mate(s). A "social helper" is a group member who increases the fitness value of the reproductive "founders" of the group by helping rear siblings. The remaining terms are self-explanatory. This scheme is an expansion of the ideas developed by Orians (1969).

Behavioral and morphological attributes of the various social systems may be further correlated by considering potential mobility. Mobility must be defined relative to the actual size of the animal. Volant and aquatic forms, regardless of size, usually are reasonably mobile with respect to resource dispersion. Nonvolant or nonaquatic forms are less mobile, but absolute mobility usually correlates with size (refer to the home range area/body size correlation, fig. 11).

Relative mobility as used here is the ability to move in terms of the animal's absolute size. A high dispersion of resources and a low potential mobility may require static defense of the resource, if the resource base can tolerate sustained exploitation. In the rating scale, an animal exhibiting a low relative mobility, such as the three-toed sloth, *Bradypus*, is given a value of 0.5. A primary herbivore that maintains a small home range and has little mobility and a relatively static distribution of resources, such as *Microtus pennsylvanicus*, is given a value of 1. By contrast, an insectivore such as the shrew, *Sorex araneus*, which moves a great deal relative to its body size in the course of foraging, receives a value of 2. Species showing great mobility in the pursuit of prey, with a correspondingly large home range, such as the Cape hunting dog, *Lycaon pictus*, are given a value of 3. Species that show seasonal migrations and utilize widely dispersed resources at different seasons of the year, such as the Cape buffalo, *Syncerus caffer*, receive a high rating of 3.2.

Reproductive strategies may be considered from two standpoints: (*a*) degree of iteroparity and (*b*) degree of reproductive investment. Iteroparity is rated on a scale of 1 to 5 for each sex. A species with an extreme tendency toward a semelparous system (e.g., *Marmosa*) will score 1.5 for females and 1.6 for males. An iteroparous species such as *Elephas* will yield a score of 4.5 for males and 5 for females.

Investment strategies are rated on a scale of 1 to 5. A score of 5 indicates a species whose reproductive strategy favors the litter's survival over that of the parent (e.g., *Sorex araneus*).

A score of 1 indicates a species that favors parental survival over the survival of offspring (e.g., *Elephas*). It should be obvious that the degree of iteroparity shown by a female and the investment indexes are mirror images, but it is useful to view the problem from these two perspectives.

A semelparous mammal has a large mean litter size and produces only one or two litters in a lifetime. It achieves the same continuity of genotype as an extremely iteroparous mammal. The iteroparous form spreads its energy for reproduction over many years (see chap. 23). All of the preceding can be summarized by the concept of "r," or rate of increase. A high r value in terms of reproduction in absolute time often reflects a semelparous tendency. As we saw before, a high r correlates to some extent with size but is negatively correlated with potential longevity (see chap. 23). Values of r are scaled from 1 to 7. The scaling is based on the mean number of young produced per female per year (see Appendix 4). A rating of 1 is equivalent to 0.25 young/♀/yr; a rating of 7 is equivalent to >30 young/♀/yr. A species with a value of 7 may approximate a semelparous mode of reproduction most strongly, whereas a value of 1 indicates a species with extreme iteroparity.

Given, then, the definition of the reproductive and investment parameters, how can we relate these to resources in the environment? I have considered three methods of measurement: seasonal flux in resources, predictability of the seasonal flux, and patch and grain of resources. Seasonal flux in resources is broken into three categories: (1) strong flux, (2) intermediate flux, and (3) relatively little seasonal flux. For each class of temporal resource distribution, the resource base may be highly predictable (a) or unpredictable (b). Thus the code 1a refers to a seasonal oscillation in resource abundance with a high predictability (see fig. 150).

The concept of patch richness and dispersion in space (grain) is borrowed from Pielou (1969). Figure 150 illustrates the five-level rating scale by plotting patch richness against the density of simultaneously available patches (as modified from Bradbury and Vehrencamp 1977*a,b*).

Seasonal availability of resources affects the population synchrony in terms of reproduction. Such synchrony may have profound effects on the mating system. The patch and grain of resources as related to mobility will determine the defensibility of classes of resources. Ultimately these environmental determinants affect the social structure. Tables 60 and 61 bring the data into a format that can be easily scanned. Figures 151 through 154 illustrate the major correlations. All these rating scales are to some extent subjective, and this presentation is not to be taken as a numerical exercise. Rather, it presents hypotheses. As we progress, one will note that the major trends are replicated in the metatherian and eutherian lines. Clearly we are perceiving a regularity in the way mammalian evolution has proceeded. There are some differences, however, which will be discussed later.

Returning to tables 58 and 59, we can make the following summary correlations. In both marsupials and eutherians, there is a strong ten-

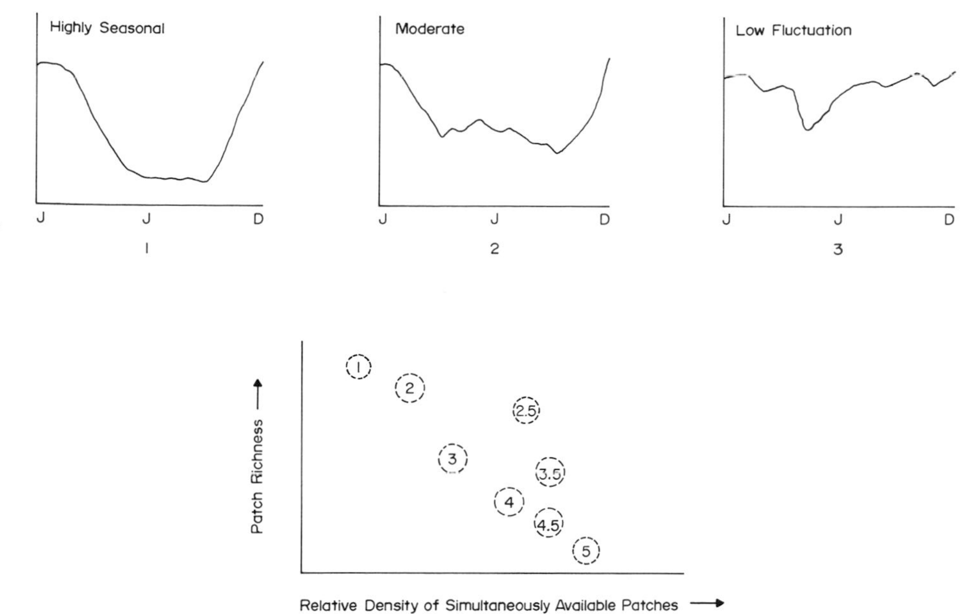

Figure 150. Upper graphs show three types of seasonal flux in resource availability as described in text. Lower graph gives the rating system of patch richness versus relative density of simultaneous available patches; 1 through 5 defines the patch richness and density of patches as employed in scoring in tables 60 and 61.

dency for polygynous mating systems. The grazing species (e.g., *Macropus parryi* and *Gazella thomsoni*) show a tendency to form larger, mobile social groupings. Complex social units are usual in refuging situations where a complex artifact is constructed (*Castor canadensis*) or where the refuge is a scarce resource (cave, sleeping tree) but will accommodate many individuals (*Myotis lucifugus, Alouatta seniculus*).

Group-living species often have an antipredator strategy based on warning of a predator's presence. In such mobile group-dwelling species as gazelles or kangaroos, the antipredator strategy may include "selfish" acts, such as moving to the center of the herd, but certain display forms may indirectly or secondarily aid other herd members (see sect. 32.6).

Foraging systems tend to be individualistic, but food sharing or "altruistic" antipredator behaviors tend to correlate with an extended rearing system based on relatives and generational overlap (*Alouatta seniculus, Macaca sinica*). Herding as a foraging strategy that may confer antipredator benefits and help in finding widely dispersed food sources but that offers little direct mutual aid is a common social tendency in grazing herbivores, both marsupials and eutherians (see also Kaufmann 1974*a,b,c*).

Tables 60 and 61 offer some general trends in the defensibility of resources. Defensibility of resources (food and refuges) is high in forms

Table 60 Some Correlations among Demographic, Ecological, and Sociological Variables for Monotremes and Marsupials

Taxon	Defense of Resources					Mobility Rating	Seasonal Flux of Food Base[a]	Distribution of Resources and Patch Richness[b]	Relative Synchrony of Reproduction	Rate of Reproduction[c] (r)	Index of Iteroparity[d]		Parental Investment Rating[e]
	Refuge	Feeding Area	Caching Area	Social Helper	Mate(s)						♂	♀	
Monotremata													
Ornithorhynchus anatinus	High	High	n.a.	n.a.	High	1.8	2a,b	2.0	Mod.	1.0	4.0	5.0	1.5
Tachyglossus aculeatus	High	Low	n.a.	n.a.	Trans.[g]	1.0	1a,b	3.5	High-Mod.	1.0	5.0	5.0	1.0
Marsupialia													
Marmosa robinsoni	High	Mod.	n.a.	n.a.	Trans.	1.0	1a,b	2.0	Mod.	6.0	1.8	1.3	4.5
Didelphis marsupialis	Mod.	Mod.	n.a.	n.a.	Trans.	1.4	2a,b, 3a	2.5	Mod.	6.0	1.7	1.4	4.3
Antechinus stuarti	Mod.	Mod.	n.a.	n.a.	Trans.	1.0	2a	3.0	Mod.	7.0	1.0	1.2	4.8
Dasyuroides byrnei	High	Mod.	n.a.	n.a.	Trans.	1.7	1b	2.0	?	6.0	2.4	3.0	2.5
Sminthopsis crassicaudata	High	Mod.	n.a.	n.a.	Trans.	1.3	1b	2.0	Mod.	6.0	2.2	2.7	2.7
Sarcophilus harrisii	Mod.	Low	n.a.	n.a.	Trans.	2.5	2a	2.5	Mod.	4.8	2.8	3.0	3.0
Perameles nasuta	High	Mod.	n.a.	n.a.	Trans.	1.7	2a	3.0	Mod.	3.0	2.0	2.1	2.8
Trichosurus vulpecula	High	High	n.a.	n.a.	High	1.4	2a	3.5	Mod.-High	3.0	2.7	3.0	3.0
Petaurus breviceps	High	Mod.	n.a.	n.a.	High	1.3	3a	2.4	Low?	3.0	2.3	2.7	2.8
Phascolarctos cinereus	Mod.	Mod.	n.a.	n.a.	Trans.	1.2	2a	4.0	Mod.	1.4	3.4	3.6	2.2
Vombatus ursinus	High	Mod.	n.a.	n.a.	Trans.	1.5	2a,b	4.0	?	1.2	3.8	4.0	1.8
Potorous tridactylus	Mod.	Low	n.a.	n.a.	Trans.	1.1	2a	3.5	?	2.4	2.0	2.4	2.0
Dendrolagus matschiei	Mod.	Mod.	n.a.	n.a.	Trans.	1.7	3a	5.0	?	1.4	3.4	3.6	2.0
Macropus giganteus	n.a.[f]	Low	n.a.	n.a.	Trans.	2.4	2a,b	4.0	Mod.	1.4	4.0	4.6	1.8
Megaleia rufa	n.a.	Low	n.a.	n.a.	Trans.	3.2	1b	4.0	Low	1.3	4.0	4.6	1.8
Macropus parryi	n.a.	Low	n.a.	n.a.	Trans.	1.9	2a	4.0	Mod.	2.0	3.8	4.2	1.8

Note: References as in table 58. For rating scales, see text.
[a] See figure 150 for scaling.
[b] See figure 150 for scaling.
[c] Rate: 1 = low; 7 = high (see text).
[d] 5 = iteroparous strategy; 1 = approximation of semelparity (see text).
[e] Parental investment in a given litter: 1 = maximum investment in mother's survival; 5 = maximum investment in litter's survival (see text).
[f] n.a. = not applicable for this species.
[g] Trans. = transitory but usually high.

Table 61 Some Correlations among Demographic, Ecological, and Sociological Variables for Eutherians

Taxon	Defense of Resources					Mobility Rating	Seasonal Flux of Food Base	Distribution of Resources and Patch Richness	Relative Synchrony of Reproduction	Rate of Reproduction (r)	Index of Iteroparity		Parental Investment Rating
	Refuge	Feeding Area	Caching Area	Social Helper	Mate(s)						♂	♀	
Edentata													
Bradypus infuscatus	n.a.	Trans.	n.a.	n.a.	?	0.5	3a	4.0	Low	1.0	5.0	5.0	1.0
Tamandua tetradactyla	High	Mod.	n.a.	n.a.	Trans.	2.8	3a	3.5	Low	1.0	4.0	4.0	1.0
Pholidota													
Manis tricuspis	High	High	n.a.	n.a.	High	2.4	3a	3.5	Low				
Macroscelidea													
Elephantulus rufescens	High	High	n.a.	n.a.	High	1.5	2a,b	3.0	Low	3.0	3.0	3.0	2.7
Rodentia													
Cynomys ludovicianus	High	High	n.a.	n.a.	High	1.2	1a	4.0	High	3.0	2.7	3.0	2.3
Tamiasciurus hudsonicus	High	High	High	n.a.	Trans.	1.5	1a	2.5	High	3.4	3.2	3.5	2.5
Tamias striatus	High	High	High	n.a.	Trans.	1.2	1a	2.5	High	3.5	2.3	2.5	3.5
Castor canadensis	High	High	High	n.a.	High	1.3	1a	3.0	High	1.8	4.0	4.0	1.2
Microtus pennsylvanicus	Mod.	Mod. ♀	n.a.	n.a.	Trans.	1.0	1a	4.0	High	5.0	1.2	1.3	4.0
M. ochrogaster	High	Mod.	n.a.	n.a.	High	1.0	1a	4.0	High	4.2	2.0	2.0	3.0
Peromyscus polionotus	High	Mod.	High	n.a.	High	1.2	1a	2.0	Mod.	4.4	2.0	2.0	3.0
P. maniculatus	Mod.	Low	Mod.	n.a.	Mod.	1.2	1a	2.0	High	4.3	1.8	1.9	3.0
Dasyprocta punctata	Mod.	High	High	n.a.	High	1.5	3b	2.0	Low	2.8	3.0	3.8	2.3
Dolichotis patagonum	High	?	n.a.	n.a.	High	1.7	1a	4.0	Mod.	2.9	3.0	3.8	2.2
Insectivora													
Sorex araneus	High	Mod.	High	n.a.	Trans.	1.4	1a	2.0	High	6.0	1.0	1.2	4.8
Talpa europaea	High	High	High	n.a.	?	0.8	1a	2.0	High	3.0	2.7	2.9	2.3
Hemicentetes semispinosus	High	Low	n.a.	n.a.	High	1.1	3b	2.0	Mod.	5.5	1.5	1.5	4.6
Carnivora													
Canis lupus	High	High	Mod.	High	High	2.8	1a	1.0	High	2.8	3.5	3.5	3.0
Nasua nasua	High	Low	n.a.	n.a.	Trans.	2.4	2b	3.5	Mod.	3.0	3.0	3.1	2.5
Panthera leo	Mod.	High	n.a.	n.a.	High	2.6	1b, 2b	2.0	Low	2.3	2.7	3.2	2.8

Table 61—*Continued*

Taxon	Defense of Resources: Refuge	Feeding Area	Caching Area	Social Helper	Mate(s)	Mobility Rating	Seasonal Flux of Food Base	Distribution of Resources and Patch Richness	Relative Synchrony of Reproduction	Rate of Reproduction (r)	Index of Iteroparity ♂	♀	Parental Investment Rating
P. pardus	Mod.	High	n.a.	n.a.	Trans.	2.4	2a,b	2.0	Mod.	2.1	2.8	3.0	2.5
Lycaon pictus	High	Mod.	n.a.	High	High	2.9	1b, 2b	1.5	Mod.	3.2	3.1	3.2	3.1
Crocuta crocuta	Mod.	High	Mod.	n.a.	High	2.6	1b, 2b	2.0	Mod.	1.6	2.7	3.5	2.5
Ursus (Euarctos) americanus	High	Low	n.a.	n.a.	Trans.	2.5	1a	2.5	High	1.8	4.0	4.1	1.4
Primates													
Microcebus murinus	High	Mod.	n.a.	n.a.	High	1.4	3b	2.0	Mod.	1.8	3.0	3.3	1.7
Saguinus oedipus	High	Mod.	n.a.	High	High	1.8	3b	2.0	Low	1.6	3.5	3.5	1.5
Cebus capucinus	High	High	n.a.	n.a.	High	2.0	2b	2.0	Low	1.2	4.0	3.8	1.0
Alouatta seniculus	High	Mod.	n.a.	n.a.	High	1.3	2b	4.0	Mod.	1.2	4.0	3.8	1.0
Macaca sinica	High	High	n.a.	n.a.	High	2.0	3b	2.0	Low-Mod.	1.4	4.0	4.0	1.0
Hylobates lar	High	High	n.a.	n.a.	High	1.8	3b	2.5	Low	1.1	4.0	4.1	1.0
Pan troglodytes	High	Mod.	n.a.	n.a.	Trans.	2.4	2b	2.0	Low	1.0	4.3	4.4	1.0
Gorilla gorilla	High	High	n.a.	n.a.	High	1.6	3b	4.0	Low	1.0	4.0	4.2	1.0
Proboscidea													
Elephas maximus	n.a.	Low	n.a.	High	Trans.	3.1	2b, 3b	5.0	Low	1.0	4.1	4.5	1.0
Perissodactyla													
Equus burchelli	n.a.	Low	n.a.	n.a.	High	3.0	2b	5.0	Low	1.3	4.0	3.8	1.2
Ceratotherium simum	n.a.	Mod.	n.a.	n.a.	Mod.	1.7	2b	5.0	Low	1.1	4.0	4.3	1.0
Artiodactyla													
Sus scrofa	High	Mod.	n.a.	n.a.	High	2.8	2b	3.0	High	3.8	3.0	2.3	2.0
Cervus canadensis	n.a.	Low	n.a.	n.a.	High	2.4	1a	4.5	High	1.7	3.0	3.3	1.5
Adenota kob	n.a.	Low	n.a.	n.a.	Mod.[a]	2.8	2b	4.5	Mod.	1.7	3.0	3.3	1.5
Gazella thomsoni	n.a.	Low	n.a.	n.a.	High	3.0	1b	4.5	Mod.	1.6	3.1	3.4	1.5
Syncerus caffer	n.a.	Low	n.a.	n.a.	Trans.	3.2	2b, 3b	4.5	Low	1.4	3.3	3.6	1.5
Ovis canadensis	n.a.	Low	n.a.	n.a.	Trans.	3.1	1a	4.5	High	1.6	3.1	3.3	1.5

Note: References as in table 59; see notes to table 60 for explanation of terms and rating scales.

[a] Forms a lek.

that are able to maintain a fixed land tenure system. This is especially true of species that have reduced mobility and tend to feed exclusively on the fixed home range (e.g., *Tamiasciurus hudsonicus*). Refuging areas that are limited in availability (sleeping trees) or constructed with great expense to the individual (complex burrows) are also often defended to a high level against conspecifics (*Tamias striatis*). The same is true of loci in the home range that are used to cache food (*Dasyprocta punctata*).

If the patch richness (in terms of available energy) of the food source tends to be high and the relative density of the patches at any time is low to moderate, and if the resources are stable through time, then if the species has a high relative mobility the denfensibility of the feeding area will be high. For example, in tables 60 and 61, nineteen species rate as having a feeding resource base of values 1 to 2; of these, seventeen show a moderate to high defense of their feeding area. Opposing this trend are species using food resources that are ubiquitous but of low energy content (leaves and grass) and that show temporal variation in their distribution. Of twenty species rated as having a resource base of the value 4 to 5, eighteen species show a low to moderate defense of their feeding areas.

The defense of mates is usually a transient phenomenon tied to the breeding season (*Cervus canadensis*), except for conditions of monogamy (*Elephantulus rufescens*) or where a year-round association occurs between a male and a group of females (*Cebus capucinus*).

Tables 59 and 61 indicate that species showing a high reproductive synchrony may opt for either a high permanent defense of a female group by a male or a transient defense by a male who chooses to court females individually. This latter case occurs most often in species where the females are not grouped, so that it is impossible for a male to monopolize a group of females. Where reproductive synchrony is low and females are grouped, male defense of a female band is possible and also maximizes the male's fitness; thus a trend toward permanent, high defense of mates may be noted (e.g., *Trichosurus vulpecula, Cebus capucinus, Macaca sinica*). Synchrony of reproduction among females of a population has some bearing on the degree of mate defense. Mate defense and its duration are profoundly affected by the fidelity mates show toward a defendable area throughout the annual cycle.

One male strategy for species exhibiting low reproductive synchrony is to defend individual females transiently. This is especially true if the feeding areas of males and females either do not overlap or cannot be defended, or if the mobility of the species is so high that a permanent defense of an area is impossible. Where a male can defend a fixed feeding area and shares it with one or more females, then female defense may be high, even with low reproductive synchrony. Mobility influences the defensibility of areas, as does the degree of congruence between male and female foraging areas. Various solutions to the problem are demonstrable, especially for the eutherians.

High defensiveness against an intruding group is often shown even if a feeding territory is incompletely defended. This tends to occur when the social unit cooperates in rearing young and is monogamous (*Saguinus oedipus, Canis lupus*).

Ecological limits, such as seasonal fluctuation in resource base or dispersion of resources in space, are the givens that shape behavioral responses through natural selection. Yet fat storage, caching food, and seasonal movements offset temporal and spatial variations in food. Some environments may be extremely restrictive and others less "demanding." At any point in time the observer is noting only a very limited sample of the ongoing processes of adaptation. A knowledge of ecological parameters may allow us to set limits for the range of behavioral variation "given a mammal," but social systems themselves can be predicted only in the most general terms. Environmental correlates of social systems do not as yet provide us with a reliable, predictive method. Post hoc arguments are useful for generating hypotheses but have as yet not yielded a predictive theory.

With the above caveats in mind, let us examine the relative mobility of a species as plotted against the feeding patch richness and availability versus the seasonal flux and predictability of food abundance (figs. 151 and 152). The simultaneous availability of many rich feeding patches may correlate with a range of population mobility forms. Where the temporal flux in patches is slight, mobility may be reduced, but where temporal flux is high, mobility may also be high. Mobility seems to invariably be high in forms with rich but widely scattered feeding patches (*Lycaon pictus*).

The herding ungulates and grazing macropods show great mobility in response to seasonal abundances of food. The social carnivores show a similar mobility that reflects their adaptation to the movements of

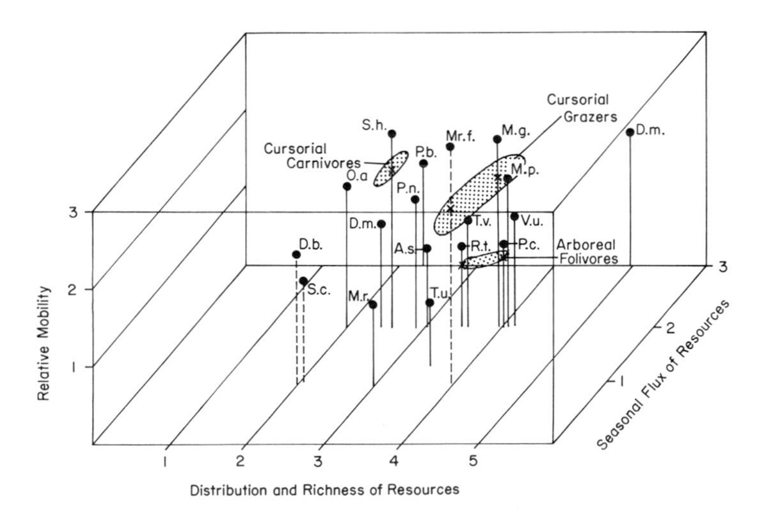

Figure 151. Three-dimensional plot of relative mobility ratings against a plot of distribution and richness of resources against seasonal flux of resources. The large cursorial grazers and carnivores have extremely high mobility. Arboreal folivores tend to show low relative mobility. Data from table 60. Letters refer to Latin binomials for the monotremes and marsupials listed in table 58.

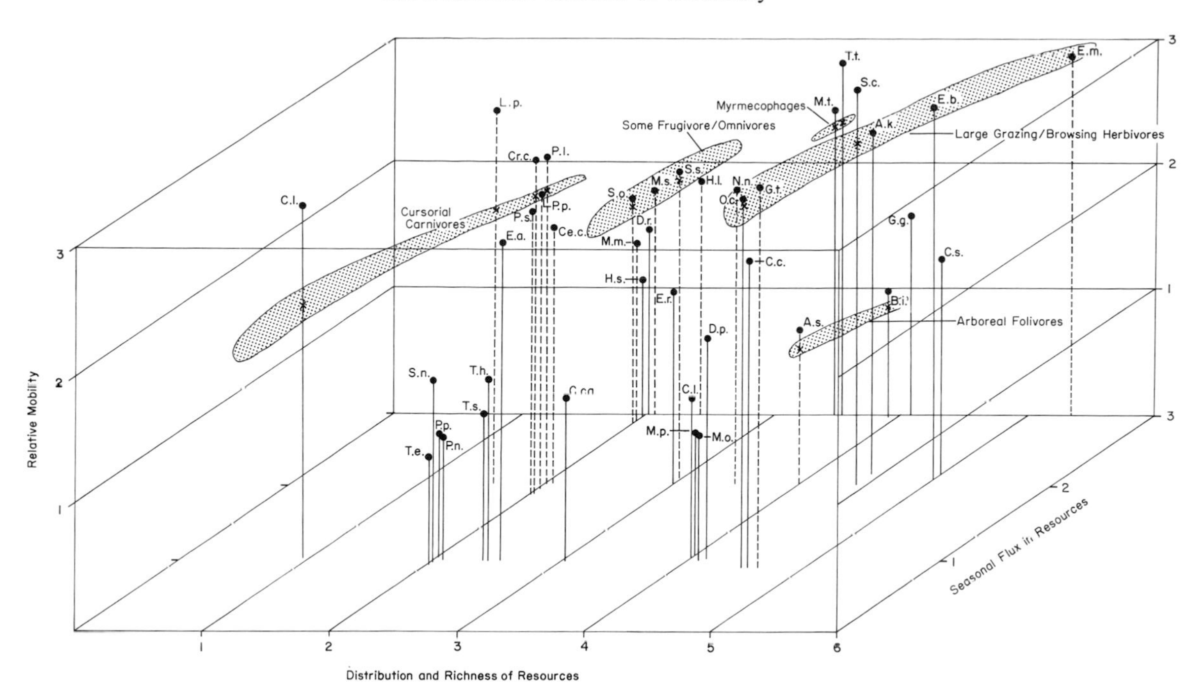

Figure 152. Three-dimensional plot of relative mobility ratings against a plot of distribution and richness of resources against seasonal flux of resources. The large cursorial grazers, carnivores, and myrmecophages have extremely high mobility. Arboreal folivores tend to show low relative mobility. Data from table 61. Letters refer to Latin binomials for the eutherians listed in table 59.

their prey. The stability of the resource base in space and time coupled with the potential mobility of the species, in part, sets the limits on the occupancy of an area. Defensibility is closely linked to the spatial stability of the resource.

Reproduction strategy has profound consequences for the form of social organization a species exhibits. A tendency toward semelparity correlates with a high reproductive potential and is tied to predictability in the abundance of food resources (*Sorex araneus, Antechinus stuarti*). In general this semelparous tendency is confined to small species with a vastly reduced complexity in social organization, though parental investment in a given litter may be very high.

Iteroparous species have low reproductive rates. The potential for a complex social structure is high if the reproductive system tends toward iteroparity. Iteroparous species are long-lived and can exist with a high generational overlap. A tendency toward iteroparity balances parental investment to favor the mother's survival over that of the offspring, but if assistance can be obtained from older siblings or relatives, thereby raising the parental fitness, then a complex society is selected for (figs. 153 and 154).

Complex social structures do not necessarily evolve in highly iteroparous reproducing forms. Qualitative differences in the degree of

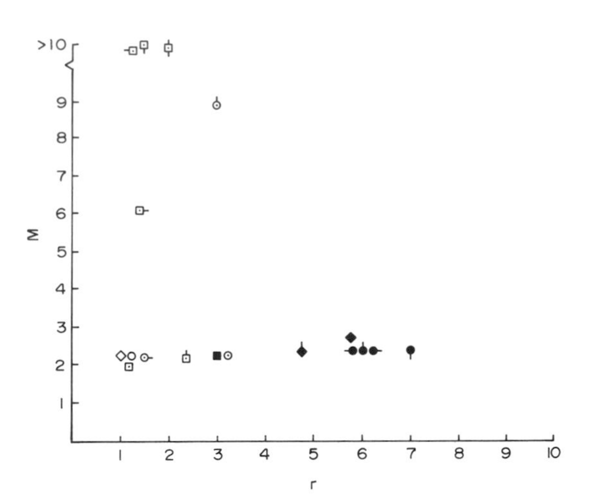

Figure 153. Sociability and reproductive rate among metatherians. Those species exhibiting low reproductive rates have the highest potential to develop complex social organizations with overlapping generations. Although the correlation is not absolute, it does suggest this relationship. Low sociability, on the other hand, may be correlated with many degrees of reproductive rate, and no simple correlation is obtainable. Ornithorhynchidae □; Tachyglossidae ◇; Peramelidae ◆ ■; Dasyuridae ●; Didelphidae ●-; Phalangeroidea ⊡ ⊙-; Macropodidae -⊡.

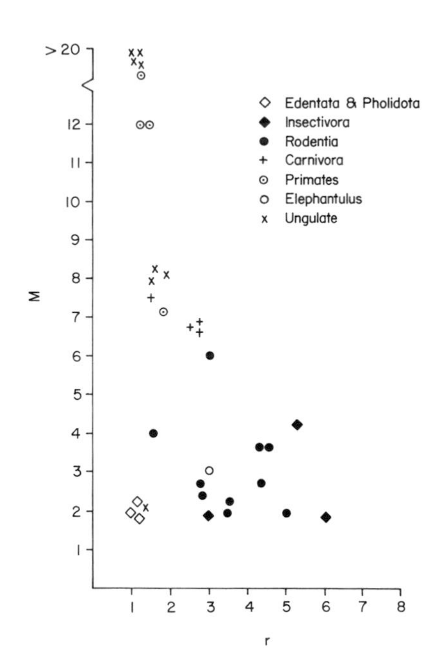

Figure 154. Sociability and reproductive rate among eutherians. Those species exhibiting low reproductive rates have the highest potential to develop complex social organizations with overlapping generations. Although the correlation is not absolute, it does suggest this relationship. Low sociability, on the other hand, may be correlated with many degrees of reproductive rate, and no simple correlation is obtainable.

social complexity and the capacity for individual recognition seem to be related to brain size (see chaps. 22 and 34). The pathway to complex social evolution in the Mammalia clearly derives from parental investment systems and the participation of older offspring. Nevertheless, mammalian evolution is characterized by selective premium on the retention of individual capacities for reproduction. The evolution of permanent neuter castes in mammalian societies must be very rare if it occurs naturally at all.

34 Synthesis

It should be clear by now, once we have developed a baseline for the evolution of mammals, that continental drift allowed certain mammalian stocks to undergo adaptive radiation in isolation from one another for considerable lengths of time. Mammals developing in Africa were swamped in the Eocene and subsequent periods by invasions from the northern continents. The South American radiations were invaded in a minor sense off and on from the Oligocene onward, but the predominant invasions and mixing of the faunas occurred in the Pliocene. Australia remained relatively isolated except that the aerial niches were occupied by the Chiroptera at an early date, followed recently by the murid rodents. The rapid spread of *Homo sapiens* to all the major continents over the past twenty thousand years probably contributed to the decline of the megafaunas, especially in Australia and South America (Martin and Wright 1967).

Where we are able to discern adaptive radiations either from observing contemporary faunas or from examining fossil records, we find convergences in antipredator behavior, mode of acquisition of food, locomotion, shelter-seeking behavior, and demographic strategies. Certain major adaptive trends are the consequences of specializing for antipredator strategies and feeding niches. We can define an arboreal folivore syndrome for tropical mammals. We can define an herbivorous fossorial syndrome, an insectivorous fossorial syndrome, a cursorial grazing syndrome, a cursorial browsing syndrome, and a cursorial carnivorous syndrome.

When we identify animals that have taken on similar feeding adaptations but have different phylogenetic origins, we find remarkable convergences in their digestive systems, metabolism, and form. Antipredator strategies, such as the evolution of the bipedal richochetal mode of locomotion, spiny coats, gliding, and speed, have—like the trends outlined for feeding adaptations—produced morphological and behavioral convergences. We can identify a small diurnal, cursorial mammal syndrome in the Macroscelidea, the dasyproctid rodents, the Tragulidae, the small Cervidae, and the cephalophine Bovidae (Rathbun 1979).

Yet, if we had knowledge of these syndromes reflecting antipredator and feeding strategies and had described the behavior and morphology of a form on one continent, we would be tempted, without having visited, to predict the occurrence of the same form on a different continent within a similar habitat. We know, however, that such convergence is never complete. First, the niches offered on the different continents are not necessarily the same in all details. The faunal complements on isolated continental masses have had different histories and, though convergence may be demonstrable, it is always relative. It appears that there exists a certain residue of phylogenetic inertia (see Wilson 1975) that constrains how far natural selection can proceed in shaping a species or a community. Why phylogenetic inertia seems to be stronger in some stocks than in others is difficult to explain. Surely it has something to do with the structure of the genome itself as well as the uniqueness of the faunal mixtures that any given population encountered at the beginning of its divergence and throughout time. Thus, knowledge derived from the past can be projected into the future only if one is mindful of such constraints. We might hope to define laws that would predict the course of evolution not only for an animal species but for our own species. We have perhaps achieved relative prediction. There are certain laws concerning the regeneration of soil, the life cycle of plants, and thermodynamics that set constraints on carrying capacity and limit how long habitats can persist. These are guidelines for the future, but they remain imperfect in predicting fine detail. Perhaps it is not too much to ask that we learn to live with grace in a system that permits only relative prediction.

The evolution of mating and rearing systems seems to be intimately tied up with the adaptations of populations to resource dispersion and predictability. Part of this adaptation complex is expressed in the demographic strategy of the species. Size and reproductive strategy are inextricably linked. The evolution of the relative brain size is in some manner also linked to demographic strategy (see sect. 23.4). As an approach to a synthesis, let us consider the two classical forms of selection, as defined by MacArthur (1972): r selection and K selection.

Regardless of the validity of r and K selection as closed categories describing the consequences of a subset of selection processes (Pianka 1972), the concepts have heuristic value and are useful in the following discussion.[1] When we look at existing species in the here and now, a K selected pattern has not necessarily been a constant throughout the phylogenetic history of the species. Rather, the history of a lineage is probably marked by alternating forms of natural selection resulting in temporary r and K trends. Furthermore, convergences in reproductive mode or demography from two entirely different genetically related stocks can be inferred from comparisons of a series of contemporary forms. Figure 155 demonstrates selection acting on a hypothetical ancestral population to produce an r selected line and a K selected line. Later in time, selection acting on the new lineages could produce a relative convergence of a K selected form from one line resembling in its demography the now less related stock. Such a set of inferences

1. A high density of competitors at the same trophic level leads to extreme K selection. It should follow, then, that extremely K selected forms are in relatively stable habitats with numerous species competing for the same food resources with similar methods of "prey" capture.

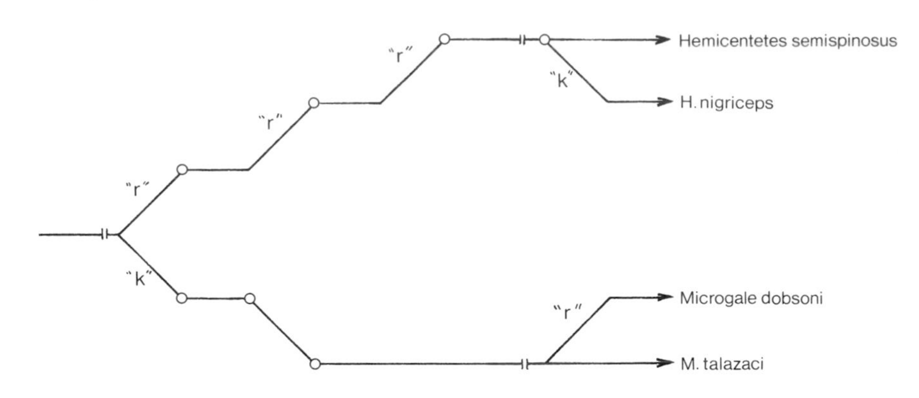

Figure 155. Hypothetical pathways of selection giving rise to two reproductive strategies for two genera of Tenrecidae. The relativity of the terms r selection and K selection are demonstrable (see text).

can be drawn from the history of speciation in the Dasypodidae and their contemporary reproductive patterns (see sect. 4.1).

Although one can demonstrate convergences in demography owing to fluctuating selective pressures, at the same time figure 155 illustrates how inertia may occur in two phylogenetic lines. This example taken from the family Tenrecidae illustrates that, although the genus *Microgale* shows an enduring trend toward K selection, some species of *Microgale* have undergone recent r selection. Conversely, although the genus *Hemicentetes* shows a long history of r selection, in recent time at least one species has undergone K selection. Thus there has been a partial convergence in reproductive mode, but the inertia of the two phylogenetic lines is evident. We would speak of the genus *Hemicentetes* as being in general r selected and of the genus *Microgale* in general as having been subjected to K selection.

The action of r selection on a population or a founder stock has profound consequences. If sustained r selection takes place, the stock soon begins to exhibit the demographic parameters of a short life-span, a large litter size, and a tendency toward semelparity (see chap. 22). The necessary consequence is a decrease in the percentage of an individual's life that is spent with the mother or older siblings in a "learning" situation. A parallel selection pressure results in a maximum genetic preprogramming of information in the nervous system (see chap. 23). This complex of events can occur only if the resource base the species exploits is predictable with respect to the average cohort. This becomes critical as the species approximates a reproduction pattern in which cohorts turn over each year. The net result of a maximum of genetic preprogramming of information in the central nervous system is a reduction in selective pressure for an increased brain size. This results in low encephalization, either through retention of a species-specific mass of neural tissue or through an outright positive selection to reduce the amount of neural tissue (see Eisenberg and Wilson 1978).

Given this state of affairs, then, if no permanent nonreproducing caste is selected for to help rear siblings, then there will be no necessary increase in the complexity of the social organization shown by the species in question. One should be wary of confusing colonies based on a close packing of distinct, noncooperative reproducing units with a communal, monogamous system. Such colonies may be the result of habitat patchiness, not of deliberate social tendencies (see fig. 142 and Eisenberg 1977).

In social insects, selection has often favored the creation of a permanent nonreproducing caste, with concomitant selection for increased fecundity and longevity of the founder female (Wilson 1975). This can lead to complex, interdependent social units without any necessary increase in neural tissue. This latter pathway, commonly pursued in the evolution of eusociality in the Hymenoptera and Isoptera, has never been a consistent pattern of selection in the class Mammalia (see Wilson 1975).*

On the other hand, a founder stock that undergoes K selection generally develops a population showing a longer relative life-span, a smaller litter size, and a tendency toward iteroparity. Such selection usually occurs when the resource base is unpredictable in a given year when the species becomes large in size, or when both conditions obtain. Two consequences can derive from this. If the trend toward iteroparity is prolonged so that a given female's reproduction extends over many years, then the predictability of resources within the environment must be averaged over the life-span of the reproducing female. The restriction of predictability from one year to the next that is so critical to an annually reproducing species is relaxed for an iteroparous species. If the environmental situation is predictable with respect to an average cohort of long-lived females, then selection can favor a maximum of genetic preprogramming of information necessary for the exploitation of the niche in question, and the percentage of the lifetime the offspring spend in a learning situation need not be increased. Thus an intermediate or low encephalization quotient can be retained or else active selection can favor a reduced mass of neural tissue. In the retention of a low encephalization quotient, then, if no permanent nonreproducing caste is selected for, which is almost always the case in mammals, then there need be no increase in social complexity.

On the other hand, if the resource base is uncertain and difficult to predict from one year to the next, and if its exploitation requires individual learning by the offspring, then selection may favor an increase in the percentage of time the young spend in learning. This will inevitably lead to a high encephalization quotient, the correlates of which generally include an increase in the capacity for homeothermy, and a long gestation with a prolonged intrauterine development of the central nervous system or a prolongation of central nervous system development after birth (see chap. 23).

Even with a high or intermediate encephalization quotient, if there is no long-term nonreproducing caste involved in the rearing of the young, then no necessary complexity in the social organization need be shown; but again colony formation may be possible. On the other hand, if a long term nonreproducing caste is selected for to help parents care for siblings, then a complex interdependent social unit can evolve, and, indeed, this seems to be the pathway by which complex interdependent social units have evolved in the class Mammalia.

Thus, high encephalization quotients, relatively long gestations and sometimes prolonged central nervous system growth after birth, a high degree of homeothermy, and a reasonably high basal metabolic rate covary positively. These characters tend to be associated with one consequence of K selection on a founder stock and seem to be prerequisites for the creation of a complex interdependent social unit; however, no single attribute of the aforementioned set of attributes necessarily in and of itself leads to or is correlated with the formation of a complex social unit. What is clear, however, is that when one does find a complex interdependent social unit in the class Mammalia, the attributes of high encephalization quotient, kin selection, mutualism, interdependency, and participation of siblings in the care of younger siblings all correlate (see fig. 156).

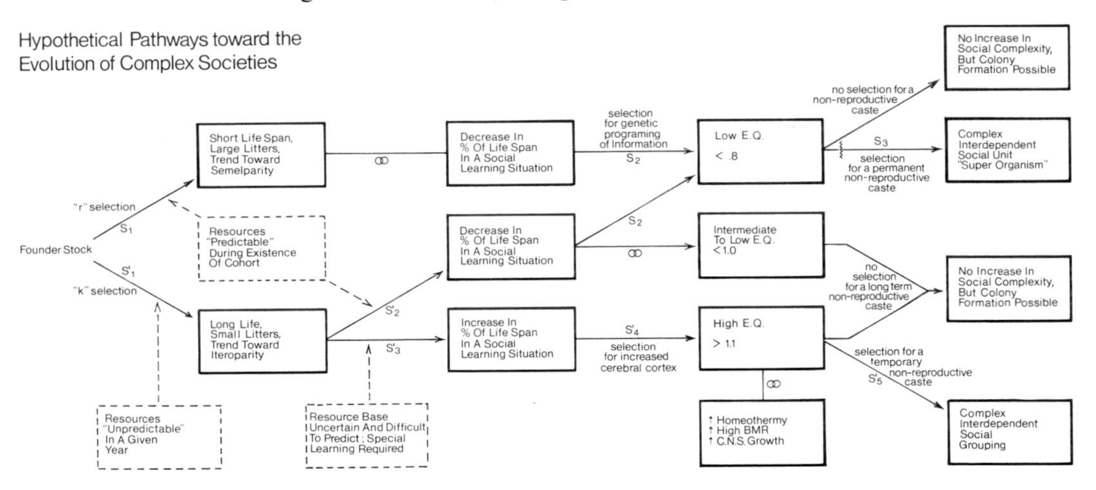

Figure 156. Consequences of different forms of selection as a result of specialization for different reproductive rates. Note that the formation of complex interdependent social groupings is one outcome of at least five different selective pressures.

ꝏ = linkage of two traits
S_1 = r selection
S'_1 = K selection
S_2 = selection for high genetically controlled programming of the CNS
S'_2 = selection for decreased association of parent and young
S_3 = selection for non-reproducing caste
S'_3 = selection for a prolonged association between parent and young
S'_4 – selection for increased volume of cerebral cortex
S'_5 = selection for delayed sexual maturation

Coda

Those who have read Darwin's *Origin of Species* (1859) know that the opening pages deal with variation. I well remember plowing through the section on the range of variation in the domestic pigeon (*Columba livia*) produced by human selection. At times I am even reminded of Darwin's words while sitting in my backyard on upper Sixteenth Street, N.W., in Washington, D.C., watching the feral pigeons feed on the leftovers from the dogs' dishes.

Darwin fully recognized that phenotypic variation is the raw material upon which selection (natural or artificial) acts. One may well ask, then, why I have based almost the entire text on averages (species, generic, and familial). The answer is clearly that I desired to point out both trends and modes through phylogenetic time and also those trends as I discern them in the present. The modes or averages are abstractions and do not represent the reality of the individual. Recognizing trends permits prediction about populations in relatively large areas, but the predictive power wanes as it is applied to individuals. Figure 157 summarizes the dilemma for the generalist. What one gains in knowledge of the broad picture, one loses in application to each specific case. Yet

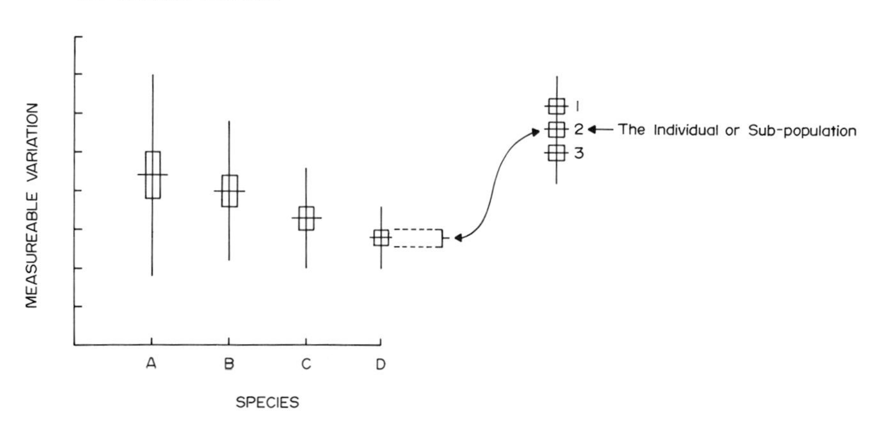

Figure 157. Hypothetical relationship between measurement of phenotypic variation for four species and the form of variation within a species' average. In the hypothetical diagram, the mean value for species D may in fact be an average value for three subpopulations that in and of themselves have statistically separable means.

a lifetime of judgments based on averages applied on successive specific cases may turn out (by the grace of God) to be respectable.

Yablokov (1974) has offered us one of the most recent treatises on variation. He has characterized his discipline as "population morphology." I consider his example a useful guide for future morphological and behavioral studies. We must not dwell only on individual case histories, nor can we be content with generalizations. The generalizations suggest questions. The discrete populations studied over time provide us with examples of the range of variation and, by proper analysis, the genesis of variation. I trust that my examples drawn on the broad landscape will suggest enough testable hypotheses to lead some of us to engage in the longitudinal studies of variation that are necessary to refine predictions.

In my final comments I would like to modify a quote from G. G. Simpson: "No morphological [or behavioral] character is inherited as such. The ultimate morphological . . . [or behavioral] . . . expression of the same hereditary factors differ markedly. The particular expression, among several possible expressions, developed in any given individual does not affect the potentialities passed on to its descendents. It certainly affects the fate of the individual and therefore helps to determine whether that individual will have descendents" (Simpson 1944, p. 30).

With this as a preface, my comments on sociobiology and future research are these. The book *Sociobiology* by E. O. Wilson and the resulting subdiscipline are creating an enormously useful synthesis of population ecology, genetics, and behavior. The hypotheses being generated by sociobiologists may be studied with great profit, and a renaissance is beginning in behavior studies.

As the genetic basis for behavior is explored in these beginning phases, simple models are constructed with the minimum assumptions—a single gene, two alleles, one-to-one correspondence between gene and behavior, and so on. The models suggest lines of investigation, and the sophisticated researcher proceeds to the field with a fresh eye and a sharp pencil. Unfortunately, the layman may be deeply influenced by the simplicity of first-order models. Worse yet, "popularizers" can exploit the unsophisticated. It is well to remember that, though mankind acknowledges the existence of natural laws that cannot be contravened, the belief in our ability to choose freely between alternatives is a psychological necessity. Even if "free will" should prove to be an illusion of our minds, it will be the role of an alien from outer space to record this. The belief in free will and individual responsibility for the consequences of choice is the absolute given for the conduct of most responsible adult human behavior. Naturally, mankind can be studied from a sociobiological perspective. What I deplore, however, is that facile generalizations have been developed by amateurs and overzealous paraprofessionals that serve no useful purpose, are self-serving, and may promote political mischief. In my 1966 monograph I speculated (pp. 74–78) on the biological bases of mankind's social and cultural institutions. I will not do so here. In 1969 the Smithsonian Institution sponsored a public symposium to explore the relevance of studies on animal behavior to the interpretation of human behavior (Eisenberg and Dillon 1972). Let the record stand. Wilson (1978) has done an admirable job of presenting his own viewpoint concerning extensions and applications.

Ultimately, the wisdom of mankind is written in the various codes of conduct that serve as the bases for moral judgment and legal systems in all organized societies. A reading of the Deuteronomic code and the Talmud or a study of English common law will indicate that such systems are open-ended and grow (albeit slowly) as social circumstances change. Generalizations about human conduct are simply that—generalizations. As such, they can be guidelines but can be neither the ultimate measure of individual cases nor a template for individual conduct in situations of real choice. There are many books that can guide the reader interested in the broad range of variation in the behavior patterns of *Homo sapiens* and the basic environmental causes underlying cultural differences (see Harris 1977 for a superb interpretation of cultural differences and their genesis). In a book like this one, *H. sapiens* is reduced to points on a graph or entries in the appendixes along with *Tenrec ecaudatus* and *Peromyscus maniculatus*. I reserve my comments on *Homo sapiens* for the future and my diaries.

Appendix 1

A Classification of the Extant Mammalia to the Level of Family.

Genera and their common names when frequently cited in text are included under each familial heading.

Taxonomic Hierarchy as Employed by McKenna (1975)

Class
    Subclass
        Superlegion
            Legion
                Sublegion
                    Infraclass
                        Supercohort
                            Cohort
                                Magnaorder
                                    Superorder
                                        Grandorder
                                            Mirorder
                                                Order
                                                    Family
                                                        Genus

A Classification of the Extant Mammalia[1]

Class Mammalia
Subclass Prototheria
Infraclass Ornithodelphia
    Order Monotremata
    Suborder Tachyglossa
        Family Tachyglossidae
            Tachyglossus Echidna
            Zaglossus New Guinea echidna
    Suborder Platypoda
        Family Ornithorhynchidae
            Ornithorhynchus Platypus
Subclass Theria
Infraclass Peramura
Supercohort Marsupialia
    Order Marsupicarnivora
        Family Didelphidae
            Caluromys Woolly opossum
            Monodelphis Short bare-tailed opossum

1. Modified from McKenna (1975) and Ride (1964). Listing is complete to family level; only genera cited in text are included below the level of family.

Marmosa Murine or mouse opossum
Philander Gray "four-eyed" opossum
Metachirus Brown "four-eyed" opossum
Didelphis Common opossum
Chironectes Water opossum
Family Microbiotheriidae
Dromiciops Monitos del monte
Family Dasyuridae
Phascogale Brush-tailed marsupial mouse
Antechinus Broad-footed marsupial mouse
Dasyurus Eastern Australian native cat
Sarcophilus Tasmanian devil
Planigale Flat-skulled marsupial mouse
Sminthopsis Narrow-footed marsupial mouse
Antechinomys Jerboa marsupial mouse
Myrmecobius Numbat
Family Thylacinidae
Thylacinus Tasmanian pouched "wolf"
Family Notoryctidae
Notoryctes Marsupial or pouched "moles"
Order Paucituberculata
Family Caenolestidae
Caenolestes "Rat" opossum
Lestoros Peruvian "rat" opossum
Rhyncholestes Chilean "rat" opossum
Order Peramelina[2]
Family Peramelidae
Perameles Long-nosed bandicoot
Echymipera New Guinean spiny bandicoot
Thylacomys "Rabbit" bandicoot
Isoodon Short-nosed bandicoot
Order Diprotodonta
Family Phalangeridae
Trichosurus Brush-tailed possum
Phalanger Cuscus
Family Petauridae
Pseudocheirus Ring-tailed possum
Petaurus Sugar glider
Schoinobates Greater gliding possum
Family Burramyidae
Acrobates Pygmy gliding possum
Cercartetus "Dormouse" possum
Burramys Mountain pygmy possum
Family Tarsipedidae
Tarsipes Honey possum
Family Phascolarctidae
Phascolarctos Koala

2. The bandicoots may be included with the "Marsupicarnivora" to create the order Polyprotodonta (Kirsch 1977).

Family Vombatidae
Vombatus Common wombat
Lasiorhinus Hairy-nosed wombat
Family Macropodidae
Macropus Greater kangaroos
Wallabia Wallaby
Setonix Short-tailed scrub wallaby
Onychogalea Nail-tailed wallaby
Dendrolagus Tree kangaroo
Bettongia "Rat" kangaroo
Potorous Broad-faced "rat" kangaroo
Supercohort Eutheria
Cohort Edentata
Order Cingulata
Family Dasypodidae
Chaetophractus Hairy armadillo
Euphractus Six-banded armadillo
Zaedyus Pichi
Priodontes Giant armadillo
Cabassous Naked-tailed armadillo
Tolypeutes Three-banded armadillo
Dasypus Long-nosed armadillo
Chlamyphorus Fairy armadillo
Burmeisteria Burmeister's armadillo
Order Pilosa
Family Myrmecophagidae
Myrmecophaga Giant anteater
Tamandua Lesser anteater
Cyclopes Fairy anteater
Family Bradypodidae
Bradypus Three-toed sloth
Choloepus Two-toed sloth
Order Pholidota (i.s.)
Family Manidae
Manis Pangolin
Cohort Epitheria
Magnaorder Ernotheria
Superorder Leptictida
Grandorder Anagalida
Order Macroscelidea
Family Macroscelididae
Macroscelides Elephant shrew
Nasilio Short-nosed elephant shrew
Elephantulus Long-eared elephant shrew
Petrodromus Forest elephant shrew
Rhynchocyon Checkered-backed elephant shrew
Order Lagomorpha
Family Ochotonidae
Ochotona Pika

Family Leporidae
Pentalagus Ryukyu rabbit
Pronolagus Rock Hare
Romerolagus Volcano rabbit
Caprolagus Hispid hare
Lepus Hares
Poelagus Scrub hare
Sylvilagus American rabbit
Oryctolagus European rabbit
Nesolagus Sumatran hare
Magnaorder Preptotheria
Order Rodentia (i.s.)
Family Aplodontidae
Aplodontia Mountain beaver
Family Sciuridae
Sciurus Tree squirrels
Syntheosciurus Grooved-tooth squirrel
Tamiasciurus Pine squirrels
Funambulus Palm squirrel
Ratufa Giant squirrel
Callosciurus Tricolored squirrel
Atlantoxerus Barbary ground squirrel
Xerus African ground squirrel
Marmota Marmot
Cynomys Prairie dog
Spermophilus (=Citellus) Ground squirrel
Tamias Chipmunk
Eutamias Western chipmunk
Petaurista Giant flying squirrel
Glaucomys New World flying squirrel
Petinomys Dwarf flying squirrel
Family Geomyidae
Geomys Pocket gopher
Thomomys Western pocket gopher
Pappogeomys Central American pocket gopher
Family Heteromyidae
Perognathus Pocket mouse
Microdipodops Kangaroo mouse
Dipodomys Kangaroo rat
Liomys Spiny pocket mouse
Heteromys Forest spiny pocket mouse
Family Castoridae
Castor Beaver
Family Anomaluridae
Anomalurus Scaly-tailed flying squirrel
Zenkerella Flightless scaly-tailed squirrel
Family Pedetidae
Pedetes Springhare

Family Cricetidae
Oryzomys Rice rat
Rhipidomys Climbing rat
Tylomys Central American climbing rat
Nyctomys Vesper rat
Reithrodontomys Harvest mouse
Peromyscus White-footed mouse or deer mouse
Ochrotomys Golden mouse
Baiomys Pygmy mouse
Onychomys Grasshopper mouse
Scotinomys Brown mouse
Sigmodon Cotton rat
Neotoma Wood rat
Rheomys Central American water mouse
Cricetulus Dwarf hamster
Cricetus Common hamster
Mesocricetus Golden hamster
Macrotarsomys Madagascan hopping mouse
Nesomys Madagascan forest rat
Eliurus Madagascan dormouse
Hypogeomys Votsotsa
Dicrostonyx Collared lemming
Lemmus Lemming
Clethrionomys Red-backed vole
Arvicola Water vole
Ondatra Muskrat
Neofiber Round-tailed muskrat
Phenacomys Heather vole
Pitymys Pine vole
Microtus Vole
Lagurus Sage brush vole
Prometheomys Long-clawed mole vole
Ellobius Mole vole
Myospalax Chinese mole rat
Beamys Long-tailed pouched rat
Saccostomus Pouched mouse
Cricetomys Pouched rat
Dendromus African tree mouse
Otomys Swamp rat
Family Gerbillidae[3]
Gerbillus Gerbil
Tatera Antelope gerbil
Meriones Jird or "gerbil"
Psammomys Sand gerbil
Family Spalacidae
Spalax Old World mole rat

3. May be considered a subfamily of the Cricetidae.

Family Rhizomyidae
Tachyoryctes African mole rat
Cannomys Lesser bamboo rat
Rhizomys Bamboo rat
Family Muridae
Micromys European harvest mouse
Apodemus Wood mouse
Mesembriomys Tree rat
Lemniscomys Striped field mouse
Thallomys Acacia rat
Rattus Old Wörld rat
Pseudomys Australian native mouse
Melomys Mosaic-tailed rat
Mus Old World mouse
Notomys Australian kangaroo mouse
Uranomys African big-toothed mouse
Hydromys Australian water rat
Phloeomys Slender-tailed cloud rat
Crateromys Bushy-tailed cloud rat
Family Myoxidae (=Gliridae)
Myoxus (=Glis) Dormouse
Muscardinus Least dormouse
Eliomys Garden dormouse
Family Platacanthomyidae
Platacanthomys Spiny dormouse
Family Seleviniidae
Selevinia Selevin's mouse, desert dormouse
Family Zapodidae
Sicista Birch mouse
Zapus Jumping mouse
Family Dipodidae
Jaculus Jerboa
Allactaga Long-eared jerboa
Family Hystricidae
Hystrix Old World porcupine
Atherurus Brush-tailed porcupine
Family Erethizontidae
Erethizon North American porcupine
Coendou Prehensile-tailed porcupine
Echinoprocta Upper Amazon porcupine
Chaetomys Thin-spined porcupine
Family Caviidae
Cavia Guinea "pigs," cavies
Kerodon Rock cavy
Galea Cuis
Microcavia Dwarf cavy or desert cavy
Dolichotis Mara or Patagonian cavy
Family Hydrochaeridae
Hydrochaeris Capibara

Family Dinomyidae
Dinomys Brannick's giant rat
Family Dasyproctidae
Agouti (=Cuniculus) Paca
Dasyprocta Agouti
Myoprocta Acouchi
Family Chinchillidae
Lagostomus Plains viscacha
Lagidium Mountain viscacha
Chinchilla Chinchilla
Family Capromyidae
Capromys Cuban hutia
Geocapromys Ground hutia
Plagiodontia Hispaniolan hutia
Myocastor Coypu[4]
Family Octodontidae
Octodon Degu
Octodontomys Chozchoz
Spalacopus Cururos
Family Ctenomyidae
Ctenomys Tuco-tuco
Family Abrocomidae
Abrocoma Chinchilla rat
Family Echimyidae
Proechimys Spiny rat
Diplomys Arboreal soft-furred spiny rat
Dactylomys Coro-coro
Echimys Spiny rat
Family Thryonomyidae
Thryonomys Cane rat
Family Petromyidae
Petromus Rock rat
Family Bathyergidae
Cryptomys Blesmole
Bathyergus Mole rat
Heterocephalus Naked mole rat
Family Ctenodactylidae
Ctenodactylus
Pectinator
Massoutiera
Felovia
} Gundis
Superorder Tokotheria
Grandorder Ferae
Order Carnivora
Family Canidae
Canis Dog, wolf, coyote, etc.
Alopex Arctic fox
Vulpes Fox
Fennecus Fennec fox

4. Usually placed in the family Myocastoridae.

Urocyon Gray fox
Nyctereutes Raccoon dog
Dusicyon South American fox
Cerdocyon Crab-eating fox
Chrysocyon Maned wolf
Speothos Guyana bush dog
Cuon Red dog, dhole
Lycaon Cape hunting dog
Otocyon Bat-eared fox
Family Ursidae
Tremarctos Spectacled bear
Selenarctos Tibetan bear
Ursus Brown bear, grizzly bear
Euarctos American "black" bear[5]
Thalarctos Polar bear
Helarctos Sun bear
Melursus Sloth bear
Family Ailuropodidae
Ailuropoda Giant panda
Family Procyonidae
Bassariscus Cacomistle
Procyon Raccoon
Nasua Coati mundi
Nasuella Mountain coati
Potos Kinkajou
Bassaricyon Olingo
Family Ailuridae
Ailurus Lesser panda
Family Mustelidae
Mustela Weasel
Vormela Marbled polecat
Martes Marten
Eira Tayra
Galictis Grison
Lyncodon Patagonian weasel
Poecilictis African banded weasel
Gulo Wolverine
Meles European badger
Arctonyx Hog badger
Mydaus Malayan stink badger
Suillotaxus Palawan stink badger
Taxidea American badger
Melogale Ferret badger
Mephitis Striped skunk
Spilogale Spotted skunk
Conepatus Hog-nosed skunk
Lutra Otter
Pteronura Amazon river otter, giant otter
Amblonyx Oriental small-clawed otter

5. Synonymized under *Ursus*.

Aonyx African clawless otter
Enhydra Sea otter
Family Viverridae
Genetta Genet
Viverricula Lesser oriental civet
Osbornictis African otter civet
Viverra Oriental civet
Civettictis African ground civet
Nandinia African palm civet
Paradoxurus Palm civet
Paguma Masked palm civet
Arctictis Binturong
Fossa Fanaloka
Hemigalus Banded palm civet
Diplogale Hose's palm civet
Cynogale Otter civet
Family Herpestidae[6]
Galidia Ring-tailed Madagascan mongoose
Suricata Meerkat
Herpestes Mongoose
Helogale Dwarf mongoose
Atilax Marsh mongoose
Mungos Banded mongoose
Liberiictis Kuhn's mongoose
Ichneumia White-tailed mongoose
Bdeogale Black-footed mongoose
Family Cryptoproctidae
Cryptoprocta Fossa
Family Hyaenidae
Proteles Aardwolf
Crocuta Spotted hyena
Hyaena Striped and brown hyenas
Family Felidae
Lynx Lynx, bobcat
Felis Smaller cats
Panthera Lion, leopard, jaguar, tiger
Neofelis Clouded leopard
Uncia Snow leopard
Acinonyx Cheetah
Puma Puma, mountain lion
Order Pinnipedia
Family Otariidae
Otaria South American sea lion
Eumatopias Stellar's sea lion
Zalophus California sea lion
Neophoca Australian sea lion

6. Separation of Viverridae and Herpestidae follows Wozencraft 1980.

Arctocephalus Southern fur seal
Callorhinus Northern fur seal
Family Odobenidae
Odobenus Walrus
Family Phocidae
Phoca Hair seal or harbor seal
Pusa Ringed seal
Halichoerus Gray seal
Erignathus Bearded seal
Monachus Monk seal
Lobodon Crab-eater seal
Hydrurga Leopard seal
Leptonychotes Weddell seal
Cystophora Hooded seal
Mirounga Elephant seal
Grandorder Insectivora
Order Tenrecomorpha
Family Potamogalidae
Potamogale Otter shrew
Micropotamogale Least otter shrew
Family Chrysochloridae (i.s.)
Chrysochloris Golden mole
Eremitalpa Desert golden mole
Family Tenrecidae
Tenrec Common tenrec
Setifer Greater hedgehog tenrec
Hemicentetes Striped tenrec
Echinops Lesser hedgehog tenrec
Oryzorictes Rice tenrec
Microgale Long-tailed tenrec
Limnogale Aquatic tenrec
Geogale Large-eared tenrec
Order Erinaceomorpha[7]
Family Erinaceidae
Echinosorex Moonrat
Erinaceus Hedgehog
Hemiechinus Long-eared desert hedgehog
Paraechinus Desert hedgehog
Order Soricomorpha[8]
Family Solenodontidae[9]
Solenodon Solenodon
Family Soricidae
Sorex Shrew
Microsorex Pygmy shrew
Neomys European water shrew

7. May be subsumed under the same order "Insectivora."
8. May be subsumed under the same order "Insectivora."
9. According to Van Valen (1967), the solenodons may be grouped with the "Tenreco-morphs."

Blarina Short-tailed shrew
Cryptotis Least shrew
Notiosorex Desert shrew
Crocidura White-toothed shrew
Suncus Oriental musk shrews
Anourosorex Mole shrew
Family Talpidae
Uropsilus Asiatic shrew mole
Desmana Russian desman
Galemys Pyrenean desman
Talpa European mole
Scaptonyx Long-tailed mole
Neurotrichus Shrew mole
Parascalops Brewer's mole
Scapanus Western mole
Scalopus Eastern mole
Condylura Star-nosed mole
Family Nesophontidae
Nesophontes[10] Nesophontes
Grandorder Archonta
Order Scandentia
Family Tupaiidae
Tupaia Tree shrew
Urogale Philippine tree shrew
Ptilocercus Feather-tailed tree shrew
Order Dermoptera
Family Cynocephalidae
Cynocephalus Flying lemur
Order Chiroptera
Suborder Megachiroptera
Family Pteropidae
Cynopterus Short-nosed fruit bat
Rousettus Rousette fruit bat
Pteropus Flying fox
Epomophorus Epauleted fruit bat
Hypsignathus Hammer-headed bat
Eidolon Straw-colored fruit bat
Eonycteris Dawn fruit bat
Macroglossus Long-tongued fruit bat
Nyctimene Tube-nosed fruit bat
Suborder Microchiroptera
Family Rhinopomatidae
Rhinopoma Mouse-tailed bat
Family Emballonuridae
Emballonura Old World sheath-tailed bat
Saccopteryx Sac-winged bat
Peropteryx Peter's sac-winged bat

10. Probably extinct.

Family Noctilionidae
Noctilio Bulldog bat
Family Nycteridae
Nycteris Slit-faced bat
Family Megadermatidae
Megaderma Old World false vampire bat
Family Rhinolophidae
Rhinolophus Horseshoe bat
Family Hipposideridae
Hipposideros Old World leaf-nosed bat
Family Mormoopidae
Pteronotus Moustached bat
Mormoops Leaf-chinned bat
Family Phyllostomatidae
Macrotus Big-eared bat
Phyllostomus Spear-nosed bat
Vampyrum False vampire bat
Anoura Geoffroy's long-nosed bat
Carollia Short-tailed leaf-nosed bat
Sturnira Yellow-shouldered bat
Artibeus Neotropical fruit bat
Glossophaga Long-tongued bat
Family Desmodontidae[11]
Desmodus Vampire bat
Family Natalidae
Natalus Funnel-eared bat
Family Furipteridae
Furipterus Thumbless bat
Family Thyropteridae
Thyroptera Disk-winged bat
Family Myzopodidae
Myzopoda Madagascan disk-winged bat
Family Vespertilionidae
Myotis Little brown bat, mouse-eared bat
Lasionycteris Silver-haired bat
Pipistrellus Pipistrelle bat
Nyctalus Noctule bat
Eptesicus Big brown bat
Tylonycteris Flat-headed bat, bamboo bat
Nycticeius Evening bat
Lasiurus Red bat
Plecotus Lump-nosed bat
Antrozous Pallid bat
Family Mystacinidae
Mystacina New Zealand short-tailed bat

11. May be considered a subfamily of the Phyllostomatidae.

Family Molossidae
Platymops Flat-headed free-tailed bat
Tadarida Free-tailed bat
Eumops Mastiff bat
Cheiromeles Naked bat
Order Primates
Suborder Strepsirhini
Family Lemuridae
Lemur Lemur
Lepilemur Sportive lemur
*Cheirogaleus** Dwarf lemur
*Microcebus** Mouse lemur
*Phaner** Forked lemur
Family Indridae
Avahi Avahi
Propithecus Sifaka
Indri Indri
Family Daubentoniidae
Daubentonia Aye-aye
Family Lorisidae
Loris Slender loris
Nycticebus Slow loris
Arctocebus Angwantibo
Perodicticus Potto
Family Galagidae*
Galago Galago
Euoticus Needle-clawed galago
Suborder Haplorhini
Infraorder Tarsiiformes
Family Tarsiidae
Tarsius Tarsier
Infraorder Platyrrhini
Family Cebidae
Aotus Night monkey
Callicebus Titi monkey
Cacajao Uakari monkey
Pithecia Saki monkey
Chiropotes Bearded saki monkey
Alouatta Howler monkey
Cebus Capuchin monkey
Saimiri Squirrel monkey
Ateles Spider monkey
Lagothrix Woolly monkey
Family Callitrichidae
Callimico Goeldi's marmoset
Callithrix Short-tusked marmoset
Cebuella Pygmy marmoset
Saguinus Long-tusked tamarin
Leontopithecus Lion tamarin

Infraorder Catarrhini
Family Cercopithecidae
Macaca Macaque monkey
Cercocebus Mangabey monkey
Papio Baboon
Theropithecus Gelada baboon
Cercopithecus Guenon monkey
Miopithecus Talapoin monkey
Erythrocebus Patas monkey
Presbytis Asiatic leaf monkeys
Nasalis Proboscis monkey
Colobus African leaf monkey
Family Pongidae
Hylobates Gibbon
Symphalangus Siamang
Pongo Orangutan
Pan Chimpanzee
Gorilla Gorilla
Family Hominidae
Homo Man (humankind)
Grandorder Ungulata
Mirorder Eparctocyana
Order Tubulidentata
Family Orycteropodidae
Orycteropus Aardvark
Order Artiodactyla
Family Suidae
Potamochoerus River hog
Sus Wild swine
Phacochoerus Warthog
Hylochoerus Forest hog
Babyrousa Babirusa
Family Tayassuidae
Tayassu Peccary
Family Hippopotamidae
Hippopotamus Hippopotamus
Choeropsis Pygmy hippopotamus
Family Camelidae
Camelus Camel
Lama Llama, guanaco
Vicugna Vicuna
Family Tragulidae
Hyemoschus Water chevrotain
Tragulus Chevrotain, mouse deer
Family Cervidae
Moschus Musk deer
Muntiacus Muntjac
Dama Fallow deer
Axis Axis or hog deer

Cervus Elk or wapiti
Elaphurus Pere David's deer
Odocoileus New World deer (mule, white-tailed, black-tailed)
Ozotoceros Pampas deer
Blastocerus Swamp deer
Hippocamelus Huemul
Mazama Brocket
Pudu Pudu
Alces Moose
Rangifer Reindeer, caribou
Hydropotes Water deer
Capreolus Roe deer

Family Giraffidae
Giraffa Giraffe
Okapia Okapi

Family Antilocapridae
Antilocapra Pronghorn antelope

Family Bovidae
Tragelaphus Bushbuck
Boocercus Bongo
Taurotragus Eland
Boselaphus Bluebuck
Tetracerus Four-horned antelope
Bubalus Indian water buffalo
Anoa Dwarf water buffalo
Bos Wild cattle
Syncerus African buffalo
Bison American bison
Novibos Kouprey
Philantomba Dwarf duiker
Cephalophus Duiker
Sylvicapra Forest duiker
Kobus Kob
Redunca Waterbuck
Hippotragus Sable or roan antelope
Oryx Oryx
Addax Addax
Damaliscus Damaliscus gazelle
Alcelaphus Hartebeest
Connochaetes Gnu
Neotragus Royal antelope
Madoqua Dik-dik
Aepyceros Impala
Litocranius Waller's gazelle
Gazella Gazelle
Saiga Saiga antelope
Naemorhedus Goral
Capricornis Serow
Oreamnos Mountain "goat"

Rupicapra Chamois
Budorcas Takin
Ovibos Musk-ox
Capra Goat
Pseudois Blue sheep
Ammotragus Barbary sheep
Ovis Sheep
Mirorder Cete
Order Cetacea
Suborder Odontoceti
Family Platanistidae
Platanista Ganges River dolphin
Inia Amazon River dolphin
Lipotes Chinese river dolphin
Pontoporia La Plata River dolphin
Family Ziphiidae
Mesoplodon True's beaked whale
Ziphius Cuvier's beaked whale
Hyperoodon Bottle-nosed whale
Family Physeteridae
Physeter Sperm whale
Kogia Pygmy sperm whale
Family Monodontidae
Delphinapterus Beluga
Monodon Narwhal
Family Delphinidae
Steno Rough-toothed dolphin
Sousa Plumbeous dolphin
Sotalia River dolphin
Tursiops Bottle-nosed dolphin
Orcinus Killer whale
Globicephala Pilot whale
Phocoena Harbor porpoise
Suborder Mysticeti
Family Eschrichtidae
Eschrichtius Gray whale
Family Balaenopteridae
Balaenoptera Rorqual or finback whale
Megaptera Humpback whale
Family Balaenidae
Eubalaena Right whale
Balaena Greenland right whale or bowhead whale
Caperea Pygmy right whale
Mirorder Phenacodonta
Order Perissodactyla
Family Equidae
Equus Zebras, horses, and asses
Family Tapiridae
Tapirus Tapirs

Family Rhinocerotidae
Rhinoceros Javan and Indian rhinoceroses
Didermocerus Sumatran rhinoceros
Diceros African "black" rhinoceros
Ceratotherium African "white" rhinoceros
Order Hyracoidea (i.s.)
Family Procaviidae
Dendrohyrax Tree hyrax
Heterohyrax Gray hyrax
Procavia Rock hyrax
Mirorder Tethytheria
Order Proboscidea
Suborder Elephantoidea
Family Elephantidae
Elephas Asiatic elephant
Loxodonta African elephant
Order Sirenia
Family Dugongidae
Dugong Dugong
Family Trichechidae
Trichechus Manatee

Note: i.s. = *incertae sedis*—of doubtful position with respect to the superior taxon.

Appendix 2

Weights and Measurements for Selected Mammalian Species

The weights and measurements recorded in this appendix were drawn from a variety of sources. They represent the range or mean weight or head and body length for an adult specimen. Where the species is strongly dimorphic, separate mean weights and measurements have been given for the male and the female. In some cases only weights are recorded; in others, only measurements. In part this may have resulted from inadequacies in the literature, but the appendix represents those standard values employed in calculations and graphs involving regression against size.

The sources for the data are diverse. We depended heavily on records from the National Zoological Park, the United States National Museum of Natural History, and, in particular, the Smithsonian Venezuela Project Data Bank established by Dr. C. O. Handley, Jr. Standard reference works were also consulted, including: Hall and Kelson 1959; Medway 1978; Kingdon 1971, 1974*a,b*; Collins 1973; Ognev 1963; Gittelman (in prep.) for the Carnivora; Harrison 1955; Napier and Napier 1967.

Numbers in parentheses were not used in the tabular computations presented in the text but are included here to complete the table.

Taxon	Weight (g)	Head and Body Length (mm)
Monotremata		
Ornithorhynchus anatinus	1,200	300–400
Tachyglossus aculeatus	4,200	350–530
Marsupialia		
Didelphis marsupialis	660–890	390–482
Marmosa robinsoni	37.5	—
Antechinus stuarti (=*flavipes*)	37	90–170
Sminthopsis crassicaudata	14	—
Antechinomys spenceri	24.2	—
Pseudantechinus macdonnellensis	43.1	—
Dasycercus cristicauda	88.8	—
Dasyuroides byrnei	89	—
Phascogale tapoatafa	157.2	—
Dasyurus hallucatus	584.4	—
D. maculatus	1,782	(350–450)
D. viverrinus	1,300	(350–450)
Sarcophilus harrisii	6,700	(525–800)
Perameles gunni	686	(200–425)
Isoodon macrourus	880	(240–490)
Trichosurus vulpecula	1,982	(320–580)
Petaurus breviceps	128	(120–320)

Taxon	Weight (g)	Head and Body Length (mm)
Vombatus ursinus	20,000	—
Bettongia lesueuri	1,500	—
Potorous apicalis	1,360	—
Megaleia rufa	30,000	—
Macropus giganteus	25,000	(800–1,000)
Insectivora[a]		
Solenodon paradoxus	800	—
Migrogale talazaci	39–61	100–125
M. dobsoni	34–35	92–114
Hemicentetes semispinosus	125–280	141–72
H. nigriceps	90–150	130+
Setifer setosus	180–270	150–220
Echinops telfairi	110–250	140–80
	180 ♀	
Tenrec ecaudatus	900 ♀	205–390
	1,600–2,400 ♂	
Echinosorex gymnurus	1,000–1,400	255–350
Erinaceus europaeus	800–1,025	217–30
Sorex araneus	10.4	79.3
S. cinereus	3–5	46–61
S. vagrans	2.5–7.5	54–66
Cryptotis parva	2.8–5.6	63–67
Blarina brevicauda	14–18	78–104
Neomys fodiens	12–18	85
Notiosorex crawfordi	5	59–61
Microsorex hoyi	2.3–3.5	51–63
Crocidura hirta	10–20	62–110
C. leucodon	6–15	47–79
C. suaveolens	5.8	67.8
C. rusulla	7–9	75
Suncus murinus	18 (30)	100–150
S. etruscus	1.8–2.4	36–50
Talpa europaea	85–110	135–43
Parascalops breweri	—	116–17
Scapanus townsendi	—	161–86
S. aquaticus	—	110–70
Edentata		
Dasypus novemcinctus	4,000	370–430
Euphractus sexcinctus	4,900	—
Chaetophractus villosus	1,800	—
Choloepus hoffmanni	9,000	600–640
Bradypus variegatus	4,000–4,500	—
Chiroptera		
Cynopterus sphinx	90	—
Rousettus leschenaulti	70	—
Pteropus geddeiri	650	—
P. giganteus	700	—
Eidolon helvum	280	150–95
Rhinopoma kinneari	31	—
Megaderma lyra (=spasma)	42.5	—
Carollia perspicillata	17.6	65
Desmodus rotundus	43	80–86
Hipposideros bicolor	8.5	—
Myotis lucifugus	9	41–54
M. myotis	34	65–85
Pipistrellus pipistrellus	6.5	40–41
P. ceylonicus	7.5	—

[a] In the classical sense.

Taxon	Weight (g)	Head and Body Length (mm)
Nyctalus noctula	26	74–79
Eptesicus fuscus	17 ♀	70
E. serotinus	28.3	48–72
Scotophilus wroughtoni	19	—
Antrozous pallidus	26	—
Plecotus townsendii	8	—
Tylonycteris robustula	8	—
Tadarida brasiliensis	12	—
Lagomorpha		
Oryctolagus cuniculus	2,000	420
Sylvilagus aquaticus	2,200 ♀	459–63
Lepus americanus	1,500 ♀	340–468
L. californicus	2,870	415–518
	2,100 ♀	—
L. timidus	3,000	580
L. europaeus	4,000–5,000	640
	3,750 ♀	
Macroscelidea		
Macroscelides proboscideus	45	110
Petrodromus tetradactylus	210	195
Rhynchocyon chrysopygus	450	383
Elephantulus rufescens	58	122
E. brachyrhynchus	—	100–135
Rodentia		
Aplodontia rufa	364	355
Sciurus carolinensis	425	220–60
S. niger	750–950	—
S. vulgaris	300	—
Tamiasciurus hudsonicus	190	(220–22)
Tamias striatus	86	137–86
Eutamias amoenus	~47	107–32
E. quadrivittatus	115	117–25
Marmota monax	5,500	318–510
Cynomys ludovicianus	1,200 ♀	283–300
Citellus columbianus	336	—
C. tridecemlineatus	145	—
	170 ♂–152 ♀	
Glaucomys sabrinus	138 ♀	148–88
G. volans	60	130–33
Dipodomys panamintinus	65	129–32
D. merriami	35	100–198
D. nitratoides	37	91–101
Liomys pictus	44	97–116
Perognathus parvus	23	71–91
P. californicus	22	87–92
Geomys breviceps (=pinetis)	(200)	153–253
Castor canadensis	24,000	820
Muscardinus avellanarius	20	80
Glis glis	92.5	180
Dryomys nitedula	92.5?	93
Eliomys quercinus	72.5	137
Tachyoryctes ruandae	250	—
Cannomys badius	325	18–20 cm
Mesembriomys gouldi	1,000	(240–350)
Mus musculus	20.5	90
Rattus norvegicus	330	250
R. rattus	187.5	210
Apodemus agrarius	20.5	108
A. sylvaticus	19.5	95

Taxon	Weight (g)	Head and Body Length (mm)
A. flavicollis	27	115
Micromys minutus	4–7	68
Mesocricetus auratus	80–100	130
Cricetus cricetus	150–250	260
Mystromys albicaudatus	99–145	—
Peromyscus leucopus	30 (22)	93–108
P. gossypinus	29	92–108
P. truei gilberti	27	—
P. yucatanicus	—	110–13
P. megalops auritus	71	—
P. crinitus	20 15.1 ♀	82–84
P. californicus	38	103–18
P. eremicus	22	77–90
P. maniculatus artemesiae	21	—
P. m. bairdii	15	—
P. m. gambelii	19	—
P. polionotus subgriseus	15	—
Baiomys taylori	7.9–9.4	53–70
Neotoma albigula	198 ♀	(207–13)
N. cinerea	271 ♀	(162–247)
Sigmodon hispidus	80–200 215♀	143–99
Onychomys leucogaster	30 ♀	(99–130)
Ochrotomys nutalli	24.7	82–97
Reithrodontomys montanus	8.5	59–80
R. megalotis	10–15	(74)
Scotinomys xerampelinus	15	77.8
S. teguina	15	78.7
Nyctomys sumichrasti	55	123–30
Ototylomys phyllotis	63.8	123–89
Tylomys nudicaudatus	280	—
Tatera indica	110	160–210
Pachyuromys duprasi	50	—
Gerbillus nanus	22	63–115
G. pyramidum	65	95–130
Psammomys obesus	212	70–300
Meriones crassus	105	111–50
M. libycus	100	117–71
M. tristrami	105	97–162
M. persicus	110	165
M. shawi	185	—
M. unguiculatus	53 ♀	—
Clethrionomys glareolus	23	108
Ondatra zibethicus	1,400 1,090 ♀	330
Arvicola terrestris	120 ♀	170
Pitymys subterraneus	17.5	98
Microtus pennsylvanicus	32 35 ♀	108
M. agrestis	30	122
M. arvalis	22	104
M. nivalis	43	130
Dicrostonyx groenlandicus	76	122–37
Sicista betulina	8	58
Jaculus orientalis	130–225	145 (N = 1)
Atherurus africanus	2,500	365–515
Hystrix cristata	20,000	—
Thryonomys swinderianus	2,000	432–584
Agouti paca	8,200	—

Taxon	Weight (g)	Head and Body Length (mm)
Dasyprocta sp.	2,700	—
Abrocoma cinerea	140	—
Octodontomys gliroides	100	—
Octodon degus	250	—
Chinchilla laniger	500	—
Lagidium boxi	1,400	—
L. peruanum	1,200	—
Lagostomus maximus	3,000	—
Cavia porcellus	1,000	—
C. aperea	500	—
Galea musteloides	400	—
Dolichotis patagona	7,856	—
Pediolagus salinicola	2,000	—
Hydrochoerus hydrochoeris	30,000	—
Erethizon dorsatum	5,000	500–560
Dinomys branickii	13000	—
Proechimys semispinosus	324–50	230
P. guairae	300	—
Ctenomys talarum	150	—
Geocapromys ingrahami	700	—
Capromys pilorides	4,300	540–55
Plagiodontia aedium	1,970	—
Myocastor coypus	6,000	—
Myoprocta pratti	1,000	—
Scandentia		
Tupaia belangeri	200	—
Primates		
Microcebus murinus	70	—
Lemur catta	2,290	—
L. fulvus	1,924	—
Galago crassicaudatus	1030	297–373
G. senegalensis	229	150–173
G. demidovii	60	125–60
Nycticebus coucang	1,230	270–300
Perodicticus potto	1,348	367–86
Arctocebus calabarensis	403	220–51
Tarsius syrichta	120	—
Cebuella pygmaea	145	130–44
Callithrix jacchus	241	—
Saguinus midas	483	—
S. oedipus	510	—
S. tamarin	483	—
Callimico goeldii	472	—
Leontopithecus rosalia	745	—
Pithecia pithecia	1,046	—
Saimiri sciureus	590	—
Cebus capucinus	2,700	444
Ateles geoffroyi	7,600	431–40
A. fusciceps	8,200–9,400*	—
Lagothrix lagotricha	5,400	—
Alouatta villosa	7,670	499–524
*Macaca mulatta**	4,370–10,659	450
Theropithecus gelada	9,830	—
Presbytis obscurus	5,400	(477–685)
Nasalis larvatus	9,873	540–723
Hylobates lar	6,800	454–73
Symphalangus syndactylus	10,300	533–42
*Gorilla gorilla**	75,000–110,000	1,070–1,600
Pongo pygmaeus	37,000	768–965

* In these strongly dimorphic species, mean weights for males and females are recorded.

Taxon	Weight (g)	Head and Body Length (mm)
Pan troglodytes	39,000	700–925
Homo sapiens	60,000	—
Fissipedia		
Canis lupus	28,000	(1,070–1,375)
C. latrans	11,000	(900)
Lycaon pictus	20,000	(762–1,015)
Cuon alpinus	17,000	(760–1,000)
Speothos venaticus	7,000	(575–750)
Vulpes vulpes	9,000	(455–865)
Alopex lagopus	6,000	(458–675)
Fennecus zerda	1,500	(357–407)
Otocyon megalotis	3,200	(460–580)
Chrysocyon brachyurus	23,000	(1,250)
Nyctereutes procyonoides	7,500	(500–550)
Ursus americanus	77,270	(1.5–1.8 m)
U. arctos middendorfi	300,000	—
U. a. arctos	105,000	(3 m)
Helarctos malayanus	80,000	(1.1–1.4 m)
Thalarctos maritimus	170,000	(2.2–2.5 m)
Mustela nivalis†	59 ♀	(114–260)
M. putorius	800	(290–460)
M. erminea	45–77	(65–325)
M. frenata†	102	(224–84)
M. vison	565–1,020	(347–79)
Lutra lutra	7,400	(630–750)
Martes martes	940	(450–580)
Taxidea taxus	6,000	(420–70)
Meles meles	5,060	(560–812)
Procyon lotor	5,270	(445–600)
Bassariscus astutus	870–1,100	(305–75)
Nasua nasua	5,000	(510–96)
Bassaricyon gabbii	2,800	(350–475)
Potos flavus	1,970	(415–575)
Ailurus fulgens	3,000–4,500	(510–635)
Ailuropoda melanoleuca	182,000	(1.2–1.3 m)
Herpestes edwardsi	1,320	(382–412)
Genetta tigrina	1,673	(490–600)
Fossa fossa	1,430–1,735	(465–520)
Eupleres goudotii	1,600	(600–700)
Galidia elegans	655–960	(340)
Mungotictis decemlineata	725–880	(325)
Cryptoprocta ferox	7,000–12,000	(610–760)
Crocuta crocuta	55,300	(1,275–1,658)
Hyaena brunnea	56,800	(1,121–1,172)
Panthera leo	151,000	(2,410–2,840)
P. pardus	39,000	(1,200–1,600)
P. onca	62,000	(1,300–1,390)
P. tigris	134,000	(1,600–3,000)
Uncia uncia	39,000	(1,070–1,300)
Neofelis nebulosa	20,000	(616–1,066)
Leopardus pardalis	12,000–14,000	(675–875)
L. tigrinus	1,600	(920–1,360)
Lynx canadensis	13,640	(626–70)
L. lynx	6,930	(735–75)
Felis nigripes	1,620	(384; N = 1)
F. margarita	2,194	(460–570)
F. sylvestris sylvestris	5,165	(470–660)
F. s. lybica	2,700	(489–632)
F. s. catus	3,260	550

† Because these are strongly dimorphic, female values are used.

Taxon	Weight (g)	Head and Body Length (mm)
Acinonyx jubatus	43,000	(1,900–2,260)
Puma concolor	43,700	(1,132–1,227)
Prionailurus bengalensis	3,270	(440–830)
Caracal caracal	7,434	(728–53)
Leptailurus serval	9,040	(670–1,000)
Pinnipedia		
Arctocephalus pusillus	120,000 ♂	—
A. tropicalis	50,000+	—
Callorhinus ursinus	55,000 ♀	1.2–1.5 m ♀, 2.2–2.5 m ♂
Eumetopias jubatus	350,000 ♀	2.2–2.8 m ♀, 3.1–4.0 m ♂
Odobenus rosmarus	560,000 ♀	3.7 m ♂
Phoca vitulina	65,000 ♀	1.5–1.9 m ♂
(Pusa) hispida	65,000 ♀	—
Pagophilus groenlandicus	140,000 ♀	—
Halichoerus grypus	?	1.7–1.9 m ♀, 3.7 m ♂
Erignathus barbatus	340,000 ♀	2.5–2.6 m ♀, 2.5–3.7 m ♂
Lobodon carcinophagus	220,000 ♀	—
Leptonychotes weddelli	370,000 ♀	—
Monachus schauinslandi	250,000 ♀	—
Cystophora cristata	270,000 ♀	2.5 m ♀, 3.3 m ♂
Mirounga lecnina	900,000 ♂	—
	420,000 ♀	—
Cetacea		
Physeter catodon	13,896,350	9.3–18 m
Tursiops truncatus	155,000	1.75–3.6 m
Phocaena phocaena	55,500	1.9 m
Balaenoptera musculus	93,869,000	24.9 m
B. physalis	50,172,680	15.5 m
B. acutorostrata	5,854,540	9.9 m
B. borealis	12,201,286	18.0 m
Megaptera novaeangliae	31,837,850	12.5 m
Tubulidentata		
Orycteropus afer	60,000	1,300
Hyracoidea		
Procavia capensis	17,000	345–574
Proboscidea		
Loxodonta africana	2,766,000	5,000
Elephas maximus	2,730,000	(5,500–6,400)
Perissodactyla		
Equus zebra	166,500	1,900
E. caballus	260,000	—
E. asinus	187,973	—
E. burchellii	219,000	—
Tapirus bairdi	185,450	1,950
T. terrestris	175,000	—
T. indicus	380,000	—
Diceros bicornis	1,081,000	3,500
Artiodactyla		
Sus scrofa	75,000	—
Potamochoerus porcus	72,270	—
Phacochoerus aethiopicus	65,000	—
Tayassu tajacu	13,640	851–85
Hippopotamus amphibius	1,277,000	—
Choeropsis liberiensis	272,000	—
Camelus dromedarius	570,000	—
C. bactrianus	450,000	—
Vicugna vicugna	50,000	—
Tragulus javanicus	1,500	(396–480)

Taxon	Weight (g)	Head and Body Length (mm)
T. napu	5,000	(500–600)
Moschus moschiferus	10,000	—
Muntiacus muntjak	18,000	—
Dama dama	39,000	—
Axis axis	46,000	—
Cervus canadensis	200,000	1,952–2,759 ♂
C. unicolor	162,000	—
Odocoileus hemionus	57,000 ♀	1,045–1,732
O. virginianus	47,730	1,188–1,732
Mazama gouazoubira	17,000	850–1,050
M. americana	20,000	1,000–1,240
Pudu mephistophiles	5,900	—
P. pudu	6,000	—
Rangifer tarandus	105,000	1,267–2,318
Giraffa camelopardalis	1,017,000	4,000
Antilocapra americana	40,000	1,156–1,294 ♂
Boselaphus tragocamelus	122,730	—
Bubalus bubalis	425,000	—
Bos grunniens	250,000	—
Bibos gaurus	702,730	—
Bos taurus	272,000	—
Syncerus caffer	447,000	2,500
Tragelaphus scriptus	28,600	1,350
Taurotragus oryx	300,000	—
Madoqua kirkii	5,300	580
Hippotragus niger	218,640	—
Kobus defassa	181,000	—
Connochaetes taurinus	219,000	—
Gazella thomsoni	17,500	—
G. granti	41,300	—
Aepyceros melampus	42,000	—
Saiga tatarica	45,000	—
Capricornis sumatraensis	120,000	(130–50)
Rupicapra rupicapra	34,000	—
Ovibos moschatus	300,000	1,939–2,332 ♂
Capra ibex	52,500	1,085–1,384
Ammotragus lervia	66,000	—
Ovis canadensis	61,360	1,089–1,803
O. (musimon) aries	30,000	—

Appendix 3

Longevity Records for Selected Mammalian Species

Longevity records were obtained in the main from the files of the National Zoological Park and from unpublished data collected by Marvin Jones. I am indebted to Mr. Jones for freely making available the results of his long-term research into records from the world's zoological parks (see also Jones 1979).

Taxon	Maximum Captive (months)	Mean Captive (months)
Monotremata		
Ornithorhynchus anatinus	168	—
Tachyglossus aculeatus	600	—
Zaglossus bruijni	231	—
Marsupialia		
Didelphis marsupialis	84	36
Marmosa robinsoni	22	—
Antechinus stuarti (=flavipes)	16	—
Sminthopsis crassicaudata	48	—
Antechinomys spenceri	38	—
Dasycercus cristicauda	72	—
Dasyuroides byrnei	78	—
Phascogale tapoatafa	36	—
Dasyurus maculatus	35	—
D. viverrinus	82	—
Sarcophilus harrisii	69	—
Trichosurus vulpecula	116	72
Thylacomys lagotis	65	—
Petaurus breviceps	126	78
Pseudocheirus peregrinus	68	—
Vombatus ursinus	312	132
Bettongia penicillata	96	—
Potorous tridactylus	78	—
Megaleia rufa	196	—
Wallabia (Protemnodon) bicolor	30	—
Onychogale frenata	—	54
Macropus giganteus	168	—
Insectivora		
Microgale talazaci	54	—
M. dobsoni	>50	—
Hemicentetes semispinosus	22	—
Tenrec ecaudatus	28	8
Edentata		
Tolypeutes tricinctus	132	—
Choloepus hoffmanni	132	—
Chiroptera		
Pteropus giganteus	205	—
Myotis sodalis	120	—

Taxon	Maximum Captive (months)	Mean Captive (months)
Pipistrellus subflavus	129	—
Eptesicus fuscus	218	—
Antrozous pallidus	77	—
Plecotus townsendii	120	—
Lagomorpha		
Oryctolagus cuniculus	156	66
Sylvilagus floridanus	117	—
Rodentia		
Sciurus carolinensis	164	108
S. niger	119	—
Ratufa bicolor	204	—
Tamiasciurus hudsonicus	96	—
Marmota marmota	164	84
Cynomys ludovicianus	120	—
Glaucomys volans	72	61
Castor canadensis	228	—
Mus musculus	36	18
Rattus norvegicus	37	30
R. rattus	84	—
Apodemus agrarius	—	24–48[a]
A. sylvaticus	—	24–48[a]
A. flavicollis	—	24–48[a]
Micromys minutus	—	48
Cricetus cricetus	30	24
Clethrionomys glareolus	—	24–36[a]
Ondatra zibethica	75	—
Arvicola terrestris	24–48[a]	—
Pitymys subterraneus	24–48[a]	—
Microtus pennsylvanicus	16	—
M. agrestis	24–36[a]	—
M. nivalis	—	24–48[a]
Jaculus orientalis	48	12
Atherurus africanus	96	—
Hystrix cristata	244	120
Thryonomys swinderianus	50	—
Agouti paca	152	—
Dasyprocta sp.	120	72
Chinchilla laniger	84	48
Cavia porcellus	72	24
Dolichotis patagona	156	60
Hydrochoerus hydrochoeris	138	—
Myocastor coypus	120	48
Primates		
Lemur macaco	252	126
Galago crassicaudatus	152	—
G. senegalensis	117	—
Nycticebus coucang	98	—
Callithrix jacchus	192	132
Pithecia pithecia	143	—
Ateles geoffroyi	216	—
A. fusciceps	288+	—
Macaca mulatta	348	180
Hylobates lar	276	—
Gorilla gorilla	312	—
Pongo pygmaeus	336	—
Pan troglodytes	444	204
Fissipedia		
Canis lupus	168	144
C. latrans	180	108

Taxon	Maximum Captive (months)	Mean Captive (months)
C. aureus	122	—
Lycaon pictus	121	—
Cuon alpinus	108	—
Vulpes vulpes	117	—
Urocyon cinereoargenteus	101	—
Alopex lagopus	168	—
Ursus americanus	311	—
U. arctos arctos	408	228
Helarctos malayanus	246	132
Thalarctos maritimus	400	192
Mustela nivalis	93	—
M. vison	120	—
Lutra lutra	144	—
L. canadensis	228	—
Martes martes	162	120
Taxidea taxus	166	132
Meles meles	168	—
Eira barbara	210	72
Galictis cuja	87	—
Mephitis mephitis	73	—
Procyon lotor	165	49
Nasua nasua	121	—
Bassaricyon gabbii	96	—
Potos flavus	233	—
Ailuropoda melanoleuca	180	—
Civettictis civetta	156	—
Genetta tigrina	150	84
Arctictis binturong	216	—
Paradoxurus hermaphroditus	180	96
Mungos mungo	96	60
Cryptoprocta ferox	204	—
Crocuta crocuta	300	144
Hyaena hyaena	288	144
Panthera leo	348	270
P. pardus	276	168
P. onca	268	168
P. tigris	234	132
Uncia uncia	104	—
Neofelis nebulosa	96	—
Leopardus pardalis	152	—
Lynx canadensis	139	73
L. lynx	193	—
Felis sylvestris catus	252	180
Acinonyx jubatus	187	72
Puma concolor	192	108
Prionailurus bengalensis	144	—
P. viverrinus	120	—
Caracal caracal	202	63
Pinnipedia		
Arctocephalus pusillus	192	—
Callorhinus ursinus	300	—
Eumetopias jubatus	228	156
Zalophus californianus	276	156
Odobenus rosmarus	>180	—
Phoca vitulina	228	—
Tubulidentata		
Orycteropus afer	116	—
Hyracoidea		
Procavia capensis	74	—

Taxon	Maximum Captive (months)	Mean Captive (months)
Proboscidea		
Loxodonta africana	544	292
Elephas maximus	684	480
Perissodactyla		
Equus zebra	307	264
E. caballus	600	300
E. asinus		600
E. burchelli	334	144
Tapirus terrestris	365	—
Diceros bicornis	271	—
Didermocerus sumatrensis	420	—
Rhinoceros unicornis	564	504
Artiodactyla		
Phacochoerus aethiopicus	182	—
Hippopotamus amphibius	594	480
Choeropsis liberiensis	475	—
Vicugna vicugna	202	144
Odocoileus hemionus	180	96
Rangifer tarandus	144	—
Giraffa camelopardalis	336	168
Boselaphus tragocamelus	252	—
Bubalus bubalis	360	—
Bos grunniens	267	—
B. taurus	360	270
Syncerus caffer	185	120
Bison bison	264	120
Tragelaphus scriptus	108	—
Taurotragus oryx	300	109
Hippotragus niger	200	—
Kobus defassa	200	108
Connochaetes taurinus	192	—
Aepyceros melampus	152	—
Oreamnos americanus	115	—
Rupicapra rupicapra	202	—
Capra hircus	216	108
Ammotragus lervia	185	96
Ovis (musimon) aries	240	150

[a] Ranges are given for species often showing a cohort turnover or where values tend toward two modes.

Appendix 4

Reproduction and Development for Selected Mammalian Species

Taxon	Gestation (days)	Age of Eye Opening (days)	Weight of Neonate (g)	Range or Mean Litter Size	Average Interbirth Interval (months)[a]	Average Litters/Year	Age at First Mating (days)	Duration of Lactation (days)	References
Monotremata									
Ornithorhynchus anatinus	—	—	—	1.3	>12	1	—	120	Fleay 1944
Tachyglossus aculeatus	—	—	—	1	>12	1	365	140	Griffiths 1968
Marsupialia									
Didelphis marsupialis	12–13	58–72	.21–.16	7	3.5	1–2	186 ♀	100	Collins 1973
Marmosa robinsoni	14	39–40	~.18	11	3	1–2	180 ♀	65	Collins 1973
Monodelphis domestica	14	—	—	7.3	4.5	2	182 ♀	—	NZP records
Antechinus stuarti (=flavipes)	23–37	62	.18	3–12	3	1–2	270 ♀	(90)	Collins 1973; Marlow 1961
Sminthopsis crassicaudata	13–16	49–50	—	10	3	1–2	115 ♀	70	Collins 1973; Godfrey and Crowcroft 1971
Dasyurus viverrinus	8–14	63	.2	6	—	1	>300 ♀	105+	Collins 1973
Sarcophilus harrisii	31	87–93	—	1–4	—	1	>700 ♀	140	Collins 1973
Perameles gunni	11	44–46	(.25)	(2.33)	2.5	4	70 ♀	(60)	Collins 1973; Lyne 1952
Isoodon macrourus	15	45	.18	4	—	—	—	63–70	Collins 1973
Thylacomys lagotis	—	—	—	1.5	—	1	—	—	Collins 1973
Trichosurus vulpecula	18	100–110	.2–.32	1.4	6	2	>300 ♀	150	Collins 1973
Petaurus breviceps	20	80	.194	1–3	5	2	<270 ♀	120	Collins 1973; Smith 1971
Pseudocheirus peregrinus	—	91	—	2	—	1	<300 ♀	160	Collins 1973
Vombatus ursinus	—	—	2	1	—	1	—	—	Collins 1973
Bettongia penicillata	—	—	—	1	—	—	—	—	Collins 1973
B. lesueuri	22	(83)	.38	1	—	—	—	—	Collins 1973; Stodart 1977
Potorous tridactylus	~38	(~94)	(.33)	1	4.5	2	<360 ♀	120	Collins 1973; Guiler 1971
P. apicalis	33	94	.33	1	—	—	—	—	Collins 1973
Megaleia rufa	32–34	144	.84	1	8	1	17 mo ♀	>300	Collins 1973
Wallabia (Protemnodon) bicolor	(35)	—	—	1	—	1	—	—	Collins 1973
Macropus giganteus	29–38	163–81	.9	1	>8	1	(18 mo)	30	Collins 1973
Insectivora									
Solenodon paradoxus	—	—	60	1–2	—	—	—	—	Eisenberg 1975*b*
Microgale talazaci	58–63	18	3.6	1–3	—	—	—	30	Eisenberg 1975*b*

Note: Numbers in parentheses were not used in the tabular computations presented in the text but are included here to complete the table.
[a] Assumes young survives after birth and the female exhibits a lactation anestrous period.

Taxon	Gestation (days)	Age of Eye Opening (days)	Weight of Neonate (g)	Range or Mean Litter Size	Average Interbirth Interval (months)[a]	Average Litters/Year	Age at First Mating (days)	Duration of Lactation (days)	References
M. dobsoni	62–64	—	4	1–2	—	—	—	—	Eisenberg 1975*b*
Hemicentetes semispinosus	57–63	8–10	6.5	2–11	3.7	2	35–40 ♀	20	Eisenberg 1975*b*
H. nigriceps	55	8	6.3	2–4	3.7	2	30–35	20	Eisenberg 1975*b*; Eisenberg and Muckenhirn 1968
Setifer setosus	65–69	13	24.7	1–5	—	1	6 mo	—	Eisenberg 1975*b*; Eisenberg and Muckenhirn 1968
Echinops telfairi	62–65	(7–9)	6	1–10 (5.8)	—	—	6 mo	~30	Eisenberg 1975*b*; Eisenberg and Muckenhirn 1968
Tenrec ecaudatus	57–63	9–14	25	1–32	3.4	1–2	6 mo	29	Eisenberg 1975*b*; Eisenberg and Muckenhirn 1968
Echinosorex gymnurus	—	—	—	1.5	—	—	—	—	Gould 1978*a,b*
Erinaceus europaeus	34–49	14–18	13.9	1–7	—	2	—	21–28	Herter 1957; Morris 1966
Parechinus micropus	—	21	—	1–2	—	—	—	—	Herter 1957
Sorex araneus	20	18–21	0.5	5.7–7.7	1.4	1.5	~280	—	Crowcroft 1957
S. minutus	≥25	—	0.25	2–4	—	1.5	—	—	Hutterer 1976
S. cinereus	—	14–17	0.28	6.7	—	—	—	—	Pruitt 1954; Forsyth 1976
S. vagrans	20	7–14	0.33	5.2	—	—	—	—	Johnston and Rudd 1957
Cryptotis parva	21–23	14	.3–.4	4–6	—	—	—	21–22	Hamilton 1944
Blarina brevicauda	21–22	—	.67–1.29	3–9	—	—	30–60	—	Blus 1971
Neomys fodiens	24	22–23	1+	3–8	—	—	—	37	Price 1953
Notiosorex crawfordi	—	—	.31–.47	—	—	—	—	—	Hoffmeister and Goodpaster 1962
Microsorex hoyi	—	—	—	5–6	—	—	—	—	Walker et al. 1968
Crocidura hirta	—	12–13	~1	—	—	—	—	17–18	Ansell 1964
C. bicolor	—	10	—	—	—	—	—	14–15	Ansell 1964
C. leucodon	28–31	12–14	.8	3–7	—	—	60–90	—	Hellwing 1973*a*
C. suaveolens	28	7–9	.42–.67	—	—	—	—	18–22	Vlasek 1972; Hanzak 1966
C. rusulla	28.5	10	.8	1–7	—	—	—	—	Vogel 1972; Hellwing 1973*b*
Suncus murinus	30.2	7–10	2–2.3	1.84	—	—	—	17–20	Dryden 1968
S. etruscus	27–28	14–16	.18–.25	2.5	—	—	—	—	Fons 1973; Vogel 1970
Talpa europaea	30–40	22	—	3.8	12	1	180	—	Godfrey and Crowcroft 1960
Parascalops breweri	28–42?	—	—	4	12	1	—	28	Eadie 1939
Scapanus townsendi	—	—	—	2–4	12	1	—	—	Ingles 1965
Scalopus aquaticus	42	—	—	2–5	12	1	—	—	Arlton 1936

Taxon	Gestation (days)	Age of Eye Opening (days)	Weight of Neonate (g)	Range or Mean Litter Size	Average Interbirth Interval (months)[a]	Average Litters/Year	Age at First Mating (days)	Duration of Lactation (days)	References
Edentata									
Dasypus novemcinctus	120 +	12–18	40–50	4	12	1	—	—	Block 1974
Tolypeutes matacus	—	21–28	85	1	—	—	—	—	Meritt 1976
Euphractus sexcinctus	60–65	22–25	95–115	1–3	—	—	—	28–35	Gucwinska 1971
Chaetophractus villosus	60	32	145	2	—	1	—	—	Encke 1965
Choloepus hoffmanni	332	0	340–400	1	18	0.5	—	~20	NZP records
Bradypus variegatus	170–80	0	230–50	1	12	1	—	<10	Montgomery and Sunquist 1978
Myrmecophaga tridactyla	190	6?	1,587	1	9	1.34	—	—	Bickel, Murdock, and Smith 1976; Hardin 1976
Tamandua tetradactyla	—	—	—	1	7	1.7	—	—	NZP records
Pholidota									
Manis tricuspis	?	—	140	1	—	—	—	—	NZP records
Chiroptera									
Cynopterus sphinx	115–25	0	12–14	1	—	—	—	—	Gopalakrishna 1969
Rousettus leschenaulti	125	0	12	1	—	—	—	—	Gopalakrishna 1969; Kulzer 1966
Pteropus geddeiri	180	0	—	1	—	—	—	—	Gopalakrishna 1969
P. niger	142	0	—	1	—	1	—	—	Gopalakrishna 1969
P. giganteus	145	0	—	1	12	1	—	—	Gopalakrishna 1969
Eidolon helvum	120	0	—	1	—	—	—	—	Gopalakrishna 1969; Mutere 1967
Rhinopoma kinneari	123[b]	?	—	—	—	—	—	—	Gopalakrishna 1969; Khajuria 1972
Megaderma lyra (=spasma)	155	0	7–8	1	—	1	—	—	Gopalakrishna 1969
Artibeus jamaicensis	—	0	—	1	5	2	200	—	NZP records
Carollia perspicillata	105–15	0	5	1	5	2	200	6 wk	Kleiman and Davis 1978
Desmodus rotundus	150	0	6	1	12	1	—	—	Turner 1975; Schmidt and Manske 1973
Hipposideros bicolor	42.5	?	1.2	1	—	—	—	—	Gopalakrishna 1969
Myotis lucifugus	55	0	2	—	12	1	—	—	Wimsatt 1966
M. myotis	53	—	—	—	12	1	—	—	
Pipistrellus pipistrellus	44	4.8	1.4	—	12	1	—	—	Kleiman 1969

[b] Includes delayed implantation time?

Taxon	Gestation (days)	Age of Eye Opening (days)	Weight of Neonate (g)	Range or Mean Litter Size	Average Interbirth Interval (months)[a]	Average Litters/Year	Age at First Mating (days)	Duration of Lactation (days)	References
P. ceylonicus	52.5	0	1.2–1.4	2	—	—	—	—	Gopalakrishna 1969
Nyctalus noctula	71.5	6.1	5.7	—	—	1	—	—	Kleiman 1969
Eptesicus fuscus	—	—	3.5	1	12	1	—	>30	Kleiman 1969
E. serotinus	—	7.8	5.8	—	12	1	—	—	Kleiman 1969
Scotophilus wroughtoni	110?[b]	—	—	2	—	—	—	—	Gopalakrishna 1969
Antrozous pallidus	63	9	3.1	—	12	1	—	33	Davis 1969
Plecotus townsendii	56	—	2.1–2.7	—	—	1	—	—	Pearson, Koford, and Pearson 1952
Tylonycteris robustula	87.5	—	1.25	—	—	—	—	—	Medway 1972
Tadarida brasiliensis	80.5?	—	—	—	12	1	270	—	Asdell 1964
Lagomorpha									
Oryctolagus cuniculus	30–37	10	38	4–15	—	—	—	~20	Flux 1967; Brambell 1944; McIlwaine 1962
Sylvilagus aquaticus	39–40	2–3	61	2.8	—	—	—	14	Hunt 1959
Lepus americanus	37	0	(50) 70–80	3–5	—	2.4	—	15	Newson 1964
L. californicus	41–47	0	69	2.3	—	4.3	—	14	Lechleitner 1959
L. timidus	44–53	0	87	1–7	—	2	—	28	Flux 1970
L. europaeus	42–44	0	123	1–6	—	3–4	—	—	Flux 1967
Macroscelidea									
Macroscelides proboscideus	76	0	6–8	1–2	2.5	—	—	—	Rosenthal 1975
Petrodromus tetradactylus	—	0	30.5	1	—	—	—	—	Rathbun 1976, 1979
Rhynchocyon chrysopygus	42	0	80	1	—	4–5	—	14	Rathbun 1976, 1979
Elephantulus rufescens	56	0	10	1–2	2	4–5	—	22	Rathbun 1976, 1979
E. brachyrhynchus	—	0	—	2	—	—	—	—	Rathbun 1976, 1979
Rodentia									
Sciurus carolinensis	44	37	16.5	2.7	—	2	—	(49–70)	Horwich 1972
S. niger	45	42	16	3	—	2	—	—	Asdell 1964
S. vulgaris	38	31	11.5	3–7	—	2	—	43–70	Mohr 1954
Tamiasciurus hudsonicus	40	26–34	7–8	3.3	12	1–2	13–15 mo	50	Dolbeer 1973; Prescott and Ferron 1978; French, Stoddard, and Bobek 1975
Tamias striatus	31	31	~2.5	3–5	3–12	1–2	—	35+	French, Stoddard, and Bobek 1975; Elliott 1978
Eutamias amoenus	32	31.5	2.6	5.25	—	1	—	—	Broadbrooks 1958

Taxon	Gestation (days)	Age of Eye Opening (days)	Weight of Neonate (g)	Range or Mean Litter Size	Average Interbirth Interval (months)[a]	Average Litters/Year	Age at First Mating (days)	Duration of Lactation (days)	References
E. quadrivittatus	31.5	31	2.8	5.25	—	1	—	—	Wadsworth 1969
Marmota monax	31–32	26–28	—	4	12	1	2 yr	—	Hamilton 1934
Cynomys ludovicianus	27–33	33–37	15	5	12	1	—	49	Koford 1957
Citellus columbianus	24	21–23	7–9	4.5	—	—	—	28	Shaw 1925
C. tridecemlineatus	28	27	2.6–4	8	—	1.5	—	29	Johnson 1931
C. citellus	25–28	28	—	6	—	—	—	—	Mohr 1954
Glaucomys sabrinus	37	31	5.8	4	—	—	—	65	Muul 1969
G. volans	39	26	3.63	3	12	1	365?	60–70	Sollberger 1943
Dipodomys panamintinus	29	17–18	4.5	3–4	—	—	—	28	Eisenberg 1963
D. merriami	33	11–12	—	2	—	—	—	20	Eisenberg 1963
D. nitratoides	32	10	4.0	2–3	—	—	—	23	Eisenberg 1963
Liomys pictus	25	19	2.5	4	—	—	—	26	Eisenberg 1963
Perognathus parvus	24	15	~1.5	5	1.5–12	1–2	—	—	Eisenberg 1967; NZP records
P. californicus	25	15	1.5	3.5	—	—	—	23	Eisenberg 1963
Geomys breviceps (=pinetis)	28	—	(5.8)	2 (1.7)	—	1.7	—	—	Wood 1949
Thomomys bottae	29	—	4.1	5.85	—	—	—	—	Scheffer 1938
Castor canadensis	128	0	340	4	12	1	2–3 yr	42	Bradt 1938; Wilsson 1971
Muscardinus avellanarius	20	17	—	6	—	1–2	—	—	Mohr 1954
Glis glis	30–32	20–22	—	5.0	—	1–2	—	—	Mohr 1954
Dryomys nitedula	24.5	21	—	3.0	—	1–2	—	—	Mohr 1954
Eliomys quercinus	23	19.5	—	5	—	1–2	—	—	Mohr 1954
Tachyoryctes ruandae	47.5	22	14	1–2	—	—	—	—	Rahm 1969
Cannomys badius	41.5	23.5	7	2.0	—	—	—	—	Eisenberg and Maliniak 1973
Mesembriomys gouldi	43–44	11	35	1–3	—	42	—	—	Crichton 1969
Mus musculus	19	13	1.5	5	—	4–6	—	—	Mohr 1954; Altman and Dittmer 1962
Rattus norvegicus	22	15	4.92	8	—	2–7	—	—	Mohr 1954; Altman and Dittmer 1962
R. rattus	22	14.5	—	—	—	2+	—	—	Mohr 1954; Altman and Dittmer 1962
Apodemus agrarius	22	12.5	—	2–8	—	3–4	—	—	Mohr 1954
A. sylvaticus	23	14	—	—	—	4–5	—	—	Mohr 1954

Taxon	Gestation (days)	Age of Eye Opening (days)	Weight of Neonate (g)	Range or Mean Litter Size	Average Interbirth Interval (months)[a]	Average Litters/Year	Age at First Mating (days)	Duration of Lactation (days)	References
A. flavicollis	24.5	14	—	3–8	—	2–3	—	—	Mohr 1954
Micromys minutus	21	8–9	—	11	—	—	—	—	Mohr 1954
Mesocricetus auratus	16	12–13	1.8	11	—	—	56–70	21	NZP records (unpublished)
Cricetus cricetus	20	14	—	11	—	—	43	—	Mohr 1954
Mystromys albicaudatus	38	—	6.5	3 (2.9)	—	—	32	—	Hall et al. 1967
Peromyscus leucopus	23	13	1.87	4.36 (5)	—	—	23.6 (21)	—	Layne 1968
P. gossypinus	23	10–18	2.2	3.7	—	—	(20) 30.2	—	Layne 1968
P. truei gilberti	26.2	14?	2.34	3.4	—	—	(21) 40[c]	—	Layne 1968
P. yucatanicus	27–31	13–21	2.5	3.5	—	—	—	—	Lackey 1976
P. megalops auritus	—	22.8	2.9–5 (3.9)	1.6	—	—	—	21	Layne 1968
P. crinitus	24	13.5	2.2	3.7	—	2.05	—	28	Layne 1968; Egoscue 1964
P. californicus	23.6	15	4.3	1–3	46 d	3–4	—	21?	Layne 1968; McCabe and Blanchard 1950
P. eremicus	21	13	2	1–4 (3.0)	—	—	—	35	Layne 1968; NZP records
P. maniculatus artemesiae	22–26	14.5	1.3–2.2	4.43	30 d	2–4	—	30.6 (25)	Layne 1968
P. m. bairdii	25	13.7	1.1–2.3	3–4	30 d	2–4	—	18	Layne 1968
P. m. blandus	22–25	14.4	1.5–2.1	3.85	—	2–4	—	25.3 (18)	Layne 1968
P. m. gambelii	23.5	13.3	1.4	3.2	—	2–4	—	18?	Layne 1968
P. polionotus subgriseus	23–24	13.7	1.1–2.2	3.35	—	3.4	—	—	Layne 1968
Baiomys taylori (?)	18?	12	1.1	2.72	—	—	—	—	Hudson 1974
Neotoma albigula	~38	15	10.9	2.2	—	—	—	30	Egoscue 1957; Schwartz and Bleich 1975
N. fuscipes	33	17	—	2–3	—	—	—	—	Linsdale and Tevis 1951
N. cinerea	27–32	14–15	—	3.5	—	—	—	21	Martin 1973
Sigmodon hispidus	27	0	7.2 (6)	5.6 (6.4)	—	—	—	20	Meyer and Meyer 1944
Onychomys leucogaster	27–32	18	—	3.7	—	—	—	33–47	NZP records (unpublished)
Ochrotomys nutalli	29–30	12.7	2.8	2.65	—	—	—	18	Linzey and Linzey 1967
Reithrodontomys montanus	23–24	8	1.0–1.3	2–4	—	—	—	—	Hall and Kelson 1959
R. megalotis	23	7–8	—	—	27 d	—	—	—	Leraas 1938
Scotinomys xerampelinus	32	17.8	2.2	2–4	—	—	—	—	Hooper 1974

[c] *Peromyscus truei trucus.*

Taxon	Gestation (days)	Age of Eye Opening (days)	Weight of Neonate (g)	Range or Mean Litter Size	Average Interbirth Interval (months)[a]	Average Litters/Year	Age at First Mating (days)	Duration of Lactation (days)	References
S. teguina	30	12–15	2.2	3	—	—	—	—	Hooper 1974
Nyctomys sumichrasti	34	16.5	4.7	2	—	—	—	—	Birkenholz and Wirtz 1965
Ototylomys phyllotis	50–52	4	10.2	2.31	—	—	—	53.6	Helm 1973, 1975
Tylomys nudicaudatus	39.2	11	—	2.5	—	—	—	40.6	Helm 1973; Tesh and Cameron 1970
Tatera indica	24	18	—	11	—	—	—	—	Eisenberg 1967
Pachyuromys duprasi	21	20.5	—	4.5	—	—	—	—	Petter 1961
Gerbillus nanus	21–23	13.5	—	3–7	—	—	—	—	Eisenberg 1967
G. pyramidum	21	18.5	2.6	4	—	—	—	—	Petter 1961
Psammomys obesus	25	15.5	—	—	—	—	—	—	Petter 1961
Meriones crassus	21	17	—	3–6	—	—	—	—	Koffler 1972
M. libycus	24	14.3	—	2–7	—	—	—	—	Petter 1961
M. tristrami	24	15.5	—	—	—	—	—	—	Petter 1961
M. persicus	22–23	16–17	—	4	—	—	—	—	Eibl-Eibesfeldt 1951*b*
M. shawi	21	18	4.8	4	—	—	—	—	Petter 1961
M. unguiculatus	23	—	2.8	5.1	—	—	—	30	NZP records
Clethrionomys glareolus	21	12	1.6	4–7 (5)	—	3–4	—	18	Asdell 1964
Ondatra zibethicus	30	11	21	8	—	4–5	—	17	Asdell 1964
Arvicola terrestris	21	9.5	10	2–7 (6.5)	—	3–4	—	—	Mohr 1954
Pitymys subterraneus	21	12.5	—	2–5	—	5–6	—	—	Asdell 1964
Microtus pennsylvanicus	21	—	2.6	4	—	—	—	14	Asdell 1964
M. agrestis	21	9.5	—	4–7	—	3–4	—	—	Mohr 1954
M. arvalis	21	9.5	1.8	8	—	3–7	—	16	Mohr 1954
M. nivalis	21	10	—	4–7	—	2	—	—	Mohr 1954
Sicista betulina	31.5	?	—	2–6	—	1	365	—	Mohr 1954
Jaculus orientalis	29	35	—	2	—	—	—	—	NZP records
Hystricomorpha									Weir 1974
Atherurus africanus	105	0	150	1	—	—	—	—	Rahm 1962
Hystrix cristata	112	0	1,000	2	—	—	—	—	NZP records
Thryonomys swinderianus	155	0	130	4	—	—	—	30	

Taxon	Gestation (days)	Age of Eye Opening (days)	Weight of Neonate (g)	Range or Mean Litter Size	Average Interbirth Interval (months)[a]	Average Litters/Year	Age at First Mating (days)	Duration of Lactation (days)	References
Caviomorpha									Weir 1974
Agouti paca	115	0	702	1.5	—	—	—	—	NZP records
Dasyprocta sp.	~120	0	200	1.5	—	—	270	20	Weir 1974
Abrocoma cinerea	115–18	0	15–18	2	—	—	—	—	NZP records
Octodontomys gliroides	99	0	15	2	—	—	—	—	NZP records
Octodon degus	90	0	14	5	—	—	—	28	NZP records
Chinchilla laniger	111	0	35	2	—	—	150–240	~49	Weir 1974
Lagidium boxi	140	0	260	1	—	—	—	—	Weir 1974
L. peruanum	140	0	180	1	—	—	—	56	NZP records
L. maximus	153	0	200	2	—	—	—	30	Weir 1974
Cavia porcellus	68	0	100	6	—	—	—	21	Weir 1974
C. aperea	61	0	60	3	—	—	—	—	Weir 1974
Galea musteloides	52	0	40	3	—	—	—	21	Weir 1974
Dolichotis patagona	93	0	432	1.5	—	—	—	—	Weir 1974
Pediolagus salinicola	77	0	193	1	—	—	—	—	NZP records
Hydrochoerus hydrochoeris	150	0	1,500	4	—	—	—	112?	Weir 1974; Trapido 1942; Ojasti 1973
Erethizon dorsatum	217	0	1,500	1	—	—	—	56	Weir 1974
Dinomys branickii	222–83	—	900	2	—	—	—	—	NZP records
Proechimys semispinosus	64.8	0	23	2.8	63 d	—	—	21	Maliniak and Eisenberg 1971
P. guairae	63	0	22	3	—	—	60 ♂–90 ♀	21	Weir 1974
Ctenomys talarum	130	0	8	4	—	—	—	35	Weir 1974
Geocapromys ingrahami	100	0	80	1	—	—	—	—	Weir 1974
Capromys pilorides	120	0	220	2	—	—	—	—	NZP records
Plagiodontia aedium	135–50	0	—	1	—	—	—	—	NZP records
Myocastor coypus	132	0	225	5	—	—	—	56	Weir 1974
M. pratti	99	0	100	2	—	—	—	14–20	Kleiman 1972*a*
Scandentia									
Tupaia belangeri	45	21.5	10	2	—	7–8	90	33–37	Martin 1968

Taxon	Gestation (days)	Age of Eye Opening (days)	Weight of Neonate (g)	Range or Mean Litter Size	Average Interbirth Interval (months)[a]	Average Litters/Year	Age at First Mating (days)	Duration of Lactation (days)	References
Primates									Ardito 1975; Leutenegger 1973
Microcebus murinus	60	2	10	2	—	—	—	45	Bourlière, Petter-Rousseaux, and Petter 1961
Lemur catta	127.5	0	88.2	1	—	—	—	—	Ardito 1975; Leutenegger 1973
L. fulvus	127.5	0	83.5	1	—	—	—	—	Ardito 1975; Leutenegger 1973
Galago crassicaudatus	133	0	46.4	2	—	—	—	—	Ardito 1975; Leutenegger 1973
G. senegalensis	122–25	0	23.2	1	—	—	365	30	Gucwinska and Gucwinska 1968
G. demidovii	—	0	7.6	1	—	—	—	—	Ardito 1975; Leutenegger 1973
Nycticebus coucang	193	0	45	1	—	1	—	90–180	Ardito 1975; Leutenegger 1973
Perodicticus potto	193	0	48.5	1	—	1	—	—	Ardito 1975; Leutenegger 1973
Arctocebus calabarensis	133	0	25.3	1	—	1	—	—	Ardito 1975; Leutenegger 1973
Tarsius syrichta	180	0	26.2	—	—	—	—	—	Ardito 1975; Leutenegger 1973
Cebuella pygmaea	145	0	16	2	—	—	—	—	Ardito 1975; Leutenegger 1973
Callithrix jacchus	144	0	27.5	2	151–56 d	—	—	180	Rothe 1977; Stevenson 1976
Saguinus midas	140	0	39	2	—	1	—	—	Ardito 1975; Leutenegger 1973
S. oedipus	140	0	40	2	—	1	—	—	Ardito 1975; Leutenegger 1973

Taxon	Gestation (days)	Age of Eye Opening (days)	Weight of Neonate (g)	Range or Mean Litter Size	Average Interbirth Interval (months)[a]	Average Litters/Year	Age at First Mating (days)	Duration of Lactation (days)	References
S. tamarin	140	0	34	2	—	1	—	—	Ardito 1975; Leutenegger 1973
Callimico goeldii	154	0	40	1	—	1	—	—	Ardito 1975; Leutenegger 1973
Leontopithecus rosalia	128	0	61	2	180 d	1–2	—	90	NZP records; Hoage 1978
Pithecia pithecia	163	0	—	1	—	—	—	—	Ardito 1975; Leutenegger 1973
Saimiri sciureus	157	0	84	1	—	—	—	—	Ardito 1975; Leutenegger 1973
Cebus capucinus	180	0	230	1	—	—	—	—	Ardito 1975; Leutenegger 1973
Ateles geoffroyi	225	0	512	1	—	—	—	330	NZP records
A. fusciceps	225	0	235	1	2–3 yr	.5	—	~330	NZP records
Lagothrix lagotricha	225	0	450	1	1.5–2 yr	.5	8 yr	9–12 mo	Williams 1967
Alouatta villosa	180–94	0	440	1	—	—	—	—	Ardito 1975
Macaca mulatta	167	0	330–600	1	—	—	—	—	Ardito 1975
Theropithecus gelada	170	0	464	1	—	—	—	—	Ardito 1975; Leutenegger 1973
Presbytis obscurus	150	0	485	1	—	—	—	—	Ardito 1975; Leutenegger 1973
Nasalis larvatus	166	0	450	1	—	—	—	7 mo	Pournelle 1967
Hylobates lar	225	0	400	1	—	—	—	—	NZP records
Symphalangus syndactylus	223	0	560	1	—	—	—	—	NZP records
Gorilla gorilla	252–55	0	2,040	1	—	—	—	14 mo	NZP records
Pongo pygmaeus	263	0	1,200	1	—	—	—	—	NZP records
Pan troglodytes	235	0	1,580	1	—	—	—	—	NZP records
Homo sapiens	270	0	3,300	1	3.5 yr	.28	—	—	
Fissipedia									Ewer 1973
Canis lupus	63	12–15	400	6.5	—	—	—	42–56	

Taxon	Gestation (days)	Age of Eye Opening (days)	Weight of Neonate (g)	Range or Mean Litter Size	Average Interbirth Interval (months)[a]	Average Litters/Year	Age at First Mating (days)	Duration of Lactation (days)	References
C. latrans	58–61	14	274	4	—	—	—	35	Bekoff and Jamieson 1975
C. aureus	60–63	~5	—	2–6	—	—	—	63	Ewer 1973
Cerdocyon thous	53	—	590	4.5	7.4	1.8	740	—	NZP records
Lycaon pictus	72–73	14	365	7	—	—	—	—	Ewer 1973
Cuon alpinus	60–70	13–15	270	2–5	—	—	—	—	Ewer 1973
Speothos venaticus	65	14–17	157	6	—	—	—	—	Ewer 1973
Vulpes vulpes	51–52	13–15	105	5–6	12	1	—	56–70	Ewer 1973
Urocyon cinereoargenteus	63	—	—	2–7	12	1	—	—	Ewer 1973
Alopex lagopus	52	14–16	75	4–10	12	1	—	—	Ewer 1973
Fennecus zerda	50–52	8	28	3	12	1	—	56–70	Ewer 1973
Otocyon megalotis	60–75	?	123	2–5	—	—	—	—	Ewer 1973
Chrysocyon brachyurus	60–65	12	431	1–3	—	—	—	105+	Brady and Ditton 1979
Nyctereutes procyonoides	59–79	9–10	75	2–8	—	—	—	56	Ewer 1973
Ursus americanus	67[d]	21	252–336	2.5	—	—	—	—	Ewer 1973
U. arctos middendorfi	70[d]	28	600	2.23	—	—	—	720	Ewer 1973
U. a. arctos	56–70[d]	44	765	2.5	—	—	—	—	Ewer 1973
Helarctos malayanus	95–96	14	225	—	—	—	—	—	Ewer 1973
Thalarctos maritimus	—	33–39	765	2.5	24	.5	2.5–4 yr	720	Ewer 1973
Mustela nivalis	35–36[d]	26–30	?	4.8	3	2	—	35	Ewer 1973; Heidt, Peterson, and Kirland 1968
M. rixosa	35–38[d]	—	1.1	4–6	3	1–2	—	—	Ewer 1973
M. putorius	41–42[d]	28	10	6–7	12	1	—	42–56	Ewer 1973
M. erminea	~28?[d]	35	—	2	12	1	—	—	Ewer 1973
M. frenata	23–24[d]	35–37	3.1	6–7	12	1	—	—	Ewer 1973
M. vison	28–30[d]	—	—	4–6	12	1	365	—	Ewer 1973
Lutra lutra	61–63[d]	35	—	2	—	—	—	—	Ewer 1973
L. canadensis	56[d]	16–19	—	2.3	—	—	—	112–40	Johnstone 1978
Martes martes	30[d]	32–36	—	3.0	12	1	—	42–46	Ewer 1973

[d] Delayed implantation corrected for; actual time from conception to birth is much longer.

Taxon	Gestation (days)	Age of Eye Opening (days)	Weight of Neonate (g)	Range or Mean Litter Size	Average Interbirth Interval (months)[a]	Average Litters/Year	Age at First Mating (days)	Duration of Lactation (days)	References
Taxidea taxus	42	—	116	2–3	12	1	—	42	Ewer 1973
Meles meles	42	10	84	1–4	—	—	—	84	Neal 1958
Spilogale putorius (western)	21	—	9.4	3–5	12	1	—	56	Ewer 1973
Mephitis mephitis	63	13–20	14	6–10	12	1	—	42–49	Ewer 1973
Procyon lotor	63	18–24	62–98	3–7 (4)	12	1	1–2 yr	70	Ewer 1973
Bassariscus astutus	60	31–34	25	3	—	—	—	120	Richardson 1942
Nasua nasua	77	5	180	1–5	—	—	—	—	Ewer 1973
Bassaricyon gabbii	73–74	≤7	51	1	—	—	—	—	Ewer 1973
Potos flavus	112–20	15	117	1	—	—	—	82+	Poglayen-Neuwall 1976; Bhatia and Desai 1972
Ailurus fulgens	130	10	195	2	12	1	570	150+	Roberts and Kessler 1979
Ailuropoda melanoleuca	140	46	104	1–2	—	—	—	—	Anonymous 1974
Civettictis civetta	72–82	1–2	680	1–4	—	2	2 yr	140	Mallinson 1973
Herpestes edwardsi	56–63	16–17	—	2–4	—	—	—	—	Ewer 1973
H. auropunctatus	70–75	10–12	25–40	2–4	—	—	365	28–35	Larkin and Roberts 1979
Suricata suricatta	77	10–12	41.5	2–5	—	—	720	58	NZP records; Larkin and Roberts 1979
Genetta tigrina	60–70	12–18	71	1–3	10	1+	—	56	Wemmer 1977
Arctictis binturong	90–92	4	284–341	2–3	8	1+	944	172+	NZP records; Aquilina and Beyer 1979
Helogale parvula	50–54	—	—	2–6	—	—	—	—	Ewer 1973
Mungos mungo	60–62	—	—	3–5	—	—	—	—	Ewer 1973
Fossa fossa	86	0	65–70	1	—	—	730	70	Albignac 1973; Larkin and Roberts 1979
Eupleres goudotii	?	0	150	1	—	—	—	60	Albignac 1973; Larkin and Roberts 1979
Galidia elegans	85	6–8	50	1	—	—	~730	56	Albignac 1973; Larkin and Roberts 1979

Taxon	Gestation (days)	Age of Eye Opening (days)	Weight of Neonate (g)	Range or Mean Litter Size	Average Interbirth Interval (months)[a]	Average Litters/Year	Age at First Mating (days)	Duration of Lactation (days)	References
Mungotictis decemlineata	90–105	0	50	1	—	—	~730	15–20	Albignac 1973; Larkin and Roberts 1979
Cryptoprocta ferox	90	12–15	100	2–4	—	—	~1,090	112	Albignac 1973; Larkin and Roberts 1979
Crocuta crocuta	110	0	1,460	1–2	—	—	—	525?	Pournelle 1965
Hyaena brunnea	90	8	680–805	1–3	—	—	—	—	
Panthera leo	110	6	1,400	1–5	24	0.5	—	231	Hemmer 1976
P. pardus	96	7	300	2–3	24	0.5	—	84	Hemmer 1976
P. onca	101	8	800	1–4	—	—	—	154	Hemmer 1976
P. tigris	103	11	1,359	1–4	1.5 yr if cubs survive	.5 if cubs survive	4–5 yr	~90	Hemmer 1976; Sankhala 1967
Uncia uncia	99	8	470	1–4	—	—	—	56–70	Hemmer 1976
Neofelis nebulosa	88	2–11	470	1–4	—	—	—	140	Hemmer 1976
Leopardus pardalis	75	10	250	1–2	—	—	—	—	Hemmer 1976
L. tigrinus	75	17	—	1–2	—	—	—	—	Hemmer 1976
Lynx canadensis	61	—	—	1–4	—	—	—	—	Hemmer 1976
L. lynx	63	6	—	1–6	—	—	—	—	Hemmer 1976
Felis nigripes	67	7	60	1–2	12	1	21 mo	—	Hemmer 1976; Leyhausen and Tonkin 1966
F. margarita	61	14	39	—	—	—	—	—	Hemmer 1976
F. sylvestris sylvestris	68	11	100	1–8	—	—	—	42–49	Hemmer 1976; Meyer-Holzapfel 1968
F. s. lybica	58	10	—	—	—	—	—	—	Hemmer 1976
F. s. catus	63	9	90	—	—	—	—	—	Hemmer 1976
F. geoffroyi	73	14	—	2.0	4.6	2	—	—	NZP records
Acinonyx jubatus	92	7	270	3–4	—	—	—	—	Hemmer 1976
Puma concolor	93	10	400	1–5	—	—	—	—	Hemmer 1976
Prionailurus bengalensis	66	10	80	1–4	—	—	—	—	Hemmer 1976
P. viverrinus	63	—	170	2–3	—	—	—	—	Hemmer 1976

Taxon	Gestation (days)	Age of Eye Opening (days)	Weight of Neonate (g)	Range or Mean Litter Size	Average Interbirth Interval (months)[a]	Average Litters/Year	Age at First Mating (days)	Duration of Lactation (days)	References
Caracal caracal	71	7	—	1–3	—	—	—	—	Hemmer 1976
Leptailurus serval	73	9	250	—	—	—	—	~112–40	Johnstone 1979
Pinnipedia									Bryden 1972
Arctocephalus pusillus	—	0	7,000	1	—	—	—	245	Bryden 1972
A. tropicalis	?	0	4,000	1	—	—	—	98	Bryden 1972
Callorhinus ursinus	273	0	5,000	1	—	—	3 yr	84	Bryden 1972
Eumetopias jubatus	270	0	20,000	1	—	—	—	84	Bryden 1972
Odobenus rosmarus	330	0	55,000	1	—	—	4–5 yr ♀ 7 yr ♂	700	Bryden 1972
Phoca vitulina	240	0	10,000	1	—	—	—	35–42	Bryden 1972
(Pusa) hispida	?	0	4,000	1	—	—	—	56–70	Bryden 1972
Pagophilus groenlandicus	240	0	11,000	1	—	—	5–8 yr ♀ 8 yr ♂	10–11	Bryden 1972
Halichoerus grypus	255	0	14,000	1	—	—	4–5 yr ♀ 6 yr ♂	14–21	Bryden 1972
Erignathus barbatus	323?	0	45,000	1	2 yr	—	6 yr ♀ 7 yr ♂	—	Bryden 1972
Lobodon carcinophagus	~255	0	25,000	1	—	—	—	35	Bryden 1972
Leptonychotes weddelli	270	0	30,000	1	—	—	—	42	Bryden 1972
Monachus schauinslandi	(330)	0	16,000	1	—	—	—	35	Bryden 1972
Cystophora cristata	—	0	12,000	1	—	—	4 yr	14–21	Bryden 1972
Mirounga leonina	240	0	40,000	1	—	—	2 yr ♀ 4 yr ♂	24	Bryden 1972
Cetacea									
Hyperoodon ampulatus	360	0	—	—	—	—	—	—	Bryden 1972
Physeter catodon	480	0	—	—	—	—	—	390	Bryden 1972
Tursiops truncatus	360	0	20,000	1	—	—	6 yr	330–480	Bryden 1972
Delphinus delphis	276	—	—	1	—	—	—	120	Bryden 1972
Globicephala melaena	435	0	—	1	—	—	—	390–488	Bryden 1972

Taxon	Gestation (days)	Age of Eye Opening (days)	Weight of Neonate (g)	Range or Mean Litter Size	Average Interbirth Interval (months)[a]	Average Litters/Year	Age at First Mating (days)	Duration of Lactation (days)	References
Phocaena phocaena	315	0	5,000–6,000	—	—	—	—	—	Bryden 1972
Eschrichtius glaucus	360	0	—	1?	2 yr	.5	—	210	Bryden 1972
Balaenoptera musculus	360	0	—	1	—	—	3.5–5 yr	210	Bryden 1972
B. physalis	370	0	—	1	—	—	3.5–5 yr	210	Bryden 1972
B. acutorostrata	300	0	—	1	—	—	—	135	Bryden 1972
B. borealis	330	0	—	1	—	—	—	150	Bryden 1972
Megaptera novaeangliae	345	0	—	1	—	—	—	315	Bryden 1972
Balaena mysticetus	280	0	—	1	—	—	—	360	Bryden 1972
Tubulidentata									
Orycteropus afer	120	0	1,870	1	—	—	—	112	Melton 1976; Sampsell 1969
Hyracoidea									
Procavia capensis	225	0	170–240	3–4	—	—	2 yr	~90	Mendelssohn 1965
Sirenia									
Trichechus manatus	385–400	0	11–27 kg	1	36	—	3–4 yr	18 mo	Hartman 1979
Proboscidea									
Loxodonta africana	660	0	—	1	—	—	—	—	Sikes 1971
Elephas maximus	628	0	90,909	1	—	—	—	—	McKay 1973
Perissodactyla									
Equus zebra	360	0	—	1	—	—	—	—	Asdell 1964
E. asinus	340	0	—	1	—	—	—	—	Asdell 1964
E. hemionus	330	0	—	1	—	—	—	—	Asdell 1964
E. burchelli	330–60	0	32 kg	1	—	—	1.5–3 yr	—	Wackernagel 1966
Tapirus bairdi	—	0	9,400	1	—	—	—	—	
T. terrestris	397	0	4,500	1	—	—	—	—	NZP records
Tapirus indicus	394	0	7,100	1	—	—	—	—	Richter 1966
Diceros bicornis	475	0	27,500	1	30	—	5–6 yr	—	NZP records
Rhinoceros unicornis	488	0	—	1	—	—	—	—	NZP records

Taxon	Gestation (days)	Age of Eye Opening (days)	Weight of Neonate (g)	Range or Mean Litter Size	Average Interbirth Interval (months)[a]	Average Litters/Year	Age at First Mating (days)	Duration of Lactation (days)	References
Didermocerus sumatrensis	—	0	—	1	—	—	—	—	
Artiodactyla									
Sus scrofa	120	0	—	3–5	—	—	1.5 yr	—	Mohr 1970
Potamochoerus porcus	120	0	—	1–4	—	—	—	—	Mohr 1970
Phacochoerus aethiopicus	171–75	0	860–910	1–4	—	—	—	—	Mohr 1970
Tayassu tajacu	112–16? 114–48?	—	—	2	—	—	—	42–56	Knipe 1957
Hippopotamus amphibius	237	0	55 kg	1	12	1	3–4 yr	—	NZP records; Dittrich 1976
Choeropsis liberiensis	194	0	4,900	1	16.7	0.9	3–5.5 yr	90	NZP records; Dittrich 1976
Camelus dromedarius	389	0	—	—	—	—	—	360	Gauthier-Pilters 1974
C. batrianus	406	—	—	—	—	—	—	—	NZP records
Vicugna vicugna	330	0	6,350	—	24	0.5	386	300+	Schmidt 1973; NZP records
Hymoschus aquaticus	120	0	—	1	—	—	—	—	Dubost 1978
Tragulus javanicus	120	0	370	1	—	—	4–5 mo	90	Davis 1966
T. napu	162	0	371–79	1	—	—	—	21?	Romer 1974
Moschus moschiferus	160	0	—	1–2	—	—	365	—	Frädrich 1966*b*
Muntiacus muntjak	—	0	1,240	1	7.6	1.5	—	—	Barrette 1975; NZP records
Dama dama	240	0	4,500	1	12	1	16 mo ♀	107	Chapman and Chapman 1975; Wayne 1967
Axis axis	210–25	0	—	1	9.4	—	—	—	NZP records
Cervus canadensis	249–62	0	—	1	12	1	—	—	NZP records
C. unicolor	240	0	14,300	1	—	—	—	—	NZP records
Elaphurus davidianus	250	0	—	1	12	1	—	—	NZP records
Odocoileus hemionus	203	0	3,000	1–2	12	1	—	21?	Linsdale and Tomich 1953
O. virginianus	204	0	—	1–2	12	1	—	150?	Taylor 1956
Mazama gouazoubira	206	0	1,300?	1	—	—	—	—	Thomas 1975
M. americana	220	0	567?	1	12	1	—	—	NZP records
Pudu mephistophiles	?	0	400	1	—	—	—	—	Hick 1969

Taxon	Gestation (days)	Age of Eye Opening (days)	Weight of Neonate (g)	Range or Mean Litter Size	Average Interbirth Interval (months)[a]	Average Litters/Year	Age at First Mating (days)	Duration of Lactation (days)	References
P. pudu	~210	0	500	1	~8	—	—	<65	Hick 1969
Alces alces	240–50	0	—	1–3	12	1	2–3 yr ♀	—	Asdell 1964
Rangifer tarandus	210–40	0	5,897–6,623	1–2	12	1	1–1.5 yr ♂	60	Lent 1966
Giraffa camelopardalis	420–50	0	47–70 kg	1	25	0.5	3–4 yr	270	Dagg and Foster 1976
Antilocapra americana	230–40	0	1.8–2.5 kg	1–2	—	—	16 mo ♀ 24 mo ♂	<60	Müller-Schwarze and Müller-Schwarze 1973
Cephalopus sylvicultor	260	0	6.5	1	12	1	—	—	NZP records
Boselaphus tragocamelus	247	0	—	1–2	—	—	—	—	
Bubalus bubalis	307	0	—	1	—	—	—	—	NZP records
Bos grunniens	258	0	18,000	1	13–27	<1	—	—	NZP records
Bibos gaurus	270	0	—	1–2	—	—	—	—	NZP records
Syncerus caffer	330	0	35,570	1	22	0.5	—	—	Mentis 1972
Tragelaphus scriptus	160	0	3,450	1	—	—	18.7 mo	—	Dittrich 1969, 1972
Tragelaphus strepsiceros	275	0	—	1	12	1	14 mo	—	NZP records
Taurotragus oryx	266	0	27,400	1	—	—	3 yr ♀ 4 yr ♂	—	NZP records
Madoqua kirkii	174	0	590	1	—	—	—	42	Dittrich 1969
Hippotragus niger	271	0	11,000	1	—	—	19.6 mo	—	Dittrich 1969, 1972
Oryx dammah	248	0	10,900	1	—	—	21.8–27 mo	—	Dittrich 1969, 1972
Kobus defassa	280	0	12,000	1	—	—	—	180–240	Dittrich 1969
Connochaetes taurinus	252	0	17,000	—	12	1	709	—	Asdell 1964; NZP records
Gazella dorcas	173	0	1,015–1,690	1	8.9	1+	21.4–22 mo	—	Dittrich 1969, 1972
G. thomsoni	145?	0	—	1	12	1	—	—	Asdell 1964
G. granti	199	0	5,050–7,310	—	—	1	—	—	Dittrich 1969
Aepyceros melampus	204	0	3,970–5,500	1	—	1	—	—	Dittrich 1969
Saiga tatarica	145	0	(3,200)	1–2	12	1	—	—	Asdell 1964
Capricornis sumatraensis	240	0	—	1–2	—	—	—	—	
Oreamnos americanus	147	0	—	1–2	—	—	—	—	Asdell 1964

Taxon	Gestation (days)	Age of Eye Opening (days)	Weight of Neonate (g)	Range or Mean Litter Size	Average Interbirth Interval (months)[a]	Average Litters/Year	Age at First Mating (days)	Duration of Lactation (days)	References
Rupicapra rupicapra	150–60	0	—	1	—	—	—	—	Asdell 1964
Ovibos moschatus	240	0	8,943–9,806	—	—	—	3–5 yr	139	Banks 1978; Oeming 1966
Capra ibex	150–80	0	—	1	12	1	—	90	Asdell 1964
Ammotragus lervia	154–61	0	3,730	1	—	—	—	—	Asdell 1964
Ovis canadensis	180	0	4,082	1	12	1	27 mo	120–50	Asdell 1964
O. aries	150	0	—	1–2	12	1	—	—	Asdell 1964

Sources: Walker et al. 1964 and subsequent revisions were used for tabulating some of the basic data in this appendix. Where other references were employed, they are cited. Unpublished records from the National Zoological Park were also utilized. Asdell 1964; Fraser 1968; and Leitch et al. 1959 were frequently consulted for additional material but are not always cited in the reference column.

Appendix 5

Metabolic Rates for Selected Mammalian Species

There are many difficulties inherent in the measurement of metabolic rate. The values given here serve as the basis for the inferences discussed in the text. These are drawn in the main from Spector, ed., *Handbook of Biological Data* (1956), and from numerous conversations with Dr. Brian K. McNab.

Taxon	Metabolic Rate (O_2/hr/g)
Monotremata	
Ornithorhynchus anatinus	0.46
Tachyglossus aculeatus	0.22
Marsupialia	
Antechinus stuarti (=flavipes)	1.53
A. maculatus	1.26
Sminthopsis crassicaudata	1.67
Antechinomys spenceri	0.98
Pseudantechinus macdonnellensis	0.63
Dasycercus cristicauda	0.52
Dasyuroides byrnei	0.87
Phascogale tapoatafa	0.81
Dasyurus hallacatus	0.51
D. maculatus	0.30
D. viverrinus	0.45
Sarcophilus harrisii	0.28
Trichosurus vulpecula	0.32
Insectivora	
Sorex araneus	7.7–8.1
S. cinereus	16.8
Blarina brevicauda	5.3
Microsorex hoyi	16.7
Edentata	
Dasypus novemcinctus	0.20
Choloepus hoffmanni	0.19
Bradypus variegatus	0.18
Chiroptera	
Pteropus poliocephalus	1.00
Myotis lucifugus	0.40
Lagomorpha	
Oryctolagus cuniculus	0.68
Rodentia	
Tamiasciurus hudsonicus	1.7
Citellus undulatus	1.2
Dipodomys merriami	1.2
Mus musculus	3.4
Rattus norvegicus	1.045
Cricetus cricetus	0.87

Taxon	Metabolic Rate (O_2/hr/g)
Peromyscus leucopus	4.4
P. eremicus	2.2
P. maniculatus bairdii	2.0
P. maniculatus	3.0
Reithrodontomys megalotis	2.5
Meriones unguiculatus	1.0
Clethrionomys gapperi	3.6
Microtus pennsylvanicus	3.2
Dicrostonyx groenlandicus	3.94
Cavia porcellus	0.58
Capromys pilorides	0.227
Primates	
Saimiri sciureus	0.83
Carnivora	
Vulpes vulpes	0.55
Alopex lagopus	0.73
Ursus americanus	0.36
Thalarctos maritimus (juvenile)	0.49
Mustela rixosa	8.15
Felis sylvestris catus	0.71
Pinnipedia	
Phoca vitulina	0.45
Cetacea	
Phocaena phocaena	0.30
Hyracoidea	
Dendrohyrax sp.	0.34
Proboscidea	
Elephas maximus	0.15
Perissodactyla	
Equus caballus	0.25
Artiodactyla	
Sus scrofa	0.11
Rangifer tarandus	0.36
Bos taurus	0.17
Bos indicus	0.14
Oreamnos americanus	0.26
Ovis aries	0.34

Appendix 6

Encephalization Quotients for Selected Mammalian Species

The values given in this appendix were calculated according to the methods outlined in the text. Brain weights were taken from the literature as outlined in table 46, with supplemental data from my own records.

Taxon	Encephalization Quotient
Monotremata	
Ornithorhynchus anatinus	.944
Tachyglossus aculeatus	.715
Marsupialia	
Didelphis marsupialis	.353–.573
Marmosa robinsoni	.716
Antechinus maculatus	.562
Dasyurus quoll	1.046
D. viverrinus	.750
Sarcophilus harrisii	.361–.419
Trichosurus vulpecula	.863
Echymipera doreyana	.599
Vombatus ursinus	.552
Onychogalea frenata	1.119 young♂
Macropus giganteus	.470 ♂
Insectivora	
Solenodon paradoxus	.545–.687
Microgale talazaci	.765
M. dobsoni	.747
M. cowani	.981
Oryzoryctes talpoides	.618
Hemicentetes semispinosus	.427–.453
Setifer setosus	.454
Echinops telfairi	.40
Tenrec ecaudatus	.345–.486
Erinaceus europaeus	.508
Sorex araneus	.621
S. minutus	.557
Blarina brevicauda	.748
Neomys fodiens	.747
Crocidura rusulla	.562
Suncus murinus	.476
S. etruscus	.447
Talpa europaea	.730–1.064
Scalopus aquaticus	1.341
Galaemys pyrannicus	1.169
Desmana moschata	.787
Edentata	
Dasypus novemcinctus	.371
Dasypus sexcinctus	.786
Euphractus sexcinctus	.862

Taxon	Encephalization Quotient
Chaetophractus villosus	.790
Zaedyus pichi	.640
Choloepus hoffmanni	1.092
Bradypus variegatus	.954–.946
Cyclopes didactylus	.907
Myrmecophaga tridactyla	.809
Tamandua tetradactyla	1.452
Pholidota	
Manis javanicus	.473? obese specimen
Chiroptera	
Cynopterus sphinx	1.118
Rousettus leschenaulti	1.062
Pteropus geddeiri	.954
Eidolon helvum	1.366
Rhinopoma microphyllum	.592
R. hardwickei	.620
Megaderma lyra (=spasma)	1.135
Carollia perspicillata	1.111
Desmodus rotundus	1.228
Myotis sodalis	.608
M. lucifugus	.519
Pipistrellus subflavus	.548
Eptesicus fuscus	.526
Antrozous pallidus	.614
Plecotus townsendii	.808
Tadarida brasiliensis	.682
Lagomorpha	
Oryctolagus cuniculus	.682
Lepus capensis	.501
Macroscelidea	
Rhynchocyon stuhlmanni	1.112
Elephantulus fuscipes	1.176
Rodentia	
Aplodontia rufa	.851
Sciurus niger	1.426
S. vulgaris	1.542–1.334
Funisciurus carruthersi	1.785
Ratufa indica	.758
Tamias striatus	1.736
Marmota marmota	.829
Perognathus parvus	1.247
Tachyoryctes splendens	.740
Mus minutoides	1.373
M. musculus	.808
Rattus norvegicus	.792
R. rattus	.835
Mesocricetus auratus	1.193
Cricetus cricetus	.564
Cricetomys emini	1.473
Mystromys albicaudatus	.710
Peromyscus leucopus	1.08 ♂, 1.23 ♀
P. californicus	.981 ♂, .930 ♀
P. eremicus	1.22 ♂, 1.23 ♀
P. maniculatus bairdii	1.07 ♂, 1.14 ♀
P. polionotus subgriseus	1.25 ♂, 1.27 ♀
Sigmodon hispidus	1.278
Gerbillus nanus	1.072
G. pyramidum	.793
Meriones crassus	.688
M. libycus	.932

Taxon	Encephalization Quotient
M. shawi	.677
Ondatra zibethicus	.624
Arvicola terrestris	.768
Microtus arvalis	.810
Spalax hungaricus	1.518
Jaculus orientalis	1.198
Atherurus africanus	1.188
Hystrix cristata	.940
Thryonomys swinderianus	.566
Agouti paca	1.086
Dasyprocta sp.	.807
Chinchilla laniger	1.342
Lagostomus maximus	.842
Cavia porcellus	.95
Dolichotis patagona	.775
Hydrochoerus hydrochoeris	.675
Erethizon dorsatum	.727
Myocastor coypus	.988
Scandentia	
Tupaia glis	1.369
Primates	
Microcebus murinus	1.638
Lemur macaco	1.652
Lemur catta	1.449
Galago senegalensis	1.782
Loris tardigradus	1.637
Nycticebus coucang	2.342
Perodicticus potto	1.535
Saguinus midas	2.426
S. oedipus	2.025
S. tamarin	1.953
Saimiri sciureus	3.758
Cebus capucinus	3.395
Ateles paniscus	2.641
A. belzebuth	2.569
Alouatta villosa	1.561
Macaca mulatta	2.342
Presbytis obscurus	1.668
Hylobates lar	3.157
Symphalangus syndactylus	2.215
Gorilla gorilla	1.402–1.68
Pongo pygmaeus	1.56–1.91
Pan troglodytes	2.449–2.18
Homo sapiens	7.33–7.69
Fissipedia	
Canis lupus	1.332 ♂, .922 ♀
Vulpes vulpes	1.885
Urocyon cinereoargenteus	1.523
Ursus arctos arctos	.908
Helarctos malayanus	2.572 (juvenile)
Mustela rixosa	1.895
M. vison	.946
Meles meles	.943
Eira barbara	1.88–2.19
Galictis cuja	1.786
Conepatus sp.	.84–1.7
Pteronura brasiliensis	1.49
Procyon lotor	1.37–1.496
Herpestes smithi	1.374
Genetta tigrina	1.414

Taxon	Encephalization Quotient	
G. pardina	1.62	
Paradoxurus hermaphroditus	1.129	
Crocuta crocuta	.9–1.14	
Hyaena hyaena	.914	
Panthera leo	.573–.829	
P. pardus	.845	
P. onca	1.357	
P. tigris	1.35	
Leopardus pardalis	1.55	
Felis sylvestris catus	1.14	
Puma concolor	.975	
Pinnipedia		
Odobenus rosmarus	1.025	
Phoca richardi	1.54	
(Pusa) hispida	1.80	
Erignathus barbatus	.79	
Cetacea		
Physeter catodon	.19–.50 ♂–♀	
Tursiops tursia	3.226	
Globicephala melaena	1.7	
Phocaena phocaena	4.9	
Balaenoptera musculus	.137–.233	seasonal fattening phenomenon decreases EQ values
B. acutorostrata	.081	
B. rostratus	.201	
Megaptera novaeangliae	.153	
Sirenia		
Trichechus manatus	.431	
Hyracoidea		
Procavia capensis	.826–.985	
Proboscidea		
Loxodonta africana	.972 ♂, 2.05 ♀	
Elephas maximus	1.62–1.72	
Perissodactyla		
Equus zebra	1.701	
E. caballus	1.068	
E. asinus	.967	
Tapirus bairdi	.872	
T. indicus	.439–.582	
Diceros bicornis	.540 ♂	
Artiodactyla		
Sus scrofa	.270	
Phacochoerus aethiopicus	.627	
Hippopotamus amphibius	.261–.392	
Camelus dromedarius	1.008	
C. bactrianus	.657	
Tragulus stanleyi	1.077	
Cervus canadensis	.902	
Odocoileus virginianus	1.055	
Rangifer arcticus	.934	
Giraffa camelopardalis	.409–.426	
Bos taurus	1.085	
Syncerus caffer	.554 $\overline{X}$ ♂ and ♀	
Bison bison	.351 ♂	
Tragelaphus scriptus	1.091–1.052	
Redunca redunca	.914	
Connochaetes taurinus	.938	
Gazella thomsoni	.950–.954	

Taxon	Encephalization Quotient
Aepyceros melampus	1.116
Capra hircus	1.084
Ovis aries	.541 ♂

Supplementary Notes

Page 75 The volume edited by Maglio and Cooke provides ample documentation of the early isolation of Africa and the early evolution of endemic stocks (V. J. Maglio and H. B. S. Cooke, *Evolution of African Mammals* [Cambridge: Harvard University Press, 1978]).

Page 82 The tripartite subordinal classification based on jaw musculature derives from Brandt, and using differences in jaw structure as a basis for subordinal classification was proposed by Tullberg (see Landry 1957).

Page 87 Sherman and his colleagues have recently analyzed the problem in *Spermophilus beldingi*. Most litters (78%) were multiply sired. Littermates of such species are not equally related. This conclusion has profound implications for kinship theory models (see J. Hanken and P. W. Sherman, "Multiple Paternity in Belding's Ground Squirrel Litters," *Science* 212[1981]:351–53).

Page 105 *Agouti* is often separated into its own family, Agoutidae.

Page 127 *Panthera* is now considered a junior synonym for the genus *Leo*.

Page 145 W. P. Luckett, *Comparative Biology and Evolutionary Relationships of Tree Shrews* (New York: Plenum, 1980), provides the latest synthesis of the data.

Page 153 P. W. Freeman, *A Multivariate Study of the Family Molossidae: Morphology, Ecology, and Evolution,* Fieldiana (Zoology), no. 7 (Chicago: Field Museum, 1981). The volume will be the standard synthesis for the family for some time to come.

Page 181 H. Marsh, ed., *The Dugong* (Townesville, Queensland: James Cooke University Press, 1981), provides a useful summary of recent dugong research.

Moeritherium is no longer considered ancestral to the Proboscidea. See V. J. Maglio and H. B. S. Cooke, *Evolution of African Mammals* (Cambridge: Harvard University Press, 1978).

Page 198 The recent volume by Gauthier-Pilters and Dagg summarizes the ecology and behavior of *Camelus dromedarius* (H. Gauthier-Pilters and A. I. Dagg, *The Camel* [Chicago: University of Chicago Press, 1981]).

Page 199 Some authors synonymize *Cervus canadensis* with the Old World red deer *C. elaphus.*

Page 218 I am using diversity in the vernacular sense; actually my usage here is equivalent to a "species richness" measure.

Page 251 Ranges reflect averages as recorded for the "macroniches" in table 43. They do not reflect ranges considered independent of trophic specializations.

Page 323 Clearly, competition and predation rates may be reduced relative to continental values on islands with a depauperate fauna.

Page 393 In the case of *Marmosa* there is no pouch, but there is a teat area surrounded by long hairs. I have used the term pouch in this paragraph to refer to a "generalized" marsupial rearing situation.

Page 397 A full discussion of play necessarily leads to considerations of behavioral development or the ontogeny of behavior. I have not specifically treated ontogeny in this volume, but an understanding of the mechanisms of neurophysiological maturation is essential for understanding behavioral development and the role of play in the adaptation of individuals. For a rigorous synthesis, the reader is urged to consult the recent volume by Robert Fagen, *Animal Play Behavior* (Oxford: Oxford University Press, 1981).

Page 441 Jarvis makes a strong argument for eusociality in *Heterocephalus.* The data are from captive colonies, and I eagerly await confirmation from a field study. (See J. Jarvis, "Eusociality in a Mammal: Cooperative Breeding in Naked Male-Rat Colonies," *Science* 212[1981]:571–73).

Page 459 The genera *Cheirogaleus, Microcebus,* and *Phaner* may be grouped in a single subfamily Cheirogalinae. Some authors in turn raise this to a familial rank. See Szalay and Delson (1979).

Some authors consider the Galaginae and Lorisinae to be subsumed under the Lorisidae. See Szalay and Delson (1979).

Glossary

This glossary contains some anatomical, evolutionary, and ecological terms that may not all be known by any given specialist who may read this work. The glossary does not pretend to be complete, but it should serve as a guideline.

Allantois One of the fundamental fetal membranes developed during the early embryology of the vertebrate egg. The allantois outpouches from the early yolk sac and is endodermal in its genesis, though it becomes covered with a layer of mesoderm during the course of its growth into the body stalk. The allantois is an important organ for gaseous exchange in the development of oviparous vertebrates.

Allogrooming Care of the body surface performed by one individual upon various parts of the body of a second individual.

Altricial Born in a very underdeveloped state. An altricial vertebrate undergoes considerable growth and development after its hatching or birth (cf. **Precocial**).

Apomorph Describes a character (morphological or behavioral) that has undergone drastic modification from a presumed earlier condition

Autogrooming Care of the body surface, such as by scratching, licking, and wiping, as performed on the actor's own body (cf. **Allogrooming**).

Basal metabolic rate The amount of energy consumed by an organism per unit of time when at rest; usually measured by O_2 consumption or CO_2 production.

Brachydont A term applied to the structure of mammalian teeth. A brachydont tooth has a large grinding surface area and a moderately high crown. It is characteristic of animals that masticate tough substances, such as fibrous plant tissue.

Bunodont dentition Molar teeth that have low, rounded cusps and are adapted for crushing.

Chorioallantoic placenta One of the placental forms developed by a fusion of the chorion and allantois that serves as the fetal membrane component in interfacing with the maternal endometrium. See also **Placenta**.

Chorion One of the fetal membranes derived from the primitive trophoblastic tissue forming a sac that encloses the embryo and all other fetal membranes.

Choriovitelline placenta A form of placentation where the chorion and yolk sac contribute to the fetal component for interfacing with the maternal endometrium. See also **Placenta**.

Cleidoic egg The egg form characteristic of higher vertebrates. The cleidoic egg was evolved as an adaptation to reproducing on land, and it thus includes a shell and a tough shell membrane. The fetal membranes are elaborated to provide a mechanism for gaseous exchange through the shell by means of the allantois and a means for embryonic nutrition during the developmental cycle within the egg by an elaboration of the yolk sac. The elaboration of the fetal membranes in mammals proceeds in the same fashion as in birds and reptiles. The cleidoic egg typifies the eggs of birds, reptiles, and monotremes

Core area That portion of a mammal's home range that is used most intensively and from which like-sexed conspecifics are usually excluded. Overt territorial defense may not be shown.

Crustacivore A species whose dietary specializations are for feeding on crustaceans, such as shrimp, crayfish, and crabs.

Cursorial Strongly adapted for running across open terrain. This adaptation typifies certain ungulates and carnivores.

Diapause A period of developmental arrest during embryogenesis when cell division ceases.

Digitigrade locomotion Walking or running on the tips of the toes as opposed to plantigrade.

Encephalization quotient A method of describing relative brain size. As defined by Jerison (1973), it is the degree to which brain size exceeds or falls below the mean brain size for animals of that weight. It is derived by comparing the regression of brain size against body weight value with an actual value for the specimen in hand.

Endothelio-chorial placenta A placenta type where the maternal circulatory system is separated from the chorion of the embryo only by the endothelial lining of the blood vessels.

Endothermy The maintenance of body temperature through the oxidation of stored food within the body.

Epipubic bones Small bones paired just opposite the pubis, characterizing monotremes and marsupials.

Epitheliochorial placenta Placentation where the epithelial lining of the maternal component of the placenta is opposed to the chorion of the embryo. Diffusion of material across the placental barrier involves passage through two epithelial layers.

Fetal membrane(s) During the early embryology of higher vertebrates, typically four separate membranes are developed as outpouchings. They come to enclose the embryo and to serve various functions in the formation of the placenta in higher mammals. They include the allantois, amnion, chorion, and yolk sac.

Filter bridge A term in biogeography indicating that, though two land masses are usually separated by a water boundary, faunal exchange via islands permits species from both land masses to exchange members. The "island bridge" is a sufficiently difficult passage so that the movement of species from one land mass to another is almost a random event.

Fossorial Adapted for burrowing and life underground.

Hemochorial placenta A placenta form where the maternal blood is in direct contact with the embryonic chorion.

Heterodont dentition Dentition where the teeth are separable into different functional classes, such as incisors, canines, premolars, and molars; it characterizes the class Mammalia.

Homeothermy The maintenance of a relatively constant body temperature throughout a defined time interval.

Homodont dentition Condition of the dentition where tooth form shows little or no differentiation into functional classes (cf. **Heterodont dentition**).

Hyoid bone A small U-shaped bone at the base of the tongue in the higher Mammalia.

Hypsodont dentition A tooth structure where the crown is extremely high. It is an adaptation for the mastication of fibrous plant substances frequently shown by grazing forms.

Infraorbital foramen An opening within the skull of some mammals anterior to the zygomatic arch (cheek bone). Its size and presence are diagnostic in the order Rodentia.

Iteroparous Reproducing in each of several consecutive years (cf. **Semelparous**).

k-Selection A process of natural selection that reduces the reproductive rate of an individual and promotes iteroparous reproduction.

Llanos Open, savanna habitat in South America.

Masseter One of the jaw muscles attaching to the zygomatic arch (cheek bone).

Monovular Producing a single egg and usually, unless twinning takes place, a single offspring during each reproductive cycle.

Myrmecophage An animal specialized for feeding on ants and termites.

Omnivore An animal that can feed on a variety of different foodstuffs; usually applied to species that have a wide range of diet, including meat, fruits, and invertebrates.

Oviparous Reproducing by laying eggs.

Ovoviviparous Retaining the egg within the reproductive tract of the female until the conclusion of incubation. The young then emerges from the reproductive tract of the female fully mobile and free from the egg membranes.

Placenta A structure composed of both a maternal and a fetal component, evolved to permit the exchange of nutritive material between the female circulatory system and the developing embryo.

Plantigrade locomotion Mode of locomotion where the weight is borne on the entire sole of the hand or foot, as opposed to digitigrade locomotion.

Plesiomorph Describes a character (morphological or behavioral) that has undergone little modification from an ancestral condition (cf. **Apomorph**).

Polyembryony A reproductive mechanism whereby several young are produced from a single egg.

Polyovular Shedding several ova during each reproductive cycle (cf. **Monovular**).

Precocial Showing advanced sensory and locomotor abilities at birth (cf. **Altricial**).

Prepenial scrotum The position of the scrotum anterior to the penis, characteristic of the Marsupialia.

r-Selection A selective process that favors a high reproductive rate and promotes a trend toward semelparous reproduction.

Ricochetal locomotion Form of locomotion in which the animal proceeds by a series of hops or jumps.

Scatter hoard Foodstuffs hoarded within a species' home range in several individual caches rather than at a single locus.

Selenodont dentition Dentition where the cusps are in the form of half-moons; this is typical of many Artiodactyla.

Semelparous Literally, reproducing once in the lifetime of an individual; as applied to the Mammalia, it refers to species that have a tendency to produce large litters and generally rear only one litter within a life cycle. Most mammals have the capacity to reproduce at least twice, however, and are said to approximate a semelparous condition.

Tribosphenic molars A condition in primitive mammals where the basic molar form exhibits three major cusps.

Triconodont molars Molar structure typified by a single extinct order of mammals, the Triconodonta, where three crowns form the basic cusp pattern of each molar. The lower molars, however, lack the shelf or talon characteristic of the derived tribosphenic condition whence all mammal molar patterns were derived (cf. **Tribosphenic molars**).

Trophoblast A portion of the cleaving cell mass that differentiates during embryogenesis of eutherian mammals to form the fetal membranes involved in placentation.

Viviparous Bearing young alive after a period of intrauterine development involving placentation and transfer of nutrients from the maternal to the fetal circulation (cf. **Ovoviviparous**).

Yolk sac placenta A form of placentation, typical of the Marsupialia, where the yolk sac is the predominant fetal membrane opposing itself to the uterine wall and providing for some nutrient exchange between the uterine glands and the developing embryo (cf. **Choriovitelline placenta**).

References

Aeschlimann, A. 1963. Observations sur *Philantomba maxwelli* (Hamilton-Smith) une antilope de la Forêt Eburnéenne. *Acta Tropica* 20:341–68.

Albignac, R. 1969*a*. Notes éthologiques sur quelques carnivores Malgaches: Le *Galidia elegans* I. Geoffroy. *Terre et Vie* 2:202–15.

———. 1969*b*. Naissance et élevage en captivité de jeunes *Cryptoprocta ferox,* Viverridés malgaches. *Mammalia* 33(1):93–98.

———. 1973. *Mammifères carnivores*. Faune de Madagascar, vol. 36. Paris: ORSTOM.

Alexander, R. D. 1974. The evolution of social behavior. *Ann. Rev. Ecol. System.* 5:325–83.

Allen, D. 1979. *The wolves of Minong*. Boston: Houghton Mifflin.

Allen, G. M. 1939. *Bats*. Cambridge, Mass.: Harvard University Press.

Allin, J. T., and Banks, E. M. 1968. Behavioural biology of the collared lemming *Dicrostonyx groenlandicus* (Traill). I. Agonistic behaviour. *Anim. Behav.* 16(2–3):245–63.

Altman, P. L., and Dittmer, D. S., eds. 1962. *Growth including reproduction and morphological development*. Washington, D.C.: Federation of American Societies for Experimental Biology.

Altmann, D. 1969. *Harnen und Köten bei Säugetiere*. Wittenburg: Ziemsen Verlag.

———. 1972. Verhaltenstudien an Mähnenwölfen. *Zool. Gart.* 41:278–98.

Altmann, S. A., ed. 1967. *Social communication in primates*. Chicago: University of Chicago Press.

Altmann, S. A., and Altmann, J. 1970. *Baboon ecology*. Chicago: University of Chicago Press.

Amoroso, C. E. 1958. Placentation. In *Marshall's physiology of reproduction,* vol. 2, ed. A. S. Parkes. 3d ed. London: Longmans Green.

Andersen, D. C. 1978. Observations on reproduction, growth and behavior of the northern pocket gopher (*Thomomys talpoides*). *J. Mammal.* 59:418–22.

Andersen, H. T., ed. 1969. *The biology of marine mammals*. New York: Academic Press.

Anderson, P. K. 1961. Density, social structure, and non-social environment in house-mouse populations and the implications for regulation of numbers. *Trans. N.Y. Acad. Sci.*, sec. 2, 23:447–51.

———. 1965. The role of breeding structure in evolutionary processes of *Mus musculus* populations. *Proceedings of the Symposium on Mutational Process,* Prague, pp. 17–21.

Anderson, P. K., and Birtles, A. 1978. Behaviour and ecology of the dugong, *Dugong dugon* (Sirenia): Observations in Shoalwater and Cleveland Bays, Queensland. *Australian Wildl. Res.* 5:1–23.

Anderson, S., and Jones, J. K., Jr. 1967. *Recent mammals of the world: A synopsis of families.* New York: Ronald Press.

Andrew, R. J. 1963. The origin and evolution of the calls and facial expressions of the Primates. *Behaviour* 20:1–109.

———. 1964. The displays of the Primates. In *Evolution and genetic biology of the Primates,* vol. 2, ed. J. Buettner-Janusch, pp. 227–309. New York: Academic Press.

[Anonymous]. 1974. On the breeding of the giant panda and the development of its cubs. *Acta Zool. Sinica* 20(2):139–47 (trans. Lois Chang, U.S. National Library of Medicine).

Ansell, W. F. H. 1964. Captivity behavior and post-natal development of the shrew *Crocidura bicolor. Proc. Zool. Soc. London* 142:132–27.

Anthony, H. E. 1916. Habits of *Aplodontia. Bull. Amer. Mus. Nat. Hist.* 35:53–63.

Apfelbach, R. 1973. Olfactory sign stimulus for prey selection in polecats (*Putorius putorius* L.). *Z. Tierpsychol.* 33:270–73.

Aquilina, G. D., and Beyer, R. H. 1979. The exhibition and breeding of binturongs as a family group at Buffalo Zoo. *Int. Zoo Yearb.* 19:185–87.

Ardito, G. 1975. *Checklist of the data on the gestation length of primates.* Torino: Istituto di Antropologia.

Arlton, A. V. 1936. An ecological study of the mole. *J. Mammal.* 17:349–71.

Armitage, K. B. 1962. Social behavior of a colony of the yellow-bellied marmot. *Anim. Behav.* 10:319–31.

Arvola, A. 1974. Vocalization in the guinea-pig, *Cavia porcellus* L. *Ann. Zool. Fennici* 11:1–96.

Arvola, A.; Ilmen, M.; and Koponen, T. 1962. On the aggressive behaviour of the Norwegian lemming (*Lemmus lemmus*) with special reference to sounds produced. *Arch. Soc. Zool. Bot. Fennicae* 17(2):80–101.

Aschoff, J. 1958. Tierische Periodik unter dem Einfluss von Zeitgebern. *Z. Tierpsychol.* 15:1–30.

Asdell, S. A. 1964. *Patterns of mammalian reproduction.* 2d ed. Ithaca: Cornell University Press.

Ashby, K. R. 1967. Studies on the ecology of field mice and voles (*Apodemus sylvaticus, Clethrionomys glareolus* and *Microtus agrestis*) in Houghall Wood, Durham. *J. Zool.* 152(4):389–513.

Asibey, E. O. A. 1974. Reproduction in the grasscutter (*Thryonomys swinderianus* Temminck) in Ghana. *Symp. Zool. Soc. London* 34:251–63.

Aslin, H. 1974. The behaviour of *Dasyuroides byrnei* (Marsupialia) in captivity. *Z. Tierpsychol.* 35:187–208.

Atwood, E. L. 1950. The life history studies of nutria or coypu in coastal Louisiana. *J. Wildl. Manage.* 14:249–65.

Augee, M. L.; Ealey, E. H. M.; and Price, I. P. 1975. Movements of echidnas, *Tachyglossus aculeatus,* determined by marking-recapture and radio tracking. *Australian Wildl. Res.* 2:93–101.

Austin, C. R., and Short, R. V. 1972. *Reproductive patterns.* Cambridge: Cambridge University Press.

Autenrieth, R. E., and Fichter, E. 1975. On the behavior and socialization of pronghorn fawns. *Wildl. Monogr.* 42:1–111.

Backhaus, D. 1961. *Beobachtungen am Giraffen in Zoologischen Gärten und Freier Wildbahn.* Brussels: Institut des Parcs National du Congo et du Ruanda-Urundi.

———. 1964. Zum Verhalten des nördlichen Breitmaulnashorns (*Diceros simus cottoni* Lydekker 1908). *Zool. Gart.* 29(3):93–107.

Baker, A. E. M. 1974. Interspecific aggressive behavior of pocket gophers *Thomomys bottae* and *T. talpoides* (Geomyidae: Rodentia). *Ecology* 55:671–73.

Baker, R. H. 1971. Nutritional strategies of myomorph rodents in North American grasslands. *J. Mammal.* 52(4):800–805.

Baker, R. J. 1967. Karyotypes of bats of the family Phyllostomatidae and their taxonomic implications. *Southwestern Nat.* 12:407–28.

———. 1970. The role of karyotypes in phylogenetic studies of bats. In *About bats,* ed. B. H. Slaughter and D. W. Walton, pp. 303–13. Dallas: Southern Methodist University Press.

———. 1973. Comparative cytogenetics of the New World leaf-nosed bats (Phyllostomatidae). *Periodicum Biol.* 75:37–45.

Baker, R. J.; Knox-Jones, J., Jr.; and Carter, D. C., eds. 1976. *Biology of bats of the New World family Phyllostomatidae, part I.* Special Publications of the Museum, no. 10. Lubbock: Texas Tech Press.

———. 1977. *Biology of bats of the New World family Phyllostomatidae, part II.* Special Publications of the Museum, no. 13. Lubbock: Texas Tech Press.

———. 1978. *Biology of bats of the New World family Phyllostomatidae, part III.* Special Publications of the Museum, no. 16. Lubbock: Texas Tech Press.

Baker, R. R. 1978. *The evolutionary ecology of animal migration.* New York: Holmes and Meier.

Bakko, E. B., and Brown, L. N. 1967. Breeding biology of the white-tailed prairie dog, *Cynomys leucurus,* in Wyoming. *J. Mammal.* 48(1):100–113.

Baldridge, A. 1972. Killer whales attack and eat a gray whale. *J. Mammal.* 53(4):898–900.

Baldwin, J. D. 1971. The social organization of a semi-free-ranging troop of squirrel monkeys (*Saimiri sciureus*). *Folia Primat.* 14:23–50.

Baldwin, J. D., and Baldwin, J. I. 1972. The ecology and behavior of squirrel monkeys (*Saimiri oerstedi*) in a natural forest in western Panama. *Folia Primat.* 18:161–84.

Baldwin, P. H.; Schwartz, C. W.; and Schwartz, E. R. 1952. Life history and economic status of the mongoose in Hawaii. *J. Mammal.* 33:335–56.

Balph, D. F., and Stokes, A. W. 1963. On the ethology of a population of Uinta ground squirrels. *Amer. Midl. Nat.* 69:106–26.

Balph, D. M., and Balph, D. F. 1966. Sound communication of Uinta ground squirrels. *J. Mammal.* 47:440–50.

Banks, D. R. 1978. Hand-rearing a musk-ox, *Ovibos moschatus,* at Calgary Zoo. *Int. Zoo Yearb.* 18:213–15.

Banks, E. M. 1964. Some aspects of sexual behavior in domestic sheep, *Ovis aries. Behaviour* 23:249–79.

———. 1968. Behavioural biology of the collared lemming *Dicrostonyx groenlandicus* (Traill). II. Sexual behaviour. *Anim. Behav.* 16(2–3):263–71.

Bannikov, A. G. 1961. Ecologie et distribution d'*Equus hemionus* Pallas: Les variations de sa limite de distribution septentrionale. *Terre et Vie,* pp. 86–100.

———. 1963. *Die Saiga-Antilope.* Wittenberg: Ziemsen.

———. 1964. Biologie du chien viverrin en U.R.S.S. *Mammalia* 28:1–39.

Bannikov, A. G.; Zhirnov, L. V.; Lebedeva, L. S.; and Fandeev, A. A. 1967. *Biology of the saiga.* Jerusalem: Israel Program for Scientific Translations.

Barash, D. P. 1973. The social biology of the Olympic marmot. *Anim. Behav. Monographs* 6(3):171–245.

Barbour, R. W., and Davis, W. H. 1969. *Bats of America.* Lexington: University Press of Kentucky.

Barfield, R. J., and Geyer, L. A. 1972. Sexual behavior: Ultrasonic postejaculatory song of the male rat. *Science* 176:1349–50.

Barkalow, F. S., Jr.; Hamilton, R. B.; and Soots, R. F. 1970. The vital statistics of an unexploited gray squirrel population. *J. Wildl. Manage.* 34:489–500.

Barnes, R. D., and Barthold, S. W. 1969. Reproduction and breeding behaviour in an experimental colony of *Marmosa mitis* Bangs (Didelphidae). *J. Reprod. Fert.,* suppl. 6:477–82.

Barnett, S. A. 1958. An analysis of social behavior in wild rats. *Proc. Zool. Soc. London* 130:107–52.

———. 1963. *The rat: A study in behavior.* Chicago: Aldine.

Barrette, C. 1975. Social behaviour of muntjac. Ph.D. thesis, Department of Biology, University of Calgary.

———. 1977*a*. Fighting behavior of muntjac and the evolution of antlers. *Evolution* 31(1):169–76.

———. 1977*b*. Some aspects of the behaviour of muntjacs in Wilpattu National Park. *Mammalia* 41(1):1–34.

Barrow, G. J. 1964. *Hydromys chrysogaster:* Some observations. *Queensland Nat.* 17:43–44.

Bartholomew, G. A. 1952. Reproduction and social behavior of the northern elephant seal. *Univ. California Publ. Zool.* 47(15):369–472.

———. 1959. Mother-young relations and the maturation of pup behavior in the Alaska fur seal. *Anim. Behav.* 7:163–71.

Bartholomew, G. A., and Cade, T. J. 1957. Temperature regulation, hibernation, and aestivation in the little pocket mouse, *Perognathus longimembris*. *J. Mammal.* 38(1):60–72.

Bartholomew, G. A., and Collias, N. 1962. Role of vocalization in the social behavior of the northern elephant seal. *Anim. Behav.* 10:7–14.

Bartholomew, G. A., and Hoel, P. G. 1953. Reproductive behavior of the Alaska fur seal, *Callorhinus ursinus*. *J. Mammal.* 34:417–36.

Bartholomew, G. A., and Hudson, J. W. 1962. Hibernation, estivation, temperature regulation, evaporative water loss and heart rate of the pigmy possum, *Cercartetus nanus*. *Physiol. Zool.* 35:94–107.

Bartholomew, G. A., and MacMillen, R. E. 1961. Oxygen consumption, estivation and hibernation in the kangaroo mouse, *M. pallidus*. *Physiol. Zool.* 34:177–83.

Bartholomew, G. A., and Rainey, M. 1971. Regulation of body temperature in the rock hyrax, *Heterohyrax brucei*. *J. Mammal.* 52(1):81–95.

Bateson, G. 1966. Problems in cetacean and other mammalian communication. In *Whales, dolphins and porpoises*, ed. K. S. Norris, pp. 569–82. Berkeley: University of California Press.

Battistini, R., and Richard-Vindard, G., eds. 1972. *Biogeography and ecology in Madagascar*. The Hague: W. Junk.

Bauchop, T. 1978. Digestion of leaves in vertebrate arboreal folivores. In *The ecology of arboreal folivores*, ed. G. G. Montgomery. Washington, D.C.: Smithsonian Institution Press.

Bauchot, R., and Stephan, H. 1966. Données nouvelles sur l'encéphalisation des insectivores et des prosimiens. *Mammalia* 30(1):160–97.

———. 1969. Encéphalisation et niveau évolutif chez les simiens. *Mammalia* 33:225–75.

Beach, F. A., and Rabedeau, R. G. 1959. Sexual exhaustion and recovery in the male hamster. *J. Comp. Physiol. Psychol.* 52:56–61.

Beck, A. M. 1973. *The ecology of stray dogs*. New York: York Press.

Beck, B. 1980. *Animal tool behavior*. New York: Garland.

Beck, U. von. 1972. Über die künstliche Aufzucht von Borstengürteltieren (*Euphractus villosus*). *Zool. Gart.* 41:215–22.

Bekoff, M. 1972. The development of social interaction, play and metacommunication in mammals: An ethological perspective. *Quart. Rev. Biol.* 47:412–34.

———. 1977. Mammalian dispersal and the ontogeny of individual behavioral phenotypes. *Amer. Nat.* 111:715–32.

———, ed. 1978. *Coyotes: Biology, behavior and management*. New York: Academic Press.

Bekoff, M., and Jamieson, R. 1975. Physical development in coyotes (*Canis latrans*) with a comparison to other canids. *J. Mammal.* 56:685–92.

Bell, H. M. 1973. The ecology of three macropod marsupial species in an area of open forest and savannah woodland in North Queensland, Australia. *Mammalia* 37(4):527–44.

Bellier, L. 1968. Contribution à l'étude d'*Uranomys ruddi* Dollman. *Mammalia* 32(3):419–47.

Berger, J. 1979. Social ontogeny and behavioural diversity: Consequences for bighorn sheep, *Ovis canadensis,* inhabiting desert and mountain environments. *J. Zool.* 118:251–66.

Bergerud, A. T. 1974. Rutting behaviour of Newfoundland caribou. In *The behaviour of ungulates and its relation to management,* ed. V. Geist and F. Walther, 1:395–436. IUCN Publications, n.s. no. 24. Morges: IUCN.

Berrie, P. M. 1973. Ecology and status of the lynx in interior Alaska. In *The world's cats,* ed. R. L. Eaton, 1:4–41. Winston, Ore.: World Wildlife Safari.

Bertram, B. C. R. 1975. Social factors influencing reproduction in wild lions. *J. Zool.* (London) 177:463–82.

Bertram, G. C. L. 1940. The biology of the Weddell and crabeater seals, with a study on the comparative behavior of the Pinnipedia. *Sci. Rept. Brit. Graham London Exped., 1934–37,* 1(1):1–139.

———. 1964. *In search of mermaids: The manatees of Guiana.* New York: T. Y. Crowell.

Bertram, G. C. L., and Bertram, C. K. R. 1973. The modern Sirenia: Their distribution and status. *Biol. J. Linn. Soc.* 5:297–338.

Bertrand, M. 1969. The behavioral repertoire of the stumptail macaque. *Bibl. Primatol.,* no. 11.

Best, T. L. 1973. Ecological separation of three genera of pocket gophers (Geomyidae). *Ecology* 54(6):1311–20.

Bhatia, C. L., and Desai, J. H. 1972. Growth and development of a hand-reared kinkajou at Delhi Zoo. *Int. Zoo Yearb.* 12:176–77.

Bickel, C. L.; Murdock, G. K.; and Smith, M. L. 1976. Hand-rearing a giant anteater at Denver Zoo. *Int. Zoo Yearb.* 16:195–98.

Bickham, J. W., and Baker, R. J. 1979. Canalization model of chromosome evolution. *Bull. Carnegie Mus. Nat. Hist.* 13:70–84.

Biggers, J. D. 1966. Reproduction in male marsupials. *Symp. Zool. Soc. London* 15:251–80.

Biggins, J. G., and Overstreet, D. H. 1978. Aggressive and nonaggressive interactions among captive populations of the brush-tail possum, *Trichosurus vulpecula* (Marsupialia: Phalangeridae). *J. Mammal.* 59(1):149–60.

Birkenholz, D. E. 1963. A study of the life history and ecology of the round-tailed muskrat (*Neofiber alleni* True) in north-central Florida. *Ecol. Monogr.* 33:255–80.

Birkenholz, D. E., and Wirtz, W. O., III. 1965. Laboratory observations on the vesper rat. *J. Mammal.* 46:181–89.

Blair, W. F. 1940. A study of prairie deer-mouse populations in southern Michigan. *Amer Midl. Nat.* 24:273–305.

———. 1941. Observations on the life history of *Baiomys taylori subater. J. Mammal.* 22:378–83.

———. 1951. Population structure, social behavior and environmental relations in a natural population of the beach mouse (*Peromyscus polionotus leucocephalus*). *Contribs. Lab. Vert. Biol., Univ. Mich.* 48:1–47.

Blasdel, P. 1955. Hunting methods of fish-eating bats, particularly *Noctilio leporinus*. *J. Mammal.* 36:390–99.

Block, J. A. 1974. Hand-rearing seven-banded armadillos, *Dasypus septemcinctus*, at the National Zoological Park, Washington. *Int. Zoo Yearb.* 14:210–14.

Blood, D. S. 1963. Some aspects of behavior of a bighorn herd. *Canadian Field Nat.* 77:77–94.

Blus, L. J. 1971. Reproduction and survival of short tailed shrews (*Blarina brevicauda*) in captivity. *Lab. Anim. Sci.* 21:884–91.

Bogan, M. A. 1972. Observations on parturition and development in the hoary bat, *Lasiurus cinereus*. *J. Mammal.* 53:611–13.

Boness, D. 1979. The social system of the gray seal *Halichoerus grypus* (Fab.), on Sable Island, Nova Scotia. Ph.D. thesis, Dalhousie University.

Bonner, J. T. 1965. *Size and cycle*. Princeton: Princeton University Press.

Borowski, S., and Dehnel, A. 1953. Augabei zur Biologie der Sorricidae. *Ann. Univ. Mariae Curie—Sklodowska,* sect. C, 7:305–448.

Bourlière, F. 1964. *The natural history of mammals*. New York: Alfred Knopf.

———. 1975. Mammals, small and large: The ecological implications of size. In *Small mammals: Their productivity and population dynamics,* ed. F. B. Golley, K. Petrusewicz, and L. Ryszkowski, pp. 1–8. Cambridge: Cambridge University Press.

Bourlière, F.; Petter-Rousseaux, A.; and Petter, J. J. 1961. Regular breeding in captivity of the lesser mouse lemur (*Microcebus murinus*). *Int. Zoo Yearb.* 3:24–25.

Bradbury, J. W. 1977. Lek mating behavior in the hammer-headed bat. *Z. Tierpsychol.* 45(3):225–55.

Bradbury, J. W., and Vehrencamp, S. L. 1976*a*. Social organization and foraging in emballonurid bats. I. Field studies. *Behav. Ecol. Sociobiol.* 1:337–81.

———. 1976*b*. Social organization and foraging in emballonurid bats. II. A model for the determination of group size. *Behav. Ecol. Sociobiol.* 1:383–404.

———. 1977*a*. Social organization and foraging in emballonurid bats. III. Mating systems. *Behav. Ecol. Sociobiol.* 2(1):1–19.

———. 1977*b*. Social organization and foraging in emballonurid bats. IV. Parental investment patterns. *Behav. Ecol. Sociobiol.* 2(1):19–31.

Bradshaw, G. V. R. 1961. A life history study of the California leaf-nosed bat, *Macrotis californicus*. Ph.D. diss., University of Arizona, Tucson.

———. 1962. Reproductive cycle of the California leaf-nosed bat, *Macrotis californicus*. *Science* 136:645–46.

Bradt, G. W. 1938. A study of beaver in Michigan. *J. Mammal.* 19:139–62.

Brady, C. A. 1978. Reproduction, growth and parental care in crab-eating foxes, *Cerdocyon thous,* at the National Zoological Park, Washington. *Int. Zoo Yearb.* 18:130–34.

———. 1979. Observations on the home range and diet of the crab-eating fox, *Cerdocyon thous.* In *Studies on vertebrate ecology in northern Neotropics,* ed. J. F. Eisenberg, pp. 161–71. Washington, D.C.: Smithsonian Institution Press.

Brady, C. A., and Ditton, M. K. 1979. Management and breeding of maned wolves, *Chrysocyon brachyurus,* at the National Zoological Park, Washington. *Int. Zoo Yearb.* 19:171–76.

Braithwaite, R. W. 1974. Behavioural changes associated with the population cycle of *Antechinus stuartii* (Marsupialia). *Australian J. Zool.* 22:45–62.

Braithwaite, R. W., and Lee, A. K. 1979. A mammalian example of semelparity. *Amer. Nat.* 113:151–55.

Brambell, F. W. R. 1944. The reproduction of the wild rabbit, *Oryctolagus cuniculus* (L.). *Proc. Zool. Soc. London* 144:1–45.

Brattstrom, B. H. 1973. Social and maintenance behavior of the echidna, *Tachyglossus aculeatus. J. Mammal.* 54(1):50–71.

Bretting, H. 1972. Die Bestimmung der Reichschwellen bei Igeln (*Erinaceus europaeus* L.) für einige Fett säuren. *Z. Säugetierk.* 37:286–310.

Broadbrooks, H. E. 1958. Life history and ecology of the chipmunk, *Eutamias amoenus,* in eastern Washington. *Misc. Publ. Mus. Zool. Univ. Mich.,* no. 103.

———. 1965. Ecology and distribution of the pikas of Washington and Alaska. *Amer. Midl. Nat.* 73:299–335.

Brodie, P. F. 1975. Cetacean energetics: An overview of intraspecific size variation. *Ecology* 56(1):152–61.

Bromlei, G. F. 1956. Girmalaiskii medved (*Selenarctos tibetanus ussuricus*). *Zool. Zhur.* 35:111–29.

Bromley, P. T. 1969. Territoriality in pronghorn bucks on the National Bison Range, Moise, Montana. *J. Mammal.* 50(1):81–90.

Bromley, P. T., and Kitchen, D. 1974. Courtship in the pronghorn *Antilocapra americana.* In *The behaviour of ungulates and its relation to management,* ed. V. Geist and F. Walther, 1:356–65. IUCN Publications, n.s. no. 24. Morges: IUCN.

Bronson, F. H. 1963. Some correlates of interaction rate in natural populations of woodchucks. *Ecology* 44:637–43.

———. 1964. Agonistic behaviour in woodchucks. *Anim. Behav.* 12:470–78.

———. 1979. The reproductive ecology of the house mouse. *Quart. Rev. Biol.* 54:265–99.

Brooks, J. W. 1954. *A contribution to the life history and ecology of the Pacific walrus.* Special Report 1. College: Alaska Cooperative Wildlife Research Unit.

Brooks, R. J., and Banks, E. M. 1973. Behavioural biology of the collared lemming (*Dicrostonyx groenlandicus* Traill): An analysis of acoustic communication. *Anim. Behav. Monogr.* 6(1):1–83.

Broom, R. 1895. Note on the period of gestation in echidna. *Proc. Linnean Soc. New South Wales* 10:576–77.

Brosset, A. 1961. Some notes on Blanford's or the white-tailed wood rat (*Rattus blanfordi*) in western India. *J. Bombay Nat. Hist. Soc.* 58:241–48.

———. 1966. *La biologie des chiroptères*. Paris: Masson.

Brower, J. E., and Cade, T. J. 1966. Ecology and physiology of *Napaeozapus insignis* (Miller) and other woodland mice. *Ecology* 47:46–63.

Brown, J. H. 1975. Geographical ecology of desert rodents. In *Ecology and evolution of communities,* ed. M. L. Cody and J. M. Diamond, pp. 315–41. Cambridge: Belknap Press of Harvard University Press.

Brown, J. H.; Reichman, O. J.; and Davidson, D. W. 1979. Granivory in desert ecosystems. *Ann. Rev. Ecol. Syst.* 10:201–28.

Brown, J. L. 1975. *The evolution of behavior.* New York: W. W. Norton.

Brown, R. Z. 1953. Social behavior, reproduction and population changes in the house mouse (*Mus musculus* L.). *Ecol. Monogr.* 23:217–40.

Bryden, M. M. 1972. Growth and development of marine mammals. In *Functional anatomy of marine mammals,* ed. R. J. Harrison, 1:1–79. London: Academic Press.

Buchanan, D. B. 1978. Communication and ecology of pithecine monkeys with special reference to *Pithecia pithecia*. Ph.D. diss., Wayne State University.

Bucher, G. C. 1937. Notes on the life history and habits of *Capromys*. *Mem. Soc. Cub. Hist. Nat.* 11:93–107.

Buchler, E. R. 1976. The use of echolocation by the wandering shrew (*Sorex vagrans*). *Anim. Behav.* 24:858–73.

Buckner, C. H. 1964. Metabolism, food capacity and feeding behavior in four species of shrews. *Canadian J. Zool.* 42:259–79.

———. 1966. Populations and ecological relationships of shrews in tamarack bogs in southeastern Manitoba. *J. Mammal.* 47:181–94.

Buechner, H. K. 1974. Implications of social behaviour in the management of Uganda kob. In *The behaviour of ungulates and its relation to management,* ed. V. Geist and F. Walther, 2:853–71. IUCN Publications, n.s. no. 24. Morges: IUCN.

Buechner, H. K.; Mackler, S. F.; Stroman, H. R.; and Xanten, W. A. 1975. Birth of an Indian rhinoceros, *Rhinoceros unicornis,* at the National Zoological Park, Washington. *Int. Zoo Yearb.* 15:160–66.

Buechner, H. K.; Morrison, J. A.; and Leuthold, W. 1965. Reproduction in Uganda kob with special reference to behaviour. *J. Reprod. Fertil.* 9:366–67.

Buechner, H. K., and Schloeth, R. 1965. Ceremonial mating behavior in Uganda kob (*Adenota kob thomasi* Neumann). *Z. Tierpsychol.* 22:209–25.

Burckhardt, D. 1958. Kindliches Verhalten als Ausdrucksbewegung im Fortpflanzungszeremoniell einiger Wiederkäuer. *Rev. Suisse Zool.* 65: 311–16.

Bürger, M. 1959. Eine vergleichende Untersuchung über Putzbewegungen bei Lagomorpha und Rodentia. *Zool. Gart.* 23:434–506.

Burghardt, G. M., and Burghardt, L. S. 1972. Notes on behavioral development of two female black bear cubs: The first eight months. In *Bears: Their biology and management,* ed. S. Herrero, pp. 207–21. Morges: IUCN.

Burrell, H. 1927. *The platypus.* Waterloo: Eagle Press.

Burrows, R. 1968. *Wild fox.* Newton Abbot, England: Daniel and Charles.

Burt, W. H. 1940. Territorial behavior and populations of some small mammals in southern Michigan. *Misc. Publ. Mus. Zool. Univ. Mich.* 45:1–58.

Burton, R. 1973. *The life and death of whales.* New York: Universe Books.

Bush, G. L. 1975. Modes of animal speciation. *Ann. Rev. Ecol. System.* 6:339–64.

Buss, D. H. 1971. Mammary glands and lactation. In *Comparative reproduction of non-human Primates,* ed. E. S. E. Hafez, pp. 315–33. Springfield, Ill.: Thomas.

Buss, I. O. 1961. Some observations on food habits and behaviour of the African elephant. *J. Wildl. Manage.* 25:131–48.

Buss, I. O., and Estes, J. A. 1971. The functional significance of movements and positions of the pinnae of the African elephant, *Loxodonta africana. J. Mammal.* 52(1):21–27.

Buss, I. O., and Smith, N. S. 1966. Observations on reproduction and breeding behavior of the African elephant. *J. Wildl. Manage.* 30:375–88.

Butterworth, B. B. 1961*a*. A comparative study of growth and development of the kangaroo rats, *D. deserti* Stephens and *D. merriami* Mearns. *Growth* 25:127–38.

———. 1961*b*. The breeding of *Dipodomys deserti* in the laboratory. *J. Mammal.* 42:413–14.

Bützler, W. 1974. Kampf- und Paarungsverhalten, soziale Rangordnung und Aktivitätsperiodik beim Rothirsch (*Cervus elaphus* L.). *Z. Tierpsychol.* 16:1–80.

Cabrera, A. 1958–61. Catologo de los mammiferos de America del Sur. *Rev. Mus. Argent. Cienc. Nat. Bernardino Rivadavia* 4(1–2):1–732.

Cadigan, F. C., Jr. 1972. A brief report on copulatory and perinatal behaviour of the lesser Malayan mouse deer (*Tragulus javanicus*). *Malayan Nat. J.* 25(2):112–16.

Calaby, J. H. 1960. Observations on the banded anteater *Myrmecobius f. fasciatus* Waterhouse (Marsupialia), with particular reference to its food habits. *Proc. Zool. Soc. London* 135(2):183–207.

———. 1968. Platypus (*Ornithorhynchus anatinus*) and its venomous characteristics. In *Venomous animals and their venoms,* ed. W.

Bücherl, E. Buckley, and V. Deulofeu, pp. 15–29. New York: Academic Press.
Calaby, J. H., and Wimbush, D. J. 1964. Observations on the broad-toothed rat, *Mastacomys fuscus* Thomas. *CSIRO Wildl. Res.* 9(2):123–33.
Caldwell, D. K., and Caldwell, M. C. 1977. Cetaceans. In *How animals communicate,* ed. T. A. Sebeok, pp. 794–808. Bloomington: Indiana University Press.
Caldwell, D. K.; Caldwell, M. C.; and Rice, D. W. 1966. Behavior of the sperm whale, *Physeter catodon* L. In *Whales, dolphins and porpoises,* ed. K. S. Norris, pp. 677–717. Berkeley: University of California Press.
Caldwell, M. C., and Caldwell, D. K. 1966. Epimeletic (care-giving) behavior in Cetacea. In *Whales, dolphins and porpoises,* ed. K. S. Norris, pp. 755–81. Berkeley: University of California Press.
Calhoun, J. B. 1956. A comparative study of the social behavior of two inbred strains of house mice. *Ecol. Monogr.* 26:81–103.
———. 1963. *The ecology and sociology of the Norway rat.* Washington, D.C.: U.S. Department of Health, Education, and Welfare, Public Health Service.
Carrick, R. 1962. Studies on the southern elephant seal, *Mirounga leonina.* IV. Breeding and development. *CSIRO Wildl. Res.* 7:161–97.
Carrick, R.; Csordas, S. E.; Ingham, S. E.; and Keith, K. 1962. Studies on the southern elephant seal. III. The annual cycle in relation to age and sex. *CSIRO Wildl. Res.* 7(2):119–60.
Carter, D. S. 1970. Chiropteran reproduction. In *About bats,* ed. B. H. Slaughter and D. W. Walton, pp. 233–46. Dallas: Southern Methodist University Press.
Cartmill, M. 1972. Arboreal adaptations and the origin of the order Primates. In *The function and evolutionary biology of Primates,* ed. R. Tuttle, pp. 97–122. Chicago/New York: Aldine-Atherton.
———. 1974. *Daubentonia, Dactylopsila,* woodpeckers and klinorhynchy. In *Prosimian biology,* ed. R. D. Martin, G. A. Doyle, and A. C. Walker, pp. 655–70. London: Duckworth.
Case, T. J. 1978. On the evolution and adaptive significance of postnatal growth rates in the terrestrial vertebrates. *Quart. Rev. Biol.* 53:243–82.
Catlett, R. A. 1961. Social behavior and bio-energetics in relation to density of house mice. Abstract of the Forty-first Annual Meeting of the American Society of Mammalogists.
Cerva, F. A. 1929. Beobachtungen an der Streifenmaus (*Sicista loriger trizona* Pet.). *Zool. Gart.* 1:390–95.
Chalmers, N. R. 1968. The visual and vocal communication of free-living mangabeys in Uganda. *Folia Primat.* 9:258–80.
Channing, A., and Rowe-Rowe, D. T. 1977. Vocalizations of South African mustelines. *Z. Tierpsychol.* 44:283–93.
Chapman, B. M.; Chapman, R. F.; and Robertson, I. A. D. 1959. The growth and breeding of the multimammate rate *Rattus (Mastomys) natalensis* (Smith) in Tanganyika Territory. *Proc. Zool. Soc. London* 133:1–9.

Chapman, D., and Chapman, N. 1975. *Fallow deer: Their history, distribution and biology.* Lavenham, England: Terence Dalton.

Charles-Dominique, P. 1972. Ecologie et vie sociale de *Galago demidovii* (Fischer 1808, Prosimii). *Z. Tierpsychol.* 9:7–41.

———. 1975. Vie sociale de *Perodicticus potto* (Primate, Lorisidés). Etude de terrain en forêt équatoriale de l'Ouest africain au Gabon. *Mammalia* 38:355–79.

———. 1977. *Ecology and behaviour of nocturnal Primates.* New York: Columbia University Press.

———. 1978. Ecology and social behaviour of the African palm civet *Nandinia binotata* in Gabon, with a comparison with sympatric prosimians. *Terre et Vie* 32:477–528.

Chasen, F. 1940. A handlist of Malaysian mammals. *Bull. Raffles Mus.* (Singapore) 15:1–209.

Chasen, F.; Kloss, N.; and Boden, C. 1929. Notes on the flying lemurs (*Galeopterus*). *Bull. Raffles Mus.* (Singapore) 2:12–22.

Chen, Pao-Lai; Tsai, K. C.; Chou, N. W.; and Fung, K. N. 1966. On the population density of mole rats in the upper Who-Tuo valley. *Acta Zool. Sinica* 18:21–27.

Chew, R. M. 1958. Reproduction by *Dipodomys merriami* in captivity. *J. Mammal.* 39:597–98.

Chew, R. M., and Butterworth, B. B. 1959. Growth and development of Merriam's kangaroo rat. *Growth* 23:75–79.

Chew, R. M., and Chew, A. E. 1970. Energy relationships of the mammals of a desert shrub (*Larrea tridentata*) community. *Ecol. Monogr.* 40:1–21.

Chivers, D. J. 1974. *The siamang in Malaya: A field study of a primate in tropical rain forest.* Contributions to Primatology, vol. 4. Basel: S. Karger.

Christie, D. W., and Bell, E. T. 1972. Studies on canine reproductive behaviour during the normal oestrous cycle. *Anim. Behav.* 20(4):621–32.

Clark, L. D. 1962. Experimental studies of the behavior of an aggressive, predatory mouse, *Onychomys leucogaster.* In *Roots of behavior,* ed. E. L. Bliss, chap. 12. New York: Harper Brothers.

Clark, M. J. 1967. Pregnancy in the lactating pigmy possum *Cercartetus concinnus. Australian J. Zool.* 15:673–83.

Clark, W. K. 1951. Ecological life history of the armadillo in the eastern Edwards Plateau region. *Amer. Midl. Nat.* 46:337–58.

Clemens, W. A. 1977. Phylogeny of the marsupials. In *The biology of the marsupials,* ed. B. Stonehouse and D. Gilmore, pp. 51–68. Baltimore: University Park Press.

Clothier, R. R. 1955. Contribution to the life history of *Sorex vagrans* in Montana. *J. Mammal.* 36:214–21.

Clough, G. C. 1963. Biology of the arctic shrew, *Sorex arcticus. Amer. Midl. Nat.* 69:69–81.

———. 1972. Biology of the Bahaman hutia, *Geocapromys ingrahami. J. Mammal.* 53(4):807–23.

———. 1974. Additional notes on the biology of the Bahaman hutia, *Geocapromys ingrahami. J. Mammal.* 55:670–71.

Clutton-Brock, J.; Corbett, G. B.; and Hills, M. 1976. A review of the family Canidae with a classification by numerical methods. *Bull. Brit. Mus. (Nat. Hist). Zool.* 29:119–99.
Clutton-Brock, T. H., ed. 1977. *Primate ecology.* London/New York: Academic Press.
Clutton-Brock, T. H., and Harvey, P. H. 1977. Primate ecology and social organization. *J. Zool.* (London) 183:1–39.
Coe, M. J. 1962. Notes on the habits of the Mount Kenya hyrax (*Procavia johnstoni mackinderi* Thomas). *Proc. Zool. Soc. London* 138:639–45.
Coehlo, A. M.; Bramblett, C. A.; and Quick, L. B. 1979. Activity patterns in howler and spider monkeys: An application of socio-bioenergetic methods. In *Primate ecology and human origins,* ed. I. S. Bernstein and E. O. Smith, pp. 175–200. New York: Garland.
Collias, N. E. 1960. An ecological and functional classification of animal sounds. In *Animal sounds and communication,* ed. W. E. Lanyon and W. N. Tavolga, pp. 368–91. Publication no. 7. Washington, D.C.: American Institute of Biological Sciences.
Collins, G. 1940. Habits of the Pacific walrus (*Odobenus divergens*). *J. Mammal.* 21:138–44.
Collins, L. R. 1973. *Monotremes and marsupials.* Washington, D.C.: Smithsonian Institution Press.
Collins, L. R., and Eisenberg, J. F. 1972. The behavior and breeding of pacaranas (*Dinomys branickii*) in captivity. *Int. Zoo Yearb.* 12:108–14.
Compoint-Monmignaut, C. 1973. Anatomie Comparée: L'encéphalisation chez les Rongeurs. *C. R. Acad. Sci., Paris* 277:861–63.
Conaway, C. H. 1958. Maintenance, reproduction and growth of the least shrew in captivity. *J. Mammal.* 39:507–12.
Constantine, D. G. 1958. Ecological observations on Lasiurine bats in Georgia. *J. Mammal.* 39:64–70.
Cope, E. D. 1880. On the genera of the *Creodonta. Proc. Amer. Phil. Soc.* 19:76–82.
Corner, E. J. H. 1945. The Durian theory or the origin of the modern tree. *Ann. Bot.,* n.s., 13:367–414.
Corner, R. W. M. 1972. Observations on a small crabeater seal breeding group. *British Antarctic Surv. Bull.* 30:104–6.
Cowan, I. McT., and Arsenault, M. G. 1954. Reproduction and growth in the creeping vole, *Microtus oregoni serpens* Merriam. *Canadian J. Zool.* 32:198–208.
Cox, B. 1974. Vertebrate paleodistributional patterns and continental drift. *J. Biogeog.* 1:75–94.
Crabb, W. D. 1948. The ecology and management of the prairie skunk in Iowa. *Ecol. Monogr.* 18:201–32.
Crandall, L. S. 1958. Notes on the behavior of platypuses (*Ornithorhynchus anatinus*) in captivity. *Handb. Zool.* 8(10):101–7.
Crespo, J. A., and De Carlo, J. M. 1963. Estudio ecologico de una poblacion de zorros colorados *Dusicyon culpaeus culpaeus* (Molina)

en el oeste de la Provincia de Neuquen. *Rev. Mus. Arg. Cien. Nat. Bs. As. (Ecol.)* 1:1–155.

Crichton, E. G. 1969. Reproduction in the pseudomyine rodent *Mesembriomys gouldii* (Gray) Muridae. *Australian J. Zool.* 17:785–97.

Crile, G., and Quiring, D. P. 1940. A record of the body weight and certain organ and gland weights of 3690 animals. *Ohio J. Sci.* 40:219–59.

Crompton, A. W., and Hiiemäe, K. 1969. How mammalian molar teeth work. *Discovery* 5(1):23–34.

———. 1970. Molar occlusion and mandibular movements during occlusion in the American opossum, *Didelphis marsupialis* L. *Zool. J. Linn. Soc.* 49:21–47.

Crook, J. H. 1966. Gelada baboon herd structure and movement: A comparative report. *Symp. Zool. Soc. London* 18:237–58.

———. 1970. The socio-ecology of Primates. In *Social behavior in birds and mammals,* ed. J. H. Crook. London: Academic Press.

Crook, J. H., and Gartlan, S. S. 1966. Evolution of Primate societies. *Nature* (London) 210:1200–1230.

Crowcroft, P. 1957. *The life of the shrew.* London: Reinhardt.

Crowcroft, P., and Row, F. P. 1963. Social organization and territorial behaviour in the wild house mouse (*Mus musculus* L.). *Proc. Zool. Soc. London* 140:517–32.

Crowcroft, P., and Soderlund, R. 1977. Breeding of wombats (*Lasiorhinus latifrons*) in captivity. *Zool. Gart.,* n.s. (Jena) 17(5):313–22.

Csordas, S. E. 1958. Breeding of the fur seal, *Arctocephalos forsteri* Lesson at Macquarie Island. *Australian J. Sci.* 21:87–88.

Cuttle, P. 1978. The behaviour in captivity of the dasyurid marsupial *Phascogale tapoatafa* (Meyer). M.Sc. thesis, Department of Zoology, Monash University.

Dagg, A. I., and Foster, J. B. 1976. *The giraffe.* New York: Van Nostrand Reinhold.

Dalby, P. L. 1975. Biology of pampa rodents, Balcarce Area, Argentina. *Publ. Mus. Michigan State Univ., Biol. Ser.* 5(3):1–271.

Dalquest, W. W. 1948. Mammals of Washington. *Univ. Kansas Publ. Mus. Nat. Hist.* 2:1–444.

Dalquest, W. W., and Orcutt, D. R. 1942. The biology of the least shrew-mole, *Neurotrichus gibbsii minor. Amer. Midl. Nat.* 27:387–401.

Daly, M., and Daly, S. 1975*a*. Socio-ecology of Saharan gerbils, especially *Meriones libycus. Mammalia* 39:289–311.

———. 1975*b*. Behaviour of *Psammomys obesus* (Rodentia, Gerbillinae) in the Algerian Sahara. *Z. Tierpsychol.* 37:298–321.

Dapson, R. W. 1968. Growth patterns in a post-juvenile population of short tailed shrews (*Blarina brevicauda*). *Amer. Midl. Nat.* 79:118–29.

Darling, F. F. 1937. *A herd of red deer.* London: Oxford University Press.

Darlington, P. J. 1957. *Zoogeography: The geographical distribution of animals.* New York: John Wiley.

Darwin, C. 1859. *The origin of species.* London: John Murray.

Dathe, H. 1963. Beitrag zur Fortpflanzungsbiologie des Malaienbaren, *Helarctos m. malayenus* (Raffl.). *Z. Säugetierk.* 28:155–62.
———. 1967. Bemerkungen zur Aufzucht von Brillenbaren *Tremarctos ornatus* (Cuv.) im Tierpark Berlin. *Zool. Gart.* (Leipzig) 34:105–33.
———. 1970. A second generation birth of captive sun bears, *Helarctos malayenus,* at East Berlin Zoo. *Int. Zoo Yearb.* 7:131.
Davenport, L. B., Jr. 1964. Structure of two *Peromyscus polionotus* populations in old-field ecosystems at the AEC Savannah River plant. *J. Mammal.* 45:95–113.
Davis, D. D. 1964. *The giant panda: A morphological study of evolutionary mechanisms.* Fieldiana: Zoology, vol. 3. Chicago: Field Museum of Natural History.
———. 1965. The lesser gymnure, *Hylomys suillus. Malayan Nat. J.* 19(2–3):147–48.
Davis, D. H. S. 1963. Wild rodents as laboratory animals and their contribution to medical research in South Africa. *South African J. Med. Sci.* 28:53–89.
Davis, J. A. 1965. A preliminary report on the reproductive behavior of the small Malayan chevrotain, *Tragulus javanicus,* at the New York Zoo. *Int. Zoo Yearb.* 5:42–44.
Davis, R. 1966. Homing performance and homing ability in bats. *Ecol. Monogr.* 36(3):201–37.
———. 1969. Growth and development of young pallid bats, *Antrozous pallidus. J. Mammal.* 50:729–36.
Davis, R. B.; Herreid, C. F.; and Short, H. L. 1962. Mexican free-tailed bats in Texas. *Ecol. Monogr.* 32:311–46.
Davis, R. M. 1972. Behaviour of the vlei rat, *Otomys irroratus* (Brants 1827). *Zool. Africana* 7(1):119–40.
Davis, W. B., and Joeris, L. 1945. Notes on the life history of the little short-tailed shrew. *J. Mammal.* 26:136–38.
Davis, W. H. 1966. Population dynamics of the bat, *Pipistrellus subflavus. J. Mammal.* 47:383–96.
Dawkins, R. 1976. *The selfish gene.* New York/Oxford: Oxford University Press.
Dawson, G. A. 1977. Composition and stability of social groups of the tamarin, *Saguinus oedipus geoffroyi,* in Panama: Ecological and behavioral implications. In *The biology and conservation of the Callitrichidae,* ed. D. G. Kleiman, pp. 23–39. Washington, D.C.: Smithsonian Institution Press.
Dawson, T. J., and Hulburt, A. J. 1970. Standard metabolism, body temperature and surface areas of Australian marsupials. *Am. J. Physiol.* 218:1233–38.
Deag, J. M., and Crook, J. H. 1971. Social behaviour and "agonistic buffering" in the wild Barbary macaque, *Macaca sylvanus* L. *Folia Primat.* 15:183–200.
de Carvalho, C. T. 1968. Comparative growth rates of hand-reared big cats. *Int. Zoo Yearb.* 8:56–59.
De Coursey, P. 1961. Effect of light on the circadian activity rhythm of the flying squirrel, *Glaucomys volans. Z. Vergl. Physiol.* 44:331–54.

Dehnel, A. 1952. The biology of breeding of common shrew *Sorex araneus* L. *Ann. Univ. M. C.-Sklodowska, Sect. C* 6:359–76.

Delibes, M. 1974. Sobre alimentacion y biologio de la gineta (*Genetta genetta* L.) en España. *Donana Act. Vert.* 1(1):143–99.

De Vore, B. I., ed. 1965. *Primate behavior: Field studies of monkeys and apes*. New York: Holt, Rinehart and Winston.

De Vos, A.; Broex, P.; and Geist, V. 1967. A review of social behaviour of the North American cervids during the reproductive period. *Amer. Midl. Nat.* 77:390–417.

De Vos, A., and Dowsett, R. J. 1966. The behaviour and population structure of three species of the genus *Kobus*. *Mammalia* 30:30–55.

Dewsbury, D. A. 1972. Patterns of copulatory behavior in male mammals. *Quart. Rev. Biol.* 47(1):1–33.

———. 1974. Copulatory behavior of California mice (*Peromyscus californicus*). *Brain, Behav. Evol.* 9:95–106.

———. 1975. Diversity and adaptation in rodent copulatory behavior. *Science* 190:947–54.

Dieterlen, F. 1959. Das Verhalten des syrischen Goldhamsters (*Mesocricetus auratus* Waterhouse). Untersuchungen zur Frage seiner Entwicklung und seiner angeborenen Anteile durch geruchsisolierte Aufzuchten. *Z. Tierpsychol.* 16(1):47–103.

———. 1960. Bemarkungen zu Zucht und Verhalten der Zwergmaus (*Micromys minutus soricinus* Hermann). *Z. Tierpsychol.* 17(5):552–54.

———. 1961. Beiträge zur Biologie der Stachelmaus, *Acomys cahirinus dimidiatus* Crezschmar. *Z. Säugetierk.* 26:1–64.

———. 1971. Beiträge zur Systemitik, Ökologie und Biologie der Gattung *Dendromus* (Dendromurinae, Cricetidae, Rodentia). *Säugetierk. Mitt.* 19:97–132.

Dijkgraaf, S. 1946. Die Sinneswelt der Fledermäuse. *Experientia* 2:438–48.

Dippenaar, N. J. 1979. Notes on the early post-natal development and behaviour of the tiny musk shrew, *Crocidura bicolor* Bocage 1889 (Insectivora: Soricidae). *Mammalia* 43:83–92.

Dittrich, L. 1969. Birth weights and weight increases of African antelopes born in the Hanover Zoo. *Int. Zoo Yearb.* 9:118–20.

———. 1972. Gestation periods and age of sexual maturity of some African antelopes. *Int. Zoo Yearb.* 12:184–87.

———. 1976. Age of sexual maturity in the hippopotamus. *Int. Zoo Yearb.* 16:171–73.

Dittus, W. P. J. 1975. Population dynamics of the toque monkey, *Macaca sinica*. In *Primates: Socio-ecology and psychology,* ed. R. H. Tuttle, pp. 125–52. The Hague: Mouton.

———. 1977*a*. The socioecological basis for the conservation of the toque monkey (*Macaca sinica*) of Sri Lanka (Ceylon). In *Primate conservation,* ed. Prince Rainier III and G. Bourne, pp. 237–65. New York: Academic Press.

———. 1977*b*. The social regulation of population density and age-sex distribution in the toque monkey. *Behaviour* 63(3–4):281–322.

Dobroruka, L. J. 1960. Einige Beobachtungen an Ameisenigeln *Echidna aculeata* (Shaw 1872). *Z. Tierpsychol.* 17:178–81.
Dobzhansky, T. 1970. *Genetics of the evolutionary process.* New York/London: Columbia University Press.
Dolbeer, R. A. 1973. Reproduction in the red squirrel (*Tamiasciurus hudsonicus*) in Colorado. *J. Mammal.* 54:536–40.
Dollo, L. 1899. Les ancêtres des Marsupiaux étaient-ils arboricoles? *Trav. Stat. Zool. Wimereux* 7:188–203.
Domning, D. 1976. An ecological model for late Tertiary Sirenian evolution in the North Pacific Ocean. *Syst. Zool.* 25:352–62.
Donaldson, S. L.; Wirtz, T. B.; and Hite, A. E. 1975. The social behaviour of capybaras, *Hydrochoerus hydrochaeris*, at Evansville Zoo. *Int. Zoo Yearb.* 15:201–6.
Doty, R. L. 1973. Reactions of deer mice (*Peromyscus maniculatus*) and white-footed mice (*Peromyscus leucopus*) to homospecific and heterspecific urine odors. *J. Comp. Physiol. Psychol.* 84:296–303.
———. 1976. *Mammalian olfaction, reproductive processes and behavior.* New York: Academic Press.
Doty, R. L., and Kart, R. 1972. A comparative and developmental analysis of the midventral sebaceous glands in 18 taxa of *Peromyscus* with an examination of gonadal steroid influences in *Peromyscus maniculatus bairdii. J. Mammal.* 53:83–99.
Douglas-Hamilton, I. 1972. On the ecology and behaviour of the African elephant. Ph.D. diss., University of Oxford.
Downhower, J. F., and Armitage, K. B. 1971. The yellow-bellied marmot and the evolution of polygamy. *Amer. Nat.* 105(946):355–70.
Drake, J. D. 1958. The brush mouse, *Peromyscus boylii*, in southern Durango. *Publ. Mus. Michigan State Univ., Biol. Ser.* 1(3):97–132.
Drüwa, P. 1977. Beobachtungen zur Geburt und Natürlichen Aufzucht von Waldhunden (*Speothos venaticus*) in der Gefangenschaft. *Zool. Gart.* n.s., 47:109–37.
Dryden, G. L. 1968. Growth and development of *Suncus murinus* in captivity on Guam. *J. Mammal.* 49(1):51–63.
———. 1969. Reproduction in *Suncus murinus. J. Reprod. Fertil.* 6:377–96.
Dryden, G. L.; Gebczynski, M.; and Douglas, E. L. 1974. Oxygen consumption by nursling and adult musk shrews. *Acta Theriol.* 19(30):453–61.
Dubost, G. 1965. Quelques renseignements biologiques sur *Potamogale velox. Biol. Gabonica* 1(3):257–72.
———. 1968. Les niches écologiques des forêts tropicales sud-americaines et africaines sources de convergences remarquables entre rongeurs et artiodactyles. *Terre et Vie* 1:3–28.
———. 1970. L'organisation spatiale et sociale de *Muntiacus reevesi* Ogilby 1839 en semi-liberté. *Mammalia* 34(3):331–56.
———. 1975. Le comportement du Chevrotain africain, *Hyemoschus aquaticus* Ogilby (Artiodactyla, Ruminantia). *Z. Tierpsychol.* 37:403–48.

———. 1978. Account on the biology of the African chevrotain *Hyemoschus aquaticus* Ogilby (Artiodactyla: Tragulidae). *Mammalia* 42:1–62.

Dubost, G., and Genest, H. 1974. Le comportement social d'une colonie de maras *Dolichotis patagonum* Z. dans le parc de Branfere. *Z. Tierpsychol.* 35(3):225–302.

Dücker, G. 1957. Farb- und Helligheitssehen und Instinkte bei Viverriden und Feliden. *Zool. Beitr.* 3:25–100.

———. 1965. Das Verhalten der Viverriden. *Handb. Zool.* 8 Band, 38 Lieferung, 19(20a):1–48.

———. 1971. Gefangenschafts beobachtungen an Pardelvollern *Nandinia binotata. Z. Tierpsychol.* 28:77–89.

Dudley, D. 1974. Paternal behavior in the California mouse, *Peromyscus californicus. Behav. Biol.* 11(2):247–52.

Dunbar, R. I. M., and Dunbar, E. P. 1974. Ecology and population dynamics of *Colobus guereza* in Ethiopia. *Folia Primat.* 21:188–208.

———. 1975. *Social dynamics of gelada baboons.* Contributions to Primatology, vol. 6. Basel: S. Karger.

Dunford, C. 1970. Behavioral aspects of spatial organization in the chipmunk, *Tamias striatus. Behaviour* 35:215–32.

Dunnett, G. M. 1956. A live trapping study of the brush-tailed possum, *Trichosurus vulpecula* Kerr (Marsupialia). *CSIRO Wildl. Res.* 1:1–18.

———. 1962. A population study of the quokka, *Setonix brachyurus* (Quoy and Gaimard) (Marsupialia). 2. Habitat, movements, breeding and growth. *CSIRO Wildl. Res.* 7:13–32.

———. 1963. A population study of the quokka, *Setonix brachyurus* (Quoy and Gaimard) (Marsupialia). 3. The estimation of population parameters by means of the recapture technique. *CSIRO Wildl. Res.* 8:78–117.

———. 1964. A field study of local populations of the brush-tailed possum, *Trichosurus vulpecula,* in eastern Australia. *Proc. Zool. Soc. London* 142:665–95.

Duplaix, N. 1980. Observations on the ecology and behavior of the giant river otter, *Pteronura brasiliensis,* in Suriname. *Rev. Ecol. (Terre et Vie)* 34:495–620.

Du Plessis, S. S. 1972. Ecology of blesbok with special reference to productivity. *Wildl. Monogr.* 30:1–70.

Eadie, W. R. 1935. *The life and habits of the platypus.* Melbourne: Stillwell and Stephens.

———. 1939. A contribution to the biology of *Parascalops breweri. J. Mammal.* 20:150–73.

Eadie, W. R., and Hamilton, W. J. 1956. Notes on reproduction in the star-nosed mole. *J. Mammal.* 37:223–31.

Ealey, E. H. M. 1967*a*. Ecology of the euro, *Macropus robustus* (Gould), in northwestern Australia. II. Behaviour, movements and drinking patterns. *CSIRO Wildl. Res.* 12:27–51.

———. 1967*b*. Ecology of the euro, *Macropus robustus* (Gould), in northwestern Australia. IV. Age and growth. *CSIRO Wildl. Res.* 12:67–80.

Eaton, R. L. 1970. Hunting behavior of the cheetah. *J. Wildl. Manage.* 34(1):56–67.

———, ed. 1973*a*. *The world's cats*. Vol. 1. *Ecology and conservation.* Winston, Ore.: World Wildlife Safari.

———. 1973*b*. *The world's cats*. vol. 2. *Biology, behavior and management of reproduction.* Winston, Ore.: World Wildlife Safari.

———. 1974*a*. *The cheetah: The biology, ecology and behavior of an endangered species.* New York: Van Nostrand Reinhold.

———, ed. 1974*b*. *The world's cats*. Vol. 3, no. 3. *Contributions to breeding biology, behavior and husbandry.* Seattle: Burke Museum, Carnivore Research Institute.

———, ed. 1976*a*. *The world's cats*. Vol. 3, no. 1. *Contributions to status, management and conservation.* Seattle: Burke Museum, Carnivore Research Institute.

———, ed. 1976*b*. *The world's cats*. Vol. 3, no. 2. *Contributions to biology, ecology, behavior and evolution.* Seattle: Burke Museum, Carnivore Research Institute.

———. 1977. Reproductive biology of the leopard. *Zool. Gart.* 47:329–51.

Eberhard, I. 1978. Ecology of the koala, *Phascolarctos cinereus* (Goldfuss) (Marsupialia: Phascolarctidae) in Australia. In *The ecology of arboreal folivores*, ed. G. G. Montgomery, pp. 315–28. Washington, D.C.: Smithsonian Institution Press.

Eberhardt, R. L., and Norris, K. S. 1964. Observations of newborn Pacific gray whales on Mexican calving grounds. *J. Mammal.* 45:88–95.

Edinger, T. 1948. Evolution of the horse brain. *Mem. Geol. Soc. Amer.* (Washington, D.C.) 25:1–177.

Egoscue, H. J. 1956. Preliminary studies of the kit fox in Utah. *J. Mammal.* 37:350–57.

———. 1957. The desert woodrat: A laboratory colony. *J. Mammal.* 38:472–81.

———. 1962. The bushy-tailed woodrat: A laboratory colony. *J. Mammal.* 43:328–36.

———. 1964. Ecological notes and laboratory life history of the canyon mouse. *J. Mammal.* 45:387–96.

———. 1972. Breeding the long-tailed *Beamys hindei* in captivity. *J. Mammal.* 53:296–302.

———. 1975. Population dynamics of the kit fox in western Utah. *Bull. Southern California Acad. Sci.* 74(3):122–27.

Egoscue, H. J.; Bittmenn, J. G.; and Petrovich, J. A. 1970. Some fecundity and longevity records for captive small mammals. *J. Mammal.* 51:622–23.

Ehrlich, S. 1966. Ecological effects of reproduction in nutria *Myocastor coypus* Mol. *Mammalia* 30:144–52.

Eibl-Eibesfeldt, I. 1950*a*. Beiträge zur Biologie der Haus- und der Ahrenmaus nebst einigen Beobachtungen an anderen Nagern. *Z. Tierpsychol.* 7(4):558–86.

———. 1950*b*. Über die Jugendentwicklung des Verhaltens eines männlichen Dachses (*Meles meles* L.). *Z. Tierpsychol.* 7:227–55.

———. 1951*a*. Beobachtungen zur Florpflanzungsbiologie und Jungendentwicklung des Eichhörnchens. *Z. Tierpsychol.* 8:370–400.

———. 1951*b*. Gefangenshafts beobachtungen an der persichen Wüstenmause (*Meriones persicus persicus* Blanford). *Z. Tierpsychol.* 8:400–423.

———. 1953. Zur Ethologie des Hamsters (*Cricetus cricetus* L.). *Z. Tierpsychol.* 10:204–54.

———. 1955. Ethologische studien am Galpagos-Seelowen, *Zalophus wollebacki* Sivertsen. *Z. Tierpsychol.* 12:286–303.

———. 1958. Das Verhalten der Nagetiere. *Handb. Zool.*, Band 8, Lieferung 12; 10(13):1–88.

———. 1970. *Ethology: The biology of behavior.* New York: Holt, Rinehart and Winston.

Einarsen, A. S. 1948. *The pronghorn antelope and its management.* Washington, D.C.: Wildlife Management Institute.

Eisenberg, J. F. 1961. Observations on the nest-building behavior of armadillos. *Proc. Zool. Soc. London* 137(2):322–24.

———. 1962. Studies on the behavior of *Peromyscus maniculatus gambelii* and *Peromyscus californicus parasiticus. Behaviour* 19:177–207.

———. 1963. The behavior of heteromyid rodents. *Univ. California Publ. Zool.* 69:1–100.

———. 1964. Studies on the behavior of *Sorex vagrans. Amer. Midl. Nat.* 72:417–25.

———. 1966. The social organizations of mammals. *Handb. Zool.*, Band 8, Lieferung 39; 10(7):1–92.

———. 1967. Comparative studies on the behavior of rodents with special emphasis on the evolution of social behavior, Part I. *Proc. U.S. Nat. Mus.* 122(3597):1–55.

———. 1968. The behavior of *Peromyscus*. In *The biology of Peromyscus*, ed. J. King, pp. 451–95. American Society of Mammalogists, Special Publication no. 2.

———. 1969. Social organization and emotional behavior. In *Experimental approaches to the study of emotional behavior,* ed. E. Tobach. *Ann. N.Y. Acad. Sci.* 159(3):752–60.

———. 1973. Mammalian social systems: Are primate social systems unique? In *Symposium of the IVth International Congress of Primatology,* 1:232–49. Basel: S. Karger.

———. 1974. The function and motivational basis of hystricomorph vocalizations. In *The biology of hystricomorph rodents,* ed. I. W. Rowlands and B. Weir, pp. 211–44. Symposium no. 34. London: Zoological Society of London.

———. 1975*a*. Phylogeny, behavior and ecology in the Mammalia. In *Phylogeny of the Primates: An interdisciplinary approach,* ed. P. Luckett and F. Szalay, pp. 47–68. New York: Plenum Press.

———. 1975*b*. Tenrecs and solenodons in captivity. *Int. Zoo Yearb.* 15:6–12.

———. 1975*c*. The behavior patterns of desert rodents. In *Rodents in desert environments,* ed. I. Prakash and P. K. Ghosh, pp. 189–224. Monographae Biologicae. The Hague: W. Junk.

———. 1976. Communication and social integration in the black spider monkey, *Ateles fusciceps robustus,* and related species. *Smithsonian Contrib. Zool.* 213:1–108.

———. 1977. The evolution of the reproductive unit in the class Mammalia. In *Reproductive behavior and evolution,* ed. J. S. Rosenblatt and B. Komisaruk, pp. 39–71. New York: Plenum Press.

———. 1978. Evolution of arboreal herbivores in the class Mammalia. In *The ecology of arboreal folivores,* ed. G. G. Montgomery, pp. 135–52. Washington, D.C.: Smithsonian Institution Press.

———. 1979. Habitat, economy and society: Some correlations and hypotheses for the Neotropical primates. In *Primate ecology and human origins,* ed. I. S. Bernstein and E. O. Smith, pp. 215–62. New York: Garland Press.

———. 1980*a*. The density and biomass of tropical mammals. In *Conservation biology,* ed. M. Soulé and B. A. Wilcox, pp. 35–56. Sunderland, Mass.: Sinauer Associates.

———. 1980*b*. Biological strategies of living conservative mammals. In *Comparative physiology: Primitive mammals,* ed. K. Schmidt-Nielson, L. Bolis, and C. Richard Taylor, pp. 13–30. Cambridge: Cambridge University Press.

Eisenberg, J. F.; Collins, L. R.; and Wemmer, C. 1975. Communication in the Tasmanian devil (*Sarcophilus harrisii*) and a survey of auditory communication in the Marsupialia. *Z. Tierpsychol.* 37:379–99.

Eisenberg, J. F., and Dillon, W. S., eds. 1972. *Man and beast: Comparative social behavior.* Washington, D.C.: Smithsonian Institution.

Eisenberg, J. F., and Golani, I. 1977. Communication in Metatheria. In *How animals communicate,* ed. T. Sebeok, pp. 575–99. Bloomington: Indiana University Press.

Eisenberg, J. F., and Gould, E. 1966. The behavior of *Solenodon paradoxus* in captivity with comments on the behavior of other Insectivora. *Zoologica* 51(1):49–58.

———. 1970. The tenrecs. A study in mammalian behavior and evolution. *Smithsonian Contrib. Zool.* 27:1–137.

Eisenberg, J. F., and Isaac, D. E. 1963. The reproduction of heteromyid rodents in captivity. *J. Mammal.* 44:61–67.

Eisenberg, J. F., and Kleiman, D. G. 1972. Olfactory communication in mammals. *Ann. Rev. Ecol. System.* 3:1–32.

Eisenberg, J. F., and Kuehn, R. E. 1966. The behavior of *Ateles geoffroyi* and related species. *Smithsonian Misc. Coll.* 151(8), no. 4683:1–63.

Eisenberg, J. F., and Leyhausen, P. 1972. The phylogenesis of predatory behaviour in mammals. *Z. Tierpsychol.* 30:59–93.

Eisenberg, J. F., and Lockhart, M. 1972. An ecological reconnaissance of Wilpattu National Park, Ceylon. *Smithsonian Contrib. Zool.* 101:1–118.

Eisenberg, J. F., and McKay, G. M. 1970. An annotated checklist of the recent mammals of Ceylon. *Ceylon J. Sci. (Biol. Sci.)* 8(2):69–99.

———. 1974. Comparison of ungulate adaptations in the New World and Old World tropical forests with special reference to Ceylon and the rainforests of Central America. In *The behaviour of ungulates and its relation to management,* ed. V. Geist and F. Walther, 2:585–602. IUCN Publications, n.s. no. 24. Morges: IUCN.

Eisenberg, J. F.; McKay, G. M.; and Jainudeen, M. R. 1971. Reproductive behavior of the Asiatic elephant (*Elephas maximus maximus* L.). *Behaviour* 38(3):193–225.

Eisenberg, J. F., and Maliniak, E. 1967. The breeding of *Marmosa* in captivity. *Int. Zoo Yearb.* 7:78–79.

———. 1973. Breeding and captive maintenance of the lesser bamboo rat, *Cannomys badius. Int. Zoo Yearb.* 13:204–7.

———. 1974. The reproduction of the genus *Microgale* in captivity. *Int. Zoo Yearb.* 14:108–10.

Eisenberg, J. F., and Muckenhirn, N. 1968. The reproduction and rearing of tenrecoid insectivores in captivity. *Int. Zoo Yearb.* 8:106–10.

Eisenberg, J. F.; Muckenhirn, N.; and Rudran, R. 1972. The relationship between ecology and social structure in primates. *Science* 176:863–74.

Eisenberg, J. F., and Redford, K. H. 1979. Biogeographic analysis of the mammalian fauna of Venezuela. In *Vertebrate ecology in the northern Neotropics,* ed. J. F. Eisenberg, pp. 46–63. Washington, D.C.: Smithsonian Institution Press.

Eisenberg, J. F., and Seidensticker, J. 1976. Ungulates in southern Asia: A consideration of biomass estimates for selected habitats. *Biol. Cons.* 10:293–308.

Eisenberg, J. F., and Thorington, R. W., Jr. 1973. A preliminary analysis of a Neotropical mammal fauna. *Biotropica* 5:150–61.

Eisenberg, J. F., and Wilson, D. 1978. Relative brain size and feeding strategies in the Chiroptera. *Evolution* 32:740–51.

———. 1981. Relative brain size in didelphid marsupials. *Am. Nat.* 118:110–26.

Eisentraut, M. 1957. *Aus dem Leben der Fledermäuse und Flughunde.* Jena: Fischer.

Eliot-Smith, G. 1898. The brain in the Edentata. *Trans. Linnean Soc. London,* ser. 2 (Zoology), 7:277–394.

Ellefson, J. O. 1974. A natural history of gibbons in the Malay Peninsula. In *Gibbon and siamang,* ed. D. Rumbaugh, 3:1–136. Basel: S. Karger.

Ellerman, J. R., and Morrison-Scott, T. C. S. 1951. *Checklist of Palearctic and Indian Mammals 1758–1946.* London: British Museum.

Elliott, L. 1978. Social behavior and foraging ecology of the eastern chipmunk (*Tamias striatus*) in the Adirondack Mountains. *Smithsonian Contrib. Zool.* 265:1–107.

Encke, W. 1965. Aufzucht von Borestengürteltieren, *Chaetophractus villosus. Zool. Gart.* 31(1/2):88–90.

Encke, W.; Gandras, A.; and Bienick, H. J. 1970. Beobachtungen am Mähnenwolf (*Chrysocyon brachyurus*). *Zool Gart.* 38:47–67.

Enders, R. K. 1935. Mammalian life histories from Barro Colorado Island, Panama. *Bull. Mus. Comp. Zool. Harvard* 78(4):385–502.

Epple, G. 1977. Notes on the establishment and maintenance of the pair bond in *Saguinus fuscicollis*. In *The biology and conservation of the Callitrichidae,* ed. D. G. Kleiman, pp. 231–38. Washington, D.C.: Smithsonian Institution Press.

Erlinge, S. 1968. Territoriality in the otter, *Lutra lutra* L. *Oikos* 19:81–89.

Errington, P. 1963. *Muskrat populations*. Ames: Iowa State University Press.

Espmark, Y. 1964. Studies in dominance-subordination relationship in a group of semi-domestic reindeer (*Rangifer tarandus* L.). *Anim. Behav.* 12(4):420–26.

———. 1969. Mother-young relations and development of behaviour in roe deer (*Capreolus capreolus* L.). *Viltrevy* [Swedish wildlife] 6(6):461–540.

Essapian, F. S. 1963. Observations on abnormalities of parturition in captive bottle-nosed dolphins, *Tursiops truncatus,* and concurrent behavior of other porpoises. *J. Mammal.* 44:405–14.

Estes, R. D. 1966. Behaviour and life history of the wildebeest (*Connochaetes taurinus* Burchell). *Nature* (London) 212:999–1000.

———. 1967. The comparative behavior of Grant's and Thomson's gazelles. *J. Mammal.* 48(2):189–209.

———. 1974. Social organisation of the African Bovidae. In *The behavior of ungulates and its relation to management,* ed. V. Geist and F. Walther, 1:166–205. IUCN Publications, n.s. no. 24, Morges: IUCN.

Estes, R. D., and Goddard, J. 1967. Prey selection and hunting behavior of the African wild dog. *J. Wildl. Manage.* 31(1):52–69.

Evans, W. E., and Herald, E. S. 1970. Underwater calls of captive Amazon manatee, *Trichechus inunguis. J. Mammal.* 51:820–23.

Ewer, R. F. 1958. Adaptive features in the skulls of African Suidae. *Proc. Zool. Soc. London* 131:135–55.

———. 1959. Suckling behavior of kittens. *Behavior* 15:146–62.

———. 1963. A note on the suckling behaviour of the viverrid, *Suricata suricatta* (Schreber). *Anim. Behav.* 11:599–601.

———. 1965. Food burying in the African ground squirrel, *Xerus erythropus* (E. Geoff.). *Z. Tierpsychol.* 22(3):321–27.

———. 1967. The behaviour of the African giant rat (*Cricetomys gambianus* Waterhouse). *Z. Tierpsychol.* 24:6–79.

———. 1968*a*. A preliminary survey of the behaviour in captivity of the dasyurid marsupial, *Sminthopsis crassicaudata* (Gould). *Z. Tierpsychol.* 25:319–65.

———. 1968*b*. *The ethology of mammals*. London: Logos Press.

———. 1969. Some observations on the killing and eating of prey by two dasyurid marsupials: The mulgara, *Dasycercus cristicauda,* and the Tasmanian devil, *Sarcophilus harrisii. Z. Tierpsychol.* 26:23–38.

———. 1971. The biology and behaviour of a free-living population of black rats (*Rattus rattus*). *Anim. Behav. Monogr.* 4(3):127–74.

———. 1973. *The carnivores*. Ithaca: Cornell University Press.

Fairbairn, D. J. 1978. Behavior of dispersing deer mice (*Peromyscus maniculatus*). *Behav. Ecol. Sociobiol.* 3(3):265–83.

Farentinos, R. C. 1972. Social dominance and mating activity in the tassel-eared squirrel (*Sciurus aberti ferreus*). *Anim. Behav.* 20:316–26.

———. 1974. Social communication of the tassel-eared squirrel (*Sciurus aberti*): A descriptive analysis. *Z. Tierpsychol.* 34:441–58.

Farner, D. S. 1965. Circadian systems in the photoperiodic responses of vertebrates. In *Circadian clocks*, ed. J. A. Aschoff, pp. 357–69. Amsterdam: North-Holland.

Fay, F. H., and Ray, C. 1968. Influence of climate on the distribution of walruses, *Odobenus rosmarus* (Linnaeus). 1. Evidence from thermoregulatory behavior. *Zoologica* 53:1–14.

Feer, F. 1979. Observations écologiques sur le néotrague de bates, *Neotragus batesi,* du Nord-Est Gabon. *Terre et Vie* 33 (2):159–240.

Fellner, K. 1968. Erste naturliche Aufzucht von Nebelpardern (*Neofelis nebulosa*) in einem Zoo. *Zool. Gart.* 35(3):105–37.

Fenchel, T. 1974. Intrinsic rate of natural increase: The relationship with body size. *Oecologia* 14:317–26.

Fiedler, U. 1974. Beobachtungen zur Biologie einiger Gerbillinen, insbesondere *Gerbillus (Dipodillus) dasyurus* (Myomorpha, Rodentia) in Gefangenschaft. II. Okologie. *Z. Säugetierk.* 39(1):24–41.

Field, R. 1978. Vocal behavior of wolves (*Canis lupus*): Variability in structure, context, annual/diurnal patterns, and ontogeny. Ph.D. diss., Baltimore: Johns Hopkins University.

Findley, J.; Harris, A.; Wilson, D.; and Jones, C. 1975. *Mammals of New Mexico*. Albuquerque: University of New Mexico Press.

Finley, R. B. 1958. The wood rats of Colorado: Distribution and ecology. *Univ. Kansas Publ. Mus. Nat. Hist.* 10:213–52.

Fisher, E. M. 1939. Habits of the southern sea otter. *J. Mammal.* 20:21–36.

Fisher, R. A. 1930. *The genetical theory of natural selection*. Oxford: Clarendon Press.

Fitch, H. S.; Goodrum, R.; and Newman, C. 1952. The armadillo in the southeastern United States. *J. Mammal.* 33:21–37.

Fitch, H. S., and Sandidge, L. L. 1953. Ecology of the opossum on a natural area in northeastern Kansas. *Univ. Kansas Publ. Mus. Nat. Hist.* 7(2):305–38.

Fleay, D. 1935. Notes on the breeding of Tasmanian devils. *Victorian Nat.* 52:100–105.

———. 1944. *We breèd the platypus*. Melbourne: Robertson and Mullens.

Fleming, T. H. 1971*a*. Population ecology of three species of Neotropical rodents. *Misc. Pub. Mus. Zool. Univ. Michigan* 143:1–77.

———. 1971*b*. *Artibeus jamaicensis:* Delayed embryonic development in a Neotropical bat. *Science* 171:402–4.

———. 1972. Aspects of population dynamics of three species of opossums in the Panama Canal Zone. *J. Mammal.* 53:619–23.

———. 1975. The role of small mammals in tropical ecosystems. In *Small mammals: Their productivity and population dynamics,* ed. F. B. Golley, K. Petrusewicz, and L. Ryszkowski, pp. 269–98. London: Cambridge University Press.
———. 1977. Growth and development of two species of tropical heteromyid rodents. *Amer. Midl. Nat.* 98(1):109–23.
Fleming, T. H.; Hooper, E. T.; and Wilson, D. E. 1972. Three Central American bat communities: Structure, reproductive cycles and movement patterns. *Ecology* 53(4):555–69.
Flux, J. E. C. 1967. Reproduction and body weights of the hare, *Lepus europaeus* Pallas, in New Zealand. *New Zealand J. Sci.* 10:357–401.
———. 1970. Life history of the mountain hare (*Lepus timidus scoticus*) in northeastern Scotland. *J. Zool.* 161:45–123.
Foelsch, D. W. 1968. Beobachtete Geburt eines Seehundes (*Phoca vitulina*). *Z. Säugetierk.* 33(1):50–55.
Fogden, M. P. L. 1974. A preliminary field study of the western tarsier, *Tarsius bancanus* Horsefield. In *Prosimian biology,* ed. R. D. Martin, G. A. Doyle, and A. C. Walker, pp. 151–67. London: Duckworth.
Fogden, S. C. L. 1971. Mother-young behaviour at grey seal breeding beaches. *J. Zool.* (London) 164:61–92.
Fons, R. 1973. Modalités de la reproduction et développement postnatal en captivité chez *Suncus etruscus* (Savi 1822). *Mammalia* 37(2):288–325.
———. 1975. Premières données sur l'écologie de la pachyure étrusque *Suncus etruscus* (Savi 1822) et comparaison avec deux autres Crocidurinae: *Crocidura russula* (Hermann 1780) et *Crocidura suaveolens* (Pallas 1811) (Insectiva, Soricidae). *Vie Milieu* 25 (2-C):315–60.
Fons, R., and Sicart, R. 1976. Contribution à la connaissance du métabolisme énergétique chez deux Crocidurinae: *Suncus etruscus* (Savi 1822) et *Crocidura russula* (Hermann 1780) (Insectivora, Soricidae). *Mammalia* 40(2):299–311.
Fontaine, R., and DuMond, F. V. 1977. The red ouakari in a seminatural environment: Potentials for propagation and study. In *Primate conservation,* ed. Prince Rainier III of Monaco and G. H. Bourne, pp. 168–236. New York: Academic Press.
Forde, E. 1934. *Habitat economy and society.* London: Methuen.
Forsyth, D. J. 1976. A field study of growth and development of nestling masked shrews (*Sorex cinereus*). *J. Mammal.* 57(4):708–21.
Fossey, D. 1972. Vocalizations of the mountain gorilla (*Gorilla gorilla beringei*) *Anim. Behav.* 20:36–53.
———. 1974. Observations on the home range of one group of mountain gorilla (*Gorilla gorilla beringei*). *Anim. Behav.* 22:568–81.
Foster, J. B. 1966. The giraffe of Nairobi National Park: Home range, sex ratios, the herd and food. *E. African Wildl. J.* 4:139–48.
Fouts, R. S., and Rigby, R. L. 1977. Man-chimpanzee communication. In *How animals communicate,* ed. T. Sebeok, pp. 1034–54. Bloomington: Indiana University Press.
Fox, M. W., ed. 1975. *The wild canids.* New York: Van Nostrand Reinhold.

Frädrich, H. 1965. Zur Biologie und Ethologie des Warzenschweines (*Phacochoerus aethiopicus* Pallas), unter Berücksichtigung des Verhaltens anderer Suiden. *Z. Tierpsychol.* 22:328–92.

———. 1966*a*. Breeding and hand-rearing warthogs *Phacochoerus aethiopicus* at Frankfurt Zoo. *Int. Zoo Yearb.* 6:200–202.

———. 1966*b*. Einige Verhaltensbeobachtungen am Moschustier (*Moschus moschiferus* L.). *Zool. Gart.* 33(1/3):65–78.

Frame, L. H., and Frame, G. W. 1976. Female African wild dogs emigrate. *Nature* (London) 263:227–29.

Frame, L. H.; Malcolm, J. R.; Frame, G. W.; and van Lawick, H. 1979. The social organization of African wild dogs (*Lycaon pictus*) on the Serengeti Plains, Tanzania, 1967–1978. *Z. Tierpsychol.* 50:225–49.

Francq, E. N. 1969. Behavioral aspects of feigned death in the opossum, *Didelphis marsupialis*. *Amer. Midl. Nat.* 81:556–68.

Franklin, W. L. 1974. The social behavior of the vicuna. In *The behaviour of ungulates and its relation to management,* ed. V. Geist and F. Walther, 1:477–87. IUCN Publications, n.s. no. 24. Morges: IUCN.

Fraser, A. F. 1968. *Reproductive behaviour in ungulates.* New York: Academic Press.

Frazer, J. F. D. 1977. Growth of young vertebrates in the egg or uterus. *J. Zool.* 183(2):189–203.

Frazer, J. F. D., and Huggett, A. St. G. 1974. Species variations in the foetal growth rates of eutherian mammals. *J. Zool.* (London) 174:481–509.

French, N. R.; Stoddard, D. M.; and Bobek, B. 1975. Patterns of demography in small mammal populations. In *Small mammals: Their productivity and population dynamics,* ed. F. B. Golley, K. Petrusewicz, and L. Ryszkowski, pp. 73–102. London: Cambridge University Press.

Friedman, H. 1955. The honey guides. *U.S. Nat. Mus. Bull.* 208:1–279.

Frisch, K. von. 1954. *The dancing bees.* London: Methuen.

Frith, H. J., and Calaby, J. H. 1969. *Kangaroos.* London: C. Hurst; New York: Humanities Press.

Fritzell, E. K. 1978. Aspects of raccoon (*Procyon lotor*) social organization. *Canadian J. Zool.* 56(2):260–71.

Fulk, G. W. 1976. Notes on the activity, reproduction and social behavior of *Octodon degus*. *J. Mammal.* 57(3):495–505.

Fuller, J. L., and DuBuis, E. M. 1962. The behaviour of dogs. In *The behavior of domestic animals,* ed. E. S. E. Hafez, pp. 415–52. Baltimore: Williams and Wilkins.

Fuller, J. L., and Thompson, W. R. 1960. *Behavior genetics.* New York: Wiley.

Gambaryan, P. P. 1974. *How mammals run.* New York: John Wiley.

Gangloff, B., and Ropartz, P. 1972. La répertoire comportemental de la genette *Genetta genetta* (Linné). *Terre et Vie* 36(4):489–560.

Gardner, A. 1977. Feeding habits. In *Biology of bats of the New World family Phyllostomatidae, part II,* ed. R. J. Baker, J. Knox-Jones, Jr., and D. C. Carter, pp. 293–350. Special Publications of the Museum 13. Lubbock: Texas Tech Press.
Gardner, B. T., and Gardner, R. A. 1969. Teaching sign language to a chimpanzee. *Science* 165:644–72.
Gaskin, D. E. 1964. Recent observations in New Zealand waters on some aspects of behaviour of the sperm whale (*Physeter macrocephalus*). *Tuatara* 12(2):106–14.
Gauthier-Pilters, H. 1959. Einige Beobachtungen zum Droh-, Angriffs-, und Kampfverhalten des Dromedarhengstes, sowie über Geburt und Verhaltensentwicklung des Jungtiers in der nodrwestlichen Sahara. *Z. Tierpsychol.* 16:593–604.
———. 1962. Beobachtungen an Feneks. *Z. Tierpsychol.* 19:440–64.
———. 1966. Einige Beobachtungen über das Spielverhalten biem Fenek (*Fennecus zerda* Zimm.). *Z. Säugetierk.* 31(5):337–50.
———. 1974. The behaviour and ecology of camels in the Sahara, with special reference to nomadism and water management. In *The behaviour of ungulates and its relation to management,* ed. V. Geist and F. Walther, 2:542–51. IUCN Publications, n.s. no. 24. Morges: IUCN.
Gautier, J.-P. 1974. Field and laboratory studies of the vocalizations of talapoin monkeys (*Miopithecus talapoin*): Structure, function, ontogenesis. *Behavior* 49:1–64.
Gautier, J.-P., and Gautier-Hion, A. 1977. Communication in Old World monkeys. In *How animals communicate,* ed. T. A. Sebeok, pp. 890–964. Bloomington: Indiana University Press.
Gautier-Hion, A. 1966. L'écologie et l'éthologie du talapoin (*M. talapoin*). *Biol. Gabonica* 2:311–29.
———. 1970. L'organisation sociale d'une bande de talapoins (*Miopithecus talapoin*) dans le nord-est du Gabon. *Folia Primat.* 12:116–41.
———. 1971. L'écologie du talapoin du Gabon. *Terre et Vie* 25:427–90.
Gebczynski, M.; Gorecki, A.; and Drozdz, A. 1972. Metabolism, food assimilation and bioenergetics of three species of dormice (Gliridae). *Acta Theriol.* 17(21):271–94.
Gee, E. P. 1953. Life history of the great Indian one-horned rhinoceros. *J. Bombay Nat. Hist. Soc.* 51:341–48.
Geist, V. 1963. On the behaviour of the North American moose (*Alces alces andersoni* Peterson 1950) in British Columbia. *Behaviour* 20(3–4):377–416.
———. 1964. On the rutting behavior of the mountain goat, *Oreamnos americanus*. *J. Mammal.* 45:551–67.
———. 1966*a*. Ethological observations on some North American cervids. *Zool. Beitr.* 12:219–50.
———. 1966*b*. The evolution of horn-like organs. *Behaviour* 27(3–4):175–215.
———. 1968*a*. On delayed social and physical maturation in mountain sheep. *Canadian J. Zool.* 46(5):899–904.

———. 1968*b*. On the interrelation of external appearance, social behaviour, and social structure of mountain sheep. *Z. Tierpsychol.* 25:119–215.
———. 1971. *Mountain sheep*. Chicago: University of Chicago Press.
———. 1974. On the relationship of ecology and behaviour in the evolution of ungulates: Theoretical considerations. In *The behaviour of ungulates and its relation to management,* ed. V. Geist and F. Walther, 1:235–47. IUCN Publications, n.s. no. 24. Morges: IUCN.
Geist, V., and Walther, F., eds. 1974. *The behaviour of ungulates and its relation to management*. IUCN Publications, n.s. no. 24, Morges: IUCN.
Genelly, R. E. 1965. Ecology of the common mole-rat (*Cryptomys hottentotus*) in Rhodesia. *J. Mammal.* 46:647–65.
Genest-Villard, H. 1972. Contribution à l'écologie et l'éthologie d'un petit rongeur arboricole, *Thamnomys rutilans,* en République Centrafricaine. *Mammalia* 36(4):543–79.
———. 1973. Données écologiques sur un Leggada des savanes intraforestières du Centre africain: *Mus oubanguii* (Rongeurs, Murides). *Mammalia* 37(2):220–31.
Gentry, R. L. 1970. Social behavior of the stellar sea lion. Ph.D. diss., University of California, Santa Cruz.
———. 1973. Thermoregulatory behavior of eared seals. *Behaviour* 46(1–2):73–94.
George, W. 1974. Notes on the ecology of gundis (F. Ctenodactylidae). *Symp. Zool. Soc. London* 34:143–60.
Gerell, R. 1967. Food selection in relation to habitat in mink (*Mustela vison*) in Sweden. *Oikos* 18:233–46.
Gessaman, J. A., ed. 1973. *Ecological energetics of homeotherms: A view compatible with ecological modeling*. Monograph Series, vol. 20. Logan: Utah State University Press.
Gewalt, W. 1967. Über den Beluga-Wal Delphinapterus leucas (Pallas 1776) im Rhein bei Duisburg. *Z. Säugetierk.* 32(2):65–86.
Gier, H. T. 1968. Coyotes in Kansas. *Kansas Agric. Exp. Stn. Bull.,* no. 393.
Gihr, M., and Pilleri, G. 1969. On the anatomy and biometry of *Stenella styx* Grey and *Delphinus delphis* L. (Cetacea, Delphinidae) of the western Mediterranean. In *Investigations on Cetacea,* ed. G. Pilleri, 1:15–65. Berne: Woldaw.
Gijzen, A., and Tijskens, J. 1971. Growth in weight of the lowland gorilla *Gorilla g. gorilla* and of the mountain gorilla *Gorilla g. beringei. Int. Zoo Yearb.* 11:183–93.
Gittleman, J. L. In preparation. Carnivore ecology and social organization. D. Phil. thesis, University of Sussex.
Glander, K. E. 1975. Habitat and resource utilization: An ecological view of social organization in mantled howling monkeys. Ph.D. diss., University of Chicago.
———. 1978. Howling monkey feeding behavior and plant secondary compounds. In *The ecology of arboreal folivores,* ed. G. G. Mont-

gomery, pp. 561–74. Washington, D.C.: Smithsonian Institution Press.

Goddard, J. 1966. Mating and courtship of the black rhinoceros (*Diceros bicornis* L.). *E. African Wildl. J.* 4:69–76.

———. 1967. Home range, behaviour, and recruitment rates of two black rhinoceros populations. *E. African Wildl. J.* 5:133–51.

———. 1968. Food preferences of two black rhinoceros populations. *E. African Wildl. J.* 6:1–19.

Godfrey, G., and Crowcroft, P. 1960. *The life of the mole*. London: Museum Press.

———. 1971. Breeding of fat tailed marsupial mouse *Sminthopsis crassicaudata* in captivity. *Int. Zoo Yearb.* 11:33–38.

Goethe, F. 1964. Das Verhalten der Musteliden. *Handb. Zool.* Band 8, Leiferung 37, 19(19):1–80.

Goffart, M. 1971. *Function and form in the sloth*. New York: Pergamon Press.

Golani, I. 1966. Observations on the behaviour of the jackal *Canis aureus* L. in captivity. *Israel J. Zool.* 15:28.

———. 1973. Non-metric analysis of behavioral interaction sequences in captive jackals (*Canis aureus* L.). *Behaviour* 44(1–2):89–112.

———. 1976. Mechanisms of motor homeostasis in mammalian display. In *Perspectives in ethology*, ed. P. P. G. Bateson and P. Klopfer, 2:69–130. New York: Plenum Press.

Golani, I., and Keller, A. 1975. A longitudinal field study of the behavior of a pair of golden jackals. In *The wild canids*, ed. M. W. Fox, pp. 303–35. New York: Van Nostrand Reinhold.

Golani, I., and Mendelssohn, H. 1971. Sequences of precopulatory behaviour of the jackal (*Canis aureus* L.). *Behaviour* 38:169–92.

Golley, F. B. 1960. Energy dynamics of a food chain of an old-field community. *Ecol. Monogr.* 30:187–206.

Goodall, J. 1965. Chimpanzees of the Gombe Stream Reserve. In *Primate behavior: Field studies of monkeys and apes*, ed. I. Devore, pp. 425–73. New York: Holt, Rinehart and Winston.

———. 1968. The behaviour of free-living chimpanzees in the Gombe Stream Reserve. *Anim. Behav. Monogr.* 1(3):165–311.

———. 1977. Infant killing and cannibalism in free-living chimpanzees. *Folia Primat.* 28:259–82.

Goodpaster, W. W., and Hoffmeister, D. F. 1954. Life history of the golden mouse *Peromyscus nuttalli* in Kentucky. *J. Mammal.* 35:23.

Goodwin, G. G., and Greenhall, A. M. 1961. A review of the bats of Trinidad and Tobago, descriptions, rabies infection, and ecology. *Amer. Mus. Nat. Hist. Bull.* 122(3):187–302.

Gopalakrishna, A. 1969. Gestation period in some Indian bats. *J. Bombay Nat. Hist. Soc.* 66:317–22.

Gordon, K. 1943. The natural history and behavior of the western chipmunk and the mantled ground squirrel. *Oregon State Monogr./Studies Zool.* 5:1–104.

Gosling, L. M. 1969. Parturition and related behaviour in Coke's hartebeest, *Alcelaphus buselaphus cokei* Gunther. *J. Reprod. Fertil.* 6:265–86.

———. 1974. The social behavior of Coke's hartebeest, *Alcelaphus buselaphus cokei*. In *The behaviour of ungulates and its relation to management,* ed. V. Geist and F. Walther, 1:488–511. IUCN Publications, n.s. no. 24. Morges: IUCN.

Gossow, H. 1970. Vergleichende Verhaltensstudien en Marderartigen. 1. Über Lautäusserungen und zum Beuteverhalten. *Z. Tierpsychol.* 27:405–80.

Gottlieb, G. O. 1950. Zur Kenntnis der Birkenmaus. *Zool. Jahrb.* 79:93–113.

Gould, E. 1965. Evidence for echolocation in the Tenrecidae of Madagascar. *Proc. Amer. Phil. Soc.* 109:352–60.

———. 1969. Communication in three genera of shrews (Soricidae): *Suncus, Blarina* and *Cryptotis*. *Comm. Behav. Biol.*, part A, 3:11–31.

———. 1970. Echolocation and communication in bats. In *About bats,* ed. B. H. Slaughter and D. W. Walton, pp. 144–62. Dallas: Southern Methodist University Press.

———. 1971. Studies of maternal-infant communication and development of vocalizations in the bats *Myotis* and *Eptesicus*. *Comm. Behav. Biol.* 5:263–313.

———. 1975. Neonatal vocalizations in bats of eight genera. *J. Mammal.* 56(1):15–29.

———. 1977. Echolocation and communication. In *Biology of bats of the New World family Phyllostomatidae, part II,* ed. R. J. Baker, J. Knox-Jones, Jr., and D. C. Carter, pp. 247–81. Special Publications of the Museum 13. Lubbock: Texas Tech Press.

———. 1978*a*. The behavior of the moonrat, *Echinosorex gymnurus* (Erinaceidae) and the pentail shrew, *Ptilocerus lowi* (Tupaiidae) with comments on the behavior of other Insectivora. *Z. Tierpsychol.* 48:1–27.

———. 178*b*. Foraging behavior of Malaysian nectar-feeding bats. *Biotropica* 19(3):184–93.

Gould, E.; Negus, N. C.; and Novick, A. 1964. Evidence for echolocation in shrews. *J. Exp. Zool.* 156(1):19–38.

Gould, S. J. 1977. *Ontogeny and phylogeny.* Cambridge: Belknap Press of Harvard University Press.

Grand, T. 1977. Body weight: Its relation to tissue composition, segment distribution, motor function. I. Interspecific comparisons. *Amer. J. Phys. Anthropol.* 47:211–40.

Grand, T., and Eisenberg, J. F. In press. Concerning the affinities of the rodent genera *Dinomys* and *Coendou*. *Säugetierk. Mitt.*

Grant, T. R. 1974. Observations of enclosed and free-ranging grey kangaroos, *Macropus giganteus*. *Z. Säugetierk.* 39:65–78.

Grassé, P. P., ed. 1955*a*. *Traité de zoologie: Anatomie, systematique, biologie*. Vol. 17 (fasc. 1). *Mammifères*. Paris: Masson.

———. 1955*b*. *Traité de zoologie: Anatomie, systematique, biologie*. Vol. 17 (fasc. 2). *Mammifères*. Paris: Masson.

Greegor, D., Jr. 1974. Comparative ecology and distribution of two species of armadillos, *Chaetophractus vellerosus* and *Dasypus novemcinctus*. Ph.D. diss., University of Arizona.

Green, R. H. 1974. Mammals. In *Biogeography and ecology in Tasmania,* ed. W. D. Williams, pp. 367–96. The Hague: W. Junk.

Green, S., and Minkowski, K. 1977. The lion-tailed monkey and its south Indian rainforest habitat. In *Primate conservation,* ed. Prince Rainier III of Monaco and G. H. Bourne, pp. 289–337. New York: Academic Press.

Greenhall, A. M. 1965. Notes on behavior of captive vampire bats. *Mammalia* 29:441–52.

Greenwood, P. J.; Harvey, P.; and Perrins, C. M. 1979. The role of dispersal in the great tit (*Parus major*): The causes, consequences, and heritability of natal dispersal. *J. Anim. Ecol.* 48:123–42.

Gregory, W. K. 1951. *Evolution emerging: A survey of changing patterns from primeval life to man,* 2 vols. New York: Macmillan.

Griffin, D. R. 1958. *Listening in the dark: The acoustic orientation of bats and men.* New Haven: Yale University Press.

Griffiths, M. 1968. *Echidnas.* New York: Pergamon Press.

———. 1978. *The biology of the monotremes.* New York: Academic Press.

Griffiths, M.; McIntosh, D. L.; and Leckie, R. M. C. 1972. The mammary glands of the red kangaroo with observations on the fatty acid components of the milk triglycerides. *J. Zool.* (London) 166(2):265–75.

Groves, C. P. 1974. *Horses, asses and zebras in the wild.* London: David and Charles, Newton Abbot.

Groves, C. P., and Kurt, F. 1972. *Dicerorhinus sumatrensis. Mammal. Species* 21:1–6.

Grubb, P., and Jewell, P. A. 1974. Movement, daily activity and home range of soay sheep. In *Island survivors: The ecology of the soay sheep of St. Kilda,* ed. P. A. Jewell, C. Milner, and J. Morton Boyd, pp. 160–94. London: Athlone.

Grundlach, H. 1968. Brutforsorge, Brutpflege, Verhaltensontogenese und Tagesperiodik beim Europaischen Wildschwein (*Sus scrofa* L.). *Z. Tierpsychol.* 25:955–95.

Grünwald, A. 1969. Untersuchungen zur Orientierung der Weiszahnspitzmause. *Z. Vergl. Physiol.* 65:191–217.

Gucwinska, H. 1971. Development of six-banded armadillos *Euphractus sexcinctus* at Wroclaw Zoo. *Int. Zoo Yearb.* 11:88–89.

Gucwinska, H., and Gucwinska, A. 1968. Breeding the Zanzibar galago, *Galago senegalensis zanzibaricus* in the Wroclaw Zoo. *Int. Zoo Yearb.* 8:111–14.

Guiler, E. R. 1970*a*. Observations on the Tasmanian devil, *Sarcophilus harrisii* (Marsupialia: Dasyuridae). I. Numbers, home range, movements, and food in two populations. *Australian J. Zool.* 18:49–62.

———. 1970*b*. Observations on the Tasmanian devil, *Sarcophilus harrisii* (Marsupialia: Dasyuridae). II. Reproduction, breeding and growth of pouch young. *Australian J. Zool.* 18:63–70.

———. 1971. The husbandry of the potoroo, *Potorous tridactylus*. *Int. Zoo Yearb*. 11:21–22.

Gunderson, H. L. 1976. *Mammalogy*. New York: McGraw-Hill.

Hafez, E. S. E.; Williams, M.; and Wierzbowski, S. 1962. The behavior of horses. In *The behavior of domestic animals*, ed. E. S. E. Hafez, pp. 370–96. Baltimore: Williams and Wilkins.

Hagga, R. 1960. Observations on the ecology of the Japanese pika. *J. Mammal*. 41(2):200–212.

Halder, U. 1976. *Ökologie und Verhalten des Banteng (Bos javanicus) in Java*. Mammalia Depicta. Hamburg and Berlin: Paul Parey Verlag.

Hall, A., III; Persing, R. L.; White, D. C.; and Rickets, R. T., Jr. 1967. *Mystromys albicaudatus* (the African white tailed rat) as a laboratory species. *Lab. Anim. Care* 17:180–88.

Hall, E. R., and Kelson, K. R. 1959. *The mammals of North America*, 2 vols. New York: Ronald Press.

Hamburg, D. A., and McCown, E. R., eds. 1979. *The great apes*. Menlo Park, Calif.: Benjamin/Cummings.

Hamilton, J. E. 1934. The southern sealion *Otaria byronia* (de Blainville). *Discov. Rep*. 8:269–318.

Hamilton, W. D. 1964. The genetical theory of social behavior, I, II. *J. Theoret. Biol*. 7(1):1–52.

———. 1970. Selfish and spiteful behavior in an evolutionary model. *Nature* (London) 228:1218–20.

———. 1971. Geometry for the selfish herd. *J. Theoret. Biol*. 31(2):295–311.

Hamilton, W. J., Jr. 1934. The life history of the rufescent woodchuck, *Marmota monax rufescens*. *Ann. Carnegie Mus*. 23:85–178.

———. 1936. The food and breeding habits of the raccoon. *Ohio J. Sci*. 36(3):131–40.

———. 1939. Observations on the life history of the red squirrel in New York. *Amer. Midl. Nat*., 22:732–45.

———. 1944. The biology of the little short tailed shrew *Cryptotis parva*. *J. Mammal*. 25:1–7.

———. 1953. Reproduction and young of the Florida wood rat, *Neotoma f. floridana* (Ord.). *J. Mammal*. 34:180–89.

Hamilton, W. J., III. 1962. Reproductive adaptations of the red tree mouse. *J. Mammal*. 43:486–504.

———. 1973. *Life's color code*. New York: McGraw-Hill.

Handley, C. O., Jr. 1966. Checklist of the mammals of Panama. In *Ectoparasites of Panama*, ed. R. L. Wenzel and V. J. Tipton, pp. 753–95. Chicago: Field Museum of Natural History.

———. 1976. Mammals of the Smithsonian Venezuela project. *Brigham Young Univ. Sci. Bull., Biol. Ser*. 20(5):1–91.

Hanney, P., and Morris, B. 1962. Some observations on the pouched rat in Nyasaland. *J. Mammal*. 43:238–48.

Hansen, R. M., and Miller, R. S. 1959. Observations on the plural occupancy of pocket gopher systems. *J. Mammal*. 40:577–84.

Hanzak, J. 1966. Zur Jungendentwicklung der Gartenspitzmaus, *Crocidura suaveolens* (Pallas 1821). *Lynx* 6:67–74.

Happold, D. C. D. 1967. Biology of the jerboa, *Jaculus jaculus butleri* (Rodentia, Dipodidae) in the Sudan. *J. Zool.* (London) 151:257–75.

———. 1970. Reproduction and development of the Sudanese jerboa, *Jaculus jaculus butleri* (Rodentia, Dipodidae). *J. Zool.* (London) 162:505–15.

Happold, M. 1976*a*. Social behavior of the conilurine rodents (Muridae) of Australia. *Z. Tierpsychol.* 40:113–82.

———. 1976*b*. Reproductive biology and development in the conilurine rodents (Muridae) of Australia. *Australian J. Zool.* 24:19–26.

———. 1976*c*. The ontogeny of social behavior in four conilurine rodents (Muridae) of Australia. *Z. Tierpsychol.* 40:265–78.

Harcourt, H. A. 1979. Social relationships among adult female mountain gorillas. *Anim. Behav.* 27:251–64.

Hardin, C. J. 1976. Hand-rearing a giant anteater at Toledo Zoo. *Int. Zoo Yearb.* 16:199–200.

Harker, J. 1964. *The physiology of diurnal rhythms*. Cambridge: Cambridge University Press.

Harlow, H. F., and Harlow, M. K. 1965. The affectional systems. In *Behavior of nonhuman Primates,* ed. A. M. Schrier, H. F. Harlow, and F. Stollnitz, 2:287–334. New York: Academic Press.

Harrington, C. R. 1968. Denning habits of the polar bear (*Ursus maratimus* Phipps). *Canadian Wildl. Serv. Rep. Ser.* 5:1–30.

Harris, C. J. 1968. *Otters: A study of the recent Lutrinae*. World Naturalist Series. London: Weidenfeld and Nicolson.

Harris, M. 1977. *Cannibals and kings: The origin of culture*. New York: Random House.

Harris, V. T. 1952. An experimental study of habitat selection by prairie and forest races of the deermouse, *Peromyscus maniculatus*. *Contrib. Lab. Vert. Biol., Univ. Michigan* 56:1–53.

Harrison, J. L. 1955. Data on the reproduction of some Malayan mammals. *Proc. Zool. Soc. London* 125:445–60.

Harrison, R. J., and Kooyman, G. L. 1968. General physiology of the Pinnipedia. In *The behaviors and physiology of pinnipeds,* ed. R. J. Harrison, R. C. Hubbard, R. S. Peterson, C. E. Rice, and R. J. Schusterman, pp. 211–95. New York: Appleton-Century-Crofts.

Harrison, R. J.; Hubbard, R. C.; Peterson, R. S.; Rice, C. E.; and Schusterman, R. J., eds. 1968. *The behavior and physiology of pinnipeds*. New York: Appleton-Century-Crofts.

Harrison, R. J., and Thomlinson, J. D. W. 1956. Observations on the venous system in certain Pinnipedia and Cetacea. *Proc. Zool. Soc. London* 126:205–33.

Hart, J. S., and Irving, L. 1959. The energetics of harbor seals in air and in water with special consideration of seasonal changes. *Canadian J. Zool.* 37:447–57.

Hartman, D. S. 1971. Behavior and ecology of the Florida manatee, *Trichechus monatus latirostris* (Harlan) at Crystal River, Citrus County. Ph.D. diss., Cornell University.

———. 1979. *The behavior and ecology of the manatee (Trichechus manatus) in Florida*. Special Publication no. 5, American Society of Mammalogists.

Hartmann, L. 1964. The behaviour and breeding of captive weasels (*Mustela nivalis* L.). *New Zealand J. Sci.* 7:147–56.

Harvey, E. B., and Rosenberg, L. E. 1960. An apocrine gland complex in the pika. *J. Mammal.* 45(2):213–20.

Harvey, P. H.; Greenwood, P. J.; and Perrins, C. M. 1979. Breeding area fidelity of the great tit (*Parus major*). *J. Anim. Ecol.* 48:305–13.

Hassenberg, L. 1965. *Ruhe und Schlaf bei Säugetieren*. Wittenberg-Lutherstadt: Ziemsen.

Hawkins, A. E., and Jewell, P. A. 1962. Food consumption and energy requirements of captive British shrews and the mole. *Proc. Zool. Soc. London* 138:137–55.

Hayden, P.; Gambino, J. J.; and Lindbert, R. G. 1966. Laboratory breeding of the little pocket mouse, *Perognathus longimembris*. *J. Mammal.* 47:412–23.

Healey, M. C. 1967. Aggression and self-regulation of population in deermice. *Ecology* 48:377–92.

Hearn, J. P., and Lunn, S. F. 1975. The reproductive biology of the marmoset monkey, *Callithrix jacchus*. In *Breeding simians for developmental biology*. *Lab Animal Handbooks* 6:191–202.

Heck, L. 1908. Echidna—Züchtung im Berliner Zoologischer Garten. *Berlin Setzber. Ges. Naturf. Fr.* 1908:187–89.

Hediger, H. 1950. Gefangenschaftsbegurt eines afrikanischen Springhasen, *Pedetes caffer*. *Zool. Gart.* 17(1/5):166–69.

———. 1958. Verhalten der Beuteltiere (Marsupialia). *Handb. Zool.* 8:1–27.

———. 1970. The breeding behaviour of the Canadian beaver (*Castor fiber canadensis*). *Forma et Functio* 2:336–51.

Hediger, H., and Kummer, H. 1961. Das Verhalten der Schnabeligel (Tachyglossidae). *Handb. Zool.*, Band 8, Lieferung 27, 10(8a):1–8.

Heerdt, P. F. van, and Sluiter, J. W. 1965. Notes on the distribution and behaviour of the noctule bat (*Nyctalus noctula*) in the Netherlands. *Mammalia* 29:463–78.

Heidt, G. A.; Petersen, M. K.; and Kirland, G. L., Jr. 1968. Mating behavior and development of least weasels (*Mustela nivalis*) in captivity. *J. Mammal.* 49(3):413–20.

Heinsohn, G. E. 1966. Ecology and reproduction of the Tasmanian bandicoots (*Perameles gunni* and *Isoodon obesulus*). *Univ. California Publ. Zool.* 80:1–96.

———. 1968. Habitat requirements and reproductive potential of the macropod marsupial *Potorous tridactylus* in Tasmania. *Mammalia* 32(1):30–43.

———. 1972. A study of dugongs (*Dugong dugon*) in northern Queensland, Australia. *Biol. Conserv.* 4:205–13.

Heinsohn, G. E., and Birch, W. R. 1972. Foods and feeding habits of the dugong, *Dugong dugon* (Erxleben), in northern Queensland, Australia. *Mammalia* 36(3):414–22.

Heinsohn, G. E., and Spain, A. V. 1974. Effects of a tropical cyclone on littoral and sub-littoral biotic communities and on a population of dugongs (*Dugong dugon* Müller). *Biol. Conserv.* 6:143–52.

Heithaus, E. R.; Fleming, T. H.; and Opler, P. A. 1975. Foraging patterns and resource utilization in seven species of bats in a seasonal tropical forest. *Ecology* 56:841–54.

Hellwing, S. 1973*a*. Husbandry and breeding of white-toothed shrews. *Int. Zoo Yearb.* 13:127–34.

Hellwing, S. 1973*b*. The postnatal development of the white-toothed shrew *Crocidura russula monacha* in captivity. *Z. Säugetierk.* 38(5):257–70.

———. 1975. Sexual receptivity and oestrus in the white-toothed shrew, *Crocidura russula monacha. J. Reprod. Fertil.* 45:469–77.

Helm, J. D., III. 1973. Reproductive biology of *Tylomys* and *Ototylomys*. Ph.D. diss., Michigan State University.

———. 1975. Reproductive biology of *Ototylomys* (Cricetidae). *J. Mammal.* 56:575–90.

Hemmer, H. 1966. Untersuchungen zur Stammesgeschichte der Pantherkatzen (Pantherinae), Teil 1. *Veröff. Zool. Staatssamml. München* 11:1–121.

———. 1968. Untersuchungen zur Stammesgeschichte der Pantherkatzen (Pantherinae). Teil 2. Studien zur Ethologie des Nebelparden *Neofelis nebulosa* (Griffiths 1821) und des Irbis *Uncia uncia* (Schreber 1775). *Veröff. Zool. Staatssamml. München* 12:155–247.

———. 1976. Gestation period and postnatal development in felids. In *The world's cats.* Vol. 3, no. 2. *Contributions to biology, ecology, behavior and evolution,* ed. R. L. Eaton, pp. 143–65. Seattle: Carnivore Research Institute, Burke Museum, University of Washington.

Hendrichs, H. 1971. Freilandbeobachtungen zum Sozialsystem des Afrikanischen Elefanten, *Loxodonta africana* (Blumenbach 1797). In *Dikdik und Elefanten,* ed. W. Wickler, pp. 77–173. Munich: Piper.

———. 1978. Die soziale organisation von Säugetierpopulationen. *Säugetierk. Mitt.* 20(2):81–116.

Hendrickx, A. G., and Houston, M. L. 1971. Prenatal and postnatal development. In *Comparative reproduction of non-human Primates,* ed. E. S. E. Hafez, pp. 334–81. Springfield, Ill.: Charles C. Thomas.

Hennig, R. 1962. Über einige Verhaltensweisen des Rehwildes (*Capreolus capreolus*) in Freier Wildbahn. *Z. Tierpsychol.* 19:223–29.

Hennig, W. 1966. *Phylogenetic systematics.* Urbana: University of Illinois Press.

Herald, E. S.; Brownell, R. L., Jr.; Frye, F. L.; Morris, E. J.; Evans, W. E.; and Scott, A. B. 1969. Blind river dolphin: First side-swimming Cetacean. *Science* 166:1408–10.

Herrero, S., ed. 1972. *Bears: Their biology and management.* Morges: IUCN.

Hershkovitz, P. 1962. *Evolution of Neotropical cricetine rodents (Muridae) with special reference to the phyllotine group.* Fieldiana: Zoology, vol. 46. Chicago: Field Museum of Natural History.

———. 1966. Catalog of living whales. *Bull. U.S. Nat. Mus.* 246:1–259.

———. 1974. A new genus of Late Oligocene monkey (Cebidae, Platyrrhini) with notes on postorbital closure and Platyrrhine evolution. *Folia Primat.* 21(1):1–35.

———. 1977. *Living new world monkeys (Platyrrhini).* Vol. 1. Chicago: University of Chicago Press.

Herter, K. 1953. Über das Verhalten von Iltissen. *Z. Tierpsychol.* 10:56–71.

———. 1957. Das Verhalten der Insektivoren. *Handb. Zool.* Band 8, Lieferung 9, 19(10):1–50.

Hewer, H. R. 1960. Behaviour of the grey seal (*Halichoerus grypus* Fab.) in the breeding season. *Mammalia* 24:400–412.

Hick, U. 1967. Geglückte Aufzucht eines Pudus (*Pudu pudu* Mol.). *Freunde des Kolner Zoo,* Heft 4 (10 Jahrgang): 111–18.

———. 1969. Successful raising of a pudu *Pudu pudu* at Cologne Zoo. *Int. Zoo Yearb.* 9:110–12.

Hickman, V. V., and Hickman, J. L. 1960. Notes on the habits of the Tasmanian dormouse phalangers *Cercartetus nanus* (Desmarest) and *Eudromicia lepida* (Thomas). *Proc. Zool. Soc. London* 135:365–74.

Hiiemäe, K. M., and Ardran, G. M. 1968. A cinefluorographic study of mandibular movement during feeding in the rat (*Rattus norvegicus*). *J. Zool. Soc. London* 154:139–54.

Hildebrand, M. 1961. Further studies on the locomotion of the cheetah. *J. Mammal.* 42:84–91.

Hill, W. C. O. 1953. Primates. In *Comparative anatomy and taxonomy.* I. *Strepsirhini.* Edinburgh: University Press.

———. 1955. *Comparative anatomy and taxonomy.* II. *Haplorhini: Tarsioidea.* Edinburgh: University Press.

———. 1957. *Comparative anatomy and taxonomy.* III. *Pithecoidea: Platyrrhini.* Edinburgh: University Press.

———. 1960. *Comparative anatomy and taxonomy.* IV. *Cebidae, Part A.* Edinburgh: University Press.

———. 1962. *Comparative anatomy and taxonomy.* V. *Cebidae, Part B.* Edinburgh: University Press.

Hill, W. C. O.; Porter, A.; Bloom, R. T.; Seago, J.; and Southwick, M. D. 1957. Field and laboratory studies on the naked mole rat, *Heterocephalus glaber. Proc. Zool. Soc. London* 128:455–514.

Hinde, R. A. 1970. *Animal behavior: A synthesis of ethology and comparative psychology.* 2d ed. New York: McGraw-Hill.

Hinton, H. E., and Dunn, A. M. S. 1967. *Mongooses: Their natural history and behavior.* Edinburgh: Oliver and Boyd.

Hirshfeld, J. R., and Bradley, D. G. 1977. Growth and development of two species of chipmunks, *Eutamias panamintinus* and *E. palmeri. J. Mammal.* 58:44–52.

Hirth, D. H. 1977. *Social behavior of white-tailed deer in relation to habitat. Wildl. Monogr.* 53:1–55.

Hladik, A., and Hladik, C. M. 1969. Rapports trophiques entre végétation et primates dans la forêt de Barro Colorado (Panama). *Terre et Vie* 1:25–117.

Hladik, C. M., and Charles-Dominique, P. 1974. The behaviour and ecology of the sportive lemur (*Lepilemur mustelinus*) in relation to its dietary peculiarities. In *Prosimian biology*, ed. R. D. Martin, G. A. Doyle, and A. C. Walker, pp. 23–39. London: Duckworth.

Hoage, R. J. 1978. Biosocial development in the golden lion tamarin, *Leontopithecus rosalia rosalia* (Primates: Callitrichidae). Ph.D. diss., University of Pittsburgh.

Hock, R. J. 1951. The metabolic rates and body temperatures of bats. *Biol. Bull.* 101:289–99.

Hoeck, H. N. 1975. Differential feeding behaviour of the sympatric hyrax, *Procavia johnstoni* and *Heterohyrax brucei*. *Oecologia* (Berlin) 22:15–47.

Hoffmeister, D. F., and Goodpaster, W. W. 1962. Life history of the desert shrew, *Notiosorex crawfordi*. *Southwestern Nat.* 7:236–52.

Hofmann, R. R., and Stewart, D. R. M. 1972. Grazer and browser: A classification based on the stomach structure and feeding habits of East African ruminants. *Mammalia* 36:226–40.

Holm, E. 1971. Contributions to the knowledge of the biology of the Namib desert golden moles, *Eremitalpa granti*. *Sci. Pap. Namib Desert Res. Sta.* 41:37–42.

Holsworth, W. N. 1967. Population dynamics of the quokka, *Setonix brachyurus*, on the west end of Rottnest Island, Western Australia. 1. Habitat and distribution of the quokka. *Australian J. Zool.* 15:29–46.

Holt, A. B.; Cheek, D. B.; Mellits, E. D.; and Hill, S. E. 1975. Brain size and the relation of the primate to the nonprimate. In *Fetal and postnatal cellular growth: Hormones and nutrition*, ed. D. B. Cheek, pp. 23–44. New York: John Wiley.

Homewood, K. M. 1978. Feeding strategy of the Tana mangabey (*Cercocebus galeritus galeritus*) (Mammalia: Primates). *J. Zool.* (London) 186:375–92.

Hoogerwerf, A. 1970. *Udjung Kulon: The land of the last Java rhinoceros*. Leiden: E. J. Brill.

Hooper, E. T. 1968. Habitats and food of amphibious mice of the genus *Rheomys*. *J. Mammal.* 49(3):550–53.

———. 1974. Reproductive patterns in the two parapatric species of the rodent genus *Scotinomys*. *Proceedings of the First International Theriological Congress, Moscow*.

Hooper, E. T., and Carleton, M. D. 1976. Reproduction, growth and development in two contiguously allopatric rodent species, genus *Scotinomys*. *Misc. Publ. Mus. Zool., Univ. Michigan*, no. 151.

Hopson, J. A. 1977. Relative brain size and behavior in Archosaurian reptiles. *Ann. Rev. Ecol. Syst.* 8:429–48.

Horner, B. E., and Taylor, J. M. 1968. Growth and reproductive behavior in the southern grasshopper mouse. *J. Mammal.* 49(4):644–61.

Hornocker, M. 1969. Winter territoriality in mountain lions. *J. Wildl. Manage.* 33(3):457–64.

Horwich, R. H. 1972. The ontogeny of social behavior in the gray squirrel (*Sciurus carolinensis*). *Z. Tierpsychol.* suppl. no. 8, pp. 1–103.

How, R. A. 1978. Population strategies of four species of Australian possums. In *The ecology of arboreal folivores,* ed. G. G. Montgomery, pp. 305–14. Washington, D.C.: Smithsonian Institution Press.

Howard, W. E. 1948. Dispersal, amount of inbreeding, and longevity in a local population of prairie deermice on the George Reserve, southern Michigan. *Contr. Lab. Vert. Biol.* 43:1–50.

Howe, R., and Clough, G. C. 1971. The Bahaman hutia, *Geocapromys ingrahami,* in captivity. *Int. Zoo Yearb.* 11:89–93.

Howell, A. B. 1920. Contribution to the life history of the California mastiff bat. *J. Mammal.* 1:111–17.

———. 1930. *Aquatic mammals.* Springfield, Ill.: Charles C. Thomas.

———. 1932. The saltatorial rodent *Dipodomys:* The functional and comparative anatomy of its muscular and osseus systems. *Proc. Amer. Acad. Arts Sci.* 67:377–536.

———. 1944. *Speed in animals.* Chicago: University of Chicago Press.

Howell, D. J. 1974. Bats and pollen: Physiological aspects of the syndrome of chiropterophily. *Comp. Biochem. Physiol.* 48A:263–76.

Howell, D. J., and Burch D. 1974. Food habits of some Costa Rican bats. *Rev. Biol. Trop.* 21:281–94.

Hrdy, S. B. 1976. Care and exploitation of nonhuman primate infants by conspecifics other than the mother. *Adv. Study Behav.* 6:101–58.

———. 1977. *The langurs of Abu: Female and male strategies of reproduction.* Cambridge: Harvard University Press.

Hubback, T. 1939. The Asiatic two-horned rhinoceros. *J. Mammal.* 20:1–20.

Huber, F. 1962. Central nervous control of sound production in crickets and some speculations on its evolution. *Evolution* 16:439–42.

———. 1963. Lokalisation und Plastizität im Zentralnervensystem der Tiere. *Zool. Anz.* (suppl.) 26:200–267.

Hudson, J. W. 1974. The estrous cycle, reproduction, growth and development of temperature regulation in the pygmy mouse, *Baiomys taylori. J. Mammal.* 55:572–88.

Huf, K. 1965. Über das verhalten des binturong (*Arctictis binturong* Raffl.). *Portugaliae Acta Biologica,* ser. A, 9(4–5):249–504.

Huggel-Wolf, H., and Huggel-Wolf, M. 1965. La biologie d'*Eidolon helvum* (Kerr) (Megachiroptera). *Acta Trop.* 22:1–10.

Huggett, A. St. G., and Widdas, W. F. 1951. The relationship between mammalian foetal weight and conception age. *J. Physiol.* 114:306–17.

Hughes, R. L.; Thompson, J. A.; and Owen, W. H. 1965. Reproduction in natural populations of the Australian ringtail possum *Pseudocheirus peregrinus* (Marsupialia: Phalangeridae) in Victoria. *Australian J. Zool.* 13:383–406.

Humphrey, S. R., and Cope, J. B. 1976. *Population ecology of the little brown bat, Myotis lucifugus, in Indiana and north-central Kentucky.* Special Publication no. 4. American Society of Mammalogists.

Hunsaker, D. 1977. Ecology of New World marsupials. In *The biology of marsupials,* ed. D. Hunsaker, pp. 95–156. New York: Academic Press.

Hunsaker, D., and Hahn, T. C. 1965. Vocalization of the South American tapir, *Tapirus terrestris*. *Anim. Behav.* 13(1):69–75.

Hunsaker, D., and Shupe, D. 1977. Behavior of the New World marsupials. In *The biology of marsupials*, ed. D. Hunsaker, pp. 279–347. New York: Academic Press.

Hunt, T. P. 1959. Breeding habits of the swamp rabbit with notes on its life history. *J. Mammal.* 40:82–91.

Husson, A. M. 1978. *The mammals of Suriname*. Leiden: E. J. Brill.

Hutterer, R. 1976. Beobachtungen zur Geburt und Jungendentwicklung der Zwergspitzmaus, *Sorex minutus* L. (Soricidae: Insectivora). *Z. Säugetierk.* 41(1):1–22.

———. 1977. Haltung und Lebensdauer von Spitzmäusen der Gattung *Sorex* (Mammalia, Insectivora). *Z. Angewandte Zool.* 23:353–69.

Hutterer, R., and Vogel, P. 1977. Abwehrlaute afrikanischer Spitzmäuse der Gattung *Crocidura* Wagler 1832 und ihre systematische Bedeutung. *Bonn. Zool. Beitr.* 28:218–27.

Huxley, J. 1932. Field studies and physiology: A correlation in the field of avian reproduction. *Nature* (London) 129:166.

———. 1972. *Problems of relative growth*. New York: Dover Publications.

Ingles, L. G. 1961. Home range and habits of the wandering shrew. *J. Mammal.* 42:455–62.

———. 1965. *Mammals of the Pacific states*. Stanford: Stanford University Press.

Innis, A. C. 1958. The behaviour of the giraffe, *Giraffa cameleopardalis*, in Eastern Transvaal. *Proc. Zool. Soc. London* 131:245–78.

Irving, L., and Hart, J. S. 1957. The metabolism and insulation of seals as bare-skinned mammals in cold water. *Canadian J. Zool.* 35:497–511.

Isaac, D. 1966. Temporal organization of activities in deer mice (*Peromyscus maniculatus*). Ph.D. diss., University of California, Berkeley.

Jacobi, E. F., and Van Bemmel, A. C. V. 1968. Breeding of Cape hunting dogs (*Lycaon pictus* Temminck) at Amsterdam and Rotterdam zoos. *Zool Gart.* 36(1/3):90–94.

Jacobs, G. H. 1978. Spectral sensitivity and colour vision in the ground-dwelling sciurids: Results from golden mantled ground squirrels and comparisons for five species. *Anim. Behav.* 26(2):409–21.

Jainudeen, M. R.; Eisenberg, J. F.; and Jayasinghe, J. B. 1971. Semen of the Ceylon elephant, *Elephas maximus*. *J. Reprod. Fertil.* 24:213–17.

Jainudeen, M. R.; Eisenberg, J. F.; and Tilakeratne, N. 1971. Oestrous cycle of the Asiatic elephant, *Elephas maximus*, in captivity. *J. Reprod. Fertil.* 27(3):321–28.

Jainudeen, M. R.; Katangale, C. B.; and Short, R. V. 1972. Plasma testosterone levels in relation to musth and sexual activity in the male Asiatic elephant, *Elephas maximus*. *J. Reprod. Fertil.* 29(1):99–103.

Jainudeen, M. R.; McKay, G. M.; and Eisenberg, J. F. 1972. Observations on musth in the domesticated Asiatic elephant (*Elephas maximus*). *Mammalia* 36(2):247–61.

Janis, C. 1976. The evolutionary strategy of the Equidae and the origins of rumen and cecal digestion. *Evolution* 30:757–74.

Jannett, F. J., Jr. 1978. The density-dependent formation of extended maternal families of the montane vole, *Microtus montanus nanus*. *Behav. Ecol. Sociobiol.* 3(3):245–65.

Jantschke, F. 1973. On the breeding and rearing of bush dogs *Speothos venaticus* at Frankfurt Zoo. *Int. Zoo Yearb.* 13:141–43.

Janzen, D. H. 1971. Seed predation by animals. *Ann. Rev. Ecol. Syst.* 2:465–92.

Jarman, M. V., and Jarman, P. J. 1973. Daily activity of impala. *E. African Wildl. J.* 11(1):75–93.

Jarman, P. J. 1974. The social organization of antelopes in relation to their ecology. *Behaviour* 58(3–4):215–67.

Jarvis, J. U. M. 1969. The breeding season and litter size of African mole rats. *J. Reprod. Fertil.* 6:237–48.

———. 1973*a*. Activity patterns in the mole rats *Tachyoryctes splendens* and *Heliophobius argenteocinereus*. *Zool. Africana* 8(1):101–19.

———. 1973*b*. The structure of a population of mole rats, *Tachyoryctes splendens*. *J. Zool.* (London) 171:1–14.

Jarvis, J. U. M., and Sale, J. B. 1971. Burrowing and burrow patterns of East African mole rats *Tachyoryctes*, *Heliophobius* and *Heterocephalus*. *J. Zool.* (London) 163:451–79.

Jenkins, F. A. 1971. Limb posture and locomotion in the Virginia opossum (*Didelphis marsupialis*) and in other non-cursorial mammals. *J. Zool.* (London) 165:303–15.

Jenkins, F. A., and Parrington, F. R. 1976. The postcranial skeletons of the Triassic mammals, *Eozostrodon*, *Megazostrodon* and *Erythrotherium*. *Phil. Trans. Royal Soc. London,* ser. B (Biol. Sci.), 273:83–103.

Jenness, R. 1974. The composition of milk. In *Lactation: A comprehensive treatise*, ed. B. L. Larson and V. R. Smith, 3:3–107. New York: Academic Press.

Jenness, R., and Studier, E. H. 1976. Lactation and milk. In *Biology of bats of the New World family Phyllostomatidae, part I,* ed. R. J. Baker, J. Knox-Jones, Jr., and D. C. Carter, pp. 201–18. Special Publication of the Museum, no. 10. Lubbock: Texas Tech Press.

Jepsen, G. L. 1970. Bat origins and evolution. In *Biology of bats,* ed. W. A. Wimsatt, pp. 1–64. New York: Academic Press.

Jerison, H. J. 1973. *Evolution of the brain and intelligence*. New York: Academic Press.

Jewell, P. A. 1966. The concept of home range in mammals. In *Play, exploration and territory in mammals,* ed. P. A. Jewell and C. Loizos, pp. 85–109. Symposiums of the Zoological Society of London, 18. New York: Academic Press.

Jewell, P. A., and Fullagar, P. J. 1965. Fertility among races of the field mouse (*Apodemus sylvaticus*) and their failure to form hybrids with the yellow-necked mouse (*Apodemus flavicollis*). *Evolution* 19(2):175–81.

Johnsingh, A. J. T. 1979. Ecology and behaviour of the dhole or Indian wild dog, *Cuon alpinus* Pallas 1811, with special reference to predator-prey relationships at Bandipur. Ph.D. thesis, Madurai University.

Johnson, G. E. 1931. Early life of the thirteen-lined ground squirrel. *Trans. Kansas Acad. Sci.* 34:282–90.

Johnston, R. E. 1975*a*. Scent marking by male golden hamsters (*Mesocricetus auratus*). I. Effects of odors and social encounters. *Z. Tierpsychol.* 37:75–98.

———. 1975*b*. Scent marking by male golden hamsters (*Mesocricetus auratus*). II. The role of the flank gland scent in the causation of marking. *Z. Tierpsychol.* 37:138–44.

———. 1975*c*. Scent marking by male golden hamsters (*Mesocricetus auratus*). III. Behavior in a seminatural environment. *Z. Tierpsychol.* 37:213–21.

Johnston, R. F., and Rudd, R. L. 1957. Breeding the salt marsh shrew. *J. Mammal.* 38:157.

Johnstone, P. 1977. Hand rearing serval at Mole Hall Wildlife Park. *Int. Zoo Yearb.* 17:218–19.

———. 1978. Breeding and rearing the Canadian otter (*Lutra canadensis*) at Mole Hall Wildlife Park, 1966–1977. *Int. Zoo Yearb.* 18:143–47.

Jolly, A. 1966. *Lemur behaviour: A Madagascar field study*. Chicago: University of Chicago Press.

———. 1972. *The evolution of Primate behavior.* New York: Macmillan.

Jones, C. 1967. Growth, development and wing loading in the evening bat, *Nycticeius humeralis* (Rafinesque). *J. Mammal.* 48(1):1–20.

Jones, M. 1979. Longevity of mammals in captivity. *Int. Zoo News*, April/May, pp. 16–26.

Jonkel, C. J., and Cowan, I. McT. 1971. The black bear in the spruce-fir forest. *Wildl. Monogr.* 27:1–57.

Jonkel, C. J.; Kolenosky, G. B.; Robertson, R. J.; and Russell, R. H. 1972. Further notes on polar bear denning habits. In *Bears: Their biology and management,* ed. S. Herrero, pp. 142–59. Morges: IUCN.

Jonkel, C. J., and Weckwerth, R. P. 1963. Sexual maturity and implantation of blastocysts in the wild pine marten. *J. Wildl. Manage.* 27(1):93–98.

Joubert, E. 1972. The social organization and associated behavior in the Hartmann zebra, *E. z. hartmannae*. *Madoqua* 6:17–56.

Jungius, H. 1970. Studies on the breeding biology of the reedbuck (*Redunca arundinum* Boddaert 1785) in the Kruger National Park. *Z. Säugetierk.* 35(3):129–46.

———. 1971. *The biology and behaviour of the reedbuck in the Kruger National Park*. Mammalia Depicta. Hamburg: Paul Parey.

Kalmbach, E. R. 1943. *The armadillo: Its relation to agriculture and game*. Austin, Tex.: Game Fish, and Oyster Commission.

Kaplan, H., and Hyland, S. O. 1972. Behavioural development in the Mongolian gerbil (*Meriones unguiculatus*). *Anim. Behav.* 20(1):147–55.

Kaufmann, J. H. 1962. Ecology and social behavior of the coati, *Nasua narica* on Barro Colorado Island, Panama. *Univ. California Publ. Zool.* 60(3):95–222.

———. 1965. Studies on the behavior of captive tree shrews (*Tupaia glis*). *Folia Primat.* 3:50–74.

———. 1974*a*. The ecology and evolution of social organization in the kangaroo family (Macropodidae). *Amer. Zool.* 14:51–62.

———. 1974*b*. Social ethology of the whiptail wallaby, *Macropus parryi,* in northeastern New South Wales. *Anim. Behav.* 22:281–369.

———. 1974*c*. Habitat use and social organization of nine sympatric species of macropodid marsupials. *J. Mammal.* 55(1):66–80.

Kaufmann, J. H., and Kaufmann, A. 1965. Observations of the behavior of tayras and grisons. *Z. Säugetierk.* 30(3):146–55.

Kavanau, J. L. 1962. Precise monitoring of drinking behavior in small mammals. *J. Mammal.* 43:345–51.

———. 1967. Behavior of captive white-footed mice. *Science* 155:1623–39.

Kawamichi, T. 1968. Winter behavior of the Himalayan pika, *Ochotona roylei*. *J. Fac. Sci. Hokkaido Univ.*, ser. 6, 16:582–94.

———. 1970. Social pattern of the Japanese pika, *Ochotona hyperborea yesoensis,* preliminary report. *J. Fac. Sci. Hokkaido Univ.,* ser. 6, Zool. 17:462–73.

———. 1971*a*. Annual cycle of behavior and social patterns of the Japanese pika, *Ochotona hyperborea yesoensis*. *J. Fac. Sci. Hokkaido Univ.,* ser. 6, Zool., 18:173–85.

———. 1971*b*. Daily activities and social pattern of two Himalayan pikas, *Ochotona macrotis* and *O. roylei,* observed at Mt. Everest. *J. Fac. Sci. Hokkaido Univ.,* ser. 6, Zool. 17:587–609.

———. 1976. Hay territory and dominance rank of pikas. *J. Mammal.* 57(1):133–48.

Kay, R. F., and Cartmill, M. 1977. Cranial morphology and adaptations of *Palaechthon nacimienti* and other Paramomyidae (Plesiodapoidea, ?Primates), with a description of a new genus and species. *J. Hum. Evol.* 6:13–53.

Kay, R. F., and Hylander, W. L. 1978. The dental structure of mammalian folivores with special reference to primates and phalangeroids (Marsupialia). In *The ecology of arboreal folivores,* ed. G. G. Montgomery, pp. 173–92. Washington, D.C.: Smithsonian Institution Press.

Kaye, S. V. 1961. Laboratory life history of the eastern harvest mouse. *Amer. Midl. Nat.* 66:439–51.

Keast, A. 1972. Comparisons of contemporary mammal faunas of southern continents. In *Evolution, mammals and southern continents,* ed. A. Keast, F. C. Erk, and B. Gloss, pp. 433–501. Albany: State University of New York Press.

———. 1977. Historical biogeography of the marsupials. In *The biology of marsupials,* ed. B. Stonehouse and D. Gilmore, pp. 69–96. Baltimore: University Park Press.
Kellogg, W. N. 1961. *Porpoises and sonar.* Chicago: University of Chicago Press.
Kelsall, J. P. 1968. *The migratory barren-ground caribou of Canada.* Ottawa: Queens Printer.
Kenagy, G. J. 1972. Saltbush leaves: Excision of hypersaline tissue by a kangaroo rat. *Science* 178:1094–96.
———. 1973*a*. Daily and seasonal patterns of activity and energetics in a heteromyid rodent community. *Ecology* 54:1201–19.
———. 1973*b*. Adaptations for leaf eating in the Great Basin kangaroo rat, *Dipodomys microps. Oecologia* (Berlin) 12:383–412.
———. 1976. The periodicity of daily activity and its seasonal changes in free-ranging and captive kangaroo rats. *Oecologia* (Berlin) 24:105–40.
Kenyon, K. W. 1960. Territorial behavior and homing in the Alaska fur seal. *Mammalia* 24:431–44.
———. 1969. *The sea otter in the eastern Pacific Ocean.* North American Fauna 68. Washington, D.C.: U.S. Department of Interior.
Kenyon, K. W., and Rice, D. W. 1959. Life history of the Hawaiian monk seal. *Pacific Sci.* 13:215–52.
Kenyon, K. W., and Wilke, F. 1953. Migration of the northern fur seal, *Callorhinus ursinus. J. Mammal.* 34:86–98.
Khajuria, H. 1972. Courtship and mating in *Rhinopoma h. hardwickei* Gray (Chiroptera: Rhinopomatidae). *Mammalia* 36(2):307–9.
Kihlström, J. E. 1972. Period of gestation and body weight in some placental mammals. *Comp. Biochem. Physiol.* 43A:674–79.
Kiley-Worthington, M. 1965. The waterbuck (*Kobus defassa* Ruppel 1835 and *K. ellipsiprimnus* Ogilby 1833) in East Africa: Spatial distribution. A study of the sexual behaviour. *Mammalia* 29:177–205.
Kilgore, D. L., Jr. 1969. An ecological study of the swift fox in the Oklahoma panhandle. *Amer. Midl. Nat.* 83(2):512–34.
Kilham, L. 1954. Territorial behavior of the red squirrel. *J. Mammal.* 35:252–53.
Kiltie, R. A. 1980. Seed predation and group size in rainforest peccaries. Ph.D. diss., Princeton University.
King, J. A. 1955. Social behavior, social organization, and population dynamics in a black-tailed prairiedog town in the Black Hills of South Dakota. *Contrib. Lab. Vert. Biol., Univ. Michigan* 67:1–123.
———. 1958. Maternal behavior and behavioral development in two subspecies of *Peromyscus maniculatus. J. Mammal.* 39(2):177–90.
———. 1961. Development and behavioral evolution in *Peromyscus.* In *Vertebrate speciation,* ed. W. F. Blair, pp. 122–47. Austin: University of Texas Press.
———. 1965. Body, brain and lens weights of *Peromyscus. Zool. Jahrb. Anat.* 82:177–88.
———, ed. 1968. *Biology of Peromyscus (Rodentia).* Special Publication no. 2. American Society of Mammalogists.

King, J. E. 1964. *Seals of the world.* London: Trustees of the British Museum (Natural History).

Kingdon, J. 1971. *East African mammals: An atlas of evolution.* Vol. 1. London/New York: Academic Press.

———. 1974*a*. *East African mammals: An atlas of evolution in Africa.* Vol. 2-A. *Insectivores and bats.* London/New York: Academic Press.

———. 1974*b*. *East African mammals: An atlas of evolution in Africa.* Vol. 2-B. *Hares and rodents.* London/New York: Academic Press.

Kinzey, W. G. 1977. Diet and feeding behavior of *Callicebus torquatus.* In *Primate ecology,* ed. T. H. Clutton-Brock, pp. 127–51. London/New York: Academic Press.

Kirkpatrick, T. H. 1965*a*. Grey kangaroo being studied. *Queensland Agric. J.* 91:454–58.

———. 1965*b*. Studies of Macropodidae in Queensland. 2. Age estimation in the grey kangaroo, the red kangaroo, the eastern wallaroo and the red-necked wallaby, with notes on dental abnormalities. *Queensland J. Agric. Anim. Sci.* 22:301–17.

———. 1965*c*. Studies of Macropodidae in Queensland. 3. Reproduction in the grey kangaroo (*Macropus major*) in southern Queensland. *Queensland J. Agric. Anim. Sci.* 22:319–28.

———. 1966. Studies of Macropodidae in Queensland. 4. Social organization of the grey kangaroo (*Macropus giganteus*). *Queensland J. Agric. Anim. Sci.* 23:317–22.

Kirmiz, J. P. 1962*a*. *Adaptation de la gerboise au milieu desertique.* Société de Publication Egyptiennes. Alexandria: S.A.E.

———. 1962*b*. *Adaptation to desert environment: A study on the jerboa, rat and man.* London: Butterworth.

Kirsch, J. A. W. 1977. The classification of marsupials. In *The biology of marsupials,* ed. D. Hunsaker, II, pp. 1–50. New York/London: Academic Press.

Kirsch, J. A. W., and Calaby, J. H. 1977. The species of living marsupials: An annotated list. In *The biology of marsupials,* ed. B. Stonehouse and D. Gilmore, pp. 9–26. Baltimore/London/Tokyo: University Park Press.

Kleiber, M. 1961. *The fire of life.* New York: John Wiley.

Kleiman, D. G. 1967. Some aspects of social behavior in the Canidae. *Amer. Zool.* 7:365–72.

———. 1969. Maternal care, growth rate, and development in the noctule, pipistrelle and serotine bats. *J. Zool.* (London) 157(2):187–212.

———. 1970. Reproduction in the female green acouchi, *Myoprocta pratti. J. Reprod. Fertil.* 23:55–65.

———. 1971. The courtship and copulatory behaviour of the green acouchi, *Myoprocta pratti. Z. Tierpsychol.* 29:259–78.

———. 1972*a*. Maternal behaviour of the green acouchi. (*Myoprocta pratti* Pocock), a South American caviomorph rodent. *Behaviour* 43(1–4):48–84.

———. 1972*b*. Social behavior of the maned wolf (*Chrysocyon brachyurus*) and bush dog (*Speothos venaticus*): A study in contrast. *J. Mammal.* 53(4):791–807.

———. 1974*a*. Patterns of behavior in hystricomorph rodents. In *Symposium on the biology of hystricomorph rodents*, ed. I. W. Rowlands and B. Weir, pp. 171–209. London: Zoological Society.
———. 1974*b*. Scent marking in the binturong, *Arctictis binturong*. *J. Mammal.* 55(1):224–27.
———, ed. 1977*a*. *The biology and conservation of the Callitrichidae*. Washington, D.C.: Smithsonian Institution Press.
———. 1977*b*. Monogamy in mammals. *Quart. Rev. Biol.* 52:39–69.
Kleiman, D. G., and Brady, C. A. 1978. Coyote behavior in the context of recent canid research: Problems and perspectives. In *Coyotes: Biology, behavior and management*, ed. M. Bekoff, pp. 163–88. New York: Academic Press.
Kleiman, D. G., and Davis, T. M. 1978. Ontogeny and maternal care. In *Biology of the New World leaf-nosed bats*, ed. R. J. Baker, pp. 387–402. Special Publication of the Museum no. 16. Lubbock: Texas Tech Press.
Kleiman, D. G., and Eisenberg, J. F. 1973. Comparisons of canid and felid social systems from an evolutionary perspective. *Anim. Behav.* 21:637–59.
Kleiman, D. G.; Eisenberg, J. F.; and Maliniak, E. 1979. Reproductive parameters and productivity of caviomorph rodents. In *Vertebrate ecology in the northern Neotropics*, ed. J. F. Eisenberg, pp. 173–83. Washington, D.C.: Smithsonian Institution Press.
Kleiman, D. G., and Racey, P. A. 1969. Observations on noctule bats (*Nyctalus noctula*) breeding in captivity. *Lynx* 10:65–77.
Klein, L. L. 1972. The ecology and social organization of the spider monkey, *Ateles belzebuth*. Ph.D. diss., University of California, Berkeley.
Klein, L. L., and Klein, D. B. 1977. Feeding behaviour of the Colombian spider monkey. In *Primate ecology*, ed. T. H. Clutton-Brock, pp. 153–80. London/New York: Academic Press.
Kleinenberg, S. E.; Yablokov, A. V.; Belkovich, B. M.; and Tarasevich, M. N. 1969. *Beluga (Delphinapterus leucas): Investigation of the species*. Translated from Russian. Jerusalem: Israel Program for Scientific Translations.
Klingel, H. 1967. Soziale Organisation und Verhalten freilebender Steppenzebras. *Z. Tierpsychol.* 24:580–624.
———. 1968. Soziale Organisation und Verhaltensweisen von Hartmann- und Bergzebras (*Equus zebra hartmannae* und *E. z. zebra*). *Z. Tierpsychol.* 25:76–88.
———. 1972*a*. Social behaviour of African Equidae. *Zool. Africana* 7(1):175–85.
———. 1972*b*. Das Verhalten der Pferde (Equidae). *Handb. Zool.* 8/10(24):1–68.
———. 1974. Soziale Organisation und Verhalten des Grevy-Zebras (*Equus grevyi*). *Z. Tierpsychol.* 36:37–70.
———. 1977. Observations on social organization and behaviour of African and Asiatic wild asses (*Equus africanus* and *E. hemionus*). *Z. Tierpsychol.* 44:323–31.

Klingel, H. 1979. Soziale organisation des Flusspferdes *Hippopotamus amphibius*. *Verh. Dtsch. Zool. Ges.*, 1:241.

Klopfer, P. H., and Gilbert, B. K. 1966. A note on retrieval and recognition of young in the elephant seal, *Mirounga angustirostris*. *Z. Tierpsychol.* 23(6):757–61.

Klopfer, P. H., and Hailman, J. P. 1967. *An introduction to animal behavior.* Englewood Cliffs, N.J.: Prentice-Hall.

Knaus, W. 1960. *Das Gamswild.* Hamburg: Paul Parey.

Knipe, T. 1957. *The javelina in Arizona.* Phoenix: State of Arizona Game and Fish Department.

Koenig, L. 1957. Beobachten über Reviermarkierung sowie Droh-, Kampf-, und Abwehrverhalten des Murmeltieres (*Marmota marmota* L.). *Z. Tierpsychol.* 14:510–20.

———. 1960. Das Aktionsystem des Siebenschläfers (*Glis glis* L.). *Z. Tierpsychol.* 17:427–505.

Koffler, B. R. 1972. *Meriones crassus*. *Mammal. Spec.* 9:1–4.

Koford, C. B. 1957. The vicuña and the puna. *Ecol. Monogr.* 27:153–219.

———. 1958. Prairie dogs, white faces and blue grama. *Wildl. Monogr.* no. 3.

Krämer, A. 1969. Soziale Organisation und Sozialverhalten einer Gemspopulation (*Rupicapra rupicapra* L.) der Alpen. *Z. Tierpsychol.* 26(8):889–965.

Kramer, G. 1952. Experiments on bird orientation. *Ibis* 94:265–85.

Kraus, C.; Gihr, M.; and Pilleri, G. 1970. Das Verhalten von *Cuniculus paca* in Gefangenschaft. *Rev. Suisse Zool.* 77:353–88.

Krebs, C. J. 1963. Lemming cycle at Baker Lake, Canada, during 1959–62. *Science* 140:674–76.

———. 1966. Demographic changes in fluctuating populations of *Microtus californicus*. *Ecol. Monogr.* 36:239–73.

———. 1970. *Microtus* population biology: Behavioral changes associated with the population cycle in *M. ochrogaster* and *M. pennsylvanicus*. *Ecology* 51(1):34–52.

Krebs, C. J.; Keller, B. L.; and Tamarin, R. H. 1969. *Microtus* population biology: Demographic changes in fluctuating populations of *M. ochrogaster* and *M. pennsylvanicus* in southern Indiana. *Ecology* 50(4):587–607.

Krebs, C. J., and Myers, J. H. 1974. Population cycles in small mammals. *Adv. Ecol. Res.* 8:267–399.

Krefeld, W. E. 1963. Bericht über Geburt und Aufzucht vom Geparden, *Acinonyx jubatus*, im Krefelder Tierpark. *Zool. Gart.* 27:177–81.

Krieg, H. 1929. Biologische Reisestudien in Südamerika. IX. Gürteltiere. *Z. Morph. Ökol. Tiere* 14(1):166–90.

Kritzler, H. 1952. Observations on the pilot whale in captivity. *J. Mammal.* 33:321–33.

Krott, P. 1960. *Der Vielfrass.* Wittenberg: Ziemsen.

Krott, P., and Krott, G. 1962. Zum Verhalten des Branbären (*Ursus arctos* L. 1758) in den Alpen. *Z. Tierpsychol.* 29(2):160–206.

Kruuk, H. 1966. Clan-system and feeding habits of spotted hyenas. *Nature* 209:1257–58.

———. 1972. *The spotted hyena: A study of predation and social behavior.* Chicago: University of Chicago Press.

———. 1976. Feeding and social behaviour of the striped hyaena (*Hyaena vulgaris* Desmarest). *E. African Wildl. J.* 14:91–111.

———. 1978. Spatial organization and territorial behaviour of the European badger, *Meles meles*. *J. Zool.* (London) 184(1):1–21.

Kruuk, H., and Sands, W. A. 1972. The aardwolf (*Proteles cristatus* Sparrman 1783) as a predator of termites. *E. African Wildl. J.* 10:211–27.

Kuehn, R. E., and Beach, F. A. 1963. Quantitative measurement of sexual receptivity in female rats. *Behaviour* 21(3–4):282–99.

Kühlhorn, F. 1936. Die anpassungstypen der Gürteltiere. *Z. Säugetierk.* 12:245–303.

Kühme, W. 1964. Über die soziale Bindung innerhalb eines Hyänenhund-Rudels. *Naturwissenschaften* 51(23):567.

———. 1965. Communal food distribution and division of labour in African hunting dogs (*Lycaon pictus lupinus*). *Nature* (London) 205:443–44.

———. 1966. Freilandstudien zur soziologie des Hyanenhundes (*Lycaon pictus lupinus* Thomas 1902). *Z. Tierpsychol.* 22:495–541.

Kuhn, H. J. 1964. Zur Kenntnis von *Micropotamogale lamottei* Heim de Balsac 1954. *Z. Säugetierk.* 29:152–73.

———. 1966. Der Zebraducker, *Cephalophus doria* (Ogilby 1837). *Z. Säugetierk.* 31(4):282–93.

Kulzer, E. 1958. Untersuchung über die Biologie von Flughunden der Gattung *Rousettus* Gray. *Z. Morph. Ökol. Tiere* 47:374–402.

———. 1962. Über die Jugendentwicklung der Angola-Bulldog fledermaus *Tadarida* (Mops.) *condylura* (A. Smith 1833) (Molossidae). *Säugetierk. Mitt.* 10:116–24.

———. 1966. Die Geburt bei Flughunden der Gattung *Rousettus* Gray (Megachiroptera). *Z. Säugetierk.* 31:226–33.

Kummer, H. 1968. *Social organization of hamadryas baboons.* Chicago: University of Chicago Press.

Kunz, T. H. 1973. Population studies of the cave bat (*Myotis velifer*): Reproduction, growth and development. *Univ. Kansas Occ. Pap. Mus. Nat. Hist.* 15:1–43.

Kurt, F. 1963. Zur Carnivorie bei *Cephalophus dorsalis*. *Z. Säugetierk.* 28(5):309–13.

———. 1968. *Das Sozialverhalten des Rehes Capreolus capreolus L.: Eine Feldstudie.* Mammalia Depicta. Hamburg/Berlin: Paul Parey.

———. 1974. Remarks on the social structure and ecology of the Ceylon elephant in the Yala National Park. In *The behaviour of ungulates and its relation to management,* ed. V. Geist and F. Walther, 2:618–34. IUCN Publications, n.s. no. 24. Morges: IUCN.

Lacher, T. E., Jr. 1979. The comparative social behavior of *Kerodon rupestris* and *Galea spixii* in the xeric caatinga of northeastern Brazil. Thesis, University of Pittsburgh.

Lachiver, F., and Petter, F. 1969. La léthargie du graphiure (*Graphiurus murinus lorraineus* Dollman 1910, Rongeurs, Glirides). *Mammalia* 33(2):165–92.

Lackey, J. A. 1976. Reproduction, growth and development in the Yucatan deer mouse, *Peromyscus yucatanicus*. *J. Mammal.* 57:638–55.

LaFollette, R. M. 1971. Agonistic behaviour and dominance in confined wallabies, *Wallabia rufogrisea frutica*. *Anim. Behav.* 19:93–101.

Lamprecht, J. 1979. Field observations on the behaviour and social system of the bat-eared fox, *Otocyon megalotis* Desmarest. *Z. Tierpsychol.* 49:260–84.

Lamprey, H. F. 1963. Ecological separation of the large mammal species in the Tarangire Game Reserve, Tanganyika. *E. African Wildl. J.* 1:63–92.

———. 1964. Estimation of the large mammal densities, biomass, and energy exchange in the Tarangire Game Reserve and the Masai Steppe in Tanganyika. *E. African Wildl. J.* 2:1–46.

Lande, R. 1979. Quantitative genetic analysis of multivariate evolution, applied to brain:body size allometry. *Evolution* 33:402–16.

Landry, S. O., Jr. 1957. The interrelationships of the New and Old World hystricomorph rodents. *Univ. Calif. Publ. Zool.* 56(1):1–118.

———. 1970. The Rodentia as omnivores. *Quart. Rev. Biol.* 45(4):351–72.

Lang, E. M. 1961. Beobachtungen am Indischen Panzernashorn (*Rhinoceros unicornis*). *Zool. Gart.* 25:369–409.

———. 1967. The birth of an African elephant, *Loxodonta africana*, at Basel Zoo. *Int. Zoo Yearb.* 7:154–57.

Lang, T. G. 1966. Hydrodynamic analysis of Cetacean performance. In *Whales, dolphins, and porpoises*, ed. K. Norris, pp. 410–32. Berkeley: University of California Press.

Larkin, P., and Roberts, M. 1979. Reproduction in the ring-tailed mongoose *Galidea elegans* at the National Zoological Park, Washington. *Int. Zoo Yearb.* 19:188–92.

Lataste, F. 1887. *Documents pour l'éthologie des mammifères*. Reprinted from *Actes Soc. Linn. Bordeaux*, vols. 40, 41, and 43. Bordeaux: Société Linné.

Laurie, A. 1978. The ecology and behavior of the greater one-horned rhinoceros. Ph.D. thesis, Cambridge University.

Laurie, A., and Seidensticker, T. 1977. Behavioral ecology of the sloth bear (*Melursus ursinus*). *J. Zool.* (London) 182(2):187–204.

LaVal, R. K. 1976. Voice and habitat of *Dactylomys dactylinus* (Rodentia: Echimyidae) in Ecuador. *J. Mammal.* 57:402–3.

Lavocat, R. 1959. Origine et affinité des rongeurs de la sous-famille des dendromurinés. *C. R. Acad. Sci. Paris* 248:1375–77.

———. 1967. A propos de la dentition des rongeurs et du problème de l'origine des muridés. *Mammalia* 31:205–16.

———. 1974. What is an hystricomorph? *Symp. Zool. Soc. London* 34:7–20.

Laws, R. M. 1956. The elephant seal. II. General, social and reproductive behavior. *FIDS Sci. Rep.*, vol. 13.

———. 1974. Behaviour, dynamics and management of elephant populations. In *The behaviour of ungulates and its relation to management,* ed. V. Geist and F. Walther, 2:513–29. IUCN Publications, n.s. no. 24. Morges: IUCN.

Laws, R. M.; Parker, I. S. C.; and Johnstone, R. C. B. 1975. *Elephants and their habitats: The ecology of elephants in North Bunyoro, Uganda.* Oxford: Clarendon Press.

Lay, D. W. 1942. Ecology of the opossum in eastern Texas. *J. Mammal.* 23:147–59.

Layne, J. N. 1954. The biology of the red squirrel, *Tamiasciurus hudsonicus lognax,* in central New York. *Ecol. Monogr.* 24:227–67.

———. 1960. The growth and development of young golden mice, *Ochrotomys nuttalli. Quart. J. Florida Acad. Sci.* 23:36–58.

———. 1968. Ontogeny of *Peromyscus.* In *Biology of Peromyscus (Rodentia),* ed. J. A. King, pp. 148–253. Special Publication American Society Mammalogists.

Layne, J. N., and Caldwell, D. K. 1964. Behavior of the Amazon dolphin, *Inia geoffrensis* (Blainville) in captivity. *Zoologica* 49:81–108.

Layne, J. N., and Glover, D. 1977. Home range of the armadillo in Florida. *J. Mammal.* 58:411–14.

LeBoeuf, B. J. 1974. Male-male competition and reproductive success in elephant seals. *Amer. Zool.* 14:163–76.

LeBoeuf, B. J.; Ainley, D. G.; and Lewis, T. J. 1974. Elephant seals on the Farallones: Population structure of an incipient breeding colony. *J. Mammal.* 55(2):370–85.

LeBoeuf, B. J.; Whiting, R. J.; and Gantt, R. F. 1972. Perinatal behavior of northern elephant seal females and their young. *Behaviour* 43(1–4):121–57.

LeBoulenge, E., and Fuentes, E. R. 1978. Some observations on the population dynamics of *Octodon degus* (Rodentia: Hystricomorpha) in central Chile. *Terre et Vie* 32:325–42.

Lechleitner, R. R. 1959. Sex ratio, age classes and reproduction of the black-tailed jack rabbit. *J. Mammal.* 40:63–81.

Lee, A. K.; Bradley, A. J.; and Braithwaite, R. W. 1977. Corticosterone levels and male mortality in *Antechinus stuarti.* In *The biology of marsupials,* ed. B. Stonehouse and D. Gilmore, pp. 209–20. Baltimore: University Park Press.

Lehrman, D. S. 1970. Semantics and conceptual issues in the nature-nurture problem. In *Development and evolution of behavior,* ed. T. R. Aronson et al. San Francisco: Freeman.

Leitch, I.; Hytten, F. E.; and Billewicz, W. Z. 1959. The maternal and neonatal weights of some Mammalia. *Proc. Zool. Soc. London* 13:11–28.

Lekagul, B., and McNeely, J. A. 1977. *Mammals of Thailand.* Bangkok: Kurusapha Ladprao Press.

Lent, P. C. 1965. Rutting behaviour in a barren-ground caribou population. *Anim. Behav.* 13:259–64.

———. 1966. Calving and related social behavior in the barren-ground caribou. *Z. Tierpsychol.* 6:701–56.

———. 1974. Mother-infant relationships in ungulates. In *The behaviour of ungulates and its relation to management,* ed. V. Geist and F. Walther, 1:14–55. IUCN Publications, n.s. no. 24. Morges: IUCN.

Leraas, H. J. 1938. Observations on the growth and behavior of harvest mice. *J. Mammal.* 19:441–44.

Leutenegger, W. 1973. Maternal fetal weight relationships in primates. *Folia Primat.* 20:280–93.

Leuthold, W. 1966. Variations in territorial behavior of Uganda Kob, *Adenota kob thomasi* (Neumann). *Behaviour* 27(3–4):214–57.

———. 1967. Beobachtungen zum Jugendverhalten von Kob-Antilopen. *Z. Säugetierk.* 32:59–63.

———. 1970. Observations on the social organization of impala (*Aepyceros melampus*). *Z. Tierpsychol.* 27:693–721.

———. 1971. Freilandbeobachtungen an Giraffengazellen (*Litocranius walleri*) im Tsavo-Nationalpark, Kenia. [Field observations on ecology and behaviour of gerenuk.] *Z. Säugetierk.* 36(1):19–37.

———. 1976. Group size in elephants of Tsavo National Park and possible factors influencing it. *J. Anim. Ecol.* 45:425–39.

———. 1977. *African ungulates: Zoophysiology and ecology 8*. New York: Springer.

Leuthold, W., and Leuthold, B. M. 1975. Parturition and related behaviour in the African elephant. *Z. Tierpsychol.* 39:75–84.

Leyhausen, P. 1956. Das Verhalten der Katzen. *Handb. Zool.* Band 8, Teil 10(21):1–34.

———. 1963. Über sudamerikanische Pardelkatzen. *Z. Tierpsychol.* 20(5):627–40.

———. 1965. Über die Funktion der Relativen Stimmungshierarchie (Dargestellt am Beispiel der phylogenetischen und ontogenetischen Entwicklung des Beutefangs von Raubtieren). *Z. Tierpsychol.* 22(4):412–94.

———. 1979. *The behaviour of cats*. New York: Garland Press.

Leyhausen, P., and Tonkin, B. 1966. Breeding the black-footed cat *Felis nigripes* in captivity. *Int. Zoo Yearb.* 6:178–82.

Leyhausen, P., and Wolff, R. 1959. Das Revier einer Hauskatze. *Z. Tierpsychol.* 16:666–70.

Lidicker, W. Z. 1966. Ecological observations on a feral house mouse population declining to extinction. *Ecol. Monogr.* 36:27–50.

———. 1975. The role of dispersal in the demography of small mammals. In *Small mammals: Productivity and dynamics of populations,* ed. K. Petrusewicz, F. B. Golley, and L. Ryszkowski, pp. 103–28. New York: Cambridge University Press.

Lillegraven, J. A. 1974. Biogeographical considerations of the marsupial-placental dichotomy. *Ann. Rev. Ecol. Syst.* 5:263–83.

Lillegraven, J. A.; Kielan-Jaworowska, Z.; and Clemens, W. A. 1980. *Mesozoic mammals*. Berkeley: University of California Press.

Lilly, J. C. 1961. *Man and dolphin*. New York: Pyramid Publications.

Lim Boo Liat. 1967. Note on the food habits of *Ptilocercus loweii* Gray (pentail tree-shrew) and *Echinosorex gymnurus* (Raffles) (moonrat) in Malaya with remarks on "ecological labelling" by parasite patterns. *J. Zool.* (London) 152(4):375–81.

Lincoln, G. A. 1971. The seasonal reproductive changes in the red deer stag (*Cervus elaphus*). *J. Zool.* (London) 163:105–23.

———. 1972. The role of antlers in the behaviour of red deer. *J. Exp. Zool.* 182(2):233–49.

Lindemann, W. 1955. Über die Jugendentwicklung beim Luchs (*Lynx l. lynx* Kerr) und bei der Wildkatze (*Felis s. silvestris* Schreb.). *Behaviour* 8:1–45.

Lindstedt, S. L. 1980. The smallest insectivores: Coping with scarcities of energy and water. In *Comparative physiology: Primitive mammals,* ed. K. Schmidt-Nielson, L. Bolis, and C. Richard Taylor, pp. 163–69. Cambridge: Cambridge University Press.

Linsdale, J. M. 1946. *The California ground squirrel.* Berkeley: University of California Press.

Linsdale, J. M., and Tevis, L. P. 1951. *The dusky-footed wood rat.* Berkeley: University of California Press.

Linsdale, J. M., and Tomich, P. Q. 1953. *A herd of mule deer.* Berkeley: University of California Press.

Linzey, A. V. 1970. Postnatal growth and development of *Peromyscus maniculatus nubiterrae. J. Mammal.* 51:152–55.

Linzey, D. W., and Linzey, A. V. 1967. Growth and development of the golden mouse, *Ochrotomys nuttalli nuttalli. J. Mammal.* 48:445–58.

Lippert, W. 1973. Zum Fortpflanzungsverhalten des Mähnen wolfes, *Chrysocyon brachyurus* Illiger. *Zool. Gart.,* n.s. (Leipzig) 43(5):225–47.

Llanos, A. C., and Crespo, J. A. 1952. Ecologia de la viscacha (*Lagostomus maximus maximus* Blainv.) en el nordeste de la provincia de Entre Rios. *Rev. Invest. Agric.* 6:289–378.

Lloyd, J. E. 1972. Vocalization in *Marmota monax. J. Mammal.* 53(1):214–16.

Lobao Tello, J. L. P., and Van Gelder, R. G. 1975. The natural history of nyala *Tragelaphus angasi* (Mammalia, Bovidae) in Mozambique. *Bull. Amer. Mus. Nat. Hist.* 155(4):321–86.

Loisa, K., and Pulliainen, E. 1968. Winter food and movements of two moose (*Alces alces* L.) in northeastern Finland. *Ann. Zool. Fennici* 5:220–23.

Loizos, C. 1966. Play in mammals. *Symp. Zool. Soc. London* 18:1–9.

Lord, R. 1961. A population study of the gray fox. *Amer. Midl. Nat.* 66:87–109.

Lorenz, K. 1961. Phylogenetische Anpassung und adaptive Modifikation des Verhaltens. *Z. Tierpsychol.* 18:139–87.

Lott, D. F. 1974. Sexual and aggressive behavior of American bison, *Bison bison.* In *The behaviour of ungulates and its relation to management,* ed. V. Geist and F. Walther, 1:382–94. IUCN Publications, n.s., no. 24. Morges: IUCN.

Loughray, A. G. 1959. Preliminary investigation of the Atlantic walrus (*Odobenus r. rosmarus* L.). *Canadian Wildl. Serv. Wildl. Manage. Bull.,* ser. 1, 14:1–123.

Low, B. S. 1978. Environmental uncertainty and the parental strategies of marsupials and placentals. *Amer. Nat.* 112:197–213.

Luckett, W. P. 1975. Ontogeny of the fetal membranes and placenta: Their bearing on primate phylogeny. In *Phylogeny of the Primates*, ed. W. P. Luckett and F. S. Szalay, pp. 157–82. New York: Plenum Press.

Luckett, W. P., and Szalay, F. S., eds. 1975. *Phylogeny of the Primates*. New York: Plenum Press.

Lusty, J. A., and Seaton, B. 1978. Oestrus and ovulation in the Casiragua *Proechimys guairae* (Rodentia, Hystricomorpha. *J. Zool.* (London) 184:255–65.

Lyall-Watson, M. 1963. A critical re-examination of food "washing" behaviour in the raccoon (*Procyon lotor* Linn.). *Proc. Zool. Soc. London* 141(2):371–93.

Lyman, C. P., and Dawe, A. R., eds. 1960. *Mammalian hibernation*. Bulletin of the Museum of Comparative Zoology, vol. 124. Cambridge: Harvard University Press.

Lyne, A. G. 1952. Notes on external characters of the pouch young of four species of bandicoot. *Proc. Zool. Soc. London* 122:625–49.

Lyne, A. G.; Pilton, P. E.; and Sharman, G. B. 1959. Oestrus cycle, gestation period and parturition in the marsupial *Trichosurus vulpecula*. *Nature* (London) 183:622–23.

MacArthur, R. H. 1970. Species packing and competitive equilibrium for many species. *Theoret. Pop. Biol.* 1:1–11.

———. 1972. *Geographical ecology: Patterns in the distribution of species*. New York: Harper and Row.

MacArthur, R. H., and Wilson, E. O. 1967. *The theory of island biogeography*. Princeton: Princeton University Press.

McBride, A. F., and Hebb, D. O. 1948. Behavior of the captive bottlenosed dolphin, *Tursiops truncatus*. *J. Comp. Physiol. Psychol.* 41:111–23.

McBride, G. 1963. The teat order and communication in young pigs. *Anim. Behav.* 11:53–56.

———. 1971. The nature-nurture problem in social evolution. In *Man and beast: Comparative sociology*, ed. J. F. Eisenberg and W. S. Dillon, pp. 35–56. Washington, D.C.: Smithsonian Institution Press.

———. 1976. The study of social organization. *Behaviour* 59:96–115.

McCabe, T. T., and Blanchard, B. D. 1950. *Three species of Peromyscus*. Santa Barbara, Calif.: Rood Associates.

McCarley, H. 1966. Annual cycle, population dynamics and adaptive behavior of *Citellus tridecemlineatus*. *J. Mammal.* 47:294–316.

McCracken, G. F., and Bradbury, J. W. 1977. Paternity and genetic heterogeneity in the polygynous bat, *Phyllostomus hastatus*. *Science* 198 (no. 4314):303–6.

McCullough, D. R. 1969. The Tule elk, its history, behavior and ecology. *Univ. California Publ. Zool.* 88:1–209.

McDougal, C. 1977. *The face of the tiger*. London: Rivington Books and Andre Deutsch.

Mace, G. M. 1979. The evolutionary ecology of small mammals. D. Phil. thesis, University of Sussex.
McGill, T. E. 1962. Sexual behavior in three inbred strains of mice. *Behaviour* 19(4):341–50.
McHugh, T. 1958. Social behavior of the American bison. *Zoologica* 43:1–40.
McIlwaine, C. P. 1962. Reproduction and body weights of the wild rabbit *Oryctolagus cuniculus* (L.) in Hawke's Bay, New Zealand. *New Zealand J. Sci.* 5:325–41.
Mack, D., and Kafka, H. 1978. Breeding and rearing of woolly monkeys at the National Zoological Park, Washington. *Int. Zoo Yearb.* 18:117–22.
McKay, G. M. 1973. The ecology and behavior of the Asiatic elephant in southeastern Ceylon. *Smithsonian Contrib. Zool.* 125:1–113.
McKeever, S. 1964. The biology of the golden-mantled ground squirrel. *Ecol. Monogr.* 34:383–401.
McKenna, M. C. 1975. Toward a phylogenetic classification of the Mammalia. In *Phylogeny of the Primates,* ed. W. P. Luckett and F. S. Szalay, pp. 21–46. New York: Plenum Press.
MacKinnon, J. R. 1971. The orang-utan in Sabah today. *Oryx* 11(2–3):141–91.
———. 1974. The behavior and ecology of wild orang-utans (*Pongo pygmaeus*). *Anim. Behav.* 22:3–74.
McLanahan, E. B., and Green, K. M. 1977. The vocal repertoire and an analysis of the contexts of vocalizations in *Leontopithecus rosalia.* In *The biology and conservation of the Callitrichidae,* ed. D. G. Kleiman, pp. 251–71. Washington, D.C.: Smithsonian Institution Press.
McLaren, I. A. 1957. Some aspects of growth and reproduction of the bearded seal, *Erignathus barbatus* (Erxleben). *J. Fish. Res. Bd. Canada* 15(2):219–27.
McManus, J. J. 1967. Observations on sexual behavior of the opossum, *Didelphis marsupialis. J. Mammal.* 48(3):486–87.
———. 1970. The behavior of captive opossums, *Didelphis marsupialis virginiana. Amer. Midl. Nat.* 84:144–69.
MacMillen, R. E., and Lee, A. K. 1969. Water metabolism of Australian hopping mice. *Comp. Biochem. Physiol.* 28:493–514.
MacMillen, R. E., and Nelson, J. E. 1969. Bioenergetics and body size in dasyurid marsupials. *Amer. J. Physiol.* 217(4):1246–51.
McNab, B. K. 1963. Bioenergetics and the determination of home range size. *Amer. Nat.* 97:130–40.
———. 1966. The metabolism of fossorial rodents: A study of convergence. *Ecology* 47:712–33.
———. 1971*a*. On the ecological significance of Bergmann's rule. *Ecology* 52(5):845–54.
———. 1971*b*. The structure of tropical bat faunas. *Ecology* 52:352–58.
———. 1974. The energetics of endotherms. *Ohio J. Sci.* 74(6):370–80.
———. 1978*a*. Energetics of arboreal folivores: Physiological problems and ecological consequences of feeding on an ubiquitous food supply.

In *The ecology of arboreal folivores*, ed. G. G. Montgomery, pp. 153–62. Washington, D.C.: Smithsonian Institution Press.

———. 1978*b*. The comparative energetics of Neotropical marsupials. *J. Comp. Physiol.* 125:115–28.

———. 1980. Food habits, energetics, and the population biology of mammals. *Amer. Nat.* 116(1):106–24.

McNally, J. 1960. The biology of the water rat *Hydromys chrysogaster* Geoffroy (Muridae: Hydromyinae) in Victoria. *Australian J. Zool.* 8:170–80.

MacPherson, A. H. 1969. *The dynamics of Canadian arctic fox populations*. Canadian Wildlife Service Report, ser. 8. Ottawa: Department of Indian Affairs and Northern Development.

Madison, D. 1978. Movement indicators of reproductive events among female meadow voles as revealed by radio-telemetry. *J. Mammal.* 59:835–43.

Malcolm, J. R. 1979. Social organization and communal rearing in African wild dogs. Ph.D. diss., Harvard University.

Maliniak, E., and Eisenberg, J. F. 1971. The breeding of *Proechimys semispinosus* in captivity. *Int. Zoo Yearb.* 11:93–98.

Mallinson, J. 1973. The reproduction of the African civet *Viverra civetta* at Jersey Zoo. *Int. Zoo Yearb.* 13:147–50.

Mallory, F. F., and Brooks, R. J. 1978. Infanticide and other reproductive strategies in the collared lemming (*Dicrostonyx groenlandicus*). *Nature* (London) 273:144–46.

Malpas, R. C. 1978. The ecology of the African elephant in Ruwenzori and Kabalega Falls National Parks, Uganda. Ph.D. thesis, Cambridge University.

Mandal, A. K., and Biswas, S. 1970. Notes on the behaviour of Szechuan burrowing shrew *Anourosorex squamipes squamipes* Milne-Edwards from Khasi Hills, Assam. *J. Bengal Nat. Hist. Soc.* 36(2):193–99.

Mann, G. 1963. Phylogeny and cortical evolution in Chiroptera. *Evolution* 17:589–91.

Mansfield, A. W. 1958. The breeding behavior and reproductive cycle of the Weddell seal (*Leptonychotes weddelli* Lesson). *FIDS Sci. Rep.*, no. 18.

Manville, R. H. 1959. The Columbian ground squirrel in northwestern Montana. *J. Mammal.* 40:26–44.

Mares, M. A. 1975. South American mammal zoogeography: Evidence from convergent evolution in desert rodents. *Proc. Nat. Acad. Sci.* 72(5):1702–6.

———. 1976. Convergent evolution of desert rodents: Multivariate analysis and zoogeographic implications. *Paleobiology* 2:39–63.

Marler, P., and Hamilton, W. J., III. 1966. *Mechanisms of animal behavior.* New York: John Wiley.

Marler, P., and Tenaza, R. 1977. Signaling behavior of apes with special reference to vocalization. In *How animals communicate*, ed. T. A. Sebeok, pp. 965–1033. Bloomington: Indiana University Press.

Marlow, B. J. 1961. Reproductive behaviour of the marsupial mouse, *Antechinus flavipes* (Waterhouse) (Marsupialia) and the development of the pouch young. *Australian J. Zool.* 9(2):203–17.

———. 1967. Mating behaviour in the leopard seal, *Hydrurga leptonyx* (Mammalia: Phocidae), in captivity. *Australian J. Zool.* 15:1–5.

———. 1975. The comparative behaviour of the Australasian sea lions *Neophoca cinerea* and *Phocarctos hookeri* (Pinnipedia: Otariidae). *Mammalia* 39(2):159–230.

Marshall, A. J. 1947. The breeding cycles of an equatorial bat (*Pteropus giganteus*) of Ceylon. *Proc. Linnean Soc.* 159:103–11.

Marshall, J. T. 1977. A synopsis of Asian species of *Mus* (Rodentia, Muridae). *Bull. Amer. Mus. Nat. Hist.*, vol. 158, art. 3.

Marshall, J. T., Jr.; Ross, B. A.; and Chantharozrong, S. 1972. The species of gibbons in Thailand. *J. Mammal.* 53:479–86.

Marshall, W. H.; Gullion, G. W.; and Scharb, R. G. 1962. Early summer activities of porcupines as determined by radio-positioning techniques. *J. Wildl. Manage.* 26:75–79.

Martin, P. 1971. Movements and activities of the mountain beaver (*Aplodontia rufa*). *J. Mammal.* 52:717–23.

Martin, P. S., and Wright, H. E., eds. 1967. *Pleistocene extinctions: The search for a cause.* Proceedings of the Seventh Congress of the International Association of Quaternary Research, Yale University. New Haven: Yale University Press.

Martin, R. D. 1966. Tree shrews: Unique reproductive mechanism of systematic importance. *Science* 152:1402–4.

———. 1968. Reproduction and ontogeny in tree shrews (*Tupaia belangeri*) with reference to their general behavior and taxonomic relationships. *Z. Tierpsychol.* 25:409–95.

———. 1972. A preliminary field study of the lesser mouse lemur (*Microcebus murinus* J. F. Miller 1777). *Z. Tierpsychol.* suppl. 9:43–89.

———. 1975. Breeding tree-shrews and mouse lemurs in captivity. *Int. Zoo. Yearb.* 15:35–41.

Martin, R. D.; Doyle, G. A.; and Walker, A. C., eds. 1974. *Prosimian biology.* London: Duckworth.

Martin, R. J. 1973. Growth curves for bushy-tailed woodrats based upon animals raised in the wild. *J. Mammal.* 54:517–18.

Mason, W. A. 1968. Use of space by *Callicebus* groups. In *Primates: Studies in adaptation and variability,* ed. P. C. Jay, pp. 200–216. New York: Holt, Rinehart and Winston.

Maynard-Smith, J. 1978. *The evolution of sex.* Cambridge: Cambridge University Press.

Mayr, E. 1963. *Animal species and evolution.* Cambridge: Belknap Press of Harvard University Press.

Mead, J. G. 1975. Anatomy of the external nasal passages and facial complex in the Delphinidae (Mammalia: Cetacea). *Smithsonian Contrib. Zool.* 207:1–12.

Mead, R. A. 1968. Reproduction in western forms of the spotted skunk (*Spilogale*). *J. Mammal.* 49:373–90.

Meagher, M. M. 1973. *The bison of Yellowstone National Park.* National Park Service Scientific Monograph Series no. 1. Washington, D.C.: Government Printing Office.

Mech, L. D. 1970. *The wolf: The ecology and behavior of an endangered species.* New York: American Museum of Natural History.

Mech, L. D.; Barnes, D. M.; and Tester, J. R. 1968. Seasonal weight changes, mortality and population structure of raccoons in Minnesota. *J. Mammal.* 49(1):63–74.

Medway, Lord. 1967. Observations on breeding of the pencil-tailed tree mouse, *Chiropodomys gliroides. J. Mammal.* 48(1):20–27.

———. 1972. Reproductive cycles of the flat-headed bats, *Tylonycteris pachypus* and *T. robustula* (Chiroptera: Vespertilioninae) in a humid equatorial environment. *Zool. J. Linnean Soc.* 51:33–61.

———. 1974. Food of a tapir, *Tapirus indicus. Malay Nat. J.* 28:90–93.

———. 1977. *Mammals of Borneo.* 2d ed. Monographs of the Malaysian Branch of the Royal Asiatic Society, no. 7. Singapore.

———. 1978. *The wild mammals of Malaya and Singapore.* 2d ed. London: Oxford University Press.

Medway, Lord, and Marshall, A. G. 1970. Roost-site selection among flat-headed bats (*Tylonycteris* spp.). *J. Zool.* (London) 161:237–45.

———. 1972. Roosting associations and flat-headed bats, *Tylonycteris* sp. (Chiroptera: Vespertilionidae) in Malaysia. *J. Zool.* (London) 168(4):463–82.

Meester, J., and Setzer, H. W. 1971. *The mammals of Africa: An identification manual.* Washington, D.C.: Smithsonian Institution Press.

Meggers, B. J. 1971. *Amazonas: Man and culture in a counterfeit paradise.* New York: Aldine.

Melton, D. A. 1976. The biology of aardvarks (Tubulidentata-Orycteropodidae). *Mammal. Rev.* 6(2):75–88.

Meltzer, A. 1967. The rock hyrax (*Procavia capensis syriaca* Schreber 1784). M.Sc. thesis, Tel Aviv University.

Mendelssohn, H. 1965. Breeding the Syrian hyrax, *Procavia capensis syriaca* Schreber 1784. *Int. Zoo Yearb.* 5:116–25.

Mentis, M. T. 1972. A review of some life history features of the large herbivores of Africa. *Lammergeyer* 16:1–89.

Menzies, J. I. 1973. A study of leaf-nosed bats (*Hipposideros caffer* and *Rhinolophus landeri*) in a cave in northern Nigeria. *J. Mammal.* 54(4):930–46.

Meritt, D. A., Jr. 1972. In nature the behavior of armadillos. In *Yearbook of the American Philosophical Society, Report for 1971–1972,* pp. 32–33. Philadelphia: American Philosophical Society.

———. 1976. The La Plata three-banded armadillo (*Tolypeutes matacus*) in captivity. *Int. Zoo. Yearb.* 16:153–56.

Merwe, M. van der. 1978. Foetal growth curves and seasonal breeding in the Natal clinging bat, *Miniopterus schreibersi natalensis. S. African J. Zool.* 14:17–21.

Meyer, B. J., and Meyer, R. K. 1944. Growth and reproduction of the cotton rat, *Sigmodon hispidus hispidus,* under laboratory conditions. *J. Mammal.* 25:107–29.

Meyer, M. P., and Morrison, P. 1960. Tissue respiration and hibernation in the thirteen-lined ground-squirrel, *Spermophilus tridecemlineatus*. In *Mammalian hibernation*, ed. C. P. Lyman and A. R. Dawe, pp. 405–21. Bulletin of the Museum of Comparative Zoology, vol. 124. Cambridge: Harvard University Press.

Meyer-Holzapfel, M. 1956. Das Spiel bei Säugetieren. *Handb. Zool.*, Band 8, Teil 19(5):1–36.

———. 1957. Das Verhalten der Bären. *Handb. Zool.*, Band 8, Teil 10(17):1–28.

———. 1968. Breeding the European wild-cat *Felis s. silvestris* at Berne Zoo. *Int. Zoo Yearb.* 8:31–38.

Michael, R. P., and Crook, J. H., eds. 1973. *Comparative ecology and behavior of Primates*. London/New York: Academic Press.

Michener, G. R. 1969. Notes on the breeding and young of the crest-tailed marsupial mouse, *Dasycercus cristicauda*. *J. Mammal.* 50:633–35.

Michielsen-Kroin, N. 1966. Intraspecific and interspecific competition in the shrews *Sorex araneus* L. and *S. minutus* L. *Archs. Néerl. Zool.* 17:73–174.

Millar, J. S. 1975. Tactics of energy partitioning in breeding *Peromyscus*. *Canadian J. Zool.* 53:967–76.

———. 1976. Adaptive features of mammalian reproduction. *Evolution* 31:370–86.

Miller, E. H. 1975. Walrus ethology. I. The social role of tusks and applications of multidimensional scaling. *Canadian J. Zool.* 53:590–613.

Miller, R. S. 1964. Ecology and distribution of pocket gophers (Geomyidae) in Colorado. *Ecology* 45(2):256–71.

Mills, M. G. L. 1978. Foraging behaviour of the brown hyaena (*Hyaena brunnea* Thunberg 1820) in the southern Kalahari. *Z. Tierpsychol.* 48:113–41.

Milton, K. 1978. Behavioral adaptations to leaf eating by the mantled howler monkey (*Alouatta palliata*). In *The ecology of arboreal folivores*, ed. G. G. Montgomery, pp. 535–49. Washington, D.C.: Smithsonian Institution Press.

Mitchell, B. L. 1965. An unexpected association between a plant and an insectivorous animal. *Puku* 3:178.

Mitchell, J. 1973. Determination of relative age in the dugong *Dugong dugon* (Müller) from a study of skulls and teeth. *Zool. J. Linnean Soc.* 53(1):1–23.

Moehlman, P. 1974. Behavior and ecology of feral asses (*Equus asinus*). Ph.D. thesis, University of Wisconsin.

———. 1979. Jackal helpers and pup survival. *Nature* (London) 277:382–83.

Moeller, H. 1975. Sind die Beutler den plazentalen Säugern unterlegen? *Säugetierk. Mitt.* 23(1):19–29.

Moeller, W. 1968. Allometrische Analyse der Gürteltierschädel: Ein Beitrag zur Phylogenie der Dasypodidae. *Zool. Jahrb.* 85(3):411–528.

Moen, A. N. 1973. *Wildlife ecology: An analytical approach*. San Francisco: W. H. Freeman.

Mohr, E. 1936. Biologische Beobachtungen an *Solenodon paradoxus* Brandt in Gefangenschaft. *Zool. Anz.* 113:177–88; 116:65–76; 117:233–41; 122:132–43.

———. 1939. Die Baum- und Ferkelratten-Gattungen *Capromys* Desmarest (sens. ampl.) und *Plagiodontia* Cuvier. *Mitt. Hamburgischen Zool. Mus. Institut in Hamburg* 48:49–118.

———. 1954. *Die freilebenden Nagetiere Deutschlands und der Nachbarlander.* Jena: Gustav Fischer.

———. 1956. Das Verhalten der Pinnipedier. *Handb. Zool.* Band 8, Teil 10(22):1–20.

———. 1957. *Sirenen oder Seekühe.* Wittenberg: Ziemsen.

———. 1963. Beiträge zur Naturgeschichte der Klappmütze, *Cystophora cristata* Erxl. 1777. *Z. Säugetierk.* 28(1):19–37.

———. 1965. *Altweltliche Stachelschweine.* Stuttgart: Franckh'sche Verlag.

———. 1970. *Wilde Schweine* Wittenberg: Ziemsen.

Möhres, F. P., and Neuweiler, G. 1966. Die Ultraschallorientierung des Grossblatt-Fledermäuse (Chiroptera-Megadermatidae). *Z. Vergl. Physiol.* 53:195–227.

Moisan, G. 1958. *La caribou de la Gaspé sie.* Montreal: Société Zoologique de Québec.

Mondolfi, E. 1974. Taxonomy, distribution and status of the manatee in Venezuela. *Mem. Soc. Ciencias Nat. La Salle* 24(97):1–85.

Montgomery, G. G. 1964. Tooth eruption in pre-weaned raccoons. *J. Wildl. Manage.* 28:582–94.

———. 1969. Weaning of captive raccoons. *J. Wildl. Manage.* 33(1):154–59.

———, ed. 1978. *The ecology of arboreal folivores.* Washington, D.C.: Smithsonian Institution Press.

Montgomery, G. G., and Lubin, Y. D. 1977. Prey influences on movement of Neotropical anteaters. In *Proceedings of the 1975 Predator Symposium,* ed. R. Phillips and C. Jonkel, pp. 103–31. Missoula: Montana Forest and Conservation Experiment Station, University of Montana.

———. 1978. Impact of anteaters (*Tamandua* and *Cyclopes,* Edentata: Myrmecophagidae) on arboreal ant populations. Abstr. 111, p. 56, American Society of Mammalogists, Abstracts of Technical Papers, 58th Annual Meeting.

Montgomery, G. G., and Sunquist, M. E. 1974. Contact-distress calls of young sloths. *J. Mammal.* 55:211–13.

———. 1975. Impact of sloths on Neotropical forest energy flow and nutrient cycling. In *Tropical ecological systems: Trends in terrestrial and aquatic research,* ed. F. B. Golley and E. Medina, pp. 69–98. New York: Springer.

———. 1978. Habitat selection and use by two-toed and three-toed sloths. In *The ecology of arboreal folivores,* ed. G. G. Montgomery, pp. 329–59. Washington, D.C.: Smithsonian Institution Press.

Moore, J. C. 1956. Observations of manatees in aggregations. *Amer. Mus. Novit.* 1811:1–24.

Moors, P. J. 1974. The foeto-maternal relationship and its significance in marsupial reproduction: A unifying hypothesis. *J. Australian Mamm. Soc.* 1:263–66.

Morgan, T. 1868. *The beaver and his works.* Philadelphia: Lippincott.

Morris, B. 1966. Breeding the European hedgehog *Erinaceus europaeus* in captivity. *Int. Zoo Yearb.* 6:141–46.

Morris, R., and Morris, D. 1966. *Men and pandas.* London: Hutchinson.

Morrison, P., and McNab, B. K. 1962. Daily torpor in a Brazilian murine opossum (*Marmosa*). *Comp. Biochem. Physiol.* 6:57–68.

Morton, E. S. 1977. On the occurrence and significance of motivation-structural rules in some bird and mammal sounds. *Amer. Nat.* 111(381):855–69.

Morton, M. L., and Tung, H. L. 1971. Growth and development in the Belding ground squirrel (*Spermophilus beldingi beldingi*). *J. Mammal.* 52:611–16.

Morton, S. R. 1978*a*. An ecological study of *Sminthopsis crassicaudata* (Marsupialia: Dasyuridae). I. Distribution, study areas and methods. *Australian Wildl. Res.* 5:151–62.

———. 1978*b*. An ecological study of *Sminthopsis crassicaudata* (Marsupialia: Dasyuridae). II. Behaviour and social organization. *Australian Wildl. Res.* 5:163–82.

———. 1978*c*. An ecological study of *Sminthopsis crassicaudata* (Marsupialia: Dasyuridae). III. Reproduction and life history. *Australian Wildl. Res.* 5:181–211.

Moynihan, M. 1964. Some behavior patterns of platyrrhine monkeys. I. The night monkey (*Aotus trivirgatus*). *Smithsonian Misc. Coll.* 135(5):1–84.

———. 1966. Communication in the Titi monkey, *Callicebus*. *J. Zool. Soc. London* 150:77–127.

———. 1970. Some behavior patterns of platyrrhine monkeys. II. *Saguinus geoffroyi* and some other tamarins. *Smithsonian Contrib. Zool* 28:1–77.

———. 1976. *The New World Primates.* Princeton: Princeton University Press.

Muckenhirn, N. A. 1972. Leaf-eaters and their predator in Ceylon: Ecological roles of gray langurs, *Presbytis entellus,* and leopards. Ph.D. diss., University of Maryland.

Muckenhirn, N. A., and Eisenberg, J. F. 1973. Home ranges and predation of the Ceylon leopard. In *The World's Cats,* ed. R. L. Eaton, 1:142–75. Winston, Ore.: World Wildlife Safari.

Mullen, D. A. 1968. Reproduction in brown lemmings (*Lemmus trimucronatus*) and its relevance to their cycle of abundance. *Univ. Calif. Publ. Zool.* 85:1–24.

Müller-Schwarze, D. 1968. Play deprivation in deer. *Behaviour* 31(1–2):144–63.

———. 1969. Complexity and relative specificity in a mammalian pheromone. *Nature* (London) 223:525–26.

Müller-Schwarze, D., and Mozell, M. M., eds. 1977. *Chemical signals in vertebrates*. New York/London: Plenum Press.

Müller-Schwarze, D., and Müller-Schwarze, C. 1973. Behavioral development of hand-reared pronghorn *Antilocapra americana*. *Int. Zoo. Yearb.* 13:217.

Müller-Using, D. 1956. Zum verhalten des Murmeltiers (*Marmota marmota* L.). *Z. Tierpsychol.* 13:135–42.

Mundy, K. R. D., and Flook, D. R. 1963. Notes on the mating activity of grizzly and black bears. *J. Mammal.* 44:637–38.

Munn, N. L. 1950. *Handbook of psychological research on the rat*. New York: Houghton Mifflin.

Murie, A. 1940. *Ecology of the coyote in the Yellowstone*. Fauna Series no. 5. Washington, D.C.: U.S. Department of Agriculture.

———. 1944. *The wolves of Mount McKinley*. Fauna of the National Parks of the U.S., Fauna Series no. 5. Washington, D.C.: National Park Service.

Murie, O. 1951. *The elk of North America*. Harrisburg, Pa.: Stackpole.

Murphy, M. R. 1973. Effects of female hamster vaginal discharge on the behavior of male hamsters. *Behav. Biol.* 9(3):367–75.

Mutere, F. A. 1967. The breeding biology of equatorial vertebrates: Reproduction in the fruit bat, *Eidolon helvum*, at latitude 0° 21′ N. *J. Zool.* (London) 153(1):153–63.

Muul, I. 1968. Behavioral and physiological influences on the distribution of the flying squirrel, *Glaucomys volans*. *Misc. Publ. Mus. Zool., Univ. Michigan*, 134:1–66.

———. 1969. Mating behavior, gestation period, and development of *Glaucomys sabrinus*. *J. Mammal.* 50(1):121.

Muul, I., and Lim Boo Liat. 1978. Comparative morphology, food habits and ecology of some Malaysian arboreal rodents. In *The ecology of arboreal folivores*, ed. G. G. Montgomery, pp. 361–69. Washington, D.C.: Smithsonian Institution Press.

Muybridge, E. 1887. *Animal in motion*, ed. L. S. Brown. Reprinted 1957. New York: Dover Publications.

Myers, J. H., and Krebs, C. J. 1971. Genetic, behavioral and reproductive attributes of dispersing field voles *Microtus pennsylvanicus* and *Microtus ochrogaster*. *Ecol. Monogr.* 41:53–78.

Mykytowycz, R. 1966*a*. Observations on odoriferous and other glands in the Australian wild rabbit *Oryctolagus cuniculus* (L.) and the hare *Lepus europaeus*. 1. The anal gland. *CSIRO Wildl. Res.* 11:11–29.

———. 1966*b*. Observations on odoriferous and other glands in the Australian wild rabbit *Oryctolagus cuniculus* (L.) and the hare *Lepus europaeus*. 2. The inguinal glands. *CSIRO Wildl. Res.* 11:49–64.

———. 1966*c*. Observations on odoriferous and other glands in the Australian wild rabbit *Oryctolagus cuniculus* (L.) and the hare *Lepus europaeus*. 3. Harder's lachrymal and submandibular glands. *CSIRO Wildl. Res.* 11:65–90.

Nader, I. A. 1968. Breeding records of the long-eared hedgehog *Hemiechinus auritus* (Gmelin). *Mammalia* 32(3):528–30.

Napier, J. R., and Napier, P. H. 1967. *A handbook of living Primates.* New York: Academic Press.
Naumov, N. P., and Lobachev, V. S. 1975. Ecology of the desert rodents of the U.S.S.R. In *Rodents in desert environments,* ed. I. Prakash and P. K. Gosh, pp. 465–590. The Hague: W. Junk.
Neal, B. J. 1959. A contribution to the life history of the collared peccary in Arizona. *Amer. Midl. Nat.* 61:177–90.
Neal, E. G. 1958. *The badger.* Harmondsworth: Penguin Books.
———. 1970. The banded mongoose, *Mungos mungo* Gmelin. *E. African Wildl. J.* 8:63–71.
———. 1977. *Badgers.* Poole, Dorset, England: Blandford Press.
Neal, E. G., and Harrison, R. J. 1958. Reproduction in the badger (*Meles meles*). *Trans. Zool. Soc. London* 29:67–130.
Negus, N. C.; Gould, E.; and Chipman, R. K. 1961. Ecology of the rice rat, *Oryzomys palustris* (Harlan), on Breton Island, Gulf of Mexico, with a critique of the social stress theory. *Tulane Studies Zool.* 8(4):95–123.
Nel, J. A. J. 1978. Notes on the food and foraging behavior of the bat-eared fox *Ocotyon megalotis.* In *Ecology and taxonomy of African small mammals,* ed. D. A. Schilitter. Bulletin no. 6. Pittsburgh: Carnegie Museum of Natural History.
Nelson, J. E. 1964. Vocal communication in Australian flying foxes (Pteropodidae: Megachiroptera). *Z. Tierpsychol.* 27:857–70.
———. 1965*a*. Movements of Australian flying foxes (Pteropodidae: Megachiroptera). *Australian J. Zool.* 13:53–73.
———. 1965*b*. Behaviour of Australian Pteropodidae (Megachiroptera). *Anim. Behav.* 13:544–57.
Nemoto, T. 1959. Food of baleen whales with reference to whale movements. *Sci. Rep. Whales Res. Inst.* (Tokyo) 14:149–290.
Nemoto, T., and Yoo, K. I. 1970. An amphipod as a food of the Antarctic sei whale. *Sci. Rep. Whales Res. Inst.* (Tokyo) 22:153–58.
Neville, M. K. 1972. The population structure of red howler monkeys (*Alouatta seniculus*) in Trinidad and Venezuela. *Folia Primat.* 17:56–86.
Nevo, E. 1969. Mole rat *Spalax ehrenbergi:* Mating behavior and its evolutionary significance. *Science* 163:484–86.
———. 1979. Adaptive convergence and divergence of subterranean mammals. *Ann. Rev. Ecol. Syst.* 10:269–308.
Nevo, E., and Amir, E. 1964. Geographic variation in reproduction and hibernation patterns of the forest dormouse. *J. Mammal.* 45:69–87.
Nevo, E.; Naftali, G.; and Guttman, R. 1975. Aggression patterns and speciation. *Proc. Nat. Acad. Sci. USA, Zool.* 72(8):3250–54.
Newsome, A. E. 1965. The distribution of red kangaroos, *Megaleia rufa* (Desmarest), plant sources of persistent food and water in central Australia. *Australian J. Zool.* 13:289–99.
Newsome, A. E., and Corbett, L. K. 1975. Outbreaks of rodents in semiarid and arid Australia. In *Rodents in desert environments,* ed. I. Prakash and P. K. Gosh, pp. 117–53. The Hague: W. Junk.

Newson, J. 1964. Reproduction and prenatal mortality of snowshoe hares on Manitoulin Island, Ontario. *Canadian J. Zool.* 10:225–67.

Neyman, P. F. 1977. Aspects of the ecology and social organization of free-ranging cotton-top tamarins (*Saguinus oedipus*) and the conservation status of the species. In *The biology and conservation of the Callitrichidae,* ed. D. G. Kleiman, pp. 39–73. Washington, D.C.: Smithsonian Institution Press.

Nicholson, A. J. 1941. The homes and social habits of the wood mouse *P. leucopus noveboracensis* in southern Michigan. *Amer. Midl. Nat.* 25:196–223.

Nikulin, P. G. 1940. Walrus: Bulletin of the Pacific Science Institute of Fisheries and Oceanography: NKRP-USSR. *Mar. Mammal. Far East* 20:21–59.

Nishiwaki, M. 1967. Distribution and migration of marine mammals in the north Pacific area. *Bull. Ocean Res. Inst.* 1:1–64.

Nissen, H. W. 1931. A field study of chimpanzees. *Psychol. Monogr.* 8:1–122.

Norris, K. S., ed. 1966. *Whales, dolphins and porpoises.* Berkeley: University of California Press.

Norris, K. S., and Prescott, J. H. 1961. Observations on Pacific cetaceans of California and Mexican waters. *Univ. California Publ. Zool.* 63:391–402.

Novick, A. 1958. Orientation in Paleotropical bats. I. Microchiroptera. *J. Exp. Zool.* 138:81–154.

———. 1963. Pulse duration in the echolocation of insects by the bat, *Pteronotus. Ergeb. Biol.* 26:21–26.

Oates, J. F. 1977. The social life of a black-and-white *Colobus* monkey, *colobus guereza. Z. Tierpsychol.* 45(1):1–60.

Oboussier, H. 1972. Evolution of the mammalian brain: Some evidence on the phylogeny of the antelope species. *Acta Anat.* 83:70–80.

O'Connell, M. A. 1979. The ecology of didelphid marsupials in northern Venezuela. In *Vertebrate ecology in the northern Neotropics,* ed. J. F. Eisenberg, pp. 73–87. Washington, D.C.: Smithsonian Institution Press.

Odum, E. P. 1971. *Fundamentals of ecology.* Philadelphia/London/Toronto: W. B. Saunders.

Oeming, A. 1966. A herd of musk-oxen, *Ovibos moschatus,* in captivity. *Int. Zoo Yearb.* 6:58–65.

O'Farrell, M. J., and Studier, E. H. 1973. Reproduction, growth and development in *Myotis thysanodes* and *M. lucifugus* (Chiroptera: Vespertilionidae). *Ecology* 54:18–30.

Oftedal, O. 1980. Milk and mammalian evolution. In *Comparative physiology: Primitive mammals,* ed. K. Schmidt-Nielson, L. Bolis, and C. Richard Taylor, pp. 31–42. Cambridge: Cambridge University Press.

Ognev, S. I. 1963. *Mammals of the U.S.S.R. and adjacent countries.* Vols. 5 and 6. *Rodents.* Jerusalem: Israel Program for Scientific Translations.

Ohsumi, S. 1971. Some investigations on school structure of sperm whales. *Sci. Rep. Whales Res. Inst. Tokyo* 23:1–25.
Ojasti, J. 1973. *Estudio biologico del chigüire o capibara.* Caracas: Fundo Nacional de Investigaciones Agropecuarias.
Olds, J. M. 1950. Notes on the hood seal (*Cystophora cristata*). *J. Mammal.* 31:450–52.
Olds, T. J., and Collins, L. R. 1973. Breeding Matschie's tree kangaroo *Dendrolagus matschiei* in captivity. *Int. Zoo Yearb.* 13:123–25.
Olivier, R. C. D. 1978. On the ecology of the Asian elephant, *Elephas maximus* Linn., with particular reference to Malaya and Sri Lanka. Ph.D. diss., Cambridge University.
Oppenheimer, J. R. 1968. Behavior and ecology of the white-faced monkey, *Cebus capucinus,* on Barro Colorado Island, C.Z. Ph.D. diss., University of Illinois, Urbana.
———. 1977. Communication in New World monkeys. In *How animals communicate,* ed. T. A. Sebeok, pp. 851–89. Bloomington: Indiana University Press.
Orians, G. H. 1969. On the evolution of mating systems in birds and mammals. *Amer. Nat.* 103(934):589–603.
Orr, R. T. 1940. *The rabbits of California.* Occasional Papers, 19. San Francisco: California Academy of Sciences.
———. 1954. Natural history of the pallid bat, *Antrozous pallidus. Proc. California Acad. Sci.* 28:165–246.
———. 1965. Interspecific behavior among pinnipeds. *Z. Säugetierk.* 30:163–70.
———. 1967. The Galapagos sea lion. *J. Mammal.* 48(1):62–70.
Osgood, W. H. 1943. A new genus of rodents from Peru. *J. Mammal.* 24:369–71.
———. 1943. *The mammals of Chile.* Zoological Series, vol. 30. Chicago: Field Museum of Natural History.
Owen-Smith, N. 1972. Territoriality: The example of the white rhinoceros. *Zool. Africana* 7(1):273–80.
———. 1975. The social ethology of the white rhinoceros *Ceratotherium simum* Burchell 1817. *Z. Tierpsychol.* 38:337–84.
Owens, M. J., and Owens, D. D. 1978. Feeding ecology and its influence on social organization in brown hyaenas (*Hyaena brunnea* Thunberg) of the Central Kalahari Desert. *E. African Wildl. J.* 16:112–36.
Packer, C. 1979. Inter-troop transfer and inbreeding avoidance in *Papio anubis. Anim. Behav.* 27:1–36.
Packer, W. C. 1969. Observations on the behavior of the marsupial *Setonix brachyurus* (Quoy and Gaimard) in an enclosure. *J. Mammal.* 50:8–21.
Pages, E. 1965. Notes sur les pangolins du Gabon. *Biol. Gabonica* 1(3):209–338.
———. 1970. Sur l'écologie et les adaptations de l'orycterope et des pangolins sympatriques du Gabon. *Biol. Gabonica* 6(1):27–92.

———. 1972*a*. Comportement aggressif et sexuel chez les pangolins arboricoles (*Manis tricuspis* et *M. longicaudata*). *Biol. Gabonica* 8(1):1–62.

———. 1972*b*. Comportement maternel et developpement du jeune chez un pangolin arboricole (*M. tricuspis*). *Biol. Gabonica* 8(1):63–120.

———. 1976. Etude éco-éthologique de *Manis tricuspis* par radio-tracking. *Mammalia* 39(4):613–41.

Paradiso, J. 1969. *Mammals of Maryland*. North American Fauna, no. 66. Washington, D.C.: U.S. Department of Interior.

Parker, P. 1977. An ecological comparison of marsupial and placental patterns of reproduction. In *The biology of marsupials*, ed. B. Stonehouse and D. Gilmore, pp. 273–86. Baltimore: University Park Press.

Parra, R. 1978. Comparison of foregut and hindgut fermentation in herbivores. In *The ecology of arboreal folivores*, ed. G. G. Montgomery, pp. 205–29. Washington, D.C.: Smithsonian Institution Press.

Patterson, B. 1965. The fossil elephant shrews (family Macroscelididae). *Bull. Mus. Comp. Zool. Harvard* 133:295–335.

Patterson, B., and Pasqual, R. 1972. The fossil mammal fauna of South America. In *Evolution, mammals and southern continents*, ed. A. Keast, F. C. Erk, and B. Glass, pp. 247–309. Albany: State University of New York Press.

Patton, J. L. 1970. Karyotypic variation following an elevational gradient in the pocket gopher, *Thomomys bottae grahamensis* Goldman. *Chromosoma* 31:41–50.

Payne, R. S., and McVay, S. 1971. Songs of humpback whales. *Science* 173:585–97.

Pearson, A. M. 1975. *The northern interior grizzly bear Ursus arctos L.* Ottawa: Canadian Wildlife Service Report Series, 34.

Pearson, O. P. 1944. Reproduction in the shrew (*Blarina brevicauda* Say). *Amer. J. Anat.* 75:39–93.

———. 1948. Life history of mountain viscachas in Peru. *J. Mammal.* 29:345–74.

———. 1959. Biology of the subterranean rodents, *Ctenomys*, in Peru. *Mem. Mus. Hist. Nat. "Javier Prado"* 9:1–56.

Pearson, O. P.; Binstein, N.; Boiry, L.; Bush, C.; di Pace, M.; Gallopin, G.; Penchaszadeh, P.; and Piantanida, M. 1968. Estructura social, distribución espacial y composición por edades de una población de Tuco-tucos (*Ctenomys talarum*). *Inv. Zool. Chilenas* 13:47–80.

Pearson, O. P.; Koford, M. R.; and Pearson, A. K. 1952. The reproduction of the lump-nosed bat (*Corynorhinus rafinesquei*) in California. *J. Mammal.* 33:273–320.

Pedersen, A. 1957. *Der Eisbär*. Wittenberg: Zeimsen.

———. 1959. *Der Eisfuchs, Alopex lagopus Linne*. Wittenberg: Ziemsen.

———. 1962. *Das Walross*. Wittenberg: Ziemsen.

Pepper, S. 1961. *World hypotheses*. Berkeley: University of California Press.

Perrin, W. F.; Coe, J. M.; and Zwiefel, J. R. 1976. Growth and reproduction of the spotted porpoise *Stenella attenuata*. *Fish. Bull.* 74:229–69.

Peters, G. 1978. Vergleichende Untersuchung zur Lautgebung einiger Feliden (Mammalia, Felidae). *Spixiana,* suppl. 1.

Peterson, R. S. 1969. Airborne vocal communication in the California sea lion, *Zalophus californianus*. *Anim. Behav.* 17:17–24.

Peterson, R. S., and Bartholomew, G. A. 1967. *The natural history and behavior of the California sea lion*. Special Publication no. 1. American Society of Mammalogists.

Peterson, R. S.; Hubbs, C. L.; Gentry, R. L.; and DeLong, R. L. 1968. The Guadalupe fur seal: Habitat, behavior, population size and field identification (124). *J. Mammal.* 49(4):665–75.

Petter, F. 1961. Répartition géographique et écologie des rongeurs désertique. *Mammalia* 24:1–219.

———. 1967. Contribution à la faune du Congo (Brazzaville) Mission A. Villiers et A. Descarpentries. Zv. Mammifères rongeurs (Muscardinidae et Muridae). *Bull. Inst. Afrique Noire* 29A:815–20.

———. 1972. The rodents of Madagascar: The seven general of Malagasy rodents. In *Biogeography and ecology in Madagascar,* ed. R. Battistini and G. Richard-Vindard, pp. 661–82. The Hague: W. Junk.

Petter, F.; Chippaux, A.; and Monmignaut, C. 1964. Observations sur la biologie, la reproduction et la croissance de *Lemniscomys striatus* (rongeurs, muridés). *Mammalia* 28:620–27.

Petter, J. J., and Hladik, C. M. 1970. Observations sur le domaine vital et la densité de population de *Loris tardigradus* dans le forêts de Ceylan. *Mammalia* 3:394–409.

Pfeffer, P. 1967. Le mouflon de Corse (*Ovis ammon musimon* Schreber 1782): Position systématique, écologie et éthologie comparées. *Mammalia* 31:1–262.

Pianka, E. R. 1972. r and K selection or b and d selection. *Amer. Nat.* 106:581–88.

Pielou, E. C. 1969. *An introduction to mathematical ecology*. New York: John Wiley.

Pilleri, G. 1959. Beiträge zur vergleichenden Morphologie des Nagetiergehirnes. *Acta Anat.* 39(suppl.):1–124.

———. 1960*a*. Beiträge zur vergleichenden Morphologie des Nagetiergehirnes. *Acta Anat.* 42(suppl.):5–34.

———. 1960*b*. Beiträge zur vergleichenden Morphologie des Nagetiergehirnes. *Acta Anat.* 42(suppl.):36–68.

———. 1961*a*. Beiträge zur vergleichenden Morphologie des Nagetiergehirnes. *Acta Anat.* 44(suppl.):5–35.

———. 1961*b*. Beiträge zur vergleichenden Morphologie des Nagetiergehirnes. *Acta Anat.* 44(suppl.):36–83.

Pilleri, G., and Knuckey, J. 1969. Behavior patterns of some Delphinidae observed in the western Mediterranean. *Z. Tierpsychol.* 26(1):48–72.

Pilters, H. 1956. Das Verhalten der Tylopoden. *Handb. Zool.,* Band 8, Lieferung 2, 19(27):1–24.

Pilton, P. E., and Sharman, G. B. 1962. Reproduction in the marsupial *Trichosurus vulpecula*. *J. Endocrinol.* 25:119–36.

Pimlott, D. M. 1967. Wolf predation and ungulate populations. *Amer. Zool.* 7:267–78.

Pirlot, P., and Stephan, H. 1970. Encephalization in Chiroptera. *Canadian J. Zool.* 48:433–44.

Pocock, R. I. 1921. The external characters and classification of the Procyonidae. *Proc. Zool. Soc. London,* pp. 389–422.

———. 1928. Some external characters of the giant panda (*Ailuropoda melanoleuca*). *Proc. Zool. Soc. London,* pp. 975–81.

Poduschka, W. 1968. Über die Wahrnehmung von Ultraschall beim Igel, *Erinaceus europaeus roumanicus*. *Z. Vergl. Physiol.* 61:420–26.

———. 1969. Ergänzungen zum Wissen über *Erinaceus europaeus roumanicus* und kritische Überlegungen zur bisherigen Literatur über europäische Igel. *Z. Tierpsychol.* 26:761–804.

———. 1970. Das Selbstbespeicheln der Igel. Film CT 1320 der Bundesstaatl. Vienna: Hptst. wiss. Kinemat.

———. 1971. Zur Kenntnis des Nordafrikanischen Erdhörnchens, *Atlantoxerus getulus* (F. Major). *Zool. Gart.,* n.s. (Leipzig) 40(4/5):211–26.

———. 1974. Augendrüsensekretionen bei den Tenreciden *Setifer setosus* (Froriep 1806), *Echinops telfairi* (Martin 1838), *Microgale dobsoni* (Thomas 1918) und *Microgale talazaci* (Thomas 1918). *Z. Tierpsychol.* 35:303–19.

———. 1977. Das Paarungsvorspiel des Osteuropäischen Igels (*Erinaceus e. roumanicus*) und theoretische Überlegungen zum Problem männlicher Sexualpheromone. *Zool. Anz.* (Jena) 199(3/4):187–208.

Poduschka, W., and Firbas, W. 1968. Das Selbstbespeicheln des Igels, *Erinaceus europaeus* Linne 1758, steht in Beziehung zur Funktion des Jacobsonschen Organes. *Z. Säugetierk.* 33(3):160–72.

Poglayen-Neuwall, I. 1962. Beiträge zu einem Ethogramm des Wickelbären (*Potos flavus* Schreber). *Z. Säugetierk.* 27(1):1–44.

———. 1966. On the marking behavior of the kinkajou (*Potos flavus* Schreber). *Zoologica* 51(4):137–41.

———. 1973. Observations on the birth of a kinkajou, *Potos flavus* (Schreber 1774). *Zoologica* 58(spring):41–42.

———. 1975. Copulatory behavior, gestation and parturition of the tayra (*Eira barbara* L. 1758). *Z. Säugetierk.* 40:176–89.

———. 1976. Zur Fortpflanzungsbiologie und Jungendentwicklung von *Potos flavus*. *Zool. Gart.,* n.s. (Jena) 46:237–83.

Poglayen-Neuwall, I., and Poglayen-Neuwall, I. 1964. Gefangenschaftsbeobachtungen an Makibären (*Bassaricyon* Allen 1876). *Z. Säugetierk.* 39(6):321–66.

———. 1976. Postnatal development of tayras (Carnivora: *Eira barbara* L. 1758). *Zool. Beitr.* 22(3):345–405.

Pohl, A. 1967. Beiträge zur Ethologie und Biologie des Sonnendachses (*Helictis personata* Gray 1831) in Gefangenschaft. *Zool. Gart.* 33(5):16–247.

Poirier, F. E. 1969. The Nilgiri langur (*Presbytis johnii*) troop: Its composition, structure, function and change. *Folia Primat.* 10:20–47.

———. 1970. The Nilgiri langur (*Presbytis johnii*) of South India. In *Primate behavior: Developments in field and laboratory research,* ed. L. A. Rosenblum, pp. 251–383. New York/London: Academic Press.

Pollock, J. I. 1974. Field observations on *Indri indri:* A preliminary report. In *Lemur biology,* ed. I. Tattersall and R. W. Sussman, pp. 287–311. New York: Plenum Press.

———. 1977. The ecology and sociology of feeding in *Indri indri.* In *Primate ecology,* ed. T. H. Clutton-Brock, pp. 38–68. London/New York: Academic Press.

Pond, C. 1978. Morphological aspects and ecological significance of fat distribution in wild vertebrates. *Ann. Rev. Ecol. Syst.* 9:519–70.

Poole, T. B. 1966. Aggressive play in polecats. *Symp. Zool. Soc. London* 18:23–44.

Poole, T. B., and Dunstone, N. 1976. Underwater predatory behaviour of the American mink (*Mustela vison*). *J. Zool.* (London) 178:395–412.

Portman, A. 1965. Über die Evolution der Tragzeit bei Säugetieren. *Rev. Suisse Zool.* 72:658–66.

Pournelle, G. H. 1952. Reproduction and early postnatal development of the cotton mouse, *Peromyscus gossypinus gossypinus. J. Mammal.* 33:1–20.

———. 1965. Observations on birth and early development of the spotted hyena. *J. Mammal.* 46:503.

———. 1967. Observations on reproductive behaviours and early postnatal development of the proboscis monkey *Nasalis larvatus orientalis* at San Diego Zoo. *Int. Zoo Yearb.* 7:90–92.

———. 1968. Classification, biology and description of the venom apparatus of insectivores of the genera *Solenodon, Neomys* and *Blarina.* In *Venomous animals and their venoms,* ed. W. Bücherl, E. Buckley, and V. Deulofeu, pp. 31–42. New York: Academic Press.

Powell, R. A. 1978. A comparison of weasel and fisher hunting behavior. Carnivore 1:28–33.

Prakash, I., and Gosh, P. K., eds. 1975. *Rodents in desert environments.* The Hague: W. Junk.

Prakash, I.; Kametkar, L. R.; and Purohit, K. G. 1968. Home range and territoriality of the northern palm squirrel, *Funambulus pennanti* Wroughton. *Mammalia* 32(4):603–12.

Prechtl, H., and Schleidt, W. 1950. Auslösender und Steuernd Mechanismen des Säugaktes I. *Mitt. Z. Vergl. Physiol.* 32:257–62.

Prescott, J., and Ferron, J. 1978. Breeding and behavior development of the American red squirrel. *Int. Zoo Yearb.* 18:125–30.

Price, M. 1953. The reproductive cycle of the water shrew *Neomys fodiens bicolor* Shaw. *Proc. Zool. Soc. London* 128:599–621.

Procter, J. 1963. A contribution to the natural history of the spotted necked otter (*Lutra maculicollis*) in Tanganyika. *E. African Wildl. J.* 1:93–102.
Provost, E. E.; Nelson, C. A.; and Marshall, D. A. 1973. Population dynamics and behavior in the bobcat. In *The world's cats,* ed. R. L. Eaton, 1:42–67. Winston, Oreg.: World Wildlife Safari.
Pruitt, W. O. 1960. Behavior of the barren-ground caribou. *Biol. Pap. Univ. Alaska* 3:1–44.
Pruitt, W. O., Jr. 1954. Notes on a litter of young masked shrews. *J. Mammal.* 35:109–10.
Pucek, Z. 1958. Untersuchungen über Nestentwicklung und Thermoregulation bei einem Wurf von *Sicista betulina. Acta Theriol.* 2(2):11–54.
Purohit, K. G.; Kametkar, L. R.; and Prakash, I. 1966. Reproduction biology and post-natal development in the northern palm squirrel *Funambulus pennanti* Wroughton. *Mammalia* 30(4):538–47.
Pye, A. 1965. The auditory apparatus of the Heteromyidae (Rodentia, Sciuromorpha). *J. Anat.* (London) 99(1):161–74.
Pye, D., and Sales, G. 1974. *Ultrasonic communication by animals.* London: Chapman and Hall; New York: Wiley.
Quilliam, T. A., ed. 1966. The mole: Its adaptation to an underground environment. *J. Zool.* (London) 149:31–114.
Quimby, D. C. 1951. The life history and ecology of the jumping mouse, *Zapus hudsonicus. Ecol. Monogr.* 21:61–95.
Rabb, G. B. 1959*a*. Toxic salivary glands in the primitive insectivore *Solenodon. Nat. Hist. Misc.* 170:1–3.
———. 1959*b*. Reproductive and vocal behavior in captive pumas. *J. Mammal.* 49(4):616–17.
Rabinovich, J. E. 1975. Demographic strategies in animal populations. A regression analysis. In *Tropical ecological systems: Trends in terrestrial and aquatic research,* ed. F. B. Golley and E. Medina, pp. 19–40. New York: Springer.
Racey, P. A. 1973. Environmental factors affecting the length of gestation in heterothermic bats. *J. Reprod. Fertil.* 19:175–89.
Radinsky, L. B. 1969. Outlines of canid and felid brain evolution. *Ann. N.Y. Acad. Sci.* 167:277–88.
———. 1973. Are stink badgers skunks? Implications of neuroanatomy for mustelid phylogeny. *J. Mammal.* 54(3):585–94.
Raedeke, K. J. 1979. Sexual selection and dimorphism in the guanaco (*Lama guanicoe*). Abst. 73, in Abstracts of Technical Papers, 59th Annual Meeting, American Society of Mammalogists.
Rahm, U. 1960. Note sur les spécimens actuellement connus de *Micropotomogale (Mesopotomogale) ruwenzori* et leur répartition. *Mammalia* 24:511–15.
———. 1961*a*. Verhalten der Schuppentiere. *Handb. Zool.* VII 10(12d):32–48.
———. 1961*b*. Beobachtungen an der ersten in Gefangenschaft gehalten *Mesopotomogale ruwenzorii. Rev. Suisse Zool.* 68(1):73–90.

———. 1962. L'élevage et la reproduction en captivité de l'*Atherurus africanus*. *Mammalia* 26:1–9.
———. 1964. Das Verhalten der Klippschliefer (Hyracoidea). *Handb. Zool.*, Band 8, Teil 10(23b):1–23.
———. 1966. Les mammifères de la forêt équatoriale de l'est du Congo. *Ann. Mus. Roy. Africain Cent.*, ser. 8, *Zool.*, 149:37–121.
———. 1967. Les muridés des environs du Lac Kivu et des régions voisines (Afrique centrale) et leur écologie. *Rev. Suisse Zool.* 74:439–519.
———. 1969. Gestation period and litter size of the mole rat, *Tachyoryctes ruandae*. *J. Mammal.* 50:383–84.
———. 1971. Oekologie und Biologie von *Tachyoryctes ruandae* (Rodentia: Rhizomyidae). *Rev. Suisse Zool.* 78:623–38.
Rainey, D. G. 1956. Eastern woodrat, *Neotoma floridana:* Life history and ecology. *Publ. Mus. Nat. Hist. Univ. Kansas* 8:535–646.
———. 1965. Observations on the distribution and ecology of the white-throated wood rat in California. *Bull. California Acad. Sci.* 64:27–42.
Ralls, K. 1969. Scent-marking in Maxwell's duiker, *Cephalophus maxwelli*. Abst. 51. *Amer. Zool.* 9:1071.
———. 1973. *Cephalophus maxwelli*. *Mammal. Spec.* 31:1–4.
———. 1975. Agonistic behavior in Maxwell's duiker, *Cephalophus maxwelli*. *Mammalia* 39(2):241–49.
———. 1976. Mammals in which females are larger than males. *Quart. Rev. Biol.* 51:245–76.
———. 1977. Sexual dimorphism in mammals: Avian models and unanswered questions. *Amer. Nat.* 111:917–38.
Ralls, K.; Barasch, C.; and Minkowski, K. 1975. Behavior of captive mouse deer, *Tragulus napu*. *Z. Tierpsychol.* 37:356–78.
Ralls, K.; Brownell, R. J.; and Ballou, J. 1980. Differential mortality by sex and age in mammals with special reference to the sperm whale. *Rep. Int. Whale Comm.* (special issue no. 2), pp. 233–43.
Ralls, K.; Brugger, K.; and Ballou, J. 1979. Inbreeding and juvenile mortality in small populations of ungulates. *Science* 206:1101–3.
Ramirez, J. F.; Freese, C. H.; and Revilla, C. J. 1977. Feeding ecology of the pygmy marmoset, *Cebuella pygmaea*, in northeastern Peru. In *The biology and conservation of the Callitrichidae*, ed. D. G. Kleiman, pp. 91–105. Washington, D.C.: Smithsonian Institution Press.
Rand, R. W. 1955. Reproduction in the female Cape fur seal, *Arctocephalus pusillus* (Schreber). *Proc. Zool. Soc. London* 124:717–40.
———. 1967. The Cape fur seal (*Arctocephalus pusillus*). 3. General behaviour on land and at sea. *Invest. Rep.*, Division of Sea Fisheries, South Africa, 60:1–39.
Randolph, P. A.; Randolph, T. C.; Mattingly, K.; and Foster, M. M. 1977. Energy costs of reproduction in the cotton rat, *Sigmodon hispidus*. *Ecology* 58:31–45.
Ransome, R. D. 1967. The distribution of the greater horse-shoe bat, *Rhinolophus ferrumequinum*, during hibernation, in relation to environmental factors. *J. Zool.* (London) 153(4):77–113.

Rasa, O. A. E. 1972. Aspects of social organization in captive dwarf mongooses. *J. Mammal.* 53(1):181–85.

———. 1973. Marking behaviour and its social significance in the African dwarf mongoose, *Helogale undulata rufula*. *Z. Tierpsychol.* 32:293–318.

———. 1977. The ethology and sociology of the dwarf mongoose. *Z. Tierpsychol.* 43:337–406.

Rathbun, G. B. 1976. The ecology and social structure of the elephant shrews, *Rhynchocyon chrysopygus* Günther and *Elephantulus rufescens* Peters. Ph.D. thesis, University of Nairobi.

———. 1979. The social structure and ecology of elephant-shrews. *Z. Tierpsychol.* 20:1–77.

Ray, C.; Watkins, W. A.; and Burns, J. J. 1969. Thc underwater song of *Erignathus* (bearded seal). *Zoologica* 54(2):79–83.

Redford, K. H. In press. The food habits of armadillos (Edentata: Dasypodidae). In *The evolution and ecology of sloths, anteaters and armadillos,* ed. G. G. Montgomery.

Reichstein, H. 1962. Beiträge zur Biologie einer Steppennagers, *Microtus (Phaeomys) brandti* (Radde 1861). *Z. Säugetierk.* 27(3):146–63.

Reig, O. A. 1970. Ecological notes on the fossorial octodontid rodent *Spalacopus cyanus* (Molina). *J. Mammal.* 51:592–601.

Renfree, M. B. 1980. Placental function and embryonic development in marsupials. In *Comparative physiology: Primitive mammals,* ed. K. Schmidt-Nielson, L. Bolis, and C. Richard Taylor, pp. 269–84. Cambridge: Cambridge University Press.

Rensch, B. 1959. *Evolution above the species level.* London: Methuen.

Rensch, B., and Dücker, G. 1959. Die Spiele von Mungo und Ichneumon. *Behaviour* 14:185–213.

Reuther, R. T. 1969. Growth and diet of young elephants in captivity. *Int. Zoo Yearb.* 9:168–77.

Reynolds, H. C. 1952. Studies on reproduction in the opossum (*Didelphis virginiana virginiana*). *Univ. California Publ. Zool.* 52(3):223–84.

Reynolds, H. G. 1950. Relation of Merriam kangaroo rats to range vegetation in southern Arizona. *Ecology* 31:456–63.

———. 1958. The ecology of the Merriam kangaroo rat (*Dipodomys merriami* Mearns) on the grazing lands of southern Arizona. *Ecol. Monogr.* 28:111–27.

———. 1960. Life history notes on Merriam's kangaroo rat in southern Arizona. *J. Mammal.* 41:48–58.

———. 1966. Abert's squirrels feeding on pinyon pine. *J. Mammal.* 47:550–51.

Reynolds, V., and Reynolds, F. 1965. Chimpanzees of the Budongo Forest. In *Primate behavior,* ed. I. DeVore, pp. 368–424. New York: Holt, Rinehart and Winston.

Rice, D. W., and Wolman, A. A. 1971. The life history and ecology of the gray whale (*Eschrichtius robustus*). Special Publication no. 3. American Society of Mammalogists.

Richard, A. 1974. Intra-specific variation in the social organization and ecology of *Propithecus verreauxi. Folia Primat.* 12:241–63.

———. 1977. The feeding behaviour of *Propithecus verreauxi.* In *Primate ecology,* ed. T. H. Clutton-Brock, pp. 72–96. London/New York: Academic Press.

———. 1978. Variability in the feeding behavior of a Malagasy prosimian, *Propithecus verreauxi:* Lemuriformes. In *The ecology of arboreal folivores,* ed. G. G. Montgomery, pp. 519–33. Washington, D.C.: Smithsonian Institution Press.

Richard, P. B. 1964. Notes sur la biologie du Dasmen des arbres (*Dendrohyrax dorsalis*). *Biol. Gabonica* 1:73–84.

———. 1967. Le déterminisme de la construction des barrages chez le castor du Rhone. *Terre et Vie* 4:339–470.

———. 1968. Inventaire des comportements individuels du castor de France (*Castor fiber* Linn.). *Rev. Comport. Anim.* 2(3):19–58.

Richard, P. B., and Vaillard, A. V. 1969. Le desman des Pyrénées (*Galemys pyrenaicus*): Premières notes sur sa biologie. *Terre et Vie* 3:225–45.

Richardson, W. B. 1942. Ring-tailed cats (*Bassariscus astutus*): Their growth and development. *J. Mammal.* 23:17–26.

Richter, W. von. 1966. Untersuchung über ageborene Verhaltensweise den Schabrackentapirs und des Flachlandtapirs. *Zool. Beitr. Inst. Univ. Erlangen-Nurenburg* 1(12):67–159.

Ride, W. D. L. 1964. A review of Australian fossil marsupials. *J. Roy. Soc. Western Australia* 74(4):97–131.

———. 1970. *Native mammals of Australia.* Melbourne: Oxford University Press.

Rijksen, H. D. 1978. *A field study on Sumatran orang-utans (Pongo pygmaeus abelii Lesson 1827).* Wageningen: H. Veenman and B. V. Zonen.

Ripley, S. R. 1967. Intertroop encounters among Ceylon gray langurs (*Presbytis entellus*). In *Social communication among Primates,* ed. S. A. Altmann, pp. 237–52. Chicago: University of Chicago Press.

———. 1970. Leaves and leaf-monkeys: The social organization of foraging in gray langurs, *Presbytis entellus thersites.* In *Old World monkeys: Evolution, systematics and behavior,* ed. J. R. Napier and P. H. Napier, pp. 481–509. New York/London: Academic Press.

Roberts, M. S. 1975. Growth and development of mother-reared red pandas, *Ailurus fulgens. Int. Zoo Yearb.* 15:57–62.

Roberts, M. S., and Kessler, D. S. 1979. Reproduction in red pandas, *Ailurus fulgens* (Carnivora: Ailuropodidae). *J. Zool.* (London) 188:235–49.

Robinson, J. F. 1977. Vocal regulation of spacing in the titi monkey, *Callicebus moloch.* Ph.D. diss., University of North Carolina.

Rodman, P. S. 1973. Population composition and adaptive organisation among orang-utans of the Kutaix Reserve. In *Comparative ecology and behavior of Primates,* ed. R. P. Michael and J. H. Crook, pp. 171–209. London/New York: Academic Press.

Rogers, L. L. 1977. Social relationships, movements, and population dynamics of black bears in northeastern Minnesota. Ph.D. diss., University of Minnesota.

Röhrs, M. 1966. Vergleichende Untersuchungen zur Evolution der Gehirne von Edentaten. I. Hirngewicht-Körpergewicht. *Z. Zool. Syst. Evolutionsforsch.*, 4:196–207.

Romer, A. S. 1966. *Vertebrate paleontology.* Chicago: University of Chicago Press.

Romer, J. D. 1974. Milk analysis and weaning in the lesser Malay chevrotain. *Int. Zoo Yearb.* 14:179–80.

Rood, J. P. 1958. Habits of the short-tailed shrew in captivity. *J. Mammal.* 39:499–507.

———. 1963. Observations on the behavior of the spiny rat *Heteromys melanoleucus* in Venezuela. *Mammalia* 27:186–92.

———. 1965*a*. Observations on population structure, reproduction and molt of the Scilly shrew. *J. Mammal.* 46:426–33.

———. 1965*b*. Observations on the home range and activity of the Scilly shrew. *Mammalia* 29:507–17.

———. 1970*a*. Ecology and social behavior of the desert cavy (*Microcavia australis*). *Amer. Midl. Nat.* 83(2):415–54.

———. 1970*b*. Notes on the behavior of the pygmy armadillo. *J. Mammal.* 51:179–80.

———. 1972. Ecological and behavioral comparisons of three genera of Argentine cavies. *Anim. Behav. Monogr.* 5:1–83.

———. 1975. Population dynamics and food habits of the banded mongoose. *E. African Wildl. J.* 13:89–111.

———. 1978. Dwarf mongoose helpers at the den. *Z. Tierpsychol.* 48:277–87.

Rosenblatt, J. S., and Lehrman, D. S. 1963. Maternal behavior of the laboratory rat. In *Maternal behavior in mammals,* ed. N. Rheingold, pp. 8–57. New York: John Wiley.

Rosenblatt, J. S., and Schneirla, T. C. 1962. The behavior of cats. In *The behavior of domestic animals,* ed. E. S. E. Hafez, pp. 453–88. Baltimore: Williams and Wilkins.

Rosenblatt, J. S.; Turkewitz, G.; and Schneirla, T. C. 1962. Development of suckling and related behavior in neonate kittens. In *Roots of behavior,* ed. E. L. Bliss, pp. 198–210. New York: Harper.

Rosenthal, M. 1975. The management, behavior and reproduction of the short-eared elephant shrew. M.A. thesis, Northeastern Illinois University.

Rosenzweig, M. L. 1966. Community structure in sympatric Carnivora. *J. Mammal.* 47:602–12.

———. 1973. Habitat selection experiments with a pair of coexisting heteromyid rodent species. *Ecology* 54:111–17.

Roth, H. H. 1964. Ein Beitrag zur Kenntnis von *Tremarctos ornatus* (Cuvier). *Zool. Gart.* 29:107–29.

Rothe, H. 1977. Parturition and related behavior in *Callithrix jacchus* (Ceboidea, Callitrichidae). In *The biology and conservation of the*

Callitrichidae, ed. D. G. Kleiman, pp. 193–207. Washington, D.C.: Smithsonian Institution Press.
Rowell, T. E. 1961. The family group in golden hamsters: Its formation and break-up. *Behaviour* 17:81–94.
———. 1972. *Social behaviour of monkeys*. London: Coe and Wyman.
Rowe-Rowe, D. T. 1978. Comparative prey capture and food studies of South American mustelines. *Mammalia* 42(2):175–96.
Rowlands, I. W., and Sadleir, R. M. F. S. 1968. Induction of ovulation in the lion, *Panthera leo*. *J. Reprod. Fertil.* 16:105–11.
Rowlands, I. W., and Weir, B. J. 1974. *The biology of hystricomorph rodents*. Symposia of the Zoological Society of London, no. 34. London: Academic Press.
Rudran, R. 1973*a*. The reproductive cycles of two subspecies of purple-faced langurs (*Presbytis senex*) with relation to environmental factors. *Folia Primat.* 19:41–60.
———. 1973*b*. Adult male replacement in one-male troops of purple-faced langurs (*Presbytis senex senex*) and its effect on population structure. *Folia Primat.* 19:166–92.
———. 1978*a*. Socioecology of the blue monkeys (*Cercopithecus mitis stuhlmanni*) of the Kibale Forest, Uganda. *Smithsonian Contrib. Zool* 249:1–88.
———. 1978*b*. Intergroup dietary comparisons and folivorous tendencies of two groups of blue monkeys (*Cercopithecus mitis stuhlmanni*). In *The ecology of arboreal folivores*, ed. G. G. Montgomery, pp. 483–503. Washington, D.C.: Smithsonian Institution Press.
———. 1979. Demography and social mobility in a red howler monkey (*A. seniculus*) population. In *Vertebrate ecology in the northern Neotropics*, ed. J. F. Eisenberg, pp. 107–26. Washington, D.C.: Smithsonian Institution Press.
Ruffer, D. G. 1965. Sexual behaviour of the northern grasshopper mouse (*Onychomys leucogaster*). *Anim. Behav.* 13:447–52.
———. 1968. Agonistic behavior of the northern grasshopper mouse (*Onychomys leucogaster breviauritus*). *J. Mammal.* 49(3):481–88.
Russell, E. M. 1970. Observations on the behaviour of the red kangaroo (*Megaleia rufa*) in captivity. *Z. Tierpsychol.* 37(4):385–404.
———. 1973. Mother-young relations and early behavioural development in the marsupials *Macropus eugenii* and *Megaleia rufa*. *Z. Tierpsychol.* 33:163–203.
Russell, J. K. 1979. Reciprocity in coati (*Nasua narica*) bands. Ph.D. thesis, University of North Carolina.
Russell, R. J. 1968. Revision of pocket gophers of the genus *Pappogeomys*. *Univ. Kansas Publ. Mus. Nat. Hist.* 16(7):581–776.
Rust, C. C. 1966. Notes on the star-nosed mole (*Condylura cristata*). *J. Mammal.* 47:538.
Saayman, G. S.; Bower, D.; and Tayler, C. K. 1972. Observations on inshore and pelagic dolphins on the southeastern Cape coast of South Africa. *Koedoe* 15:1–24.

Saayman, G. S., and Tayler, C. K. 1973. Some behaviour patterns of the southern right whale (*Eubalaena australis*. *Z. Säugetierk.* 38(3):172–83.

Sacher, G. A. 1959. Relation of lifespan to brain weight and body weight in mammals. In *Colloquia on aging,* ed. G. E. W. Wolstenholme and M. O'Connor, pp. 115–33. London: CIBA Foundation, Churchill.

Sacher, G. A., and Staffeldt, E. F. 1974. Relation of gestation time to brain weight for placental mammals: Implications for the theory of vertebrate growth. *Amer. Nat.* 108:593–615.

Sadleir, R. M. F. S. 1965. The relationship between agonistic behaviour and population changes in the deermouse, *Peromyscus maniculatus*. *J. Anim. Ecol.* 34:331–52.

———. 1966. Notes on reproduction in the larger Felidae. *Int. Zoo Yearb.* 6:184–87.

———. 1969. *The ecology of reproduction in wild and domestic mammals.* London: Methuen.

St. Girons, M. C. 1962. Notes sur les dates de reproduction en captivité du fenec. *Z. Säugetierk.* 27:181–84.

Sale, J. B. 1965*a*. The feeding behaviour of rock hyraxes (genera *Procavia* and *Heterohyrax*) in Kenya. *E. African Wildl. J.* 3:1–18.

———. 1965*b*. Gestation period and neonatal weight of the hyrax. *Nature* (London) 205:1240–41.

Sampsell, R. N. 1969. Hand rearing an aardvark, *Orycteropus afer,* at Crandon Park Zoo, Miami. *Int. Zoo Yearb.* 9:97–99.

Sanborn, C. C., and Watkins, R. 1950. Notes on the Malay tapir and other game animals in Siam. *J. Mammal.* 31:430–33.

Sandegren, F. E. 1970. Breeding and maternal behavior of the Steller sea lion (*Eumetopias jubata*) in Alaska. M.S. thesis, University of Alaska, College.

Sandegren, F. E.; Chu, E. W.; and Vandevere, J. E. 1973. Maternal behavior in the California sea otter. *J. Mammal.* 54(3):668–80.

Sankhala, K. S. 1967. Breeding behaviour of the tiger, *Panthera tigris,* in Rajasthan. *Int. Zoo Yearb.* 7:133–47.

Sauer, E. G. F., and Sauer, E. M. 1971. Die Kurzohrigen Elefantenspitzmaus in der Namib. *Namib und Meer* 2:5–43.

———. 1972. Zur Biologie der Kurzohrigen Elefantenspitzmaus. *Z. Kölner Zoo* 15(4):119–39.

Scammon, C. M. 1874. *The marine mammals of the northwestern coast of North America, described and illustrated, together with an account of the American whale fishery.* San Francisco: John H. Carmany.

Schaffer, W. M., and Reed, C. A. 1972. *The co-evolution of social behavior and cranial morphology in sheep and goats (Bovidae, Caprini).* Fieldiana, Zoology, no. 61. Chicago: Field Museum of Natural History.

Schaller, G. B. 1961. The orang-utan in Sarawak. *Zoologica* 46(2):73–82.

———. 1963. *The mountain gorilla: Ecology and behavior.* Chicago/London: University of Chicago Press.

———. 1967. *The deer and the tiger.* Chicago: University of Chicago Press.
———. 1968. Hunting behaviour of the cheetah in the Serengeti National Park, Tanzania. *E. African Wildl. J.* 6:95–101.
———. 1969. Food habits of the Himalayan black bear (*Selenarctos thibetanus*) in the Dodrigam Sanctuary, Kashmir. *J. Bombay Nat. Hist. Soc.* 66:156–59.
———. 1972. *The Serengeti lion: A study of predator-prey relations.* Chicago/London: University of Chicago Press.
———. 1977. *Mountain monarchs: Wild sheep and goats of the Himalaya.* Chicago: University of Chicago Press.
Schaller, G. B., and Hamer, A. 1978. Rutting behavior of Pere David deer, *Elaphurus davidianus*. *Zool. Gart.* 48:1–15.
Scheffer, T. H. 1931. Habits and economic status of the pocket gophers. *Tech. Bull. U.S. Dept. Agric.* 224:1–26.
———. 1938. Breeding records of Pacific coast pocket gophers. *J. Mammal.* 19:220–24.
Scheffer, V. B. 1958. *Seals, sea lions and walruses.* Stanford: Stanford University Press.
Scheffer, V. B., and Slipp, J. W. 1944. The harbor seal in Washington State. *Amer. Midl. Nat.* 32:373–416.
Schenkel, R. 1948. Ausdrucks-Studien am Wölfen. *Behaviour* 1:81–129.
———. 1966. On sociology and behaviour in impala (*Aepyceros melampus suara* Matschie). *Z. Säugetierk.* 31(3):177–205.
Schenkel, R., and Lang, E. M. 1969. Das Verhalten der Nashörner. *Handb. Zool.* 19(25):1–56.
Schenkel, R., and Schenkel-Hulliger, L. 1969*a*. The Javan rhinoceros in Udjung Kulon. *Acta Trop.* 26:97–135.
———. 1969*b*. *Ecology and behavior of the black rhinoceros.* Mammalia Depicta. Berlin: Paul Parey.
Schevill, W. E., ed. 1974. *The whale problem: A status report.* Cambridge: Harvard University Press.
Schevill, W. E., and Watkins, W. A. 1965. Underwater calls of *Trichechus* (manatee). *Nature* (London) 205:373–74.
Schlichte, H.-J. 1978. The ecology of two groups of blue monkeys (*Cercopithecus mitis stuhlmanni*) in an isolated habitat of poor vegetation. In *The ecology of arboreal folivores,* ed. G. G. Montgomery, pp. 505–17. Washington, D.C.: Smithsonian Institution Press.
Schloeth, R. 1956. *Zur Psychologie der Begegnung zwischen Tieren.* Leiden: E. J. Brill.
———. 1961. Das Sozialleben des Camargue-Rindes. *Z. Tierpsychol.* 18:574–627.
Schmid, B. 1938. Psychologische Beobachtungen und Versuche an einem jungen männlichen Ameisenbären (*Myrmecophaga tridactylus* L.) *Z. Tierpsychol.* 2:117–26.
Schmidt, C. R. 1973. Breeding seasons and notes on some other aspects of reproduction in captive camelids. *Int. Zoo Yearb.* 13:387–90.
———. 1976. Verhalten einer Zoogruppe von Halsband-Pekaris (*Tayassu tajacu*). Ph.D. diss., University of Zurich.

Schmidt, U., and Manske, U. 1973. Die Jugendentwicklung der Vampirfledermäuse (*Desmodus rotundus*). *Z. Säugetierk.* 38:14–33.

Schmidt-Nielson, K. 1964. *Desert animals: Physiological problems of heat and water.* Oxford: Clarendon Press.

Schneider, K. M. 1936. Zur Fortpflanzung, Aufzucht und Jugendentwicklung des Schakbachen-tapirs. *Zool. Gart.* (Leipzig) 8:83–96.

———. 1954. Vom Baumkanguruh (*Dendrolagus leucogenys* Matschie). *Zool. Gart.* 21(1–2):63–106.

Schöenberner, D. 1965. Beobachtungen zur Fortpflanzungsbiologie des Wolfes, *Canis lupus*. *Z. Säugetierk.* 30:171–78.

Schoenholzer, L. 1959. Beobachtungen über das Trinkverhalten bei Zootieren. *Zool. Gart.*, n.s., 24:346–434.

Scholander, P. F. 1940. Experimental investigations on the respiratory function in diving mammals and birds. *Hvalradets Skrift. Norske Videnskaps-Akad.* (Oslo) 22:1–131.

Schomber, H. W. 1963. Beiträge zur Kenntnis der Giraffengazelle (*Litocranius walleri* Brooke 1878). *Säugetierk. Mitt.* 11:1–44.

Schonewald, C. 1977. Temporal patterns of courtship in a caviomorph rodent, the green acouchi (*Myoprocta pratti* Pocock). Ph.D. diss., University of Maryland.

Schramm, P. 1961. Copulation and gestation in the pocket gopher. *J. Mammal.* 42:167–70.

Schultz, W. 1967. Der Schwertwal—*Orcinus orca* (Linnaeus 1758). *Z. Säugetierk.* 32(2):90–103.

Schultze-Westrum, T. G. 1965. Innerartiliche Verstandigung durch Düfte beim Gleitbeutler *Petaurus breviceps papuana* Thomas (Marsupialia: Phalangeridae). *Z. Vergl. Physiol.* 50:151–220.

———. 1969. Social communication by chemical signals in flying phalangers. In *Olfaction and taste,* ed. C. Pfaffmann, pp. 268–77. New York: Rockefeller University Press.

Schumacher, U. 1963. Quantitative Untersuchungen an Gehirnen mitteleuropäischer Musteliden. *J. Hirnforsch.* 6:137–63.

Schusterman, R. J. 1967. Perception and determinants of underwater vocalization in the California sea lion. In *Les systèmes sonars animaux,* ed. R. G. Busnel, pp. 535–617. Jouy-en-Josas-78, France: Laboratoire de Physiologie Acoustique.

Schusterman, R. J., and Bailliet, R. F. 1971. Aerial and underwater visual acuity in the California sea lion (*Zalophus californianus*) as a function of luminance. *Ann. N.Y. Acad. Sci.* 188:37–46.

Schusterman, R. J., and Dawson, R. G. 1968. Barking, Dominance and territoriality in male sea lions. *Science* 160:434–36.

Schusterman, R. J.; Gentry, R.; and Schmook, J. 1967. Underwater sound production by captive California sea lions, *Zalophus californianus*. *Zoologica* 52(1):21–24.

Schwartz, O. A., and Bleich, V. C. 1975. Comparative growth in two species of wood rats, *Neotoma lepida intermedia* and *Neotoma albigula venusta*. *J. Mammal.* 56:653–66.

Scott, J. P., and Fuller, J. L. 1965. *Genetics and the social behavior of the dog.* Chicago: University of Chicago Press.

Scott, W. B. 1937. *A history of land mammals in the western hemisphere*. New York: Macmillan.

Sebeok, T. A., ed. 1977. *How animals communicate*. Bloomington/London: Indiana University Press.

Seidensticker, J. C., IV; Hornocker, M. G.; Wiles, W. V.; and Messick, J. P. 1973. Mountain lion social organization in the Idaho primitive area. *Wildl. Monogr.* 35:1–60.

Seitz, A. 1955. Untersuchungen über angeborene Verhaltensweisen bei Caniden Beobachtungen an Marderhunden. *Z. Tierpsychol.* 12:463–89.

Seton, E. T. 1953. *Lives of game animals*. Vols. 1–4. Boston: Charles T. Brandford.

Severaid, J. H. 1956. The natural history of the pikas. Ph.D. diss., University of California, Berkeley.

Shadle, A. R. 1943. The play of American porcupines (*Erethizon d. dorsatum* and *E. epixanthum*. *J. Comp. Psychol.* 37:145–50.

———. 1946. Copulation in the porcupine. *J. Wildl. Manage.* 10:159–62.

Shapiro, J. 1949. Ecological and life history notes on the porcupine in the Adirondacks. *J. Mammal.* 30:247–57.

Sharman, G. B. 1959. Marsupial reproduction. *Monogr. Biol.* 8:332–68.

———. 1963. Delayed implantation in marsupials. In *Delayed implantation*, ed. A. C. Enders, pp. 3–14. Chicago: University of Chicago Press.

———. 1970. Reproductive physiology of marsupials. *Science* 167:1221–28.

Sharman, G. B., and Calaby, J. H. 1964. Reproductive behaviour in the red kangaroo, *Megaleia rufa*, in captivity. *CSIRO Wildl. Res.* 9(1):58–85.

Sharp, W. M., and Sharp, L. H. 1956. Nocturnal movements and behavior of wild raccoons at a winter feeding station. *J. Mammal.* 37(2):170–77.

Shaw, W. T. 1925. Breeding and development of the Columbian ground squirrel. *J. Mammal.* 6:106–13.

Sheldon, W. G. 1975. *The wilderness home of the giant panda*. Amherst: University of Massachusetts Press.

Sheppard, D. H. 1969. A comparison of reproduction in two chipmunk species (*Eutamias*). *Canadian J. Zool.* 47:603–8.

Sherman, P. W. 1977. Nepotism and the evolution of alarm calls. *Science* 197:1246–53.

Shillito, J. F. 1963. Observations on the range and movements of a woodland population of the common shrew, *Sorex araneus*. *Proc. Zool. Soc. London* 140:533–46.

Siebenaler, J. B., and Caldwell, D. K. 1956. Cooperation among adult dolphins. *J. Mammal.* 37:126–28.

Sikes, S. K. 1971. *The natural history of the African elephant*. London: Weidenfeld and Nicolson.

Simpson, C. D. 1968. Reproduction and population structure in greater kudu in Rhodesia. *J. Wildl. Manage.* 32(1):149–62.

Simpson, G. G. 1940. Mammals and land bridges. *J. Wash. Acad. Sci.* 30:137–63.

———. 1944. *Tempo and mode in evolution*. New York: Columbia University Press.
———. 1945. The principles of classification and a classification of the mammals. *Bull. Amer. Mus. Nat. Hist.* 85:1–350.
———. 1959. The nature and origin of supraspecific taxa. *Cold Spring Harbor Symp. Quant. Biol.* 24:255–72.
———. 1965. *The geography of evolution*. Philadelphia: Chilton Books.
———. 1970. *The Argyrolagidae, extinct South American marsupials*. Bulletin of the Museum of Comparative Zoology, vol. 139. Cambridge: Harvard University Press.
———. 1975. Recent advances in methods of phylogenetic inference. In *Phylogeny of the Primates,* ed. W. P. Luckett and F. S. Szalay, pp. 3–19. New York: Plenum Press.
———. 1980. *Splendid isolation*. New Haven: Yale University Press.
Sinclair, A. R. E. 1977. *The African buffalo*. Chicago: University of Chicago Press.
Slaughter, B. H., and Walton, D. W. 1970. *About bats*. Dallas: Southern Methodist University Press.
Slijper, E. J. 1958. Das Verhalten der Wale (Cetacea). *Handb. Zool.*, Band 8, Teil 19(14):1–32.
———. 1962. *Whales*. London: Hutchinson and Company.
Smith, C. C. 1968. The adaptive nature of social organization in the genus of tree squirrels *Tamiasciurus*. *Ecol. Monogr.* 38:31–63.
———. 1970. The coevolution of pine squirrels (*Tamiasciurus*) and conifers. *Ecol. Monogr.* 40:349–71.
———. 1977. Feeding behaviour and social organization in howling monkeys. In *Primate ecology,* ed. T. H. Clutton-Brock, pp. 97–126. London/New York: Academic Press.
Smith, J. D. 1975. Comments on flight and the evolution of bats. In *Major patterns in vertebrate evolution,* ed. M. K. Hecht, P. C. Goody, and B. M. Hecht, pp. 427–37. New York/London: Plenum Press.
———. 1976. Chiropteran evolution. In *Biology of bats of the New World family Phyllostomatidae, part I,* ed. R. J. Baker, J. Knox-Jones, Jr., and D. C. Carter, pp. 49–69. Special Publications of the Museum 10. Lubbock: Texas Tech Press.
Smith, M. 1973. *Petaurus breviceps*. *Mammal. Spec.* 30:1–4.
Smith, M. H. 1966. The evolutionary significance of certain behavioral, physiological and morphological adaptations of the old field mouse, *Peromyscus polionotus*. Ph.D. diss., University of Florida, Gainesville.
Smith, M. J. 1971. Breeding the sugar-glider *Petaurus breviceps* in captivity, and growth of pouch young. *Int. Zoo Yearb.* 11:26–28.
Smith, M. S. R. 1966. Injuries as an indication of social behavior in the Weddell seal (*Leptonychotes weddelli*). *Mammalia* 30:241–47.
Smith, W. J. 1977. *The behavior of communicating*. Cambridge: Harvard University Press.

Smith, W. J.; Smith, S. L.; Oppenheimer, E. C.; deVilla, J. G.; and Ulmer, F. A. 1973. Behavior of a captive population of black-tailed prairie dogs: Annual cycle of social behavior. *Behaviour* 46:189–220.

Smuts, G. L. 1975. Home range sizes for Burchell's zebra *Equus burchelli antiquorum* from the Kruger National Park. *Koedoe* 18:139–46.

Smythe, N. 1970*a*. On the existence of "pursuit invitation" signals in mammals. *Amer. Nat.* 104(938):491–94.

———. 1970*b*. Relationship between fruiting seasons and seed dispersal methods in a Neotropical forest. *Amer. Nat.* 104:25–35.

———. 1978. The natural history of the Central American agouti (*Dasyprocta punctata*). *Smithsonian Contrib. Zool.* 257:1–52.

Snyder, D. P. 1978. *Populations of small mammals under natural conditions*. Special Publications Series, vol. 5. Pittsburg: Pymatuning Laboratory of Ecology, University of Pittsburg.

Sody, H. J. V. 1959. Das Javanische Nashorn. *Z. Säugetierk.* 24:109–241.

Soholt, L. F. 1973. Nest-building in *Dipodomys merriami* at different ambient temperatures. *J. Mammal.* 54(4):997–98.

———. 1977. Consumption of herbaceous vegetation and water during reproduction and development of Merriam's kangaroo rat, *Dipodomys merriami*. *Amer. Midl. Nat.* 98:445–57.

Sollberger, D. E. 1943. Notes on the breeding habits of the eastern flying squirrel. *J. Mammal.* 24:163–73.

Somers, P. 1973. Dialects in southern Rocky Mountain pikas, *Ochotona princeps*. Lagomorpha. *Anim. Behav.* 21(1):124–37.

Sorenson, M. W. 1962. Some aspects of water shrew behavior. *Amer. Midl. Nat.* 68:445–62.

———. 1970. Observations on the behavior of *Dasycercus cristicauda* and *Dasyuroides byrnei* in captivity. *J. Mammal.* 51:123–31.

Soulé, M., and Wilcox, B., eds. 1980. *Conservation biology*. Sunderland, Mass.: Sinauer.

Southern, H. N. 1948. Sexual and aggressive behavior in the wild rabbit. *Behavior* 1:173–94.

Southwick, C. E. 1955. Regulatory mechanisms of house mouse population: Social behavior affecting litter survival. *Ecology* 36:627–34.

Sowls, L. K. 1974. Social behavior of the collared peccary *Dicotyles tajacu*. In *The behaviour of ungulates and its relation to management,* ed. V. Geist and F. Walther, 1:144–65. IUCN Publications, n.s. no. 24. Morges: IUCN.

Sowls, L. K., and Phelps, R. J. 1968. Observations on the African bushpig *Potamochoerus porcus* Linn. in Rhodesia. *Zoologica* 53(3):75–105.

Spain, A. V., and Heinsohn, G. E. 1973. Cyclone associated feeding changes in the dugong (Mammalia: Sirenia). *Mammalia* 37:678–80.

———. 1975. Size and weight allometry in a North Queensland population of *Dugong dugon* (Müller) (Mammalia: Sirenia). *Australian J. Zool.* 23:159–68.

Spector, W. S., ed. 1956. *Handbook of biological data*. Philadelphia: W. B. Saunders.

Spurway, H., and Haldane, J. B. S. 1953. The comparative ethology of vertebrate breathing, I. *Behaviour* 6:8–34.

Stains, H. J. 1956. The raccoon in Kansas. *Misc. Publ. Univ. Kansas Mus. Nat. Hist.* 10:1–76.

Stanley, M. 1971. An ethogram of the hopping mouse, *Notomys alexis*. *Z. Tierpsychol.* 29:225–58.

Stanley, S. M. 1979. *Macroevolution: Pattern and process*. San Francisco: W. H. Freeman.

Starck, D. 1956. Primitiventwicklung und plazentation der Primaten. In *Primatologia,* ed. H. Hofer, A. H. Schultz, and D. Starck, 1:723–886. Basel: S. Karger.

Starrett, A., and Fisler, G. F. 1970. Aquatic adaptations of the water mouse, *Rheomys underwoodi*. *Los Angeles County Mus. Contrib. Sci.* 182:1–14.

Start, A. N., and Marshall, A. G. 1976. Nectarivorous bats as pollinators of trees in West Malaysia. In *Tropical trees: Variation, breeding and conservation,* ed. J. Burley and B. T. Styles, pp. 141–50. London: Academic Press.

Stebbings, R. E. 1966. A population study of bats of the genus *Plecotus*. *J. Zool.* (London) 150:53–75.

Stehlin, H. G., and Schaube, S. 1951. Die trigonodontie der Simplicidentaten Nager. *Schweiz. Palaeont. Abh.* 67:1–385.

Steiner, A. L. 1970. Etude descriptive de quelques activités et comportements de base de *Spermophilus columbianus columbianus* (Ord.). *Rev. Comp. Anim.* 4:23–42.

———. 1971. Play activity of Columbian ground squirrels. *Z. Tierpsychol.* 28:247–61.

———. 1973. Self- and allo-grooming in some ground squirrels (Sciuridae): A descriptive study. *Canadian J. Zool.* 51:151–61.

Stephan, H.; Bauchot, R.; and Andy, O. J. 1970. Data on size of the brain and of various brain parts in insectivores and primates. In *Primate brain,* ed. C. R. Noback and W. Montagna, pp. 289–97. New York: Appleton.

Stephan, H., and Pirlot, P. 1970. Volumetric comparisons of brain structures in bats. *Z. Zool. Syst. Evol. Forsch.* 8:200–236.

Stevenson, M. F. 1976. Maintenance and breeding of the common marmoset with notes on hand-rearing. *Int. Zoo Yearb.* 16:110–16.

Stine, C. J., and Dryden, G. L. 1977. Lip-licking behavior in captive musk shrews, *Suncus murinus*. *Behaviour* 62(3–4):298–314.

Stirling, I. 1969. Ecology of the Weddell seal in McMurdo Sound, Antarctica. *Ecology* 50(4):573–86.

Stodart, E. 1966*a*. Management and behavior of breeding groups of the marsupial *Perameles nasuta* Geoffroy in captivity. *Australian J. Zool.* 14:611–23.

———. 1966*b*. Observations on the behaviour of the marsupial *Bettongia lesueuri* (Quoy and Gaimard) in an enclosure. *CSIRO Wildl. Res.* 11:91–99.

———. 1977. Breeding and behaviour of Australian bandicoots. In *The biology of marsupials,* ed. B. Stonehouse and D. Gilmore, pp. 169–79. Baltimore/London: University Park Press.

Stonorov, D., and Stokes, A. W. 1972. Social behavior of the Alaska brown bear. In *Bears: Their biology and management,* ed. S. Herrero, pp. 232–43. Morges: IUCN.

Storer, T. I. 1962. *Pacific rat ecology: Report of a study made on Panape and adjacent islands 1955–1958.* Bulletin no. 225. Honolulu: Bishop Museum.

Storm, G. L. 1965. Movements and activities of foxes as determined by radio tracking. *J. Wildl. Manage.* 29:1–13.

Storm, G. L.; Andrews, R. D.; Phillips, R. L.; Bishop, R. A.; Siniff, D. B.; and Tester, J. R. 1976. Morphology, reproduction, dispersal and mortality of midwestern red fox populations. *Wildl. Monogr.* 49:1–82.

Strahan, R., and Thomas, D. E. 1975. Courtship of the platypus, *Ornithorhynchus anatinus. Australian Zool.* 18(3):165–78.

Strickland, D. L. 1967. Ecology of the rhinoceros in Malaya. *Malayan Nat. J.* 20(1–2):1–17.

Struhsaker, T. T. 1967*a*. Social structure among vervet monkeys (*Cercopithecus aethiops*). *Behaviour* 29:83–121.

———. 1967*b*. Ecology of vervet monkeys (*Cercopithecus aethiops*) in the Masai Amboseli Game Reserve, Kenya. *Ecology* 48:891–904.

———. 1967*c*. Behavior of elk (*Cervus canadensis*) during the rut. *Z. Tierpsychol.* 24:80–114.

———. 1969. Correlates of ecology and social organization among African cercopithecines. *Folia Primat.* 11:80–118.

———. 1975. *The red colobus monkey.* Chicago: University of Chicago Press.

———. 1976. A further decline in numbers of Amboseli vervet monkeys. *Biotropica* 8(3):211–14.

———. 1977. Infanticide and social organization in the redtail monkey (*Cercopithecus ascanius schmidti*) in the Kibale Forest, Uganda. *Z. Tierpsychol.* 4:75–84.

Stuewer, F. W. 1943. Raccoons: Their habits and management in Michigan. *Ecol. Monogr.* 13(2):203–57.

Sugiyama, Y. 1967. Social organization in Hanuman langurs. In *Social communication among Primates,* ed. S. A. Altmann, pp. 221–36. Chicago: University of Chicago Press.

———. 1968. Social organisation of chimpanzees in the Budongo Forest. *Primates* 9:225–58.

———. 1973. The social structure of wild chimpanzees: A review of field studies. In *Comparative ecology and behaviour of Primates,* ed. R. P. Michael and J. H. Crook, pp. 375–410. London/New York: Academic Press.

Sukhanov, V. B. 1974. *General system of symmetrical locomotion of terrestrial vertebrates and some features of movement of lower tetrapods.* Translated from Russian. Washington, D.C.: Smithsonian Institution and National Science Foundation.

Sunquist, M. E. 1979. The movements and activities of tigers (*Panthera tigris tigris*) in Royal Chitawan National Park, Nepal. Ph.D. diss., University of Minnesota.

Sunquist, M. E., and Montgomery, G. G. 1973. Activity patterns and rates of movement of two-toed and three-toed sloths (*Choloepus hoffmanni* and *Bradypus infuscatus*). *J. Mammal.* 54:346–54.

Sussman, R. W. 1975. A preliminary study of the behavior and ecology of *Lemur fulvus rufus* Audebert 1800. In *Lemur biology,* ed. I. Tattersall and R. W. Sussman, pp. 237–59. New York: Plenum Press.

———. 1977. Feeding behaviour of *Lemur catta* and *Lemur fulvus*. In *Primate ecology,* ed. T. H. Clutton-Brock, pp. 1–36. London/New York: Academic Press.

Sussman, R. W., and Raven, P. H. 1978. Pollination by lemurs and marsupials: An archaic coevolutionary system. *Science* 200:731–36.

Suzuki, A. 1969. An ecological study of chimpanzees in a savanna woodland. *Primates* 10(2):103–48.

Svihla, A. 1935. Development and growth of the prairie deermouse, *Peromyscus maniculatus bairdi. J. Mammal.* 16:109–15.

Szalay, F. 1977. Phylogenetic relationships and a classification of the eutherian Mammalia. In *Major patterns in vertebrate evolution,* ed. M. K. Hecht, P. C. Goody, and B. M. Hecht, pp. 315–74. New York/London: Plenum Press.

Szalay, F., and Delson, E. 1979. *Evolutionary history of Primates*. New York: Academic Press.

Taber, F. W. 1945. Contribution on the life history and ecology of the nine-banded armadillo. *J. Mammal.* 26:211–26.

Talmage, P. V., and Buchanan, G. P. 1954. The armadillo (*Dasypus novemcinctus*): A review of its natural history, ecology anatomy and reproductive physiology. *Rice Inst. Pamphlet, Monogr. Biol.* 41(2):1–135.

Tamsitt, J. R., and Valdivieso, D. 1965. Reproduction of the female big fruit-eating bat, *Artibeus lituratus palmarum,* in Colombia, *Caribbean J. Sci.* 5:157–66.

Tarling, D., and Tarling, M. 1975. *Continental drift: A study of the earth's moving surface*. Garden City, N.Y.: Anchor Press/Doubleday.

Tate, G. H. H. 1951. Results of the Archbold expedition no. 65: The rodents of Australia and New Guinea. *Bull. Amer. Mus. Nat. Hist.* 97:183–430.

Tattersall, I., and Sussman, R. W., eds. 1975. *Lemur biology.* New York: Plenum Press.

Tavolga, M. C. 1966. Behavior of the bottlenose dolphin (*Tursiops truncatus*): Social interactions in a captive colony. In *Whales, dolphins and porpoises,* ed. K. S. Norris, pp. 718–30. Berkeley: University of California Press.

Tavolga, M. C., and Essapian, F. A. 1957. The behavior of the bottlenosed dolphin (*Tursiops truncatus*): Mating, pregnancy, parturition and mother-infant behavior. *Zoologica* 42:11–31.

Tayler, C. K., and Saayman, G. S. 1972. The social organisation and behaviour of dolphins (*Tursiops aduncus*) and baboons (*Papio ursinus*): Some comparisons and assessments. *Ann. Cape Provincial Mus. (Nat. Hist.)* 9(2):1–49.

Taylor, J. M. 1961. Reproductive biology of the Australian bush rat, *Rattus assimilis*. *Univ. California Publ. Zool.* 60(1):1–66.

Taylor, J. M., and Horner, B. E. 1970. Reproduction in the mosaic-tailed rat *Melomys cervinipes* (Rodentia: Muridae). *Australian J. Zool.* 18:171–84.

Taylor, W. P. 1935. *Philippine land mammals*. Manila: Government of the Philippine Islands, Department of Agriculture and Commerce.

———, ed. 1956. *The deer of North America*. Harrisburg, Pa.: Stackpole.

Teleki, G. 1973. *The predatory behaviour of wild chimpanzees*. Lewisburg, Pa.: Bucknell Univesity Press.

Tembrock, G. 1957. Das Verhalten des Rotfuchses. *Handb. Zool.*, Band 8, Leiferung 9(10:15):1–20.

Temple-Smith, P. D. 1974. Seasonal breeding biology of the platypus *Ornithorhynchus anatinus* with special reference to the male. Ph.D. thesis, Australian National Museum, Canberra.

Tenaza, R. R. 1975. Territory and monogamy among Kloss gibbons (*Hylobates klossii*) in Siberut Island, Malaysia, *Folia Primat.* 24:60–80

Tener, J. S. 1965. *Muskoxen in Canada*. Ottawa: Queen's Printer.

Terborgh, J. 1974. Preservation of natural diversity: The problems of extinction prone species. *BioScience* 24(12):715–22.

———. 1975. Faunal equilibria and design of wildlife preserves. In *Tropical ecological systems: Trends in terrestrial and aquatic research,* ed. F. B. Golley and E. Medina, pp. 369–80. New York: Springer.

Terman, C. R. 1961. Some dynamics of spatial distribution within seminatural populations of prairie deermice. *Ecology* 42:288–302.

Terwilliger, V. J. 1978. Natural history of Baird's tapir on Barro Colorado Island, Panama Canal Zone. *Biotropica* 10(3):211–20.

Tesh, R. B. 1970. Observations on the natural history of *Diplomys darlingi*. *J. Mammal.* 51:199–202.

Tesh, R. B., and Cameron, R. V. 1970. Laboratory rearing of the climbing rat, *Tylomys nudicaudatus*. *Lab. Anim. Care* 20:93–96.

Thenius, E. 1969. Stammesgeschichte der Säugetiere. *Handb. Zool.* 8:1–369, 369–722.

Thenius, E., and Hofer, H. 1961. *Stammesgeschichte der Säugetiere*. Berlin: Paul Parey.

Thiede, U. 1973. Zur Evolution von Hirneigenschaften mitteleuropäischer und südamerikanischer Musteliden. II. Quantitative Untersuchungen an Gehirnen südamerikanischer Musteliden. *Z. Säugetierk,* 38(4):208–15.

Thiessen, D., and Yahr, P. 1977. *The gerbil in behavioral investigations*. Austin: University of Texas Press.

Thomas, J. A., and Birney, E. C. 1979. Parental care and mating system of the prairie vole, *Microtus ochrogaster. Behav. Ecol. Sociobiol.* 5(2):171–86.

Thomas, W. 1975. Observations on captive brockets *Mazama americana* and *M. gouazoubira. Int. Zoo Yearb.* 15:77–78.

Thompson, D'Arcy W. 1917. *On growth and form.* Cambridge: Cambridge University Press.

———. 1942. *On growth and form: A new edition.* Cambridge: Cambridge University Press.

Thomson, J. A., and Owen, W. H. 1964. A field study of the Australian ringtail possum *Pseudocheirus peregrinus* (Marsupialia: Phalangeridae). *Ecol. Monogr.* 34:27–52.

Tinbergen, N. 1951. *The study of instinct.* London: Oxford University Press.

Tomilin, A. G. 1967. *Mammals of the USSR and adjacent countries.* Vol. 9. *Cetacea.* Translated from Russian 1967. Jerusalem: Israel Program for Scientific Translations.

Topachevskii, V. A. 1976. *Fauna of the USSR: Mammals: Mole rats, Spalacidae* (Akademiya Nauk SSSR Zoologicheskii Inst. Novaya Seriya no. 99, 1969). Translated from Russian 1976. Washington, D.C.: Smithsonian Institution and National Science Foundation; New Delhi: Amerind Publishing Company.

Trapido, H. 1942. Gestation period, young and maximum weight of the isthmian capybara, *Hydrochoerus isthmius* Goldman. *J. Mammal.* 30:433.

Trapp, G. R. 1978. Comparative behavioral ecology of the ringtail and gray fox in southwestern Utah. *Carnivore* 1(2):3–32.

Trebbau, P., and van Bree, P. J. H. 1974. Notes concerning the freshwater dolphin *Inia geoffrensis* (de Blainville 1817) in Venezuela. *Z. Säugetierk.* 39(1):50–57.

Trivers, R. L. 1971. The evolution of reciprocal altruism. *Quart. Rev. Biol.* 46(4):35–57.

———. 1972. Parental investment and sexual selection. In *Sexual selection and the descent of man, 1871–1971,* ed. B. Campbell, pp. 136–79. Chicago: Aldine.

———. 1974. Parent-offspring conflict. *Amer. Zool.* 14(1):249–64.

Trivers, R. L., and Willard, D. E. 1973. Natural selection of parental ability to vary the sex ratio of offspring. *Science* 179:90–92.

Troughton, E. 1967. *Furred animals of Australia.* 9th ed. Sydney: Angus and Robertson.

Troyer, W. A., and Hensel, R. J. 1964. Structure and distribution of a Kodiak bear population. *J. Wildl. Manage.* 28:769–72.

Tschanz, B.; Meyer-Holzapfel, M.; and Bachmann, S. 1970. Das Informationssystem bei Braunbären. *Z. Tierpsychol.* 27:47–72.

Tulloch, D. G. 1969. Home range in feral water buffalo, *Bubalus bubalis. Australian J Zool.* 17:143–52.

Turnbull-Kemp, P. 1967. *The leopard.* Cape Town: Howard Timmins and Tri-Ocean Books.

Turner, D. C. 1975. *The vampire bat*. Baltimore/London: Johns Hopkins University Press.
Tuttle, R. 1975. Parallelism, brachiation and hominoid phylogeny. In *Phylogeny of Primates*, ed. W. P. Luckett and F. S. Szalay, pp. 447–80. New York: Plenum Press.
Tyler, S. J. 1972. The behaviour and social organization of the New Forest ponies. *Anim. Behav. Monogr.* 5(2):87–194.
Tyndale-Biscoe, H. 1973. *Life of marsupials*. New York: American Elsevier.
Tyndale-Biscoe, H., and Smith, R. F. C. 1969*a*. Studies on the marsupial glider *Schoinobates volans* (Kerr). 2. Population structure and regulatory mechanisms. *J. Anim. Ecol.* 38:637–50.
———. 1969*b*. Studies on the marsupial glider *Schoinobates volans* (Kerr). 3. Response to habitat destruction. *J. Anim. Ecol.* 38:651–59.
Ullrich, W. 1966. Beobachtungen zur Biologie des Arni (*Bubalus arnee fulvus* Blanford) in Assam. *Zool. Gart.* 32(4):146–58.
VanCuylenberg, B. W. B. 1977. Feeding behaviour of the Asiatic elephant in southeast Sri Lanka in relation to conservation. *Biol. Conserv.* 12:33–54.
Van Gelder, R. 1953. The egg-opening technique of a spotted skunk. *J. Mammal.* 34:255–56.
Van Heel, W. H. Dudok. 1966. Navigation in Cetacea. In *Whales, dolphins and porpoises*, ed. K. S. Norris, pp. 597–606. Berkeley: University of California Press.
Van Lawick, H. 1971. Golden jackals. In *The innocent killers*, ed. J. and H. Van Lawick-Goodall, pp. 105–48. Boston: Houghton Mifflin.
Van Lawick-Goodall, J. 1968. A preliminary report on expressive movements and communication in the Gombe Stream chimpanzees. In *Primates: Studies in adaptation and variability*, ed. P. C. Jay, pp. 313–74. New York: Holt, Rinehart and Winston.
———. 1971. *In the shadow of man*. New York: Dell Publication Company.
Van Lawick-Goodall, J., and Van Lawick, H. 1971. *The innocent killers*. Boston: Houghton Mifflin.
VanTien, D. 1966. Données sur la biologie du petit écureuil rayé de Hainan (*Callosciurus swinhoei hainanus*) au Nord-Vietnam. *Z. Säugetierk.* 31(6):478–79.
Van Valen, L. 1965. Treeshrews, primates and fossils. *Evolution* 19:137–51.
Van Valen, L. 1967. New Paleocene insectivores and insectivore classification. *Bull. Amer. Mus. Nat. Hist.* 135:219–84.
Van Wigjngaarden, A., and Van de Peppel, J. 1964. The badger, *Meles meles* (L.), in the Netherlands. *Lutra* 6(1–2):1–60.
Varona, L. S. 1974. *Catologo de los mamiferos viventes y extinguides de las Antillas*. Havana: Academia de Ciencias de Cuba.
Vaughn, T. A. 1978. *Mammalogy*. 2d ed. Philadelphia: W. B. Saunders.
Veenstra, A. J. 1958. The behaviour of the multi-mammate mouse, *Rattus (Mastomys) natalensis*. *Anim. Behav.* 6:195–206.

Vehrencamp, S. L.; Stiles, F. G.; and Bradbury, J. W. 1977. Observations on the foraging behavior and avian prey of the neotropical carnivorous bat, *Vampyrum spectrum*. *J. Mammal.* 58(4):469–79.

Venables, U. M., and Venables, L. S. V. 1957. Mating behaviour of the seal *Phoca vitulina* in Shetland. *Proc. Zool. Soc. London* 128:387–99.

Verheyen, R. 1954. *Monographie éthologique de l'hippopotame (Hippopotamus amphibius Linne): Exploration du Parc National Albert.* Brussels: Institut des Parcs Nationaux du Congo Belge.

Verschuren, J. 1957. *Ecologie, biologie, et systématique des Chiropteres*. Vol. 7. *Exploration du Parc National de la Garancha—Mission H. de Saeger.* Brussels: Institut des Parcs Nationaux du Congo Belge.

Verts, B. J. 1967. *The biology of the striped skunk.* Urbana: University of Illinois Press.

Vesey-Fitzgerald, D. F. 1966. The habits and habitats of small rodents in the Congo River catchment region of Zambia and Tanzania. *Zool. Africana* 2(1):111–22.

Vincent, R. E. 1958. Observations on red fox behavior. *Ecology* 39:755–57.

Vlasek, P. 1972. The biology of reproduction and post-natal development of *Crocidura suaveolens* Pallas 1811 under laboratory conditions. *Acta Univ. Carol. Biol.* 1970:207–92.

Vogel, P. 1969. Beobachtungen zum intraspezifischen Verhalten der Hausspitzmaus (*Crocidura russula*). *Rev. Suisse Zool.* 76:1079–86.

———. 1970. Biologische Beobachtungen an Etruskerspitzmäusen *Suncus etruscus* (Savi 1822). *Z. Säugetierk.* 35:173–85.

———. 1972. Vergleichende Utersuchung zum Ontogenesemodus Einheimischer Soriciden (*Crocidura russula, Sorex araneus* und *Neomys fodiens*). *Rev. Suisse Zool.* 79:1201–1332.

———. 1974. Kälteresistenz und reversible Hypothermie der Etruskerspitzmaus (*Suncus etruscus,* Soricidae, Insectivora). *Z. Säugetierk.* 39:78–88.

———. 1976. Energy consumption of European and African shrews. *Acta Theoriol.* 21:195–206.

von Bonin, G. 1937. Brain-weight and body-weight of mammals. *J. Gen. Psychol.* 16:379–89.

Vorhies, C. T., and Taylor, W. P. 1940. Life history and ecology of the white throated wood rat (*Neotoma albigula albigula*) in relation to grazing in Arizona. *Univ. Agric. Exp. Sta. Tech. Bull.* 86:455–529.

Vorontsov, N. N. 1960. The ways of food specialization and evolution of the alimentary system in Muroidea. In *Symposium theriologicum,* ed. J. Kratochvil, pp. 360–77. Brno: Ceskoslovenska Akademic Ved.

Vose, H. M. 1973. Feeding habits of the western Australian honey opossum *Tarsipes spenserae*. *J. Mammal.* 54:245–46.

Vosseler, J. 1929. Beitrag zur Kenntnis der Fossa (*Cryptoprocta ferox* Benn.) und iher Fortpflanzung. *Zool. Gart.* (Leipzig), n.s., 2:1–9.

Waage, J. K., and Montgomery, G. G. 1976. *Cryptoses choloepi:* A coprophagous moth that lives on a sloth. *Science* 193(4248):157–58.

Wackernagel, H. 1966. Grant's zebra, *Equus burchelli boehmi* at Basle Zoo: A contribution to breeding biology. *Int. Zoo Yearb.* 6:38–41.

Wadsworth, C. E. 1969. Reproduction and growth of *Eutamias quadrivittatus* in southeastern Utah. *J. Mammal.* 50:256–61.

Wagner, H. O. 1961. Die Nagetiere einer Gebirgsabdachung in Südmexiko und ihre Beziehungen zur Umwelt. *Zool. Jahrb. Syst. Bd.* 89:177–242.

Walker, E. P.; Warnick, F.; Hamlet, S. E.; Lange, K. I.; Davis, M. A.; Uible, H. E.; and Wright, P. F. 1964. *Mammals of the world.* 1st ed. Baltimore: Johns Hopkins University Press.

———. 1968. *Mammals of the world.* 2d ed. Baltimore: Johns Hopkins University Press.

Walther, F. 1960. Verhaltensweisen im Paarungszeremoniell des Okapi (*Okapi johnstoni* Sclater 1901). *Z. Tierpsychol.* 17:188–210.

———. 1964. Zum Paarungsverhalten der Sömmeringgazelle (*Gazella soemmeringi* Cretzschmar 1826). *Zool. Gart.* 29:145–60.

———. 1965. Psychologische Beobachtungen zur Gesselschaftshaltung von Oryx-Antilopen (*Oryx gazella beisa* Rüpp.). *Zool. Gart.* 31(1/2):1–58.

———. 1966. Zum Liegenverhalten des Weisschwanz-gnus (*Connochaetes gnou* Zimmerman 1780). *Z. Säugetierk.* 31:1–16.

———. 1968. *Verhalten der Gazellen.* Wittenberg: Ziemsen.

———. 1969. Flight behaviour and avoidance of predators in Thomson's gazelle (*Gazella thomsoni* Guenther 1884). *Behaviour* 34(3):184–221.

Ward, G. D. 1978. Habitat use and home range of radio-tagged opossums *Trichosurus vulpecula* (Kerr) in New Zealand lowland forest. In *The ecology of arboreal folivores,* ed. G. G. Montgomery, pp. 267–87. Washington, D.C.: Smithsonian Institution Press.

Waring, G. H. 1966. Sounds and communication of the yellow-bellied marmot (*Marmota flaviventris*). *Anim. Behav.* 14:177–83.

———. 1970. Sound communications of black-tailed, white-tailed and Gunnison's prairie dogs. *Amer. Midl. Nat.* 83(1):167–85.

Waser, P. M. 1975. Monthly variations in feeding and activity patterns of the mangabey, *Cercocebus albigena*. *E. African Wildl. J.* 13:249–63.

———. 1976. *Cercocebus albigena:* Site attachment, avoidance and intergroup spacing. *Amer. Nat.* 110:911–35.

Watson, R. M. 1965. Observations on the behavior of young spotted hyaena (*Crocuta crocuta*) in the burrow. *E. African Wildl. J.* 3:122–23.

Watton, D. G., and Keenleyside, M. H. A. 1974. Social behaviour of the arctic ground squirrel, *Spermophilus undulatus*. *Behaviour* 50:1–2.

Wayne, P. 1967. Artificial rearing of roe deer and fallow deer, *Capreolus capreolus* and *Dama dama* at Norfolk Wildlife Park. *Int. Zoo Yearb.* 7:168–71.

Weber, M., and Abel, O. 1928. *Die Säugetiere.* Vol. 2. Jena: Gustav Fischer.

Webster, D. B. 1962. A function of the enlarged middle ear cavities of the kangaroo rat, *Dipodomys*. *Physiol. Zool.* 35:248–55.

Wegener, A. 1929. *The origin of continents and oceans*. English translation of the 4th ed. London: Methuen.
Weir, B. J. 1970. The management and breeding of some hystricomorph rodents. *Lab. Anim.* 4:83–97.
———. 1973. Another hystricomorph rodent: Keeping casiragua (*Proechimys guariae*) in captivity. *Lab. Anim.* 7:125–34.
———. 1974. Reproductive characteristics of the hystricomorph rodents. *Symp. Zool. Soc. London* 34:265–301.
Wells, M. E. 1968. A comparison of the reproductive tracts of *Crocuta crocuta, Hyaena hyaena* and *Proteles cristatus*. *E. African Wildl. J.* 6:89–95.
Wemmer, C. M. 1977. Comparative ethology of the large spotted genet (*Genetta tigrina*) and some related viverrids. *Smithsonian Contrib. Zool.* 239:1–93.
Wemmer, C. M., and Collins, L. R. 1978. Communication patterns in two phalangerid marsupials, the gray cuscus (*Phalanger gymnotis*) and the brush possum (*Trichosurus vulpecula*). *Säugetierk. Mitt.* 26(3):161–72.
Wemmer, C. M.; Collins, L.; Beck, B.; and Leja, B. In press. Sociology of Pere David's deer and its implications for captive management. In *Pere David's deer: Its biology and captive management*. New York: Garland Press.
Wemmer, C. M.; VonEbers, M.; and Scow, K. 1976. An analysis of the chuffing vocalization in the polar bear (*Ursus maritimus*). *J. Zool.* (London) 180:425–39.
West-Eberhard, M. J. 1975. The evolution of social behavior by kin selection. *Quart. Rev. Biol.* 50:1–33.
Western, D. 1979. Size, life history and ecology in mammals. *African J. Ecol.* 17:185–205.
Wetzel, R. M. 1977. The Chacoan peccary, *Catagonus wagneri* (Rusconi). *Bull. Carnegie Mus. Nat. Hist.* 3:1–36.
Wetzel, R. M., and Mondolfi, E. 1979. The subgenera and species of long-nosed armadillos, genus *Dasypus* L. In *Vertebrate ecology in the northern Neotropics*, ed. J. F. Eisenberg, pp. 72–126. Washington, D.C.: Smithsonian Institution Press.
Wharton, C. H. 1950. Notes on the life history of the flying lemur. *J. Mammal.* 31:269–73.
———. 1957. *An ecological study of the kouprey, Novibos sauveli (Urbain)*. Monograph no. 5, Institute of Science and Technology. Manila: Manila Bureau of Printing, Republic of the Philippines.
Whitaker, J. O. 1963. A study of the meadow jumping mouse *Zapus hudsonicus* in central New York. *Ecol. Monogr.* 33:215–54.
Whitaker, J. O., and Mumford, R. E. 1972. Ecological studies on *Reithrodontomys megalotis* in Indiana. *J. Mammal.* 53(4):850–61.
Wickler, W. 1961. Ökologie und Stammesgeschichte von Verhaltensweisen. *Fortschr. Zool.* 13:303–65.
Wickler, W., and Seibt, U. 1976. Field studies on the African fruit bat *Epomophorous wahlbergi* (Sundevall), with special reference to male calling. *Z. Tierpsychol.* 40:345–76.

Williams, E., and Scott, J. P. 1953. The development of social behavior patterns in the mouse in relation to natural periods. *Behaviour* 6:35–64.

Williams, L. 1967. Breeding Humboldt's woolly monkey *Lagothrix lagothricha* at Murrayton Woolly Monkey Sanctuary. *Int. Zoo Yearb.* 6:86–89.

———. 1968. *Man and monkey.* Philadelphia/New York: J. B. Lippincott.

Willis, E. O. 1974. Populations and local extinctions of birds on Barro Colorado Island, Panama. *Ecol. Monogr.* 44:153–69.

Wilson, A. C.; Carlson, S. S.; and White, T. J. 1977. Biochemical evolution. *Ann. Rev. Biochem.* 46:573–639.

Wilson, D. E. 1970. Ecology of *Myotis nigricans* on Barro Colorado Island, Panama. *J. Zool.* (London) 163:1–14.

———. 1973*a*. Reproduction in neotropical bats. *Period. Biol.* 75:215–17.

———. 1973*b*. Bat faunas: A trophic comparison. *System. Zool.* 22(1):14–29.

Wilson, D. S. 1980. *The natural selection of populations and communities.* Menlo Park, Calif.: Benjamin/Cummings.

Wilson, E. O. 1975. *Sociobiology: The new synthesis.* Cambridge: Belknap Press of Harvard University Press.

———. 1978. *On human nature.* Cambridge: Harvard University Press.

Wilson, E. O., and Willis, E. O. 1975. Applied biogeography. In *Ecology and evolution of communities,* ed. M. L. Cody and J. M. Diamond, pp. 522–34. Cambridge: Belknap Press of Harvard University Press.

Wilson, F. J., and Clarke, J. E. 1962. Observations on the common duiker *Silvicapra grimmia* Linn., based on material collected from a tsetse control game elimination scheme. *Proc. Zool. Soc. London* 138:487–97.

Wilson, S. C. 1973*a*. Juvenile play of the common seal *Phoca vitulina* with comparative notes on the gray seal *Halichoerus grypus. Behaviour* 48:37–60.

———. 1973*b*. Mother-young interactions in the common seal *Phoca vitulina vitulina. Behaviour* 48:23–36.

Wilson, S. C., and Kleiman, D. G. 1974. Eliciting play: A comparative study (*Octodon, Octodontomys, Pediolagus, Phoca, Choeropsis, Ailuropoda*). *Amer. Zool.* 14:341–70.

Wilsson, L. 1968. *My beaver colony.* New York: Doubleday.

———. 1971. Observations and experiments on the ethology of the European beaver (*Castor fiber* L.). *Viltrevy Swedish Wildl.* 8(3):115–266.

Wimsatt, W. A. 1960. An analysis of parturition in Chiroptera, including new observations on *Myotis lucifugus. J. Mammal.* 41:183–200.

———. 1970*a*. *Biology of bats.* Vol. 1. New York: Academic Press.

———. 1970*b*. *Biology of bats.* Vol. 2. New York: Academic Press.

———. 1977. *Biology of bats.* Vol. 3. New York/London: Academic Press.

Wing, L. D., and Buss, I. O. 1970. Elephants and forests. *Wildl. Monogr.* 19:1–92.

Winn, H. E., and Olla, B. L. 1979. *Behavior of marine animals*. Vol. 3. *Cetaceans*. New York: Plenum Press.

Winn, H. E., and Perkins, P. J. 1976. Distribution and sounds of the minke whale with a review of mysticete sounds. *Cetology* 19:1–12.

Wirtz, J. H. 1967. Social dominance in the golden-mantled ground squirrel, *Citellus lateralis chrysodeirus* (Merriam). *Z. Tierpsychol.* 24(3):342–51.

Wirtz, W. O. 1968. Reproduction, growth and development, and juvenile mortality in the Hawaiian monk seal. *J. Mammal.* 49:229–38.

Wirz, K. 1950. Zur quantitativen Bestimmung der Rangordung bei Säugetieren. *Acta Anat.* 9:134–96.

Wodinsky, J. 1977. Hormonal inhibition of feeding and death in octopus: Control by optic gland secretions. *Science* 198:948–51.

Wood, A. E. 1931. The phylogeny of the heteromyid rodents. *Amer. Mus. Novit.* 501:1–20.

———. 1935. Evolution and relationships of the heteromyid rodents with new forms from the Tertiary of western North America. *Ann. Carnegie Mus. Nat. Hist.* 24:73–262.

———. 1955. A revised classification of the rodents. *J. Mammal.* 36:165–87.

———. 1965. Grades and clades among rodents. *Evolution* 19(1):115–30.

———. 1974. The evolution of the Old World and New World hystricomorphs. In *The biology of hystricomorph rodents,* ed. I. W. Rowlands and B. J. Wier, pp. 21–60. New York/London: Academic Press.

Wood, D. H. 1970. An ecological study of *Antechinus stuarti* (Marsupialia) in a Southeast Queensland rainforest. *Australian J. Zool.* 18:185–207.

Wood, J. E. 1949. Reproductive patterns of the pocket gopher (*Geomys breviceps brazensis*). *J. Mammal.* 30:36–44.

Wood, W. S. 1927. Habits of the flying lemur (*Galeopterus penninsulae*). *J. Bombay Nat. Hist. Soc.* 32:372.

Woodward, S. L. 1979. The social system of feral asses (*Equus asinus*). *Z. Tierpsychol.* 49:304–16.

Woolard, P.; Vestjens, W. J. M.; and MacLean, L. 1978. The ecology of the eastern water rat, *Hydromys chrysogaster,* at Griffith, N.S.W.: Food and feeding habits. *Australian Wildl. Res.* 5:59–73.

Woolley, P. 1966. Reproduction in *Antechinus* spp. and other dasyurid marsupials. In *Comparative biology of reproduction in mammals,* ed. I. W. Rowlands, pp. 281–94. London/New York: Academic Press.

Wozencraft, W. C. 1980. The phylogenetic status of civets and mongooses. Paper presented at the sixtieth annual meeting of the American Society of Mammalogists, Kingston, R.I.

Wright, P. C. 1978. Home range, activity pattern, and agonistic encounters of a group of night monkeys (*Aotus trivirgatus*) in Peru. *Folia Primat.* 29:43–55.

Wünschmann, A. 1966. Einige Gefangenschaftsbeobachtungen an Breitstirn-Wombats (*Lasiorhinus latifrons* Owen 1845). *Z. Tierpsychol.* 23(1):56–71.

———. 1970. *Die Plumpbeutler (Vombatidae)*. Wittenberg: Ziemsen.

Yablokov, A. V. 1974. *Variability of mammals*. Trans. from Russian by L. Van Valen. New Delhi: Amerind Publishing Company.
Yeaton, R. I. 1972. Social behavior and social organization in Richardson's ground squirrel (*Spermophilus richardsonii*) in Saskatchewan. *J. Mammal.* 53(1):139–48.
Yerkes, R. M. 1943. *Chimpanzees: A laboratory colony*. New Haven: Yale University Press.
Yerkes, R. M., and Yerkes, A. W. 1929. *The great apes*. New Haven: Yale University Press.
Yoeli, M.; Alger, N.; and Most, H. 1963. Tree rat, *Thamnomys surdaster surdaster,* in laboratory research. *Science* 142:1585–86.
Young, S. P. 1958. *The bobcat of North America*. Harrisburg, Pa.: Stackpole.
Young, S. P., and Goldman, E. A. 1944. *The wolves of North America*. New York: Dover.
———. 1946. *The puma: Mysterious American cat*. Washington, D.C.: American Wildlife Institute.
Zannier, F. 1965. Verhaltensuntersuchungen an der Zwergmanguste (*Helogale undulata rufula*) im Zoologischen Garten Frankfurt am Main. *Z. Tierpsychol.* 22:672–95.
Zara, J. L. 1973. Breeding and husbandry of the capybara *Hydrochoerus hydrochoeris* at Evansville Zoo. *Int. Zoo Yearb.* 13:137–39.
Zeeb, K. 1961. Der freie Herdensprung bei Pferden. *Wienen Tierarz. Monat.* 48:90–102.
Zimen, E. 1971. *Wölfe und Königspudel, Vergleichende Verhaltensbeobachtungen*. Munich: R. Piper.
———. 1976. On the regulation of pack size in wolves. *Z. Tierpsychol.* 40:300–341.
Zippelius, H.-M. 1972. Die Karawanenbildung bei Feld- und Hausspitzmaus. *Z. Tierpsychol.* 30:305–20.

Technical Index

Italic numbers refer to illustrations

Index of Common Names